W9-AMW-164

EARTH UNDER SIEGE

From Air Pollution to Global Change

Richard P. Turco

with a Foreword by Carl Sagan

Oxford   New York
OXFORD UNIVERSITY PRESS
1997

Oxford University Press

Oxford   New York
Athens   Auckland   Bangkok   Bogota   Bombay
Buenos Aires   Calcutta   Cape Town   Dar es Salaam
Delhi   Florence   Hong Kong   Istanbul   Karachi
Kuala Lumpur   Madras   Madrid   Melbourne   Mexico City
Nairobi   Paris   Singapore   Taipei   Tokyo   Toronto

and associated companies in
Berlin   Ibadan

Copyright © 1997 by Oxford University Press, Inc.

Published by Oxford University Press, Inc.,
198 Madison Avenue, New York, New York, 10016
http://www.oup-usa.org

Oxford is a registered trademark of Oxford University Press.

All rights reserved. No part of this publication may be reproduced,
stored in a retrieval system, or transmitted, in any form or by any means,
electronic, mechanical, photocopying, recording, or otherwise,
without the prior permission of Oxford University Press.

Library of Congress Cataloging-in-Publication Data

Turco, Richard P., 1943-
Earth under siege : from air pollution to global change / Richard P. Turco
p.   cm.
Includes bibliographical references and index.
ISBN 0-19-507286-3. — ISBN 0-19-507287-1 (pbk.)
1. Air—Pollution. 2. Climatic changes. I. Title.
TD883.T85   1995   363.73'92—dc20   94-11635

9 8 7 6 5 4 3 2 1

Printed in the United States of America
on acid-free paper

Contents

Foreword

The Earth is surrounded by a halo of atmosphere, about as thick (compared to its diameter) as the coat of shellac is on a large schoolroom globe. And yet, the lives of all of us sensitively depend on the integrity of this thin layer of air. Our technology has now achieved formidable, if not awesome, powers. We are pouring all sorts of materials, in enormous quantities, into this protective, life-supporting layer—often without a clue as to what the long-term consequences will be. Depletion of the ozone layer, global greenhouse warming, nuclear winter, acid rain, and ground-level pollution are only a few of the developing or potential hazards that have been identified. Doubtless, new perils will be uncovered in coming years.

These dangers are compounded by the fact that citizens and policy makers often have only a limited understanding of what the issues are. There are those who believe that the world is coming to an end every time a potential hazard is announced, and others who stolidly insist that there are no serious environmental dangers no matter what the scientists say; or that the Earth will heal itself.

We urgently need a comprehensive and comprehensible book on the dangers to the environment to which someone with no more than a high school education can turn. That need has now been supplied in the present book by Richard Turco, who has himself made important research contributions to our understanding of these leading environmental dangers. I wish every college student, every lawmaker, and every concerned citizen would delve into it.

Carl Sagan
Director, Laboratory for Planetary Studies
Cornell University
Ithaca, New York

Preface

Air pollution—on local, regional, and global scales—stands as one of the most important problems of the modern technological age. All the peoples of the Earth are affected by air pollution, although most are only vaguely aware of the dangers. Few people outside a small cadre of professional scientists and engineers understand the fundamental principles that relate to air pollution: the sources of the numerous pollutants that plague our world, their dispersion through the environment, the chemical and physical processes that produce and modify these compounds, the toxic effects of air pollutants in humans and other species, and the measures that have been taken and could be taken in the future to reduce or eliminate specific offending pollutants. This book is meant to provide the foundation for such an understanding among university students. In fact, the book is directed to all students, regardless of their background in mathematics, physics, and chemistry, although I assume they have had high-school college-preparatory course work. The text itself is written so that students in the arts, social sciences, law, medicine, and other professional endeavors should be able to grasp the underlying principles and follow the scientific arguments.

It is my belief that solutions to existing and future environmental problems must be forged by an *informed* corps of professionals in business and policy-making positions. Scientists can gather data, interpret their meaning, and even sound warning bells on impending environmental disasters, but sensitive government administrators, business leaders, and concerned citizens must provide the impetus for action.

This book evolved from a series of lectures that I developed for a course in air pollution at the University of California, Los Angeles. My excitement at the prospect of educating hundreds of students on the facts and myths of air pollution and other issues related to "global change" led me to create a broad course dealing with such issues on all spatial and temporal scales. I also recognized the opportunity for using the topic as a means to teach non-science majors a wide range of basic scientific concepts, from chemical reactions to the mechanisms that control the climate. I have attempted to present these concepts in the framework of simple models of the physical world, gently introducing students to the physical laws that govern our everyday lives. Many students who will be future leaders seem to resist attempts to teach them science and an understanding of the natural order; this should worry those of us who will eventually be led by them. It is my sincere wish that the entire core of our educated leadership in government, business, and the professions be knowledgeable about and sensitive to issues involving air pollution and the global environment in general.

I must express special thanks to a number of people who contributed to this book. F. Marty Ralph, a graduate student at UCLA, first transcribed my lectures in air pollution into a detailed set of class notes for my students. Ralph expended a great effort to ensure that the notes offered an accurate representation of the facts presented in those lectures. The notes, continually expanded and improved over a period of several years, have been adopted by several instructors at UCLA and other campuses. Indeed, some of the material contained in this work was introduced by others who taught the course. It became clear to me, however, that students at an introductory level require much more detail and clarity of exposition. Accordingly, this textbook represents a complete rewriting of the original notes to conform more closely in depth and style to my original lecture materials and commentary, and to include further thoughts I have had on the subject. I have expanded the dialogue, for example, into areas that, at the time, lay beyond the scope of a single-quarter course consisting of some twenty 1-hour lectures. The material in this first edition of the text has also been updated where possible from the final edition of the precursor class notes.

Jeffrey Lew, a postdoctoral researcher in cloud physics and lecturer at UCLA, is responsible for all the illustrations and graphics in this book. His enormous skill and creativity in producing understandable and informative artwork from my often cryptic and cluttered view-foils has made possible a comprehensible presentation of this diverse and difficult material. Lew also created the overall layout for the book. It is fair to say that without his contributions, this book would have been, if not impossible, of significantly lesser quality.

I have appreciated the enthusiasm and patience of the editors at Oxford University Press, particularly Joyce Berry, but also Irene Pavitt and Elyse Dubin, and their efforts in preparing the manuscript for publication. I had not realized how much work must go into producing a readable and accurate textbook of several hundred pages. The Oxford editors, and those colleagues whom I avoided during the 5-year gestation of this book, receive my apologies for any apparent neglect on my part. Several able assistants spent hours digging out facts for me at the university library, and Deborah McDonald deserves special mention in this regard. I also thank a large contingent of friends and fellow scientists for their encouragement, suggestions, concern, and scientific contributions that made this book possible. More important, I thank all the hardworking researchers who are seeking solutions to the daunting issues discussed in this text.

Among those to whom I owe special thanks are Brian Toon (NASA Ames Research Center) who, in typically exhaustive fashion, pored over every sentence and paragraph in an earlier version; Donald Stedman (University of Denver), who provided me with ideas for exciting and illuminating class demonstrations; Joseph Casmassi (South Coast Air Quality Management District), who taught this class at UCLA with flair for many quarters; Ralph Cicerone (University of California at Irvine), who continually encouraged me by adopting draft versions of this book for his courses; Jack Winchester (University of Florida at Tallahassee), who provided the initial impetus for me to write this text; and my students Sybil Anderson and Katja Drdla, who read and commented frankly on the contents. My friend and colleague Carl Sagan, who generously provided the foreword, has been especially supportive over the years and has provided a role-model that all writers of science should emulate. I particularly thank my wife, Linda, for her contributions to the clarity and accuracy of the text, for her rendering of the cover art, and for her perseverance during all those evenings and weekends that I sat at my computer rewriting what had already been revised several times over.

Los Angeles
August 1996

R. P. T.

EARTH UNDER SIEGE

From Air Pollution to Global Change

1

Introduction

1.1 The Cronus Syndrome

Mythological tales often reflect events that occur in everyday life, although in exaggerated or exalted forms. Nevertheless, valuable lessons might be learned by considering the proposed behavior of the gods and goddesses in compromising situations. Selfish or arrogant acts associated with strength and power could be avoided, and generous or virtuous behaviors would be adopted.

In Greek mythology, Uranus was the god of the sky, and Gaia, his wife, was the goddess of the Earth. Gaia, or Mother Earth, gave birth to 12 children, six male and six female, who were the Titans. Cronus was their lastborn son. The Titans were forced by Uranus, who feared their intentions, to live within the confines of the Earth. But Cronus, ambitious and angry, turned on his father and castrated him. Wounded and impotent, Uranus retreated to the far reaches of the universe, and Cronus rose to supremacy. Fearing an end similar to Uranus's, Cronus thought it best to swallow his own children (begat by another Titan, his sister Rhea). However, one of his offspring, Zeus, escaped being eaten through a deception, when Rhea substituted a large stone wrapped in infant's clothes for the child. Zeus later assumed the throne of Mount Olympus and destroyed Cronus and the other Titans in a shower of lightning bolts that made the Earth shudder and ignited global fires that consumed the forests and laid the land to waste. Out of this devastation arose a new and gentler human society.

This mythology may be seen as an analogue to the present despoliation of the Earth's environment. It might be called the **Cronus syndrome**. Humans are literally the children of the Earth, or Gaia, having evolved over billions of years from humble origins in the organic brines of the primal oceans. Thus, in a sense, we are modern Titans, and unfortunately, we have retained many of the unsavory aspects of that legacy. In our quest for personal gain and our neglect of the natural order, we have effectively disfigured our father, the sky, Uranus. In many places, the air is unfit to breathe; it reeks of foul odors; and it reflects the dull hues of industrial wastes. Having flaunted our powers by defiling the air, we are now engaged in the piecemeal destruction of the entire global environment—often, ironically, in the name of "economic development." In the process, we are swallowing our children's future. Unfortunately, it may take something like Zeus's thunderbolts to sweep aside the Titans and lead a return to a state of existence in closer harmony with nature. We can only hope that before disaster strikes, actions necessary to preserve a livable environment will be taken.

The self-destructive tendencies of modern society were recognized by Chief Seattle, leader of the Duwamish and Suquamish tribes of Puget Sound, Washington. He lived the natural life of a Native American and for a time adopted Western customs. From that unique perspective, Chief Seattle noticed that Western society had begun to lose touch with nature. The modern world, in many ways, had lost respect for the environment. The potential consequences are clearly expressed in a quote attributed to him:

> Whatever befalls the earth befalls the sons of the earth. Man did not weave the web of life; he is merely a strand in it. Whatever he does to the web, he does to himself.

Indeed, the human species is the only organism that treats its environment as a disposable commodity and thus must consciously act to preserve it.

1.2 On the Quality of Life

There was a time in the recent past when the citizens of the United States and other countries were, for the most part, unaware of or indifferent to the declining quality of the air they breathed, the water they drank, and the land they lived on. Greater problems—jobs, food, and housing—occupied their waking thoughts. Of course, there were visionaries, like John Muir, who foresaw the widespread destruction of the natural environment as a consequence of selfish plundering of resources and unregulated industrialization. Today, by contrast, most of us are keenly aware of pressing environmental issues: urban air pollution, acid rain, global ozone depletion, and greenhouse warming. A 1988 study by the World Health Organization concluded that most of the world's 1.8 billion city dwellers were breathing air of unacceptable quality. More than 1.1 billion people are subjected to excess sulfur dioxide, and 1.2 billion to high levels of smoke and dust. The situation is particularly bad in Third World countries, where government priorities have focused on economic development rather than a clean environment. Nevertheless, major urban centers even in developed nations often find themselves smothered by a pall of irritating smog. The widely recognized home of urban air pollution is Los Angeles, with its cloudless skies, millions of automobiles, and tenacious foul air.

There is no unique definition of environmental pollution. The kinds and effects of pollutants are so diverse that they defy generalization. In the broadest sense, the pollution of air, water, and land might be considered to be any change from their natural state that is caused by human activities. However, this definition is so broad that even trivial changes might be defined as "polluting." Many normal human animal functions might be considered to create "pollution" (for example, bathing in a stream or defecating in a forest). These actions are themselves purely natural when people are regarded as an integral part of the environment. Moreover, a broad definition of pollution implies a detailed knowledge of the natural, or pristine, state of the environment *as well as* the normal variations that occur over time. It is important to separate natural variability from changes induced by human activities. A drift from a perceived "pristine" state of the environment may represent nothing more than a natural long term change in a highly complex system. Pollution may have nothing to do with it. Distinguishing natural variability from the effects of pollution often requires careful measurements recorded over a long time span. To detect the signals of pollution, data may need to be analyzed using sophisticated statistical techniques. When data of sufficient quality are not available, interpretations can become muddled and controversial.

A more restrictive definition of pollution identifies it as the presence in the environment of any substance, no matter what the source, that affects human health or well-being, or the well-being of any other specific organism exposed to the pollutant. To be useful, this definition requires that a threshold concentration, or a level of exposure, be specified for every potential environmental pollutant. The threshold level is used to identify substances that are merely nuisances without causing serious effects. For example, house dust may occasionally make you sneeze, but there are no serious proposals to outlaw it. The idea of a strict "threshold" level of exposure is difficult to justify, because the sensitivity of individual persons to any pollutant can vary over a wide range.

In practice, pollution is identified by its potential or actual impact on human health, structures and artifacts (for example, ancient marble statuary), or aesthetic sensibilities (for example, the clarity of air in national parks). Adverse effects on the environment at large—aside from effects on people—have only recently become valid criteria for identifying pollution hazards. Even so, there still seem to be far too many government leaders and officials who cannot see the value in protecting—from acid rain, ultraviolet radiation, and other threats—the majority of species on our planet that do not vote.

1.3 Global Change and Preservation

The term *global change* has achieved wide popularity. Global change refers to recent worldwide changes in the environment related primarily to increasing concentrations of airborne pollutants. A natural protective layer of "ozone" in the stratosphere is under attack from artificial compounds manufactured for use in air conditioners and refrigerators. The problem has become so severe that a "hole" recently appeared in the ozone lying over the South Pole. The Earth's climate seems to be warming beneath a thickening blanket of carbon dioxide emitted from power plants and cars. Superhurricanes and freakish weather may be associated with the heating.

Global change is really a jigsaw puzzle whose pieces correspond to changes occurring on smaller "local" and "regional" scales. A local change may affect your town or even your backyard. A regional change may affect a state or even a small country. Furthermore, when the global ozone layer or the climate system changes, the effects are not the same everywhere. There are distinct patterns of local and regional effects. In addition, the pollution that causes global change itself originates from a mosaic of local and regional sources. These pollutant emissions can aggregate over the entire planet, from top to bottom and end to end. Automobiles in Los Angeles, power plants in Beijing, cooking fires in India, rice paddies in Japan, steel mills in Russia, personal computers in Germany—all may contribute to global pollution.

Our planet is constantly changing, in any case. Continents drift thousands of miles over tens of millions of years; earthquakes are the local, often devastating, episodic result of this motion. Species of plants and animals disappear and new ones appear over millions of years; evolution is a dynamic, ceaseless process occasionally punctuated by colossal events of extinction. The climate also drifts from state to state, periodically falling into ice ages separated by pleasant warm periods. The oceans and atmosphere are in continuous agitation, varying from day to day and place to place. Each element of the environment—the atmosphere, oceans, continents, and "biosphere"—exhibits natural variations with a wide spectrum of time spans over which fluctuations occur. Global change is necessarily a combination of all these natural fluctuations, plus the changes induced by humans. To determine the components of change associated with human activities, the natural fluctuations must somehow be identified and subtracted. As we already noted, extensive data sets are usually needed to separate natural from human-induced environmental change. Looking back over the history of the Earth, the quantity of specific detailed information that can be recovered from geological and fossil records and used to decipher global change in the past diminishes rapidly as time recedes.

For logical reasons, the study of global change has focused on recent times. We can directly measure variations in the composition of the present atmosphere and in the state of the world climate, and we have a relatively large environmental database for the past 50 years or so of Earth history. It is important, nevertheless, to determine the extent to which present-day conditions have drifted from the norms of the past 10,000 years, the Holocene epoch during which human civilization developed. Climate studies of the past few million years, the period over which the human species evolved, are also of great interest. Global change may be better understood when viewed in the context of variations that have occurred over long spans of time.

The impact of global environmental change on living organisms is of paramount concern. In particular, potential effects on human society and its "quality of life" must be ascertained. The time scales of interest span decades to centuries. Over these time intervals, the change that can be tolerated is quite limited. How fast can change occur without seriously disrupting society and degrading health? What kinds of change are acceptable to preserve economic structures and standards of living? What sacrifices *must* be made now to preserve a habitable, happy planet for the future? Is it perhaps even possible to improve the state of the global environment through prudent long-term engineering projects spurred by fears of global change?

Environmental pollution and global change involve complex scientific, socioeconomic, and political issues. The problems are not intractable, however. Those people who are concerned about the well-being of their family and the life-sustaining environment around them should become familiar with the basic principles of environmental science. This science requires understanding the way the climate system works, where air pollution comes from, and how the ozone layer behaves, among other things. The atmosphere plays a central role in almost all the environmental problems that we face today. Accordingly, there are good reasons to focus in this book on air—its properties and behavior. We will discuss the origin of air and the evolution of its composition; interactions among the atmosphere, oceans, land, and the biosphere; the role of the atmosphere in global climate change; and implications of recent changes in atmospheric composition. After studying the following chapters, readers should be confidently armed with a broad—but today essential—understanding of the global environment.

1.4 Methodology for Study

This book is divided into three parts that are further divided into chapters focusing on specific topics. Part I introduces the atmosphere and the fundamental scientific principles that are useful in studying environmental pollution. The evolution of the Earth is also outlined so as to provide a backdrop for current pollution issues and to introduce the Earth as a system of interacting physical, chemical, and biological elements. Part II discusses local and regional environmental problems of air pollution, including smog and acid rain. We explain in detail concepts related to the human impacts of pollution in this part of the book. Part III is devoted to a survey of global-scale, long-term environmental problems—such as climate change and stratospheric ozone depletion—associated with anthropogenic pollution, particularly carbon dioxide and chlorofluorocarbon emissions. Extensive discussions of biogeochemical cycles and the Earth's climate system are contained in this part of the book.

Each chapter includes a reading list, covering work in both scientific journals and the more popular media. Further inquiry is encouraged and can be accommodated on any level of detail in these references. At the end of each chapter are review questions and problems designed to highlight important points and to stimulate independent thinking on the part of the reader. The material in each chapter builds on earlier sections. But individual chapters and the three parts themselves may be treated as independent units of study if sufficient introductory and supporting information is supplied by the instructor.

This book has two appendices. Appendix A describes scientific notation and common mathematical operations for students unfamiliar with these concepts. Appendix B outlines a series of experiments that demonstrate some of the interesting physical and chemical phenomena introduced in the book.

Throughout the text are references to similar or related material in other sections of the book. To some degree, this material is redundant, which seemed to be unavoidable in a text covering such a wide range of topics. Certain subjects, for example, arise in different contexts and must therefore be discussed at different points in the manuscript. Smog, for example, is rightly a distinct topic of particular interest to those who live in or around cities, but in many respects, urban air pollution is the cause of regional and global environmental problems. This common connection among local, regional, and global pollution is emphasized in several places. Likewise, the chemical reactions that generate smog also can be found in the stratosphere and in other locales. The basic chemistry thus may need to be recapitulated for clarity or altered in form to fit each context. Climate is another broad and diverse subject that appears throughout the text. There are connections to be drawn between local air pollution and global climate change, the evolution of the Earth and of climate, the theory of climate change and observations of greenhouse warming, between chemistry and climate, climate change and global environmental engineering, and so on.

Owing to the difficulty in organizing such a far-ranging subject as "environmental pollution" into a unique, completely logical sequence of subtopics, the presentation given here is somewhat disjointed. The number of possible arrangements of the material in this text is enormous. Obviously I have used my own judgment to decide on the order of presentation. That judgment is based on my experience in the lecture hall, where the outline for this text evolved. The cross-referencing and correlation of related topics should aid, rather than distract, the reader.

A detailed index is included. Longer rather than shorter indexes are more likely to be useful to students exploring new topics. Accordingly, the indexing here is repetitive where necessary to help in tracking down a specific subject. Boldface type is also used in the text to emphasize key terms and concepts.

The most difficult section of the book is Part I. This collection of material is meant as a resource for students and instructors alike to learn or teach basic physical and chemical principles. Here these principles are laid out on a level that should be understandable to entry-level college students who are not scientists or engineers. Toward this end, I have incorporated everyday experiences and events to illustrate the behavior and properties of the natural world. But this approach is not always successful, and some students may still find particular sections or concepts beyond their comprehension. Instructors must be sensitive to the capabilities of their charges. In using this book as a resource, instructors should carefully prepare their corresponding classroom lectures, distribute supplementary notes, and provide additional reading material, such as the articles listed at the end of each chapter. Selective reading assignments and homework exercises from the text can be designed to ease students through the most difficult parts. Most of the later chapters can be read with only passing reference to Chapter 3, the most technical chapter. Chapter 2 is recommended as an introduction to Part II, and Chapter 4 as preparatory material for Part III.

The book is also formulated with a historical perspective. Students are introduced to some of the key practitioners of the art of science over the centuries. Many of the most notable scientists of the past have contributed to our current understanding of the environment. Science is shown to be a process of discovery and learning that eventually leads to the "truth," although the path is often tortuous and subject to temporary detours set up by human weaknesses. Indeed, the history of environmental science and technology includes unconscionable acts and judgmental errors that have led to the current state of the environment's degradation. In one sense, this book is a confession of past environmental misdeeds and crimes, offered to victims present and future; it is a testament of the environmental legacy of earlier generations to those yet unborn. In another, more hopeful, sense, the text is a catalogue of remarkable discoveries and activities that may yet save our fragile planet.

Suggested Readings

Caldicott, H. *If You Love This Planet: A Plan to Heal the Earth.* New York: Norton, 1992.

Commoner, B. *Making Peace with the Planet.* New York: Pantheon Books, 1990.

Complete Guide to Environmental Careers. Washington, D.C.: Island Press, 1993.

Ehrlich, P. *The Machinery of Nature.* New York: Simon and Schuster, 1986.

Ehrlich, P. and A. Ehrlich. *Healing the Planet.* New York: Addison-Wesley, 1991.

Gore, A. *Earth in the Balance: Ecology and the Human Spirit.* New York: Penguin Books, 1992.

Harte, J. *The Green Fuse: An Ecological Odyssey.* Berkeley: University of California Press, 1993.

Nesbit, E. *Leaving Eden: To Protect and Manage the Earth.* Cambridge: Cambridge University Press, 1991.

Schneider, S. *Global Warming: Are We Entering the Greenhouse Century?* San Francisco: Sierra Club Books, 1989.

Shabecoff, P. *A Fierce Green Fire: The American Environmental Movement.* New York: Farrar, Straus & Giroux, 1993.

Part I
Fundamentals

The first part of this book presents many of the basic principles that apply to the science and technology of the environment. The science of the environment is just the classic natural science of the past century in the context of modern environmental issues. It is Thoreau with a computer rather than a pen. It is the study of the world around us and how it works; the collection of everyday events that we are most familiar and comfortable with. It is the wind, rain, waves, lightning, and clouds. It is the smell of a forest or the seaside. It is a top spinning or a vortex swirling in a drain, smoke rising over a fire, brilliantly colored sunsets, bubbles in beer. All these phenomena can be understood on the basis of fundamental concepts developed over the centuries by straightforward observation and experimentation.

The science of the environment is built on a foundation of basic physics and chemistry. These two subjects, along with mathematics, form the backbone of all scientific inquiry and understanding. Knowledge of basic effects provides the inquisitive mind with powerful tools for reasoning and discovery. Accordingly, a number of ideas and concepts—from the laws that govern the behavior of gases to the rules that apply to global chemical cycles—are laid out in the following chapters. Events from common experience are invoked when possible to show the essential connections between "physics" and "chemistry," and the real world.

The real world is nevertheless infinitely more complex than any experiment ever carried out in a physics or chemistry laboratory. For the most part, that complexity is caused by the presence, in countless numbers, of living organisms. You cannot see the solution of a jigsaw puzzle by looking at the jumbled pieces that fall out of the box. Instead, you must sort through the bits, picking out the ones that look alike, and painstakingly assembling the individual pieces that are most closely related. After a while, subsections of the puzzle become recognizable, and these larger elements then begin to fit together into a grander scheme, until the entire picture is complete. You do not have to assemble every piece of the puzzle to find out what the subject is. Depending on how perceptive you are, you can discern the scene long before the puzzle is finished (if you have not looked at the cover of the puzzle box, of course).

Environmental science is like a puzzle. First you must have the basic pieces that, in certain combinations, define how specific elements or processes of the environment work. Then you must fit the pieces together to formulate the elements or processes, which themselves represent building blocks in the overall structure of the environment. Finally, you must assemble all the elements that compose a particular environmental setting. An environmental problem is almost always recognized long before the puzzle is complete. Accordingly, debates may arise over different interpretations of the fragmentary image that has been created and the importance of the missing information.

For example, suppose you wish to understand the cause of precipitation in a certain region. You first should identify the basic physical laws that control the behavior of water as a vapor and in its condensed states in the atmosphere and you should also describe the motions of air using simple laws. By assembling these basic concepts into a descriptive model for a cloud, the source of rain in convective clouds becomes apparent. Better yet, you may be able to predict quantitatively the occurrence of rain using the cloud model. Before any serious use, however, the model should be tested by comparing its predictions with observations of real clouds. The determination of patterns and amounts of rainfall requires that the cloud model then be placed in the context of local and regional weather systems. The numbers and sizes of the clouds and their locations can then be estimated using larger-scale information. The interactions between the clouds, and the sum of their effects on regional precipitation, can be investigated to complete the picture.

Chapters 2, 3, and 4 provide the basic science building blocks for the environmental analysis we wish to pursue in later chapters. These tools, if mastered, should prove to be useful in thinking, in other regards, about the environment around you.

2

Air: The Medium of Change

2.1 What Is Air?

The Earth's atmosphere is a mixture of gases that we call *air*. The gases that make up the atmosphere exist as **molecules**, the smallest unit of a particular material that carries the properties specific to that material. Molecules are arrangements of atoms, the smallest units of matter that define the elements, the basic building blocks of the universe (Section 3.1). The word *atmosphere* derives from the Greek *atmo* (vapor) plus *spherios* (sphere): hence, the *vapor sphere*, which the ancient Greeks perceived as a basic element of nature. Air usually contains a number of small particles and often is filled with clouds of condensed water. Although from our vantage point on the surface, the atmosphere is seemingly endless in extent, in reality it is a delicate veil. From space it is seen as a thin, luminous layer of vapor hugging the surface.

Air—clean and clear, fresh and cool—nurtures life on Earth and protects it from the deadly environment lying beyond it. It fulfills this role in several important ways: Atmospheric gases participate in essential life processes, providing oxygen for animal respiration and carbon dioxide for plant respiration; air molecules filter out deadly ultraviolet radiation emitted by the sun before it can reach the ground and kill living organisms; and the atmosphere offers a vast repository for pollutants generated by human activities, as well as the means to render those pollutants less harmful. Because air is the medium in which winds blow, it has the capacity to disperse and dilute air pollutants quickly over a large area, reducing their local effects. This is a double-edged sword, however, because much larger regions—perhaps even the entire planet—may be affected by local pollution sources. Because air is the medium in which clouds and precipitation form, it has the ability to cleanse itself by washing out, or scrubbing, air pollutants to the ground. Again, this can be a disadvantage, for example, in the case of acid rain, in which clouds are responsible for cleaning the atmosphere but, at the same time, contaminating sensitive ecological systems with corrosive acids.

Of the three principal repositories for pollutants in the environment—the atmosphere, the oceans, and the land—the atmosphere is the most susceptible to degradation. The atmosphere has the smallest mass by far. Climbers who reach the summit of Mount Everest (about 29,000 feet, or roughly 9 kilometers, above sea level) find themselves surrounded by rarefied air and in need of pressurized oxygen (about 70 percent of the atmosphere lies below them). A typical commercial jetliner cruises at an altitude of 40,000 feet (about 13 kilometers), above roughly 80 percent of the atmospheric mass. Although the atmosphere of Earth stretches hundreds of kilometers beyond that height, the sensible atmosphere may be considered to end at an altitude of roughly 60 miles (about 100 kilometers). By this height, more than 99.999 percent of the atmosphere has been traversed. The air is so thin at 100 kilometers—just one-millionth the density at the ground—that one would suffocate within minutes if suddenly exposed to the atmosphere there.

The maximum horizontal extent of the Earth's atmosphere is the circumference of the planet, a distance of about 40,000 kilometers (24,000 miles). It follows that the ratio of the vertical depth to the horizontal extent of the Earth's atmosphere is roughly 1:400 (Figure 2.1). If the depth-to-length ratio of a typical 30-foot swimming pool were similar, the water in the pool would be about 1 inch deep—hardly enough to get your toes wet! In fact, most of the air is compressed into a layer one-tenth as thick—about 10 kilometers. At this scale, the relative thickness of the Earth's atmosphere can be compared with the thickness of the skin on an apple: not an impressive shield.

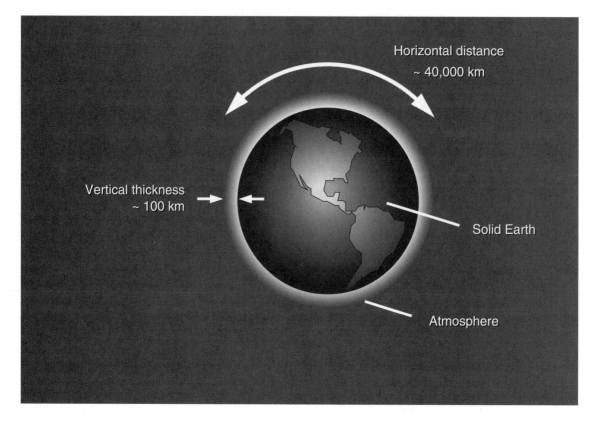

Figure 2.1   The overall structure of the Earth's atmosphere and its relationship to the solid Earth (however, not to true scale, since the atmosphere would be reduced to merely the thickness of a heavy line in the drawing).

2.1.1 SENSING AIR

We know that we are surrounded by air. We sense it in a hundred ways, even if we do not understand exactly what it is or why we feel its presence. The ancient Greeks obviously were aware of air and elevated it as one of only four basic elements of which all substances were composed. Like water, air is a fluid, although much less dense. Nevertheless, as a fluid, air can flow in great currents, creating the winds that blow from zephyrs to gales. There are numerous tangible and intangible manifestations of air.

• *Tactile*: the feel of the wind on your face and in your hair, the cool freshness of a breeze on your skin, the rush of hot air when you open the oven door.
• *Visual*: the pure blue color of the sky, iridescent sunsets, clouds bobbing and flowing in the wind, whitecaps over waves, the fading of mountain ridges with distance, the haziness of a polluted day.
• *Olfactory*: the distinct aromas of forests and oceans, the fresh smell of ozone after lightning, fetid odors of garbage and industrial waste, smells of flowers, buses, and friends.
• *Auditory*: whistling of the wind through a window pane, thunderclaps and sonic booms, the rustling of trees.

Air has many other effects on our everyday lives. Air causes friction if we try to move quickly through it, so that running and bicycling are more difficult than they would be on a planet without air (although without air to breathe, any activity would be impossible). Our comfort is controlled by the temperature and humidity (or moisture content) of air. Air is also filled with ions—microscopic particles that carry an electric charge, either positive or negative. Some researchers believe that a negative charge can make you feel good and a positive charge, irritable (the opposite of what you might think).

If you are an incipient scientist, you will be wondering how you can translate casual observations of the atmosphere into real, quantitative measurements of the composition and properties of air.

Table 2.1   The composition of air (by number of molecules)

		Fraction of air	
Constant gases	N_2 (nitrogen)	78.08%	
	O_2 (oxygen)	20.95%	
	Ar (argon)	0.93%	
	Ne (neon)	18.2 ppmv* (= 0.00182%)	
	He (helium)	5.2 ppmv	
	Kr (krypton)	1.14 ppmv	
	Xe (xenon)	0.09 ppmv	
Variable gases	H_2O (water vapor)	4%	(maximum, in the tropics)
		0.00001%	(minimum, at the South Pole)
	CO_2 (carbon dioxide)	~365 ppmv	(increasing ~0.4% per year, and by about 35% since 1800)
	CH_4 (methane)	~1.8 ppmv	
	H_2 (hydrogen)	~0.6 ppmv	
	N_2O (nitrous oxide)	~0.31 ppmv	
	CO (carbon monoxide)	~0.09 ppmv	
	O_3 (ozone)	~0.4 ppmv	
	CF_2Cl_2 (fluorocarbon 12)	~0.0005 ppmv	

* ppmv indicates "parts per million volume"

The history of that endeavor is a long and fascinating story, taken up in the next section. The study of air provided the foundation for the development of scientific thought itself. But don't worry. You need not be a scientist to enjoy and appreciate the atmosphere. Painters, poets, and writers all have eloquently recorded fascinating phenomena seen in the air. By the time you have finished this book, you too will have a much deeper appreciation and understanding of the many natural events constantly on display around you.

2.1.2 THE BASIC INGREDIENTS

Below an altitude of about 100 kilometers (Figure 2.1), the atmosphere consists of a relatively homogeneous, or uniform, mixture of gases, referred to as the **homosphere**. Table 2.1 lists the most abundant chemical constituents in the homosphere. Nitrogen (N_2) is the dominant component, comprising about 78 percent of the air molecules. Oxygen (O_2) is the second most common gas, accounting for about 21 percent of the air. Thus about 99 percent of the air in the homosphere is composed of N_2 and O_2. Next

in line in terms of quantity are argon (Ar), about 1 percent, and water vapor (H_2O), very roughly 1 percent, but highly variable in concentration. Beyond these gases, the other components of the atmosphere—even carbon dioxide (CO_2)—are measured in parts per million or less (Section A.1).

We introduce chemical notation at this point for convenience. That is, we use the symbol N_2 to represent nitrogen gas. Later, we will discover that this chemical nomenclature tells us a lot about a compound. In the case of N_2, it tells us that each molecule of nitrogen gas contains two nitrogen atoms. Nitrogen (N) is an element, the simplest chemical unit found in nature. Compounds are combinations of elements in fixed proportions (nitrogen gas, for example, is a compound of nitrogen consisting of two nitrogen atoms per molecule). (Chemical nomenclature, which provides a shorthand for writing chemical compounds and equations, is discussed in subsequent sections [for example, Section 3.3.1].

The composition of the atmosphere is much richer than is suggested by Table 2.1. The actual number of trace (that is, "minor," or low-concentration) constituents reaches into the thousands. Some of these compounds will be discussed in other chapters of this book, but most will not even be mentioned, which is just as well. At this level of study, you should concentrate on the basic composition of the atmosphere (the "major" constituents) and the key trace compounds that cause environmental problems, avoiding the confusion of keeping track of countless secondary compounds.

The gases that make up the atmosphere may also be characterized as either *constant*, or invariant, gases or *variable* species. Constant gases are those that always have the same mixing ratio in the homosphere over time scales of decades to centuries. The trace gases neon, helium, and krypton fall into this category (Table 2.1). Imagine that you have captured a bottle of air containing exactly 1 million gas molecules. The sample may be collected at any point in the homosphere. Out of this collection of molecules, you would find about one krypton atom, five helium atoms, and 18 neon atoms (Table 2.1). The same numbers would always be found, no matter where or when you took the sample. These gases are invariant.

In contrast, variable gases are found to have different relative concentrations in different parts of the homosphere at different times. For example, water vapor can have concentrations as high as 0.04 (4%) of all air molecules in the tropics, where it is hot and moist, and as low as 0.0000001 (0.00001%) in the extremely cold, dry air at the South Pole in winter. Another important variable gas is carbon dioxide (which has the chemical abbreviation CO_2). Its variation is most notable with changes in the seasons. During the spring and summer, plants, which consume carbon dioxide during photosynthesis and convert it to organic matter, absorb CO_2 at a faster rate than they do during the fall and winter, when they are dormant. Accordingly, CO_2 concentrations are lower in the summer than in the winter. This effect is seen in both the Northern and Southern Hemispheres—although the CO_2 seasonal cycle is "out of phase" by six months between the Hemispheres (which means summer occurs in the Southern Hemisphere during winter in the Northern Hemisphere, spring during fall, and so on). The amplitude of the seasonal cycle is much smaller in the Southern Hemisphere than in the Northern Hemisphere. Although about one-third of the Earth's surface is covered by land (the rest being open oceans and ice), in the Southern Hemisphere only one-fifth of the total area is available for vegetation. Hence there are fewer plants and a less noticeable seasonal variation in CO_2 in that hemisphere.

Carbon dioxide also exhibits a long-term increasing trend in its concentration that has been associated with the burning of fossil fuels (coal, petroleum, and natural gas) and deforestation. This behavior of carbon dioxide, and its consequences for the global environment, are discussed in greater detail in Chapter 12.

Table 2.1 lists a few additional common atmospheric constituents such as methane, nitrous oxide, and ozone in the order of their abundance. These substances will be discussed later in this book. Note that the concentrations of certain atmospheric gases are very low, yet those compounds are still considered to be important. In other words, one must be cautious in discounting the environmental significance or effects of a compound simply on the basis of its concentration. Examples of this rule will become apparent as you continue.

2.1.3 THE BASIC PROPERTIES

To study the behavior and properties of the atmosphere and understand the processes that control air pollution and its environmental impacts, we should become familiar with a few simple physical concepts and parameters. These all relate to the properties of gases, with particular application to the atmosphere.

Temperature

Temperature measures the amount of internal energy, or heat, that a substance holds; the more energy there is, the higher the temperature will be. This energy is manifested as the movement of the atoms and molecules comprising the substance (the speed of the molecules in a gas, for example). The higher the temperature is, the faster the atoms and molecules are moving. That is all very logical. But what is amazing is that the random motions of the countless individual molecules in a complex material like air, which is composed of hundreds of different species or compounds, can be described by a *single* quantity, the temperature of a gas (usually simply denoted, T). If we know the temperature of the gas, we know almost everything of practical value about the motions of the gas molecules and their energy content.

Thermometers are instruments that measure temperature. The first crude thermometers were invented in the early seventeenth century. All these

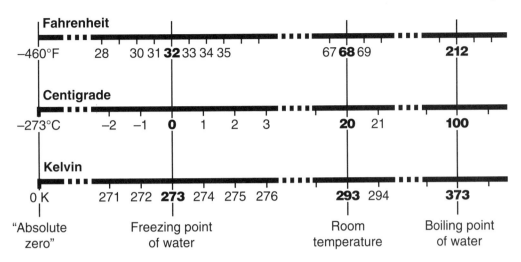

Figure 2.2 The three principal scales of thermometry. The Fahrenheit temperature scale is the oldest and most popular in English-speaking countries. The centigrade (Celsius) and kelvin scales are used by scientists today. These latter two scales are offset by about 273 degrees centigrade (Celsius).

early instruments were designed using the simple observation that most materials, including gases and liquids, expand when they are heated. However, it required considerable ingenuity to calibrate these devices to yield accurate temperature measurements. The three most common temperature scales are illustrated in Figure 2.2. The **Fahrenheit** scale was invented by the Polish-born physicist Gabriel D. Fahrenheit (1686–1736) in the early eighteenth century. It uses two fixed references of temperature: The lower temperature point was achieved by mixing together equal weights of snow and common table salt, or sodium chloride. That temperature is 0°F (note that the symbol ° means **degrees**). The upper-reference temperature was determined by placing the thermometer in the armpit of a healthy male (note the chauvinism of that time). Today, that temperature, considered the average normal temperature of a healthy adult, is 98.6°F.[1] On the Fahrenheit scale, pure water freezes into ice at 32°F and, at sea level, boils at 212°F. The **centigrade**, or **Celsius**, temperature scale was first clearly defined in 1742 by the Swedish astronomer Anders Celsius (1701–1744). The centigrade scale today uses as reference points the temperature at which pure water freezes into ice (0°C), and the boiling point of pure

water at sea level (100°C). Interestingly, when Celsius first introduced his scale, he reversed the temperatures, calling the boiling point 0°C!

On the Fahrenheit and centigrade scales, the temperature difference between the freezing and boiling points of water is divided into either 180 or 100 equal units, or degrees, respectively (note that *centi* means "one-hundredth"). The Fahrenheit scale is the one in common use in Great Britain and many of its former colonies (including the United States). However, scientists prefer the simpler centigrade temperature scale.

The conversion between Fahrenheit and centigrade temperatures follows a very simple rule: To convert from °F to °C, subtract 32°F and multiply by the fraction 5/9. To convert from °C to °F, multiply °C by 9/5 and add 32°F. Written as equations, these rules are simply:

$$°C = \left(°F - 32°\right) \times \frac{5}{9}$$

$$°F = °C \times \frac{9}{5} + 32° \tag{2.1}$$

Physicists and chemists usually use the **kelvin** scale of temperature, named in honor of the famous Irish physicist Lord Kelvin (Sir William Thomson, [1824-1907]). It is an absolute scale for the range of temperatures found in nature. The reference temperature is called **absolute zero**, or 0 K (note that in the kelvin system, the ° symbol is often omitted). Absolute zero is equivalent, in precise terms,

---

1. Fahrenheit originally set the upper-reference temperature as exactly 96°F. Later, when the freezing and boiling points of water were adopted as the reference points (armpits went out of vogue), the Fahrenheit scale was recalibrated, and everybody's temperature rose about 2.5 degrees.

to –273.16°C (that is, somewhat more than 273 degrees centigrade below freezing), or –459.69°F. Absolute zero is a strange place to be. At absolute zero, all motion, even that of the atoms in molecules, is frozen out—there is no heat left to be drawn away; it is as cold as it can possibly get. No experiment has ever reached 0 K because it is so difficult to extract every last bit of heat from a substance, but some experiments have reached temperatures close to 0.1 K! A simple relationship exists between the centigrade and kelvin scales: They are merely shifted by 273.16 degrees, or symbolically:

$$°C = K - 273.16 \qquad (2.2)$$

In all the following discussions, the kelvin or centigrade temperatures will be used, unless Fahrenheit is appropriate for other reasons.

The average speed at which an air molecule travels through space is determined by the temperature of the air (as well as the mass of the molecule). Indeed, the average molecular velocity is just the speed of sound in the gas, about 360 meters per second in air at the ground. The distance from a lightning stroke can be estimated by counting the seconds before thunder is heard. Each second is equal to 360 meters (about 1100 feet).[2] As the temperature of a gas increases, so does the molecular speed and the speed of sound. If the temperature of two samples of different gases is the same but one gas is composed of smaller, lighter molecules, then the speed of sound will be greater in that gas. Have you ever wondered why your voice sounds "squeaky" after you have inhaled gas from a helium-filled balloon? It is because helium gas is much lighter than air (remember, helium is used in balloons to make them buoyant, or lighter than air—more on buoyancy later) and the sound vibrations from your vocal cords move with greater speed and resonate differently in your throat and mouth with helium present rather than air.

Pressure

Pressure and force are related concepts. Pressure can be thought of as force per unit area, or total force divided by the surface area over which the force acts.

2. The light from the lightning flash travels at the speed of light, which is 300 million meters per second, nearly 1 million times faster than sound. At any normal distance, the light flash arrives "instantaneously," and the sound plods along behind.

One material exerts pressure on another at the boundary, or surface, separating those materials. **Force** is the agent of action. When you pitch a baseball, you exert a continuous force on the ball until the instant it leaves your hand. Force is the agent, acting over time, that accelerates the ball and directs it toward the strike zone. When you push against a wall, the situation is more complicated. You exert a force—and therefore pressure—on the wall, but it does not move. That is because the building, to which the wall is connected, exerts an equal and opposite force on the wall sufficient to keep it stationary.

Gases and liquids also produce force and pressure. If you have ever been toppled in heavy surf or stood in the howling wind of a hurricane, you will appreciate the strength of this force. Even when a fluid is at rest, it exerts force and pressure. If a gas is confined in a container, the bombardment of the surface of the container by fast-moving gas molecules creates a force on the wall. As you might suspect, the pressure force increases as the number of molecules—or their concentration—bombarding the surface increases. Pressure also increases as the speed of the impinging molecules—or as we now understand it, their temperature—increases.

In the case of the atmosphere, the pressure exerted by air at any point is actually a measure of the weight of the air above that point. The upward pressure of the atmosphere must exactly balance the downward force caused by the weight of the overlying air. Gravity acting on the overlying air produces the force, which is just the weight of the gas. Imagine that the pressure is too low to support the overlying air. Then the air will be compressed, its concentration will increase, and its pressure will build up until the weight above is balanced. The same effect occurs in water, except that liquids are not very compressible, so they effectively resist being squeezed by building up their internal force (pressure) with little volume contraction. Because liquids are much denser and heavier than gases, pressure builds up rapidly with depth under water. A 10-meter-deep column of water weighs as much as the entire column of air above it in the homosphere: 100 kilometers in depth.

The pressure of the atmosphere at sea level is 14.7 pounds per square inch (of surface area), or about 10 metric tons per square meter. The surface area of your body is roughly 1 square meter, so the atmosphere is pressing against you with a total force of 10 tonnes! Why aren't you crushed? Because the fluids and solids in your body are exerting that same pressure against the atmosphere. As the air squeezes against you, your body compresses slightly to balance that pressure, as explained for the case of water. Your lungs do not collapse because your mouth (or nose) is always open (somewhat), allowing air to rush in (or out) to equalize the pressure inside your lungs with that of the atmosphere.

An aquanaut diving at a depth of 10 meters while attempting to breathe through a snorkel connected to the surface would experience a crushing force of about 1 tonne on his lungs (which would probably collapse as a result) (Figure 2.3). That is why divers must breathe highly pressurized air at such a depth and must use special control valves that accurately match the internal air pressure with the external water pressure. At the other extreme, an astronaut in space who is not wearing a pressurized suit but is breathing air from a tank pressurized to 1 atmosphere (the normal pressure of air at sea level) would experience an explosive force of about 1 tonne within their lungs (which would most likely burst). So, an astronaut's suit must be pressurized and the suit itself must be extremely strong to withstand the explosive forces caused by 1 atmosphere of pressure.[3]

You can experience the power of atmospheric pressure very simply by sucking on a paper cup (or fast food drink container). The experiment is sketched in Figure 2.4. By forcing your muscles to expand your lungs, you increase the volume of your lungs. The quantity of gas trapped in the cup and lungs is fixed. Hence the gas must expand with the increasing volume, reducing its concentration. But as the concentration in the cup decreases, so does the pressure in the cup. The pressure of the atmosphere just outside the cup remains constant—nothing in the experiment has caused it to change. The result is a pressure difference between the inside and the outside of the cup, which translates directly into a net force pushing inward on the cup. Eventually, this pressure difference (net force) becomes large enough to collapse the cup. Actually, atmospheric pressure is more than powerful enough to crush a metal can (refer to the demonstration of the pressure force in Appendix B).[4]

3. Actually, astronauts can breath pure oxygen, which comprises about one-fifth of air, at a reduced pressure of one-fifth of an atmosphere, which relaxes the strength requirements on the space suit.

4. A sealed bag of peanuts brought onto an airplane will swell up as the plane climbs in altitude and the pressure in the cabin

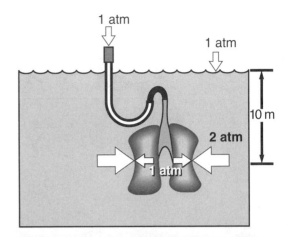

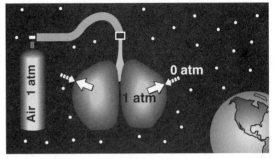

Figure 2.3 The effects of pressure, for someone submerged in 10 meters of water and for someone in space. In both cases, the internal pressure of the lungs is assumed to be maintained at 1 atmosphere. In space, there is no external pressure, and the internal pressure creates an outward force. Under water, the external pressure always exceeds 1 atmosphere, producing a net crushing force.

Volume

Volume can be thought of simply as a piece of space. In practical situations, any volume is occupied by substances (gases, liquids, and solids). To obtain the volume of a room, we multiply its length by its width by its height. If all these dimensions are measured in units of meters (Appendix A), then the volume calculated has units of (meters) × (meters) × (meters), or **cubic meters**, or **meters cubed** = m^3. In discussing air pollution and the global environment, we will be interested in volumes that range in size from small (the capacity of your lungs, about one-thousandth of a cubic meter) to large (the

drops. The air trapped in the bag tries to expand to equalize the pressure inside and outside. A bag of potato chips in your backpack similarly expands as you hike to the top of a mountain. Unfortunately, the calories do not evaporate.

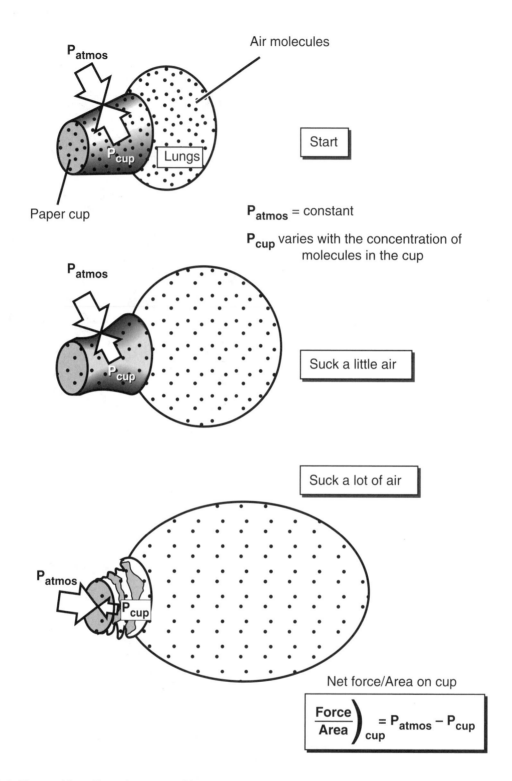

Air molecules

Patmos

Pcup

Lungs

Paper cup

Start

Patmos = constant

Pcup varies with the concentration of molecules in the cup

Patmos

Pcup

Suck a little air

Suck a lot of air

Patmos

Pcup

Net force/Area on cup

$$\left.\frac{\text{Force}}{\text{Area}}\right)_{cup} = P_{atmos} - P_{cup}$$

Figure 2.4 The crushing effect of pressure. After a paper cup is placed over the mouth and the nose is held shut, the lungs are expanded, drawing air from the cup and decreasing the pressure in the cup relative to the atmosphere, until the cup is crushed by the external force. The concentration of air molecules trapped in the cup and lungs is indicated by the density of dots: The number of dots is constant, although the change in the volume of the "lungs" shown is greatly exaggerated.

space filled by the sensible atmosphere, roughly 10 million trillion cubic meters).

Density

The *density* of a gas, or indeed any liquid or solid, is simply a measure of the mass of that material present in a specified volume of space. The units of density are mass/volume (for example, kilograms/meter cubed, or kg/m^3). If M is the mass (kg), V the volume (m^3), and ρ the density (kg/m^3), then the relationship between these quantities is

$$\rho = \frac{M}{V} \qquad (2.3)$$

The element mercury (13,600 kg/m^3) has a much higher density, than water (1,000 kg/m^3). Thus a bottle of mercury is much heavier than the same bottle filled with water. When the bottle holds only air (1.3 kg/m^3), it is much lighter yet (in fact, the weight of the bottle is all you can sense). All the air in an average room weighs about 40 kilograms, or 90 pounds. If the room were filled with water, the weight would be about 30 metric tons. Here and throughout the text, **metric tons**, or **tonnes**, will be the adopted unit for large amounts of mass. One metric ton equals 1000 kilograms. The standard **English ton**, or **ton**, equals 2000 English pounds; 1 metric ton is about 2200 English pounds—pretty close to an English ton.

If you have studied science in the past, you will be aware that when discussing density, we used the quantities mass and weight interchangeably. **Mass** is an absolute measure of the amount of a material and is the product of the density and volume of the material. **Weight** is the measure of the force of gravity acting on a given mass. At the surface of the Earth, gravity has a certain value that is nearly constant everywhere on the planet. So measuring the weight of, or gravitational force acting on, a body is equivalent to measuring its mass. In our discussions, we will not differentiate between mass and weight unless there is a critical ambiguity in the usage.

The density of a substance can be determined by weighing a sample of the substance and dividing that by the volume or size of the sample. For gases, the resulting density can be expressed in other ways, which sometimes are more convenient. Two other useful measures of gas density are the **concentration** and the **mixing ratio**.

Concentration

Imagine that in a particular sample of gas, rather than weighing the gas itself to determine the total mass of the molecules in it, we instead count the number of molecules in the sample and divide that total number by the volume of the sample. We have then calculated the gas concentration, having units of number of molecules per volume (number/volume, or molecules/cubic meter = number/m^3). If N is the total number of molecules in the sample and V is the volume, then the concentration, n, is

$$n = \frac{N}{V} \qquad (2.4)$$

There is also a simple relationship between the density and concentration. Every molecule has a well-defined mass. When each molecule in a gas has mass, m (kg), the gas density is proportional to the concentration as

$$\rho = nm \qquad (2.5)$$

Obviously, the mass of a single molecule is very small, but it is not zero, and the number of molecules is usually enormous (a shorthand representation of very large and very small numbers in "**scientific notation**" is discussed below.

Mixing Ratio

A quantity that is closely related to the concentration and density of a gas is its mixing ratio (Section A.1). A typical sample of air contains many species, as we have seen. The mixing ratio defines the relative amount of a specific gas in a bulk sample of air. It may be specified in terms of either the relative mass (the mass mixing ratio) or the number of molecules (number, or volume, mixing ratio). Suppose, for example, that we measure the concentration of carbon dioxide molecules in a sample of air, n_{CO_2} (molecules/m^3) and we know the total concentration of air molecules in the sample, n_A (molecules/m^3). The mixing ratio, or fraction, of CO_2, r_{CO_2}, is then

$$r_{CO_2} = \frac{n_{CO_2}}{n_A} \qquad (2.6)$$

Calculating this fraction, one would find that r_{CO_2} is about 0.000365. Notice that if we multiply this ratio by some large number—say 1 million—we will obtain a reasonable number, like 365. Another way of stating this is that there are 365 CO_2 molecules for every 1 million air molecules. Because the ratio is taken relative to the total concentration of air molecules, r is usually expressed as the mixing ratio "by volume." The carbon dioxide mixing ratio is thus 365 **parts per million by volume** (**ppmv**). The relationship between the concentration of molecules and their volume may not be obvious to you. Suffice it to say at this point that in a normal mixture of gases, each molecule, regardless of type, occupies the same volume of space. Mixing ratios by concentration or number is then analogous to mixing ratios by volume (occupied by the gas molecules).

The same arguments leading to Equation 2.6 can be used if densities are substituted for concentrations. The density of carbon dioxide is related to the mass of a CO_2 molecule, m_{CO_2}, and the CO_2 molecular concentration by

$$\rho_{CO_2} = m_{CO_2} n_{CO_2} \qquad (2.7)$$

The density of air is a known quantity and can be related to the total concentration of air molecules by an expression like Equation 2.7. This mixing ratio would be stated as so many parts by *mass*, rather than by volume (Section A.1). The mass mixing ratio of CO_2 is about 550 **parts per million by mass** (**ppmm**).

There are other practical ways to think about mixing ratios. When you concoct a recipe, you are mixing substances in specific proportions, or mixing ratios. By adding an ounce of salt to a pound of flour, you have created a very unhealthy batter with a salt mass-mixing ratio of 1/17. Watch out for hypertension! Life is full of useful mixing ratios that you may not even recognize.

The most useful aspect of the mixing ratio is its constancy for any constituent of air that has a moderately long lifetime and can therefore be uniformly mixed throughout the atmosphere. Thus carbon dioxide has a mixing ratio of about 365 ppmv everywhere in the homosphere. There is no need to measure separately its concentration at each altitude. Only *one parameter* is needed to define completely the entire global distribution of CO_2.

Large and Small Numbers

Scientific discussion may use numbers that are either very large or very small. Examples are the total number of molecules in the Earth's atmosphere, 10^{44}, and the fraction corresponding to a mixing ratio of one **part per trillion by volume** (**pptv**), 10^{-12}. To obtain a feeling for the small fraction that 1 pptv represents, consider that this is equivalent to 1 cent out of 10 billion ($10,000,000,000), or roughly 1 drop of sweetener to 1.5 million gallons of lemonade (a *very* sour cooler, indeed).

If we had to write out the number representing the number of molecules in the atmosphere, it would read 100,000,000,000,000,000,000,000,000, 000,000,000,000,000,000. That is, the digit 1 followed by a string of 44 zeros! The number representing one part per trillion would read 0.000000000001. Because these longhand numbers are clumsy and impractical, in the rest of this text we use **scientific notation**. (This notation is described in Section A.1, where useful information on units of measure and conversion factors between different systems of units is also provided). In scientific notation, we would write the product of the two long numbers simply as, $(1 \times 10^{44}) \times (1 \times 10^{-12}) \times (1 \times 10^{32})$. This is obviously much more compact than writing out the actual numbers themselves. Scientific notation is particularly useful when multiplying and/or dividing large and small numbers.

An amazing fact emerges when dealing with the global environment. Even very small percentages of a specific material in a very large domain can translate into huge amounts of that material. In the Los Angeles area, for example, about 3 million acre-feet of water (about 4 billion metric tons) are used every year. Imagine that there are contaminants in the water with mixing fractions of only one part per billion by mass (ppbm, or 10^{-9}). Certainly this is a small amount. But is it? If the contaminant were lead, there would be enough to make 1 million bullets; if insecticide, enough to fill 5 million cans of bug spray; and if mercury, enough to fill 4 million rectal thermometers.[5] These are impressive quantities. We will frequently come across such ponderous numbers in the following discussions.

2.2 A Short History of Discovery

The history of the discovery of air and its composition covers many centuries and cultures. It has been

---

5. These figures were estimated by Dr. Warren B. Crummett of the Dow Chemical Company.

a slow, painstaking process filled with misinterpretations and errors, but also with brilliant insights and discoveries. The original scientific ideas about air can be traced to the Greek philosophers, with the most recent fundamental work occurring at the end of the nineteenth century. In this section, we summarize some of the key discoveries regarding the composition and physical properties of air. The roster of scientists and events described here is far from exhaustive. The discussion, however, illustrates the nature of the scientific process, which uses theories and hypotheses to explain observations and guide new experiments designed to test a theory or suggest alternative concepts.

2.2.1   THE AIR REVEALED

In their leisure, the Greek philosophers pursued their deep curiosity of the natural world. Empedocles, near the beginning of the fifth century B.C., proposed that all material things in nature were composed of four eternal and unchanging **elements**: earth, water, fire, and air.[6] These were things that could be easily sensed and seemed universal. Everything in the physical world was assumed to be composed of mixtures of the four elements. Earth could be felt and provided the basic materials for building temples and growing food. Water sustained life and cleansed the environment; the seas stretched to the ends of the Earth. Fire was crucial to ancient life and, at the same time, unique and mysterious: Why did things burn? Fire, it was thought, must be present in those materials that are flammable and must be released when they burn. Air sustained the winds and clouds; indeed, it must have been a ghostly substance to the ancients, who were not able to determine very well its real properties. Accordingly, the early Greeks considered air to be one uniform substance.

Aristotle (384–322 B.C.) was a truly brilliant thinker in many ways. He adopted Empedocles's four-element theory. As a scientist, he deduced that water vapor was a distinct component of air. He reasoned that water must evaporate from the Earth's surface, travel upward as a vapor, and then condense into clouds, from which the water fell back as rain.

He realized that rain could not be continually supplied from the sky above and that the water falling from clouds was indistinguishable from the water in lakes and streams. Water, being a basic element, must be conserved and therefore had to be recycled. Aristotle was able to theorize that water must take at least two forms: liquid and vapor. In the vapor state, it followed, water must be a partial and variable constituent of air.

Fixed, Mephitic, and Vitiated Air

It was many centuries after Aristotle before additional progress was made in understanding the composition of the atmosphere. In seventeenth century Europe, physicians and natural scientists were becoming interested in air and undertook many novel experiments. Fire was often used as a means of processing air to separate the components, and the nature of fire itself was a subject of controversy. John Mayow (1641–1679) determined in 1676 that air consisted of at least two distinct components, one of which sustained fire and life and the other of which did not. He named the former *fire-air* and wrote that this air was necessary for respiration. Much later, fire-air was quantitatively defined as oxygen (O_2).

An early explanation of the major findings in chemistry to that time was the phlogiston theory, first advanced by Johann Joachim Becher and formalized by George Ernst Stahl around 1700. The theory proposed that a substance called **phlogiston** (from the Greek for "to set fire") permeated all combustible materials. During combustion, the phlogiston was released, and the burned residue, being depleted of phlogiston, would not sustain combustion (or life, as it turned out). This erroneous theory remained the basis for much scientific investigation and debate until the end of the eighteenth century.

Joseph Black (1728–1799) was a chemistry professor who experimented with treatments for urinary stones, or calculi. Black and others believed that the stones, which are calcified mineral deposits, could be dissolved by certain potions that remained undiscovered. Nevertheless, in his experiments, begun in 1752, Black noted that by heating a compound, known today as magnesium carbonate, a heavy gas was produced that neither supported a flame nor sustained animal life. The gas was carbon dioxide (CO_2). Black, however, named it fixed air, because it seemed to be attached, or "fixed," to certain natural

6.   The modern understanding of the chemical elements was developed by Robert Boyle in 1661, as discussed later in this section.

compounds—at the time called calxes—and was released when these compounds were heated or burned. The calxes, he reasoned, were formed by the natural presence of fixed air in the atmosphere. Black also developed a technique for selectively capturing and measuring carbon dioxide from gas samples using common lime, a calcium compound that reacts with CO_2 to form calcium carbonate.

The discovery of nitrogen was made soon thereafter by Daniel Rutherford (1749–1819). He burned materials in air until nothing more would sustain combustion. Then he removed the carbon dioxide (fixed air) using Black's technique. Rutherford found that most of the air remained. When tested for its ability to sustain life, the gas failed miserably. Rutherford therefore called it mephitic air (nitrogen) after the mephitis, a noxious or foul exhalation from the earth. Today we call it nitrogen (N_2). Rutherford's work is the earliest published research identifying nitrogen as a component of air and clearly differentiating it from Black's fixed air. Rutherford also believed that phlogiston, the mysterious substance long associated with fire, was a component of mephitic air, although he offers no quantification on this point. Phlogiston was assumed to be emitted during combustion, and its accumulation in a closed volume extinguished flames; that is, phlogiston was antifire. The effect was demonstrated by placing a candle in a closed glass vessel. Soon the candle flame was snuffed out. The assumed cause was a buildup of phlogiston, but the real cause was a depletion of fire-sustaining oxygen.

Henry Cavendish (1731–1810) had a long and distinguished career studying the composition of air. In 1766, he published a paper describing "factitious airs" capable of being freed from materials and made "elastic." He isolated several such airs. Together with the work of Black on carbon dioxide, these experiments showed that air was a complex mixture of compounds. Cavendish also experimented with nitrogen, much as Rutherford did, although the details of these results were not published. Later he contributed to the discovery of trace compounds in air by devising a method for removing nitrogen from Rutherford's mephitic air. Although mephitic air was mostly nitrogen, it still held small quantities of other gases. Today, we know that the residue in fact consists mainly of argon, although that discovery came much later.

Joseph Priestley (1733–1804) could both preach and do science, but he is remembered mainly for the latter. Among his interests in life was a desire to understand the properties of air. Priestley developed a technique to distill the fire-bearing component of air. He burned mercury in air, forming mercuric oxide, which contains oxygen extracted from the air. By heating the mercuric oxide in an evacuated container, the oxygen could be released again in an almost pure state, with a concentration about five times that of natural air (which is about 20 percent oxygen [Table 2.1]). Recall that earlier researchers had *removed* the fire-air through combustion and studied the remaining "fixed" and "mephitic" airs. Priestley's approach was fundamentally different. In this new gas, flammable materials burned with an intense brightness. In 1774, Priestley realized he had isolated the source of combustion, and named it *dephlogisticated air* (later to be named oxygen [O_2] by Lavoisier). Priestley reasoned that since flames burned so brightly in his distilled air samples, it must initially be totally free of phlogiston.

It happens that two other notable scientists of Priestley's time had also isolated oxygen from air: Carl Scheele in 1773 and Antoine-Laurent Lavoisier in 1789. Scheele experimented with fire-air and the residue, which he called *vitiated air* (essentially Rutherford's mephitic air). Scheele also believed in phlogiston but, unlike Priestley, thought his fire-air contained it.

Lavoisier and Modern Science

Antoine-Laurent Lavoisier (1743–1794) placed the chemistry of air on a sound scientific basis, although he is not recognized as making any fundamental discoveries regarding the composition of air. Remarkably, he was the first scientist to point out that all substances can exist in any one of three states—solid, liquid, or gas—depending on the temperature and pressure of the substance. He originally named Priestley's gas *eminently respirable air*, but later called it *oxygen*, from the Greek for "begetter of acids," because he showed that oxygen is present in the common acids of nitrogen, sulfur, and phosphorus (Section 3.3.4 and Chapter 9). In 1777, Lavoisier first announced, in his *Memoirs on Combustion in General*, that fire, long associated with the ethereal substance phlogiston, is actually caused by reactions between oxygen and other elements. He, with the famous mathematician Pierre Laplace, demonstrated that animal respiration involves the consumption of oxygen from air, with the release of carbon dioxide and heat energy; that is, respiration is a form of "slow combustion." Accordingly, Lavoisier is connected with many of the aspects of chemistry that are relevant to air pollution and the environment.

Lord Rayleigh (William Strutt [1842–1919]) discovered argon (Ar) with Sir William Ramsay in 1894 and named it *inactive air* because of its chemical inactivity (later, it was labeled argon, from the Greek for "idle or lazy"). Today we know that argon is the most common of a family of very inert gases referred to as **noble gases** (in part because they are immune to reaction with oxygen and cannot be burned under any circumstances).[7] Since argon comprises only about 1 percent of air and is chemically inactive, its discovery required unusually precise measurements. In fact, in 1795 Cavendish had isolated a residue of mephitic air that could not be oxidized no matter how much sparking the gas was exposed to. Unlike Cavendish's chemical approach to the problem, Rayleigh and Ramsay identified this residue by carefully determining the density of mephitic air and comparing it with the density of pure nitrogen prepared from ammonia (a nitrogen compound often used in household cleaning solutions). The pure nitrogen was slightly lighter, and they deduced that the mephitic air must contain a small amount of a heavier gas—argon. The discovery was met with skepticism, however, because the chemists of that day could not believe that a genuinely new element had gone undiscovered for so long. Even objective scientists sometimes fall victim to denial when faced with an embarrassing oversight. But Lord Rayleigh prevailed and was awarded the Nobel Prize in physics in 1904.

Rayleigh was an extremely resourceful physical scientist who made numerous fundamental contributions to the understanding of sound, light, and gases that were far more important than his discovery of argon. For example, Rayleigh also discovered hydrogen in the atmosphere and provided the first explanation of the blueness of the sky (Section 3.2).

Although Sir William Ramsay (1852–1916) is known for the discovery of argon with Lord Rayleigh, he also identified a number of other inert noble gases—helium (He), neon (Ne), krypton (Kr), and xenon (Xe)—in the atmosphere (Table 2.1). Ramsay's technique was very clever: He literally froze the air. At very low temperatures ($-57°C$ for carbon dioxide,

$-183°C$ for oxygen, and $-196°C$ for nitrogen), the common atmospheric gases condense. However, the noble gases do not liquefy until even lower temperatures ($-186°C$ for Ar, $-246°C$ for Ne, and $-269°C$ for He). Ramsay was thus able to separate the gases by means of cryodistillation. This creative work won him the Nobel Prize in chemistry in 1904 (the same year Rayleigh won the prize in physics).

The discovery of **ozone** (O_3) in the upper atmosphere was made by John Hartley in 1881. In the late 1830s, German chemist Christian Friederich Schonbein (1799–1868) had identified a new compound, ozone, which he produced in various laboratory experiments using oxygen. Schonbein noted the distinctively sweet smell of ozone and probably named the gas for the Greek *ozien* (to smell). Hartley later experimented with samples of ozone to determine how it absorbs light. In 1881, Hartley identified the most important absorption band in ozone (named appropriately enough the **Hartley band**), which controls ultraviolet radiation in the Earth's atmosphere (Sections 3.2 and 13.4). It occurred to Hartley that if ozone were present in the atmosphere in significant amounts, the spectrum of sunlight reaching the ground would reveal telltale signs of ozone absorption in his new band (in 1880, spectroscopic work in the visible spectrum by M. J. Chappuis also demonstrated that ozone was present in air). Hartley made the important discovery that the ultraviolet limit of the sunlight spectrum measured by M. A. Cornu at the ground in 1879 coincided exactly with his absorption band. From these observations Hartley deduced that a relatively thick layer of ozone must reside in the upper atmosphere (Chapter 13).

Some 90 years later, concern about depletion of the Earth's ozone layer—which protects life against the deadly ultraviolet radiation of the sun—launched an era of global environmental action. We also mention that Hartley's discovery is one of the first instances in which an atmospheric constituent was measured quantitatively through **remote sensing**, in this case by determining the absorption of sunlight. That is, the ozone that Hartley discovered was not in his laboratory; it was 20 kilometers over his head!

The early discoveries regarding the composition of the atmosphere were painstakingly slow. Three centuries of scientific inquiry led to the identification of only about a dozen constituents (Table 2.1). In the early years, these were called airs of various sorts (fire-air, mephitic air, and so on). The more precise chemical characterization of the constituents of air

---

7.   Originally, chemists used the term *Greek ozien* to describe very inert and unreactive metals, particularly gold and platinum.

required the development of the atomic theory of chemistry and the law of proportions of elements in compounds by John Dalton and others in the late eighteenth century. Reflecting on the history of the composition of air, however, the sequence of discovery is very logical. First to be identified were the major constituents: nitrogen (78%), oxygen (21%), and water vapor (up to several percentage points). The most reactive constituents were also subject to isolation and early study, particularly oxygen (21%) and carbon dioxide (~0.0003%). The last constituents to be uncovered were the inert trace gases, beginning with argon (1%), the most abundant, followed by helium, neon, and so on.

The pace of discovery of the atmosphere's composition has accelerated in the latter half of this century, with the advance of technology and growth of interest in the environment, and the training of armies of scientists and technicians with the skills to carry out sophisticated atmospheric experiments. Unlike the early pioneers of natural science and chemistry, today's researchers are equipped with precision instruments and can draw on an enormous reservoir of accumulated data.

2.2.2 THE MECHANICS OF AIR

In parallel with the discoveries concerning the chemical composition of the atmosphere, advances were made in understanding the behavior of gases and mixtures of gases like air. These discoveries regarding the physical properties of air were as important as those regarding its composition, because they allowed chemists to carry out very accurate measurements of gas samples. A few of the key discoveries are summarized next.

Boyle and Charles

Robert Boyle (1627–1691), an English chemist, and other natural scientists of his time experimented with the physical behavior of gases held in containers. A sample of gas was placed in a closed vessel with a plunger, for example, and the relationship between the force, or pressure, on the plunger and the volume of the confined gas was measured. These early researchers—we might think of them as the original physicists—found that three parameters were critical to their experiments: the temperature of the gas, the volume of the gas, and the pressure exerted

by the gas. Their discoveries defined the relationships among these basic properties of gases (Section 2.2.1). The quantitative relationships will be given later (Section 3.1.1).

Boyle—considered by some historians of science to be the "father of chemistry"—determined in 1662 that the pressure exerted by a fixed amount of gas confined to a chamber varies inversely with the volume of the chamber (that is, the volume occupied by the gas) if the temperature is held constant during the experiment. For example, if the volume of a sealed chamber is halved by inserting a plunger, the pressure within the chamber is doubled. The change in pressure can be measured in terms of the force, or weight, required to hold the plunger in. If the plunger, on the other hand, is retracted, so that the final volume of the container is twice the original volume, the pressure will be one-half the starting value. The quantitative relationship between pressure and volume is known as **Boyle's law**. The experiment is illustrated in Figure 2.5(a).

Boyle's law is a logical result, in accord with our earlier discussion of pressure. If the volume of the container is reduced, the concentration of the gas in the container will increase, and so, in proportion, will the pressure on the walls of the container. Boyle, in fact, believed that gases were composed of tiny particles, or atoms that behaved like tiny coiled springs and rebounded from the walls of a container. This kind of simple conceptual picture may help you visualize the mechanics of a gas and its pressure effect.

Jacques Charles (1746–1823), a French physicist, discovered in 1787 that the volume of a fixed quantity of gas varied in proportion to the temperature of the gas if the pressure of the gas were fixed. Like Boyle, Charles performed a series of careful experiments to derive this relationship, known as **Charles's law**.[8] Figure 2.5(b) sketches a simple experiment that demonstrates this law.

Charles's curiosity about the relationship between the volume of a fixed amount of gas (and thus its density) and the temperature of the gas arose from his interest in ballooning. Charles may be thought of as the first hot-air ballooning specialist, pointing out many concepts still used in modern ballooning. He also thought of the idea of using hydrogen gas—the

---

8. John Dalton, who was studying the mechanical behavior of common gases at the same time as Charles, independently discovered the relationship between gas volume and temperature that later became known as Charles's law.

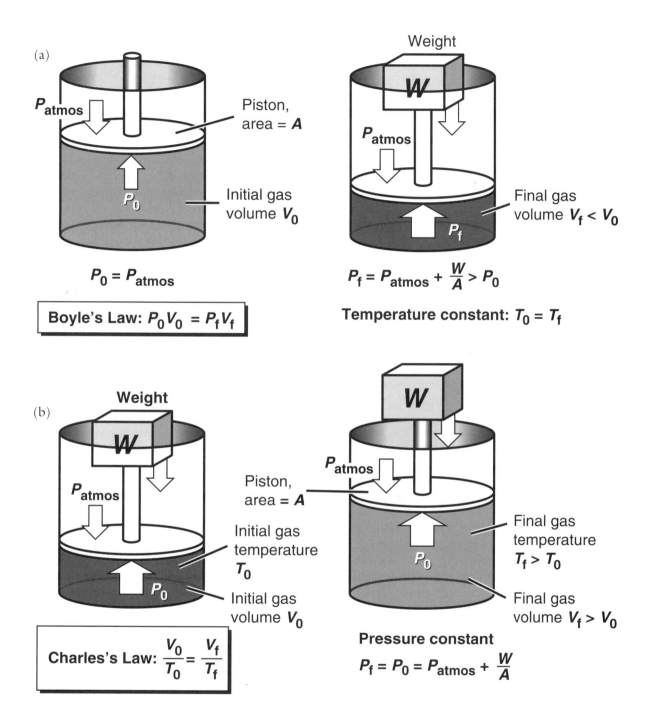

Figure 2.5 The simple gas experiments by Boyle (a) and Charles (b). In these experiments, the piston is assumed, for simplicity, to have negligible weight. The applied external weight, W, produces a pressure (force per unit area) given as W/A, where A is the area of the piston. In Boyle's experiment, the temperature is held constant (any temperature may be used). In Charles's experiment, the pressure is held constant by maintaining an appropriate force on the piston.

lightest natural gas—in balloons to replace hot air. That practice ended tragically with the explosion of the dirigible *Hindenberg* over the Lakehurst Naval Air Station in Lakehurst, New Jersey, on May 6, 1937. Nowadays, large blimps are filled with helium, the second lightest natural gas. Being a noble gas, helium will not burn and so is safe to use in large quantities.

Dalton and Atoms

John Dalton (1766–1844) was another English chemist who dabbled in gases. In 1793 he discovered that the total pressure of a mixture of different gases was simply the sum of the pressures of each of the component gases, for a constant temperature and volume of the gases. The quantitative relationship is known as **Dalton's law of partial pressures**. Again, such a relationship seems intuitive. If we add different gases one at a time to a container, the total concentration of gas in the container at each step will increase, and so will the total pressure. Moreover, the same result is expected regardless of the order in which the gases are added, as long as the same amount of each gas eventually ends up in the container. If we had originally held the gas samples in identical containers, we could have measured the pressure of each gas in its container. If we then transferred all the gases into one container and measured the total pressure, it would equal the sum of the pressures of each component measured separately. This is what Dalton deduced quantitatively.

During the period of discovery of the physical behavior of gases, ideas about the very nature of matter itself were being developed. In Greece in the sixth century **B.C.**, Anaximenes of Miletus concluded that water was the only basic element. Empedocles later proposed four elements: water, earth, fire, and air. It was also widely believed that matter was continuous in nature; that is, however finely a material was subdivided, it would continue to display the same continuous properties. This perception was in part the result of the impracticality, using even the sharpest and most precise tools, of dividing matter until it is no longer indivisible. Long before that point, the pieces would become invisible! Nearly two millennia passed before the true mechanical nature of matter and the existence of distinct **elements** and discrete **atoms** became widely accepted. Robert Boyle was an early advocate of this so-called "mechanical theory of matter." This theory assumes the existence of simple microscopic units of matter,

or atoms. The atoms consist mainly of empty space lying between the dense nucleus and a swarm of orbiting electrons (see Section 3.1). In a gas, the atoms comprise only a minuscule volume of the total space occupied by the vapor.

Boyle established the modern usage of the word *element*, describing elements as "primitive and simple, or perfectly unmingled bodies." By the mid-nineteenth century, about 60 true elements had been discovered. Dalton played a major role in categorizing these elements. He had found that every element in its pure form had a particular weight differing from that of all the other elements. Dalton made the first crude arrangement of these elements in a table according to their weight. Indeed, as Dalton noted, each of the known elements could be assigned a unique **"atomic number"** based on its relative weight. The heavier the element, the larger its atomic number (see in addition Section 7.3.1, where the relation of atomic number to the radioactivity of elements is discussed).

Later, in 1869, the Russian chemist Dmitri I. Mendeleyev undertook the task of organizing the known elements more systematically into groups and families with similar characteristics, thereby establishing the modern **periodic table** of the elements. Mendeleyev noted that the properties of the elements seem to repeat in regular patterns as the atomic number of the element increases. Such properties, for example, include the degree of reactivity with certain reagents, ability to conduct electricity, and so on. Elements can therefore be divided into subgroups in various ways based on their observable behavior. Mendeleyev cleverly noted that the atomic number could be used to arrange the elements neatly into rows and columns. First, the elements are listed sequentially starting at atomic number one. Next, this long sequence is broken into a series of segments at special points along the chain. Finally, the segments are arranged as rows placed one above the other in a table. Mendeleyev's genius was in locating the break points in such a way that, in each resulting column, all of the elements have roughly similar characteristics. Moreover, the atomic numbers of the elements in a given row differ by values of 2, 8, 18, or 32 from those in adjacent rows.[9] Recognition of these fixed numerical relationships had a substantial influence on the development of atomic theory and

---

9. The actual periodic table of the elements is somewhat more complicated in form, because of the greater number of elements (111 discovered so far) that need to fit into the basic

modern physics. Indeed, such a simple scheme that allows one to arrange all of the basic building blocks of matter throughout the universe is quite remarkable.

Dalton made other important contributions to both chemistry and natural science. He was the first to deduce that elements combined in simple proportions to form compounds, and that the combination corresponding to any pure compound is always the same. For example, in carbon dioxide, the ratio of carbon to oxygen is *always* one-to-two (1:2). Even though, by Dalton's time, many basic elements had been identified, the nature of chemical reactions that occurred between different elements and of compounds that such reactions produced remained uncertain. Because compounds that could be decomposed and analyzed for their elemental basis were always found to have fixed relative amounts of specific elements, Dalton argued convincingly that the elements themselves must be composed of indistinguishable microscopic units, or atoms, consistent with the mechanical theory of matter (as pointed out earlier). He further concluded that the atoms of each element also had a fixed weight. Hence, the weight of any compound could be figured immediately on knowing its elemental composition.

These discoveries led directly to the modern language of chemistry, in which the proportions of the elements in a compound are expressed explicitly through chemical formulas. For example, carbon dioxide, or CO_2, (Table 2.1), consists of one atom of carbon and two atoms of oxygen; the proportion of carbon to oxygen is 1:2. For methane, with one carbon atom and four hydrogen atoms, the chemical formula is CH_4. The composition is *always* this way in methane, a C:H ratio of 1:4. A more complete discussion of chemical nomenclature is in Section 3.3.1.

Dalton was also one of the first brilliant meteorologists of vision. In his classic work, *Meteorological Observations and Essays* (1793), he offered an explanation of the aurora borealis and speculated about the cause of variations in barometric pressure. He also elucidated his ideas on the evaporation of water vapor into air, which was the seed for the atomic theory of matter. In that same work, Dalton described a theory of the trade winds proposed earlier by George Hadley (Section 2.4.2). Obviously, Dalton's interests in atmospheric science were far-ranging. His contributions to the study of air were both fundamental and profound.

2.3 The Structure of the Atmosphere

The atmosphere has not only a certain composition but also a certain structure and certain motions. Those will be discussed in the following sections. A fundamental question that arises is why there is an atmosphere at all. In Section 2.1.3, we introduced the concepts of temperature and pressure and noted that air is composed of molecules moving, on average, at the speed of sound. The molecules are continuously colliding with one another and, at the ground, cannot travel more than a ten-millionth of a meter before hitting another air molecule. These countless collisions produce the tenuous fluid that we sense (in its bulk state) as air.

Imagine that you can carry a spherical container of air far into space on a rocket and that you can quickly remove the containment, leaving a ball of air free in space. What will happen next? The air molecules in the ball are instantaneously moving in all directions because of thermal agitation. However, about half those at the edge of the ball, where the container had held them an instant before, are moving outward at the speed of sound. But there is no gas or other obstruction in space to inhibit their free flight; in one second those air molecules will have shot some 350 meters away (assuming that the air is initially at the temperature of Earth's surface). The ball of air literally explodes into space and is gone in seconds!

Our atmosphere is precisely like a ball of gas exposed to free space. Why doesn't it explode away from the Earth?

---

pattern. The top three rows of the table have been split because they contain only 2, 8, and 8 elements, respectively, whereas the next three rows contain 18 elements each. The seventh row is incomplete because the heaviest elements detected fall short of the end. Elements above 92 do not normally appear in nature, but can be artificially generated up to atomic number 111. In addition, the elements from 58 to 71 and 90 to 103 form two distinct sequences of fourteen elements each (note that 14 is just the difference between 32 and 18, two of the basic numerals separating the elements in adjacent rows in the table); when these two subsequences are placed one over the other (outside of the main table, although they may be inserted neatly in rows six and seven, respectively), the atomic numbers of each pair of adjacent elements differ by 32 units. A perceptive individual might conclude that the magic numbers 2, 8, 18, and 32 are intimately related to the internal structure of atoms themselves and express a more fundamental universal law of nature. Indeed, this was the brilliant feat of quantum mechanics, the science of phenomena at atomic scales established by Max Planck, Niels Bohr, Max Born, Werner Heisenberg, Louis de Broglie, Paul Dirac, and Erwin Schrödinger, among others.

Three important factors determine the overall structure and stability of our atmosphere: the total amount of gas present, the temperature of the gas, and the gravity of the Earth. It is the last factor that prevents the atmosphere from leaving us at high speed. If you drop a billiard ball, it will fall to the ground and stay there. Air molecules experience the same force of gravity. But, the molecules do not fall to the surface and stay there like a billiard ball because they are continuously agitated by the thermal motions of other air molecules. The gas is thus pushed up against the force of gravity. If the atmosphere were at a temperature of absolute zero, the thermal motions of the molecules would cease, and the atmosphere would be pulled to the surface by gravity and deposited there as solid nitrogen oxygen ice. If the air temperature were raised, on the other hand, the thermal agitation would become more violent, and the air would push farther upward against gravity. But at what temperature would the atmosphere begin to boil away from the Earth?

Individual air molecules (or atoms) can escape to space only if they are high enough and hot enough: high enough that they have a clear shot at escaping from the atmosphere to space without hitting another air molecule and hot enough that they have sufficient velocity to escape the gravity of Earth. In the latter regard, an escaping molecule is much like a ball thrown into the air. The ball always arcs downward and falls back to the ground. It happens that if you could throw a ball with a speed of about 11 kilometers per second (you would have a lot of offers from the major leagues), the ball would have a chance of escaping Earth's gravity to wander around the solar system. Molecules at the top of the atmosphere—where the air is so thin that collisions rarely occur—are still restrained by gravity, like the ball. That is, unless a molecule has the **escape velocity**—about 11 kilometers per second—it will always fall back into the atmosphere. Fortunately, at the temperatures found in the Earth's upper atmosphere, very few molecules or atoms have the escape velocity. The global atmosphere is therefore stable and will remain so for eons.

Small particles suspended in the atmosphere are also agitated by collisions with air molecules. These collisions help keep the particles suspended. The random movement of the particles resulting from collisions is referred to as **Brownian motion** (Section 3.1.2). All objects suspended in a gas are subject to Brownian motion. Even a billiard ball is subject to

such random thermal motion, although the movement is imperceptibly small. When two objects of about the same size collide, both can rebound at about the same velocity. The game of billiards is based on this principle. But microscopic air molecules hitting a billiard ball cannot move it perceptibly (try moving a bowling ball by tossing beans at it). Thus objects much larger than an air molecule are not greatly deflected during collisions, and their Brownian motion is small.

2.3.1 How Much Air Is There?

We have now deduced that because of Earth's gravity, the atmosphere and the underlying planet represent a closed system with respect to the major constituents of air. That is, the total amounts of these materials remain fixed. Of course, over very long periods of time, even very slow processes can cause significant cumulative changes in the atmosphere—for example, the accretion of extraterrestrial material carried by meteoroids and comets, or the leaking of certain light gases from the top of the atmosphere into space. (These processes will be discussed in Chapter 4 in relation to the evolution of the Earth's atmosphere.)

In studying the atmosphere of Earth, in which many of the important processes governing the state of the environment occur, it is useful to understand how much material is actually present in the atmosphere, the oceans, and the solid Earth. The total mass of the atmosphere, including all constituents, is approximately 5×10^{18} kilograms (kg), or 5×10^{15} tonnes (metric tons). (See Appendix A for a discussion of units.) One part per billion by mass (ppbm) is equivalent to 5 million tonnes, and one part per trillion by mass (pptm), 5,000 tonnes. Hence, even relatively small perturbations of the global atmosphere require large amounts of material, but not as much as is required to pollute the oceans.

The mass of the oceans is 1.4×10^{21} kg, or 1.4×10^{18} metric tons. The oceans are therefore about 300 times more massive than the atmosphere. Obviously, most of the ocean mass consists of water. The oceans hold most of the water on the Earth (a large amount is tied up in mineral hydrates in the Earth's crust, although this quantity is uncertain). The surface temperature of Earth is cool enough for water to condense and remain liquid, a fact that is crucial to the evolution of the atmosphere, the global environment, and life (Chapter 4).

The "solid" Earth—the lithosphere, mantle, and core—has a combined mass of about 6×10^{24} kg, or 6×10^{21} metric tons. (See Section 4.1.1 for additional information.) This is roughly 1.2 million times the mass of the atmosphere and 4000 times the mass of the oceans.

It is important to recognize that the atmosphere is the most vulnerable component of the total Earth system consisting of land, oceans, and air. The mass of the atmosphere, although huge by ordinary standards, is trivial compared with the mass of the Earth as a whole. From space, this impression is vivid. Astronauts and cosmonauts alike describe the sense of fragility they feel when viewing our thin veil of blue air against the black depths of space. We must not be fooled into thinking that there is so much air we cannot spoil it.

2.3.2 Temperature Profiles

The atmosphere is not characterized by a single temperature. In fact, the temperature changes from place to place, with time (particularly with season), and with height through the atmosphere. A discussion of all of the factors that determine the temperature of air is beyond the scope of this book, but we will review in later sections the general factors that affect temperature. These include the position and orientation of the Earth with respect to the sun, the presence of clouds, the time of day, and so on. It is important to note, however, that the temperature distribution shows a number of regularities that define the fundamental structure of the atmosphere. These regularities are most clearly revealed by averaging measurements of temperature in certain ways.

As you may be aware, temperature can vary minute to minute everywhere on the planet. In your room, in a car, at the beach, the temperature is constantly changing. There are so many measurements of temperature that might be collected that no computer would be able to handle them. Thus to be of any use, temperature and other data must be processed for interpretation. Usually, averages are taken over time and space. We can, for example, define a seasonally averaged temperature at any point, over a state, or over a hemisphere; or we can calculate the average temperature of the entire planet over a decade or century. These are examples of the measures of temperature used by meteorologists and climatologists. Your doctor measures the temperature at one time and

point in your body (under your tongue, for instance) as a means of determining your overall health (and remember Fahrenheit's use of the human armpit). Climatologists take the temperature of the Earth to determine its state of well-being.

Atmospheric temperatures depend most strongly on altitude and season, at a given geographical location. Figure 2.6 shows the average vertical distribution of temperature in the homosphere over the whole planet. The various regions of the atmosphere are named according to this temperature profile. The transition from one region to another is determined by the rate of change of temperature with height. When the temperature decreases with increasing altitude, the rate of change is referred to as the **temperature lapse rate**. (A physical explanation of the temperature lapse rate is in Section 3.1.1, and its importance to atmospheric dispersion and air pollution is discussed in Section 5.3.) When the temperature is observed to decrease with height, it can be inferred that the region is unstable; strong convective, or vertical, motions develop that rapidly mix air between the levels and create turbulence. A convectively unstable region also generally exhibits weather.

The lowest region of the Earth's atmosphere is the **troposphere** (from the Greek *tropo* [turning, changing]) (Figure 2.6). Here the temperature decreases with increasing altitude, and the air is typically unstable. This situation occurs because sunlight heats the surface and produces buoyancy in the air just above the surface (just as happens in a hot-air balloon). The result is **convection**, or vertical motion, which redistributes the heat from the surface throughout the atmosphere. The temperature still drops with altitude, however, as explained in Chapter 3. The troposphere contains 90 percent of the total mass of the atmosphere. All the weather that we are familiar with occurs here.

In the atmosphere, when warm air overlies cooler air, a **temperature inversion** is said to exist. Another more specific way to define a temperature inversion is to say that the temperature remains constant or increases as we ascend in the atmosphere. At the bottom of and within a temperature inversion, the air is quite stable against vertical motions, unlike the situation in an unstable region. The **stratosphere** of Earth is such a stable region (Figure 2.6). *Strato* is Latin for "stratified". The stratosphere indeed represents a temperature inversion that is global in extent (since the stratosphere

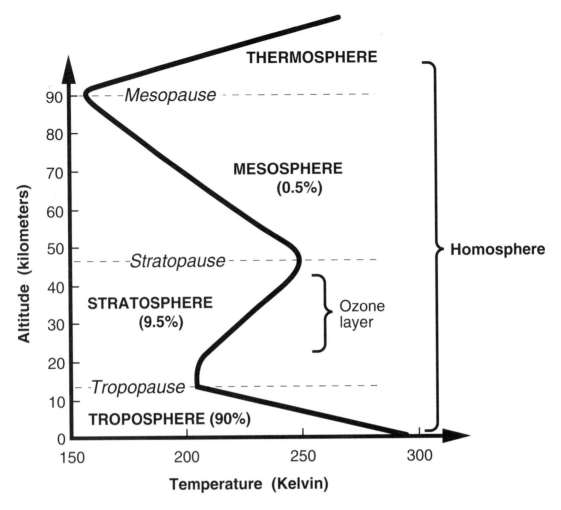

Figure 2.6 The average vertical temperature structure of the atmosphere. This temperature structure defines the different regions of the homosphere, from the troposphere at the surface to the thermosphere above it. The temperature varies by about a factor of two over the vertical range shown, from 150 to 300 kelvin. The fraction (in percent) of the total mass of the atmosphere contained in each well-defined layer is indicated in parentheses.

exists everywhere on Earth). The height that separates the troposphere and the stratosphere—we can think of it as a surface or membrane that envelops the entire planet—is called the **tropopause** (that is, the height at which the troposphere "pauses"). The tropopause is variable but is usually found at 10 to 16 kilometers above the surface. The tropopause is higher in the tropical latitudes than at the polar latitudes, because the heating of the surface is greater in the tropics; therefore, the buoyancy of the air is greater and the convection is stronger, and so the tropospheric mixing penetrates to greater altitudes.

The stratosphere contains about 9.5 percent of the total atmospheric mass—that is, only about 10

percent of the mass held in the troposphere, but more than 90 percent of the remaining mass of the atmosphere. The stratosphere also contains the natural ozone layer, which absorbs dangerous ultraviolet sunlight (Section 2.2.1; see also Section 13.4). The absorption of this sunlight causes the stratosphere to warm and explains the development of a temperature inversion and the tropopause. Because the stratosphere is very stable against vertical mixing, gases and particles introduced into this region can remain there for up to several years. The ozone layer extends through the entire depth of the stratosphere but is concentrated in its lower half.

The stratosphere extends to a height of about 50 kilometers. At this point, the temperature begins to decrease again because the ozone concentration is so low that the heating by solar ultraviolet radiation begins to drop off. This region of decreasing temperatures is called the **mesosphere**. Of course, the height of demarcation between the stratosphere and mesosphere, which occurs at roughly 50 kilometers, is named the **stratopause**. The mesosphere holds less than 1 percent of the total atmospheric mass. The mesosphere is the region where small meteors burn up: You may have seen these as "shooting stars" on a romantic evening. The lowest layers of the Earth's ionosphere, consisting of highly electrified air, also are in the mesosphere. These layers occasionally interfere with radio and television reception on an otherwise stress-free weekend.

The mesosphere is capped by the **thermosphere**. In this unusual, rarefied region of the atmosphere, the air is heated by X rays and other deadly energetic radiation from the sun (incidentally, preventing this radiation from reaching the ground and causing environmental havoc). The altitude that separates the mesosphere and thermosphere is aptly named the **mesopause**, which is usually located about 85 kilometers above the Earth's surface. The thermosphere is highly ionized and is responsible for long-distance radio communications: Radio waves are reflected from this region and can bounce between the ground and the thermosphere all the way around the world.

The air layers located above the stratosphere are so tenuous that their influence on the global environment and the composition of air is, as far as we can tell, negligible. Accordingly, these regions, although interesting and unique in their own right, are not discussed further in this text.

2.3.3 THE STRATIFICATION OF THE ATMOSPHERE

The pressure of the atmosphere and the density of air are closely related. Both decrease steadily with increasing altitude above the ground. At the surface of Earth, the average atmospheric pressure is about 1013 millibars (mb). (Appendix A explains the commonly used units of pressure). Typically, the surface pressure does not vary by more than a few percentage points between the extremes of normal weather and hurricanes. Pressure, however, does fall off very rapidly with height, to about 100 mb at 13 kilometers (the tropopause) and 1 mb at about 50 kilometers (the stratopause). Atmospheric density follows a similar behavior. The profiles of pressure and density are shown in Figure 2.7.

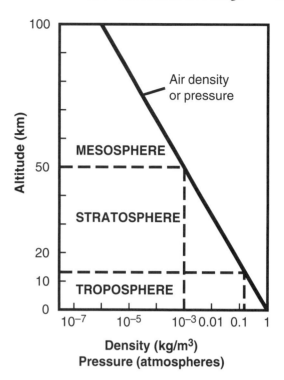

Figure 2.7 The profile of air density in the Earth's homosphere. The density falls off with height in a uniform way, because the force of gravity pulls the air toward the surface. It is difficult to distinguish the different atmospheric regions, as defined by the temperature profile, in the density profile. The curve of atmospheric pressure is almost identical to that for density, as indicated in the figure.

As explained earlier, gravity is responsible for the vertical structure of the atmosphere, for holding atmospheric gases close to the Earth. The decreases in pressure and density can be related to a distance called the **scale height** of the atmosphere. This is the vertical distance over which the pressure or density decreases by a factor of about 2.7. The scale height is inversely related to gravity: The greater the force of gravity is, the smaller the scale height and the more compressed the atmosphere will be.

2.4 Air in Motion

The wind systems on our planet are driven by three principal factors: the heating caused by sunlight at the Earth's surface, the rotation of the Earth, and the friction of the atmosphere that dissipates motion. Additional important factors include the evaporation and condensation of water, which involves

latent heat (that is, the energy that must be added to water to make it evaporate and that is released when the water condenses, which explains why your body sweats when it needs to cool down or why you can feel so cold when you are wet) and **radiant heat** emitted by the atmosphere and the surface of Earth (Section 3.2.1). The warming of land and ocean surfaces by sunlight induces vertical motions associated with buoyancy (this was mentioned when discussing the tropospheric temperature lapse rate). Buoyancy occurs when air is heated at a fixed pressure, because its density decreases as the temperature increases (Charles's Law—recall his interest in hot-air balloons). Warm air tends to rise, and cool air to sink, creating relative atmospheric motions. Any relative motion can produce **friction** or **drag** that retards the motion and dissipates the energy of motion, called **kinetic energy**[10] (feel the resistance or drag as you attempt to pull your hand through water, for example). These forces and effects are rather obvious. But it is not so obvious how the rotation of the Earth can affect the winds.

The **Coriolis force** (named for the French engineer[11] who first defined the effect) occurs when movement is attempted on a rotating object (like the planet Earth). Anyone riding a carrousel can easily experience the Coriolis force. If you try to walk from the outer edge toward the center of the carrousel on a direct path, you will be pushed sideways: to the right if the carrousel is moving clockwise, and to the left if counterclockwise. From your frame of reference on the carrousel, you are simply trying to walk a straight line. But your friends standing next to the carrousel in a more stable frame of reference, the ground, will see you spiraling inward. The differences between these apparent motions are related to the force you feel, which is trying to keep you on the spiral.

Try another experiment on the moving carrousel. Throw a ball from the edge toward the center. The ball doesn't travel a straight line, but swerves sharply away to the side! If the carrousel is turning clockwise, the ball will swerve to the left. This is in the opposite direction to the force you experienced in attempting to walk toward the center. Obviously, the errant flight of the ball is related to the absence of the force you had felt. Oddly enough, there appears to be an *apparent* force on the ball causing it to deviate from a straight line, and this is called the Coriolis force. There are more sophisticated explanations of the Coriolis effect based on the relationships between velocity and acceleration in a rotating frame of reference, but we need not bother with such detail here. Your everyday experience demonstrates the Coriolis effect (if not, go take a carrousel ride).

Varying combinations of all these effects drive the wind systems on Earth like an enormous engine. The winds can be roughly considered on two scales: local and regional winds that compose weather, and large-scale wind systems and jets that systematically move air around the entire planet.

2.4.1   Local Winds and Weather

Our sense of experience of atmospheric winds occurs on small scales. We feel the wind around us; we see clouds in the sky; and we can view a storm front approaching. These are manifestations of weather, or meteorological phenomena on (relatively) small scales. Local winds and weather are important to air pollution. (Chapter 5 describes some specific events that affect air quality and the dispersion and removal of air pollutants.)

The study and prediction of weather are developing into a fairly accurate science. In the past, weather forecasters looked to almanacs and to the skies to make short-term and long-term predictions. A few even relied on the pain levels in arthritic joints to see into the future (that ache in your leg, for example, may be related to a change in temperature or humidity or pressure, or all of these). Today, much weather analysis is still done the "old-fashioned way" using weather maps and charts, as well as experience and intuition. But more and more research and forecasting is carried out with sophisticated models that represent numerically the dynamics of the atmosphere. Moreover, the advent of satellite observation systems has reaped volumes of data that can be used in real time, or with hindsight, to guide the development and application of forecast models.

Weather forecasts are essential to pollution forecasts. Air pollution, particularly suffocating urban air pollution, is strongly controlled by weather. As the

---

10.   The heat energy connected with the mechanical motion of molecules through space is also called kinetic energy. However, *all* motion involves kinetic energy, whereas heat energy may also be carried in the form of radiation or may be partly stored as potential energy, say within vibrating molecules. The precise distinction among these various forms of energy is not necessary for the discussions in this book.

11.   Gaspard G. de Coriolis (1792–1843) was a French physicist and engineer who made early studies of the motion of bodies in rotating frames of reference.

forecasting of weather has become more reliable, the prediction of pollution levels has grown more feasible. Calculating air pollution is much more complex than calculating weather patterns, as is easily appreciated when one realizes that the weather provides only one element of the air pollution equation. One must also understand the sources of pollutants, the amounts emitted into the air, their chemistry, the processes that remove them, and the impacts of myriad pollutants on humans and the environment.

High and Low Pressure

The pressure of air at the surface of the Earth is not constant,[12] but changes continuously with changes in the weather. Large and small waves propagating through the atmosphere constantly modify the amount of air overhead and thus change the local pressure (or weight) of the atmosphere. Large pools of air can form as a result of the global circulation of the atmosphere. Huge domes of cloudless air form high-pressure systems. The high pressure is created as air subsides within a stable mass of air. High pressure brings good weather—clear skies and sunny days. When the air is unstable (see Section 5.3 for a complete discussion of atmospheric stability), low-pressure systems can develop. Rising unstable air creates the centers of low pressure. Hurricanes and typhoons are examples of low pressure gone amok. Tornadoes provide the extreme example of low pressure and its potentially damaging effects. Bad weather is associated with low pressure—clouds, thunder, lightning, rain, and wind.

A **barometer** can be used to measure atmospheric pressure. Evangelista Torricelli invented the mercury barometer in 1643.[13] It is a glass tube, sealed at one end, that is filled with mercury (a metal that is fluid at room temperature); the tube is inverted, with the open end in a pool of mercury exposed to the atmosphere. The weight of the mercury causes it to fall within the glass tube, creating a vacuum at the top of the tube. The height of the mercury in the glass enclosure above the pool depends on the air pressure exerted on the pool. The greater the air pressure is, the higher the column of mercury will be. One atmosphere of pressure can sustain a column of mercury about 760 millimeters in height.

Torricelli discovered that the height of the mercury column varied from day to day. He was apparently the first to attribute this behavior to changes in atmospheric pressure. It was not long before it was noticed that when the weather was fine, the height of the mercury column was high and that when the weather was poor, the column was low. The first weather forecasts were based on these simple observations of atmospheric pressure using a barometer. If the column is rising, good weather is expected. If the column is falling, break out the umbrellas. Many people keep simple barometers around to know what to wear each day.

2.4.2 GLOBAL WIND SYSTEMS

The global wind systems can distribute air pollution over the entire planet. These winds often reach speeds of several hundred kilometers per hour and therefore transport air over great distances in short order. Two kinds of global winds are distinguished: the **meridional circulation**, which represents average air motions oriented primarily from north to south (or varying with latitude), and the **jet streams**, which, being oriented mainly from east to west, are **zonal** in character (or encircle the Earth at constant latitude). The atmosphere is a complex three-dimensional fluid that can sustain numerous modes of oscillation and many types of waves. Waves associated with buoyancy oscillations, for example, are called **gravity waves**. These waves are seen locally as aligned cloud streaks and clear air in high-altitude

---

12. The pressure measured at sea level. Of course, as noted in Section 2.3.3, atmospheric pressure also changes with height above sea level. In fact, the change in pressure with height is one way to measure altitude (and this is done in aircraft, for example). At any given point on the Earth (and thus at a fixed height above sea level), the pressure invariably varies as the atmosphere drifts overhead.

13. Evangelista Torricelli (1608–1647) was a physicist and mathematician who studied the behavior of fluids and contributed to the development of geometry. Torricelli's law determines the speed at which a liquid flows out of a tank through an aperture. During the last few months of Galileo's life, Torricelli acted as his assistant. Galileo had suggested an experiment to create a **vacuum**—that is, a space completely devoid of matter in any form, solid, liquid or vapor. Torricelli's exploration of Galileo's idea led to the invention of the barometer. Blaise Pascal (1623–1662), another physicist of that period, used Torricelli's barometer to determine

that atmospheric pressure falls off with height, by having his brother-in-law carry a barometer up a mountainside. Pascal is also known for his law concerning pressure, which states that pressure exerted anywhere on a confined fluid (a liquid or gas) is transmitted uniformly throughout the fluid. All hydraulic systems are based on this principle.

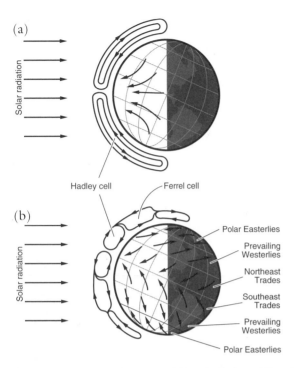

(a)

Solar radiation

Hadley cell   Ferrel cell

(b)

Solar radiation

Polar Easterlies

Prevailing
Westerlies

Northeast
Trades

Southeast
Trades

Prevailing
Westerlies

Polar Easterlies

Figure 2.8 The development of the principal meridional (latitude-oriented) circulation of the Earth's atmosphere. (a) if the Earth did not rotate on its axis and always faced the sun in one direction, the circulation would be dominated by strong rising motions in the tropics on the daylit side and subsidence at the poles and on the night side of the planet. (b) the rotation of the Earth and the effect of the Coriolis force lead to a complex pattern of circulation cells.

cirrus clouds. On a global scale, waves associated with the meandering of the midlatitude jet stream are called **planetary waves** (or **Rossby waves**, after the early pioneer in the study of this phenomenon[14]), which stretch around the entire planet.

The basic meridional structure of the atmospheric circulation is illustrated in Figure 2.8. Two cases are shown, which emphasize the effect of the Coriolis force. Because sunlight is most intense at low latitudes (the tropics, near the equator) and least intense at high latitudes (toward the North and South Poles) the heating of the Earth by the sun is concentrated at the lower latitudes. One consequence of this pattern of solar-energy deposition is the creation of a permanently warm and humid tropical belt encircling the Earth. It's a nice place to go on vacation. But the

tropical heating also fuels the most important circulation of the atmosphere, the **Hadley circulation** (named for its discoverer).[15] Figure 2.8(a) shows the conditions that would prevail if the Earth were assumed to always face the sun in one direction (or, effectively, to rotate only once on its axis during each orbital revolution around the sun). The buoyant air in the tropics would rise straight up, moving toward the poles in the upper troposphere, settling downward to lower altitudes in the polar regions and on the dark side of the planet, and flowing back toward the tropics near the surface. The Earth's circulation would then be dominated by one enormous **Hadley cell**, and the weather and climate would be quite different from what we presently enjoy.

Figure 2.8(b) shows the actual meridional circulation features of the atmosphere. The Hadley cell is still preeminent. In this circulation, the rising motion is concentrated in a narrow band near the equator, which is called the **Intertropical Convergence Zone** (**ITCZ**). The ITCZ is characterized by intense thunderstorms and heavy rainfall. The ITCZ migrates back and forth across the equator as the seasons change. It is always found in the tropical zone of the summer hemisphere (that is, the Northern Hemisphere in July and the Southern Hemisphere in January). The ITCZ is one of the most powerful engines in the climate system, converting the latent heat of evaporated water to dynamic motions of convection. The resulting rains bring an abundance of life to the tropical regions.

At higher latitudes, the atmospheric circulation reverses direction in a pattern referred to as the **Ferrel cell**. The motion of air downward from higher altitudes toward the surface over a large area is referred to as **subsidence**. In the common branch of the Ferrel and Hadley cells, subsidence in the subtropics creates semipermanent areas of high pressure and fair weather, with low rainfall. Southern California is one area that both benefits and suffers from these highs (benefits from the continual pleasant weather but suffers from the resulting drought and air pollution). The rising motions in the poleward branch of the Ferrel cell are associated with the

14.   Carl-Gustov Rossby (1898–1957), a Swedish meteorologist, founded in 1928 the first Department of Meteorology in the United States at the Massachusetts Institute of Technology.

15.   George Hadley (1685–1768) was an English physicist and meteorologist who first described the cause of the trade winds and the related meridional circulation bearing his name. Hadley was first trained as a lawyer but was more fond of physics. His accomplishments in devising a theory for the atmospheric circulation were largely unknown until John Dalton acknowledged them in 1793, well after Hadley's death.

unstable weather and precipitation common at mid- to high latitudes. Finally, there is a polar cell at high-latitudes that is characterized by rising motions in its common branch with the Ferrel cell and by general subsidence over the pole. The polar subsidence is accentuated by the attenuated solar heating at high latitudes.

Highways in the Sky

Figure 2.9 shows the global wind systems in more detail. The Coriolis force is responsible for the development of a subtropical jet, which is commonly referred to as the jet stream. As air flows poleward in the upper tropopause in the Hadley cell, it is deflected and accelerated eastward by the Coriolis force to form the jet stream. Note that winds blowing from west to east are called **westerlies** and those blowing from east to west are **easterlies**. That is, the direction from which a wind comes provides its name (just as children are given the name of their parents).

In the jet stream, air velocities can reach 400 kilometers per hour (with respect to the ground). The jet streams, coursing in the upper troposphere,

Figure 2.9   The prinicpal wind systems and jets in the Earth's atmosphere from a three-dimensional perspective. The main atmospheric circulation systems are indicated by heavy arrows. Each of the meridional circulation cells is associated with an easterly or westerly jet. The surface winds in the southwest Pacific and Indian oceans' basins are identified by single arrows. (From G. Meehl, "The Tropics and Their Role in the Global Climate System," *Geographical Journal* 153 [1987]: 21)

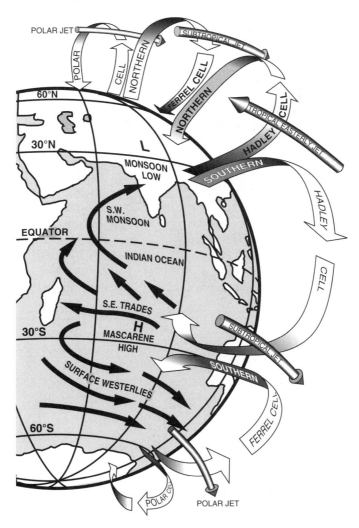

were discovered during World War II. American bombers heading for Japan from Pacific bases occasionally found themselves flying at great speed and making no headway; they were unwittingly bucking the powerful jet stream flowing against them. Pilots soon learned how to avoid these concentrated wind streams, and the war progressed. The jet stream is also responsible for steering the weather systems that move from west to east across the midlatitudes. For example, on the west coast of the United States, storms hit the southern and central regions of California only when the jet stream has moved south of its normal position over northern California and Oregon.

Two jet streams are found in each hemisphere: the subtropical jet just mentioned and the polar jet, which occurs at higher latitudes of 50 to 60°. Both these zonal flows are westerly. Another jet is formed in the tropics where the Hadley cell bifurcates, or divides, into a northern and a southern branch (Figure 2.9). This jet is not strongly controlled by Coriolis forces because of its location near the equator, and it can be either easterly or westerly, the direction being determined by other atmospheric constraints.

The winds at the surface of the Earth are of particular importance to human activities. In ancient times, these **prevailing winds** determined the feasibility of crossing the oceans under sail and carrying on profitable trade: hence the common name, **trade winds**. The trade winds are a consequence of the pattern of meridional circulation shown in Figures 2.8(b) and 2.9. These winds represent the return flows in the lower atmosphere associated with the Hadley circulation, as first described by George Hadley. At tropical latitudes, the trades are generally easterly, converging toward the ITCZ (Figure 2.8[b]). In the midlatitudes, the trade winds are generally westerly. We all know that weather arrives from the west on the prevailing winds. At high latitudes, weak easterlies are typical.

From this discussion, we can see that the atmosphere has certain fixed general features that characterize its circulation. The Hadley cell is a principal, constant feature. The prevailing winds also are reliable. In the United States, for example, air tends to move from west to east at all altitudes, although at different speeds. Pollution, therefore, tends to move toward the east. Thus acid rain is seen to be a problem eastward of Ohio and other midwestern states that dish out the pollutants causing the acidity. Before a pollutant can become a global problem, it must be carried by the major circulation systems over the entire Earth. At jet stream speeds of several hundred kilometers per hour, this can happen within a week, although the actual time required to mix a pollutant into all the nooks and crannies of the global atmosphere is closer to 2 years.

Questions

1. Can you list some of the ways that you perceive the atmosphere through different senses? Use examples that are not discussed in the text.
2. Would you rather be trapped in a room filled with mephitic air, fixed air, phlogisticated air, dephlogisticated air, fire-air, or vitiated air? Can you explain the relationships among these different kinds of airs?
3. If all the nitrogen and oxygen gas were removed from the atmosphere but the other gases remained, by what factor (approximately) would the pressure of the atmosphere at the ground be reduced?
4. A television picture tube has no air in it; rather, a nearly pure vacuum exists in the glass enclosure. Explain what happens when you hit the tube, causing the glass to shatter. How does the pressure of the atmosphere come into play?
5. Explain why it takes about 1 hour longer to fly nonstop from New York to Los Angeles than the other way around.
6. You have a clean, empty gas can in the trunk of your car. While at home, in the mountains, you open the can to check if it is empty and then seal it tightly. Later, as you are driving to the seashore, you hear a metallic crunching sound coming from the trunk. What is causing this noise?
7. You are riding on a Ferris wheel. At the exact top of the ride, with the wheel rotating briskly, you accidentally drop a soda pop bottle from the side. Does the bottle fall directly toward the center of the Ferris wheel below you? What "apparent" force is coming into play in this situation? What difference in the flight of the bottle would you notice if the wheel had been stopped at the top?

Problems

1. You are 10 meters under water and feel a certain pressure on your body from the weight of the water and air above you. If you dive 10 meters deeper, to 20 meters, what is the pressure you

will experience relative to what you felt at 10 meters?

2. Someone tells you that the bottled water you have been drinking contains one part per trillion by volume of arsenic, which is poisonous. How many glasses of water must you drink to ingest the equivalent of one glass of arsenic? (Certainly, much less than this would kill you!)

3. The wind in the upper troposphere blows from west to east with a velocity of 200 meters per second. How long would it take a balloon floating at this altitude to travel around the globe (assuming that the balloon is at the equator)?

4. Your doctor tells you that your body temperature is 22°C. Are you sick (your body temperature should be within a few degrees of 98.6°F)? The doctor says your temperature is 310 Kelvin. Do you feel better?

Suggested Readings

Dalton, J. *Meteorological Observations and Essays.* London, 1793.

Graedel, T. and P. Crutzen. *Atmospheric Change: An Earth System Perspective.* New York: Freeman, 1993.

Hartley, W. N. *Air and Its Relations to Life.* New York: Appleton, 1875.

Ingersoll, A. "The Atmosphere." *Scientific American* **249** (1983): 162.

Lavoisier, A. *Memoirs on Combustion in General.* 1777.

Ramage, C. "El Niño." *Scientific American* **254** (1986): 76.

Schaefer, V. and J. Day. *Field Guide to the Atmosphere.* Boston: Houghton Mifflin, 1981.

Scorer, R. and A. Verkaik. *Spacious Skies.* London: David and Charles, 1989.

Snow, J. "The Tornado." *Scientific American* **250** (1984): 86.

Strutt, J. W. (Lord Rayleigh) and W. Ramsay. "Argon, a New Constituent of the Atmosphere." *Philosophical Transactions of the Royal Society* **186A** (1895): 187.

Tricker, R. *The Science of Clouds.* New York: Elsevier, 1970.

Wallace, J. and P. Hobbs. *Atmospheric Science: An Introductory Survey.* New York: Academic Press, 1977.

3

Basic Physical and Chemical Principles

The fundamental laws that govern the natural world and that also bear on the problem of environmental pollution are described in this chapter. The treatment is not exhaustive by any means. The point is to introduce basic principles and concepts that provide a deeper understanding of the complex environmental issues to be taken up in the second and third parts of this book. Some equations are presented to define quantitative relationships between key physical parameters that determine the state of the environment. An equation provides a shorthand description of the effect that a change in a basic physical quantity (the temperature of a gas, say) has on another measurable quantity (the pressure exerted by the gas, say). The mathematical analysis in this book is restricted to the simplest arithmetic and algebraic operations (Appendix A). Accompanying each equation are words and diagrams, which by themselves should be adequate to mastering the underlying concepts. The mathematical equations in the text are used mainly to illustrate physical relationships and solve straightforward problems.

Chemical equations are more symbolic in nature than mathematical equations are. Chemical equations represent a condensed linguistic form used by scientists to describe the interactions among different substances brought into contact with one another. This compact and handy language of chemistry is discussed in Section 3.3.

The general topics discussed in this chapter are divided into three main areas: the physics of gases in the atmosphere, atmospheric radiation, and air chemistry.

3.1 The Mechanical Behavior of Gases and Particles

All matter, including the atmosphere, is made up of atoms, which are usually associated into molecules.

Atoms are the simplest building blocks comprising all the materials familiar to us (refer to Section 2.2.2). There are different kinds of atoms, each of which corresponds to a different element. A pure **element** is composed only of atoms of that element, which are distinct from the atoms of any other element. A block of pure gold should thus have only gold atoms in it. Likewise, given any quantity of a pure element, the smallest mote that it can be subdivided into, no matter the effort, is one atom of that element (for example, pure gold can be decomposed only into gold atoms, not into any smaller element).[1] There are about 92 or so naturally occurring elements. No other stable kinds of matter can be formed (refer to Section 7.3.1 for a discussion of the stability of natural and manmade elements). Accordingly, everything in the universe is composed of a relatively small number of basic ingredients—the natural elements.

Atoms themselves are composed of smaller "sub-atomic" particles; that is, **electrons**, **protons**, and **neutrons**. Every atom has an equal number of protons and electrons, unless it is has been "ionized," or had one or more electrons added or removed. Protons carry a single unit of positive charge, and electrons carry a single unit of negative electric charge. Because electrons and protons come in matched pairs, normal atoms are electrically neutral (that is, the negative and positive charges offset each another). Protons are clustered in the **nucleus** of the atom, and the electrons swarm around the nucleus in confined orbits. Within the nucleus, neutrons are mixed with the protons to provide "glue" that holds

---

1. The transmutation of one element into another was the goal of alchemy, which was pseudo-chemistry practiced prior to the Renaissance. The primary hope of the alchemists was to create gold from the heavier element, lead, or other "base" metals. In fact, elements can be transformed by radioactive decay (Section 7.3) or bombardment by highly energetic sub-atomic particles, the latter of which is both difficult and economically impractical.

the positive charges together in a tight cluster (as the name suggests, neutrons have no electric charge—they are neutral). The number of neutrons is roughly the same as the number of protons, but can be more or less. We could visualize an atom as a microscopic solar system with the sun (nucleus) at the center and the planets (electrons) in orbits around the nucleus.[2] In recent years, it was discovered that protons, electrons, and neutrons may themselves be composed of even smaller fundamental particles, called **quarks**. However, in the discussions here, we can limit our consideration to atoms and molecules, and avoid dealing with these strange entities.[3]

Every atom of a given element *has the same number of protons* in its nucleus. For example, every atom of gold has 79 protons; of course, there would be 79 electrons circling each gold nucleus as well. An atom of hydrogen, with only one proton and no neutrons in its nucleus, and one orbiting electron, is the simplest element that can be arranged. Uranium, with 92 protons and 146 neutrons, is the largest of the common elements. Each element has its own atomic number, which is simply equal to the number of protons in its atomic nucleus. The arrangement of elements into a periodic table based on atomic numbers is described in Section 2.2.2. However, at the time the periodic table was assembled, the internal structure of atoms was unknown. It has since been determined that the chemical properties of the elements are controlled by the electrons present in the outer orbits of the atom. The number and positions of these "**orbitals**" is fixed by the total number of electrons, or protons, in the atom, and hence the atomic number (see also Section 3.3.2).

As noted earlier, the number of neutrons in an atom can vary even for the same element (the number of electrons and protons, of course, does not vary). Atoms of an element with different numbers of neutrons are referred to as **isotopes** of that element. For example, an atom of normal hydrogen, H, has

one proton and no neutron, but there also is a stable isotope with one proton and one neutron, **deuterium**, and another isotope with two neutrons, **tritium** (tritium is a major ingredient of the so-called hydrogen bomb). Many isotopes are unstable and decompose through **radioactive decay** (Section 7.3.1; see also Section 8.2).

A **molecule**[4] is a stable combination of two or more atoms of the same or different elements. Molecules can be thought of as the fundamental building blocks of compounds consisting of two or more elements, or as stable arrangements of atoms of a single element (for example, O_2 as the principal form of pure oxygen). The atoms in molecules are held together by relatively strong forces associated with **chemical bonds** between the atoms. The chemical bonds involve electron interactions between the outermost orbitals of the constituent atoms. These are basically the same interactions that control chemical reactions (Section 3.3.2). Molecules can have properties that are quite different from those of the pure elements from which the molecules are made. The properties of water, for example, are amazingly (and importantly) different from the properties of pure hydrogen and oxygen (both of which are colorless, odorless gases under normal conditions). This critical property of compounds—of exhibiting characteristics and behavior quite different from their constituent elements—is what provides the variety in nature that allows life to exist.

In considering the mechanical behavior of gases, the fact that a molecule is a stable entity implies that it may be simply treated as an object with a certain mass and size, much like a microscopic billiard ball. In fact, this is a useful picture of a molecule that is commonly adopted by physicists. This "hard sphere" model is certainly adequate for the present discussions.

3.1.1 GAS LAWS AND HYDROSTATICS

Air is a gas mixture consisting mainly of molecules, with some atoms. The atoms are mainly noble gases, like argon. Molecules of oxygen and nitrogen dominate the composition of air. The mechanical

2. A proton or a neutron is about 1800 times more massive than an electron. An atom consists mainly of empty space, just like the solar system. Most of the mass is concentrated in the nucleus, which occupies merely a fraction of 10^{-15} of the volume of the atom; the sun occupies a fraction of 10^{-12} of the volume of the solar system defined by the orbit of Pluto.

3. Physicist Murray Gell-Mann introduced the concept of quarks in 1964 to provide a simpler framework for explaining the large number of sub-atomic particles discovered up to that time. According to the theory, quarks have flavor, "up" and "down" or "strange," and color, red, green, or blue. These weird particles with other unusual properties elegantly help to explain how the sub-atomic world works.

4. Amedeo Avogadro (1776–1856) coined the term *molecule* as the smallest unit of a gas. A unit to measure the number of molecules in a specific mass of a material is **Avogadro's number**, 6.02×10^{23} molecules, a very large number indeed, considering that the amount involved is measured in grams. One gram is about the weight of a penny.

behavior of the molecules in a gas determines the bulk properties of that gas. The laws that govern the motion of molecules in the gas or vapor phase were first quantified by James Maxwell[5] and Ludwig Boltzmann.[6] They determined the range of values of the random velocities, or speeds, that individual molecules have in a gas, and expressed the probability that a molecule will have a given velocity in terms of a simple **distribution function**. It turns out that many of the interesting properties of gases can be derived from this velocity distribution function.

The most probable velocity of a gas molecule can be shown to depend on only the mass of the molecule and the temperature of the gas. Each atom of an element has a well-determined mass, which is usually expressed as the **atomic weight** in **atomic mass units**, or **amu**. Hence the mass of any molecule of known elemental composition is readily determined as the sum of the masses of the atoms that compose the molecule. The temperature of a gas can be measured with a thermometer. The most probable molecular speed, $\hat{v}$, is then given by the relation

$$\hat{v} = \sqrt{\frac{2k_B T}{m}} \qquad (3.1)$$

where T is the temperature (kelvin), m is the mass of a molecule, and k_B is a universal physical constant known as **Boltzmann's constant**. (The constants are defined in Section A.3; the $\sqrt{\ }$ operation is defined in Section A.4.)

The Maxwell-Boltzmann velocity distribution function implies that the average, or most probable, velocity of a molecule in a gas is directly proportional to the square root of the temperature

of the gas and inversely proportional to the square root of the mass of the molecule.

In Chapter 2, we discussed the velocity of the molecules in a gas but did not quantify it. We now see that the most probable velocity is easy to calculate; it is proportional to the square root of the temperature. The speed of sound in a gas like air is also proportional to this velocity; and so the speed of sound varies as the square root of the temperature. Sound travels faster in warmer air because the molecular velocity is greater. Communication across the backyard fence is quicker in summer than in winter. Likewise, the higher the temperature is, the greater the force with which molecules will strike the wall of their container and the greater the pressure that the gas will exert on the wall (Section 2.1.3).

The Ideal Gas Law

An ideal gas is one that behaves "ideally," that is, it obeys the **ideal gas law**, which is a physical relationship among the temperature, molecular concentration, and pressure of a gas. The ideal gas law is fundamental to the physics and chemistry of gases. And it is practical, since air is an ideal gas. Under normal conditions, all gases can be treated as ideal. The pressure exerted by a gas on a surface is related to the molecular motion of the gas molecules. When a molecule collides with a surface, it produces a force on the surface (a minuscule force, to be sure). That force multiplied by the time duration of the collision is called the **impulse**. The mass and velocity of the molecule determine the force exerted on, and the impulse imparted to, the surface. The temperature and mass of the molecule determine its speed (Equation 3.1). Even though this all seems to be pretty messy, the relationship among the various parameters turns out to be straightforward.

Impulse is related to another property of mass in motion, its momentum. The **momentum** of an object is just its mass multiplied by its velocity, or mv. Momentum expresses the persistence of movement, or **inertia**, of a body in motion. A freight train traveling at 50 miles per hour has a lot more momentum than does a car traveling at the same speed, because the train is much more massive. Impulse is the force that must be exerted over time to stop the moving object. The same impulse accelerates the object from rest to its present velocity. Thus impulse represents a change in momentum.

5. James Clerk Maxwell (1831–1879) was born in Edinburgh. He had been shy since he was a boy, when he was given the nickname Dafty. Nevertheless, Maxwell possessed great insights into the natural world. He studied color vision and produced the first color photograph. He earned a prize for his studies of the dynamic motions of Saturn's rings. One of his greatest contributions was to define the distribution of velocities of the molecules in a gas, later known as the Maxwell-Boltzmann distribution. Maxwell also made fundamental contributions to the theory of electromagnetic radiation (Section 3.2).

6. Ludwig Boltzmann (1844–1906) was born in Vienna. Boltzmann elaborated Maxwell's work on the kinetic theory of gases and was the nineteenth century's champion of the "atomistic-mechanical" concept of matter. The stress of the continuing controversy over the validity of the atomistic-mechanical viewpoint may have led to his early death by suicide in 1906, on the threshold of discoveries in physics that vindicated many of his ideas.

A confined gas continuously bombards the surface of its container. Every time a molecule ricochets from the surface, its direction of travel (and so its momentum) changes, and a force (and impulse) is exerted on the surface. The pressure of the gas is just the average force exerted on a unit area of the surface from the continuous bombardment by gas molecules.

The **flux** of molecules impinging on a surface (that is, the number of molecules hitting a unit area of the surface in a unit-time interval) is proportional to the density, or concentration, of the gas multiplied by the molecular velocity. In more formal terms, the average flux is $nv/4$.[7] The pressure exerted by the gas on the surface is related to the rate at which the momentum is transferred by the molecules to the surface, or the total impulse (including the rebound effect of molecules from the surface).[8] The total impulse is the momentum change of an individual molecule multiplied by the flux of molecules, or $2mv \cdot nv/4$. Using Equation 3.1 for the velocity, it turns out that the pressure, concentration, and temperature of the gas are related as

$$p = nk_B T \qquad (3.2)$$

This relationship is known as the ideal gas law.

The ideal gas law states that the pressure of a gas varies in direct proportion to the concentration of gas molecules and to the temperature of the gas. The proportionality constant is Boltzmann's constant.

Another way to determine the relationship among pressure, concentration, and temperature is to measure it. This, in fact, was the approach of the classical chemists, as recounted in Section 2.2.2. Those early experimental results can be shown to agree with the later molecular description of gas behavior expressed by the Maxwell-Boltzmann distribution function.

The Gas Laws of Boyle and Charles

In Section 2.2.2, we described the experiments of Robert Boyle and Jacques Charles. Basically, Boyle discovered that the volume and pressure of a fixed

quantity of gas are inversely related, *when the temperature is fixed*. Mathematically, this relationship can be expressed as **Boyle's law**:

$$p \propto \frac{1}{V} \quad \text{or} \quad pV = \text{constant} \qquad (3.3)$$

where the symbol $\propto$ indicates a proportionality, with the exact equality defined by a specific constant. Note that if p and V are inversely proportional, their product is equal for any state of the gas:

$$p_1 V_1 = p_2 V_2 = p_3 V_3 = \dots \qquad (3.4)$$

where the string of dots indicates "and so on."

Boyle's law states that the pressure of a fixed amount of gas varies inversely with the volume of (the container holding) the gas when the temperature of the system is constant. Equivalently, the law states that the product of the pressure and volume of a fixed amount of gas at a fixed temperature is invariant.

Charles discovered a direct relationship between the volume and temperature of a fixed quantity of gas *when the pressure is fixed*. This is expressed as **Charles's law**:

$$V \propto T \quad \text{or} \quad \frac{V}{T} = \text{constant}$$

$$\frac{V_1}{T_1} = \frac{V_2}{T_2} = \frac{V_3}{T_3} = \dots \qquad (3.5)$$

Charles's law states that for a fixed amount of gas held at a constant pressure, the volume (occupied by the gas) varies in direct proportion to the temperature. Equivalently, the law states that the quotient of the temperature and volume of a fixed amount of gas held at a constant pressure is invariant.

Before proceeding to combine Boyle's and Charles's laws, first note that these experiments are carried out with a fixed quantity of gas trapped in a closed container. This quantity can be expressed as the total mass of gas, M, or as a specific total number of gas molecules, N. The mass and number of molecules are also related as

$$N = \frac{M}{m} \qquad (3.6)$$

where m is the mass of an individual molecule (a fixed mass for each type of molecule). Thus N, M, and m all are constant in the experiments of Boyle and Charles.

---

7. The factor of one-fourth results when the direction of motion of the molecules is taken into account.

8. The force exerted by a molecule rebounding from a surface is related to the *change* in the momentum of the molecule. The change in momentum for a perfectly elastic collision is twice the initial momentum in the direction perpendicular, or normal, to the surface.

The concentration of gas molecules, n, is related to N and the volume of the container holding the gas, V, as

$$n = \frac{N}{V} \qquad (3.7)$$

Obviously, since V may vary in these experiments, so can n, even though the mass and the number of molecules are fixed. In fact, V and n can be thought of as interchangeable parameters, since N is constant (Equation 3.7).

With this understanding, you can perform combined experiments. For example, imagine doubling the pressure on the gas in a container with the temperature fixed, thereby halving its volume (Equation 3.3 and Figure 2.5[a]), and, while maintaining the higher pressure, doubling the temperature of the gas and thereby doubling its volume (Equation 3.5 and Figure 2.5[b]). Although the volume is now the same as the initial volume, both the pressure and the temperature are *doubled*. After a little thought, you can deduce from this experiment that, at a fixed volume, pressure must be directly proportional to temperature, or $p \propto T$. Moreover, it follows from these elementary considerations that Boyle's and Charles's laws can be combined to yield the relationship

$$p = c''T = c'\frac{T}{V} = c\frac{NT}{V} = cnT \qquad (3.8)$$

where the c'', c', and c are numerical constants determined for experiments in which N and V, or N only, or neither N nor V is held constant, respectively. Equation 3.8 is in fact the ideal gas law, noting that the final constant of proportionality, c, must in reality be Boltzmann's constant, k_B.

Dalton's Law of Partial Pressures

In Boyle's and Charles's experiments, the total number of gas molecules was fixed. Dalton designed a different series of experiments, in which the number of molecules in a container was varied while the volume and temperature were held constant. Dalton also mixed together different gases in the container and observed the effect on the total pressure exerted by the gas. Dalton's variable was N. For a mixture of gases, the total number of molecules is the sum of the number of each kind:

$$N = N_1 + N_2 + N_3 + \cdots \qquad (3.9)$$

If each component were placed in the container separately (at constant V and T), the pressure exerted by that component would be given, as expected, by the combined gas law (Equation 3.8), which at constant temperature may be expressed as

$$p_i = c_i n_i \qquad (3.10)$$

where the subscript i refers to a particular gas species. However, Dalton proved that the constant c_i is the same for all gases, and for mixtures of nonreactive gases,[9] a fact later explained on the basis of molecular kinetics through the work of Avogadro, Maxwell, Boltzmann, and others. Thus Dalton derived the relationship

$$p = p_1 + p_2 + p_3 + \cdots$$
$$= (n_1 + n_2 + n_3 + \cdots)k_B T \qquad (3.11)$$

The ideal gas law has been applied to obtain the final relationship in Equation 3.11. The pressures contributed by the individual component gases—that is, p_1, p_2, and so on—are called the **partial pressures** of the components in the mixture.

Dalton's law of partial pressures states that for a fixed temperature and volume (container), the total pressure of a mixture of gases is the sum of the partial pressures that each of the individual component gases would have if placed alone in the same volume (container).

As a consequence of Dalton's law, a complex mixture of gases like air may be treated as a single medium for mechanical and dynamical analysis (however, even trace amounts of water vapor, which lead to the formation of clouds, must be taken into account [Section 3.1.3], and radiatively and chemically active trace gases must be treated in detail [Sections 3.2 and 3.3]).

Buoyancy

The weight of a volume of gas depends strictly on its density, or the concentrations of the specific molecules making up the gas. Although Dalton showed

---

9. The condition that the gases be non-reactive is necessary so that complicating chemical transformations do not occur during the experiments. The gases should also not be condensable, like water vapor, which can condense as water on the surfaces of the container.

that gases at the same concentrations and temperatures have the same pressure, the weights of equal volumes of different gases at the same pressure can differ for two reasons:

1. The molecules of different gases differ in mass;
2. The density of a gas held at a fixed pressure is affected by the temperature of the gas.

Recalling that $\rho = nm$ (where ρ is the density [Equation 2.5]), it is easy to see that if p and T, and therefore n, are fixed, ρ is simply proportional to m. For a mixture of gases, the density is the sum of the densities of the components, or

$$\begin{aligned} \rho &= \rho_1 + \rho_2 + \rho_3 + \cdots \\ &= n_1 m_1 + n_2 m_2 + n_3 m_3 + \cdots \end{aligned} \quad (3.12)$$

If the pressure and composition of the gas is fixed but the temperature can change, the density varies inversely with the temperature (from Equations 2.5 and 3.2 with p fixed),

$$\begin{aligned} \rho &\propto \frac{1}{T} \\ \rho T &= \text{constant} \quad (3.13) \\ \rho_1 T_1 &= \rho_2 T_2 \end{aligned}$$

When an object is suspended in a fluid that is denser than the object, the fluid exerts a force on the object that causes it to rise. This force is referred to as **buoyancy**.[10] Because of buoyancy, wood—with a density of about 500 kilograms per cubic meter—floats on water, which has a density of about 1000 kg/m^3. An inflated raft remains afloat because of buoyancy. Bubbles rise in a glass of champagne for the same reason. The buoyant force is directly related to the difference in the densities of the fluid and the object. The force is also proportional to the volume of the object that is displacing the fluid. It is easy to hold a tennis ball under water but very difficult to keep a large beach ball submerged. In another simple interpretation, the buoyant force is simply the *difference* between the weight of the submerged object and the weight of the fluid displaced by the object.

10. The famous Greek mathematician and scientist Archimedes (ca. 280–212 B.C.) first derived the law of buoyancy, also known as the *Archimedes principle*, by observing ships afloat on the Mediterranean Sea. Among many other things, he invented the "Archimedes screw," an ingenious device for lifting water without the need for buckets. The war machines he designed delayed the fall of the city-state Syracuse to the Romans, soon after which he was slain.

If the object is denser than the fluid, the object will sink; it has **negative buoyancy**. Most people are slightly buoyant in water (referring to body density, not personality) and can float in quiet repose.

The law of buoyancy states that any material or object immersed in a fluid will tend to rise through the fluid if the fluid density is greater than the material density. The force associated with buoyancy is the difference between the weight of the displaced fluid and the weight of the immersed material.

Air is a fluid with a density at the ground of about 1 kg/m^3. Objects that are lighter than air rise in the atmosphere. A helium-filled balloon floats upward, since helium gas molecules are roughly seven times lighter than air molecules. Similarly, hot air is less dense than cold air: hence the success of the hot-air balloon. When sunlight is absorbed at the ground, the surface becomes warmer and heats the air just above it. This warmer air becomes buoyant, or unstable, and rises in a process called **convection**. When convection is very intense, thunderstorms can result.

We can think of parcels of air that are warmer or colder than their surroundings as "objects" subject to buoyant forces. The concept of an **air parcel** is useful in thinking about atmospheric processes. A parcel may be of any size, with imaginary boundaries and particular values of temperature, density, composition, and so on. Parcels move through the atmosphere subject to various forces and undergo physical and chemical transformations, some of which are described later. All these gas laws apply to these air parcels, as if they were confined within a weightless, perfectly elastic film.[11] (See Section 5.3 for additional discussion and applications of the "air parcel" concept.)

Adiabatic Cooling and Heating

The term **adiabatic** literally means "without the loss or gain of heat." In an adiabatic process involving a parcel of a gas like air, no energy in the form of heat is exchanged between the parcel and its surroundings. The parcel is thermally insulated. For example, when a parcel of gas is allowed to expand or is compressed over a short time span, the process is usually adiabatic and little or no heat is transferred

11. Such a film might be approximated by a soap bubble, although that is so fragile it would be difficult to carry out many useful experiments on gas properties.

between the gas and its neighboring environment.[12] Under the circumstances of an adiabatic process, a very specific relationship exists between the pressure and the temperature of the gas as it undergoes expansion or contraction. That relationship is called the **adiabatic gas law**, and the process is **adiabatic expansion** or **compression**. In general, as a gas expands from a higher pressure to a lower pressure, it cools, or experiences **adiabatic cooling**. As a gas is compressed from a lower pressure to a higher pressure, it warms, or experiences **adiabatic heating**. This seems contrary to the logical notion that if no heat is transferred with an isolated parcel of gas undergoing expansion or contraction, the temperature (which is just a measure of the heat content of the molecules in the parcel) should remain constant. The explanation lies in the fact that the expansion of a gas requires that some work be expended by the gas during the expansion process to push aside the surrounding gas to make room for the expansion; this "work" energy can be drawn only from the internal heat reservoir of the gas itself, since no other source of energy is available. The gas thus becomes cooler as it expands. Likewise, the compression of a gas requires that work be done on the gas to force it into a smaller space, adding to its internal heat reservoir and heating the gas.

If you have ever used your thumbnail to release air from an inflated tire, you may recall that the digit felt cold afterward. If you have vigorously pumped up a bicycle tire and then touched the valve, it would have felt warm. The first experience demonstrates adiabatic expansion and cooling. The air in the tire is at a greater pressure than the air in the surrounding atmosphere. When the valve is opened, the air rushes out of the tire under pressure, expands at the valve opening, and cools adiabatically. When a tire is inflated, the compression of the air by the piston in the pump warms the air, and this adiabatic heating can be felt through the metal of the valve. Air released from an inflated balloon also cools noticeably.

The law of adiabatic expansion or compression states that any gas will cool that is allowed to expand freely from a higher pressure to a lower pressure without the transfer of external energy to the gas. Similarly, a gas will heat if compressed from a lower to a higher pressure in the absence of a transfer of energy from the gas.

12. Heat could transfer by means of a radiative process (Section 3.2) or by the conduction of heat through the boundary separating the gas from its surrounding medium.

The pressure in the Earth's atmosphere decreases with altitude. If an air parcel near the ground is carried upward through the atmosphere, it will experience a decreasing external pressure. The air inside the parcel will respond quickly to this external pressure change, with the internal pressure closely tracking the external pressure. However, according to Boyle's law (Equation 3.3), as the pressure of the parcel decreases, its volume increases. This increase in volume represents an adiabatic expansion, and the temperature of the parcel decreases accordingly. The opposite response occurs, of course, if the air parcel is subsiding through the atmosphere from a region of lower pressure to one of higher pressure. These situations are illustrated in Figure 3.1.

Adiabatic expansion explains the overall decrease in temperature with increasing altitude in the troposphere (Figure 2.6; see also Section 5.3.1). The rate at which temperature decreases with increasing altitude is called the temperature **lapse rate**. For the case under discussion (that is, an adiabatic process in dry air), the **dry adiabatic lapse rate** applies, which has a value of $10°C/km$. When the air is moist or has a large **relative humidity**, water may begin to condense as the air cools. The relative humidity is defined as the ratio of the partial pressure of water vapor in air to the equilibrium vapor pressure of water over a liquid surface.[13] Conversely, if the air warms again, the condensed water can evaporate. The excess energy that is released (or absorbed) during condensation (or evaporation) is called the **latent heat of condensation** (or **latent heat of evaporation**). Latent heat associated with atmospheric water vapor condensation and evaporation is one of the principal energy sources driving the climate system (Section 12.1.2).

Whenever water condenses or evaporates in an air parcel, the pure adiabatic assumption must be modified. A new lapse rate must be defined, which accounts for the latent heat effect to either warm or cool the parcel. This is referred to as a **saturated adiabatic lapse rate**. Over the entire planet, the average temperature lapse rate is about $6.5°C/km$

13. High humidity leads to the formation of dew and rain. When the humidity is high, the air is filled with water vapor, which can condense on any cool surface—for example, on the windows of an air-conditioned house or a cold bottle of beer. The "vapor pressure" defines the amount of water vapor that air can hold before it becomes "saturated." Once saturated, liquid water (or ice, at very low temperatures) forms in the air as haze or fog or dew or clouds. (See also Section 12.1.2, where the concept of a "vapor pressure" is discussed.)

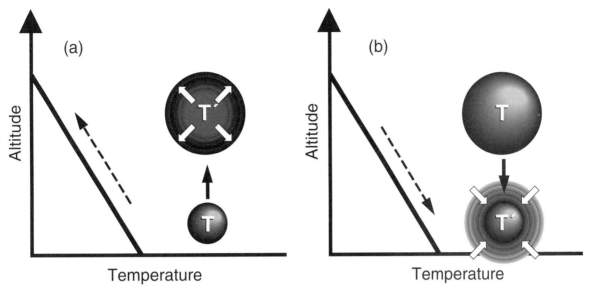

Figure 3.1 The response of an air parcel to vertical displacement in the atmosphere. The parcel is assumed to have an initial temperature T. We imagine it to be a weightless, perfectly elastic balloon. (a) The parcel starts at the ground and expands and cools by adiabatic expansion as it rises. (b) A parcel aloft descends toward the surface, warming by adiabatic compression. The behavior of the temperature of the air in the parcel as a function of the altitude of the parcel is also illustrated. This is referred to as the *adiabatic temperature profile*.

(Figure 3.2). The variation from a dry adiabatic lapse rate (10°C/km) is due to the condensation of moisture in rising air parcels.

For the average temperature lapse rate, the air at the top of a mountain 10,000 feet high would be about $20°$ centigrade colder than the air at the ground. This is why mountaintops can be snowcapped all year around and why the snow line moves up and down a mountain with the changing seasons. The temperature lapse rate is related to the stability of an air mass, which is discussed in Section 5.3.2.

Hydrostatic Balance

The atmosphere is in "hydrostatic" balance. This simply means that the pressure at *every* altitude is just that needed to balance the total weight of the air above that height (actually, the weight per unit area). Weight is the result of gravity, the force of mutual attraction between the Earth and an object above the Earth (sounds romantic, doesn't it?). Anything with mass creates gravity. The greater the mass of an object is, the stronger its gravity will be.[14] The Earth

is very massive, and forcefully attracts the molecules of air closer to the ground. Thermal motions agitate the molecules away from the ground. The balance is hydrostatic.

The weight of any object, W, is its mass, M, multiplied by the **gravity constant**,[15] g (Appendix A):

$$W = Mg \qquad (3.14)$$

In the atmosphere, if M is the total mass of the air above an area at some height (with M measured in kilograms per square meter, kg/m^2), then the hydrostatic law relating M to the pressure at that height may be stated as

$$p = Mg \qquad (3.15)$$

Because the atmospheric density decreases steadily with altitude, the total mass above each altitude, M, also decreases, causing the pressure to fall off with increasing altitude (Figure 2.7).

The law of hydrostatic balance states that the pressure at any height in the atmosphere is equal

---

14. Gravity is the force discovered by Sir Isaac Newton (1643–1727) in 1665 when he supposedly saw an apple fall to the ground and realized that the moon should likewise fall toward the Earth. Newton had fled to the English countryside to avoid the bubonic plague in London. In his spare time that year, he invented calculus,

the advanced form of mathematics that is most useful even today, and discovered the colors of the spectrum in sunlight.

15. The gravity constant is actually the acceleration associated with the force of gravitational attraction.

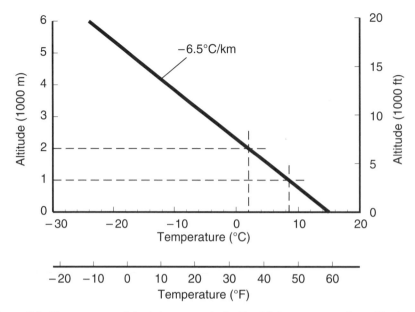

Figure 3.2   The average adiabatic lapse rate in the Earth's lower atmosphere. The lapse rate is very constant at about 6.5°C per kilometer of height, or 4°F per 1000 feet.

to the total weight of the gas above that level, when expressed as a weight per unit area (for example, kilograms per square meter).

The mass burden, M, is related to the air density and the **scale height** of the atmosphere, H, by the simple equation

$$M = \rho H \qquad (3.16)$$

The scale height is a measure of the effective "thickness" of the atmosphere if it had a constant density with altitude, and it has an average value of roughly 10 kilometers. The scale height is related to other now familiar parameters as follows:

$$H = \frac{k_B T}{mg} \qquad (3.17)$$

Notice that by substituting Equations 3.16, 2.5, and 3.17 in Equation 3.15, the ideal gas law, Equation 3.2, can be recovered.

3.1.2   PARTICLES IN SUSPENSION

Small particles suspended in the atmosphere are called **aerosols**. Aerosols are an extension in size and complexity of individual molecules. Even a very small particle is composed of millions of distinct molecules. For that reason, aerosols have properties and behavior different from those of gases. For example, aerosols can settle out of the atmosphere with substantial **sedimentation velocities**, or **fall speeds**, whereas molecules do not fall out in this manner. Take a handfull of beach sand and throw it into the air; most of the grains will fall out almost immediately. However, launch some fine powdery soil from your garden and a puff of dust will probably remain in the air long after the larger debris has dropped out. That fine dust will drift away and may take hours to settle to the ground. The differences between fall speeds of fine particles and those of coarse ones is quite obvious.

Cigarette smoke consists of particles so small that they do not effectively fall out of the air. They must be removed by other processes. One of these is **dry deposition**. When a smoke particle comes into contact with a wall or another surface, it readily sticks to that surface. Molecules are not typically this sticky. Thus smoke tends to foul draperies and haze mirrors by adhering to their surfaces. Aerosols can also **coagulate**, or stick together, when they collide with one another. Coagulation has two important effects. First, it directly reduces the number of particles in the air. Second, the coagulated particles are larger than the original aerosols and can settle out more quickly.

Aerosols have three kinds of motion. Normally, atmospheric winds and related air movements (convection, advection, and turbulence) provide the main modes of transport, although large particles are also subject to sedimentation, as we noted. Finally, the smallest aerosols experience **Brownian motion**.[16] Brownian motion is an apparent random, jerky movement of very small particles caused by fluctuations in the rate at which they are bombarded by molecules in the surrounding gas. The aerosols appear to take a **"random walk"** through space, a motion that has been compared to the staggering walk of a drunk (Section 5.2.1). Brownian motion is responsible for the coagulation of small particles, which represents an important loss process for these particles.

Gas-particle interactions are important for some aerosols. **Primary aerosols** can be formed by mechanical means, such as wind lifting dust from the ground or smoke belching from industrial facilities. Aerosols can also form by the condensation of vapors; these are called **secondary aerosols**. For example, emissions of sulfur gases from power plants form sulfuric acid in the atmosphere, which can condense as a sulfate aerosol. Similarly, many organic compounds generated photochemically in smog can condense on particles. The aerosols then can become part of the overall pollution problem.

As a result of their mechanical behavior, aerosols represent an important aspect of air pollution. These small particles do not readily fall out of the atmosphere and thus may remain suspended for days or weeks and be transported over great distances. If injected directly into the stable stratospheric layers, aerosols can remain aloft for several years, causing effects on a global scale. In smoggy air, aerosols are inhaled and may be deposited deep in the lungs, posing a significant health hazard. Our primary interest in aerosols relates to their effects on atmospheric radiation and climate and to their role as pollutants, which will be discussed in later sections. Aerosols also contribute to the chemistry of the atmosphere and the formation of clouds (Section 3.1.3). Further, aerosols are important industrial materials: From talcum powder to tints for paints, from cake flour to kitchen cleanser, aerosol technology is present everywhere in our lives (but thankfully there is more to life than small particles!).

16. Named for Scottish botanist Robert Brown (1773–1858), who discovered the effect in 1827 while observing under a microscope the apparently random movements of particles suspended in water.

3.1.3 CLOUDS AND PRECIPITATION

Clouds and precipitation represent an entire field of study connected to meteorology. The first observers of the sky wondered about clouds, catalogued them, and related their appearances to changes in the weather. Clouds provide a pleasant diversion on a lazy summer afternoon. They create beauty at sunset. And clouds are essential to the pleasant climate and bountiful life on our planet. (The climatic effects of clouds are referred to in Chapter 11; their general effects on atmospheric radiation and light are summarized in Section 3.2.)

It is clear after years of scientific investigation that clouds are composed of water droplets and ice crystals condensed upon small particles, or aerosols (Section 3.1.2). Not all aerosols act as condensation centers, however. A subgroup of the total aerosol population called **cloud condensation nuclei**, (**CCN**) are responsible for forming clouds. Typical CCN concentrations are 100 to 1000 per cubic centimeter (100 to $1000/cm^3$). The total aerosol concentration can be 10 times larger.

Because water vapor is relatively plentiful in the atmosphere (up to a large proportion of air [Table 2.1]) and readily condenses, cloud droplets and crystals can grow to be 10 to 1000 times larger than typical aerosols. Although the basic physical processes of cloud droplets are similar to those of aerosols, differences arise because of the greater mass of a cloud drop. The fall speeds of cloud droplets are slow enough that clouds appear to be stationary when seen in the sky. To form precipitation, millions of these cloud droplets must **coalesce** to form a precipitation drop, or **hydrometeor** (literally, a falling object composed of water). **Precipitation** consists of a stream of hydrometeors (in the form of droplets or ice crystals).

You may have observed the processes of coalescence and precipitation on the windshield of your car. In a dense fog or in light drizzle, a car's windshield becomes covered with thousands of tiny water droplets that stick to the glass by surface tension (these small drops have too little weight to slide down the surface). Some of the droplets grow larger by colliding with droplets falling from the air or by coalescing with neighboring droplets on the windshield. Apparently at random, one droplet begins to grow rapidly by collecting up the surrounding droplets (such a random process occurring over time is called a **stochastic process**). This larger droplet now has sufficient weight to begin sliding down the glass. As it does so, it engulfs many more droplets, growing even more rapidly and accelerating down the window. A clear streak is left as this "precipitation" drop sweeps up all the droplets in its path.

There are many forms of natural precipitation. Drizzle is the gentlest form of precipitation, consisting of very small (millimeter size) water droplets. Ordinary rain is the most common form of precipitation, in which droplets can approach a centimeter in diameter. The precipitation of ice occurs in the form of **hail**, large spherical ice balls consisting of layers of frozen water; **graupel**, or granular hail formed when frozen particles collect together; and **snow**, consisting of open, lacy, crystalline ice structures. Hailstones occasionally grow as large as golf balls and, in rare events, the size of baseballs. These large, dense ice particles are associated with violent thunderstorms with powerful convective cells. Hail can cause significant damage to crops, not to mention the finish on your car. Intricate snowflakes, falling gently on a cold winter evening, are marvelous to inspect closeup (it is true that no two snowflakes are identical!).

Intensely convective clouds produce heavy precipitation and strong electrification in the form of lightning and thunder. Cloud and precipitation droplets tend to become electrically charged during the evolution of a storm (the exact process that causes this charging has not been identified). Because different-size particles hold different charges and fall at different speeds, the positive and negative charges are separated over time, with the negative charge accumulating at the base of the cloud and the positive charge building up at the top of the cloud and at the ground. This charge-separation process creates voltages that eventually become so large (several hundred million volts) that an electrical discharge occurs between the ground and the cloud or within the cloud itself. These atmospheric "sparks," or **lightning strokes**, can be kilometers in length and carry tens of thousands of amperes of electricity.[17]

The sudden release of electrical energy and heating of the air along the lightning stroke channel (to temperatures as high as 30,000 kelvin) creates shock waves that, at a distance, become the rumbling sound of **thunder**. Lightning heats the air along the stroke channel so strongly that it causes significant changes in the composition of the air. For example, lightning generates ozone from oxygen; the ozone may be noticed as a sweet, or "fresh," aroma after a lightning storm.

Cleansing the Atmosphere

All forms of precipitation have a critically important function in maintaining a clean environment. Precipitation scavenges and removes most pollutants from the atmosphere. This occurs by two processes: **rainout** and **washout**. Rainout occurs when gases and aerosols present in clouds are incorporated into the cloud particles, which later coalesce to form rain. The pollutants are thereby carried to the ground and generally do not reenter the atmosphere afterward. In washout, gaseous and particulate pollutants below a cloud are collected and removed by the falling raindrops and snowflakes. The air looks and smells cleaner after a rainstorm in part because of the removal of pollutants, but also because the air brought in by a storm is usually cleaner.

3.2 Radiation and Energy

Before describing the effects of radiation in the Earth's atmosphere, we must explain the basic principles governing radiation fields. *Radiation*, in the sense it is used here, is really a shorthand way of saying **electromagnetic radiation**. The term radiation also is used to describe the emissions from radioactive material (Radioactivity is discussed in Chapter 7 in a separate context). The theory of

---

17. Electric current is measured in amperes. One ampere (amp) is equivalent to the flow of 6.2×10^{18} electrons per second past a fixed point. A typical light bulb uses about 1 amp of current at 110 volts. The ampere is named for Andre-Marie Ampère (1775–1836), the French physicist who founded the science of electromagnetism. Electrical current is driven by an electrical potential, which is measured in volts. The electricity in a home is close to 110 volts. The volt is named for Italian physicist, Alessandro Giuseppe Volta (1745–1827). Volta constructed the first electrical battery. He followed the lead of Luigi Galvani, who had discovered that two different metals inserted into the muscle

of a frog produced an electric current. A debate arose when Volta claimed he could get the same effect without the frog. Volta won the argument, to the good fortune of frogs everywhere. Napoleon was so impressed with the discovery that he made Volta a duke. Volta was also clever enough to isolate and identify methane gas for the first time. The electrical power dissipated in a circuit is calculated as the product of the current flowing in the circuit and voltage applied. If the current is given in amperes and the voltage in volts, the product is the power in watts (that is, joules per second [Section A2]). In a lightning stroke, the peak power output can reach 10 trillion watts.

electromagnetic radiation, developed by James Maxwell[18] in 1864, is one of the great successes of classical physics. This elegant theory deals precisely with forms of radiation ranging from the gamma rays emitted by an exploding supernova, to the diagnostic X-rays used by doctors and dentists, to the light of the sun, to the heat radiating from a fireplace, to radio and television signals. All these are forms of electromagnetic radiation, and all are governed by the same laws.[19] The only difference between them is their frequency, or wavelength.

Electromagnetic radiation can be thought of as waves moving through space at the speed of light (the fastest that anything can go, 300,000 kilometers in 1 second!). There are many different kinds of waves in nature; all have an associated **wavelength**. Figure 3.3 illustrates the meaning of wavelength for several different types of waves. The speed at which traveling waves, like electromagnetic radiation, move is usually determined by the medium through which the waves are propagating, and can usually be taken as a constant (the speed of light, $c = 300,000,000$ [or 3×10^8] m/sec in air). The **frequency** of the wave is usually fixed by its source of energy; for example, the radiation emitted by a molecule has the frequency of internal oscillation of the molecule. Once the speed and frequency of a wave are fixed, the wavelength, λ, is automatically determined by the simple relationship

$$\lambda = \frac{c}{f} \qquad (3.18)$$

where f is the frequency. Frequency has units of inverse time and is measured in **hertz**,[20] or vibrations/sec. Wavelength, as expected, has units of

18. James Clerk Maxwell elegantly elucidated the nature of electromagnetic radiation in his brilliant *Treatise on Electricity and Magnetism* (1873). All the properties of electromagnetic waves and light were summarized by Maxwell in a few simple equations, the famous "Maxwell equations."

19. Maxwell's, and earlier, theories of light invoked a mystical substance—the "ether"—which acted as the medium for the propagation of light waves. The ether permeated all space, was weightless, and, as it turned out, did not exist. Fortunately for Maxwell, his theory was robust enough to survive the evaporation of the ether.

20. Heinrich Rudolph Hertz (1857–1894) was a German physicist who performed early experiments with radio waves to test Maxwell's theory of electromagnetic radiation. In 1887, he confirmed that light and heat radiation, like radio waves, are a form of electromagnetic radiation. His nephew, Gustav Hertz, received the Nobel Prize in physics in 1925 for his studies of the interactions between electromagnetic radiation and atoms. Such interactions produce photochemical reactions (Section 3.3).

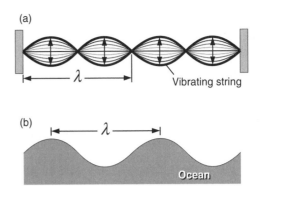

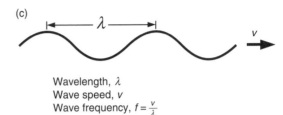

Wavelength, λ
Wave speed, v
Wave frequency, $f = \frac{v}{\lambda}$

Figure 3.3 Different types of waves: (a) standing waves, such as those occurring when a guitar string is plucked; (b) ocean waves, which are initially driven by winds but subsequently propagate because of gravity; and (c) traveling waves, which are responsible for sound and light, and generally move at a fixed speed (the speed of sound or the speed of light). The wavelength, λ, in each case can easily be recognized in the drawings.

length, although the wavelength can vary over many **orders of magnitude** (that is, by factors of 10, upward or downward), from very small values (billionths of a meter) for gamma rays to very large values (tens of meters) for radio waves.

In most common materials, the wavelength of electromagnetic radiation is inversely proportional to the frequency of the radiation. The proportionality constant is the speed of light in that material. Equivalently, for electromagnetic radiation traveling through space or a gas such as air, the wavelength multiplied by the frequency equals the speed of light in free space (the highest velocity possible in nature).

The energy of the radiation depends on the wavelength (or frequency). This dependence is given by another simple relation,

$$E = hf = \frac{hc}{\lambda} \qquad (3.19)$$

where E is the energy carried by a single unit of electromagnetic radiation and h is the constant named after the famous physicist Max Planck, who discovered this relationship.[21] (Physical constants are given in Appendix A.) The smallest unit of electromagnetic radiation has been measured and is called a **photon**. One photon carries an amount of energy given by Equation 3.19, which depends on the frequency of the radiation. Photons can be thought of as fundamental *particles* of radiation; radiant energy cannot be subdivided into units smaller than photons.

Packets of photons make up waves, and groups of waves comprise a beam, or field, of radiation. All **electromagnetic fields** actually consist of coupled electric and magnetic waves or fields (hence the term "electromagnetic"). The electric and magnetic fields are so tightly coupled that it is not necessary to differentiate between them in most discussions. Accordingly, only the electric field need be mentioned. A photon can be thought of as carrying both electric and magnetic field energy.[22]

The law of photons states that the smallest increment of energy that is carried in a beam of radiation having a given frequency and/or wavelength is proportional to the frequency of the radiation or inversely proportional to the wavelength of the radiation. In the first case, the proportionality constant is Planck's constant; in the second case, it is Planck's constant multiplied by the speed of light.

Spectrum of Radiation

Beams of radiation are generally composed of waves of different wavelengths. The distribution of energy

in the beam, or the concentration of photons, varies with the wavelength; that variation is called the **spectrum** of the radiation. The spectrum can be displayed as a graph showing the **intensity** (or energy in the beam) of the radiation versus the wavelength. Figure 3.4, for example, shows the spectrum of sunlight, including the common colors of light that we perceive. The spectral regions and units of wavelength of interest here are defined in Table 3.1 and in Appendix C.

The key spectral regions are **visible radiation** (which we can sense with our eyes), **ultraviolet (UV) radiation** (which we cannot see but which is responsible for tanning our skin), and **infrared (IR) radiation** (which, although invisible, may be felt as heat). As the names suggest, *ultra*violet refers to radiation with frequencies above those at the violet (or extreme blue) end of the visible spectrum, and *infra*red to radiation with frequencies below those at the red end of the visible spectrum. The wavelengths of the radiation in these spectral regions vary in the opposite sense from the frequency. The higher the frequency of a photon is, the shorter its wavelength will be; that is, frequency and wavelength are inversely related (as defined in Equation 3.19). Infrared wavelengths are therefore longer than visible wavelengths (frequencies are lower), and ultraviolet wavelengths

Figure 3.4 The basic spectrum of sunlight. The intensity of sunlight peaks in the visible part of the spectrum and decreases in the ultraviolet region, at shorter wavelengths, and in the infrared region, at longer wavelengths. Within the visible region, the spectrum can be further subdivided into the primary colors of light: blue, green, and red. The visible region actually contains a continuous spectrum of colors ranging from deep red at one end to violet at the blue end.

21. Max Karl Ernst Ludwig Planck (1858–1947), a German physicist, discovered the discrete nature of radiation and contributed to the advance of modern quantum physics, for which he won the Nobel Prize in physics in 1918.

22. Magnetism is a fascinating and important basic property of matter. Magnetism results from the motions of charged particles, usually the orbiting electrons in atoms or the electrical currents traveling in a wire or through an ionized vapor, as in a neon lightbulb or a lightning stroke. The apparently fixed magnetism of certain minerals and metals, as in a common magnet, is actually caused by the electron motions in the atoms of the materials that make up the magnet. The discovery of natural magnetism by the Chinese, perhaps more than 4500 years ago, led to the invention of the compass. This allowed navigation over vast expanses of open ocean. The Earth itself is an enormous magnet due to powerful electric currents flowing in the metallic molten core, which act as a large electric circuit deep inside the planet. The list of famous scientists who have studied magnetism over the past two centuries includes Charles-Augustin Coulomb, Joseph Priestley, Andre-Marie Ampère, Michael Faraday, James Clerk Maxwell, and Heinrich Hertz.

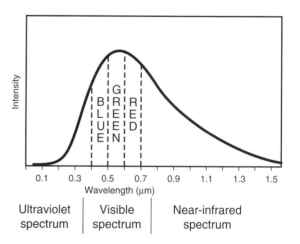

Table 3.1 Wavelength units and spectral regions

Wavelength unit	Measure	Spectral region
Angstrom (Å)	$1 \text{ Å} = 1 \times 10^{-10}$ meters	X ray (< 1 Å)
Nanometer (nm)	$1 \text{ nm} = 1 \times 10^{-9}$ meters	Ultraviolet (0.1–400 nm) Visible (400–700 nm)
Micrometer (μm) (Micron)	$1 \text{ } \mu\text{m} = 1 \times 10^{-6}$ meters	Near infrared (~0.7–5 μm) Thermal infrared (~5–100 μm)

Note: The following abbreviations hold for the spectral regions of interest: ultraviolet = UV; infrared = IR. Also see Appendix C.

are shorter (their frequencies are higher). To put the size of a typical wavelength of light in perspective, a sheet of paper or a human hair is about 100 **micrometers**, or **microns** (μm), thick, and the wavelengths of visible radiation are ~0.4 to 0.7 μm. The common, often confusing, nomenclature used to describe radiation is discussed in Appendix C.

3.2.1 SUNLIGHT AND HEAT

The enormous energy generated by nuclear reactions deep within the sun is eventually emitted to space mainly as light. The spectrum of sunlight (the **solar spectrum**) is illustrated schematically in Figure 3.4. The spectrum actually consists of a continuum of wavelengths. In the visible part of the spectrum, this continuum can be displayed using a prism that disperses the different colors of light into a continuous rainbow of colors. Each color has a specific fixed location within the spectrum.

The Primary Colors

Three colors of light are considered to be **primary colors**: blue (with wavelengths of about 0.45 μm), green (0.55 μm), and red (0.65 μm). All other colors can be produced using combinations of the primary colors (in other words, by mixing together light waves of the primary wavelengths, or frequencies, in specific proportions). In fact, when the three primary colors of light are combined in equal proportions, the resulting light is white. The combination of primary colors in pairs produces specific intermediate colors: Green and red produce yellow; blue and red, produce magenta; and blue and green produce cyan. Direct sunlight is somewhat yellow in color because

of the slightly greater intensity of the green and red components. Clouds appear white because they scatter all three primary colors with roughly equal strength (Section 3.2.2).

The color we perceive in a substance is the result of the combination of the wavelengths of light that are transmitted, reflected, or emitted by that substance. The perception of color involves the relative sensitivity to different wavelengths of the human sensor of light, the retina of the eye, and the rules applied by the interpreter of color, the brain. Optical illusions can occur when our brains attempt to interpret ambiguous light signals. Sometimes we see shadows moving in the dark; occasionally lights seem to float mysteriously in the air. Color is also interpreted by the brain using certain rules that may become confused. For example, the three primary colors in combination appear white. An object appears blue in color because it reflects blue light and absorbs red and green light. If you stare at a bright blue object for a minute and then look at a white background, the "afterimage" of the object will appear yellow in color, which is the combination of the *complementary* colors green and red. The blue appears to be absent because your retina, in response to the brightness of the blue object, has turned down the blue sensors on which the image of the object fell. When you look away from the object, these sensors take several minutes to recover. Your brain can interpret only those signals it receives from your eye, and thus "sees" only the green and red light reflected from the white background. Similar effects are seen in the afterimages of objects or lights of other colors.

Nature rarely produces pure colors. There is no single blue wavelength; indeed, there are countless frequencies of light that fall within the blue portion of the visible spectrum and thus appear blue to the eye. Likewise, any practical color of light consists of countless frequencies, among which some are dominant and define the principal coloration. Scientists have recently developed methods to obtain spectrally pure radiation (that is, having only a very narrow range of frequencies or wavelengths) using laser beams and other devices. On rare occasions, natural light is spectrally pure—for example, during certain auroral displays at high latitudes. More often, natural radiation is very diverse spectrally, like sunlight.

The intensity of sunlight peaks at a wavelength of about 550 **nanometers** (**nm**), or 0.55 micrometers or microns [μm]), which is at the center of the visible spectrum (Table 3.1). Vision in animals has evolved to use this radiation advantageously. Similarly, green plant photosynthesis is activated by plentiful visible radiation. The solar intensity in the ultraviolet spectrum (on the short wavelength end of the solar spectrum) is much smaller than in the visible region. Only about 1 percent of all the solar energy is in the form of UV radiation. At the other end of the solar spectrum, at longer wavelengths in the **near infrared** (near-IR) region, a considerable fraction of the total solar energy is found—about 50 percent, in fact. Although plants cannot utilize this energy and we cannot see by it, the near-IR solar energy is important to the energy balance and climate of the Earth (Chapter 11).

The spectrum and intensity of sunlight reaching the Earth are relatively constant over time. Accordingly, solar radiation may be considered a fundamental "constant" of the environment. The detailed spectrum of sunlight impinging on the Earth is illustrated in Figure 3.5. Note that the solar radiation reaching the ground is substantially modified by the presence of an atmosphere, owing to scattering and absorption by air molecules and clouds (Section 3.2.2). Although the solar intensity at the top of the atmosphere is relatively fixed over long spans of time, the radiation (and energy) absorbed by the Earth can be modulated on much shorter time scales by changes in atmospheric composition and cloudiness. (The radiative energy balance is treated in detail in Chapter 11.)

Solar radiation actually varies slightly over long time spans (Section 11.6.2). Also note that over the history of the solar system—about 4.5 billion years—the intensity of sunlight may have increased by as much as 50 percent. Although these long-term variations are not relevant to recent concerns about global climate

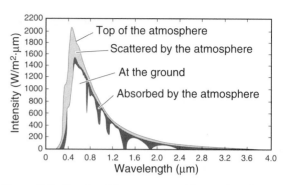

Figure 3.5 Details of the spectrum that reaches the top of the Earth's atmosphere and penetrates to the surface. The outer envelope is the full intensity of sunlight that one would encounter in space. The inner curve is lower because the Earth's atmosphere scatters some of the radiation back to space, particularly at short wavelengths. The shaded region below the inner curve indicates those regions of the spectrum where atmospheric water vapor, carbon dioxide, and ozone molecules absorb the sunlight, further reducing its penetration. The units of solar intensity are often expressed as watts per meter squared per micron of wavelength (w/m²-μm). (From J.N. Howard, J. I. F. King, and P. R. Gast, "Thermal Radiation," *Handbook of Geophysics*, [New York: Macmillan, 1960], Chapter 16, p. 15)

change, they have been critical to controlling the climatic and environmental history of Earth and the conditions under which life evolved (Section 4.2).

It is both interesting and important that the properties of sunlight (intensity and spectrum) are largely determined by the temperature of the sun's upper atmosphere. All material objects that have a temperature above absolute zero emit radiation. The sun is such an object. You are such an object. An ice cube also emits radiation (although it is not very intense). Before discussing sunlight further, therefore, we will discuss the nature of this so-called blackbody radiation.

Blackbody Radiation

What a mysterious name. What could a "blackbody" actually radiate? If you were told the sun is a blackbody, would you believe it? Max Planck, to whom we referred earlier, first described the properties of blackbody (**Planckian**) radiation. All objects emit (and absorb) electromagnetic radiation. Any object that has a temperature above absolute zero (0 kelvin) emits some radiation. In fact, a blackbody emits some radiation at *all* wavelengths *all* the time! However, the intensity of the radiation at a particular wavelength may be so low that it is completely negligible. Blackbodies at different temperatures also have different colors: A blackbody is red at 1000 kelvin, yellow at 3000 K, white at 6000 K, pale blue at 9000 K, and brilliant sky blue at 90,000 K.

If an object has a temperature below about 1000 K, it will emit most of its radiation at wavelengths that are too long to be detected visually (infrared radiation). Still, these objects emit radiation that can be felt as heat (for example, when you bring your hand close to a room radiator or near an electric oven burner) and can be detected by special infrared sensors (or "night vision" scopes). Objects that are 1000 K or hotter emit large amounts of visible light, which can be seen and felt. Examples of light emitters are a red-hot poker, the coals in your barbecue, the tungsten filament of a light bulb, and the sun itself (you can see sunlight and also turn your face toward the sun and feel its warmth). Both heat radiation and light are electromagnetic radiation. Both obey Maxwell's electromagnetic equations. Both can be emitted as blackbody radiation.

Planck's radiation law states that *all* the properties of the radiation emitted by a blackbody (spectrum, intensity, total energy, and so forth) are determined once the temperature of the blackbody is known (Planck's constant, Boltzmann's constant, and the speed of light also enter into the calculation but are well known).

There is a law in radiation physics called **Kirchhoff's law,**[23] which relates the emission and absorption properties of matter. This law states that the **emissivity**, ε, and **absorptivity**, α, are equal at every wavelength, and for any material (including gases), or

$$\varepsilon(\lambda) = \alpha(\lambda) \qquad (3.20)$$

A blackbody by definition emits radiation perfectly; that is, its emissivity is exactly 1.0 at all wavelengths. According to Kirchhoff's law, a blackbody is therefore also a perfect absorber, with an absorptivity of 1.0 at all wavelengths. Herein lies the origin of the term, *blackbody*. It is a perfectly black body in the sense that it absorbs *all* the radiation that falls on it, like a black cape. In reality, perfect

23. Gustav R. Kirchhoff (1824–1887) was a German physicist who defined the relationship between the absorption and emission of radiation by matter. He was also a pioneering spectroscopist who discovered the elements cesium and rubidium using a spectroscopic technique.

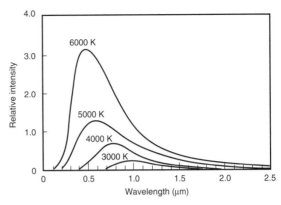

Figure 3.6 The blackbody, or Planckian, radiation spectrum. The intensity varies with wavelength in a smooth and relatively simple manner. The shape and position of the spectrum depend on the temperature. In general, the lower the temperature is, the greater the wavelength of the peak intensity and the lower the overall intensity of the radiation will be. For an object at a temperature of 6000 K, the intensity peaks at about 0.5 microns, similar to sunlight.

blackbodies are very rare. But in many common situations, objects may be approximated as blackbodies—for example, the visible light emitted by the sun or the heat emitted by everyday objects at room temperature.

Kirchhoff's law states that the way a substance absorbs the radiation impinging on it is identical to the way it emits radiation by thermal excitation. In the case of a blackbody, the absorption and emission efficiencies achieve their maximum values at all wavelengths.

Figure 3.6 shows the **blackbody radiation spectrum.** Notice that the spectrum for a 6000 K blackbody closely resembles the solar spectrum shown in Figure 3.5. It is also clear that the spectrum and the total amount of emitted radiation (given by the area under the blackbody curve at a specific temperature) depends sensitively on the temperature. A red-hot iron poker emits a lot of radiation and heat. The human body, a much cooler radiator, emits in total about as much power as does a 60-watt light bulb (not very impressive).

All the properties of a blackbody, including its emission spectrum and energy output, can be defined once the temperature of the object is known. Moreover, the properties of the radiation from a blackbody are completely independent of its composition, size, shape, or any other physical aspect. These amazingly simple properties of blackbodies were originally deduced by Planck, who also derived the

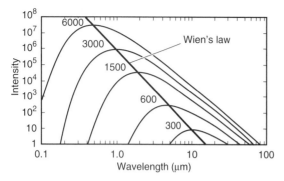

Figure 3.7 Blackbody radiation spectra as a function of temperature (kelvin), over the entire range of temperatures relevant to environmental studies. The values are displayed here on a log-log graph, so that both the wavelength and intensity scales are greatly compressed and cover many orders of magnitude. (From P. R. Gast, Air Force Cambridge Research Laboratory, McGraw Hill (1967). Appendix B of Revision of Chapter 22 of the *Handbook of Geophysics and Space Environments*, Air Force Survey in Geophysics #199, Office of Aerospace Research, USAF, Bedford, Mass. Template only.)

mathematically precise form of the intensity spectrum shown in Figure 3.6. The expression for Planck's spectral distribution function is somewhat complicated and is not given here. However, its smooth form is evident in Figure 3.6. The similarity between the blackbody spectra at different temperatures is more evident in Figure 3.7. Note that any blackbody can be made to emit any of these spectra of radiation simply by changing its temperature.

Two important properties of blackbody radiation are the wavelength of the most intense emission and the total power radiated by the object per unit area. The first property is defined quantitatively by **Wien's displacement law**,[24] which states that

$$\lambda_p = \frac{2900}{T} \, (\mu m) \qquad (3.21)$$

where λ_p is the wavelength (in micrometers) at which the peak emission intensity occurs, and T is the temperature of the blackbody in degrees kelvin. This relationship is quite useful. For example, if we know the wavelength of the peak emission of a star, we can determine its temperature from Equation 3.21 as $T^* = 2900/\lambda_p$. If we would like to know the wavelength of the peak radiation for an

object at room temperature, $T \cong 290$ K, then Equation 3.21 tells us it would be at a wavelength of about 10 μm, which is in the thermal infrared region. Ordinary objects such as chairs and walls radiate heat primarily near this wavelength. In Figure 3.7, Wien's displacement law is represented by the straight line that intersects the peak of each blackbody curve for a different temperature. Notice that the wavelength of the peak is "displaced" as the temperature changes.

Wien's displacement law states that the wavelength at which the blackbody emission spectrum is most intense varies inversely with the blackbody's temperature. The constant of proportionality is Wien's constant (about 2900 kelvin-μm).

The **power** emitted by a blackbody is the *energy per unit time* emitted in the form of electromagnetic radiation. Power can be specified in units of joules per second, or watts (J/sec, or W [Section A.2]). The **Stefan-Boltzmann law** determines the total blackbody power. According to this law, the total power emitted at *all* wavelengths per unit area of surface of the blackbody is

$$F_B = \sigma_B T^4 \qquad (3.22)$$

Here, F_B is the power or energy flux (watts per square meter per second, W/m²-sec), and σ_B is the **Stefan-Boltzmann**[25] **constant** (Appendix A). Equation 3.22 in fact is equal to the total area under the blackbody spectrum (for example, in Figure 3.6). Amazingly, this area turns out to have a relatively simple dependence on temperature. The power emission or energy flux is given per unit area; the total power emitted by an object is just F_B multiplied by the surface area that is free to radiate effectively in all directions (assuming a fairly smooth surface). This makes sense: A large hot object gives off more heat than a small hot object does.

Equation 3.22 states that the power emission of a blackbody is proportional to the "fourth power" of its temperature (Section A.4). That is, if the temperature of the object doubles, the energy it emits per unit area per second will increase by a factor of $2 \times 2 \times 2 \times 2 = 16$; if the temperature triples, the energy increases by a factor of $3 \times 3 \times 3 \times 3 = 81$. This fourth-power relation provides a very strong

24. Wilhelm Wien (1864–1928) was a German physicist who won the Nobel Prize in physics in 1911 for his discovery of the law named after him in 1893.

25. Joseph Stefan (1835–1893) was an Austrian physicist who conducted fundamental research on thermal radiation and the kinetics of gases.

dependence of emitted power on temperature. In another way of looking at it, if you wish to double the power output of a heating element, you need to increase its temperature by only about 20 percent.

The Stefan-Boltzmann law states that the total power (energy per unit time) emitted by a blackbody, per unit surface area of the blackbody, varies as the fourth power of the temperature. The constant of proportionality is the Stefan-Boltzmann constant.

The Wien and Stefan-Boltzmann laws allow one to determine easily the wavelength and power of the radiation emanating from an object, knowing only its temperature. In climate studies of the Earth, this information is fundamental.

Heat Versus Light

Recalling that the sun is a blackbody with a temperature of 6000 K, it is easy to deduce from Wien's law that the spectral intensity of sunlight should peak at about 0.5 μm. The Earth has a temperature closer to 300 kelvin (actually, about 255 K [Chapter 11]) and should therefore radiate energy at roughly 10 μm. The relative spectra of sunlight and terrestrial blackbody radiation, or **Earthglow**, are illustrated in Figure 3.8 (also refer to Appendix C).

The spectra as shown in Figure 3.8 accurately reflect the fact that at the Earth, the total energy flux in sunlight and in Earthglow are about the same (that is, the areas under the respective emission curves in Figure 3.8 are roughly equal). However, from the Stefan-Boltzmann law we know that the power output of an object per unit area is proportional to T^4. Since the sun's temperature is about 6000 K and the Earth's is about 300 K, the solar output should be about 160,000 times greater than that of the Earth. In fact, it is. The reason that the solar intensity

Figure 3.8 The relative spectra of sunlight and Earth's blackbody radiation (referred to as terrestrial radiation or Earthglow). The spectral regions of the emissions are seen to be quite distinct, with little overlap of spectra.

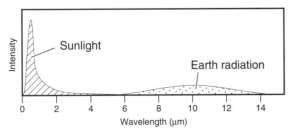

is so much lower when measured *at the Earth* is that the solar radiation has spread out over a much larger (spherical) area in traveling from the sun to the Earth. It is simple to correct for this effect by multiplying the solar spectrum found at the surface of the sun (described using Planck's function) by a factor that accounts for the increased area that the sun illuminates at a distance:[26]

$$\left(\frac{R_s}{D_{es}}\right)^2 \qquad (3.23)$$

where R_s is the radius of the sun (~700,000 km) and D_{es} is the distance of the Earth from the sun (~150,000,000 km). The term defined in Equation 3.23 is about 1/45,000, smaller than but comparable to the factor of 1/160,000 expected on the basis of estimated radiation temperatures alone. The difference between these numbers can be fully reconciled through a careful consideration of the emission temperatures of the sun and the Earth, and of the reflectivity of the Earth, referred to as its **albedo** (Sections 3.2.2, 11.3.1, and 11.6.5).

3.2.2 SCATTERING AND ABSORPTION

Radiation interacts with matter in a number of important ways. The principal types of interactions are illustrated in Figure 3.9. Scattering and absorption are generally treated as microscopic, or atomic scale, phenomena. Reflection, refraction, and diffraction are usually treated as macroscopic processes that occur when extended objects interact with light. Microscopic processes may be analyzed using a combination of atomic theory and Maxwell's equations. The macroscopic phenomena may be formally treated using Maxwell's electromagnetic radiation theory, although in many practical situations the simpler classical **ray theory** may be applied. In this case, light

26. This effect of the weakening of the intensity with distance from a source is common to both light and sound. A campfire looks dim from a distance. A shout is muffled. Nevertheless, the campfire is emitting the same amount of light no matter where you, the observer, happen to be standing. But the total amount of light energy (number of photons) falling on your eye, or the intensity of the light, decreases with distance from the fire, so that your ability to sense it is reduced. In fact, the intensity decreases inversely with the square of the distance from the source, for both light and sound (but less so if the light or sound is *focused*, to concentrate the energy in one direction, as in the beam of a flashlight).

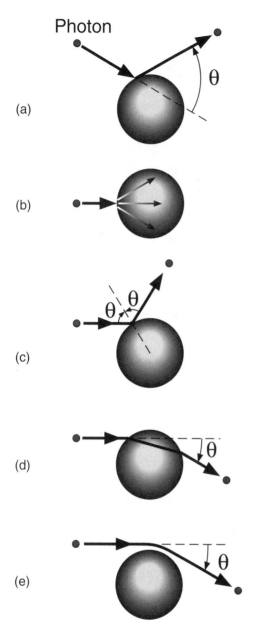

Photon

(a)

(b)

(c)

(d)

(e)

Figure 3.9 The principal interactions between light and matter involve (a) scattering, in which photons of light rebound in all directions from a molecule or particle after colliding with it; (b) absorption, in which the energy carried by a photon is deposited in a material, producing an effect such as heat; (c) reflection, in which a beam of photons bounces from an extended surface at an angle equal to the incident angle of the beam; (d) refraction, in which the direction of a beam of photons is changed by a specific amount when it enters and/or exits a material; and (e) diffraction, in which a beam of photons is deflected from its path by the presence of an object, without actually hitting the object.

is assumed to travel as rays or beams that interact with objects according to straightforward geometric laws. The laws of reflection and refraction are two such relationships. Ray theory is convenient for describing many common atmospheric optical phenomena (for example, rainbows and halos [Section 3.2.3]), but it is inadequate to describe the interaction of radiation with gases and aerosols (small particles).

In the atmospheric context, scattering and absorption involve the processes by which electromagnetic radiation interacts with molecules and small particles. The scattering and absorption properties of gases are determined by their atomic structure and, for particles, by their composition and shape as well. This subject is quite complex in its details, and so we will not develop it here except in a historical and practical context. One way to look at these interactions on a microscopic scale is to imagine particles of light, or photons, colliding with the electrons revolving in orbits around the nuclei of atoms. Photons can bounce off these electrons in various directions (scattering). Or they can knock the electrons into a new orbit or atomic state at a higher energy, the difference in energy between the original and final states being gained from the photon itself (absorption). The photon may be reemitted if the electrons relax into their original states. The electron states in molecules may be classified according to the configuration of the many electrons around the atomic nuclei, the vibration of these nuclei under a restoring force created by the electron orbits (like two masses connected by a spring), and the rotation of the molecules around their center of mass (like a spinning top).

The fact that photons can be scattered in many different directions can be demonstrated by throwing stones at a tree. Depending on the angle at which an individual stone hits the trunk, it can be deflected backward, to one side, or, merely grazing the edge, in a forward direction. After throwing many stones, a pattern might become obvious in the scattering. This would represent a **scattering phase function**, which describes the directional properties of the scattering in a particular case.

In the case of scattering, light energy is conserved. The total amount of energy in the radiation field is not diminished or enhanced, although, the direction in which the radiation moves may be altered. On the other hand, when absorption occurs, energy is removed from the radiation field and is converted to another form (for example, the internal vibration of a molecule). This energy may later be reemitted as

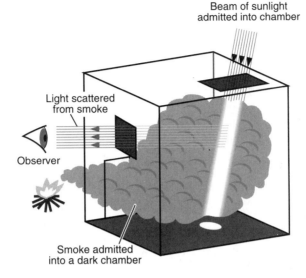

Figure 3.10 Leonardo da Vinci's experiment to demonstrate the origin of the blue sky. He admitted wood smoke and sunbeams into a small, otherwise dark room and observed the light scattered from the smoke against a dark background. The color of the scattered light was noticeably blue.

radiation in a process known as **fluorescence**. Fluorescence, which is stimulated by irradiating an object with an external light beam, is different from blackbody radiation, which is generated from the internal heat of an object.

Why Is the Sky Blue?

Before the twentieth century, one of the great mysteries of natural science and meteorology was the cause of the sky's blueness. Although the Greeks pondered this question, they remained puzzled. The first cogent explanation did not appear until the turn of the sixteenth century. Leonardo da Vinci[27] was an incredibly talented artist (his Mona Lisa still haunts

27. Leonardo da Vinci (1452–1519), an Italian painter, sculptor, anatomist, biologist, architect, geologist, engineer, and physicist, believed that natural phenomena exhibited a logic and order that could be detected by the senses. Leonardo was particularly sensitive to the geometric relationships in nature and believed that all form, inorganic and organic, was the result of forces that could be described precisely and compactly as a set of harmonious laws. Leonardo studied gravity, the conservation of energy, and friction in mechanical motion. His interest in the anatomy and flight of birds led to extended research on the properties of air. He carried out hydrological studies to determine the laws of currents in water and compared these motions with the movement of air. In his later years, Leonardo had dark visions of a global apocalypse; being an accomplished designer of armaments, perhaps he foresaw the invention of weapons of mass destruction and the "nuclear winter" described in Chapter 14.

us with her mischievous smile, and the *Last Supper* is one of the most famous paintings in history), inventor (he designed the first flying machines after studying the anatomy of birds, invented the first parachute, and designed a vehicle powered only by springs), and scientist (he originally identified friction as a universal force that impedes mechanical motion and on observing fossils in the mountains cleverly deduced that over eons the land must have risen out of the sea by means of tectonic motions). Leonardo also devised a simple experiment to show that the color of the sky is caused by the scattering of sunlight from small particles suspended in the atmosphere (Figure 3.10).

The scientific deduction concerning the color of the sky is eloquently described in Leonardo's own words:

The blueness of the atmosphere—it is not intrinsic color, but—is caused by warm vapor evaporated in minute and insensible atoms on which the solar rays fall, rendering them luminous. If you produce a small quantity of smoke from dry wood and the rays of sun fall on this smoke, and if you place (behind it) a piece of black velvet on which the sun does not fall you will see the black stuff appear of a beautiful blue color. Water, violently ejected in a fine spray and in a dark chamber where sunbeams are admitted produces then blue rays. Hence it follows as I say, that the atmosphere assumes this azure hue by reason of the particles of moisture which catch the rays of the sun.

Leonardo's observations and experiments on the scattering of solar radiation were essentially correct, although his work did not reveal why the scattered light was blue, rather than, say, red, or the identity of the particles that actually scattered the light. A quantitative description of this phenomenon did not come until the nineteenth century. Like other great scientists in history, Leonardo created a new idea by observing natural events in everyday life—occurrences that are usually overlooked or taken for granted by others—and connecting them with possible causes and effects through simple conceptual models and experiments.

It was the prolific William Strutt (Lord Rayleigh), known for his identification of argon (Section 2.2.1), who discovered the true origin of the blue sky. The effect, he deduced, is due to the scattering of sunlight by air molecules, not particles, through a process that now bears his name, **Rayleigh scattering**. When molecules such as the N_2 and O_2 in air are exposed to the electromagnetic fields of light waves, the molecules become **polarized** [28] and oscillate at the same frequency as the light. This oscillating polarization in turn radiates energy like an antenna, creating the scattered radiation field. Rayleigh constructed a clever argument to show that the intensity of the scattered radiation, I, depends on the inverse fourth power of the wavelength, or

$$I \propto \frac{1}{\lambda^4} \qquad (3.24)$$

Rayleigh also deduced the scattering phase function for this case, referred to as the **Rayleigh phase function**, and explained why the sky light has a particular polarization.[29]

The Rayleigh scattering law produces sky coloration in the manner shown in Figure 3.11. As noted earlier (Section 3.2.1), sunlight, and white light in general, is composed of roughly equal amounts of the three primary colors: blue, green, and red. Each

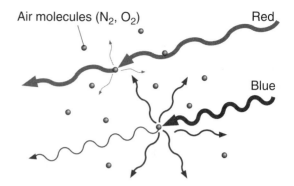

Figure 3.11 Light scattering by air molecules. According to the Rayleigh scattering law, blue light, which has shorter wavelengths than red light, is scattered more effectively by air molecules. Hence the clear sky illuminated by the sun takes on the blue color of the scattered light.

of these colors has a different wavelength: blue, ~0.45 μm; green, ~0.55 μm; and red, ~0.65 μm. According to Rayleigh's theory, these wavelengths are scattered with different efficiencies; Equation 3.24 can be used to define the relative scattering efficiencies. Blue light is scattered about 2.2 times more effectively than green light and 4.4 times as effectively as red light. The sky is blue, then, because the scattered light that we perceive as sky light is greatly enriched in blue photons, compared with green and red ones.

The Rayleigh scattering law states that the effectiveness of a gas molecule in scattering radiation of a given wavelength varies inversely as the fourth power of the wavelength.

Particles and Clouds

Particles suspended in air also scatter radiation. Aerosol particles and cloud droplets are much more effective at scattering light than air molecules are. The scattering (and absorption) of radiation by particles is usually classified by comparing the size of

---

28. In particular, the electric field component of the light wave produces a slight displacement of the negatively charged electrons in relation to the positively charged nucleus of the molecule, creating a small electric charge separation that is referred to as *electric polarization* (meaning, in this case, electric-charge polarization).

29. Polarization of light refers to the way that a light wave is oriented as it moves through space. Light polarization and electric-charge polarization are related, but distinct, effects. Generally, a light wave has two components of polarization that point in orthogonal directions, like crossed swords. When the light wave

is scattered, one component of the polarization may be scattered more strongly than the other. You can study the polarization of scattered skylight using a pair of sunglasses. The lenses of polarized sunglasses are designed to transmit light waves of one polarization, while blocking light rays of the other polarization. Hold the glasses up to the blue sky on a clear day, and slowly rotate one lens as you look through it; the sky will darken and brighten as the lens turns, indicating the degree of polarization of the light scattered from the atmosphere. Moving the lens to a different part of the sky leads to a change in the range of brightness observed through the lens, indicating a variation of polarization across the sky.

the particles, a, with the wavelength of the scattered radiation, calling this parameter $x = 2\pi a/h$. When x is much smaller than unity ($x \ll 1$), Rayleigh scattering theory applies. In the case of small particles, this is called **Tyndall**[30] **scattering**, although the λ^{-4} dependence on wavelength is the same as in Rayleigh scattering. When $x \gg 1$, the regime of ray optics applies, which is discussed in Section 3.2.3.

The Tyndall scattering law states that the total intensity of radiation scattered by a particle whose size is small compared with the wavelength of the radiation varies inversely as the fourth power of the wavelength. The Tyndall and Rayleigh laws are identical in this regard.

For intermediate values of x, a new theory must be used, called the **Mie**[31] **scattering** theory. Mie's theory provides an exact solution of Maxwell's equations for electromagnetic radiation impinging on a uniform spherical particle. Details of the theory are beyond the scope of this book, and application of the theory requires extensive computer calculations. Though impractical when first published in 1908, Mie theory now provides the basis for most studies of atmospheric radiation involving aerosols and cloud droplets.

Aerosols tend to scatter radiation preferentially in the forward direction. The forward direction refers to the direction in which the radiation is traveling. Both Rayleigh and Tyndall scattering result in equal amounts of forward and backward scattered energy. However, as the particles become larger, Mie scattering solutions demonstrate that most of the deflected radiative energy remains concentrated in the forward direction. The consequences of this effect can be seen in common optical effects, such as the brightness of a haze viewed from different directions with respect to the sun. Looking toward the sun, the haze appears much brighter because of the forward-scattered light.

Light "Extinction" and "Optical Depth"

Some materials both scatter and absorb impinging radiation. Many gas molecules, and numerous types of particulate material have this property. In such cases, it is often practical to define the extinction associated with a substance. **Extinction** is the sum of scattering and absorption, and so extinction represents the *total* effect of a material on radiation passing through the material. The effect of scattering, absorption, or extinction by a material can be expressed in terms of the scattering, absorption, or extinction **cross section**, respectively, for that material. These cross sections are basic properties of the material but usually vary with the wavelength of the radiation. The cross section can be expressed as an effective area that a single molecule or particle of the material presents as a target (m^2/molecule) to the photons of radiation or as the total area presented by a certain mass of the material (m^2/kg). The scattering cross section of individual molecules is typically very small, in the range of 1×10^{-21} m^2 to 1×10^{-28} m^2. Of course, the effect of a material on a beam of radiation depends on the total amount of material present. In this case, the amount is most conveniently measured along the radiation beam from the source of the light to the point of observation. The situation is pictured in Figure 3.12.

The amount of a gas or particle cloud along a beam of radiation can be specified as the number of molecules or particles in a column of unit cross-sectional area along the beam, N (Figure 3.12), or as the total mass of the material, M, in such a column. This **column concentration**, or **mass column**, respectively, is related to the number concentration of the molecules or particles along the beam multiplied by the length of the beam through the material:

$$N = nL; \quad M = mnL = \rho L = mN \quad (3.25)$$

where L is the beam length and ρ is the gas density (Section 2.1.2). N would be specified as molecules per unit area (for example, #/m^2), and M would be the mass per unit area (for example, kg/m^2) along the path of interest. The column number or mass may be referred to as the **path length** of the gas in question.

---

30. John Tyndall (1820–1893) was an Irish scientist who studied the interactions of radiation with small particles and deduced a law of scattering similar to Rayleigh's law for molecules. Tyndall was perhaps the first to popularize science and was renowned for his lucid and insightful lectures (which he gave in part to supplement his modest salary as a professor). His interest in aerosols led him to study airborne microorganisms and methods of sterilization using heat radiation.

31. Gustav Mie (1868–1957) was a German physicist who in 1908 published the first detailed theory describing the interaction of electromagnetic radiation with perfectly spherical homogeneous particles of an arbitrary size. Mie's theory could explain theretofore inexplicable phenomena such as the "glory" (associated with intensity peaks in the backscatter phase function) and "blue moons" (caused by the anomalous extinction of monodispersed cloud droplets).

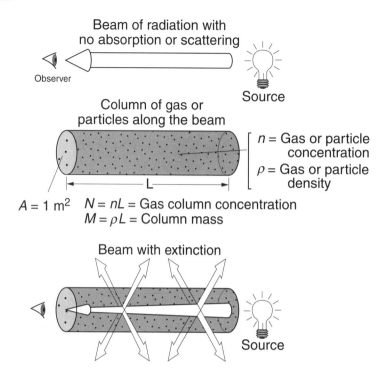

Figure 3.12 The effect of a gas or cloud of particles on a beam of radiation. The gas or cloud is represented as a column of material between the source of radiation and the observer. If the column has a unit cross-sectional area, the number of gas molecules or particles in the column, n, is called the *column concentration*. The optical depth, τ, corresponds to the column concentration multiplied by the cross section, σ, of a single molecule or particle. The optical depth for extinction is the most convenient measure of the effect on the radiation beam. The optical depths for absorption and scattering may be calculated separately for specific applications. The mass column may be substituted for the column concentration when the cross sections per unit mass are specified.

The **optical depth** of the material, τ, is defined as the product of the cross section, σ, and the column amount of the material. The optical depth is a **dimensionless number**; that is, it has no units, but is just a number. Scattering, absorption, and extinction optical depths may be defined. The relationships among these various parameters are (see also Figure 3.12):

$$\sigma_e = \sigma_s + \sigma_a$$
$$\tau_s = \sigma_s N \; ; \; \tau_a = \sigma_a N$$
$$\tau_e = \sigma_e N = \sigma_s N + \sigma_a N = \tau_s + \tau_a$$

(3.26a)

where the subscripts e refer to extinction, s to scattering and a to absorption. Although these relations have been written in terms of the column concentration N, they can just as easily be expressed in terms of the mass column, M, as long as the cross sections, σ, are appropriately defined. Indeed, there is a simple connection between the cross sections corresponding to number, σ_e, and those corresponding to mass, σ_{eM}

$$\sigma_{sM} = \frac{1}{m} \sigma_s;$$
$$\sigma_{aM} = \frac{1}{m} \sigma_a;$$
$$\sigma_{eM} = \frac{1}{m} \sigma_e$$

(3.26b)

The units of the cross sections, σ_M, are area per unit mass (m^2/kg, for example).

The optical depth is useful as a measure of the probability that a beam of light will be affected when passing through a medium (like air). If τ has a value much less than 0.1, the light beam will be essentially unaffected. As the optical depth increases to about 0.1 to 0.5, the fraction of the beam affected is roughly equal to τ. When τ is much greater than 1 ("unity"), most of the light in the beam is affected.

Note that τ varies directly with the concentration of the gas, n, the length of the path, L, and the cross section, σ. If any one of these parameters increases, τ will increase also, and if any one decreases, so will τ.

The optical depth can also be specified as a local property of a gas in terms of the **scattering, absorption**, and **extinction coefficients**. These coefficients are defined as the *optical depth per unit length*, and they have units of *inverse* length (for example, m^{-1}). The *product* of, say, the scattering coefficient for an atmospheric gas and the distance along a path through the gas yields the scattering optical depth for that path. Because they depend on the local concentration of the gas, the scattering, absorption, and extinction coefficients are not basic molecular properties of the gas, as are the cross sections. The coefficients are simply defined as

$$\varepsilon_s = n\sigma_s \; ; \; \varepsilon_\alpha = n\sigma_a \; ; \; \varepsilon_e = n\sigma_e$$
$$\varepsilon_s = \rho\sigma_{sM} \; ; \; \varepsilon_\alpha = \rho\sigma_{aM} \; ; \; \varepsilon_e = \rho\sigma_{eM} \qquad (3.27)$$
$$\tau_s = \varepsilon_s L \; ; \; \tau_\alpha = \varepsilon_a L \; ; \; \tau_e = \varepsilon_e L$$

The various parameters discussed here can be summarized as follows: The cross sections for scattering, absorption, and extinction are fundamental molecular properties of gases and other materials that determine their interaction with electromagnetic radiation. The cross sections can be given for a single molecule or a specific mass of material. The scattering, absorption, and extinction coefficients represent the fundamental properties of a gas expressed in terms of its local concentration or density. The optical depth (for scattering, absorption, or extinction) describes the overall effect of a gas on radiation moving through the gas along a specific path. For each parameter, extinction is the sum of the scattering and absorption.

Coloration and Contrast

The effects of gas and particle scattering on the light of the sky, for a range of particle sizes, is summarized in Figure 3.13. Imagine a beam of pure white light incident on the atmosphere. Selective scattering at certain wavelengths (due to Rayleigh scattering) depletes the direct radiation at those wavelengths. The transmitted light is therefore enriched in the complementary wavelengths. If blue wavelengths are depleted, the complementary wavelengths of green and red are relatively enhanced, adding a yellow tinge to the light. That is why the sun normally appears somewhat yellowish in color. The scattered light, of course, takes on the color of the scattered wavelengths, which is blue in the case of Rayleigh scattering. When clouds are present, the scattering is **neutral**; that is, the scattering efficiency of cloud droplets is independent of wavelength. Hence both the scattered *and* the transmitted light contain equal amounts of the three primary colors and so appear white (Figure 3.13). You may have seen the sun at the edge of a cloud or through a fog bank, looming as a pure white disk because of the neutral transmission of the water droplets.

Table 3.2 indicates the coloring effects of several molecular and particulate absorbers. **Absorbers** affect atmospheric color by removing certain wavelengths from the light illuminating a given material. Both the scattered and the transmitted light are depleted in that color. If blue wavelengths are absorbed, for example, the transmitted *and* the scattered light will appear yellowish to reddish in color (the precise color we see depends on a number of factors, including the optical depth of the material, the angles of illumination and observation, the background brightness, and so on). If all wavelengths are absorbed with about the same efficiency, the transmitted and the scattered light again will appear white. Note, however, that gray is a shade of white and that black, as a hue, is the absence of all color. A cloud of soot thus appears black because it evenly absorbs all the colors of white light and scatters only a small fraction of the impinging light at any wavelength. It is this reduction in reflected white light, when contrasted against a bright background such as the sky, that creates the visual perception of black. However, if you were to scatter light from a cloud of soot in a darkened chamber, the soot would appear bright, because of the contrast with the darker background.

Contrast is the difference in the brightness and color of objects or materials that are adjacent to or overlap one another. The greater the contrast is, the easier it is for the eye to discriminate among the objects. Thus a red object against a red wall may be difficult to see. Animals, both hunters and prey, have evolved natural coloration and patterns to blend in with their environments, to reduce their contrast against the background scene. The army uses camouflage paint on tanks to hide them. Contrast is also important to defining the **visibility** of the atmosphere (Section 6.5). Obviously, it is easier to see an object at a distance with high contrast against the

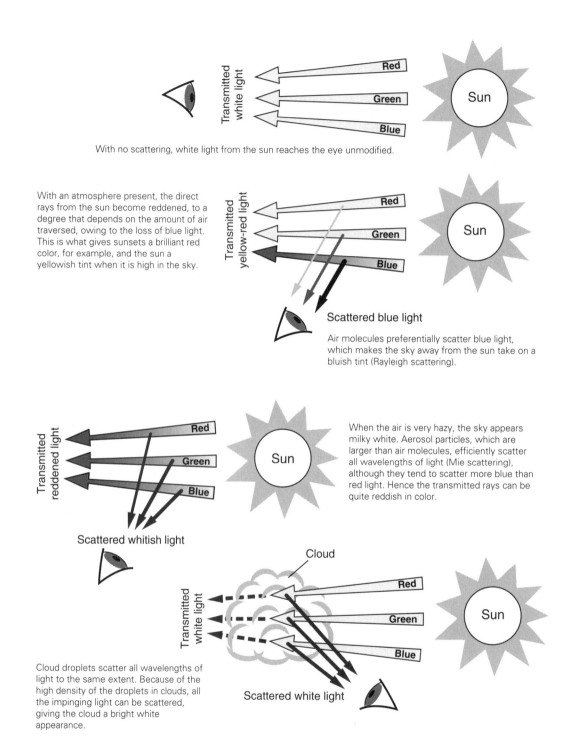

With no scattering, white light from the sun reaches the eye unmodified.

With an atmosphere present, the direct rays from the sun become reddened, to a degree that depends on the amount of air traversed, owing to the loss of blue light. This is what gives sunsets a brilliant red color, for example, and the sun a yellowish tint when it is high in the sky.

Scattered blue light

Air molecules preferentially scatter blue light, which makes the sky away from the sun take on a bluish tint (Rayleigh scattering).

When the air is very hazy, the sky appears milky white. Aerosol particles, which are larger than air molecules, efficiently scatter all wavelengths of light (Mie scattering), although they tend to scatter more blue than red light. Hence the transmitted rays can be quite reddish in color.

Scattered whitish light

Cloud droplets scatter all wavelengths of light to the same extent. Because of the high density of the droplets in clouds, all the impinging light can be scattered, giving the cloud a bright white appearance.

Scattered white light

Figure 3.13 The effects of air molecules and airborne particles (aerosols and clouds) on the transmission and scattering of sunlight, and the color of the sky. The effect of each material on the three primary colors of light is shown. The overall impression of color and brightness depends on the relative depletion or enhancement of the primary colors in the incident and scattered beams, and the contrast of the scattered light against the background illumination.

Table 3.2 Coloring effects of atmospheric materials due to scattering and absorption of light[a]

		Substance				
		Air	Clouds	Haze	Nitrogen dioxide	Soot
Color of light	Blue	Strong scattering	Strong scattering	Strong scattering	Strong absorption	Strong absorption
	Green	Moderate scattering	Strong scattering	Moderate scattering	Very weak absorption	Strong absorption
	Red	Weak scattering	Strong scattering	Moderate scattering	No effect	Strong absorption
	Tint[b]	Blue	White	Gray	Brown	Black

[a]The dominant effect, either scattering or absorption, for each substance and each primary color of light is indicated. The relative magnitude of the effect is indicated by "Strong," "Moderate," "Weak," or "Very Weak."

[b]The apparent color tint of the material as seen by an observer for illumination by sunlight is indicated (the tint can change with the conditions under which an observation is made).

background than one with low contrast. In the atmosphere, contrasts between light and dark areas create the gradations of shading and brightness that we see in clouds on a sunny afternoon, the grayness of an overcast day, and the blackness of the soot billowing from an oil fire.

It should be apparent from these discussions that the prediction and interpretation of color in the sky can be a tricky business.

3.2.3 COMMON OPTICAL EFFECTS

The ray theory of light treats radiation as a beam moving in a straight line. The beam, or ray, can be deflected, however, in several ways: reflection, refraction, and diffraction (Figure 3.9). According to the classic **ray theory of optics**, reflection, refraction, and diffraction phenomena may be expressed in terms of simple laws. Reflection and refraction effects are illustrated in Figure 3.14.

Reflection and Refraction

The **law of reflection** is a simple fact of everyday life. According to this law, the light ray that bounces from a smooth interface between two media (for example, the surface of a still pool of water; the two media in this case are the water and the air over the water) has

the same angle relative to the surface as does the incident ray. The relationship between the angle of incidence, θ_i, and the angle of reflection, θ_r, is

$$\theta_r = \theta_i \qquad (3.28)$$

Mirrors conform to this simple relationship. The law of reflection also holds for curved surfaces as long as the radius of curvature of the surface is much greater than the wavelength of the reflected light. In other words, the reflection law applies to very small sections of a surface that, at the length scales of light waves, appears to be flat.[32] In Figure 3.14, reflection is represented by the incident and reflected rays drawn at the same angle from the vertical line that is perpendicular to the interface.

The law of reflection states that for the reflection of light from a smooth surface, the angle of incidence equals the angle of reflection.

The **law of refraction** is more complicated and involves the **index of refraction** of a material. Every bulk material has an index of refraction, which measures the speed at which light can propagate through that medium relative to the speed of light in

32. An example of this phenomenon of "local flatness" is easily demonstrated by rowing to the middle of a small lake. The lake appears perfectly flat. But in reality, it has the same curvature as the Earth, with a radius of curvature equal to the radius of the Earth, about 6400 kilometers.

free space, c. If the index of refraction is η, then the velocity of light in the medium, v, is given by

$$v = \frac{c}{\eta} \qquad (3.29)$$

Since the index of refraction is always greater than 1, the velocity of light is always reduced in a medium (compared with free space). The reduction in speed is negligible for gases; the index of refraction for air, for example, is 1.0001. Water, on the other hand, has an index of refraction of 1.33, and glass, 1.5. When the indices of refraction of two adjacent media are different, the light rays are bent in crossing from one medium into the other. The angle by which the transmitted ray is deviated from the incident ray depends on the angle of incident ray on the interface and the ratio of the indices of refraction. The situation is defined geometrically in

Figure 3.14 for the conditions that apply roughly to refraction in glass. The **angle of refraction**, or **transmission**, θ_t, is measured relative to the perpendicular to the interface, not from the incident ray. The light ray bends toward the perpendicular to the interface in passing into a medium of higher index of refraction (for example, from air into water or glass). The light ray bends away from the perpendicular when passing into a medium of lower index of refraction (Figure 3.14). This fact is expressed as **Snell's law**.

According to Snell's law, **a light ray traveling in one medium, and being incident on the interface with a second medium, upon transmission into the second medium is deviated from its original direction to a new angle of refraction, θ_t, that is smaller than the original angle of incidence, θ_i, if the index of refraction of the second medium is greater than that of the first,**

Figure 3.14   The laws of reflection and refraction at the interface between two bulk media. The two circles are drawn so that the ratio of their diameters equals the ratio of the indices of refraction of the two media. Solid lines represent rays that originate above the interface; dashed lines represent rays that originate below the interface; and dotted lines represent rays projected by the observer's brain. Squares refer to the situation when the original object lies above the interface; triangles, when the object is below. Reflection is described by rays that remain in the same medium after rebounding from the interface. (the filled figure is the original object; the striped figure, the reflected image; and the speckled figure, the image projected by the observer's brain to the opposite side of the interface.) Refraction is described by rays that cross the interface between the media. (The filled figure is the original object; the open figure, the refracted image; and the shaded figure, the projected image for the refracted ray.) Notice that in each case, the reflected and refracted rays must intercept the circles corresponding to the index of refraction in their respective media at the same distance from the vertical center line (that is, along the vertical dashed line that is fixed by the reflected ray once the angle of incidence is chosen); these are points at which the images appear to lie.

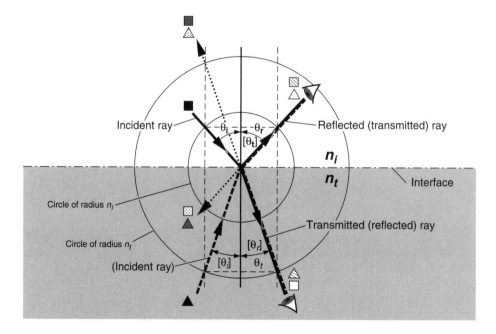

or to a larger angle if the index of refraction is smaller.

The law of refraction is easily demonstrated using a pencil and a bowl of water. When you dip the pencil into the water at various angles, the pencil will appear to be shortened, bent, or some combination of these distortions, owing to the refractive effect of water. Because the brain has no built-in correction for the refractive effect, it always assumes that light travels in straight lines (it accepts the simplest laws of physics and experience). Accordingly, the brain attempts to place the image of an object in a position along the direction that the light from the object entered the eye. This "projection" effect is illustrated in Figure 3.14, in which the eye is seen to place images along straight lines in places where no object actually exists; nevertheless, the brain "sees" the object there. In the pencil experiment (the triangles in Figure 3.14), the object appears compressed and closer to the surface than it actually is. Similar distortions can be seen in a fishbowl or a swimming pool.

Diffraction and Dispersion

Light rays that pass close to an object but do not intercept it may still be deviated from a straight path because of **diffraction** (Figure 3.9). The angle of deviation, called the **diffraction angle**, is a simple function of the wavelength, λ, of the radiation and the size of the object, d:

$$\theta_d = \frac{\lambda}{d} \qquad (3.30)$$

Diffraction is an interesting effect that has few practical consequences in studying the atmosphere, except for its important role in limiting the performance of optical instruments.

The index of refraction of common materials such as water, glass, and air changes with wavelength. This property is called **dispersion** (in the sense that white light can be dispersed into its color spectrum by such materials). The change in the index of refraction over the entire visible spectrum may be very small. In water, blue light has an index of refraction of 1.338 and red light, 1.332. Even so, this difference is large enough to create rainbows, which are basically enormous prismatic displays of sunlight. The reason that dispersion in a material can separate colors is related to the phenomenon of light refraction. The angle of

refraction depends on the index of refraction in such a way that the larger the index is, the smaller the angle of refraction will be. Thus for a beam of light passing through a glass prism, the red rays of light are deviated less than the blue rays are, and the beam displayed at a distance is separated according to color (wavelength).

Rainbows, Halos, Sunsets, and Twilights

Rainbows are beautiful. We stop and wonder at them. There is a pot of gold at the end of a rainbow, but we never seem to be at the end. On rare occasions, two concentric rainbows appear in the sky. What is the cause of these enjoyable phenomena? Very simply, rainbows are the result of the reflection and refraction of sunlight by raindrops. The effect of a single drop on a light ray is illustrated in Figure 3.15(a). The refracted and reflected rays happen to be concentrated at a certain angle, which is determined by the index of refraction of water. This is the rainbow angle of about 42 degrees from the direction of the sun, or 138 degrees from the **antisolar point**, which is the point opposite the sun along a line from the sun through the observer (Figure 3.15[b]). This angle is independent of the size of the raindrops (as long as they are large enough to form a rainbow), which you can demonstrate by expanding or shrinking the image of Figure 3.15(a) on a photocopying machine; the geometry remains the same regardless of the size.

The different wavelengths of white light (sunlight) are refracted, and therefore deviated, at different angles by the drop because of the wavelength dependence on the index of refraction of water. The rays at the red end of the spectrum are deviated the least, and those at the violet (or blue) end, the most. The color spectrum of the rainbow is formed in the order of red, orange, yellow, green, blue, and violet. In the common, or primary, rainbow, the red color is on the *outside* of the bow; the violet, on the inside. This may seem opposite to expectation, inasmuch as red light is deflected the least by raindrops. The effect can be understood by realizing that the smaller the angle that a ray of a specific color is deflected in the backward direction, the farther from the antisolar point that the color will appear to be concentrated.

The rainbow is seen as a prismatic arc of color in the sky; every point on the arc is 42 degrees from the antisolar point. The raindrops causing the rainbow

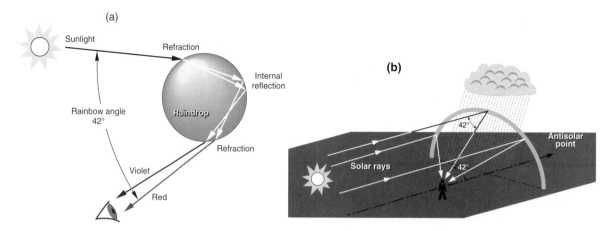

Figure 3.15   (a) The deflection of sunlight rays by a raindrop. The differences in the deviations of the red and violet rays are exaggerated for clarity. The angle of the rainbow is determined by the index of refraction of water. The apparent widths of the rainbow and the color separation are determined by the dispersion of the index of refraction. (b) The rainbow arc is composed of rays of sunlight deflected from countless raindrops to the observer.

actually deflect sunlight in many directions. However, one sees *concentrated* light from the drops only at an angle of 42 degrees from the antisolar point. Put another way, only one observer will see the particular bright rays deflected in that person's direction. In this sense, every rainbow you see is a personal experience. No one else can horn in on your rainbow! To observe a rainbow, the sun must lie behind you, as you probably already know from experience. Moreover, the sun must be relatively low in the sky (at an elevation of less than 42 degrees) for the rainbow to appear above the ground. Segments of rainbows appear when the rainfall in the distance deflecting the sunlight is patchy or localized. Have you ever tried to chase a rainbow? It is a frustrating quest indeed, since the rainbow will *always* remain 42 degrees away from you.

A number of variations on the common rainbow may be observed on rare occasions. These phenomena include the secondary rainbows, which lie outside the primary bow and have their color sequence reversed; red bows, which are colored by a reddened sun at twilight; vibrating rainbows which are caused by raindrops oscillating because of thunder; and fog bows, which consist of hazy white arcs associated with the reflection of sunlight from fog droplets.

Halos are similar to rainbows in that they are caused by the refraction of sunlight. In the case of halos, the refractors are tiny ice crystals suspended in the atmosphere, often in cirrus clouds. Ice condenses in specific crystalline shapes, mainly with hexagonal conformations. These hexagonal crystals refract light, as shown in Figure 3.16. Here the deflection is in the

forward direction, with the greatest intensity occurring at an angle of about 22 degrees (the common halo). The observer views the halo looking toward the sun. Refraction leading to the halo is caused primarily by flat plate hexagonal crystals. Although these crystals have random orientations as they fall through the atmosphere, the majority become oriented with their flat faces parallel to the ground.[33] Accordingly, halos that are low in the sky have, corresponding to these highly oriented crystals, two bright points referred to as **sundogs** (perhaps because they faithfully accompany the sun as it moves across the sky) (Figure 3.16).

Halos also differ from rainbows in that the red arc is seen on the inside of the halo and the blue on the outside (although it is often difficult to discern the colors in a halo, which typically appears as a brownish ring around the sun). The common halo can always be distinguished by its fixed angle of 22 degrees from the sun. There are many less common, oddly shaped ice crystals and hence ways in which light can be refracted through ice clouds, leading to a variety of halo phenomena. In spectacular displays, a dozen halos and arcs may be juxtaposed in the sky.

When the sun sets in the evening (or rises in the morning), the sky can explode in brilliant color. The effects are due to the scattering and refraction of sunlight. The atmosphere is a refractive medium, although the refractive effect is small and great

33.   Objects falling through air tend to stabilize in a specific orientation so that the drag (wind resistance) on the object is maximized. Therefore, when you drop a piece of paper, it tends to fly horizontally.

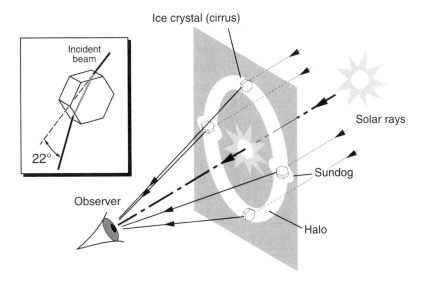

Figure 3.16 The formation of a common halo through the refraction of sunlight by ice crystals. Inset: The refraction of a ray of light through a hexagonal crystal occurs at a deflection angle of about 22 degrees; different wavelengths are deflected at slightly different angles owing to the dispersion of the index of refraction of ice. The observed halo consists of light refracted through countless ice crystals with specific orientations.

distances are required to produce a noticeable deflection (the refractive index of air, 1.00001, is just about as close to 1, and thus no deflection, as you can get). Nevertheless, astronomers are concerned about refraction and must correct for it in careful measurements. The most obvious manifestation of atmospheric refraction is the flattening of the sun's disk when it is near the horizon. The geometry is given in Figure 3.17. The refractive effect of air is such that when the image of the sun is just seen touching the horizon, the sun is actually already fully below the horizon.

Another common atmospheric refractive effect is the **mirage**. In a mirage, light rays are refracted as they pass through adjacent layers of air in which the density is strongly modified by temperature gradients, usually associated with either very warm or very cold air masses. Looking down a long asphalt road on a hot summer day, with the air shimmering from the heat, fractured images and reflections of cars can be seen appearing and disappearing as in a wavy mirror. On a desert, images of distant objects may appear to be close by, and nearby objects may become invisible. These images may seem like hopeful hallucinations to those lost in the desert, where chasing mirages can be deadly.

The twilight glow that we enjoy in the evening (it also occurs in the morning, but we are not usually as attentive then) is produced by the scattering and absorption of sunlight in the atmosphere. Figure 3.18 shows the locations of the principal twilight colors in the sky, and the origins of the rays contributing to the palette. The red skies at the horizon result from the wavelength dependence of the extinction of sunlight (when direct solar radiation is not present, considerable scattered and refracted sunlight is available in the early twilight sky). Blue rays are strongly attenuated by Rayleigh scattering in the dense air near the horizon, and the red rays suffer much less extinction and thus are prominent. As one looks higher in the sky above the horizon, the colors shift toward the blue end of the spectrum. The air is thinner at high altitudes, and the blue rays can penetrate farther through the atmosphere before being scattered, to brighten the evening sky. The red rays also penetrate but are too weakly scattered to affect the color at that height. At higher altitudes, the air becomes so thin that even the scattered blue light is too faint to be seen.

The **purple glow** actually originates in the lower stratosphere and is the result of two effects (Figure 3.18[b]). First, the scattering of the sun's rays in

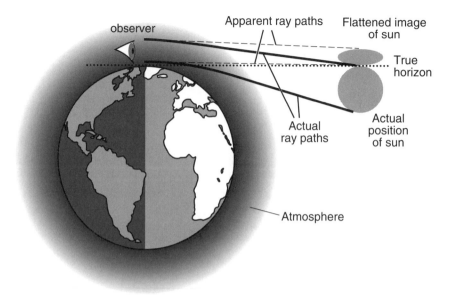

Figure 3.17   The flattening of the sun's image when it lies near the horizon is caused by atmospheric refraction. When traveling to an observer, the light rays originating at the bottom of the sun's disk must pass through a denser layer of air than do the rays from the top of the disk. The lower rays are thus more highly refracted. In reconstructing the image, the observer's brain traces straight paths that tend to raise the bottom of the sun to a greater extent than the top. The sun's image is flattened accordingly.

the lower stratosphere is enhanced by tiny sulfuric acid particles that are always present there (these particles, surrounding the Earth, comprise the so-called **Junge, or sulfate aerosol, layer**).[34] Second, the rays scattered by these aerosols also pass through the ozone layer, which absorbs some of the green light. After scattering from the aerosols, the remaining blue and red light combines to form the faint iridescent purple arc that caps the twilight on clear evenings (Figure 3.18[a]).

3.3   Chemistry and the Environment

In considering the mechanical properties of a gas like air, it is sufficient to treat the gas molecules as simple microscopic spheres that collide with one another under thermal agitation. In considering the radiative properties of a gas, the molecules must be further characterized according to their basic internal structure to determine their specific vibrational and rotational motions. When studying the chemistry of gases, beyond the simplest processes, the interactions between the atoms in a molecule and between molecules undergoing collisions must also

34.   Named for German scientist Carl Junge, who first collected stratospheric sulfate particles in 1960 while searching for radioactive debris from nuclear test explosions.

be defined. Before proceeding, however, we will present the fundamental chemical nomenclature used in this book. It is designed to enable students with little formal training in chemistry to understand basic chemical formulas, which provide a useful shorthand notation for the chemical processes occurring in the atmosphere.

The elements from which most atmospheric compounds are made are listed in Table 3.3. Note that out of some 90 natural elements, only a handful are important to atmospheric chemistry (except in special cases, some of which are discussed elsewhere in this book). In fact, only the first four elements in Table 3.3 are the essential ones. Nevertheless, the number of stable compounds that can be generated from these elements is staggering, particularly when biologically produced organic compounds are included. Literally thousands of compounds are found in natural and polluted air. We will focus on a few dozen key species.

3.3.1   Symbols and Terminology

All the elements and chemical compounds made up of these elements have specific chemical names and formulas. For example, the molecule consisting of two oxygen atoms is written as O_2, which is

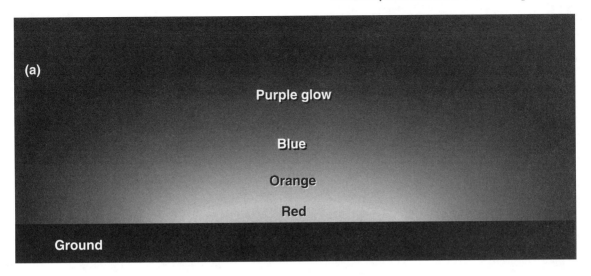

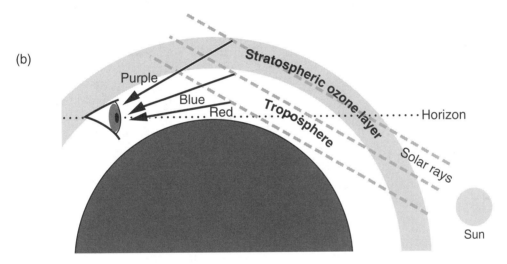

Figure 3.18   (a) Colors of the twilight sky. The various arcs and colors have related, although distinct, origins. The reddish colors near the horizon are caused by atmospheric Rayleigh extinction, the blue colors higher up by Rayleigh scattering, and the purple glow, by ozone absorption and aerosol scattering in the stratosphere. (b) Ray paths for red, blue, and purple twilight. The angles and distances are exaggerated for clarity. The stratospheric aerosol layer reflects the light seen as the "purple glow."

pronounced "oh-two." Note that the number of oxygen atoms in an "oh-two" molecule is shown by the subscript following the chemical symbol for the oxygen atom. The common name for "oh-two" is oxygen, a more formal chemical name is molecular oxygen, and a chemically precise name is dioxygen. To a chemist, all these names mean the same thing. They are like nicknames—for example, in a descending order of decorum, of a name, say, Arthur: Arthur, Art, Artie, and "the Big A." When chemists discuss chemical reactions or processes, they often use the literal or common names because they are convenient. We will do likewise, because it is not sensible

to be concerned here with the precision that formal chemical nomenclature offers. Table 3.4 provides several examples of the different levels of nomenclature for important compounds.

Compounds consisting of several elements can be represented by chemical symbols that precisely define their elemental composition. For example, N_2O is a molecule made up of two nitrogen atoms and one oxygen atom. Br_2 is a molecule consisting of two bromine atoms. (The nature of chemical compounds is discussed in Section 3.3.2.)

Different gases are often capable of reacting spontaneously with one another, causing a mutual trans-

Table 3.3   Most Common Elements in the Environment

Element (chemical symbol)		Atomic weight (atomic mass units, amu)	Electrons (number)
Hydrogen	(H)	1	1
Carbon	(C)	12	6
Nitrogen	(N)	14	7
Oxygen	(O)	16	8
Fluorine	(F)	19	9
Sulfur	(S)	32	16
Chlorine	(Cl)	35	17
Bromine	(Br)	80	35

formation into new species. In the study of such reactions, the names of compounds allow the reactions to be expressed literally. For example, the important chemical reaction

$$NO + O_3 \rightarrow NO_2 + O_2 \qquad (3.31)$$

becomes a statement of action, with the following variations:

1. En-oh plus oh-three go to en-oh-two plus oh-two.
2. En-oh reacts with oh-three to form en-oh-two and oh-two.
3. Nitric oxide (NO) reacts with ozone (O_3) to produce nitrogen dioxide (NO_2) and oxygen (O_2).

The first statement is simply phonetic yet clear in meaning. The second statement uses obvious chemical terms such as reacts, but is no more complex. The last version uses the common names of the substances involved. Obviously, there are other ways to state the same thing. But the beauty of chemical nomenclature is that *any* reaction may be stated in a compact, unambiguous form that everyone else can understand immediately. From another view, chemical equations such as Equation 3.31 are simply statements of fact, equivalent to common statements like "Mike and Margie went to the party with Stu and Kate" or "Michael and Margaret brought a gift for Stuart and Katherine." Indeed, by putting a number of statements together in a sequence, a more complex chemical process, or dialogue, may be constructed that conveys a clear plot and action.

Chemical equations also are statements of mass conservation (but not energy conservation). For example, the numbers of N and O atoms in Equation

3.31 are exactly balanced on both sides of the arrow (there are two nitrogen atoms and four oxygen atoms both before *and* after the reaction). In any chemical reaction, the number or mass of every element involved is conserved.

A chemical reaction is a quantitative statement describing the interaction between two materials (or compounds or species), the reactants, that leads to the generation of different materials, the products. The reactions are usually written using chemical nomenclature, in which materials are represented in terms of their simplest molecular structure.

A chemical process is the overall chemical transformation of one initial set of species (reactants) into another set of species (products) through a well-defined sequence of chemical reactions.

Common chemical reactions can involve two or three reactants and two or three products. An example of a chemical process is the following sequence of reactions for the oxidation of nitric oxide (NO) by ozone (O_3):

$$\begin{aligned} NO + O_3 &\rightarrow NO_2 + O_2 \\ NO_2 + O_3 &\rightarrow NO_3 + O_2 \\ NO_2 + NO_3 &\rightarrow N_2O_5 \end{aligned} \qquad (3.32)$$

At each step of the process a specific chemical reaction occurs between the reactants to produce products. At each step, the mass (and number) of the elemental atoms are conserved. The overall process can be summarized as follows:

$$NO + O_3 \rightarrow \cdots \rightarrow N_2O_5 + O_2 \qquad (3.33)$$
(overall process, non-quantitative)

Table 3.4   Chemical Nomenclature

Compound	Names
O_2	Oh-two (literal name)
	Oxygen (common name)
	Molecular oxygen (formal name)
	Dioxygen (strict nomenclature)
N_2O	En-two-oh (literal name)
	Laughing gas (popular name)
	Nitrous oxide (formal name)
	Dinitrogen monoxide (strict nomenclature)
NH_3	En-aitch-three (literal name)
	Ammonia (common name)
	Ammonia (formal name)
	Nitrogen trihydride (strict nomenclature)

where the final product species are indicated. Equation 3.33 is not quantitative, however, because the conservation of the elements is not shown. A more specific quantitative statement of the overall process is

$$2NO + 3O_3 \rightarrow \cdots \rightarrow N_2O_5 + 3O_2 \quad (3.34)$$
(overall process, quantitative)

which now conserves the elements exactly (in the chemical equation for a reaction or process, the number preceding the symbol for a substance indicates the number of molecules of that substance involved in the reaction or process). In our simple nomenclature, process Equation 3.34 reads: Two en-oh react with three oh-three in an overall process to form one en-two-oh-five and three oh-two. *Voilà*.

3.3.2 PROPERTIES OF COMMON SUBSTANCES

Gases have specific chemical properties related to their atomic structure (refer to the introduction to Section 3.1). A molecule composed of two or more atoms has a definite structure, or arrangement of atoms. The structures of some common atmospheric gases are shown in Table 3.5. Among the important physical properties of gases are their color, odor, and toxicity to humans. Their chemical properties are determined by the specific reactions in which the

gases participate and the rate coefficients for those reactions. Some of the key reactions for the atmosphere will be discussed in later sections.

An important aspect of gas behavior is the atomic structure of molecules and the dependence of physical and chemical properties on the arrangement and bonding of atoms in molecules. Why is oxygen (O_2) so different from ozone (O_3)? Why is nitrous oxide (N_2O) chemically inert, whereas nitrogen dioxide (NO_2) is reactive? The answers lie in the bonding of the atoms composing these molecules. Figure 3.19 shows the detailed atomic structures of hydrogen and oxygen atoms and of their important compound, water. Hydrogen, oxygen, and water have widely differing properties although they are made of the same H and O. These differing properties can be traced to the electrons circling the atomic nuclei. The electrons are held in fixed **orbitals** that may hold up to two or eight electrons each (in the case of the lighter elements of interest here). Moreover, electrons in an orbital like to be paired with a partner of opposite **spin** (spin is an intrinsic property of an electron, like its negative charge, which may be thought of as the rotation of the electron on its axis; only two spin orientations are possible, however, with the rotation axis pointed either up or down).

When all the orbitals are filled with paired electrons, an atom or a molecule tends to be very unreactive. The best examples of this behavior are found in the noble gases—helium, neon, and

Table 3.5   Molecular Structures of Some Common Gases

Molecule	Symbol	Structure	Properties
Nitrogen	N_2	N—N	Chemically inert in air, transparent
Oxygen	O_2	O—O	Breathable, source of ozone
Ozone	O_3	O∕ \O∖ O	Reactive oxygen form, crested in smog
Carbon dioxide	CO_2	O—C—O	Stable product of combustion
Carbon monoxide	CO	C—O	Odorless, colorless, toxic gas
Hydrogen	H_2	H—H	Explosive with oxygen
Water vapor	H_2O	H∕ O \H	Polar molecule, forms clouds and fog
Nitric oxide	NO	N—O	Produced by engines, reacts with ozone
Nitrogen dioxide	NO_2	O—N—O	Brown gas common in smog
Nitrous oxide	N_2O	N—N∖ O	Laughing gas, stable in the atmosphere
Methane	CH_4	H \ / H C H / \ H	Swamp gas, biogenic origins
Hydrogen chloride	HCl	H—Cl	Strong acid if mixed in water
Sulfur dioxide	SO_2	O—S—O	Acrid odor, forms sulfuric acid
Hydrogen sulfide	H_2S	H—S—H	Smell of rotten eggs, highly toxic gas

argon—each of which has completely filled electron orbitals, and each of which is chemically inert. As Figure 3.19 shows, atoms with partly empty orbital shells try to fill them by accepting electrons from other atoms while pairing any odd electrons. In the process of sharing electrons in orbitals, **chemical bonds** are established between atoms to form stable molecular compounds.

In Figure 3.19, the oxygen atom in water accepts one electron from each of the hydrogen atoms, thus completing its outer orbital shell. The hydrogen atoms are left with empty shells. This bonding configuration is very stable, and water vapor exhibits little chemical reactivity. Moreover, because the H_2O molecule is bent (Table 3.5), the electrical charge is separated internally with the O atom becoming a center of negative charge and the H atoms becoming centers of positive charge. This charge distribution creates a permanent electric dipole moment in the water molecule, which allows water to condense easily to form clouds. In water, the H atoms are **electron donors**, and the O atoms are **electron acceptors**. Different elements are stronger or weaker donors or acceptors. Every molecular compound has a distinct pattern of electron sharing that determines the properties of that compound.

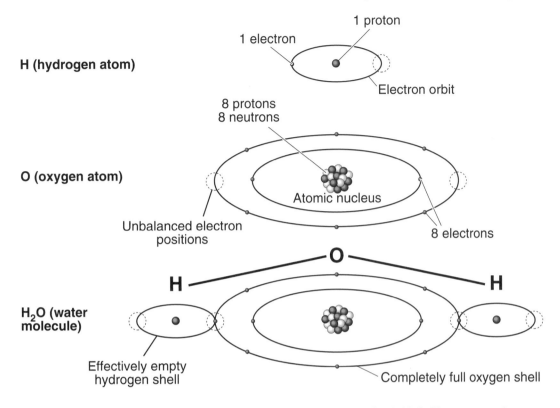

Figure 3.19   The electronic structures of H and O atoms and the atomic bonding in H_2O. Electrons are shown as small spheres in orbits around the atomic nuclei. The atoms are bonded into molecules between the electron orbitals. Each orbital prefers to be completely filled with electrons or, to a lesser degree, empty.

The atomic nuclei play no substantial role in the chemical properties of atoms and molecules except to carry the positive charge of the protons (equal in number to the orbiting electrons) that holds the electrons within orbitals through **Coulomb attraction**.[35] The isotopes of an element, which involve different numbers of neutrons in the nucleus of the atom, have a negligibly small influence on the chemistry of interest here.

Organic and Inorganic Compounds

Chemists usually distinguish between **organic** and **inorganic** compounds, although several definitions are possible:

1. An organic compound is any substance containing atoms of the element carbon and an inor-
ganic compound contains no carbon. Examples of organic compounds according to this definition are CO_2 and CH_4, and examples of inorganic compounds, are H_2O and NO_2.

2. An organic compound is any substance containing carbon and hydrogen atoms that are chemically bound in as stable molecules. Inorganic compounds have no carbon hydrogen bonds. Examples of organic compounds by this definition are CH_4, CH_2O, and C_3H_8, and examples of inorganic compounds are CO, CO_2, and H_2O.

3. An organic compound is any substance related to or derived from living organisms that contains carbon and hydrogen. All other compounds are inorganic. Examples of organic compounds under this definition are CH_4 and amino acids, and examples of inorganic compounds are CO_2 and manufactured natural gas.

35.   The attractive force between positive and negative electrical charges was quantified by French physicist Charles-Augustin Coulomb (1736–1806) in 1785. **Coulomb's law** states that the force between two charges is proportional to the product of the magnitudes of those charges (measured in coulombs of charge, of course) and is inversely proportional to the square of the distance between the charges. Coulomb also studied windmills and silk cloth. Pure Coulomb forces cannot explain the stability of atoms and must be modified according to the rules of the "quantum theory" of matter, which allows only certain discrete electron states to exist. These states make up the electron "orbitals."

We will use the second definition of organic and inorganic in this book. The first definition is too strict, as, for example, in treating carbonate minerals as organic rather than inorganic. The last definition is too loose and has a "touchy-feely" ring to it. The second definition seems just right.

3.3.3   THE MECHANISMS OF CHEMICAL REACTIONS

Chemical reactions occur through the interactions of colliding molecules, because the molecules have frequent, close, and, indeed, violent encounters as a result of their thermal agitation. In any gas mixture, the molecules are continually bombarding one another: Each air molecule at sea level is hit approximately 10 million times a second. In most of the collisions, the molecules simply rebound without any change in their molecular structure, but occasionally, the molecules form an **activated complex** during the collision. In the activated complex, the atoms of the colliding species form transient chemical bonds. The bonds hold the reactants together for a longer period, during which the atoms of the reactants may form new, permanent chemical bonds with other atoms in the complex. The reorganized complex then flies apart into new species, or products of reaction. If new chemical bonds are not formed in the activated complex, the original colliding molecules will reappear as the complex flies apart.

Figure 3.20 illustrates two of the most important kinds of molecular encounters. In Figure 3.20(a), a **binary reaction**, or **two-body reaction**, is shown schematically. In Figure 3.20(b), a **ternary reaction**, or **three-body reaction** is shown. In a binary reaction, two reactant species are transformed into two product species. In a ternary reaction, two reactant species associate to form a new product species through a three-way collision involving a **third-body**, denoted M. The key point here is that a chemical reaction differs from a simple molecular collision in that the chemical species is transformed during a reaction. Thus in the reaction shown in Figure 3.20(a), reactant species A and B are transformed into the product species C and D. In the ternary reaction depicted in Figure 3.20(b), the reactants A and B combine to form the more complex molecule AB. If we chose to be more detailed in describing the pathways by which atoms and molecules can interact chemically, we could define several other specific kinds of reactions. But to understand the chemistry discussed in this book, only the two types of reactions shown in Figure 3.20 are required. The most common type of reaction by far is the simple binary reaction (Figure 3.20[a]).

A chemical reaction is the transformation of two or more interacting molecules through the intermediary phase of an activated complex, in which existing chemical bonds are broken and new bonds are formed, in which the chemical bonds define the molecular species before and after the reaction.

Figure 3.20   The fundamental mechanisms of chemical and photochemical processes. (a) A binary, or two-body, reaction between reactant molecules A and B yields the product molecules C and D. (b) A ternary, or three-body, reaction, in which reactant molecules A and B combine to form the product molecule AB in a threeway collision with another air molecule, M. (c) The photodissociation of molecule AB into products, following the absorption of a photon of radiation. In each panel, the colliding reactant molecules form an activated complex, AB*, in which the molecules are held together by a temporary bond, labelled in (a).

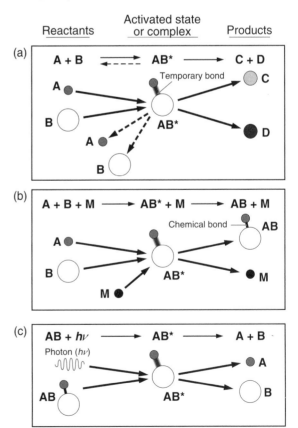

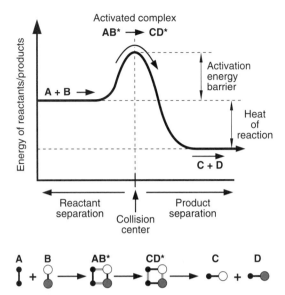

Figure 3.21 The potential energy diagram for a binary reaction, A + B → C + D. The reactants are shown to the left and the products to the right. The distances between the reactants (A and B) and products (C and D) are measured relative to the collision center for the reaction. For example, moving to the left from the collision center in the graph indicates an increasing separation of the reactants A and B. Thus the reaction occurs from left to right in the diagram, along the potential energy curve. The molecular strcture of the species involved and the chemical bonds—permanent and temporary—that exist at each stage of the reaction are depicted at the bottom of the figure. Chemical systems always seek to reach the minimum potential energy state, which in this case corresponds to the products C + D.

The mechanism of a binary chemical reaction is further explained in Figure 3.21 in terms of the **chemical potential energy** of the reaction of species A and B to form species C and D. This potential energy is a property that each chemical species inherits when it is formed. Every molecule of a given species has the same potential energy, which is known or readily measurable. In any mixture of different gases, chemical and physical interactions between the gases work to *minimize* the total potential chemical energy of the system. In the process, potential chemical energy is converted to energy of motion or heat (or **kinetic energy**). If a process raises the average potential chemical energy of the gases, that energy must be drawn from the kinetic energy pool, or the heat, already in the system. Nature resists this conversion.

The combined potential energy of the reactants (A and B) remains constant as long as A and B are separated from each other (that is, they are not interacting). As A and B approach each other in a collision, however, they face an energy barrier against forming an activated complex. The barrier is caused by the normal repulsion of two molecules or atoms that are pushed close together (try pushing two volleyballs together). During some of the collisions, the molecules travel up the barrier to form an activated complex, AB*. In the activated complex, chemical bonds can be established between A and B. Then, when the activated complex flies apart (to lower its potential energy), it can follow one of two paths. It can separate into the original species, A and B, or it can permanently rearrange the chemical bonds and dissociate into the new products, C and D. The second path represents a chemical transformation. Note that because the path to C and D is much farther downhill than the path back to A and B (Figure 3.21), the total potential energy of C + D is lower than that for A + B. The extra potential energy is released during the reaction as heat. The **heat of reaction** is simply the potential energy difference between the reactants and the products. When this energy difference is positive (the overall reaction is downhill), the reaction is said to be **exothermic** (that is, heat is generated during the reaction). If the reaction path is uphill overall, the reaction is **endothermic** (that is, heat must be absorbed from the environment to make the reaction go).

Imagine riding a bicycle along a level road. With no head wind, it is no sweat. But a hill is ahead. You now have to work vigorously to build up a head of steam to power up the slope. You are expending energy and plenty of sweat. But if you cannot reach the peak, you will roll backward down the hill. Hit the brakes! At the summit, however, you need only the slightest push to begin the long, easy downhill coast into the lowland beyond. You are regaining potential energy and building up speed as you head downhill. The speed will carry you some distance along the flatland below (until friction and wind slow you down again).

The additional energy that the colliding molecules must have to reach the top of the reaction hill (Figure 3.21) is called the **activation energy** for the reaction. The energy of activation is drawn from the thermal motion (kinetic energy) of the molecules. Some or all of this energy may reappear as kinetic energy when the colliding molecules separate into products. The activation barrier acts to inhibit a chemical reaction. If the barrier is higher, the probability is lower that the colliding molecules will reach the top. The hotter the gas is, the greater the speed of the molecules will be, and the more likely it is they will get to the top and be "activated."

The overall change in the potential chemical energy of the system (either exothermic or endothermic) determines whether a reaction will occur in the atmosphere. Practically all reactions of atmospheric importance are exothermic reactions. Endothermic reactions, which absorb thermal energy from the environment, rarely occur in the atmosphere because of its low temperature (compared, for example, with the temperatures found in flames, chemical explosions, and the atmospheres of stars).

Reaction Rates and Coefficients

The rate at which a chemical reaction occurs is described by a **reaction rate coefficient**, k. The reaction rate coefficient can be thought of as a collision kernel for the reactants to produce the products. The presence of a substantial activation energy implies that the rate coefficient is very sensitive to temperature. The warmer the reacting gases are, the larger the rate coefficient will be. For binary (two-body) reactions, k has units of cm^3/sec-molecule; for ternary (three-body) reactions, the units are cm^6/sec-molecule^2. Of course, endothermic reactions have very small rate coefficients at normal atmospheric temperatures.

The rate of a chemical reaction is obtained by multiplying the rate coefficient by the concentrations of the reacting gases. The **chemical reaction rate** is the number of reaction events occurring in a unit volume of air per unit time. Consider a binary reaction:

$$A + B \rightarrow C + D \qquad (3.35)$$

(for example, Figure 3.20[a]). The chemical rate for this reaction is given by the simple formula

$$R_{ij} = k_{ij} n_i n_j \qquad (3.36)$$

where k_{ij} is the rate coefficient for the particular reaction between species i and j, the concentrations of the reacting species are n_i and n_j, and R_{ij} is the chemical reaction rate. When the gas concentrations are measured in molecules/cm^3, the chemical rate has the dimensions of molecules/cm^3-sec,

where molecules refers to the number of each molecule produced in a unit volume of air per second. The total effect of chemistry on a given constituent of air is determined by the sum of the chemical rates for all the processes that produce or destroy that constituent, including any processes that involve radiation (see the section Sunlight and Photodissociation).

When reactant molecules are present at low concentrations, the likelihood of their collision is small, and the reaction rate (Equation 3.35) is reduced accordingly. On the other hand, if the rate coefficient for a reaction is very small—say, because of a high activation barrier—the reaction rate may be negligible, even though the reactant concentrations are high. For example, the chemical reaction rate of ozone with nitric oxide (NO) is much greater than its reaction rate with water vapor, even though water vapor is 1 million times more abundant in the lower atmosphere. The reaction between ozone and water has a large activation barrier, which prevents it from occurring under atmospheric conditions. Such reactions may proceed only at very high temperatures. Thus although N_2 and O_2 do not react at all in outdoor air (the reaction has a large activation barrier and also is endothermic), in an internal-combustion engine, the temperatures can become so great as to push the reactants over the barrier.

Molecular Ménage à Trois

When two species associate to form a more complex molecule, physics demands that another molecule be present to carry away excess energy. The third molecule is the third-body. A third-body does not actively participate in the chemical reaction, and is not chemically transformed during the reaction. Its identity therefore is not important. In air, nitrogen and oxygen are the most common third-bodies, as they are the most abundant gases. A three-body reaction is usually written as

$$A + B + M \rightarrow AB + M \qquad (3.37)$$

Figure 3.20(b) illustrates the mechanism of a three-body reaction. A third-body can be thought of as "mediating" a reaction, and so it is represented as M. The chemical rate of a three-body reaction is given by

$$R_{ij} = k_{ij} n_i n_j n_M \qquad (3.38)$$

where n_M is the concentration of third-body molecules.

Sunlight and Photodissociation

When a molecule absorbs a photon of solar radiation, it may dissociate into two or more product species. This process is referred to as **photodissociation**. Photodissociation can be viewed as occurring through the formation of an activated complex when a photon of radiation collides with the molecule. The activated complex breaks apart into the product species, as happens during a chemical reaction. The photodissociation process is depicted in Figure 3.20(c) for the case

$$AB + h\upsilon \rightarrow A + B \qquad (3.39)$$

In photodissociation, the fragment species carry away as chemical potential energy most of the energy of the absorbed photon (the rest can appear as the kinetic energy of the fragments). Subsequently, this potential energy drives exothermic chemical reactions that further alter the composition of the atmosphere. Sunlight is therefore continuously charging the atmosphere with chemical energy. The chemistry of the atmosphere driven by sunlight is referred to as **photochemistry**.

Photodissociation is the cleavage of a molecule into two or more (smaller) atomic or molecular fragments through the absorption of radiant energy.

The photodissociation rate, R_l, of a molecular species, l, is calculated as follows:

$$R_l = J_l n_l \qquad (3.40)$$

where the **photodissociation coefficient**, J_l, depends on the spectrum of sunlight and the absorption properties of the molecule, and J_l has units of inverse time, or sec^{-1}. The **photodissociation rate**, R_l, is the number of molecules l that are dissociated by radiation in a unit volume of air each second. Most photodissociation involves ultraviolet radiation, although some molecules may be weakly photodissociated by visible radiation. No atmospheric photodissociation occurs at infrared wavelengths.

The stability of a molecule in the atmosphere is often determined by its photodissociation coefficient. If the coefficient is large ($\geq 10^{-6}$/sec), the molecule is **photolytically unstable**; that is, the species can be quickly destroyed by exposure to sunlight. The atmospheric lifetimes of such molecules range from a few seconds to a few days, at most. If the photodissociation coefficient is small ($\leq 10^{-7}$/sec), the molecule is generally **photolytically stable**. Such long-lived molecules include all of the major atmospheric constituents and many air pollutants. The lifetime of a species is not determined by photodissociation alone, however. A chemical reaction can be the primary loss process for some species, whereas physical removal (for example, by rainout) is the dominant sink in other cases.

3.3.4 Basic Chemical Reactions

Some of the basic reactions that characterize the chemistry of the atmosphere are described in this section. The specific effects and consequences of these reactions and processes are discussed in later chapters.

Reactions in Gases

Among the most abundant air constituents are N_2, O_2, H_2O, N_2O, and CO_2. Nitrogen has *no* photochemistry of interest in the lower atmosphere. The N_2 molecule is too stable to be photodissociated by the wavelengths of solar radiation that can penetrate below about 100 kilometers. On the other hand, the temperatures in internal-combustion engines are high enough to cause nitrogen and oxygen to react as follows:

$$N_2 + O_2 + \xrightarrow{\quad Heat \quad} NO + NO \qquad (3.41)$$

In Equation 3.41, **nitrogen oxides** are generated initially as nitric oxide, which is later converted to nitrogen dioxide. The nitrogen oxides NO and NO_2 are often lumped together and called **NO_x**. The chemistry of NO_x is discussed later.

Molecular oxygen can be photodissociated by sunlight in the stratosphere and mesosphere, producing oxygen atoms,

$$O_2 + h\upsilon \text{ (sunlight)} \rightarrow O + O \qquad (3.42)$$

The oxygen atoms then quickly associate with another oxygen molecule in a three-body reaction as follows:

$$O + O_2 + M \rightarrow O_3 + M \quad (3.43)$$

Equation 3.43 is one of the most important reactions in the atmosphere, inasmuch as ozone is generated as a byproduct. Almost all of the ozone in the stratosphere, which protects the surface of the Earth from hazardous ultraviolet radiation, is directly produced by reaction (3.43).

Once ozone is formed, it is readily photodissociated by

$$O_3 + h\upsilon \rightarrow O^* + O_2 \quad (3.44)$$

The oxygen atom produced in Equation 3.44 is in an energetically excited state; we will refer to it as **excited atomic oxygen**. The excited oxygen atoms can be calmed by collisions with air molecules to become normal oxygen atoms:

$$O^* + M \rightarrow O + M \quad (3.45)$$

However, some of the excited oxygen atoms react with other constituents. For example, O* can react with water vapor,

$$H_2O + O^* \rightarrow OH + OH \quad (3.46)$$

Equation 3.46 produces one of the most important minor chemical constituents found in the atmosphere, the **hydroxyl radical** (OH). OH is called a **radical** because it has an unpaired electron in its outermost electronic orbital. Recalling that unpaired electrons are lonely and would like to join an electron from another atom or molecule (Section 3.3.2), it follows that radicals are quite reactive species.

A **minor constituent** of the atmosphere is one whose concentration is much smaller than the major constituents like N_2, O_2, H_2O, and CO_2. Minor constituents can have mixing fractions of parts per million by volume (ppmv) to parts per trillion by volume (pptv) or less.

OH is the main chemical "scavenger" in the atmosphere. Like a goat scavenging through unwanted debris and rubbish, OH reacts with a variety of compounds that would otherwise accumulate in, and pollute, the atmosphere. Thus, OH reacts efficiently with most hydrocarbons, hydrogen sulfide, carbon monoxide, and many other toxic and undesirable chemicals. We will refer often to OH in the following discussion.

The excited oxygen atoms produced by ozone photodissociation can also react with nitrous oxide to generate nitric oxide:

$$N_2O + O^* \rightarrow NO + NO \quad (3.47)$$

Reaction (3.47) is particularly important in the stratosphere. The nitrogen oxides participate in other reactions of note:

$$NO + O_3 \rightarrow NO_2 + O_2 \quad (3.48)$$
$$NO_2 + OH + M \rightarrow HNO_3 + M \quad (3.49)$$

The second reaction in this sequence forms nitric acid from nitrogen dioxide and hydroxyl radicals. This is a key process both for cleansing NO_x from the atmosphere and forming acid rain.

Ozone Catalysis

Certain chemical species have an especially powerful effect on ozone. One molecule of such a species is capable of destroying hundreds or thousands of ozone molecules. These species can participate in a repetitive reaction sequence with ozone without being destroyed in the process. Such chemical species are called **catalytic agents**, and the reaction sequences through which they destroy ozone are referred to as **catalytic cycles**. A few generic reaction sequences define the effect of catalysts on ozone. The following simple reaction sequence, for example, involves nitric oxide as the catalytic agent and is central to the problem of stratospheric ozone depletion:

$$NO + O_3 \rightarrow NO_2 + O_2 \quad (3.48)$$
$$O_3 + h\upsilon \rightarrow O + O_2 \quad (3.50)$$
$$\underline{NO_2 + O \rightarrow NO + O_2} \quad (3.51)$$

$$\text{(net)} \ 2O_3 \rightarrow 3O_2 \quad (3.52)$$

In particular, the occurrence of Equations 3.48, 3.50, and 3.51 in sequence results in the destruction of two ozone molecules (which are converted to three oxygen molecules), and the nitric oxide molecule that initiated the sequence is recovered at the end of the sequence. This is precisely the definition of a chemical catalyst.

An ozone catalyst is a chemical species that reacts with ozone molecules, leading to ozone

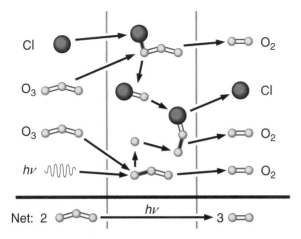

Figure 3.22 The catalytic reaction cycle of chlorine with ozone. The initial reactants are shown on the left, and the final products on the right. The intermediate steps and activated complexes for the reaction cycle also are given. Note that Cl is regenerated at the end of the cycle (as a product) and that ozone is converted to oxygen. The overall reaction is summarized at the bottom of the figure.

destruction, although the catalyst itself is not consumed in the overall reaction process.

In the atmosphere, the key ozone catalysts are NO, OH, Cl, and Br. Each of these species could be substituted for NO in the reaction sequence, with the same result. Chlorine and bromine (the most abundant halogens in the atmosphere, except for fluorine) are the most potent catalysts. One chlorine atom can destroy hundreds of thousands of ozone molecules. The actual process of chlorine catalysis of ozone is represented in Figure 3.22. (Ozone destruction by various catalytic agents is discussed at length in Chapter 13).

Chemistry of Smog

To illustrate a more complex atmospheric process, consider the formation of photochemical smog (that is, Los Angeles-style air pollution). The series of reactions involved can be summarized as

$$OH + RH \rightarrow R\bullet + H_2O \quad (3.53)$$
$$R\bullet + O_2 + M \rightarrow RO_2\bullet + M \quad (3.54)$$
$$RO_2\bullet + NO \rightarrow RO\bullet + NO_2 \quad (3.55)$$
$$NO_2 + h\upsilon \rightarrow NO + O \quad (3.56)$$
$$O + O_2 + M \rightarrow O_3 + M \quad (3.43)$$

In the process shown, the hydrocarbons, indicated as RH—for example, emitted from an oil refinery—react with the universal "scavenger," OH. The reaction removes, or **abstracts**,[36] a hydrogen atom to form water vapor and an **organic radical**, represented by the symbol, R•. The organic radical can be composed of many atoms and have a complex molecular structure. It is not necessary, however, for us to identify the specific radical.

Of more general importance are the subsequent reactions of R• (Equations 3.54 and 3.55) which, in the presence of nitric oxide, create NO_2. The formation of smog usually is initiated by emissions of nitric oxide from road vehicles powered by internal-combustion engines. Accordingly, both essential ingredients of smog—hydrocarbons (RH) and NO—can originate from the same source. The RH in this case consists of incompletely burned gasoline, or "fugitive" fuel vapors. Sunlight is also required to produce ozone, through Equations 3.56 and 3.43. Finally, it is important to have an **initiator** available to start the smog reaction sequence, and the ubiquitous scavenger OH serves that purpose.

The overall process of smog formation can be summarized as

$$RH + NO + h\upsilon \xrightarrow{O_2} \cdots \rightarrow O_3 + NO_2 + HC \quad (3.57)$$

In this representation, RH is the initial reactive hydrocarbon that reacts to form organic radicals, R•, and the principal initiating reactant is OH. The organic radicals subsequently react with oxygen and nitric oxide to generate ozone. The products of smog formation also include a variety of photolytically stable hydrocarbon products, which are indicated as HC in Equation 3.57. Some of these compounds are toxic air pollutants and others condense to form photochemical haze. Interestingly, although OH scavenges the emitted reactive (and unhealthful) hydrocarbons from the atmosphere, another consequence is the formation of unsightly (and unhealthful) smog as a by-product.

Carbon monoxide (CO) is a key component of smoggy air. CO, a colorless, odorless, and poisonous gas, is one of the pollutants that is regularly moni-

36. Hydrogen abstraction is a common chemical process for organic compounds with hydrogen-carbon bonds. A hydrogen atom, with only one orbital electron, readily combines with a variety of other chemical species. The abstraction of hydrogen by the OH radical (yielding water vapor) is an important process in organic chemistry.

tored in urban areas as an indicator of air quality. The primary source of CO in tainted air is the combustion of carbon based fuels, particularly gasoline. The atmospheric chemistry of carbon monoxide is amazingly simple. One reaction is sufficient to describe its behavior in the atmosphere. The loss of CO is through the reaction

$$CO + OH \rightarrow CO_2 + H \qquad (3.58)$$

That's all there is to it. Notice that here OH plays the good-guy role of an air pollution scavenger. Equation 3.58 converts the CO generated during incomplete or inefficient fuel combustion to carbon dioxide, the normal product of complete combustion. The hydrogen atom created in Equation 3.58 can act as a radical, $R\bullet$, and hook up with an oxygen molecule to form the hydroperoxy radical, HO_2. This hydrogen species is an important link in the catalysis of ozone by OH (Section 13.2.2), and in the formation of hydrogen peroxide, an important atmospheric oxidizing agent (Section 9.2.2).

The shorthand representation of photochemical smog formation by Equation 3.57 illustrates how a complex chemical process can be expressed using simple chemical notation. In this way, the critical reactants and products, and other critical factors contributing to the process, can be quickly identified. (The problem of smog in cities is discussed in depth in Chapter 6.)

Acids and Bases

Two important classes of compounds are **acids** and **bases**. Both types of compounds can be powerfully corrosive and unpleasant. When dissolved in water, acids produce positively charged hydrogen ions, whereas bases produce negatively charged **hydroxide (hydroxyl) ions**. The molecular structures and aqueous[37] components of some strong acids and bases are shown in Figure 3.23. The most common **strong acids** in the atmosphere are sulfuric acid (H_2SO_4), nitric acid (HNO_3), and hydrochloric acid (HCl). All three of these acids are found in clouds

37.   An **aqueous solution** refers to the dissolution of a particular substance, the **solute**, in water, the **solvent**. Sugar dissolved in a glass of water is an aqueous solution of glucose. Common table salt dissolved in water is an aqueous saline solution. Not all substances dissolve in water. Thus oil normally does not mix with water, but may be churned into a **colloidal dispersion** of very fine oil droplets suspended in the water. A colloidal dispersion is not a solution.

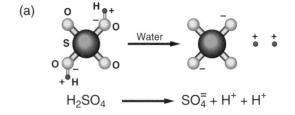

(a)

$$H_2SO_4 \longrightarrow SO_4^= + H^+ + H^+$$

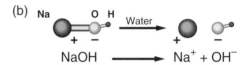

(b)

$$NaOH \longrightarrow Na^+ + OH^-$$

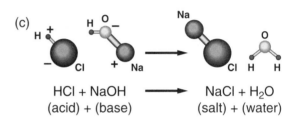

(c)

$$HCl + NaOH \longrightarrow NaCl + H_2O$$
(acid) + (base)   (salt) + (water)

Figure 3.23   The molecular structure and chemistry of acids and bases. (a) The molecular structure of sulfuric acid (a strong acid) and its dissociation in aqueous solution (that is, when dissolved in water) to form a hydrogen ion (H^+). (b) The structure and dissociation of sodium hydroxide (a strong base) in water to form the hydroxide ion (OH^-). (c) How an acid (hydrochloric acid, HCl) and a base (sodium hydroxide, NaOH) react to neutralize each other, forming a salt (NaCl) and water.

and fog. Two important **weak acids** are carbonic acid (H_2CO_3) and sulfurous acid (H_2SO_3). Carbonic acid is produced when carbon dioxide dissolves in water. Since the atmosphere is filled with CO_2, clouds are naturally composed of a carbonic acid solution. Commercial "carbonated water" is produced by forcing CO_2 to dissolve in water under pressure, and so it is also weakly acidic. Strong acids differ from weak acids in that they tend to become completely dissociated in water, thus releasing copious hydrogen ions, which are the reactive components of the acid solution. Weak acids only partially dissociate into hydrogen ions. Strong bases like sodium hydroxide (NaOH) are not normally found in natural environments. A weak base commonly found in the environment, however, is ammonium hydroxide (NH_4OH), which results when ammonia gas (NH_3) dissolves in water. Certain minerals may also act as weak bases in neutralizing acids in rain and lake water. Calcium carbonate ($CaCO_3$) and calcium hydroxide ($Ca[OH]_2$), or "slaked lime," are two abundant minerals that have mildly basic properties.[38]

Acids and bases act to neutralize each other. Hence, equivalent amounts of a strong acid and a strong base mixed (carefully!) together produce a **salt** and water. The salt consists of the **anion**[39] of the acid (that is, the negatively charged ion fragment of the acid in solution (Figure 3.23[a]) and the **cation** of the base (the positively charged fragment [Figure 3.23(b)]. The salts produced by neutralization are not corrosive and are often extremely compatible with the environment. For example, the salt of ammonium hydroxide and sulfuric acid is ammonium sulfate, a fertilizer for plants (but please do not fertilize your houseplants with sulfuric acid). In clouds, acids are often neutralized by bases derived from ammonia gas emissions or by basic minerals carried on dust particles. Nevertheless, the atmosphere is, on balance, acidic in general. (In Chapter 9, the acidity associated with precipitation is thoroughly surveyed.)

The two most abundant atmospheric acids, H_2SO_4 and HNO_3, are formed by chemical reactions in air. These acids are not directly injected into the atmosphere. But the gases from which they are produced, SO_2 and NO, respectively, are emitted in large quantities. A reaction sequence that generates sulfuric acid is

$$SO_2 + OH + M \rightarrow HSO_3 + M \quad (3.59)$$
$$HSO_3 + O_2 \rightarrow SO_3 + HO_2 \quad (3.60)$$
$$SO_3 + H_2O + M \rightarrow H_2SO_4 + M \quad (3.61)$$

Notice that the conversion of SO_2 to H_2SO_4 is initiated by the hydroxyl radical. The intermediate sulfur radical, HSO_3, is very short lived and can otherwise be ignored. Similarly, SO_3 is very quickly converted into sulfuric acid. In fact, Equations 3.60 and 3.61 occur so quickly that the overall chemical transformation of SO_2 to H_2SO_4 can be simply represented as

$$SO_2 + OH \xrightarrow{O_2, H_2O, M} \cdots \cdots \rightarrow H_2SO_4 \quad (3.62)$$

The formation of nitric acid from NO proceeds in accordance with reactions (3.48) and (3.49), discussed earlier:

$$NO + O_3 \rightarrow NO_2 + O_2 \quad (3.48)$$
$$NO_2 + OH + M \rightarrow HNO_3 + M \quad (3.49)$$

For both nitric and sulfuric acids, the key initiating species—aside from the gaseous precursors, SO_2 and NO—is OH. Thus even as OH is removing the toxic gases SO_2 and NO from the atmosphere, the process of cloud and rain acidification is being set in motion. (See Chapter 9 for a detailed discussion of the chemistry of acid rain.)

Questions

1. You may know that opposite electrical charges attract each other (positive and negative) and that like charges repel (positive and positive, or negative and negative). This force of attraction or repulsion is called the **Coulomb force**. Atoms are composed of a positively charged nucleus made up of protons and neutrons surrounded by a cloud of negatively charged electrons swirling in orbits around the nucleus. Fortunately, this arrangement is extremely stable for most elements. How might the Coulomb force play a role in stabilizing atoms, allowing the electrons and nucleus to remain associated? In the solar system, the gravitational force of attraction between two masses plays the same role in holding the planets to the sun so that they do not spin off into space. In the nucleus of the atom, the protons are closely packed together in the nucleus (except for hydrogen, which has only one proton). However, the Coulomb force acts powerfully to push these protons apart. What else do you know about the nucleus that might explain its stability against disintegration?

2. You have just filled a balloon very rapidly with air using a pump. You quickly tie the end and release the balloon. It seems to float freely before slowly settling to the floor. Can you

---

38. Common lime is actually calcium oxide, CaO. When exposed to moisture, lime absorbs water of hydration, forming calcium hydroxide. That is, $CaO + H_2O$ becomes $Ca(OH)_2$ The "thirst" of the lime has thus been "slaked."

39. An anion is so named because it migrates toward a positive electrode, or *anode*, placed in a solution. Obviously, an anion carries a negative charge. Similarly, a cation moves to the negative electrode, or *cathode*, and must carry a positive charge. The ions thus take their names from the electrode to which they are attracted. Note the possible confusion, since the charge on the ion is actually opposite in sign to the charge on its destination electrode.

explain what might be going on in terms of the common gas laws?

3. You release a helium-filled balloon. It rises rapidly at first because its density is lighter than that of the surrounding air, but eventually it stops rising. What factors might be preventing the balloon from going higher? Do you think the properties of the balloon material itself may have an effect? How?

4. You fill a bottle with air at room temperature (290 K) and seal it with a strong metal cap. You place the bottle in the freezer and cool it to 260 K. What has happened to the density of the air in the bottle? Next, you heat the bottle in warm water (not too hot!) to about 320 K. What is the change in the air density? In your analysis, ignore the small expansion or contraction of the bottle caused by heating or cooling, respectively.

5. You are sipping champagne at a party and notice the bubbles on the glass. Some of the bubbles seem to grow larger, absorb their neighbors, and rise quickly to the surface (and tickle your nose). What processes are occurring in the champagne that may explain your observations (based on the properties and processes of gases and particles discussed in this chapter)?

6. Try to imagine light in the form of waves traveling through space. Contrast this concept with the familiar waves traveling across the ocean. Now try to visualize a light beam consisting of discrete particles (photons) moving through space. Compare this concept with firing BBs at a target in a shooting gallery—both the BBs and the photons carry energy from one point to another in a straight line. Finally, see if you can visualize the ocean waves as discrete particles. This is more difficult! Why do you think that is?

7. You can experiment with the intensity of light using a candle in a darkened room. Try to read a newspaper by the candlelight; close to the candle it is fairly easy. How far away from the candle can you move before the print is impossible to read? Generally, the farther you move from the candle, the less intense the illumination from the candle will be. Can you explain this effect? The human eye is extremely sensitive to light. Sitting at a distance where you can just read by candlelight, switch on the room lights for a minute and then switch them off again. You will not be able to read the newsprint because your eyes have responded to the brighter room light by narrowing the pupils (the circular opening you can see in the iris of each eye), allowing less of the available light to reach the retina (your eyes also make some other physiological adjustments). However, after only a second or two, your eyes become reacclimated to the dark, and you can read again. Why is it more difficult to read the paper with your pupils contracted under the same conditions of illumination?

8. Can you explain why a lake has a clear blue color on a cloudless day and a gray color on an overcast day?

9. You are trying to catch fish in a pond with a small net. You can see the fish through the surface of the water, but when you attempt to scoop them up with the net, you always come up empty handed. Explain how the refraction of light may be affecting your success. If you had to survive by spearing fish in a stream for food, how would refraction be relevant, and how might you compensate for it?

10. Describe some of the effects that the Earth's atmosphere has on sunlight passing through it. Formulate your answer in terms of common effects that you have witnessed when looking at the sky.

11. Explain why a cloud of nitrogen dioxide might appear brownish in color, and a cloud of soot, black. Under what conditions might the soot cloud appear whitish or bluish?

12. What are the key factors that determine the magnitude of a chemical rate coefficient and thus the speed of a chemical reaction? What are the effects of temperature, reactant partial pressures, the height of the activation barrier, and the heat of reaction on the rate coefficient?

13. What would happen to the composition of the atmosphere if nitrogen (N_2) and oxygen (O_2) could react very rapidly to form NO (that is, if the rate coefficient for this reaction were large)? Would any oxygen be left to breathe?

14. Suppose that the NO emissions from automobiles in a polluted city could be reduced to zero. What effect would that have on the levels of photochemically generated ozone?

15. A lake on your property in the country is becoming overly acidic from polluted rains that fall in the area. In the shed, you have the following chemicals: sodium hydroxide (lye), ammonium hydroxide, and calcium hydroxide

(Ca[OH]$_2$). Which of these compounds might you decide to sprinkle on the lake to reduce its acidity?

Problems

1. Show, using Equations 3.16, 2.5, and 3.17, that the hydrostatic law expressed by Equation 3.15 is exactly equivalent to the ideal gas law, given by Equation 3.2.

2. You want to use a laser beam to measure the distance between two points. You do this by bouncing the light beam of a laser off a target at the distant point and record the time it takes for the light to return to the starting point. By making extremely careful measurements, you determine this time to be 2.66 seconds. How far away is the target point? Do you have any idea what the target might be? If the wavelength of the radiation from the laser is 0.4 micrometer, how many wavelengths are between you and the target?

3. Imagine that the surface area of the sun decreases to one-fourth its present value (owing to contraction). What must the temperature of the smaller sun be for the Earth to receive the same amount of energy as it currently does? Assume that the current temperature of the sun is 6000 K and that no other factors change (for example, the orbit of the Earth).

4. Consider the following reaction between two gases, X and Y:

$$X + Y \rightarrow U + V$$

You know that one molecule of Y weighs exactly twice as much as one molecule of X. In your laboratory, you precisely react 1 gram of X with just enough Y to consume all of X. How many grams of Y have you used in this experiment? What is the combined weight of the products U + V? Suppose you then react the U generated in the first reaction with more X through the process,

$$X + U \rightarrow V + V$$

How much additional X must you use (in grams)? How many grams of V do you end up with?

Suggested Readings: The material in this chapter is quite diverse, and many articles and books on the subject are available. A few selected works, chosen on the basis of readability, content, or historical significance, are listed below.

Bohren, C. *Clouds in a Glass of Beer: Simple Experiments in Atmospheric Physics.* New York: Wiley, 1987.

Brou, P., T. Sciascia, L. Linden, and J. Lettvin. "The Colors of Things." *Scientific American* 255 (1986): 84.

Hellemans, A. and B. Bunch. *The Timetables of Science: A Chronology of the Most Important People and Events in the History of Science.* New York: Simon and Schuster, 1988.

Humphreys, W. J. *Physics of the Air.* New York: McGraw-Hill, 1940.

Jeans, J. H. *The Dynamical Theory of Gases.* London: Cambridge University Press, 1925.

Lavenda, B. "Brownian Motion." *Scientific American* 252 (1985): 70.

Leicester, H. *The Historical Background of Chemistry.* New York: Dover, 1971.

Scott, A. "The Invention of the Balloon and the Birth of Chemistry." *Scientific American* 250 (1984): 126.

4

The Evolution of Earth

To understand the nature of air pollution and the impacts it may have on society and the planet, we must first establish the baseline for a "clean" or natural (or pristine) atmosphere. For example, the "natural" atmosphere may refer to the conditions that existed before modern civilization developed (that is, the atmosphere some 10,000 years ago). More practically, we might hope to reclaim the cleaner environment of our youth (recognizing that some of us are older than others and that reality can fade with age). Knowledge of Earth's past atmospheric conditions allows us to judge more reasonably if the future environment will be acceptable, or even livable. It is essential to understand that the atmosphere and life coevolved. The presence of an atmosphere of a favorable composition allowed life to flourish; the explosion of life forces unalterably changed the atmosphere. We could not survive in the atmosphere that existed in the early history of the Earth. Legions of living things have processed and altered the early air, filling it with oxygen and making it breathable for us. In this chapter, we will explore some of this long and complex history to place the present atmospheric state in perspective.

How can we determine the history of the atmosphere? There were no scientists recording information about the air thousands or millions of years ago. Instead, we must rely on **proxy records** of the past. Proxies are bits of information that have been preserved over time in the form of physical or chemical artifacts. The fossils of ancient organisms molded into sedimentary rocks are a prime source of knowledge about the evolution and distribution of life on the Earth. Geologic formations themselves are critical to reconstructing the history of the Earth and the formation and movements of continents. Methods of assigning dates to fossils and sediments must be available to establish the chronology of events that shaped the Earth and its life-forms. Geologists and paleontologists have found clever ways to date rocks over billions of years by measuring the rates of decay of radioactive elements. Other proxies can take on any number of strange forms, from the pollen of ancient plants to the urine residue of long-dead pack rats. Another invaluable source of information has been the glaciers covering Antarctica and Greenland. The ice sheets form when snowfall accumulates and, through melting and refreezing, forms a porous granular ice sheet called a firn. As the firn consolidates into ice under increasing pressure of continuing snowfalls, air in the pores becomes trapped as bubbles in the ice. The ice in the Antarctic ice sheet is more than 200,000 years old. By drilling a core from deep in the ice sheet, the trapped air can be recovered and analyzed for chemically stable constituents like carbon dioxide, methane, and nitrous oxide. Air from even earlier epochs of the Earth's history may be encapsulated in the resinous sap from trees, which hardens into amber (which is also valued as jewelry).

The overall evolution of the Earth and its atmosphere is subdivided into a number of epochal events, each of which had a major impact on the eventual composition of air, and thus on the global environment:

- The accretion of the Earth and its primitive atmosphere from the primordial solar nebula.
- The differentiation of the interior of the planet and the associated outgassing of volatile materials.
- The chemical era of abiotic photochemical transformation of the primordial atmosphere to form the organic molecules from which life could spring.
- The microbial era during which the first simple life-forms evolved, proliferated, and forever modified the atmosphere and environment.

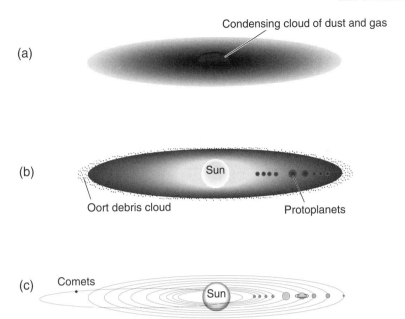

Figure 4.1 The formation of the solar system. (a) In the earliest stage, a dense cloud of dust and gas coalesced as the primordial Solar nebula. (b) Later, the sun and planets condensed as clearly defined objects. (c) Finally, most of the residual dust and gas was swept up, with only the asteroids and comets remaining.

- The geologic era in which the physical reconfiguration of oceans and continents caused major deviations in the evolution of life and the atmosphere.
- The recent age, when humans appeared with the intelligence to exploit fully all of the resources of nature, thereby inheriting the capability to alter significantly the global atmosphere and environment.

Importantly, during all of this change, life and the natural environment evolved together and therefore co-exist in a more-or-less harmonious state. If the environment is threatened by pollution, or other destructive agents, then so is life.

In the story that follows, information from a variety of proxies has been assembled to create a more or less consistent picture of a world in continuous evolution. The picture is fuzzy, however, because proxies alone are inadequate to describe fully the long and complex history of an entire planet. In looking at the history of the Earth, we are peering through a foggy window seeing only shadowy figures. Nevertheless, even shadows from the past may hold important lessons for the future.

4.1 The Origin of the Earth

The story of the Earth begins with the formation of the universe some 15 billion to 20 billion years ago. The expanding gases, consisting mainly of hydrogen from the original "**big bang**," cooled (adiabatically perhaps) and began to condense into the first stars and galaxies. The nuclear furnaces burning in the cores of these early stars were so hot that hydrogen nuclei fused into heavier elements. Occasionally, one of these first stars exploded in a spectacular supernova, spewing its elemental guts over the entire galaxy. The heavier elements, now condensed as mineral and ice particles, were mixed into galactic gas clouds. The dust and gases in these clouds later coalesced under the influence of gravity to form second-generation stars. The sun, our home star, which is about 4.6 billion years old, may be a third- or fourth-generation star.

The formation of our solar system (the sun and the planets) is depicted in Figure 4.1. The planets, including Earth, condensed from the solar nebula about 4.5 billion years ago under the influence of gravity. The new planets rapidly grew in size by gathering up the residual dust and smaller

planetesimals that also condensed from the solar nebula. Not all the planetesimals and dust were collected, however. A large number of large rocky asteroids remain in orbits between Mars and Jupiter. In addition, a swarm of icy comets resides in the **Oort cloud** beyond Pluto (the outermost planet). Occasionally, an asteroid or a comet passes near the Earth and, every now and then, collides with our planet. Today, impacts with large objects are very rare; still, about 10,000 metric tons of fine meteoritic particles enter the upper atmosphere each year as the "shooting stars" seen in the sky on clear evenings. In the early days of the Earth, collisions with massive bodies were frequent and devastating. These impacts churned up the surface, allowing gases from the interior to escape into the atmosphere.

4.1.1 EARLY EVOLUTIONARY PHASES

In its infancy, the atmosphere on Earth consisted of light, volatile gases captured from the primordial solar nebula, principally hydrogen (H_2) and helium (He). As the planet began to cool, a variety of volatile gases were plentiful in the atmosphere, including the noble gases helium, neon, and argon. When powerful nuclear reactions ignited in the collapsing center of the solar nebula caused the sun to turn on as a star, a stiff solar wind blew through the solar system, sweeping away the early atmosphere of volatile gases. Accordingly, Earth is observed to be greatly depleted of noble gases, relative to the composition of the sun.

The second stage of the evolution of the Earth's atmosphere followed within a few hundred million years. This epoch consisted of the **planetary differentiation** and **outgassing** of the interior. Planetary differentiation occurred as the densest elements (for example, iron [Fe] and nickel [Ni]) settled toward the center of the planet under the influence of gravity, forming the core.[1] During this process of settling, gravitational potential energy is converted to heat. The early composition of the Earth also included many radioactive elements. As these decayed to more stable states, their radioactive energy caused further heating of the Earth's interior. The gravitational and radioactive heating kept much of

the interior of the Earth molten. This fluid interior is called the **mantle**. The excess heat escaped from the interior through massive lava flows, which also released gases such as water vapor and carbon dioxide at the surface. Although the outer solid Earth, or **lithosphere**, is considerably cooler today, heat from the interior still escapes to the surface as lava during volcanic eruptions and at midoceanic ridges.[2]

The processes that controlled the early evolution of the Earth's atmosphere are shown in Figure 4.2. The bombardment of planetoids and other rocky and icy debris added heavy elements (such as iron and silicates) and **volatiles** (including nitrogen, water, carbon dioxide, methane, and ammonia) to the Earth. The volatiles were particularly important to the development of the early atmosphere. Most of the volatiles were added to the atmosphere from the interior of the Earth by outgassing associated with volcanic activity. The lightest, most volatile gases, hydrogen and helium, could escape to space by a process called **Jeans's escape**.[3] This escape rate has been slow enough over time that most of the primordial hydrogen still trapped in the atmosphere after its infancy remains today, mainly in the form of water (H_2O).

Water is the most important volatile on Earth. Soon after the planet began to cool, water condensed to form the extensive seas that still cover most of the surface. Those seas have remained for the entire history of the Earth. In fact, ours is the only planet in the solar system with surface oceans of water. The other **terrestrial planets**,[4] Mars and

1. A heavy fluid settles through a lighter one. Thus water poured into a container of oil sinks to the bottom of the container because water has a greater density than oil does; similarly, mercury can easily sink through water. Recall the principle of **buoyancy** discussed in Chapter 3.

2. The outermost solid part of the Earth, composed of common rocks and minerals, is the **crust**; the crust is the upper crystallized layer of the lithosphere. The mantle is more fluid than the lithosphere, and the core has a molten outer layer but a solid center.

3. Sir James Hopwood Jeans (1877–1946) was an English physicist who, along with Rayleigh, Thompson, and other English scientists of the late nineteenth century, conducted fundamental studies of the mechanics of gases and of electromagnetic radiation. Jeans first described how gases could escape the gravitational attraction of a planet by virtue of their thermal velocity if the gas molecules were high enough in the atmosphere and hot enough in temperature. In essence, the molecules are then like small rockets launched from the upper atmosphere into space. If a molecule has at least the "escape velocity" (Section 2.3), it can leave the Earth forever.

4. The terrestrial planets, Venus, Earth, and Mars, are the second, third, and fourth planets from the sun, respectively, and are solid, rocky worlds quite different from the gaseous giant outer planets. Earth and Venus are called "sister" planets because they are so similar in size and composition. Yet the surface temperature of Venus is about 750 kelvin, almost hot enough to melt lead. No oceans can exist under these conditions. On Mars, the surface temperature is about 220 kelvin, and any water that might be

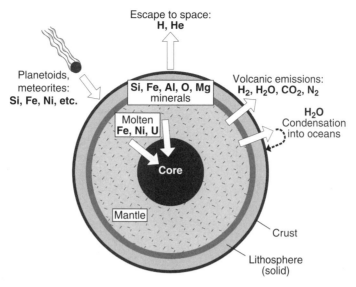

Figure 4.2 The differentiated structure of the early Earth and the principal geophysical and geochemical processes that contributed to the evolution of the atmosphere.

Venus, have completely dry surfaces. Oceans, it turns out, were essential to the evolution of life and ultimately to the development of the current state of the global environment.

Notice that a critically important volatile, oxygen (O_2), was not present in the early atmosphere. Indeed, all of the elemental oxygen in the solar nebula was originally chemically bound in compounds such as CO_2 and H_2O. Thus no free molecular oxygen was available for the atmosphere. In fact, the early atmosphere was in a highly **reduced** chemical state; that is, it was dominated by compounds that can react with and deplete free oxygen. On the other hand, we know that free molecular oxygen is now plentiful in the atmosphere and is essential to present-day life. (The fascinating story of oxygen is told in Section 4.2.)

The inventories of key volatiles on the Earth, as they were originally accreted when the Earth formed and remain today, are summarized in Table 4.1. Each of the three dominant volatiles—H_2O, CO_2

and N_2—has a different primary **reservoir**. Water is concentrated in the ocean reservoir, carbon dioxide in the solid sedimentary rock reservoirs (mainly as carbonate deposits), and nitrogen in the atmospheric reservoir (as nitrogen gas). (The movement of materials between these reservoirs is discussed in this chapter and in Chapter 10.) The cycling of volatiles, in addition to the presence of free oxygen (not an original volatile), controls most of the environmental conditions on the Earth.

Figure 4.3 depicts the movement of volatile materials through the Earth's reservoirs. The principal reservoirs are the atmosphere, oceans, and solid sediments of the lithosphere. The time scales for the processes shown in Figure 4.3 to cycle materials through the reservoirs are hundreds of millions of years. You have to be patient when you are dealing with an entire planet. Even at this slow rate, materials in the upper layers have been recycled many times over the history of the Earth. The atmosphere is the most tenuous and dynamically active of the reservoirs. Materials can be filtered through air rather rapidly. The quantities of volatiles in the atmosphere are strongly influenced by transfers with the ocean and land reservoirs on relatively short time scales (only hundreds to thousands of years for carbon dioxide, for example, but up to several million years for oxygen). The most inert volatiles—the noble gases and nitrogen—are concentrated in the atmosphere, and the amounts are quite invariant

present is frozen in the ground—again, no oceans. At different times in recent history, people have believed that life existed on Mars and Venus. Astronomers have seen "canals" on Mars, implying intelligent "builders." Others have speculated on the swampy conditions that must exist beneath the dense clouds that permanently enshroud Venus and stuporously go on to conclude that "dinosaurs" probably roam there. Direct, close-up investigations of both worlds by spacecraft clearly show neither "builders" nor "dinosaurs."

Table 4.1   Volatiles on the Earth[a]

Volatiles	Total Quantity (grams)	Atmosphere (grams)	Ocenas (grams)	Sediments (grams)
H_2O	1.6×10^{24}	1.7×10^{19} (0.001%)	1.4×10^{24} (88%)	1.9×10^{23} (12%)
CO_2	2.4×10^{23}	2.5×10^{18} (0.001%)	0.4×10^{20} (0.06%)	2.4×10^{23} (99.9%)
N_2	4.9×10^{21}	3.9×10^{21} (79.5%)	2.2×10^{19} (0.5%)	1.0×10^{21} (20%)
Cl	3.1×10^{22}	5.0×10^{12} (0%)	2.6×10^{22} (84%)	5.0×10^{21} (16%)
S	5.2×10^{21}	5.0×10^{12} (0%)	1.2×10^{21} (23%)	4.0×10^{21} (77%)

[a] The amounts are given as the mass of H_2O, CO_2, N (nitrogen-atom equivalents), Cl (chlorine-atom equivalents), or S (sulfur-atom equivalents).
*Source: J. C. G. Walker, *Evolution of the Atmosphere* (New York: Macmillan, 1977).

over geologic eras.[5] The geologic cycles of several volatiles of importance to the present state of the environment are discussed in Chapter 10.

4.1.2 BOX MODELS FOR EARTH RESERVOIRS

The amount of a material in a major reservoir, such as the atmosphere or the oceans, depends on the rates at which that material is added to and removed from the reservoir. In some cases, we simply want to know the total amount of material in a reservoir. In other situations, we want to know the concentration of the material in the reservoir or to understand the processes that control the amount. To deal with all these problems, we have developed the important concept of a **box model**, in which we imagine that the reservoir is literally a "box" into which a material of interest can be added or removed in measurable amounts at controllable rates. For example, imagine your bathtub as a reservoir, or "box," for water (if

5. The geologic time scale has five primary intervals: eras, periods, ages, epochs, and times, in the order of decreasing length. Geologic eras are characterized by major differences in the dominant life-forms (microbes versus mammals). Geologic periods are punctuated by major extinction events, when 50 percent or more of all species may disappear.

you are lucky enough to have a tub, since most of us shower these days). Water can be added to the tub by turning on the faucet, and it can be removed by opening the drain. If you close the drain and turn on the faucet to a certain position, the tub will begin to fill at a constant rate. You can easily measure the rate at which water is added (so many quarts per minute, say). Now open the drain. If the drain is large enough and the faucet is not turned on too far, the level of water (and so the amount of water in the tub) may begin to fall. If you could vary the opening of the drain, you could stabilize the amount of water in the tub. That is, with the faucet and drain in fixed positions, the level of water in the tub would become stationary. You would have achieved a **steady state** condition.

A "steady state" is exactly what the term implies; the state of the system is invariant, or "steady," in time. A system in steady state is still a *dynamic*, not a fixed, system. In the example of the tub, only the level of the water is invariant, or steady. Water itself continues to pour into and drain out of the tub (at equal rates in the steady state situation). Moreover, the water in the tub is continuously changing, even though the *total amount* of water remains constant. In other words, water is cycled through the tub even in a steady state.

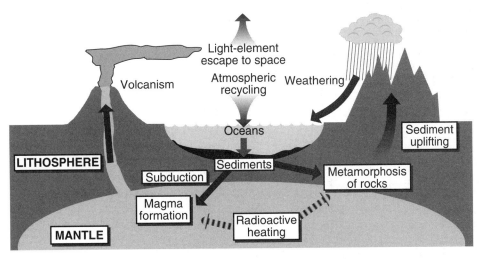

Figure 4.3 The main processes that recycle material through the Earth's reservoirs and thereby influence the global environment. Transport and transformation in the atmosphere, absorption and settling in the oceans, and subduction and volcanism in the lithosphere all are involved.

Figure 4.4 illustrates a simple box model. The magnitude of the source, S, and the residence (or lifetime), τ, of the material in the box determine the total quantity in the box:

$$Q = S\tau \qquad (4.1)$$

The source is typically specified in units of mass per unit time (for example, tonne/sec). The lifetime, τ, then must have consistent units (seconds in this case). In this case, the amount of material in the box can be characterized in terms of its total mass. This amount may be given in any quantitatively equivalent units. When we discuss the carbon dioxide cycle, we can thus use units of kilograms of CO_2, or equivalent kilograms of carbon atoms, C, or the total number of CO_2 molecules, or the number of moles of C, or any other convenient measure of the quantity of carbon dioxide. We can also use different basic units for Q, S and τ, provided that we multiply Equation 4.1 by an appropriate conversion factor, c, so that

$$Q = cS\tau \qquad (4.2)$$

would have the same units on both sides of the equation. Thus if Q is given in kilograms, S in grams per second, and τ in seconds, c must have a value of

$$c = \frac{1\ \text{kg}}{1000\ \text{g}} = 0.001$$

The concentration of material in the box is related to the volume, V, or size, of the box (assuming that the material is uniformly distributed throughout the box):

$$q = \frac{Q}{V} = \frac{S\tau}{V} \qquad (4.3)$$

The box has a volume that is defined in relation to the reservoir of interest. For the entire Earth, the atmospheric reservoir is essentially one continuous volume. Although the world's oceans are physically separated to a great extent, they can also be aggregated into one reservoir or box. If we are concerned with local air pollution, the box can consist of an urban airshed; if indoor pollution is our problem, the relevant box can be a room or a building. The box model can be used to represent any enclosure that holds a material (even if it leaks).

In steady state balance, we must have

$$S = L \qquad (4.4)$$

That is, the source must equal the **sink**, or loss or destruction rate, L. If we can measure the *steady state* concentration in the box, q, and know the residence time, τ, and the size of the box, V, then we can determine the source or sink rate for the material in the box:

$$S = L = \frac{qV}{\tau} = \frac{Q}{\tau} \qquad (4.5)$$

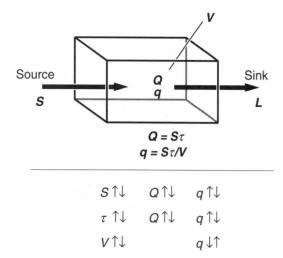

$$Q = S\tau$$
$$q = S\tau/V$$

$$
\begin{array}{ccc}
S\uparrow\downarrow & Q\uparrow\downarrow & q\uparrow\downarrow \\
\tau\uparrow\downarrow & Q\uparrow\downarrow & q\uparrow\downarrow \\
V\uparrow\downarrow & & q\downarrow\uparrow
\end{array}
$$

Figure 4.4 The box model. The box, or reservoir, is envisioned to have a fixed volume, V, with material source(s), or rate(s) of addition to the box, given by S, and material sink(s), or rate(s) of removal from the box, given by L. The total amount of material in the box is Q, and the concentration of material in the box is q. The average residence time, or lifetime, of the material in the box is given by the time constant, τ.

Suppose that while fishing in the middle of a lake, your rowboat springs a leak. The water flows through the breach at a constant rate. You begin to bail out the water with an empty beer can. If the leak is slow, you may be able to bail fast enough to stabilize the amount of water in the boat. Congratulations. You have achieved a steady state; that is, the rate at which you are dumping water over the side exactly equals the rate at which the leak is letting it back in. Soon, we hope, someone will come by and rescue you. If the leak were much larger, however, you would have to abandon ship, for no matter how fast you moved the beer can, the level of water would continue to rise. You are not in a steady state. **In a steady state system, either the source or the sink, or both, must adjust to establish a mass balance for the reservoir.**

The key parameters for the box model are summarized below:

Q = total quantity in the box; such as tonnes,
V = volume of the box; such as cubic meters,
q = concentration in the box; such as tonnes per cubic meter,
S = source per unit time; such as tonnes per day,
L = loss per unit time; such as tonnes per day,
t = residence time in the box; such as days.

Using these parameters, the box model equation can be written in a number of equivalent forms, including [from Equations 4.1 and 4.3]

$$Q = qV = S\tau \qquad (4.6)$$

In general, if the box parameters vary, the amount of material, Q, in the box also will change. For example, if the source, S, increases with all other factors remaining constant, Q and q must also increase. Similarly, if the residence time increases, so shall Q and q. On the other hand, an increase in the volume causes a dilution of the material in the box and a decrease in q; Q, the total amount in the box, is not affected in this case (Equation 4.6).

Consider the atmosphere as a global reservoir for carbon dioxide represented by a simple box model. For this example, we can take the volume of the atmosphere to be $5 \times 10^{18}\,\mathrm{m^3}$. The residence time of CO_2, about 10 years, is determined over short time scales by its rate of assimilation by plants during photosynthesis. Moreover, the concentration of CO_2 in the atmosphere has been measured to be about 365 ppmv. Thus we have q, V, and τ in this case. Of course, we must convert units:

$$
\begin{aligned}
365\ \mathrm{ppmv}\ CO_2 &= 365 \times 10^{-6} \\
&\quad \times 0.001\ \mathrm{tonne\text{-}air/m^3} \\
&\quad \times \frac{44}{29} \\
&= 55 \times 10^{-8}\ \mathrm{tonne}\ CO_2/m^3 \\
&= 15 \times 10^{-8}\ \mathrm{tonne}\ C/m^3
\end{aligned}
$$

The factor of $44/29$ converts the mass of an air molecule to the mass of a carbon dioxide molecule. Now it is easy to calculate the total mass of carbon dioxide in the atmosphere as 55×10^{-8} tonne-$CO_2/m^3 \times 5 \times 10^{18}\,\mathrm{m^3} = 2750 \times 10^9$ tonnes CO_2, or 2750 **gigatonnes** ($1\ \mathrm{Gt} = 10^9$ tonnes) of CO_2. The total mass of carbon would be about 750 Gt. Moreover, from Equation (4.6), the source (and sink) rate to the atmosphere must be roughly 2750 Gt-CO_2/10 yr = 275 Gt-CO_2/yr, or about 75 Gt-C/yr. From a few simple measurements and facts, we can specify to a great extent the behavior of carbon dioxide in the atmosphere.

In real environmental reservoirs (boxes), there can be many sources and sinks, and the reservoirs may not be homogeneous. In addition, the reservoir system may not be in a steady state, or equilibrium, which complicates the interpretation of box parameters. In this book, specific applications of the box model are discussed in the context of these limitations.

4.1.3 THE PREBIOTIC ATMOSPHERE

The atmosphere that existed before the appearance of life forms was controlled by chemical and physical processes. This early period, lasting perhaps several hundred million years, can be referred to as the **chemical era**. Because of the elemental composition of the solar nebula, the atmosphere at this time was reducing in its chemical activity. No free oxygen or, consequently, ozone existed. Accordingly, ultraviolet sunlight would be present throughout the atmosphere to initiate photochemical processes involving gases such as water vapor (H_2O), carbon dioxide (CO_2), methane (CH_4), ammonia (NH_3), and hydrogen (H_2). In particular, ammonia would have been photochemically unstable during this period and would have decomposed to form nitrogen by the overall process

$$2NH_3 + \text{Sunlight } (h\nu) \rightarrow \cdots \rightarrow N_2 + 3H_2 \tag{4.7}$$

There is some controversy about the fraction of present-day atmospheric nitrogen that is primordial N_2 and the fraction that was generated from primordial ammonia. Over the entire history of the Earth, nitrogen has no doubt been present in the atmosphere in roughly the same amount as it is today. The implications of nitrogen levels for other evolutionary processes are minor.

In discussing the quantity of a gas in the atmosphere at different times, it is convenient to scale that amount to the **present atmospheric level, (PAL)**. In these units, 1 PAL is just the present amount in the atmosphere, 2 PAL is twice as much, and so on. In the case of nitrogen, the atmosphere has apparently always held about 1 PAL. In the case of oxygen, the situation has been quite variable, with the early atmosphere holding much less than 1 PAL.

Geochemical Cycles

Figure 4.3 shows that the early atmosphere was dominated by geochemical and geophysical processes involving the land and oceans. The condensation of water into oceans on Earth was responsible for three critical processes:

1. The evaporation of water from the oceans and its subsequent recondensation into clouds and precipitation established a vigorous **hydrological cycle**. The hydrological cycle provides energy to drive the atmospheric circulation, supplies water to an otherwise parched land, and causes **weathering** of the soil and rocks, which recycles uplifted mineral sediments and volcanic lava emissions into the oceans.

2. The oceans themselves became one of the key reservoirs of volatiles in the Earth system and especially provided a means of regulating the Earth's climate, in several ways—by generating reflective water clouds and snowfields through the hydrological cycle, by limiting the buildup of carbon dioxide in the atmosphere through CO_2 conversion to limestone sediments (Sections 10.2.4 and 12.2.3), and by acting as a reservoir where *heat* could be stored, to act as a temperature-regulating flywheel for the climate system (Section 11.5).

3. The oceans provided the first living organisms with a haven from the deadly ultraviolet radiation of the sun that penetrated the early oxygen-poor atmosphere and provided a nurturing environment for life to evolve during the microbial era and beyond.

The oceans provided an ideal medium for the formation of mineral sediments that were later consolidated into sedimentary rocks. The sediments consisted of silicates (sand) washed from the land or carried long distances by winds over the oceans (the aeolian dust) and the detritus of organisms living in the oceans. The conversion of carbon dioxide to carbonate sediments in the oceans is particularly important. The formation of calcium carbonate, or limestone, is the most common. The process begins with the dissolution of CO_2 in the oceans to form carbonate ions, CO_3^{2-}, in solution. This chemical transformation is described by the following reactions:

$$CO_2 \text{ (gaseous)} \overset{H_2O}{\Longleftrightarrow} H_2CO_3 \text{ (aqueous)}$$

$$H_2CO_3 \text{ (aqueous)} \Longleftrightarrow H^+ + HCO_3^-$$

$$HCO_3^- \Longleftrightarrow H^+ + CO_3^{2-} \tag{4.8}$$

The symbol $\Longleftrightarrow$ indicates a reaction in an **aqueous solution** (that is, in water) that proceeds rapidly in each direction without any activation barrier and thus achieves **chemical equilibrium**. In an equilibrium chemical process, the rates of the chemical reactions are equal in both directions.

The carbonate ions in solution may then combine with calcium ions that have been weathered from rocks at the land surface and carried to the oceans by rivers and streams. The calcium carbonate reaction is

$$Ca^{2+} + CO_3^{2-}$$
$$\xrightarrow[\text{Aqueous solution}]{} CaCO_3 \text{ (precipitate)}$$

(4.9)

The calcium carbonate that precipitates at the floor of the ocean produces sediments of limestone. After life evolved, marine organisms began to utilize the available calcium and carbonate ions to construct hard shells of calcium carbonate, for skeletal strength and protection (clams and oysters have such shells, and so do many microscopic organisms). When the organisms died, their shells settled to the deeper ocean and were deposited in the sediments, later to be compressed and consolidated into sedimentary limestone rocks. Over a very long time, the limestone sediments are recycled by metamorphosis and uplifting (Figure 4.3).[6] The recycled sediments are exposed to weathering, and the mineral cycle is completed.

Abiotic Organic Synthesis

The secondary atmosphere that developed as a result of these geochemical and geophysical mechanisms contained large amounts of reduced gases such as H_2, N_2, and CH_4. These simple gases cannot directly combine to form the complex organic molecules of life. However, when mixtures of such gases are exposed to high energies—such as those associated with lightning discharges, ultraviolet radiation from the sun, and high-energy impacts of meteors—

6. *Metamorphosis* refers to the transformation of the limestone into other materials at the high temperatures and pressures that exist deep in the Earth. For example, at high temperatures, limestone is decomposed into carbon dioxide (and calcium oxide). The highly pressurized CO_2 gas is expelled explosively in volcanic eruptions. You can simulate such eruptions by shaking a bottle or can of soda before opening it, which causes the dissolved carbon dioxide to form bubbles under pressure in the soda.

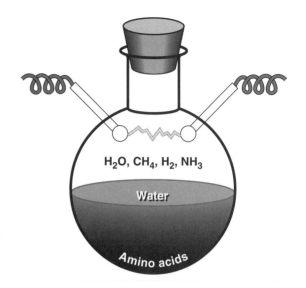

Figure 4.5 The Miller–Urey experiment to create the building blocks of life. Simple primordial gases were sealed in a flask with excess water and exposed to an electrical arc. The organic compounds formed collected in the "microocean" in the flask.

reactions can occur that lead to more complex species. In 1953, Stanley Miller and Harold Urey set up a simple experiment, illustrated in Figure 4.5, in which H_2O, CH_4, NH_3, and H_2 were sealed in a flask and subjected to an electrical discharge (to simulate lightning). The result of the Miller and Urey experiment was a brown soup rich in complex organic molecules at the bottom of the flask. Even though little squiggley living things did not pop out of the soup, the experiment was remarkable in generating **amino acids**, the complex building blocks of proteins. Proteins are essential components of all living cells. This work proved that the basic material for evolution could be formed by physical and chemical processes in an abiotic world. Later experiments by Carl Sagan and others showed that the same result could be obtained by ultraviolet irradiation of primordial gas mixtures. Even the irradiation of simple mixtures of CO and water vapor resulted in the formation of complex hydrocarbons.

One can imagine the organic materials produced in the atmosphere settling into the oceans, creating an organic-rich chemical soup. Over hundreds of millions of years, the ceaseless chemical interactions between these compounds led to the first self-replicating molecules, the precursors of **deoxyribonucleic acid** (**DNA**), the molecule carrying the genetic code. These extraordinarily complex organic molecules are very sensitive to ultraviolet radiation; they are easily dissociated and destroyed by solar radiation between 250 and 300 nanometers (nm). Fortunately, in primitive oceans, the earliest lifeforms would have found refuge beneath layers of murky organic debris deposited from the sky that served as both the spring of life and its food supply. It has recently been suggested that life formed beneath a thick layer of ice covering the primordial oceans. Ice could have protected microbes evolving in deeper unfrozen waters from harsh solar rays. Ice is also an excellent thermal insulator (igloos are constructed of ice). Heat was generated when large meteors collided with the Earth and fell into the deep oceans. The heat would have been held in as the ice resolidified. The continual bombardment of the young Earth by comets and asteroids would have guaranteed a regular supply of energy to fuel evolution.

The moment in time and the chemical means of formation of the first living organism remain unknown. To get a feeling for the number of natural "experiments" that were going on at the time, consider the Miller–Urey experiment. Miller and Urey had sealed in their flask about 10^{23} molecules and irradiated them for perhaps 10^{6} seconds. The number-time product for the experiment was then about 10^{29}. By comparison, the early atmosphere contained about 10^{44} molecules, and the experiment continued for about 10^{16} seconds (300 million years) before life appeared. This number-time product was 10^{60}, a factor of 10^{31} greater than in the Miller–Urey experiment! If every human that ever lived were to run the experiment in a 1 cubic meter bottle for 50 years, nature would have still done the experiment 1,000,000,000,000,000 as many times. It is very unlikely that you will create life in a test tube by sparking gases.

It has been argued that life may have been established on Earth by organisms carried here on comets from other parts of the solar system or galaxy. In this scenario, life would have had to evolve elsewhere, perhaps on cold interstellar dust grains bathed in starlight. The organisms would also have had to survive a meteoric entry through the atmosphere and a fiery impact on the surface. But even if they were devoid of life, comets may have contributed substantial inventories of organic molecules to the early environment. It is known that complex organic compounds are synthesized in space. The first life on Earth, shivering in deep frigid ocean waters, may

have sprung from this cosmic manna. Later, life may have been extinguished by impacts of huge objects from space (Section 4.3), which themselves refertilized the oceans.

4.2   The Coevolution of the Environment and Life

Following the abiotic geochemical era, the new chemical processes of life became a major influence on the evolution of the atmosphere and other elements of the environment. The period during which life evolved in the form of microscopic organisms (microorganisms) is referred to as the **microbial era**. Biochemical processes can be extremely efficient because enzymes and other specialized materials in cells can mediate complex chemical transformations, acting like microscopic factories. Moreover, living organisms have the means for reproducing rapidly. Hence living organisms multiply to exploit effectively an available resource, including an atmospheric gas that can be assimilated. The simplest living organisms are the **prokaryotes**. Bacteria and green algae are included in this category. These single-celled creatures do not even have internal organ structures, such as a cell nucleus. Nevertheless, they are designed, as all life-forms, to absorb nutrients from the environment in order to create energy for reproduction and other life functions. The prokaryotes evolved over a period of 3 billion years before more complex one-celled and multicelled eukaryotic organisms (**eukaryotes**, microorganisms) developed. Eukaryotic cells have internal organelles and all the bells and whistles of life; we are constructed from such cells.

The biochemical processes that initially affected the atmosphere were fairly simple and over time grew more and more sophisticated. This is exactly the course that would be expected. The following section gives a thumbnail sketch of the evolution of biological metabolism and its effects on the atmosphere.

4.2.1   THE EVOLUTION OF LIFE PROCESSES

Some of the key developments in the long-term evolution of life are shown in Figure 4.6. All living things require energy and draw it from their environment to carry out the tasks needed to survive and reproduce. Energy may be absorbed directly as light

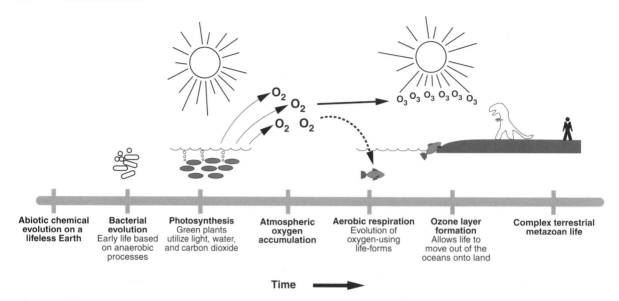

Figure 4.6 Milestones in the coevolution of life and the environment. Living organisms have become increasingly complex over time, and biological processes have had a major impact on the composition of the atmosphere. Evolution, in turn, has responded to changes in the atmosphere's composition by producing life-forms that could utilize atmospheric gases.

or heat, or it may be generated by chemical reactions occurring within the organism itself. Life-forms have over time developed ingenious methods of exploiting every conceivable source of energy in the environment. These methods range from relatively simple to quite complex. Naturally, one would expect the simplest energy-generating schemes to have appeared early in the evolution of life and to have later increased in sophistication.

In the earliest environment, organic molecules created by abiotic synthesis offered the first microbes a steady source of nutrients. These organisms were likely to use **fermentation** as a source of energy for biosynthesis (that is, to fuel their metabolism and fabricate organic body mass). Fermentation is a process that transforms an available organic compound into other compounds, thereby releasing excess chemical energy. A common example of bacterial fermentation is the conversion of sugar to ethyl alcohol (**ethanol**). The overall chemical process is

$$C_6H_{12}O_6 \rightarrow \cdots \rightarrow 2C_2H_5OH + 2CO_2$$
$$\text{sugar} \rightarrow \cdots \rightarrow \text{ethyl alcohol}$$
$$+\text{carbon dioxide}$$
$$(4.10)$$

Note that carbon dioxide is generally released during fermentation, which is what gives beer and

champagne their effervescence.[7] Fermentation is one of the simplest biochemical processes, but it is not very efficient at generating energy because the reactants and products have similar chemical potential energies. Fermentation developed in response to the abiotic source of organic matter. With organisms growing and dying, however, an additional source of organic "foodstuff" was present later in the microbial era.

The biosynthesis of living organic matter from existing organic molecules is referred to as **chemical heterotrophy**. Fermentation is a heterotrophic metabolic process. Modern organisms that eat other organisms are heterotrophs. Humans are heterotrophs, eating plants and animals in great quantities (with Americans perhaps the most prolific heterotrophs of all).

The next step in the evolution of bacterial chemistry could have been the development of **chemical autotrophy**. In this process, microbes transform

7. Baking bread is another use of fermentation. To leaven the bread, or make it "rise," living cells of a fermenting fungus, *Saccharomyses cerevisiae*, are added to the dough as yeast. Sugar is included in the batter for the fungi to feed on. They eat the sugar and generate carbon dioxide and ethanol (ethyl alcohol). The CO_2 puffs up the bread dough, which must be periodically beaten down, or "sized." Salt may be added to control the growth of the fungi. You might get inebriated if you ate raw dough, but baking drives off the alcohol. And the extra fungi? You eat them. They are high in protein and very nutritious (some health enthusiasts eat yeast as a dietary supplement).

inorganic materials into organic matter (recall our definitions of *organic* and *inorganic* in Chapter 3). For example, common **methane bacteria** use carbon dioxide and hydrogen gas to synthesize methane and generate energy:

$$CO_2 + 4H_2 \rightarrow \cdots \rightarrow CH_4 + 2H_2O \quad (4.11)$$

Chemical autotrophy is important because it provides a new source of organic matter for the environment. The quantities of inorganic molecules available as food for bacteria are much greater than the quantities of organic molecules created by abiotic processes. For example, hydrogen gas was probably plentiful in the primordial atmosphere as a result of volcanic outgassing and the low levels of oxygen (which would react vigorously with hydrogen).

One of the most abundant sources of energy in the environment is sunlight. As soon as microbes could evolve chromophores to effectively absorb photons and use their energy to catalyze chemical reactions, the important process of **photosynthesis** developed. Photosynthesis is the production of organic matter by living organisms using the energy of sunlight. The first **bacterial photosynthesizers** probably used organic molecules as a source of carbon for biosynthesis; therefore, they were heterotrophic organisms. Later in the evolution of photosynthesis, organisms learned to use carbon dioxide as a source of carbon and gases like hydrogen sulfide as sources of hydrogen for biosynthesis. The process of **autotrophic bacterial photosynthesis** developed, involving such processes as

$$CO_2 + H_2S \xrightarrow{h\nu} \cdots \rightarrow \text{``}CH_2O\text{''} + SO_4^{2-} \quad (4.12)$$

where the symbol "CH_2O" indicates a "unit" of organic matter formed by biosynthesis. Equation 4.12 represents the basic life process of the purple bacteria that currently live near volcanic vents rich in carbon dioxide and sulfide emissions. Photosynthesis offered a powerful energy source that allowed life to exist on previously indigestible nutrients. Even so, the supplies of reduced compounds, such as hydrogen sulfide and hydrogen, for photosynthesis were very limited.

Green plant photosynthesis most likely developed after bacterial photosynthesis became established. Photosynthesis using water as a reactant is considerably more complex than other forms of bacterial photosynthesis. Originally it was carried out by microscopic algae that had developed the ability to use water rather than H_2 or other reduced compounds in biosynthesis. The overall process of photosynthesis, which can be classified as **photoautotrophy** is,

$$CO_2 + H_2O \xrightarrow{h\nu} \cdots \rightarrow \text{``}CH_2O\text{''} + O_2 \quad (4.13)$$

There are three critical observations to be made concerning photosynthesis. First, the energy to drive the process is derived from sunlight. Accordingly, an organism can directly harness the power of sunlight, rather than depend on chemicals in the environment that must first be energized by sunlight through photochemical processes. Second, the food for this process consists of water and carbon dioxide, two of the most plentiful compounds in the environment. These photosynthetic organisms thus are not limited in their growth potential by the lack of basic foodstuffs. Third, one of the products of green plant photosynthesis is molecular oxygen, and so for the first time in the history of the Earth, a potent biological source of free oxygen came onto the scene.

With the accumulation of oxygen in the atmosphere caused by the proliferation of green plant photosynthesis, **aerobic respiration** would have become a favorable biological option. In the earlier stages of evolution, bacteria developed that did not require oxygen to "breathe"; instead, they sniffed the air for hydrogen and other gases then available. Indeed, oxygen was as toxic to these organisms as hydrogen sulfide is to us. Organisms that do not depend on oxygen for energy are said to be **anaerobic**.[8] You can imagine the pickle these microbes found themselves in when green plant photosynthesis became popular. As the air filled with oxygen, these poor fellows had to retreat into environments remote from air and oxygen. Today, in fact, anaerobic bacteria can be found only in oxygenfree mud at the bottom of ponds and marshes, in festering biomass, and in the guts of larger animals. (Their Dantean fate is discussed in Section 4.2.2.)

---

8. The terms *aerobic* and *anaerobic* are common today in the health and exercise industry. When you exercise aerobically, you are doing so at a level of exertion at which your breathing maintains a sufficient supply of oxygen to your muscles to avoid burnout. If you push yourself harder into the anaerobic regime, your body will require more oxygen than you can take in, you will pant heavily, and your oxygen-starved muscles will become fatigued and useless. This often happens to me on the way to the refrigerator.

Aerobic respirators use oxygen to "burn" various fuels to produce energy. For example, the **hydrogen bacteria** burn hydrogen gas:

$$2H_2 + O_2 \rightarrow \cdots \rightarrow 2H_2O \quad (4.14)$$

and the **sulfur bacteria** burn hydrogen sulfide:

$$2H_2S + O_2 \rightarrow \cdots \rightarrow S_2 + 2H_2O \quad (4.15)$$

Of course, with green plant photosynthesizers producing a large volume of biomass and releasing copious amounts of oxygen, it is advantageous for organisms to evolve that can eat the green plants and use the available free oxygen to generate energy. The aerobic respiration process corresponding to Equation 4.13 is

$$\text{"CH}_2\text{O"} + O_2 \rightarrow \cdots \rightarrow CO_2 + H_2O \quad (4.16)$$

The respiration of plants and other biological matter is a heterotrophic process. Humans oxidize the tissues of plants and animals in order to generate energy. When organic material, such as rotting vegetable matter, decays, the process may involve heterotrophic, aerobic, or anaerobic bacteria. Alternatively, the conversion of organic compounds to carbon dioxide (and water) by direct oxidation—for example, by combustion—is called **mineralization**.

Another important biogeochemical function performed by a special class of microbes is **nitrification**. Here, ammonia, or other reduced nitrogen, can be oxidized to nitrate, for example, by the steps

$$2NH_3 + 3O_2 \rightarrow \cdots \rightarrow$$
$$2NO_2^- + 2H^+ + 2H_2O \quad (4.17)$$

$$2NO_2^- + O_2 \rightarrow \cdots \rightarrow 2NO_3^- \quad (4.18)$$

Nitrification most likely evolved with the buildup of oxygen in the atmosphere and may have been contemporaneous with aerobic respiration. The nitrate generated is used by many modern plant species as a valuable nutrient. When you feed your favorite potted plant to make it green and healthy, you are probably adding a nitrate fertilizer. In the ancient environment, the sources of **fixed nitrogen** for nitrate formation were abiotic photochemical production of ammonia, and the **nitrogen-fixing bacteria**, which produced fixed nitrogen for their own

metabolism. Nitrifying bacteria are quite ancient and are important today in the creation of "natural" fertilizers.[9]

Just as there were nitrifying organisms that evolved in the presence of oxygen and reduced nitrogen, **denitrifying** bacteria evolved to exploit the new sources of oxidized fixed nitrogen. This was the last of the great microbial innovations in metabolic chemistry.

4.2.2 ANCIENT ORGANISMS AND GREENHOUSE GASES

During evolution, synergistic relationships were created between organisms that shared local ecosystems. One particularly important relationship involves the **food chain**, in which one organism might have been food for another, which in turn was food for a third, and so on. Food chains may be closed within a **biome**, or ecological community. In this case, certain nutrients are cycled among groups of species. Over time, the species may become fully dependent on one another. Two simple biological food cycles involving primitive microbes are shown in Figure 4.7. In the anaerobic photosynthesis process, bacteria use sunlight to convert CO_2 and H_2S to sulfate and organic biomass (the purple and green sulfate bacteria perform this function). When these microbes die, a different organism utilizes the sulfate and organic material to produce metabolic energy through sulfate reduction, releasing CO_2 and H_2S. The nutrient cycle is completed by these steps. Although the actual biological cycle, referred to as the **sulfuretum**, involves other organisms and is quite complex, the essential elements are described in Figure 4.7.

The biological cycle consisting of green plant photosynthesis and aerobic respiration also is shown in Figure 4.7. Here, algae collect sunlight as an energy source to metabolize H_2O and CO_2 into organic tissues and free oxygen. Subsequently, the oxygen is used by other organisms in aerobic respiration, which converts the organic matter back to H_2O and CO_2 while releasing energy. This key biological cycle—carried out in many ways almost

9. For example, a farmer may alternate plantings of wheat and alfalfa on a field, since the wheat requires nitrate to flourish, and the alfalfa, which is leguminous, holds nitrogen-fixing microorganisms in nodes on its roots, thus nitrifying the soil in which it is planted.

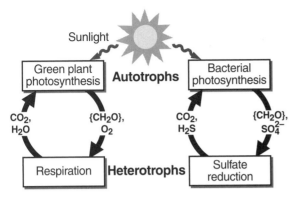

Figure 4.7 Two closed biological cycles. (a) An anaerobic cycle in which autotrophs utilize sunlight to generate biomass and heterotrophs utilize the metabolic by-products to assimilate the biomass for energy. (b) The key biological cycle of green plant photosynthesis and aerobic respiration. The photosynthesizers and respirators are locked in a mutually supporting, continuous life cycle. (From J. C. G. Walker, *Evolution of the Atmosphere* [New York: Macmillan, 1977]. Based on Figure 1–13, p. 43.)

everywhere on Earth—has dominated the biosphere for more than a billion years.

Many of these biological cycles involve the transfer of simple gases between organisms. For example, aerobic respiration/green plant photosynthesis involves the important gases CO_2 and O_2. Methanogenesis generates CH_4; denitrification releases N_2O. The CO_2, CH_4, and N_2O are **greenhouse gases** that play a role in determining the climate of Earth (Sections 11.4, 12.1, 12.2, and 12.3; also see Sections 10.2, 14.2, and 14.3). Notice that all these gases are by-products of microbial metabolism.

The evolution and establishment of microbes in the environment have been critical to determining the trace composition of the atmosphere for the past 2 billion to 3 billion years. In some cases, microscopic organisms have become very efficient at generating greenhouse gases, in part because other organisms utilize these gases and recycle their nutrient value through biological cycles. If such complex microbial processes and interrelationships had not evolved over billions of years, the atmosphere, currently so favorable to life, would not exist. To a large degree, microbes still control the flow of nutrients and greenhouse gases through the atmosphere, oceans, land, and biosphere. However, an interloper has entered the scene—*Homo sapiens*. Humans are beginning to compete with microbes as sources of

greenhouse gases, altering the intricate microscopic machinery of life that took so long to evolve in synergy with the physical world.

4.2.3   PHOTOSYNTHESIS AND THE OZONE LAYER

The earliest microbial life very likely appeared in the oceans, and very few organisms could venture out onto the land. In an atmosphere without oxygen, the direct ultraviolet rays of the sun would stream to the surface and scorch any living cells exposed to it. As marine green algae pumped up the atmospheric level of oxygen, the ozone layer was formed. Ozone is effective at filtering ultraviolet radiation (Chapter 13). With a more benign environment on land, plants and animals that evolved in the seas could begin to move out onto the shores and mud flats of the world's oceans. This environment, rich in nutrients and largely unoccupied, would see a rapid expansion of new species. Vegetation, in particular, thrived in soils that held and recycled nutrients efficiently.

The Oxygen Story

Because oxygen is so important to our existence, we will summarize the history of atmospheric oxygen. As we already stated, the primordial and early atmospheres contained very little free molecular oxygen (O_2). The evolution of the sophisticated process of green plant photosynthesis 2 billion to 3 billion years ago created a steady source of free oxygen for the atmosphere. Over the first 500 million years or so, the oxygen could have reacted with reduced minerals, particularly reduced **ferrous** iron in the world's oceans. During this period, extending from about 2 billion to 3 billion years ago, there was a transition in the ancient sediments from banded reduced iron deposits to red oxidized (**ferric**) iron beds, produced when iron minerals are weathered in an oxygen-rich atmosphere.

The rate of increase in the level of atmospheric oxygen over the period from about 1 billion to 2 billion years ago is uncertain. The mechanisms that control the total amount of oxygen in the atmosphere are complex and involve the carbon cycle as well (Section 10.2.4). At higher oxygen concentrations (perhaps 30 percent or more of air), vegetation becomes explosively combustible,[10] and so combustion would

10. You may recall a high-school chemistry experiment in which a glowing reed, thrust into a test tube filled with pure oxygen, bursts into flames.

act to limit the buildup of oxygen. If oxygen concentrations are higher, aerobic organisms more thoroughly scavenge organic debris produced by photosynthesis before the debris can be trapped in sediments, and the production of free oxygen must be accompanied by the long-term burial of organic detritus. For a number of reasons, therefore, it appears that the atmospheric level of oxygen has been relatively constant for the last 500 million to 1 billion years.

The development of the ozone layer in the Earth's atmosphere is illustrated in Figure 4.8 as a function of oxygen level in PALs. The abundance of atmospheric oxygen changed so slowly over time that the total amount (Figure 4.8[a]) and vertical distribution (Figure 4.8[b]) of ozone would have remained in photochemical equilibrium with oxygen. The amount of ozone required to protect life at the surface depends, of course, on the specific organisms involved. Early life may have been more resistant to ultraviolet radiation than life in general is today. Accordingly, a billion years ago, as little as 0.01 to 0.1 PAL of O_2 may have been sufficient to produce an effective ozone shield. At 0.1 PAL of O_2, 80 to 90 percent of today's ozone may have been present. Note that the curves in Figure 4.8 suggest that the current levels of ozone are robust against substantial variations in the oxygen abundance (which in reality is not at all likely [Chapter 10]). Indeed, the modern threat of ozone-layer depletion is not related to a drop in the oxygen level, but to an increase in the concentrations of trace chemicals that can react with ozone catalytically (Section 3.3.4).

4.3 The Mass Extinction of Life

Life on Earth has, in general, evolved through a poorly understood process of speciation (that is, the formation of new species of life derived from and related to existing species) punctuated by massive extinctions of species.[11] No species have ever been observed in the process of speciation (although viruses mutate often enough to be caught at it occasionally). Whereas speciation takes place over periods of tens of thousands of years, extinction can apparently occur

in just a few months or years. The remnants of both processes are captured in the fossil record.

4.3.1 FOSSIL HISTORY

Fossils, or preserved images of once living things, are a proxy for the diversity and complexity of life-forms that existed at earlier periods in the Earth's history. The **fossil record** is imperfect on three accounts:

Figure 4.8 The relationship between the ozone layer and the oxygen level of the atmosphere. (a) The total ozone in the atmosphere (as a fraction of the present atmospheric level, or PAL) versus the total atmospheric oxygen (O_2) concentration (also in PAL). (b) The distribution of ozone with altitude for different amounts of oxygen in PALs. (From Kasting, J. and T. M. Donahue, "Evolution of Atmospheric Ozone," *Journal of Geophysical Research* 85 [1980]: 3255. Based on Figures 1 & 2, p. 3259)

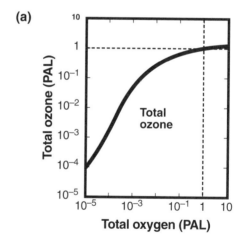

(a)

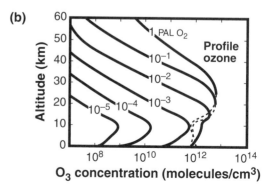

(b)

---

11. Life-forms are grouped according to heritage and biological similarities—with increasing specialization—into the categories of kingdom, phylum, class, order, family, genus, and species.

1. The record is incomplete at best and may not accurately represent the range of life-forms present at any given time or the precise pathways of speciation over time.
2. Fossils provide little information about biochemical or physiological processes, since soft tissues are usually annihilated over the ages.
3. The dating of fossils is inaccurate, with the inaccuracy increasing as we move farther back toward the origin of life.

Nevertheless, the fossil record is the most valuable tool we have for investigating our origins. Figure 4.9 depicts the principal geologic eras and periods and shows the rate of extinction at the species level over the past several billion years. A small **background extinction rate** is usually seen in the fossil data, which is consistent with an average duration of any single species of about 3 million years. Most striking, however, are the occasional **mass extinctions** of life, which appear as sharp peaks in the extinction rate in Figure 4.9. One of the most fascinating intellectual pursuits of science is searching for the cause of these mass extinctions. An event of particular interest is the extinction of the dinosaurs roughly 65 million years ago at the boundary between the Cretaceous and Tertiary periods.

4.3.2   THE DINOSAURS: A LESSON IN LONGEVITY

The dinosaurs were a group of large reptiles, often reaching extraordinary size, that dominated the land and seas for several hundred million years. Then suddenly, about 65 million years ago, they all disappeared. Their departure occurred over a period of time that, by geologic measures, was a blink of the eye. Explanations have ranged from slow changes in the climate, to which they could not adopt, to the evolution of small carnivorous mammals that ate their eggs. None of these explanations has been satisfactory. The extinctions at this time involved more than 75 percent of all species and ranged across all families of organisms (although with a considerable degree of selectivity).

In 1980, Luis Alvarez[12] and his co-workers proposed that the massive extinction was caused by the impact of a large meteor. As evidence for the impact, they presented measurements of unusually high concentrations of the rare element iridium in a sedimentary layer of clay coincident with the geologic horizon (or date) of the extinction. Iridium is known to be greatly enriched in meteors, compared with that in normal rocks and sediments. The constant background influx of iridium to the Earth carried by small meteors provides a useful geologic marker for the rate at which sediments are deposited. This was exactly the idea Alvarez had when he measured the iridium content of the clay layer that his son, a geologist, had dug up at the Cretaceous–Tertiary (or **K–T**) boundary. What Alvarez found puzzled him; the 1-centimeter-thick layer had 100 times the expected background concentration of iridium. He concluded that the iridium had been delivered by an enormous meteor 10 kilometers in diameter traveling at a speed of 10 kilometers per second. When such a meteor hits land, a crater 100 kilometers across is created, and an enormous quantity of soil and dust is blown into the atmosphere. Alvarez concluded that the clay layer itself was formed by the ejected iridium-laced dust as it settled over the entire planet.

Searches around the world have confirmed the global distribution of an iridium-rich clay layer. Quartz crystals have also been found that could have been formed only at the tremendous pressures generated by a meteor impact. A potential ancient site of the impact has been pinpointed on the Yucatán Peninsula, where geologists have found the remnants of a 100-kilometer-diameter crater dated 65 million years old. Alternative explanations for the extinctions have been put forward, of which the most reasonable assumes that volcanic eruptions were responsible. The best contender, however, is the impact.

The mechanism of the extinction is shown in Figure 4.10. The number of devastating effects produced by a large meteor impact is impressive. The explosive power of the impact would have been equivalent to the detonation of 100 million megatons of nuclear explosives.[13] The heat flash and shock waves

---

12. Luis Alvarez (1911–1988) won the Nobel Prize in physics in 1968 for his work using a bubble chamber to detect new atomic particles. Alvarez was a prolific scientist and inventor and was instrumental in designing the first hydrogen bomb during

World War II. In 1980, Alvarez and a team of scientists, including his son Walter, a geologist, proposed the meteor impact theory to explain the demise of the dinosaurs.

13. The energy released by a nuclear explosion is measured in **megatons**, MT. One megaton of nuclear explosive is equivalent to 1 million tons of TNT, or about 4×10^{15} joules of energy. An energy of 100 million MT is equivalent to 100 trillion tons of TNT, or 4×10^{23} joules.

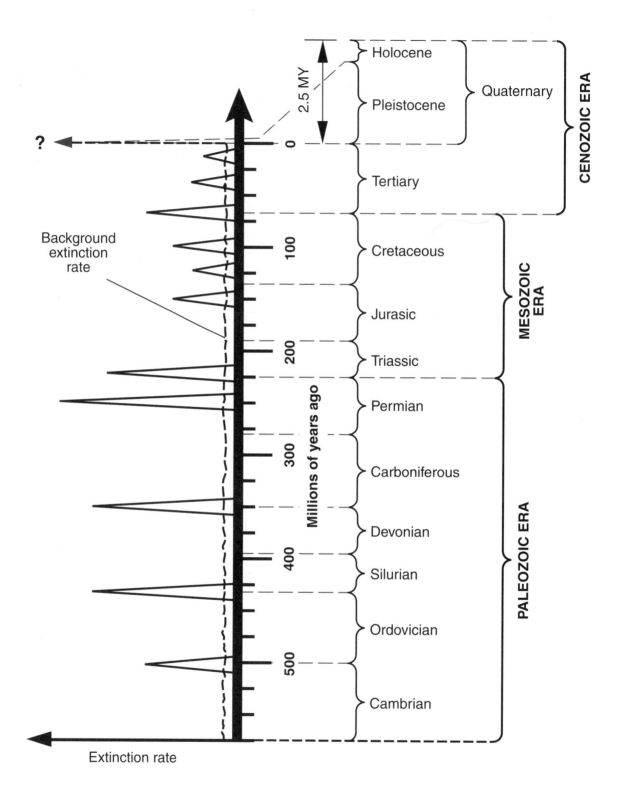

Figure 4.9 The extinction rate of species over time. Both the chronology of the major geologic periods and the relative extinction rate as a function of time are shown. Major extinction events have occurred periodically over the history of the Earth, and these events fall at major geological boundaries.

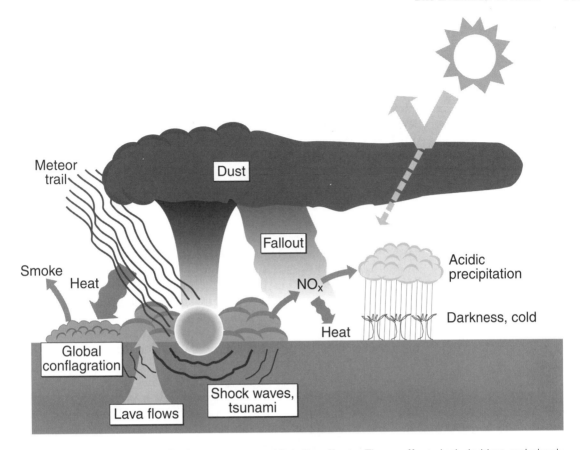

Figure 4.10 The impact of a large meteor and its aftereffects. These effects include blast and shock waves, impulsive heating of the surface by radiation, production of NO and acid rain, generation of dust and smoke clouds, and sudden severe shifts in the climate.

alone would flatten and burn everything within several hundred kilometers. Evidence in the form of soot and charcoal deposits also suggests major worldwide fires in the wake of the impact. The fires might have been ignited by the incandescent trail of the meteor and by heat radiated as the atmosphere was compressed by ejected dust settling from high altitude.

Other major global effects would be associated with the gases and dust generated by the impact that would affect the radiation balance and chemistry of the global atmosphere. The extensive dust and smoke clouds would block sunlight from the surface and create conditions similar to a **nuclear winter** (Section 14.2.1). This darkening and cooling of the land have been called **meteorite winter**. A meteorite winter would last for a year and have particularly devastating effects on large cold-blooded animals such as the dinosaurs. In addition, the severe heating associated with the meteor impact would generate large quantities of nitrogen oxides, NO_x. The NO_x would be chemically transformed into nitric acid

(Section 3.3.4), creating intense acid rains. This pulse of acidity might have had its most significant effect on organisms living in the seas, particularly those creatures bearing shells made of calcium carbonate. The shells may have dissolved, leaving these poor critters bare against the cruel elements.

As previously mentioned, new information places the location of the meteor impact on the Yucatán Peninsula in Mexico. Here, thick layers of sediments called **evaporites** can be found. The evaporites are composed largely of calcium sulfate, which is formed when seawater evaporates from an enclosed ocean basin. The Yucatán evaporites were deposited long before the meteor disaster of 65 million years ago. The impact that generated huge amounts of dust also pulverized and vaporized the evaporite and this might have released hundreds of millions of tonnes of sulfur dioxide into the atmosphere. The sulfur would soon form a dense layer of sulfuric acid particles encircling the Earth for many years, causing additional climate change. (See the description of the effects of volcanic eruptions on the climate in Sections 11.6.4 and 11.7.2; see also Section 14.3.3.) In addition to the sulfur dioxide, large amounts of carbon dioxide may have been liberated by the meteor impact. Calcium carbonate, one of the most common crustal minerals, can be decomposed by the heat and pressure of an impact, releasing the carbon dioxide that formed it. If the bedrock were primarily carbonate rocks, enough CO_2 could have been released to cause a century of sweltering heat following the brief meteorite winter (Section 12.4). Finally, it is likely that all of these other major perturbations would have led to a severe depletion of the ozone layer, lasting perhaps for decades.

All or none of these effects may have occurred at the K–T boundary. The best physical data support the idea that a violent meteor impact did trigger physical and chemical events that wiped out the dinosaurs and most other species. The exact sequence of events will probably never be known.

The Death Star

Mass extinction seems to occur periodically (for example, note the apparent regularity of extinction events over the last 250 million years in Figure 4.9). Careful scrutiny of the data reveals that a cyclic process with a period of about 26 million years can account for all the major extinction episodes in the fossil record. The origin of a cyclic process must involve the mechanical motions of the solar system. An intriguing candidate for the causative agent is the **death star**, or **nemesis**. A nemesis would be a dark companion to the sun, too small for its internal nuclear fire to be ignited. Nevertheless, a nemesis would be gravitationally bound to the sun, orbiting the sun roughly every 26 million years.

As a nemesis in its orbit passed through the Oort cloud outside the solar system, numerous comets would be scattered gravitationally toward the inner solar system (Figure 4.1). These comets, careening toward the sun, would bombard the inner planets. Such an attack of killer comets is not certain, however. Every now and then, the Earth could get off without a hit. Indeed, such a sporadic character to the comet showers is consistent with the fossil record, which does not show a mass extinction *every* 26 million years (Figure 4.9).

The existence of a dark companion star to the sun has not been proved. A nemesis may not exist.

Instead, periodic extinction of life on Earth may be the result of natural fluctuations in populations, of periodic climatic change or some other environmental or biological factor still unrecognized. Even so, a nemesis is a sweet concept. It is physically plausible and explains the major facts in a simple way. Scientists generally favor simplicity over complexity in explaining events (an important exception is the intellectual contentiousness over small points that keeps scientists busy and rigorous). The extinction of the dinosaurs is one of the greatest scientific mysteries of recent geologic times. Reliable evidence is scarce, however, and clever analysis must be applied to connect the clues. The tale of a meteor impact induced by a death star is a wonderful story woven from many individual threads of fact; and if it is still a thin fabric, it is not so full of holes as to leave one cold against the drafts of doubt.

4.3.3   GODDESS GAIA AND HOMEOSTASIS

The theory of a sudden and violent end of the dinosaurs and many other species has as its counterpoint the concept of Gaia. The Gaia hyposthesis, first formulated by James Lovelock and Lynn Margulis, holds that life and the physical environment are closely coupled and in tune and that each responds to changes in the other in such a manner as to maintain a quiescent state favorable to life. Lovelock discovered Gaia while studying the atmospheres of other planets. He observed that the atmosphere of Mars, for example, appeared to be controlled by physical processes and photochemistry. Indeed, martian air has no substantial amounts of oxygen or methane or other gases associated with biological activity. Earth, on the other hand, has an atmosphere filled with oxygen as well as unusual quantities of methane and nitrous oxide, all formed by life processes. Moreover, the atmospheric composition is essential to controlling the climate, which is ideal for life.

Lovelock saw in these observations a synergism—almost a cooperative alliance—in which life shapes its own physical and chemical environment. No one, not even Lovelock, would go so far as to say that this shaping is carried out consciously. But natural forces and processes would have to exist that maintain the whole organism, Mother Earth, or Gaia, in an equilibrium state. The Earth would then be in a state of **homeostasis**. Any agent or event that disturbed the

Sunlight

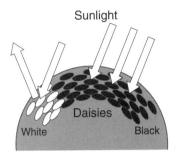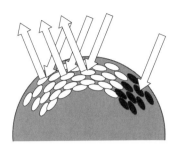

Figure 4.11 A Gaian world populated by white and black daisies. The white daisies thrive on warm temperatures, and the black daisies prefer cold temperatures. The white daisies reflect sunlight and act to cool the climate of "daisy world"; the black daisies absorb sunlight and warm the climate.

status quo would be strongly resisted and counteracted. The environment and life would be locked in a robust stability against change. Nature could be seen to react in its own behalf.

This sort of homeostatic behavior is common in living organisms and colonies of organisms. For example, if your body becomes suddenly chilled, "goose bumps" appear on your skin, signaling the withdrawal of blood from the surface to reduce the loss of body heat. By contrast, if you become overheated, sweat freely flows onto your skin where it evaporates, absorbing latent heat and cooling you down. If the intensity of light increases or decreases, the pupils in your eyes automatically adjust so that you can see perfectly. Even lowly bacteria respond to heat, light, and other physical stimuli in an attempt to maintain their cheerful environment. If environmental conditions are forced too far out of balance, however, adequate adjustments may not be possible, and the organism will die.

In a Gaian world, each organism among the vast diversity of life contributes to maintaining the environment for all. A few organisms are freeloaders, of course, riding the easy wave of the moment. Other species may take on major responsibility for controlling the state of the world. As the properties of the oceans, the continents, the atmosphere, and the climate drift over time, species regularly find themselves unable to cope with the change and so disappear. New species evolve that like the situation better. Life is more closely associated with states of quasi-equilibrium than with strict homeostasis.

Lovelock provided a simple example of the type of interactions between the physical and biological world that would be found on a Gaian world. He conjured up a **daisy world**, as illustrated in Figure 4.11. In a daisy world, there are two types of daisies:

white daisies, which like warm weather, and black daisies, which like cold weather. In addition, the white daisies very efficiently reflect the sunlight that shines on the daisy world, and the black daisies just as efficiently absorb the radiation. In other words, white daisies increase the reflectivity, or albedo, of the daisy world, whereas black daisies decrease the albedo. As explained in Section 11.6.5, an increase in reflectivity causes a planet to cool because less sunlight energy is absorbed and converted to heat; conversely, a decrease in reflectivity causes a planet to warm.

Imagine the daisy world initially in a state in which there are many more black than white daisies (first panel in Figure 4.11). The sunlight hitting the daisy world is effectively absorbed by the profusion of black daisies, and the daisy world is quite warm as a result. But the white daisies thrive in the warm climate and begin to reproduce copiously. Slowly the black daisies, weakened by the heat, are pushed out. Further imagine that the white daisies continue to proliferate and soon dominate the planet (middle panel in Figure 4.11). Now, because the albedo is large, daisy world begins to cool. The colder climate allows the black daisies to make a comeback. If the system is allowed to evolve for a time, it should reach an equilibrium state in which both white and black daisies grow in just the right numbers to maintain a climate suitable for both. In this kind of system, there is a **negative feedback**. A negative feedback is a stabilizing mechanism, which limits the excursions that may occur in a system. In the case of the daisy world, the feedback involves an interaction between the temperature sensitivity of each species and its effect on the climate through reflectivity. For each type of daisy, an increase in population is limited by a self-induced change in the temperature.

At this time, no clear evidence exists for extremely complex Gaian worlds, such as the Earth would have to be. Only one or two mechanisms have been suggested that link biological activity to a proximate stabilizing feedback process (Section 14.3.3), and none of these mechanisms has been proved. The strong likelihood of major meteor impacts, and smaller ones as well, over the history of the Earth implies their greater influence on the course of evolution than the possible homeostatic conditions that might develop during the intervals between impacts. It also happens that homeostasis implies a robust environment, one that can adjust to gradual natural change. Perhaps, then, Gaia should be abandoned as a natural philosophy, lest those who would overdevelop the Earth be encouraged to proceed.[14]

4.4   The Coevolution of Intelligence and Pollution

Of all the organisms that have lived on Earth, the microbes have had the greatest influence on the atmosphere, in terms of basic metabolic chemical transformations. The metazoans, although far more impressive in size, strength, and speed, have played a minor role. The enormous numbers of microbes occupying every imaginable ecological niche explains their importance. Only recently has a metazoan species developed the capacity to damage the global environment single-handedly—*Homo sapiens*.[15] Our power to cause worldwide destruction derives not so much from our number as from our

unprecedented intelligence. We have learned how to redirect the forces of nature, altering the air, land, and waterways in our quest to harvest the resources around us. We have genetically modified plants to yield their fruits more reliably. We have domesticated numerous animal species to labor for us and to serve as our dinner. We have produced countless corrosive chemicals and other artificial materials to support a sophisticated technology-based urban civilization.

At the present time, biologists and ecologists are recording unprecedented rates of extinction among a number of vertebrate families. The demise of dozens of species of birds has been recorded in recent years. An alarming die-off of frogs, toads, salamanders, and other amphibians led to an urgent national investigation in 1990. The cause of these extinctions is not fully understood. It is apparent that human development and the expansion of civilization into once isolated and pristine environments may be responsible for many extinctions. The dodo (*Didus ineptus*), a pigeonlike flightless bird the size of a large turkey, once roamed fearlessly on the island of Mauritius. The dodo, now extinct, was a victim of early human exploration and senseless slaughter. The bison of the American West were nearly eradicated in massive executions by gunfire for sport. The rhinoceros is being hunted into extinction for the value of its horn, which is believed in some cultures to have aphrodisiacal properties. The mere act of confiscating and cultivating extensive land areas, particularly in tropical regions, ensures the demise of countless plant and animal species. The clear-cutting of tropical rain forests irreversibly destroys sensitive ecosystems that nurture hundreds of thousands of different species. The wholesale **anthropogenic** pollution of rivers, lakes, and estuaries; the widespread application of agricultural pesticides and herbicides; and the global contamination of the atmosphere and modification of the climate all place unprecedented stress on the environment and the living things that depend on the environment, including humans (Figure 4.12).

4.4.1   POPULATION AND TECHNOLOGY

The number of humans populating the Earth is a mere 5 billion or so compared with quadrillions of microorganisms. The total human "biomass" is less than 0.1 percent of all the biomass on Earth. Technology allows humans to proliferate and thrive at the

---

14.   The general issues of the robustness of the environment against increasing human pollution, the extent to which it can be poisoned before irreversible destruction occurs, and the responsible actions that could be taken to prevent serious damage are discussed at several points in this book; for example, in relation to urban smog (Section 6.6), greenhouse climate warming (Section 12.4), and global environmental engineering (Chapter 14).

15.   The species *Homo sapiens* (meaning "wise man") emerged during the Pleistocene epoch, perhaps 1 million to 2 million years ago, and probably in Africa. The direct ancestors of *Homo sapiens* are unknown, but the lineage seems to follow a line from a series of apelike creatures beginning with *Australopithecus* (southern ape) dating from roughly 3 million to 5 million years ago, followed by *Homo habilis* (handy man), who may have appeared 2 million to 3 million years ago, and leading to *Homo erectus* (upright man, with a relatively small brain), who evolved some 1 million to 2 million years ago. The more modern forms of man are the Neanderthals (*Homo sapiens neanderthalensis*), who developed nearly 200,000 years ago, and Cro-Magnon, or modern man (*Homo sapiens sapiens*), who emerged in sub-Saharan Africa about 50,000 years ago.

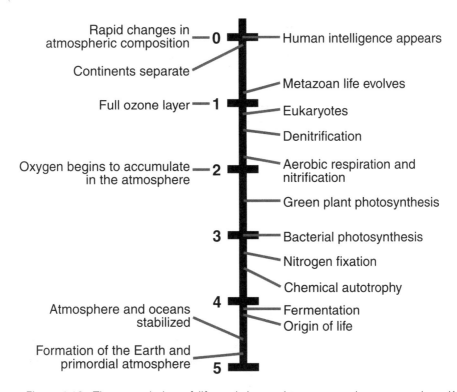

Figure 4.12 The coevolution of life and the environment to the present time. Key evolutionary advances are indicated to the right of the time line, and critical developments in the physical environment are on the left. (Chronology from J. C. G. Walker, *Evolution of the Atmosphere* [New York: Macmillan, 1977])

expense of other species. If the only supply of food for human survival were derived from natural ecosystems—that is, if we had to "live off the land"—the world could support perhaps several hundred million people. There are 10 times that number at present, and the human population is expected at least to double in the future. To support all these people, industry and technology have developed to produce the food and other goods and services demanded by all these souls. Humans, particularly in the "developed" countries, are voracious consumers of manufactured goods. Expectations of high lifestyles and standards of living have spread with time, leading to enormous demands on ecosystem services, with little attention given to the preservation or sustainability of the environment.

Often, a technological innovation that solves a specific problem related to human services introduces unforeseen pollution hazards. The discovery that coal could be burned for cooking and heating was certainly a great advance in human comfort and technology, ultimately fueling the Industrial Revolution. However, one side effect was suffocating air

pollution that befouled the air and impaired health throughout Europe for a hundred years (Chapter 6). Those who used coal were warmer, but those who dug it suffered debilitating black-lung disease. Buildings that were more comfortable to work in had facades defaced by a layer of sooty smoke. The invention of chlorofluorocarbons (CFCs) for the purposes of refrigeration and air conditioning displaced several dangerous toxic gases previously used in those applications and made cheap comfort and safe food available to everyone. But it was soon discovered that the CFCs also were depleting the ozone layer, possibly threatening all life on the planet (Chapter 13). Those miraculous (and profitable) compounds were suddenly potential despoilers of the global environment.

In the past decade, carbon dioxide generated as a by-product of fossil fuel (coal, oil, and natural gas) combustion has become a major environmental issue. It has been proposed that CO_2 emissions will cause a significant warming of the Earth's climate, with serious implications for human society (Chapter 12). But carbon dioxide is fundamental to our way of life. Beginning with the Industrial Revolution, the use of fossil fuels for energy production and industrialization has grown exponentially. To reduce or eliminate CO_2 emissions to preserve the climate and global environment would entail nothing short of a complete overhaul of human society.

Carbon dioxide is not the only human-generated compound that may affect the climate. Our activities are producing excess methane, nitrous oxide, CFCs, and other gases that can lead to global warming (Chapter 12). Such revelations often come long after the offending technology is well entrenched or serious damage has been done. In addition, businesses that profit from polluting activities may suppress information that might reveal such a problem, or they may vigorously resist any attempts to limit their polluting activities even after a problem is discovered. So what else is new in the world?

An intriguing question is whether the human species, given its propensity for denial, selfishness, tribalism, and shortsightedness, can survive the environmental challenges of the next century. Are we likely to destroy the global ecosystems that sustain us, deplete all accessible resources, and overpopulate the globe? Are cockroaches and weeds the chosen inheritors of the Earth? Or can we, through education and action, sustain a livable planet for all?

Questions

1. Explain the difference between abiotic and biological chemical synthesis in the evolution of the environment. What were the sources of energy for these processes?

2. Describe the various possible roles of comets in the evolution of the atmosphere.

3. Why does the Earth have so little helium (He) relative to silicon (Si), compared with the relative amounts in the sun (that is, why is the Earth depleted in He relative to Si, compared with the sun)?

4. Identify the principal milestones in the evolution of life in regard to the use of atmospheric gases available at different times for bioassimilation. What specific roles did H_2, CO_2, and O_2 play? Give the chronology for the utilization of these atmospheric compounds by living organisms.

5. Discuss the sequence of events that might have killed the dinosaurs about 65 million years ago.

Could such events occur today? If the dinosaurs had not suddenly become extinct, what might the world be like now?

6. Name the principal reservoirs of volatiles in the Earth system, and identify the processes that transfer volatiles among the reservoirs. Sketch a diagram of the reservoirs and the pathways for transfer.

7. Nitrogen gas (N_2) is hardly soluble in water. Accordingly, at any time very little atmospheric nitrogen is normally dissolved in the oceans. Imagine a genetically engineered organism that thrives in the oceans and uses nitrogen to form a hard, insoluble "nitrate" shell. Explain how such an organism, if it existed, might affect the atmospheric abundance of nitrogen.

Problems

1. An hourglass is a kind of "box model." The sand in one-half the glass bulb falls into the other half in a fixed time, t (when the hourglass is inverted). If you design the bulb so that it holds M grams of sand, determine the average loss rate, L, of sand from the upper bulb (when in use). Assume that t = 2 min and M = 40 g. If you redesign the hourglass to have the same timing but to use 80 grams of sand, what will be the new value of the loss rate, L? For this new bulb design—assuming that L is independent of the amount of sand in the bulb—how much sand must be added to make a 3-minute-egg timer?

2. In our discussion of the box-model concept, we talked about the carbon dioxide balance of the atmosphere. The natural steady state quantity of CO_2 in the atmosphere was estimated to be about 750 Gt-C (gigatonne of carbon), and the lifetime of CO_2 to be about 10 years. Suppose, in this simple one-box model for atmospheric CO_2, an additional *source* of CO_2 of 7.5 Gt-C/ year is introduced, with the other factors unchanged. Compute the new steady state atmospheric burden of carbon dioxide, in Gt-C. What is the percentage increase in the CO_2 concentration? Do you see any problems with this model?

3. You have a laboratory chamber that has a volume of 1 cubic meter (1 m^3), the size of a very large garbage can. You fill the chamber with perfectly dry air (that is, nitrogen and

oxygen only). Then you place in the chamber a small dish holding 1 gram of water (which is exactly 1 cubic centimeter of water, since the density of water is 1 g/cm^3). After a time, the water evaporates from the dish into the chamber. Determine the concentration of water vapor in the chamber (in units of mass per volume). If you want to produce a water vapor concentration one-half as large, how many grams of water must you put into the chamber? What is the *maximum* amount of water that you can add to the chamber, and what is the corresponding concentration of water in the chamber?

Suggested Readings

"What Caused the Mass Extinction?" *Scientific American* **263** (1990): 76.

Alvarez, L.W. Alvarez, F. Asaro, and H. Michel. "Extraterrestrial Cause for the Cretaceous-Tertiary Extinction." *Science* 208 (1980): 1095.

Broecker, W. "The Ocean." *Scientific American* 249 (1983): 146.

Cloud, P. *Oasis in Space: Earth History from the Beginning.* New York: Norton, 1988.

Coleman, G. and W. Coleman. "How Plants Make Oxygen." *Scientific American* 262 (1990): 50.

Ericson, D. B. and G. Wollin. *The Deep and the Past.* New York: Knopf, 1964.

Herschel, J. "On the Astronomical Causes Which May Influence Geological Phenomena." *Transactions of the Geological Society,* 2nd ser., vol. **3** (1830): 293.

Kuiper, G. (ed.). *The Atmospheres of the Earth and Planets.* Chicago: University of Chicago Press, 1949.

Lovelock, J. *Gaia: A New Look at Life on Earth.* Oxford: Oxford University Press, 1979.

McPhee, J. *Assembling California.* New York: Farrar, Straus and Giroux, 1993.

———. *Basin and Range.* New York: Farrar, Straus and Giroux, 1981.

———. *In Suspect Terrain.* New York: Farrar, Straus and Giroux, 1983.

———. *Rising from the Plains.* New York: Farrar, Straus and Giroux, 1983.

Meyers, N. (ed.). *Gaia: An Atlas of Planetary Management.* New York: Anchor Books, Doubleday, 1993.

Miller, S. and L. Orgel. *The Origins of Life on Earth.* Englewood Cliffs, N.J.: Prentice-Hall, 1974.

Miller, S.L. and H. Urey. "Organic Coumpound Synthesis on the Primitive Earth." *Science* 130 (1959): 245.

Muller, R. *Nemesis.* New York: Weidenfeld and Nicolson, 1988.

Nance, D., T. Worsley, and J. Moody. "The Supercontinent Cycle." *Scientific American* 259 (1988): 72.

Newell, N. "Crises in the History of Life." *Scientific American* 208 (1963): 76.

Olson, J. "The Evolution of Photosynthesis." *Science* 168 (1970): 438.

Raup, D., J. Sepkoski, et al. "Testing for Periodicity of Extinction." *Science* 241 (1988): 94.

Redfield, A. "The Biological Control of Chemical Factors in the Environment." *American Scientist* 46 (1958): 205.

Sagan, C., W. R. Thompson, R. Carlson, D. Gurnett, and C. Hord, "A Search for Life on Earth from the Galileo Spacecraft." *Nature* 365 (1993): 715.

Schneider, S. and P. Boston (eds.). *Scientists on Gaia.* Cambridge, Mass.: MIT Press, 1991.

Schneider, S. and R. Londer. *The Coevolution of Climate and Life.* San Francisco: Sierra Club Books, 1984.

Siever, R. "The Dynamic Earth." *Scientific American* 249 (1983).

Walker, J. *Evolution of the Atmosphere.* New York: Macmillan, 1977.

Weinberg, R. "The Molecules of Life" [special issue]. *Scientific American* 253 (1985): 48.

Wilson, E. *The Diversity of Life.* Cambridge, Mass.: Harvard University Press, 1992.

Part II
Local and Regional Pollution Issues

Environmental hygiene concerns the pollution of the air, water, and land. Of these three media, air is the most important in spreading the greatest number of undesirable compounds far and wide. Water pollution may cause serious health hazards. In many places, people have taken to drinking bottled "spring water" to avoid the bad taste of the water distributed to kitchen taps. For our discussion, however, the contamination of water supplies—surface waters and deeper aquifers—is a secondary concern. Rather, we address the overriding influence of air pollution on health and on the global environment.

The study of air pollution can be subdivided into more specialized topics in a number of ways. In this book, we will consider air pollution issues that apply to different scales of size. Problems that are "local" are more personal in nature. They may affect us or our neighbors the most directly. On regional scales, larger urbanized areas, such as southern California and the Ohio River valley, can be affected. Regional issues may extend to areas covering many states or perhaps even nations. A new important class of air pollution issues has emerged that affects the entire planet. These global-scale environmental problems include depletion of the stratospheric ozone layer and warming of the world's climate by carbon dioxide emissions. Such issues will be taken up in Part III.

Part II deals with air pollution on local and regional scales. Examples of "local" pollution include fumes emitted by individual industrial sites, or the pollution of the air inside a home by radon gas seeping from the soil. The term *local* can refer to a town or neighborhood or, extending to smaller scales, our personal air space, both at home and work. Incidents may expose us to tobacco smoke, asbestos fibers, and a variety of dangerous vapors. Accordingly, we will discuss indoor air pollution and occupational pollution hazards. The smog that enshrouds many large cities such as Los Angeles and Mexico City may be treated as local air pollution, although when aggregates of urban zones are considered (for example, as in southern California), the problem assumes regional dimensions. Acidic precipitation in the northeastern United States asso-ciated with power plants in the Midwest is clearly a regional air pollution issue.

The extent and effects of air pollution are controlled by a number of factors that vary from one pollutant to another. For example, although radon gas can be unhealthy when confined inside a house where it accumulates, most of the radon gas released into the global atmosphere, where it is dispersed by winds, is considered harmless. Similarly, the carbon monoxide found in automobile exhaust can be a serious health hazard for someone trapped on a gridlocked highway. That same carbon monoxide may later affect the chemistry of the atmosphere over an entire polluted continent. Carbon monoxide contamination in remote regions of the Earth far from polluted cities is unusual. On the other hand, when the stratospheric ozone layer is depleted by the chlorofluorocarbons used in refrigerators and air conditioners, life everywhere on the planet is threatened.

Local air pollution often may be associated with a single, well-defined source. Regional air pollution occasionally may be caused by isolated sources, but typically involves numerous distinct sources, each of which contributes significantly to the problem. Global air pollution is related to the accumulation of emissions from countless pollutant sources distributed around the world. These sources are often difficult to identify individually (a specific automobile or refrigerator). Nevertheless, estimates of aggregated pollutant emissions often can be deduced from industrial data. Degradation of the ozone layer by chlorofluorocarbons is related to the emissions of a dozen chemical compounds from hundreds of millions of air conditioners, refrigerators, dry-cleaning establishments, semiconductor manufacturing plants, and so on. The formation and transport of photochemical smog are another example of the cumulative effect of multitudinous small sources of pollution. Although smog is characteristically dominated by the exhaust emissions of road vehicles, its severity in a region can be affected by paint vapors, hair spray, and even the hydrocarbons emitted by trees, to name just a few identified sources of smog-forming pollutants.

In Chapter 5, we consider the atmospheric dispersion of air pollutants on local and regional scales, since this determines the manner in which pollution is spread from source regions (for example, smokestacks or freeways) and is eventually diluted and dissipated. In Chapter 6, we discuss common urban smog, with an emphasis on "photochemical" smog, such as that found in Los Angeles and other major cities around the world. We look at the toxicity and health effects of the most prevalent air pollutants, as well as a simple method for assessing exposure to pollution. Indoor air pollution—involving radon, formaldehyde, cigarette smoke, and other potentially toxic compounds—is the subject of Chapter 8. Finally, the important regional problem of acid rain is dealt with in Chapter 9.

5

Sources and Dispersion of Pollutants

Air pollution is not stationary. It does not sit where it is formed. Rather, it visits other places, carried on the winds across state lines and national borders. Polluted air produced in Czechoslovakia migrates to Austria. Sulfur dioxide emitted by power plants in Ohio falls as acid rain in New York. Because of this easy mobility, it is essential to understand the relationship between the motions of the atmosphere and the distribution of pollutants. We must not only determine the degree to which air quality is degraded, but also identify the sources and devise measures to control them.

5.1 The Source of the Problem

The sources of pollution are almost literally countless, but they are usually divided into generic classes or types—for example, power plants and automobiles. In California, the South Coast Air Quality Management District, one of the most highly organized and effective pollution-control agencies in the world, regularly identifies and fines polluters. The roster of offenders, which is made public, recently listed a bed-frame manufacturer (emissions from paint solvents), a toy maker (failure to implement a mandatory ride-sharing plan for employees), a meat renderer (excessive odors and fugitive vapors), a stereo manufacturer (illegal use of finish coatings), a wallboard producer (uncontrolled dust emissions), a demolition firm (inadequate handling procedures for asbestos), an oil refinery (violation of pollution permits), a computer manufacturer (emissions from unregistered equipment), and an aircraft-lubricant plant (solvent venting from degreasing operations). Everyday household products contribute to air pollution as well. The solvent in hair spray releases hydrocarbons; as much as 98 percent of the spray evaporates into the atmosphere. Attempts to refor-

mulate hair spray using water as a solvent have proved to be tacky at best. In some places, products such as perfume, deodorants, and glass cleaners are being forced to meet increasingly strict emission standards.

Industrial plants are a major source of pollutants. Many facilities burn coal and oil for energy. The emitted smoke, particularly from coal-fueled plants, contains sulfur, hydrocarbons, and numerous particulate by-products. A typical coal-burning industrial boiler can emit, for every kilowatt-hour of energy generated, 1 kilogram of particulates, 2 kilograms of sulfur dioxide, and 1 kilogram of nitrogen oxides. Almost every type of manufacturing produces air pollutants: oil refineries, iron and steel foundries, cement plants, wood and paper mills, food-processing facilities, textile mills, chemical-processing and allied firms, and printing and publishing companies. The kinds, amounts, and effects of some of the most important pollutants—sulfur, nitrogen oxides, organic compounds, heavy metals (elemental, not musical types), and radioactivity—are discussed in subsequent chapters.

5.1.1 WHAT TO CALL POLLUTANTS?

Pollutants can be subdivided into primary and secondary types. The pollutants emitted directly into the atmosphere from the sources just identified are called **primary pollutants**. Another entire class of pollutant compounds is produced as the secondary products of chemical reactions involving primary pollutants. These **secondary pollutants** are not directly emitted into the atmosphere but are generated there over time (Section 6.2). Pollutants can be further subdivided into gaseous (vapor) and particulate (aerosol) types. Thus **primary gaseous pollutants** include direct emissions of nitric oxide (NO) and carbon monoxide (CO) from automobiles. **Primary**

111

particulate pollutants are, among other things, the smoke from power-plant stacks and the dust raised by construction vehicles. **Secondary gaseous pollutants** are usually derived from primary gaseous emissions. The best example here is ozone, which is photochemically generated in smog from the primary pollutants emitted by automobiles (Section 6.2).

Secondary particulate pollutants, on the other hand, are typically produced from primary and secondary gaseous emissions (often with a primary particle as a substrate, or core). For example, some organic vapors emitted as primary pollutants, and many products of chemical reactions involving these vapors, condense into a haze of fine aerosol particles. Sulfur emitted in the form of sulfur dioxide is readily oxidized into sulfuric acid (Section 3.3.4), which may also form an acidic aerosol (Sections 6.5.3 and 9.4).

Pollutants leaking from industrial facilities are often referred to as "fugitive" emissions. Dust raised by farm machinery is "fugitive" dust. Soot from city buses is "fugitive" soot. It appears that much of the pollution bedeviling us has willfully decided to escape from otherwise clean factories and vehicles. Fugitive pollutants are on the 10-most-wanted list of the Environmental Protection Agency.

A multitude of compounds released into the atmosphere are highly toxic. Among these primary toxic gases are chlorine, in the form of hydrochloric acid, ammonia, and common pesticides. Indeed, most of the gaseous pollutants mentioned are toxic at high concentrations (Chapter 7). Radioactivity is a pollutant that is dangerous in very small quantities. Radioactive materials are frequently found attached to particles, which may be inhaled. (The ways in which we are exposed to primary and secondary gaseous and particulate-borne pollutants are reviewed in Chapter 7.)

5.1.2 DISTRIBUTED AND POINT SOURCES

Pollutant sources can be distinguished by the characteristics of the emissions—the type of materials, amounts released per day, and spatial extent of the emitting region. **Point sources** can be thought of as very localized. For example, a smokestack is a point source of smoke. The tailpipe of an automobile is a point source of carbon monoxide. Similarly, a leaking gas pipeline is a point source. An entire industrial facility may also be considered as a point source, particularly if our interest lies in the effect of the

emitted pollutants over a much larger region around the facility. **Distributed sources** of pollutants cover a wider area. In this case, the pollutant emissions may originate from an extended or a continuous source—a collection of rice fields is such a source—or from a concentrated cluster of point sources—the collective carbon dioxide emissions from the vehicles on a freeway, for example.

The characterization of a source as either a point or a distributed source can depend on how one chooses to aggregate the individual emissions that compose the source and on the spatial scales of interest. For example, the emissions from individual automobiles are of concern in the effort to control smog. Thus in some states such as California, the vehicular emissions of pollutants like carbon monoxide, hydrocarbons, and nitrogen oxides are tested regularly. The tests ensure that the source of pollutants from each car and truck falls below a specified standard. Unfortunately, as happens in many programs of this kind administered by authorities, people have found ways to cheat on the tests. A recent investigation in California showed that the actual reductions in hydrocarbon and nitrogen oxide emissions achieved through smog checks (followed by mandatory engine adjustments) were just one-third of the reductions necessary to meet emission standards. Nevertheless, mandatory testing of vehicle exhaust is an essential tool in reducing air pollution.

For the purpose of evaluating pollution in a whole city, however, the exact emissions from each vehicle on the road are not relevant. What is needed, in practice, is the total aggregate of emissions from all operating vehicles hour by hour over regions that measure several miles across. The aggregated emissions must be determined statistically. Thus the emissions of a "typical" automobile are determined by experimenting with a sample of ordinary cars. Next, the number of automobiles on the road, their speed, and their locations are estimated using urban traffic models. Once the number, locale, and operating conditions of vehicles of all types (cars, trucks, motorcycles, buses) are fixed for a certain time of day, the geographical distribution of the total emissions can be calculated. In this case, millions of individual point sources have been aggregated into an average distributed source.

Errors may be introduced in several ways. First, the individual sources are not really known, so a "typical," or statistical mean, source is used. The typical emission values may be based on only a few measurements taken under conditions that do not represent actual road conditions. In 1991, for example, it was discovered that the hydrocarbon emissions of automobiles in the Los Angeles area had been underestimated by a factor of two to four. The error arose because exhaust measurements were carried out on stationary test stands using well-tuned vehicles, whereas drivers on the road tend to "gun" their accelerators frequently and run their cars poorly tuned. Moreover, the engine-testing program, which is supposed to bring all cars into compliance with emission standards, was not as effective as hoped. A second major source of uncertainty is related to the assumed traffic patterns. In California, for example, data are regularly collected on freeway traffic using emplaced sensors, but there is very little information about the surface streets. Despite these limitations, the monitoring of vehicular and other sources of pollution has evolved into a high technology enterprise, which is crucial to pollution abatement assessment.

5.1.3 SIZE SCALES OF DISPERSION

The size of a pollutant source determines how the pollutant will be dispersed throughout the atmosphere. Similar considerations apply to pollutants dumped into oceans, lakes, and rivers or onto soils. However, the focus here is on air pollution and the processes that control its spread and removal. The source size is important because the movement of air takes place on many scales of size. These different scales of motion are discussed in the following major section.

As one might expect, spatially small sources of pollution are dispersed initially by small-scale motions in the vicinity. Large sources are dispersed by large-scale motions. Point emissions are at first diverted by local currents of air, like the smoke from a cigarette. As the pollutant spreads out, it is affected by larger motions of the atmosphere, including the prevailing winds that blow across the land and the updrafts in clouds. The movement of air occurs on all scales of size, from microscopic swirls to waves as long as a continent. Every source of pollution is dispersed. Indeed, the pollutants from any source, no matter how small, can eventually be dispersed over the entire planet. The time scale for such complete global dispersion is several years, assuming that the pollutant is not removed from the atmosphere or chemically transformed during the time of dispersal.

The chlorofluorocarbons (CFCs [Section 13.5.2]), for example, have the necessary characteristics and are found everywhere on Earth in about the same mixing ratio.

The dispersion of pollutants that are easily removed from the atmosphere or are chemically unstable is limited by their residence time in the atmosphere. Many pollutants are soluble in water, and so they can dissolve in the water droplets that make up clouds and be carried to the surface by rainfall. This scrubbing by clouds and precipitation is a primary natural mechanism for keeping air clean (Section 3.1.3). Soluble pollutants generally have an atmospheric residence time of a few days. In this time, pollutants may travel the distance of a country. Many pollutants react to form other compounds. For instance, carbon monoxide (CO) reacts efficiently with the hydroxyl radical (OH [Section 3.3.4]). The average lifetime of CO against a reaction with OH is about 1 month, in which time, carbon monoxide emissions can travel thousands of kilometers.

The dispersion of secondary pollutants also depends on the time required for their production from primary emissions. Ozone is a principal secondary pollutant in photochemical smog (Chapter 6). Because ozone can be generated in a matter of hours from primary pollutants, it is both a local and a regional problem. If ozone required several days to form, it would not be poisoning many of the world's major cities: The primary pollutants would have been so widely dispersed that the resulting ozone levels could not reach the very high levels found in smog.

5.2 The Dispersion of Pollutants

Pollutants are dispersed through a variety of processes involving the motions of air. For our purposes, these processes may be divided into the general categories of diffusion, convection, and advection. These processes are depicted in Figure 5.1 and are discussed next.

5.2.1 DIFFUSION AND TURBULENCE

Molecules and very small particles move in still air by colliding with air molecules (Section 3.1.1). The molecules of a pollutant in air are in constant motion and naturally drift from place to place. But, the

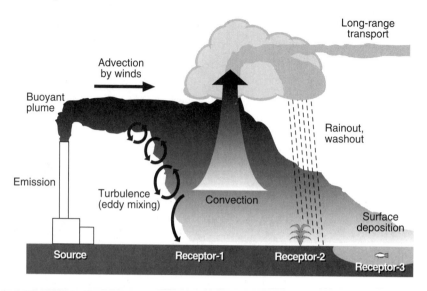

Figure 5.1 The various transport and dispersion processes for pollutants. The processes occur on small and large scales and affect "receptors" both near pollutant sources and at great distances from them.

average distance that each molecule moves is constrained by the frequency of its collisions with other molecules. Diffusion is the average motion of a molecule as a result of these collisions. Molecules in the air at sea level are subject to about 1 billion collisions per second! It is difficult to make progress under these conditions. Imagine yourself at a crowded rock concert; if you need to get to the bathroom, it can take a long time to bump your way through the crowd. If a mob decides to rush the stage, you are along for the ride (if you are lucky enough to avoid being trampled), because you cannot "diffuse" out of the crowd fast enough.

A Drunken Walk

The motion of a molecule through a gas can be described by analogy with a "drunken walk" (Figure 5.2). Picture a drunk leaning against a light post. He decides to head home. But for each step forward, he also takes one back or sideways. The drunk staggers randomly from side to side and back and forth. By this random sequence of steps, he moves slowly away from the lamppost, although in no particular direction. He is diffusing. The jerky movement of a small particle suspended in a liquid, called Brownian motion, is a similar random process (Section 3.1.2). You may have dropped a tablet of dye in a cup of water to color Easter eggs. If the water is still and is not stirred, the tablet will quickly dissolve, but the dye will remain near it. The dye molecules must diffuse

through the water. Diffusion is a very slow process over ordinary distances.

The first law of diffusion states that the distance that a gas molecule travels in air from its initial position increases in proportion to the square root of the time elapsed.[1]

Turbulence is the apparently random churning of fluids like air and water. This complex motion can usually be decomposed into individual vortices, or eddies. An eddy may be envisioned as a small swirl, like a wisp of smoke from a cigarette or a small vortex observed in a pond. In a fluid that is agitated, all sizes of eddies can exist, from fairly large ones down to microscopic sizes at which molecular diffusion becomes important. In fact, larger eddies are continuously breaking down into smaller eddies because of friction between the airstreams in the eddies. The combination of all the eddies acting simultaneously produces the effect called *turbulence*.

Turbulence can be generated by any process that induces motion in a fluid like the atmosphere. Buoyancy (Section 3.1.1), caused by instability in the atmosphere (Section 5.3), creates vertical motions (convection) that may decompose, or cascade, into turbulence. Whenever two wind streams collide, some of their energy appears as turbulence. Wind

1. The square root is explained in Section A.4. Because of the square root dependence of diffusion, the average distance traveled by a molecule is much shorter than if it maintained a constant speed in one direction at all times.

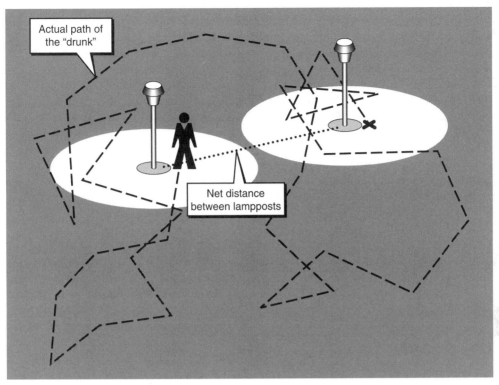

Figure 5.2 The staggering "walk" of a drunk. When traveling from one lamppost to another, a drunk may actually travel a much greater distance than the separation between the posts. The path shown was generated by assuming that all steps have equal length but random directions determined by the spin of a pen on a flat, smooth, surface.

passing over a rough surface, and thus diverted by the structure of the surface, has some of its momentum converted to small turbulent motions. As a warm plume pushes into the atmosphere from a smoke-stack, friction between the rising core of the plume and the ambient air at the interface between the plume and the atmosphere sets up small eddies that become turbulence. You can watch the eddies and turbulence develop in a stream of smoke blown into the air. But if the air is still, the turbulence will quickly die away. The energy in the turbulent motion is largely converted to heat by frictional dissipation. A ball rolling along the floor or a spinning top soon slows down and stops. At night and with low winds, there is very little turbulence in the air. Above an active thunderstorm or in high surface winds over rugged terrain, the turbulence can be intense enough to toss a jumbo jet around like a toy.

You can witness turbulence in a glass snowscape. These fascinating childhood toys consist of a small model—perhaps a charming wintry scene or Frosty the Snowman—covered with snowflakes held in a glass sphere filled with water. When you shake the glass, the snow churns up, producing a blizzard. Stop

shaking, and the snow settles to the bottom. By shaking the glass, you force the water inside the sphere to move. The motion can be very turbulent, causing the snowflakes to swirl around randomly and filling the entire glass with snow. The harder you shake the glass, the more extreme and erratic the motion of the snowflakes will be. When you stop shaking, at first the snow organizes into more uniform streaks, because the small scale turbulence dies away quickly. The remaining larger-scale motion consists of the steady currents set up in the fluid, which carry the snowflakes along in streamers. These currents soon die away, and the snowflakes settle out of the still water.[2] The turbulence in the water is very effective at mixing the snowflakes throughout the glass.

Turbulence also mixes the atmosphere quite efficiently. Unlike the flight of a molecule, which is interrupted every billionth of a second, a turbulent

2. The rate at which the flakes settle to the bottom of the glass depends on their size and weight. They are purposely made large enough to fall out in a few seconds. The smaller the flakes are, the longer they will remain suspended. Too small, and the glass will become permanently milky. The behavior of the flakes is similar to that of aerosol particles in the atmosphere (Section 3.1.2).

eddy can carry a pollutant at the speed of a typical wind for a distance equal to the size of the eddy. The pollutant molecule may hitch a ride on one eddy after another, moving in different directions with the varying airstreams. The motions of the eddies are random, like molecular motions or Brownian motion. A pollutant molecule caught up in the eddies would therefore appear to stagger along, like the drunk. Likewise, the average distance traveled obeys the first law of diffusion, but the mean speed of the turbulent diffusion is much greater than for molecular diffusion; that is, turbulent diffusion moves molecules over a distance much faster. At the ground, the time to travel a fixed distance through the atmosphere is a million times faster for turbulent diffusion than for molecular diffusion. The local rate of turbulent diffusion depends on the local intensity of the turbulence. The more agitated an air mass is, the more numerous and larger the eddies and the more rapid the turbulent mixing and dispersion will be.

Diffusing Toward Uniformity

A pollutant tracer will be diffused from one point to another if the pollutant is not uniformly mixed in the atmosphere. If the pollutant is present everywhere in the same mixing ratio, then at any point, equal numbers of pollutant molecules will be diffusing in all directions; the local concentration is therefore unaffected. Similarly, the turbulent exchange, or mixing, of air masses that have equal abundances of a pollutant cannot change its overall distribution. It follows that diffusive processes transfer a pollutant only when there is a variation in its mixing ratio with location. In fact, diffusion acts to reduce that variation.

The second law of diffusion states that because of diffusion, the spatial distribution of a trace gas in air becomes more uniform as time elapses.

In the case of molecular diffusion, the transport of a tracer (pollutant) occurs as a result of gradients in its concentration from place to place. In a region of high pollutant concentration, more pollutant molecules diffuse out than diffuse back in from the adjacent regions of lower concentration. A net amount of the pollutant flows out of the denser region. This effect can be seen in the cloud of color diffusing from a dye tablet. If you want to dye Easter eggs, you need to stir the water, create some turbulence, and mix the dye through the glass well before adding the eggs. Turbulence also causes clean air to be entrained into a plume, diluting the pollution there. Of course,

pollutants are simultaneously mixed from the plume into the clean air outside. Eventually, turbulent mixing will transport pollutants from a smokestack plume into the cleaner air surrounding it (Section 5.4.1).

You can see the effect of turbulence on pollutant distributions in a number of common situations. If you sit in a closed room with a person smoking, you may wish to move as far away from him as possible. Unfortunately, turbulence is working against you. The smoke spiraling from the cigarette is soon mixed throughout the room. You can smell it and feel it burning your eyes. There is no escape. But be still. The more agitated you become, the more turbulence you will generate, and the faster the smoke will reach you.

Turbulence is a complex phenomenon. It is difficult to express mathematically or to represent in models used to predict pollution concentrations and distributions. Thus the action of turbulent diffusion is usually ignored or only roughly approximated in pollution forecasts. Nevertheless, turbulence is an important characteristic defining the state of the atmosphere. In a closed room, the air seems quite still. But the spread of cigarette smoke shows that the air is nonetheless in motion. In a storm, turbulence can rip the wings from an aircraft. Generally, we classify as turbulence those motions that are smaller than the phenomenon of interest. So clear-air turbulence can jar and shake an airplane, but a downdraft can take it all the way to the ground. In other words, turbulence is the sum of all the small winds and eddies that we do not wish to, or need to, describe explicitly.

5.2.2   Convection and Lofting

Convection is vertical motion driven by buoyancy (Sections 3.1.1 and 3.1.3). The buoyancy is caused by solar heating of the surface and by the latent heat release resulting from the condensation of moisture in the rising air column. The upward velocities in strong convective cells may reach 50 meters per second or more. At these speeds, air can travel from the surface to 10 kilometers height in about 3 minutes. Convective transport of pollutants is important for several reasons. First, convection lofts pollutants away from the surface, where they otherwise could be in contact with people, plants, and animals (Figure 5.1). Second, in the rising convective column, precipitation may form and wash out the soluble pollutants, especially particulates. Third, pollutants that escape rainout are injected into upper air levels with stronger prevailing winds; these winds are strong enough to disperse the pollutants over great distances.

Of course, convection has side effects. The pollutants that immediately return to the ground in raindrops may themselves be detrimental. Such pollutants can contaminate local soils or form strongly acidic runoff. Pollutants that escape into the upper-level winds may later contribute to regional and global pollution problems, such as acid rain (Chapter 9) and stratospheric ozone depletion. Clearly, distributing local pollution over a larger region in the hope of diluting its effects is not always a smart move.

The Mixed Layer

Convection is a localized phenomenon. It is most spectacular in the huge thunder cells that produce lightning, thunder, and hailstones. On less intense and smaller scales, convective cells represent the largest sizes in the spectrum of turbulent eddies. Convective cells and eddies are driven primarily by surface heating during the day. The convective eddies, and all the smaller eddies comprising turbulence, work together to create a well-mixed layer of air at the ground during the day. Within this "mixed layer," churned by turbulence, pollutants can become uniformly dispersed rather quickly. The height (or depth) of the mixed layer depends on the amount of heating at the surface and on the temperature stability of the atmosphere (Section 5.3). The more heating there is, and the less stability there is, the higher the mixed layer will be. In extreme cases—a humid summer day in Florida, for example—the mixed layer can extend upward to the tropopause (Section 2.3.2). In southern California during the winter, by contrast, the mixed layer may be only a few hundred meters deep. Nevertheless, even this modest mixing layer is critical to diluting the pollutant concentrations of smog.

Pollutants emitted into the mixed layer can be dispersed throughout the depth of the layer. This reduces the concentrations at the ground and, in many cases, helps to limit the health impacts of air pollution. At night, in cold weather, and over the oceans, the mixed layer tends to be shallow. In each of these instances, the buoyancy generated at the surface, which drives convection, turbulence, and mixing, is relatively weak. The result can be high

concentrations of pollutants trapped near the surface and, accordingly, air pollution alerts.

5.2.3 ADVECTION AND LONG-RANGE TRANSPORT

The strongest winds blow horizontally around the Earth. (Section 2.4 describes the principal wind systems.) Typically, at the surface these winds move fastest in the east-west direction. At middle latitudes, the prevailing winds flow from west to east (westerlies), and in the tropics, the trade winds blow generally from the east (easterlies). It is also true that surface winds can change suddenly in direction and speed whenever a weather front passes, but the average winds are more zonal (longitudinal) in direction. The horizontal motion of the atmosphere is referred to as **advection**, and the prevailing winds are known as **advective winds**.

Advection is the process responsible for the long-range transport of pollutants downwind from sources (Figure 5.1).[3] Advective winds may have velocities of up to 400 kilometers per hour. Such velocities are rarely seen at the surface, where typical wind speeds are perhaps 10 to 20 kilometers per hour. The fastest winds are found aloft, in the jet streams. If pollutants reach these wind currents, they will be carried great distances. Even in the lower atmosphere, winds can transport pollutants across a continent in a week or so. Horizontal transfer by turbulence or convection is negligible compared with the lateral displacements associated with advection. As noted earlier, convection and turbulence play a key role in mixing pollutants to levels where they can be swept away by advective airstreams.

Advection has two functions in dispersing pollutants: Advection removes offending materials to a distance from the source and acts to dilute the pollutants. In general, faster winds generate more intense mixing associated with turbulence. Thus advective winds can enhance dilution by turbulent mixing. Since winds often vary dramatically in strength and direction over time at a given location, they tend to distribute pollutants in complex, unpredictable patterns. An observer downwind of a pollution source may experience high pollutant concentrations at

---

3. Downwind from a source means away from the source in the direction the wind is blowing. Thus pollutants emitted from a source are carried downwind. An observer standing downwind of the source can detect the pollutants, but an observer upwind cannot.

certain times but not at others. The twists and gyrations of the winds determine the level of exposure to pollutants at points distant from sources. Even so, one would not choose to live downwind of a major source of pollution if reasonable housing were available upwind. In many cities and towns, the poorest neighborhoods are downwind of factories and industrial zones. This is not by accident, but by economic selection. The areas upwind, usually in a direction dictated by the prevailing winds, have cleaner air and thus higher land values. The next time you seek housing, check out the local sources of pollution and the direction of the prevailing winds. It could save you a lot of headaches, literally.

Sources and Receptors

Advection carries pollutants from their sources to distant points. Eventually, the pollutants are either transformed into other compounds (harmless ones, or secondary pollutants) or deposited at the surface. They may be deposited through scavenging by rain, by falling onto surfaces (dry or wet), or being inhaled or ingested by living organisms. The place where the pollutant ends up is referred to as the **receptor site** (Figure 5.1). The study of sources and receptors is important to evaluating the effects of air pollution. One would not like to be a major receptor of airborne toxic waste, for example.

Consider Los Angeles as a source of a variety of pollutants, including particulates that reduce visibility (Section 6.5). The Los Angeles basin can be treated as a local source of these particles. Many of the particles are deposited locally in the regional forests (as well as the residents' lungs). Even so, a large mass of particulates escapes from the basin, carried by prevailing westerlies into the high-desert areas east of Los Angeles. Once in the dry desert environment, the fine aerosol particles can remain suspended for hundreds of miles, eventually reaching the national monuments located in the Four Corners region (Utah, Colorado, Arizona, and New Mexico). Visibility at receptors like the Grand Canyon may be degraded by these particles carried over long distances (Section 6.5).

As pollutants are transported, they are continuously lost by deposition and may be replenished by local sources along the way. In the case of Los Angeles's pollution traveling to the deserts, there are only small additions from cities and towns. But large power plants are located in the remote sections of the southwestern United States, far away from most people. These facilities generate electricity to keep Los Angeles and other major urban areas running. The electrical power is carried hundreds of miles by high-voltage towers and lines stretching across the pristine desert. The power plants emit substantial quantities of airborne pollution, including particulates, carbon dioxide, and sulfur from coal burning. It has been suggested that these operations, together with large ore smelters, are in fact the principal sources of visibility-degrading aerosols in the region.

In determining the sources of the pollution found at a specific receptor site, care must be taken to account for all the potential contributors along the path of the air masses reaching the receptor. One technique used to unravel the history of an air parcel compares the relative amounts of specific pollutants in the air. For example, if the major source of the pollution is a power plant, the vapors and particles will have a distinct composition. Each source has a "fingerprint" that can be used to identify it. If the source were primarily automobiles in Los Angeles, a different pollutant fingerprint might be expected. Problems arise with this scheme for allocating blame, however. When several sources of pollution are intermingled, it is often difficult to determine the individual contributions of each. The variability in the pollution reaching a receptor, associated with variations in the sources themselves and the vagaries of meteorology, complicates the interpretation. Imprecision in measurements of minute quantities of the most unusual tracers that could pinpoint a source—for example, a rare metal—make interpretations fuzzy. And polluters, in general, like to hide behind uncertainty.

5.3 Temperature Inversions

As we already noted, the ability of the atmosphere to mix pollutants vertically depends on the temperature structure of the lower atmosphere. The normal temperature structure in the troposphere is a decreasing temperature with increasing height (Section 2.3.2). A temperature inversion, or inversion layer, is usually defined as the layer of the atmosphere in which the temperature is *increasing* with height. The implication of this structure is that warm air is lying over the colder air just below. That configuration is very stable and not prone to the vertical transfer of air.

The concept of an **air parcel** is useful in the following discussions. An air parcel is a small volume of air held in a weightless, perfectly elastic or stretchable, heat-insulated balloon. We all have batted around balloons at birthday parties. They float nearly weightless through the air.[4] Our balloon is *exactly* weightless, and it stretches so easily that the pressure inside it is always exactly equal to the pressure outside.[5] If the pressure outside decreases, then the balloon will expand until the pressures are equalized. The expansion is adiabatic; that is, there is no exchange of heat between the inside and outside of the balloon. The balloon is insulated to ensure that no heat is exchanged. (A more comprehensive explanation of adiabatic processes is given in Section 3.1.1.)

An air parcel can be used as an experimental tool. Once we create such a parcel at the ground, we can perform various experiments with it. We can move the parcel around and see how it responds. We can fill it with different kinds of air at different temperatures and see whether the behavior of the parcel changes. Two important features of such a parcel of air are that (1) the pressures inside and outside are always the same no matter where we move it and (2) the parcel

is small enough for the air inside to have uniform properties throughout, including the temperature, density, and humidity.

5.3.1   TEMPERATURES IN THE LOWER ATMOSPHERE

The changes of temperature with height in the lower atmosphere are conveniently described by a single parameter, the lapse rate. Examples of temperature profiles with different lapse rates are illustrated in Figure 5.3. The **temperature lapse rate**, is literally the rate at which the temperature lapses, or decreases, as the height increases above the ground. The lapse rate varies with the amount of moisture in the air. Figure 5.3 depicts the range of lapse rates typically found in the atmosphere. Two important lapse rates are the dry adiabatic lapse rate and the average global lapse rate.

To investigate the properties of these lapse rates, we can use our experimental air parcels. Here, specially designed air parcels will be moved up and down through the atmosphere to see how they respond to the changing conditions of temperature and pressure. As a parcel rises, it always experiences decreasing atmospheric pressure (Section 2.3.3) and so the parcel expands adiabatically. If we move the parcel back down, the external (atmospheric) pressure will increase, and the parcel will compress adiabatically. Because no heat has been transferred to or from the parcel, it will end up in exactly the same state after such a cycle of rising and falling. However, because the air in the parcel has experienced adiabatic expansion and compression, its temperature has changed with time along the path.

The Dry Adiabatic Lapse Rate

In the case of the **dry adiabatic lapse rate**, the air in our parcel is assumed to be perfectly dry. We can dehydrate the air before we inflate our parcel or go to the driest desert on Earth and collect a bottle of highly dessicated air.[6] In either case, the result will be nearly the same. We begin the experiment with our dry parcel at the ground. We also make sure the parcel is at the same temperature as the air at the ground, by heating the dry air to the proper temperature before

---

4. The behavior of the balloon depends on the gas that you fill it with. If a light gas such as helium (He), with a low molecular weight, is used, the density of the gas in the balloon will be lower than the density of the surrounding air (Section 2.2.1); hence, the balloon is buoyant and rises quickly to the ceiling (Section 3.1.1). If you pump ordinary air into the balloon, the densities of the gas inside and outside the balloon will be similar; the balloon has neutral buoyancy and floats, slowly falling to the floor under the weight of the rubber skin. If you blow up the balloon yourself, you will exhale carbon dioxide into the balloon. Since CO_2 molecules are heavier than air molecules, the gas density in the balloon will be greater than that of air, and the balloon may sink more quickly.

5. In a real balloon, the rubber skin is stretched tightly and exerts a force on the air in the balloon. This compresses the air inside slightly compared with the air outside and increases its pressure. The extra pressure (or force per unit area) just compensates the elastic force exerted by the rubber skin. If the balloon and surrounding air are at the same temperature, the density inside the balloon will also be greater (based on the ideal gas law in Section 3.1.1). In the case of a perfectly stretchable balloon, no force is exerted on the gas in the balloon as it either expands or contracts, so the pressure difference between the inside and outside is always zero. If there is a temperature difference between the inside and the outside, however, then the density will change, although the pressure will remain the same (because the pressure is fixed by the atmospheric pressure). In this case, a warmer temperature in the balloon implies a lower density, which can make it buoyant. (The physical properties of air are discussed in Section 2.2.2.)

6. The driest air at the Earth's surface actually is found in the coldest regions of Antarctica. The air there is so cold—as low as −90°C—that almost all the moisture is frozen out.

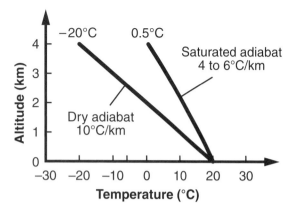

Figure 5.3 Temperature profiles in the lower atmosphere corresponding to two temperature "lapse rates." The dry adiabatic lapse rate (for very dry air) is about 10°C per kilometer of altitude. When the air is naturally humid with water vapor, the lapse rate is smaller, as small as 4°C per kilometer. The average global lapse rate is about 6.5°C per kilometer.

inflating the parcel. We place one thermometer inside the parcel and another outside, to measure the relative temperatures for comparison.

In our first experiment, we move the parcel up and down and record the temperature inside the parcel. That temperature is displayed in Figure 5.3 as the dry adiabat. Notice that the temperature changes with height at an almost uniform rate of about 10°C per kilometer of altitude (10°C/km). Moving upward, the temperature decreases at this rate, and moving downward, it increases at this rate. Moving upward, the parcel of air expands and cools adiabatically; moving downward, it contracts and warms. When air is dry, the adiabatic lapse rate holds over the whole depth of the lower atmosphere. Even if the air is moist, it will still follow the dry adiabat as long as water does not condense (see the following section).

The lapse rate is defined more precisely as the change in temperature divided by the change in height, or

$$\Gamma = -\frac{\Delta T}{\Delta z} = -\frac{T_2 - T_1}{z_2 - z_1} \quad (5.1)$$

Here the lapse rate is represented by the Greek letter gamma, Γ. The temperature is T, and the height, z. Equation 5.1 is simply the change in temperature measured at two altitudes, z_1 and z_2, divided by the change in height between those levels. A negative sign is used here to define the rate of decrease with increasing height as a positive number (that is, the

lapse rate as defined gives the *decrease* in temperature corresponding to a 1 kilometer increase in altitude; accordingly, if the temperature *increases* with altitude, the lapse rate is *negative*).[7] For example, along the measured dry adiabat in Figure 5.3, $T_1 = 20°C$ at $z_1 = 0$ km (the ground), and $T_2 = 0°C$ at $z_2 = 2$ km. Plugging these numbers into Equation 5.1 yields the value of the dry adiabatic lapse rate, Γ_d:

$$\Gamma_d = -\frac{0 - 20}{2 - 0} = -\frac{-20}{2} = +10°C/km \quad (5.2)$$

The dry adiabatic lapse rate always has a value of 10°C/km. An isolated parcel of dry air cools by 10°C for every 1 kilometer it rises. By the same reasoning, the parcel warms by 10°C for every 1 kilometer it descends.

Mountaintops are cold because of the adiabatic lapse rate. As you climb a mountain, you encounter parcels of air that, some time in the recent past, were probably near the ground. To be at your height now, the parcels had to rise. In doing so, they cooled. If your local environment is dry, as it would be in the desert, the cooling most likely followed the dry adiabatic lapse rate in Figure 5.3. At several kilometers above the ground, the mountainsides would be frozen, warmed only by the sun during daylight. The incongruous sight of snowcapped peaks above a baking desert floor is explained by the temperature lapse rate of the lower atmosphere. The temperature lapse rate is determined by adiabatic expansion and compression.

The opposite effect can occur when cold air travels down mountain slopes. The air can warm considerably. You can understand the principle by thinking about your idealized air parcels. In southern California, for example, air frequently descends from the high deserts into the coastal regions. The air is compressed and warms, producing the famous Santa Ana winds. These hot, dry winds are often a prelude to fierce brushfires in the coastal mountains.

Saturated Lapse Rates

In many places, at most times, the air is not dry but moist. Moisture can reduce the temperature lapse

---

7. The lapse rate is often defined with the opposite arithmetic sign, then giving the temperature *change* for a 1 kilometer increase in altitude. The convention adopted here is to have a positive lapse rate correspond to a profile of decreasing temperature. Obviously, the two definitions are related by just a minus sign.

rate because when the moisture becomes saturated, it condenses, releasing **latent heat**. As an air parcel rises and cools (along a dry adiabat), the saturation point may be reached at some elevation. This condensation is most commonly seen as clouds, but also as fog and haze. (The phase changes of water vapor are discussed at greater length in Section 12.1.2.)

Latent heat can warm an air parcel from the inside, slowing its rate of cooling by means of adiabatic expansion. As an air parcel rises above its saturation level, water condensation and latent heat release prevent the air from cooling as fast as a dry parcel does. Accordingly, the slope of the temperature line in Figure 5.3 is less steep, and the rate of temperature falloff is smaller. The **saturated adiabatic lapse rate**, Γ_s, corresponds to the case in which the air inside a parcel is initially saturated at the ground. *Any* cooling leads to water vapor condensation and latent heat release. A rising parcel of saturated air cools more slowly than does an unsaturated, or dry, air parcel.

The **global average temperature lapse rate** in the troposphere is roughly 6.5°C/km. This represents a state of the atmosphere lying between perfectly dry (Γ_d = 10°C/km) and saturated (Γ_s = 4°C/km) conditions. This atmospheric state does not exist everywhere on the globe, nor does it actually exist at very many individual places. Rather, the global average lapse rate would represent the mean state of the troposphere if it could be geographically averaged. There are a number of reasons for the variation with location: the amount of sunlight warming the region, the humidity of the air, the topography and prevailing winds, and the state of the surface (wet, snow covered, and so forth). The main factor, however, is the local humidity. At a fixed location, the lapse rate also changes with time. The global average lapse rate is chosen to account for all the local variations in time and space.

At any specific location and time, the temperature lapse rate normally has a value lying between the dry adiabatic lapse rate and the saturated adiabatic lapse rate, the two extremes. We will refer to the instantaneous temperature structure as the **environmental lapse rate**. In other words, the environmental lapse rate, Γ_e, is the rate of change in temperature with increasing height in the local atmosphere at the time of interest. If we are carrying out some tests using an idealized air parcel, the temperature of the environment outside the parcel can be measured by a thermometer placed next to the parcel. The environmental lapse rate describes the change in the measured temperature with altitude. Because the atmosphere is highly variable, Γ_e is also variable. Moreover, the environmental lapse rate itself can change with altitude. The variations in lapse rate occur for many reasons. Typically, one finds air masses with different meteorological histories overlying one another. The air masses have had different heating rates and humidity and thus show different temperature lapse rates. In these cases, the environmental temperature profile can be represented by a sequence of air layers stacked one atop the other. Each layer is chosen so that the lapse rate is nearly constant, although it may differ from the lapse rates in adjacent layers above and below.

The environmental lapse rate can be determined by simple instrumental observations. In fact, the temperature profile is one of the most fundamental meteorological parameters used in interpreting the behavior of the atmosphere. Measurements of temperature versus altitude are referred to as soundings. Soundings are usually made using helium-filled balloons carrying a set of standard instruments, including a thermometer (for temperature), barometer (for pressure) and hygrometer (for moisture). With these data, the temperature lapse rate as a function of altitude can be calculated.

The dry adiabatic, saturated and global average lapse rates represent basic ideal states that are rarely seen in the real atmosphere, although the natural environment is generally constrained to behave within the limits defined by these basic states. The actual temperature lapse rate, or segments of the temperature profile in distinct layers, often can be reasonably approximated using the basic states. Accordingly, the atmospheric behavior discussed in subsequent sections will frequently refer to these basic lapse rates.

Inverted Temperature Profiles

When a temperature inversion exists in the atmosphere, the temperature can actually increase with altitude. This is, of course, contrary to the adiabatic law of expansion and compression for an air parcel, which determines the basic temperature lapse rates in the lower atmosphere. The most common causes of temperature inversions are described in Section 5.3.3. In general, temperature inversions require heating of the air aloft or advection of warm air masses over cold ones. In these cases, the environmental lapse rate may be *negative*; that is, $\Gamma_e < 0°C/km$, with the temperature increasing with increasing height. (See

Section A.4 for a description of the "inequality" symbol, <.)

5.3.2 ATMOSPHERIC STABILITY

Vertical motions are essential to maintaining the structure and motion of the atmosphere and to dispersing and removing air pollutants. Without vertical mixing and convection, the air near the surface would become stagnant and intolerably polluted. Most vertical motion is induced by the instability of air near the surface.

Stability and instability are important concepts in nature. Stability guarantees that a system will not spontaneously fly apart or disintegrate. Atoms are stable: The nucleus remains intact, and the electrons stay bound to it (Section 3.1). If atoms were not fundamentally stable, we would be continuously transmuting into different materials and forms. Buildings are stable because of construction principles developed over many centuries. If buildings regularly collapsed, we would not go into them very often. The steering on an automobile is stable owing to the design of the axle mounting. If wheels were prone to suddenly turning sideways, cars would not be very popular.

Figure 5.4 provides a simple example of a stable and an unstable system. In a stable system (Figure 5.4[a]), restoring forces keep the system in a particular state. In the case shown, the ball lies at the bottom of a depression. From our experience, we know that this is a stable configuration. Nudge the ball, and it will return to its original position after rolling back and forth a few times through the low spot. Gravity and friction are the forces that produce this stable state (along with the physical rigidity of the depression itself). However, gravity can work both ways. Try to steady yourself on a taut wire, balance a quarter on its edge, or perch a book on a straw. Each of these situations is similar to the experiment depicted in Figure 5.4(b). The balancing act, in each case, can be done. Each object can be placed in a state of apparent stability. Unfortunately, the slightest nudge, vibration, or breeze causes the object to careen away, the system to disassemble. Gravity pulls the precariously balanced objects from their unstable positions, seeking to find a new stable configuration. The quarter topples flat, the book crashes to the table, and you fall on your keester!

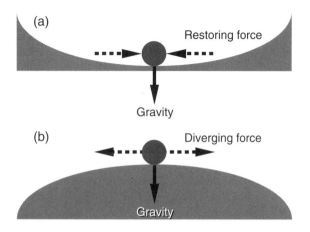

Figure 5.4 The concept of stability. (a) In a stable configuration, a ball remains at the bottom of a valley; any attempt to move the ball to the side, away from the point of stability, results in an opposing force (in this case, owing to gravity) that restores the ball to its original position. (b) However, if the ball lies at the peak of a hill, it is in an unstable condition, and any displacement of the ball away from the peak will result in a diverging force that pushes it farther from its original position. In a stable state, any displacement of the ball is resisted; in an unstable state, any displacement of the ball is amplified.

"Chaos" results when a system has a number of possible stable states, but switches unpredictably between these states from time to time. The simplest chaotic system has two stable states, and the system switches erratically from one to the other (like a friend with a volatile temper). The Earth's atmosphere has many stable states. Among these, one can think of specific weather patterns as a kind of reproducible state of the atmosphere. (The chaotic behavior of the climate is discussed in Section 11.6.1.) The weather that accompanies each pattern is fairly predictable for a period of several days. Over longer times, however, the weather becomes unpredictable because the atmosphere can switch to another pattern. We are never sure which pattern will be next. The weather is chaotic over a period of months and years.

In the main, the atmospheric structure is very stable. Our air does not blow away into space because gravity holds it obstinately to the Earth (Section 3.1.1). The winds blow regularly, if not erratically at times. Rains come. Climate holds its course (Chapter 11). Nevertheless, atmospheric instability is essential to maintaining the global environment. It induces life-giving rain and cleans the air. The origin of atmospheric instability lies in the normal patterns of solar energy deposition and dissipation, and their connection with the temperature structure of the lower atmosphere.

Relation to the Lapse Rate

Most of the solar energy received by the Earth is deposited at the surface where it heats the land and oceans. That heating causes water to evaporate into the air. The hot land surface also directly warms the air in contact with it. As a result, the air near the surface is generally warm and moist. What happens next depends on the local temperature profile.

Stability in the atmosphere is associated with the absence of vertical motion, and instability induces vertical motion. The latter effect is caused by the natural tendency of warm air to rise. You can feel warm air rising over the burner on your stove. The hot smoke from a candle flame shoots straight up in a still room. Similarly, warm parcels of air near the ground tend to rise. Such parcels are inherently unstable, although they may be stabilized by a temperature inversion. When a layer of warm air lies below a layer of cold air, an unstable situation generally exists. The layers can spontaneously overturn and mix until the tendency for the instability is removed. The original instability was related to the temperature profile, with the presence of a strong temperature lapse rate (very warm air at the surface and very cold air aloft). The induced vertical motion and mixing reduce the lapse rate by bringing cooler air toward the surface. However, it will still be warmer near the ground than at higher altitudes. A more typical lapse rate is set up: The atmosphere is stabilized, and vertical motions cease.

The atmosphere is always in a transient state. When the surface is continuously heated or cooled, the atmosphere cannot achieve a permanent, or perhaps even a temporary, stable state. The lower air layers are constantly overturning and mixing. These processes establish the local environmental temperature lapse rate. The state of the atmosphere at a certain place and time is reflected by the lapse rate. Because atmospheric stability is closely related to overturning and vertical motion, stability can be assessed from the lapse rate, which is established in part by vertical mixing. We can investigate stability using idealized air parcels, watching their behavior as we release them into different layers of the atmosphere. In the case of air pollution, we are most interested in the fate of air parcels released at the ground. When the parcels contain pollutants, we hope they will rise quickly and disappear.

A few of the possible responses of an air parcel released into the lower atmosphere are shown in Figure 5.5. In each case, an idealized air parcel is assumed, as defined at the beginning of Section 5.3. The initial properties of the air in the parcel can be changed to suit our needs. For many experiments, the air parcel is simply a puff of local air. In other cases, it may be a heated parcel of gas emitted by a smokestack. We might also want to see how a dry air parcel acts in a humid environment, or a moist parcel in a dry environment. We can warm or cool the test parcel to suit our needs. We are in control.

The atmosphere is **neutrally stable** when parcels of local air that are displaced vertically simply remain where they are left. This situation is shown in Figure 5.5(a). Such air parcels have the same characteristics as the air around them. The situation of neutral stability is often found in the mixed layer, or boundary layer, near the Earth's surface. In a well-mixed layer of air, one might expect that all the parcels have the same characteristics. Then the parcels can be moved around without greatly disturbing the local state of the atmosphere. The value of the corresponding environmental lapse rate is determined by the moisture content of the air near the surface.

In the case of a "stable" atmosphere, a local air parcel also stays where it is placed, but the parcel returns to its original position whenever it is displaced vertically. If the parcel is moved upward, it will move down when released. If it is moved downward, it will respond with an upward motion. This behavior is typical of air in a temperature inversion layer. Under these conditions, a parcel that is displaced vertically and released tends to oscillate about its original position.

Inversions and Stability

Consider Figure 5.5(b) in which a normal temperature lapse rate in the lower atmosphere is capped by a temperature inversion (indicated by the increasing temperature with height). The temperature of an air parcel near the surface that rises follows the solid-dashed line. Below the inversion, the parcel is free to move about and is neutrally stable. This is a well-mixed layer. As the parcel is raised into the inversion, it continues to cool by adiabatic expansion even as the environmental temperature increases. The parcel is therefore colder than its

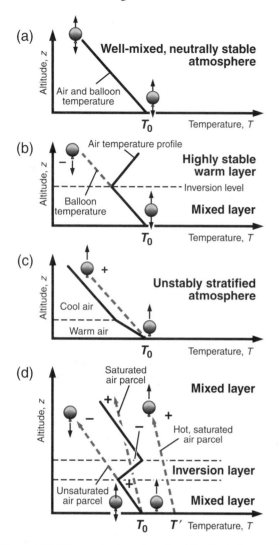

Figure 5.5 The vertical motion of an idealized air parcel in the atmosphere. The air parcel can be thought of as contained in a balloon that has no weight and is perfectly elastic. (a) In a well-mixed atmosphere, the adiabatic cooling (or heating) of an ascending (or descending) air parcel follows the background temperature lapse rate. The motion is neutrally stable, since the vertical displacement is neither resisted nor amplified. (b) If a temperature inversion exists above the ground, the vertical motion of the air parcel will be strongly resisted when it reaches the inversion. Above the inversion, an air parcel originating below has negative buoyancy. (c) If the air parcels near the ground are warmer than the air higher up, they may be unstable. As they rise, they remain warmer than the background atmosphere and experience a positive buoyancy that accelerates them upward. (d) If a temperature inversion layer exists, normal air parcels will rise to the inversion and stop, experiencing negative buoyancy. But if the surface air is sufficiently warm and moist, it can penetrate the inversion and permanently leave the lower mixed layer.

environment. At each altitude, the air parcel has the same pressure as the surrounding air. The density of air at a constant pressure varies inversely with temperature (ideal gas law). The colder air in the air parcel is therefore denser (heavier) than the air around it. The denser air experiences negative buoyancy and sinks (Section 3.1.1). Unless the parcel is forcibly held above the inversion, it will drop back into the lower atmosphere.

A strong barrier against the vertical intermingling of air masses exists when a temperature inversion sits over a mixed layer. The simple demonstration in Figure 5.5(b) shows that air in the mixed layer cannot easily penetrate an overlying inversion. In this case, the lower atmosphere is capped, and so pollution emitted near the surface is trapped. In one sense, the term *atmospheric stability* can imply the presence of a temperature inversion layer that inhibits vertical motions.

In some instances, colder air may lie over a warm surface layer. The situation is illustrated in Figure 5.5(c). An air parcel holding a normal amount of moisture might follow the dashed temperature line as it rises and cools. Even though the parcel gets colder, it remains warmer than the ambient air around it. Because the air in the parcel is warmer and less dense (lighter), it experiences positive buoyancy. The buoyancy pushes the parcel upward. For the conditions shown, the parcel remains lighter and continues to rise. The atmospheric structure is unstable. Clearly, parcels of air rise spontaneously and mix with the upper level air, warming it until the lapse rate is more neutrally stable.

Occasionally, we are interested in the motion of a moist air parcel through the lower atmosphere. The parcel may represent a puff of pollution from a smokestack, for example. In Figure 5.5(d), an environmental temperature profile is shown with a mixed layer at the surface, an inversion layer at intermediate heights, and a second mixed layer above the inversion. A local air parcel at the ground, with initial temperature, T_0, follows the solid-dashed line as it is lifted. The buoyancy is neutral, since the layer is well mixed. At the base of the inversion, however, the parcel begins to experience negative buoyancy. If it is not pushed upward, it will stop rising. The mixed layer is strongly capped.

Suppose that we add water vapor to this parcel until it is saturated. Now when the parcel is released at the ground, it immediately feels a positive buoyant force and is lofted along a saturated adiabatic profile (the dashed line from T_0). Recall that this occurs because the water vapor in the parcel is condensing, releasing latent heat, and providing some additional warmth for the parcel as it rises. The saturated adiabat, nonetheless, intersects the inversion at a higher altitude, as shown. At this point, the parcel is subject to negative buoyancy and stops in the inversion layer. The parcel has gotten a boost from the moisture, but not enough to break through the inversion.

If the parcel is now also heated to a temperature $T' > T_0$ and additional water vapor is injected to keep the air saturated, it can easily penetrate the inversion (Figure 5.5[d]). In this case, the parcel still cools adiabatically as it rises (it must) but, owing to the original addition of heat, stays warmer than the environment during its flight. Likewise, a hot, humid chimney plume may also rise farther into an inversion layer than an ambient air parcel would. When a fire is burning, we can watch the hot smoky plume rise far into the atmosphere. These plumes are slowed down as they entrain cooler ambient air from the environment and eventually take on the characteristics of the local environment. Entrainment is the result of the friction between the rising air column and the surrounding air. The friction causes turbulence that mixes together the two air masses at their boundary.

Stable and Unstable Motions

The lower atmosphere can be either stable and calm or turbulent and well mixed. In general, the overall stability of an air mass depends on the amount of heating at the surface, the surface wind speed, and the thermal stability of the overlying air. If an air mass is very dry, then parcels of that air will follow a dry adiabat as they move up and down (Figure 5.3). Artificially heated parcels, however, would rise spontaneously in this case. The strong thermals and dust devils over deserts are the result of superheating air near the surface under the glare of the noonday sun. Parcels saturated with water vapor would also accelerate upward, driven by latent heat release. Such parcels of air might come from smokestacks, for example.

Air parcels with specific initial characteristics (temperature, moisture content) may be inserted into layers of air with quite different characteristics. For example, they may be inserted at the ground—into the boundary layer—or at the interface between the mixed layer and an overlying inversion layer. The inserted air parcel may encounter a range of tempera-

ture lapse rates that represent a variety of possible states of stability of the atmospheric layer (Figure 5.5). A dry parcel inserted into an environment with an "average" lapse rate experiences a stabilizing (negative) buoyant force, which resists its vertical motion. Figure 5.3 shows why this is so. The dry lapse rate is much larger than the saturated (or average) lapse rate, and so a dry parcel cools more rapidly as it moves upward and becomes heavier.

If the local lapse rate approaches a saturated adiabat, the atmosphere is usually less stable. Heating at the surface can trigger convection and cloud formation. The latent heat of condensation forces buoyant parcels upward. Under these conditions, almost any parcel that is released at the surface is carried upward in turbulent eddies and convective cells. Any negative buoyancy the parcel might experience would be overwhelmed by the vertical winds and mixing. Under these conditions, any parcel that happens to be around near the surface will be taken for a ride.

The transport and dispersion of atmospheric pollution are controlled to a larger extent by the stability of the ambient atmosphere than by the stability of the polluted parcel itself in the background environment. For example, the initial behavior of a smokestack plume is determined by the properties (temperature, velocity, and other factors) of the polluted gas parcels leaving the stack. The local and regional transport and dispersion of the effluents, however, depend on atmospheric conditions such as stability and wind profile.

The presence of a stable atmosphere is not necessarily good news for those concerned about exposure to air pollution. A stable atmosphere implies less mixing and dilution of pollutants. For those who had hoped their smoke might rise harmlessly into the upper atmosphere, a low-level temperature inversion means the smoke will be trapped to annoy the neighbors. During the winter, we enjoy a roaring blaze in the fireplace. But if the air is cold and stable, the smoke from countless fireplaces will accumulate. The pollution may reach intolerable levels, and the authorities may step in to regulate our fireplaces.

5.3.3 Large-scale Inversions

The stability and mixing of air near the surface are controlled to a great degree by the presence of temperature inversions in the lower atmosphere. These inversions typically are very extensive, blanketing

(a)

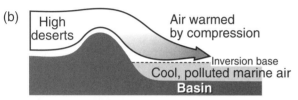

(c)

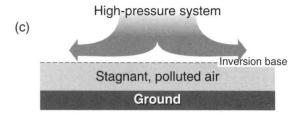

(b)

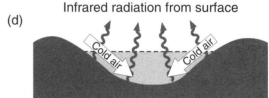

(d)

Figure 5.6 Common situations leading to inversion layers. (a) Cool marine air moves in and undercuts the warm air over land, resulting in a marine inversion. (b) The flow of air from a high plateau descends and warms by adiabatic compression, creating a cap above a cooler mass of air in a basin below the highlands. (c) When a stationary high-pressure weather system is present, air subsides over a large area, producing clear skies and a temperature inversion near the surface. (d) A radiation inversion forms when the ground efficiently radiates heat in the evening, allowing cold, dense air to pool at the surface below the warmer overlying atmosphere (the atmosphere is a much less efficient radiator of heat than the land is).

entire cities or regions. Figure 5.6 shows several ways that inversions are formed. In each case, warm air is produced over a layer of cooler air in contact with the surface. (The interactions of such inversions with local meteorology and topography to create pollution events are discussed in Sections 5.4 and 5.5.)

Marine Inversions

If you live near a coastline, you have probably experienced a marine inversion layer. Near cool oceans, such as in the western United States and western Europe, a **marine inversion** can turn a hot summer day at the beach into a cool, foggy, overcast shiver. The cause of marine inversions is depicted in Figure 5.6(a). The ocean is typically quite cool compared with the land, particularly at middle latitudes where the heating by sunlight is not strong. Air sitting over cold water also tends to be cool. Moreover, a great deal of water evaporates from the oceans. This moisture readily condenses in the marine mixed layer (or marine boundary layer) into stratus clouds and fog. Particularly on the western coastlines of continents, the marine air is blown inland by prevailing winds. In addition, the relatively intense heating of the coastal plain generates

strong onshore sea breezes, which also draw marine air inland.

Sea and land breezes are created by the contrasting temperature between the land and the oceans (or other large bodies of water, such as lakes). During the day, sunlight heats the land more than the adjacent ocean.[8] Warm air rises over the land and is replaced by cool marine air flowing in near the surface. The warm air can circulate out toward the ocean above the marine boundary layer, setting up a temperature inversion in the coastal region. As a result, pollutants emitted into the marine air layer over land are trapped near the ground. In the evening, the land cools faster than the oceans. A reverse circulation may be set up, with cool air moving from the land over the water. Such a land breeze can last

8. Land has a small effective **heat capacity** compared with the ocean, as explained in Section 11.5.1. In general, land heats up rapidly during the day as the heat of the sun is absorbed, and it cools off quickly at night as the heat is lost by radiation. The oceans, on the other hand, remain at a constant temperature throughout the day. The sunlight absorbed at the ocean surface is efficiently mixed through a 100-meter depth that is churned by winds (the ocean mixed layer). There is no equivalent mixing process to carry heat into soil. The difference in temperature between land and oceans usually has a significant daily variation, with the land hotter than the nearby ocean during the day and cooler at night.

through the night. (The effect of the sea/land breeze on air pollution in the Los Angeles basin is described in Section 5.5 [Figure 5.11].)

Regional Subsidence Inversions

When air flows over an obstacle such as a mountain range or blows from a high plateau and descends into a lower basin, a **regional subsidence inversion** can be created. Figure 5.6(b) pictures such an inversion. Subsidence is an important factor. As we noted earlier, air that descends heats by adiabatic compression. This descending air therefore creates a hot wind, like the Santa Ana winds that often blow through Los Angeles. The air is typically very dry. It might be dry because of its source over land, or it might have been desiccated as it rose over a mountain barrier, cooled, and lost water vapor by condensation and precipitation. As it is compressed, the dry air heats up along a dry adiabat (Figure 5.3). This produces the greatest warming for a given altitude change, which explains the unusually high temperatures of these winds. Similar hot winds that blow in other parts of the world are the *Chinook* (eastern Rocky Mountains) and *foehn* (United States and Europe).

If the descending air encounters colder air at the surface, it may be unable to push aside this denser air. The warm winds may then spread out above the surface layer, producing a temperature inversion. In other cases, strong subsidence winds fill a region with dry warm air that later cools at the surface by radiation, creating a radiation inversion. In southern California, weak Santa Ana winds often blow from the northeast across the mountain ranges to the north of the Los Angeles basin. These warm winds can trap colder denser marine air that is often present in the basin. These warm winds are usually associated with a high-pressure system.

High-pressure Inversions

A large-scale subsidence inversion (Figure 5.6[c]) can form when a stationary high-pressure system settles over a region. For example, during the summer season, the southern California coastal zone lies under the eastern edge of the Pacific high. This high-pressure system is a downward branch of the tropical Hadley circulation. Within the dome of high pressure, very dry air slowly subsides as it circulates clockwise about a center located off the coast of California. The air parcels brought to the Los Angeles basin thus arrive from the north and northeast. Subsidence compresses and warms the dry air as it descends into the lower troposphere. The dome of warm air then effectively traps the layer of cool marine air in the coastal region. A relatively shallow marine layer of several hundred meters to 1 kilometer in depth is formed. Separating the two air masses is a strong temperature inversion layer, which persists almost continuously during the spring and summer half of the year and appears occasionally at other times of the year. The temperature inversion suppresses convection and mixing, allowing pollutants to build up in the lower layer (Section 5.5).

Large weather systems that create conditions leading to high-pressure subsidence are fairly common. In some regions, these weather patterns regularly generate unusually intense hot winds, such as the *khamsin* (Egypt), *simoon* (Asia), and *sirocco* (North Africa). When the subsiding air encounters a cool, stable layer near the surface, a strong temperature inversion can be set up, as in the Los Angeles area.

Radiation Inversions

A **radiation inversion** (Figure 5.6[d]) is created when heat is rapidly lost from the surface by thermal radiation. (The properties of thermal [heat] radiation are reviewed in Section 3.2.1, and the radiative properties of the atmosphere are discussed in Section 11.4.) At night, the ground and the air just above the ground can cool off by radiating their heat energy while the air higher up remains warm. This occurs because air, particularly dry air, is a poor heat radiator, whereas land is an excellent radiator. The air just above land can lose heat by coming into direct contact with the surface. However, the air must be mixed downward to the surface. Recall (Section 5.2.2) that at night, convection dies away and the turbulent mixing of the lower atmosphere is shut off. The result is a very shallow layer of cold air that forms at the surface.

You can experience the formation of a radiation inversion on almost any summer night in the desert. Desert air is extremely dry. During the day, the soil and the atmosphere over a great depth are heated by the sun. Heat is carried from the surface into the atmosphere by intense mixing in the lower air layers. When the sun sets in the evening, the source of heat is suddenly turned off. The desert sands quickly radiate their heat and cool down. Sand is an excellent heat insulator, and the warmth deeper in the sand cannot reach the surface fast enough to maintain its temperature. The air just above the surface is immediately chilled, losing its heat to the ground. The air temperature can drop from daytime highs above 100°F to 40°F in a few hours. Standing on the desert floor, you can literally feel the chilling effect as the sun sets memorably in a purple glow.

The cold surface and warm upper air layers constitute a temperature inversion. Such radiation inversions are typically very close to the ground and very strong. The inversion can be strongest at sunrise, after a long night of surface cooling and deepening of the warm inversion layer. Under dry, clear weather conditions, such an inversion can occasionally be seen in the Los Angeles basin at first light in the morning. The pollution from the automobiles of early-bird commuters appears as a dense layer of smog hugging the ground beneath the shallow inversion. As the sun rises and warms the surface, the inversion is broken down because of increased mixing. The high concentrations of pollutants are dissipated.

If the area where the cooling occurs is hilly or mountainous, then the cold air forming on ridges and plateaus will flow into valleys and depressions. In certain regions, this flow creates fierce winds, called **katabatic winds**. The bora winds along the Adriatic coast of Yugoslavia and the mistral winds, which roar down the Rhône valley in France to the Mediterranean Sea, are two seasonal katabatic wind systems. In more common cases, the cold air tends to pool in low-lying areas. Pollutants released into these pools—such as the smoke from fireplaces—can build up to high concentrations.

Radiation inversions are most likely to form on clear nights with low winds. Clear skies are necessary for the surface radiation to escape; overlying clouds or fog absorb thermal radiation and effectively radiate it back to the surface. Calm winds prevent the mixing of warm air at higher altitudes downward to the surface.

A Day in the Life of an Inversion

When a temperature inversion exists over the oceans, the inversion does not vary significantly from day to night. The ocean surface temperature does not respond to solar heating during the day, and so there is no significant forcing from the surface to modify the mixed layer depth. Similarly, the ocean surface temperature does not cool as it emits radiation during the night, because the ocean has a high heat capacity (Section 11.5.1). The inversion height, accordingly, does not drop.

Over land, the situation is quite different. The inversion layer and, at the same time, the turbulent mixed layer (or boundary layer) experience a large diurnal cycle.[9] During the day, solar energy absorbed by land rapidly heats the surface and generates strong turbulence and mixing. The mixing erodes the stable base of the overlying inversion. The height of the inversion layer rises as turbulence and mixing continue to nibble away from below. The vertical thickness of the temperature inversion is narrowed. In the evening, without solar heating, the surface cools radiatively, and the turbulence subsides. The warm air above the cool surface layer becomes an extension of the inversion layer. The temperature inversion deepens above the colder surface layer.

5.4   Plumes of Pollution

Local pollution from industrial sources can be controlled in several ways. The total mass of the pollutant emissions can be reduced. The polluting facilities can be placed at favorable locations, from which their emissions are transported to less populated areas. The heights of the chimneys used to expel the pollutants can be raised so that less of emitted material is mixed back to the surface. Finally, the temperature of the emissions can be raised to encourage the polluted air parcels to rise higher into the atmosphere. None of these control measures, except the first, results in less pollution being emitted into the environment. Rather, they are designed mainly to decrease the local concentrations of pollutants so that fewer people will complain about them.

Wherever pollutants are emitted from a source into the atmosphere continuously over a period of time, a plume is formed. Within the plume, the concentrations of pollutants may be unhealthful or even hazardous. The plume carries pollutants from their source to various "receptors." During transit, the primary pollutants may be diluted, and secondary pollutants may appear. The concentrations of the

---

9. A *diurnal cycle* is simply a 24-hour, or daily, cycle based on the apparent movement of the sun across the sky. Sunlight and the heating produced by it are said to have a diurnal cycle.

pollutants at various receptor sites determine the effectiveness of the emission scheme. Accordingly, the properties and behavior of tracer plumes in the atmosphere under different meteorological conditions must be understood. Such knowledge allows effective emission control policies to be developed and implemented.

5.4.1 SMOKESTACK PLUMES

The polluted air rising from a chimney, and the trail it leaves downwind, constitutes a smokestack plume. The behavior of the plume effluents is of interest to the engineers who design the stacks and to the people who live near them. Two of the factors that influence the dispersion of smokestack plumes are the local atmospheric stability and the winds (Figure 5.1). The stability determines the rate of vertical mixing and dilution of the plume. The wind controls the distance that the pollution can travel and the areas that will be affected. The transport of a plume by winds is related to advection and long-range transport, as outlined in Section 5.2.3.

The local mixing of smokestack plumes is controlled by turbulence and convection. The degree of such mixing depends on the local temperature profile and lapse rate. For the purposes of our discussion, the corresponding atmospheric state may be described as stable, neutral, or unstable. As we stated earlier, a negative lapse rate, or temperature inversion (increasing temperature with increasing height), is very stable. Little turbulence or convection occurs. In moist unstable air, with a small lapse rate, intense convection and turbulent mixing may occur. If the air layer in which the plume is emitted is stable, pollutants may be inhibited from mixing either up or down. In unstable air, the plume can be shuttled by large eddies in a looping fashion. When neutrally stable, the air motions spread the plume in a uniform manner, forming a cone as it moves away from the stack.

Several common plume configurations are illustrated in Figure 5.7. The examples are probably familiar to you. From the point of view of pollution at ground level, the situations in Figure 5.7(b) and (e) are the most favorable. In these cases, the pollutants are held aloft for a long distance while they are being diluted. On the other hand, the plume behaviors depicted in Figure 5.7(c) and (d) are the worst. The pollution is trapped under an inversion in (c)

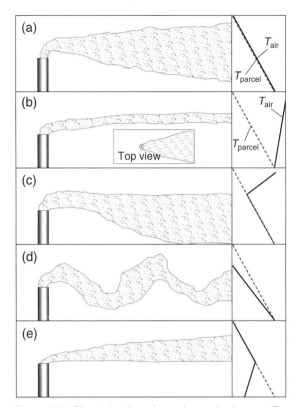

Figure 5.7 The behavior of smokestack plumes. The evolution of the plume depends on the local winds and atmospheric stability (or temperature lapse rate) and the temperature and humidity of the smoke exiting the stack. (a) When the atmosphere is neutrally stable, the plume cones, or spreads out in all directions as it travels from the stack. (b) If the atmosphere is stable, the plume cannot spread vertically but can disperse horizontally, producing a fan. (c) Fumigation occurs when stack emissions are dispersed to the ground by the overturning of the atmosphere below an inversion layer. (d) Looping occurs in an unstable atmosphere where upward and downward motions in large turbulent eddies are equally likely. (e) The injection of stack emissions above a stably stratified layer results in a lofting of the emissions. Next to each panel is a schematic of the environmental lapse rate, Γ_e (solid line), and parcel lapse rate, Γ_p (dashed line). (From David H. Slade, ed., *Meteorology and Atomic Energy 1968* [Springfield, VA: U.S. Atomic Energy Commission Office of Information Services, 1968], Figure 2.49, p. 59)

and slowly mixes downward, fumigating the surface. The situation in (d) can be the most serious for surface air pollution. Here, because the vertical transport in eddies is so rapid, high concentrations of pollutants can reach the surface near the source before being much diluted. At the same time, turbulence causes the plume to wander, sparing any one receptor

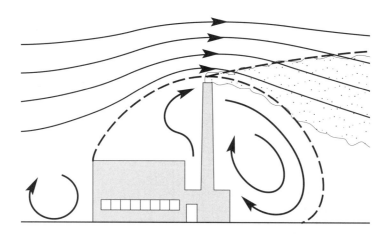

Figure 5.8 The dispersion of an individual plume depends on the micrometeorology in the vicinity of the source. The small-scale wind circulation induced by the local terrain and nearby structures can strongly influence the distribution of effluents.

site from continuous exposure. Farther away, the plume is rapidly dispersed by the vigorous mixing.

In addition to the stability of the atmosphere and wind speed and direction, several other factors influence the rate of dispersion of smokestack pollution. The exit velocity of the fumes, the temperature of the effluents, and the height of the stack are important parameters. Higher chimneys generally produce less local pollution, as the effluents take longer, on average, to reach the surface from greater heights. In the meantime, they travel farther from the source and become more dilute by the time they settle to the ground. The previous discussion of air parcel behavior stated that heated air parcels are likely to rise higher in the atmosphere under neutrally stable conditions and may even penetrate inversions in some circumstances. Accordingly, the warmer the emissions from a stack are, the longer they are likely to remain aloft. Greater exit velocities from the stack have a similar, but smaller, effect.

The dispersion of pollution from smokestacks is also affected by topography. Mountains, hills, and valleys can deflect the large-scale flow of air, creating wind channels, turbulence, and inversions. High concentrations of pollutants may accumulate against foothills downwind of a smokestack. The heating and cooling of mountainsides cause upslope winds during the day and downslope winds at night, which redistribute pollution emissions. Even the buildings around a facility can create air circulation that influences the initial dispersion of smokestack fumes. Figure 5.8 shows how such conditions might develop.

Buildings divert the prevailing wind and generate eddies of the same size as the buildings. These eddies can immediately bend a plume toward the ground, creating intense pollution at street level. In isolated areas, surface pollution might be diluted with upper level air in this way.

5.4.2 Ground Plumes

A ground plume, or cloud, is a special case of a pollution plume. Many pollutants are emitted at ground level. The sources are numerous, including automobiles, garbage dumps, and industrial holding ponds. Agricultural pollutants are usually released at the surface, and dust is raised by vehicles from the ground. Occasionally an accident releases a highly toxic compound into the atmosphere. Pesticide spills, acid leaks, and other poisonous nightmares are reported almost daily. The toxic clouds from these incidents spread as a ground plume. Figure 5.9 depicts the expected behavior of such a plume under normal atmospheric conditions. Notice that in this case, the highest pollutant concentrations are found at the surface. Even as the cloud drifts over a substantial distance, the surface-level pollution level may remain high. Accordingly, ground plumes involving toxic substances can pose a serious hazard over a large area.

The dispersal rate of a ground plume depends on the atmospheric stability and surface winds, like smokestack plumes. In an unstable atmosphere, the pollutants can be rapidly mixed vertically, thus substantially diluting their surface concentrations. If convection is occurring, the pollutants may be sucked from the surface to high altitudes relatively quickly. Stiff winds also tend to disperse the plume quickly by advection and induced mixing. The most serious threat develops when a strong, low-altitude temperature inversion is present. The toxic compounds are trapped beneath the inversion and remain at dangerous concentrations over a much larger area than if vertical mixing were active. Toxic emissions at night or in the early morning can be trapped in a very narrow layer at the surface beneath a strong, low-altitude temperature inversion. Such a layer might form, for example, as a radiation inversion.

5.4.3 URBAN HEAT ISLANDS

The construction of large cities has fundamentally changed the landscape in many regions of the world. Where forests and fields once dominated the countryside, vast expanses of concrete and asphalt now cover the land. These materials can absorb more heat and hold it much longer than trees and grass do. So it is not surprising that large urban areas are noticeably warmer than surrounding rural areas. The dif-

ference in temperature can be 6°C or more, creating an **urban heat island** effect. The heat island effect has been measured in towns as small as a large shopping mall, to the major cities of the world.

The effect on the local circulation is indicated in Figure 5.10. On a still day, the warmer air over a city rises, and the cooler air from the countryside moves into the city from all directions. The circulation produces an urban plume of pollution. The heat island effect has the advantage that pollutants in the city are raised up, as by a chimney, and may be carried off by the prevailing winds aloft.

The excess warmth of cities has several causes. As we mentioned, the concrete and asphalt in a city absorb and store a large amount of heat. This heat reservoir maintains the elevated temperature of the urban heat island. We all know that an asphalt street becomes so hot on a sunny summer day you cannot walk on it in your bare feet. In the evening, the streets remain warm long after the ground has cooled. Lizards take advantage of this effect. You can see them lounging on a rock in the late afternoon. Occasionally they find highways a bit too inviting (squish—oops!). Forests and fields also have a built-in thermostat that is absent in cities. When they get hot, the soil and vegetation release water in the form of evaporation into the atmosphere. This helps keep

Figure 5.9 A ground plume is similar in behavior to a smokestack plume but is most significantly influenced by the stability of the atmosphere near the surface. A plume at the ground tends to expand vertically and horizontally as it moves away from the source. The plume expands faster when the atmosphere is unstable and turbulent. (From Warren B. Johnson, Ralph C. Sklarew, and D. Bruce Turner, "Urban Air Quality Simulation Modeling," in *Air Pollution*, vol. 1: *Air Pollutants: Their Transformation and Transport,* ed. A. Stern [New York: Academic Press, 1976], Figure 2, p. 515)

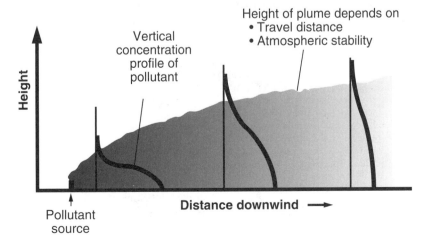

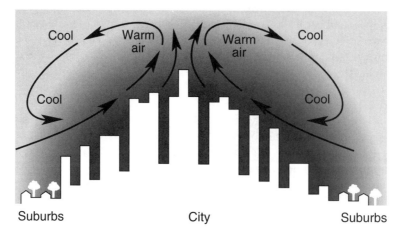

Figure 5.10 The development of a local circulation pattern over an urban heat island. Cities tend to warm up faster than the surrounding countryside during the day and then remain warmer at night. The warmer urban area induces convection over the city, affecting the local winds and weather patterns and generating a massive urban pollution plume.

the surface cool.[10] In cities, the evaporation of water is limited by pipes and sewers. Excess energy is also generated in cities. All the electrical power used, all the gasoline burned, and all the fuel consumed by industry are converted to heat. Air conditioning cools the inside of your apartment or office, but the heat removed is dumped outdoors, along with the excess heat generated by the compressor. All these factors and others make cities hotter, smoggier places to live than the suburbs are.

5.5   Regional Dispersion of Pollutants

The previous section focused on the small-scale (local) dispersion of pollution in the form of individual plumes. These plumes can be as large as a small city. Many pollution problems involve large urban complexes or even entire regions. The Los Angeles basin is a major regional conglomeration of cities and

counties with the worst air pollution in the United States. Worldwide, however, there are many cities with equally bad or even worse air pollution—Cairo, Mexico City, Santiago, and Tokyo, to name a few. Los Angeles will be used as a reference for studying regional air pollution and, later in this book, smog. Los Angeles is chosen not because it has the most severe examples of air pollution, but because, there the air pollution has been so widely investigated, is so well documented, and has been so intensively subjected to measures for control.

Los Angeles is topographically and meteorologically interesting from the point of view of air pollution. Sitting on a western coastline, surrounded by mountain ranges, and subject to regional temperature inversions most of the time, Los Angeles has most of the features that contribute to air pollution.

5.5.1   In Coastal Zones

The weather in coastal zones usually has a strong component of the sea/land breeze. The formation of a marine inversion and its relationship to the sea/land breeze were discussed in Section 5.3.3. In places like Los Angeles, the sea breeze, along with mountain winds, dominates the local circulation during episodes of heavy pollution. The cool, moist air in the marine layer initially has a temperature and

10.   The trick is in the latent heat of evaporation of water. It takes a certain amount of energy to change liquid water into water vapor; this is called the *latent heat of evaporation*. Water can evaporate rapidly when you boil it on a stove because the energy supplied by the stove is easily converted to latent heat of evaporation. The latent heat drawn from surface materials when water evaporates produces a cooling effect at the surface. We perspire during exercise because our bodies are attempting to cool us down by evaporating sweat. When water vapor condenses in the atmosphere (as fog, clouds, and rain), the same latent heat is released, warming the air wherever condensation occurs (Section 5.2.1).

humidity characteristic of the ocean surface above which the air has traveled for many days. This air soon is heated by the land and injected with a variety of gaseous and particulate pollutants. Figure 5.11 shows measurements of the polluted cloud of air over the Los Angeles basin on a typical summer day. The pollution is indicated by the particulate content of the air in this case, although it would look almost the same if carbon monoxide or ozone concentrations were shown. Notice that the smog layer is nearly completely contained below the marine inversion layer, which is capped by a large-scale subsidence inversion.

The regional pollution plume in Figure 5.11 is moving inland at the time shown (early afternoon), pushed from the ocean side by the sea breeze, and pulled from the land side by upslope mountain winds. By late afternoon, the smoggy air will engulf Riverside and Redlands. The movement of the air pollution resembles a massive ground cloud (Figure 5.9) confined by the inversion layer from above. The smog front moves like a weather front across the landscape, engulfing all before it. This amorphous plume, however, is actually the aggregate of thousands of individual plumes from point sources and distributed sources, all commingled and cooked by the sun. The further evolution of a coastal plume of this sort is greatly affected by the coastal mountains and other topographical features.

5.5.2 NEAR MOUNTAIN BARRIERS

Mountains can block the movement of regional plumes, and the western coasts of North America and South America are lined with mountains. With the prevailing winds blowing off the ocean, the mountains make a perfect trap for coastal pollution. Western cities like Los Angeles huddled against the picturesque mountains are prone to pollution stagnation. Indeed, the smog is usually so dense, and the corresponding visibility so low (Section 6.5), that the mountains are rarely seen even by local residents.

Figure 5.12 shows the evolution of the regional air pollution plume over the Los Angeles basin, three principal stages in the development, movement, and dissipation of the plume. The entire process is very complex (even without considering the actual composition and chemistry of the smog). The plume is controlled by the atmospheric temperature structure (lapse rates and inversions), the diurnal cycle of solar heating and the corresponding response of the mixed layer, the distribution of pollutant sources, the sea/land breeze cycle, and the regional effects of coastline and mountains.

In the early stage of development, the polluted air accumulates in a narrow stable boundary layer. The impinging sunlight warms the layer, induces a sea breeze, and generates turbulence that causes the

Figure 5.11 A layer of pollution in the Los Angeles basin at midday. The air pollutants are trapped and concentrated below a coastal inversion layer. The pollution cloud moves inland during the day, pushed by onshore sea breezes. The relative concentration of light-scattering aerosols is indicated by the shading, from high concentrations (dark shading) to low concentrations (light shading). The observations were made in July 1973, during the early afternoon. (Data from the Environmental Protection Agency, 1973)

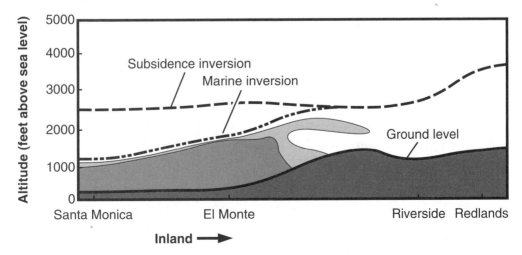

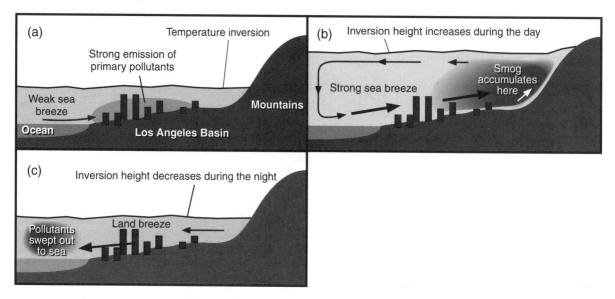

Figure 5.12   The life cycle of a pollution episode in the Los Angeles basin. The abundance and persistence of the pollution are controlled by the interaction among the numerous pollutant sources, the coastal breezes, and the local topography. (a) In the morning, pollutants are emitted in the coastal regions but are confined near the surface by the low morning inversion layer. (b) By midafternoon, ozone and other secondary pollutants have formed and move inland under the influence of strong sea breezes induced by the heating of the land and inland mountains. The inversion layer is also pushed up by the surface heating and increased turbulence in the boundary layer. (c) In the evening, the ground cools, and the inversion level drops. Pollutants left above the inversion may be swept away by upper-level prevailing winds, and the pollution trapped below the inversion may be blown to the coastal regions by the offshore land breeze.

boundary layer to expand upward. Sunlight also initiates the formation of photochemical smog (Chapter 6). The sea breeze and upslope winds transport the plume inland, where it stagnates against the mountain barriers. In the evening, radiative cooling of the surface causes the inversion level to drop and creates downslope winds and land breezes that can reposition the plume over the coastal region. Much of the daily pollution can escape the basin by leaking through mountain passes or being swept away at higher altitudes as the inversion layer deepens (pollution that had mixed upward is stabilized in the inversion layer and is subsequently advected away by the winds associated with the large-scale subsidence).

Predicting the formation, movement, and dispersion of air pollution is obviously a challenging problem. To date, no one can do it with any precision. Teams of scientists working on the biggest computers regularly make fairly accurate regional forecasts. Predicting the movement of pollutants—ranging from the smoggy palls that hang over many large cities, to dense ground plumes of highly toxic compounds escaping from ruptured storage vessels—requires comprehensive simulations of local and regional meteorology. We are still working on that.

Questions

1. Describe four processes that disperse pollutants in the atmosphere. Which process is least effective in dispersing pollutants? Which process may be most effective in diluting local pollution? Pollution on a regional scale?

2. Explain why you might want to increase the temperature of the gases emitted from a chimney in order to reduce the concentrations of pollutants near the ground downwind of the chimney.

3. When an idealized air parcel rises and expands adiabatically, which of the following characteristics of the air inside the parcel remain unchanged: (a) temperature, (b) pressure, (c) density, (d) total mass of air, (e) the concentration of a specific tracer in the air, (f) the mixing ratio of that tracer, or (g) the density of that tracer? Explain your choices.

4. What is meant by the term *temperature lapse rate*? Flying in an airplane, you notice that as your altitude increases, the temperature outside also increases. What is the atmospheric condition you are experiencing? Describe the corresponding temperature lapse rate.

5. You happen to live near a chemical plant manufacturing a deadly volatile pesticide that is stored in huge tanks nearby. What measurements could you take on a daily basis to ensure that if an accident occurred at the plant, you would be able to assess your personal danger in time to evacuate your home?

6. Why is a temperature inversion inherently stable against vertical motions in the atmosphere?

7. Describe the events leading to the formation of a radiation inversion. Why are such inversions rarely seen along western coastlines?

8. You wish to have a nice evening fire at home in your fireplace, but you do not want to disturb your neighbor with smoke. What information would you need to ensure that your neighbor will not have to choke on your chimney smoke?

9. You stir up a bowl of water with an eggbeater and then stop. Just after you stop, the water is agitated, but this turbulence quickly dies away. You whip a bowl of cream with the same eggbeater and stop. What difference in the response of the whipped cream from that of the water do you expect to see when you stop? Can you explain the reason for the difference?

10. You open a smelly can of tuna fish in the kitchen, and your dozing cat immediately dashes in hoping for the juice. Did the cat get wind of the meal because the molecules of tuna fish aroma diffused into the living room from the kitchen? Because the molecules were carried there by turbulent mixing of the air? Is something else going on?

Suggested Readings

Battan, L. *The Unclean Sky.* Garden City, N.Y.: Doubleday, 1966.

Gleick, J. *Chaos: Making a New Science.* New York: Penguin Books, 1987.

Howard, L. *Climate of London Deduced from Meteorological Observations.* 3rd ed. London: Harvey and Darton, 1833.

Landsberg, H. *The Urban Climate.* New York: Academic Press, 1981.

Stern, A., ed. *Air Pollution.* Vol. 1: *Air Pollutants: Their Transformation and Transport,* 3rd ed. New York: Academic Press, 1976.

6

Smog: The Urban Syndrome

Cities in developing and developed countries alike are being smothered by soupy clouds of tainted air. As civilization continues its industrial development and population expands, the production and emission of unhealthful compounds into the environment accelerate. In some places, restraints are placed on polluters. But there are too many places where the laws discouraging pollution are weak or the enforcement of laws is halfhearted. The serious polluters escape to these places, or they relieve themselves of polluting waste under cover of darkness. When economic factors are considered, pollution control takes a back seat to business development and jobs. Millions of subsistence farmers leave the land and flock to cities in search of menial work in hazardous air. Facilities and infrastructure lag behind the needs of these masses. Manufacturing is done under primitive conditions in which pollution abatement is not a priority. Developed countries export the most hazardous industries to the Third World. The eastern European nations of the former Soviet bloc struggle with inefficient, highly polluting facilities to earn a marginal living. Sophisticated technological fabrication introduces new pollutants with unknown long-term effects. Minute quantities of dangerous compounds are detected in the atmosphere, water, and soil.

Urban air pollution, or **smog**, is not a recent problem or the most serious environmental threat that we now face. However, pollution in cities affects hundreds of millions of people, leading to illness and general malaise. Santiago, Chile, for example, is one of the smoggiest cities on Earth. Santiago has a coastal mountain topography extremely well suited to trapping smog (Section 5.5). Almost two decades of communist rule with disregard for controls on development and pollution led to the predicament. The largest polluters in the city are privately owned buses, once encouraged by the government in order to provide cheap transportation. The noxious diesel

fumes emitted by the unregulated fleet of 11,000 buses has created smog so dangerous that children are often kept indoors to prevent their breathing the air. Only recently have new policies been adopted to remove the oldest buses from service, but Santiago still remains shrouded in a pall of bad air.

6.1 The History of Smog

When small groups of humans roamed the Earth as hunter-gatherers, there were no cities and little air pollution. To be sure, there were many natural sources of gases and particles that would be considered pollutants if emitted from industrial facilities today. Organic vapors from trees, smoke from vegetation fires, sulfur fumes from volcanic vents, and fetid vapors from swamplands all cause natural air pollution. Unless one is unlucky enough to live next to a stinking swamp or an erupting volcano, however, natural pollution is generally a minor bother. Rather, the pollution we are concerned with nowadays, which evolved with human industry and technology, is dense and dangerous and, for millions of us, unavoidable.

When agriculture developed into a major human enterprise, villages and towns sprang up as permanent residences for most people. As these embryonic cities grew, the waste from human activities and its disposal became a serious problem. Even when the Spanish explorer Juan Rodríguez Cabrillo first set anchor in October 1542 in San Pedro bay near the present site of Los Angeles, the air was clouded by the smoke from Native American campfires. He named the harbor the Bay of Smoke. Although the ravaging effects of air pollution were known well before that time, the problem had not yet been systematically described. In ancient Rome, the blackening of buildings by smoke from wood fires was noted in passing by the chronicler Horace (b. 65 b.c.). Seneca (b. 3 b.c.), Nero's tutor, noticed that his health improved markedly once he left the "oppressive fumes and culinary odors" of Rome.

Coal may have been used as a fuel by the Chinese as early as 1000 b.c. It was identified and used in Europe by at least around A.D. 1200. The early natural scientist Moses Maimonides wrote of the poor quality of air in cities even in the twelfth century:

> Comparing the air of cities to the air of deserts and arid lands is like comparing waters that are befouled and turbid to waters that are fine and pure. In the city, because of the height of its buildings, the narrowness of its streets, and all that pours forth from its inhabitants and their superfluities . . . the air becomes stagnant, turbid, thick, misty and foggy . . . If there is no choice in this matter, for we have grown up in cities and have become accustomed to them, you should select from the cities one of open horizons . . . endeavor at least to dwell at the outskirts of the city.
>
> . . .[I]f the air is altered ever so slightly, the state of the Psychic Spirit will be altered perceptibly. Therefore you find many men in whom you can notice defects in the actions of the psyche with the spoilage of the air, namely, that they develop dullness of understanding, failure of intelligence and defect of memory.[1]

The extensive burning of coal did not begin until the early eighteenth century, with the discovery of a process for making coke (another solid form of coal) and coal gas.[2] Nevertheless, Eleanor, queen consort of King Henry III of England, reportedly complained around 1250 about the pollution created by the burning of coal. King Edward I, the son of Henry III, later (ca. 1300) issued a proclamation against the use of coal (presumably around the palace where it upset his wife, another Eleanor): "Be it known to all within the sound of my voice, whosoever shall be found guilty of burning coal shall suffer the loss of his head."

The use of coal accelerated despite Edward's dissatisfaction with its side effects. In his classic work

Fumifugium, or *The Inconvenience of the Air and Smoke of London Dissipated* (1661), John Evelyn described the air quality in seventeenth-century English cities:

> It is this horrid smoake which obscures our church and makes our palaces look old, which fouls our cloth and corrupts the waters, so as the very rain, and refreshing dews which fall in the several seasons, precipitate to impure vapour, which, with its black and tenacious quality, spots and contaminates whatever is exposed to it.
>
> . . .[I]t is evident to every one who looks on the yearly bill of mortality, that near half the children that are born and bred in London die under two years of age (a child born in a country village has an even chance of living near forty years). Some have attributed this amazing destruction to luxury and the abuse of spiritous liquors. These, no doubt, are powerful assistants; but the constant and unremitting poison is communicated by the foul air, which, as the town still grows larger, has made regular and steady advances in its fatal influence.[3]

In fact, the use of coal was the backbone of the **Industrial Revolution**, which began in mid-eighteenth century England. The rapid expansion of manufacturing based on steam energy generated from coal combustion was the early hallmark of the revolution. **Steam** is nothing more than heated water vapor. James Watt invented an engine that could be driven by steam, which expanded into piston chambers, much as burning gasoline expands in the cylinders of an internal-combustion engine. Naturally, the levels and extent of the accompanying air pollution also rose dramatically.

6.1.1 AIR POLLUTION AND POETS

In the social and business culture of the Industrial Revolution, which fostered sweat shops and child exploitation, the environmental effects of industrial pollutants could be completely disregarded. Degeneration in the quality of life was noticeable to all, however. Over the years, writers and poets have captured the feeling of grayness and depression during this period. William Shakespeare himself serendipitously expressed an early opinion on the

---

1. Moses Maimonides (b.1135 in Cordoba, Spain; d. 1204 in Egypt) was a Jewish philosopher, jurist, and physician. He wrote a classic code of Jewish law, *The Guide of the Perplexed*. He contributed to both science and religion in his lifetime. Quotation is from V. Goodhill, "Maimonides—Modern Medical Relevance," *Transactions of the American Academy of Opthalmology*.

2. When coal, especially bituminous coal, is heated in the absence of air, the volatile components are given off as gases, consisting of methane, hydrogen, and carbon monoxide. The residual solid, coke, is almost pure carbon. Coal gas can be employed as a fuel and in the past was used for illumination.

3. John Evelyn (1620–1706) was a country gentleman who chronicled life in England during his life. He wrote 30 books, treatises, and discourses on various subjects, including forest silviculture and stamps, and kept a diary, later published, covering the last 50 years of his life.

state of the atmosphere and sky in English cities before the Industrial Revolution: "This most excellent canopy, the air, look you, this brave o'erhanging firmament, this majestical roof fretted with golden fire, why, it appears no other thing to me than a foul and pestilent congregation of vapours" (*Hamlet*, Act 2).

The great poet Percy Bysshe Shelley,[4] who lived during the early period of the Industrial Revolution, was very explicit in his description of London at the time:

> Hell is a city much like London—
> a populous and smoky city.

> *Peter Bell the Third*, Part III, Stanza I

Similarly, William Morris (1834–1896) contrasted two visions of London:

> Forget six counties overhung with smoke,
> Forget the snorting steam and piston stroke,
> Forget the spreading of the hideous town;
> Think rather of the pack-horse on the down,
> And dream of London, small, and white, and clean.

> Prologue to *The Earthly Paradise*

By the mid-nineteenth century, the Industrial Revolution had spread to the rest of Europe and the United States. The building of machines and the introduction of new technologies and materials picked up speed, and the degradation of the environment did as well. Listen to James B. Dollard (1872–1936) from the countryside in Scotland:

> I'm sick o' New York City an' the roarin' o' the thrains
>   That rowl above the blessed roofs an' undernaith the dhrains;
> Wid dust an' smoke an' divilmint I'm moidhered head an' brains,
>   An' I'm thinkin' o' the skies of ould Kilkinny!

> *"Ould Kilkinny!"*

But every dark cloud has its silver lining. So William Henry Davies (1871–1940) wryly noted the pleasing side effects of air and water pollution:

> What glorious sunsets have their birth
>   In cities fouled by smoke!
> This tree—whose roots are in a drain—
>   Becomes the greenest oak!

> *"Love's Rivals"*

One of the leading thinkers of the twentieth century, Buckminster Fuller,[5] grasped the true nature of modern-day pollution of the environment by recognizing that "pollution is nothing but resources we're not harvesting."

6.1.2 LONDON SMOG

Beginning in the mid-nineteenth century and extending through the first half of the twentieth, the major cities of Europe and the United States experienced episodes of choking air pollution associated with the burning of coal to generate heat and energy. Thousands of people died as a result of exposure to these toxic palls. The most serious event occurred in London in December 1952. The British Isles were capped by a large-scale temperature inversion (Section 5.3.3) and blanketed in dense fog. For five days, from December 5 to 9, air pollutants accumulated in the Thames River valley in stagnant air. About 4000 excess deaths were attributed to the inhalation of smoke, sulfurous particles, and soot mixed with fog.[6] Most of the victims suffered respiratory and heart failure. Thousands of others, especially asthmatics and people with bronchitis and other respiratory ailments, were left gasping for oxygen. The very young and very old were most vulnerable.

London had experienced killer pollution events in December 1873 (1150 dead), January 1880, February 1882, and December 1891. All these tragedies had one thing in common: The deadly conditions were precipitated by the combination of a stagnant fog and the smoky emissions of coal. In 1905, Harold Antoine des Voeux, a medical doctor, first

---

5. Richard Buckminster Fuller (1895–1983) never formally completed college although he was nonetheless accomplished at architecture, engineering, philosophy, cartography, and poetry. He invented the geodesic dome, the only known structural design that can be scaled to any size without collapsing. Following the death of his daughter at the age of 4 from influenza and polio, he dedicated his remaining years to designing environmentally safe and efficient technologies and industries. He engineered the first streamlined car with omnidirectional steering (you could park sideways) and fully surrounding bumpers (like a "bumper car"), the Dymaxion. His 1943 design promised 40 to 50 miles per gallon of gasoline, but the car was never produced. Fuller felt that a "comprehensive and anticipatory" approach to design, demonstrated by the Dymaxion, could solve the world's problems of housing, hunger, transportation, and pollution.

6. The term *excess deaths* refers to the additional number of fatalities counted above the number expected under otherwise normal conditions. The latter number is determined using statistics on mortality rates during normal periods.

4. Percy Shelley's wife, Mary Wollstonecraft Shelley, wrote the famous novel *Frankenstein* in 1816 (Section 11.6.4).

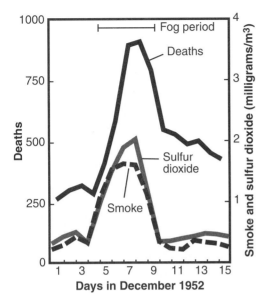

Figure 6.1 Measurements of the concentrations of smoke and sulfur dioxide during the great London smog episode of December 1952. The concentrations are shown on a daily basis during the episode. The "excess" deaths each day in London attributed to exposure to the smog also are indicated. (From Helmut E. Landsberg, *The Urban Climate* [New York: Academic Press, 1981], Figure 10.5, p. 237)

used the term smog, which combines the words smoke and fog to describe the dark palls he observed hanging over many British towns. The term became popular when he published a report in 1911 on a killer smog episode in Glasgow, Scotland, which in 1909 killed 1063 residents. We will refer to this type of urban pollution as **London smog**.

As the Industrial Revolution spread, so did the killer London smog episodes. A notable nasty pall developed in the Meuse Valley, Belgium, from December 1 to 5, 1930, resulting in 63 deaths and general misery for many others. The first major smog episode in the United States occurred in Donora, Pennsylvania, from October 26 to 31, 1948. In this incident, 20 excess deaths were recorded, and 43 percent of the population fell ill. Poza Rica, Mexico, suffered 22 deaths in November 1950 when hydrogen sulfide gas escaped from a natural gas facility under temperature inversion conditions similar to those that cause London smog. New York City has had a number of serious smog events, including one from November 12 to 22, 1953, that engulfed the entire metropolitan area with less severe but very widespread effects. In the November 24 to 30, 1966, tragedy, 168 people were recorded as victims of

smog in the New York area. In all these incidents, thousands of persons became seriously ill but recovered. The long-term damage to their health and the health of other millions exposed to the pollution can never be assessed.

The typical London smog results from the accumulation of smoke from coal burning. This smoke has a high sulfur content and leads to the production of high concentrations of sulfuric acid in fog droplets. These acidic particles, along with high densities of smoke, inhibit the normal functioning of the lungs. The symptoms include chest constriction, difficulty in breathing, headache, nausea, vomiting, and eye, ear, nose, and throat irritation. Figure 6.1 shows the buildup of smoke and sulfur dioxide (the precursor to sulfuric acid [Section 6.5.3 and Chapter 9]) during the great 1952 London smog. The episode corresponded to an extended period of fog, during which the death rate (excess deaths per day) soared. Effective legislation to control smoke emissions throughout Great Britain was passed in 1956.

6.1.3 LOS ANGELES SMOG

Today the term *smog* is used to describe another type of air pollution experienced in many cities around the world. This smog is not derived mainly from smoke and fog; rather, the emissions of automobiles and other vehicles are the primary cause. This different kind of smog forms when the meteorological conditions are right—that is, in stagnant air capped by a strong temperature inversion and illuminated by plenty of sunlight. Cities like Los Angeles are ideal for the production of this type of smog. In addition to the temperature inversions and sunlight (Section 5.5), Los Angeles is overrun with cars. The mixture of ingredients that come from automobiles reacts in the presence of sunlight to create high ozone concentrations and brown haze. Accordingly, this pollution is known as **photochemical smog**. We will also refer to it as **Los Angeles smog**, since this was the place where such smog was originally identified.

Los Angeles developed an early love affair with the automobile, installing the country's first automatic traffic signals in 1922. The potential problems associated with photochemical smog began to appear there in the early 1940s. The haze already common in the Los Angeles basin started to thicken. Catalina Island, off the coast, and the majestic San Bernardino and San Gabriel Mountains disappeared from view more frequently. Agricultural crops began to show signs of damage, particularly the bronzing of foliage, which was most noticeable on parsley. The rubber in car tires and tubing showed premature aging and cracking. According to newspaper accounts, citizens were weeping, sneezing, coughing, and complaining. By 1947, the problem was considered serious enough to set up the first Los Angeles Air Pollution Control District.

Scientific studies revealed that some component of the polluted air was damaging crops. Although the compound could not be positively identified, it was proved not to be sulfur dioxide (the key ingredient of London smog). In 1951, Arie Haagen-Smit, working in Los Angeles, showed through laboratory simulations that mixtures of hydrocarbon vapors and ozone smell the same as the smog does, and cause leaves to bronze in the same way. He suggested in 1952 that ozone actually forms in air containing hydrocarbons and nitrogen oxides. Further experiments carried out soon thereafter confirmed the Haagen-Smit theory of photochemical smog formation.[7]

In fact, ozone had been detected in urban air during the second half of the nineteenth century. Christian Schonbein, who first identified ozone as a form of oxygen, also designed a crude ozone detector. The instrument, called an *ozonometer*, consisted of a specially treated paper strip that reacted with ozone, causing its color to bleach out. The amount of fading determined the ozone concentration, which Schonbein calibrated in his laboratory. These novel ozonometers soon appeared all around Europe. One of them, situated near Paris, provided a 30-year record of ozone concentrations between 1876 and 1907. During this period, the average ozone concentration in that locale was about 10 parts per billion by volume (ppbv). That amount is comparable to the abundance of ozone in "clean," or natural, tropospheric air today. By the mid-twentieth century, average ozone concentrations had increased in rural Europe to 20 to 30 ppbv, still below the average concentrations of 100 to 200 ppbv measured in heavily polluted air. Of course, Schonbein never figured out how ozone was actually formed in

the atmosphere. More sophisticated analytical instruments and some imaginative thinking on the part of Haagen-Smit were needed.

The recognition of the cause of smog in Los Angeles did not stem the tide of bad air. After World War II, in fact, with population and industry in the area booming, air pollution soared. By 1958, Los Angeles was experiencing 219 days with stage-1 smog alerts (Section 6.2.2). At such concentrations, air pollution is considered to be hazardous to health.

6.2   Primary and Secondary Pollutants

In Section 5.1.1, we defined several categories of pollutants. The gaseous compounds are referred to as primary if they are directly emitted into the atmosphere and secondary if they are generated in the atmosphere (from primary pollutants). Normally, gases and aerosols are differentiated as well (gaseous and particulate pollutants). Urban smog is usually characterized by heavy concentrations of both primary and secondary pollutants and particulates as well as vapors. The most offending gaseous components are usually irritants or toxicants. The particulates are irritating and unhealthful and also degrade visibility. In addition, soupy palls over polluted cities, caused mainly by particulates, can have a significant negative psychological impact when they persist over entire seasons.

6.2.1   THE BASIC INGREDIENTS

The principal sources of primary pollutants for London smog are coal-burning power plants. In many parts of the world, residents burn coal, peat, wood, or dung for heating and cooking. When burned inefficiently, these fuels generate large amounts of smoke and organic fumes, and their high sulfur content leads to high levels of sulfur dioxide. The result can be acid fog (Section 6.5.3) and, at a greater distance, acid rain (Chapter 9). In most cases, there is a substantial contribution to local smog from smoke, sulfur, and other pollutants from power plants, oil refineries, fireplaces, and so on. In many of the most polluted cities, however, the major contribution to the overall smog problem is the formation of photochemical smog.

A major source of primary pollutants for urban photochemical smog is automobile exhaust. Com-

---

7. Arie Haagen-Smit is considered to be the "father of photochemical smog." He not only discovered the cause of the Los Angeles air pollution, but also campaigned vigorously to see that tough rules to control smog were put into effect. Although strongly resisted by special interests and hindered by public apathy, he eventually prevailed.

bined with intense sunlight and strong inversions, these vehicle emissions can lead to a dense toxic smog. The cities suffering the most from photochemical smog—Los Angeles, Mexico City, Santiago, Denver—all have favorable demographics, topography, and weather for photochemical smog. Without the chemical ingredients in the exhaust from internal-combustion engines, the density and frequency of smog would be much lower.

The internal-combustion gasoline-fueled engine operates at temperatures that are high enough to generate nitrogen oxides from the nitrogen and oxygen in air (Equation 3.41). The resulting NO_x is a key ingredient of smog. These engines also produce a large amount of carbon monoxide in the fuel-rich environment of the combustion chamber. If the hydrocarbon fuel (gasoline, diesel) were mixed with exactly the right amount of oxygen (in air) and burned to completion, the only significant by-products would be carbon *di*oxide (CO_2) and water (H_2O). This is called **stoichiometric combustion**. Engines that are not perfectly tuned emit a fraction of the fuel as carbon *mon*oxide (CO) and unburned reactive hydrocarbons (RH). The atmosphere already has very large concentrations of CO_2 and H_2O and so the vehicular sources are a small perturbation on the background.[8] The amounts of CO, NO, and RH emitted are large, however, compared with the background concentrations.

The main pollutants emanating from vehicle exhaust pipes are thus carbon monoxide, nitrogen oxides (NO and NO_2), and reactive hydrocarbons. Measures have been taken to reduce the emissions of these offending compounds. Catalytic converters are placed in the exhaust stream (usually in the muffler) to convert CO to CO_2. The reactive hydrocarbon emissions are controlled by keeping the engine tuned to burn fuel most efficiently (stoichiometrically). The nitrogen oxides can be reduced by designing engines to run at lower temperatures and by placing catalysts in the exhaust stream to convert NO and CO to N_2 and CO_2. In California, engines must be periodically tested to see that they meet the emission standards.

The reactive hydrocarbons (RH) have many sources besides automobile exhaust. Gasoline vapors

that escape from gas pumps and automobile tanks or that evaporate from storage or processing facilities contribute to smog-forming RH. Organic vapors from solvents used in painting and as cleaning agents are an important source. Many household products—for example, hair spray, pesticides, and cleaning fluids—emit some reactive hydrocarbons. Even cooking releases hydrocarbons that can form smog. Similarly, living plants emit hydrocarbons that are reactive. These natural sources of organic compounds are a normal part of the global carbon cycle (Section 10.2.4). Even in the absence of anthropogenic RH emissions, the background concentrations of hydrocarbons in many places are high enough to create substantial smog from NO emissions alone. Experiments have shown that many tree species—such as the common liquidambar, which adds color to an otherwise seasonless California autumn—are heavy emitters of hydrocarbons, mainly isoprene.[9] Other species of trees, such as the crape myrtle, camphor, and certain pines and cedars, emit low amounts of RH and are environmentally safer for cities.[10]

In addition to the primary gaseous pollutants, a rogue's gallery of particulates are injected into the atmosphere by vehicles, industry, and other sources. These primary aerosol particles have a number of important effects: They directly degrade visibility, one of the key problems related to smog (Section 6.5). The particles commonly contain dangerous toxins, which can be carried deep into the lungs with the particles (Chapter 7). The primary aerosols also act as a collection site for other substances that, if inhaled, can be unhealthful. Primary particles are generated from a variety of sources, including smoke from combustion processes and dust raised from roadways by vehicles. The dust may consist of silicate minerals, particles of rubber from car tires, and countless other materials.

8. The accumulation of the carbon dioxide emitted by hundreds of millions of vehicles contributes to the global buildup of atmospheric CO_2 over a period of many decades. This is the carbon dioxide greenhouse warming problem discussed in Chapter 12.

9. Among the tree species that emit reactive hydrocarbons in the largest amounts are the carrotwood and liquidambar. Two carrotwood trees can emit as much RH in a day as an automobile on a commute to work. One estimate places the total hydrocarbon emissions from trees in the Los Angeles basin at about 150 tonnes daily. That represents roughly 10 percent of the total primary emissions of hydrocarbons in Los Angeles.

10. Trees are always pleasant to have around and help control pollution. They are generally very pleasing to look at and smell good. Trees provide shade on hot days; by evaporating water, they also tend to cool the adjacent area. The leaves of trees collect smog particles and dust, making the air cleaner. Trees provide a home for birds, squirrels, and other lovable creatures and a place for kids to hang out. Without trees, cities would be ugly, gray, sterile prisons.

Vehicles release important specific pollutants in addition to CO, NO, and RH. Most fossil-derived fuels contain sulfur as a contaminant. Gasoline has a relatively small amount of sulfur, typically 0.1 percent or less by weight. By comparison, coal may have a high percentage of sulfur by weight, and raw petroleum contains 1 percent sulfur on average. When petroleum is refined into gasoline, most of the sulfur is removed. And when gasoline is burned, the residual sulfur is emitted primarily as sulfur dioxide, which is later converted to sulfuric acid and sulfates (Sections 3.3.4 and 6.5.3). These materials condense onto the primary particles already injected into the atmosphere.

Diesel engines use a cheaper, lower grade of fuel and run at higher temperatures than gasoline engines do. These engines are notorious soot generators. The soot consists of black greasy carbonaceous material in the form of very small particles.[11] Especially when cold started or accelerated, diesel-powered vehicles spew large amounts of this soot into the air. It can be seen as a dark cloud coming from trucks, buses, and some cars. The soot fouls everything it lands on. Many dangerous organic chemicals are attached to the soot particle surfaces and can be ingested with the soot (Chapter 7). Indeed, soot emissions are likely to be one of the key urban air pollution issues for the next decade (Section 6.6).

Until recently, gasoline contained trace amounts of lead as an additive to make engines run smoother. The lead is emitted as particulate that can be inhaled or ingested. The effects of lead ingestion include brain damage and memory loss, and so in most countries, gasoline is no longer produced with lead.

The quantities of primary pollutants released into the Los Angeles basin each day currently amount to about 5000 metric tons (tonnes) of CO, 1100 tonnes of NO, 1400 tonnes of reactive hydrocar-

bons, 130 tonnes of SO_2 and 1100 tonnes of particulates. Of these amounts, all stationary sources (power plants, industries) account for roughly 33 percent of the CO and NO emissions, 50 percent of the RH source, 66 percent of the SO_2 output, and up to 90 percent of the particulate release. Mobile sources (automobiles, trucks, buses) emit the rest. These primary emissions will be discussed further in later sections.

Smog Formation and By-Products

The basic chemistry of smog formation is described in Section 3.3.4. The key initiating reaction involves the hydroxyl radical (OH) and the reactive hydrocarbons (RH). The overall process of smog generation can then be summarized as[12]

$$RH + OH + NO \rightarrow \overset{h\nu+O_2}{\cdots} \rightarrow O_3 + NO_2 + HC \quad (6.1)$$

Both sunlight and oxygen are essential to this process, and of course, oxygen is readily available in air. Sunlight is also abundant in places like Los Angeles. The products shown are the principal secondary products composing smog, but there are many intermediate chemical species not shown. These species usually have lower concentrations and short atmospheric residence times.

The fate of carbon monoxide is relatively simple. CO reacts with hydroxyl and is converted to carbon dioxide:

$$CO + OH \rightarrow CO_2 + H \quad (6.2)$$

The atmospheric residence time of CO depends on the concentration of OH. In heavily polluted air, OH concentrations can be greatly elevated over the background abundances. Thus although carbon monoxide may have a residence time of about 1 month in the free atmosphere, in heavily polluted air it may be only a few days. Nevertheless, that is long enough for most of the CO to disperse in the winds.

11. When fossil fuels are burned with too little oxygen (or "fuel-rich"), soot is produced. This condition occurs in engines that are not properly tuned or when the accelerator is quickly depressed, flooding the engine with fuel. At flame temperatures, excess unburned organic vapors can lose their hydrogen as they condense to form a high-carbon solid much like graphite in microstructure. Toxic organic compounds generated during combustion tend to condense on the soot surfaces as the exhaust vapors cool. Soot is also produced in a fireplace or candle flame. Every now and then, you should have your chimney swept to remove the accumulated soot, which may burn. You can generate soot in a candle flame by placing a cold spoon in the flame. The spoon draws off some of the heat, cooling the flame so that the soot is not burned in the outer regions of the flame zone.

12. Equation 6.1 is somewhat different from the overall smog-formation process defined in Chapter 3, Equation 3.57. In the latter case, the key ingredients are shown as reactive hydrocarbons, nitric oxide, and sunlight ($h\nu$), with oxygen as an intermediate ingredient. Equation 6.1 is another representation of the same process that recognizes the significance of the initiating chemical reaction involving OH.

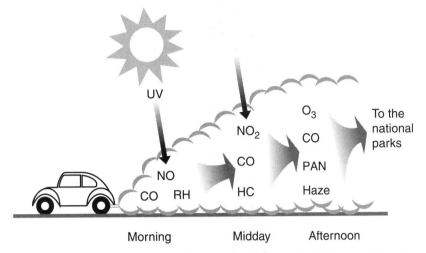

Figure 6.2 Automobiles are the primary cause of photochemical smog. The primary emissions from automobiles, NO, CO, and reactive hydrocarbons (RH), are converted by sunlight to NO$_2$, ozone, and a variety of other pollutants during the day.

The lifetime of carbon monoxide is also long enough that its concentration may be used to track the spread of a polluted air mass.

The key products of the smog-formation process are identified in Equation 6.1. These secondary pollutants include ozone (O$_3$), nitrogen dioxide (NO$_2$), and a variety of hydrocarbons (lumped together as HC). In addition, the secondary particulates generated in smog include small droplets of concentrated sulfuric acid (H$_2$SO$_4$) and haze aerosols composed of low-volatility hydrocarbons that tend to condense.

The ozone is not formed directly but is a result of nitrogen dioxide production. The relevant reactions are

$$RO_2\bullet + NO \rightarrow RO\bullet + NO_2 \qquad (6.3)$$

$$NO_2 + h\nu \rightarrow NO + O \qquad (6.4)$$

$$O + O_2 + M \rightarrow O_3 + M \qquad (6.5)$$

Here, the RO$_2\bullet$ is created by reactions of RH, as described in Section 3.3.4. The nitrogen dioxide is easily photodissociated by sunlight (Equation 6.4). On average, an NO$_2$ molecule is broken apart by photodissociation once every 2 minutes or so. The nitric oxide left behind (Equation 6.4) may later react with ozone, reclaiming an oxygen atom originally detached from nitrogen dioxide by means of photodissociation:

$$NO + O_3 \rightarrow NO_2 + O_2 \qquad (6.6)$$

The sequence of Equations 6.4, 6.5, and 6.6 establishes a balance among the concentrations of NO, NO$_2$, and O$_3$ during the day.

The two most prevalent secondary pollutants in smog are nitrogen dioxide and ozone. Both of these gases are hazardous in high concentrations and must be monitored and controlled in smog-prone cities. Many other unhealthful compounds may form in smog. Hundreds of different compounds may be produced by the reactions of RH, for example. One of the most important of these is peroxyacetylnitrate (PAN). PAN is a long-lived substance that can irritate eyes and lungs. It is formed by the combination of organic radicals from RH and nitrogen dioxide, both of which are plentiful in smog.

Figure 6.2 depicts the stages in the development of smog from primary pollutant emissions. The evolution of smog has a specific timetable, which is described in Sections 6.3 and 6.4. The development of toxic secondary pollutants usually peaks in the afternoon when ozone, PAN, and photochemical haze attain large concentrations.

The Deposition of Pollutants

Most gases and aerosols that come into contact with a surface stick to it. Their sticking efficiency varies from gas to gas. Generally, the more reactive a gas is, the more easily it will stick to a surface. Different surfaces attract the same gas at different rates. Leaves and other vegetation with large surface areas are usually very good pollutant scavengers. The air in a park has less pollution because of the trees there (except for natural emissions of hydrocarbons, which are less significant in adding to the local pollution).

The rate of removal of pollutants to a surface can be estimated using a **dry deposition velocity**, the effective speed at which the molecules of a gas move toward a surface as the gas is scavenged. If the deposition velocity is zero, the gas will not interact with the surface and thus will not deposit on the surface. The largest deposition velocities are on the order of 1 centimeter per second (1 cm/sec, about 0.33 inch each second). That might not sound very fast, but at that rate an entire kilometer of atmospheric depth can be scrubbed clean by surface deposition in one day.

The time, t, required to clean out a layer of air can be simply estimated if the deposition velocity, v_{dep}, and the thickness of the layer, h, are known. In this case,

$$t = \frac{h}{v_{dep}}$$

$$\text{or} \quad h = \frac{t}{v_{dep}} \tag{6.7}$$

For example, if the deposition velocity of a pollutant is 0.1 cm/sec and the boundary mixed layer is 300 meters thick, the pollutant will be deposited on the surface in about 3×10^5 seconds, or a few days ($[300 \text{ m} \times 100 \text{ cm/m}]/0.1 \text{ cm/sec} = 3 \times 10^5$ sec). The dry deposition velocity usually increases with the gas's reactivity. The reactivity of common air pollutants increases in the order CO, NO, O_3, NO_2, SO_2. In the case of carbon monoxide, for example, the dry deposition velocity ranges from about 0.001 to 0.005 centimeter per seconds. At these velocities, only about 1 meter of air can be cleansed of CO during an evening. Ozone has a dry deposition velocity of roughly 0.1 to 0.5 cm/sec. Accordingly, 100 meters of air can be cleared of ozone in one night by this process. Nocturnal radiation inversions often are this narrow.

6.2.2   CLEAN AND DIRTY AIR

Whether air is "clean" or "dirty" is often a matter of opinion. If that was all there was to it, we could simply argue with one another about the state of the environment and its effects on our health. Fortu-

nately, a set of **standards** has been established to indicate the amounts of pollutants in air and provide guidelines for the likely effects of that pollution on health and well-being. "Clean" air is generally air in which pollution concentrations fall below the standard. The air is not pristine; it may not even be healthful for you to breathe over a lifetime. But it is clean enough that the long-term risk of serious disease is acceptably low.[13] Above the standard, air pollution is likely to cause long-term health effects in many people. Although unacceptable, it is held to be tolerable until a cleanup is possible. At some level, air pollution becomes very unhealthful and even hazardous. At such levels, well above the standard, a smog "episode" occurs, and a smog "alert" may be sounded. Children are kept indoors, and their activities are limited so that they will breathe as little of the polluted air as possible. People suffering from asthma and emphysema head for an oxygen tank.

Exposure to pollutants can cause a variety of physical symptoms. Each person has a different response to smog, which depends on the concentrations of the pollutants, the duration of exposure, and individual sensitivity to each pollutant. Over the years, tests conducted in laboratories and studies of people exposed to pollutants—in their normal life or accidentally in high concentrations—have contributed to the definition of standards for judging the severity of air pollution. (The physiological effects of exposure to many common pollutants are discussed in Chapters 7 and 8.)

In general, serious health effects may result either from short exposure to high concentrations of a pollutant, *or* from long exposure to lower concentrations. In the case of smog, the short-term exposure is used to establish criteria for determining air quality and to control pollution levels. This is done because sensitive people can quickly experience severe effects of exposure to high levels of ozone, sulfur dioxide, particulates, and other common components of smog. Accordingly, the concentrations of certain key "criteria" pollutants are monitored in cities like Los Angeles. In all cases, the concentrations are averaged over a specific period of time. For example, the concentration of ozone may be measured continuously

---

13.   The definition of acceptable risk is itself a risky business. Some of the factors that determine the risk to health caused by exposure to pollutants are the duration of the exposure, efforts to mitigate the exposure and its impacts, and the quantitative physiological responses to the pollutants involved. (These factors are discussed further in Chapter 7.)

Table 6.1 Pollution alert level[a]

Carbon Monoxide

California/Federal

Standard	20 ppm	(1-hr avg.)
	9.3 ppm	(8-hr avg.)
Stage 1	40 ppm	(1-hr avg.)
	20 ppm	(12-hr avg.)
Stage 2	75 ppm	(1-hr avg.)
	35 ppm	(12-hr avg.)
Stage 3	100 ppm	(1-hr avg.)
	50 ppm	(12-hr avg.)

Federal

Standard	35 ppm	(1-hr avg.)
	9.5 ppm	(8-hr avg.)
Stage 1	15 ppm	(8-hr avg.)
Stage 2	30 ppm	(8-hr avg.)
Stage 3	40 ppm	(8-hr avg.)

Ozone

California

Standard	0.09 ppm	(1-hr avg.)
Stage 1	0.20 ppm	(1-hr avg.)
Stage 2	0.35 ppm	(1-hr avg.)
Stage 3	0.50 ppm	(1-hr avg.)

Federal

Standard	0.12 ppm	(1-hr avg.)
Stage 1	0.20 ppm	(1-hr avg.)
Stage 2	0.35 ppm	(1-hr avg.)
Stage 3	0.50 ppm	(1-hr avg.)

Nitrogen Dioxide

California/Federal

Standard	0.25 ppm	(1-hr avg.)
Stage 1	0.60 ppm	(1-hr avg.)
	0.15 ppm	(24-hr avg.)
Stage 2	1.20 ppm	(1-hr avg.)
	0.30 ppm	(24-hr avg.)
Stage 3	1.60 ppm	(1-hr avg.)
	0.40 ppm	(24-hr avg.)

Federal

Standard	0.053 ppm	(annual avg.)

[a]All units are parts per million by volume of air.
* Data from the South Coast Air Quality Management District, Los Angeles, California.

Table 6.2   Standards for air pollutants[a]

Pollutant	California standard	Federal standard	Time of exposure
Sulfur dioxide	0.25 ppmv	————	1-hour avg.
		0.50 ppmv	3-hour avg.
	0.05 ppmv	0.14 ppmv	24-hour avg.
	————	0.03 ppmv	annual avg.
Particulate matter (PM_{10})[b]	50 $\mu g/m^3$	150 $\mu g/m^3$	24-hour avg.
	30 $\mu g/m^3$	50 $\mu g/m^3$	annual avg.
Sulfates	25 $\mu g/m^3$	————	24-hour avg.
Lead	1.5 $\mu g/m^3$	————	30-day avg.
	————	1.5 $\mu g/m^3$	90-day avg.
Hydrogen sulfide	0.03 ppmv	————	1-hour avg.
Vinyl chloride	0.01 ppmv	————	24-hour avg.

a.   The concentrations are given in parts per million by volume (ppmv) for gases and micrograms per cubic meter of air ($\mu g/m^3$) for particulates.

b. PM_{10} refers to the total particulate matter (aerosols) in the form of particles with sizes smaller than 10 micrometers (μm) in diameter. Particles of this size are also called respirable suspended particles (RSP) because they tend to lodge in the lungs and bronchial tubes (Section 6.5.1).

*Data from the South Coast Air Quality Management District, Los Angeles, California.

during the day. The actual measurements occur, say, every minute. But these minute-by-minute values are averaged each hour to obtain the ozone concentration that is used to determine air quality. Other pollutants may be averaged over periods of 1 hour (carbon monoxide, nitrogen dioxide, sulfur dioxide), 3 hours (sulfur dioxide), 8 hours (carbon monoxide), 12 hours (carbon monoxide), and 24 hours (sulfur dioxide, nitrogen dioxide, particulates, lead, sulfate, vinyl chloride). There also are standards for long-term exposure based on annual average concentrations (nitrogen dioxide, sulfur dioxide, and particulates), and occupational exposure to airborne toxins (Chapter 7).[14]

The standards for carbon monoxide, nitrogen dioxide, and ozone in the atmosphere are summarized in Table 6.1. Additional standards for sulfur dioxide and other common pollutants are listed in Table 6.2. In all cases, the standards are meant to define the levels of exposure to pollutants that might be considered relatively safe for an average person in good health. This is a judgment call. No one would choose to be exposed to hazardous compounds at any level. Economic and social forces influence where people live and the jobs they hold, factors that are at least as important to determining personal levels of pollution exposure as are the standards for pollutant concentrations.

The different averaging times for pollutants relate to different sets of standards in different places, for assessments of different kinds of health effects and for different monitoring capabilities. For example, California has its own air quality standards, which differ somewhat from the federal standards. In fact, the California standards are tougher. Throughout the 1980s the federal government dragged its feet on controlling air pollution. California stepped in to take the lead in defining rational air pollution policy by establishing stiffer standards for emissions and fines for offenders. California's rules have been widely

14.   The standards for exposure to pollutants on the job fall within the purview of the Occupational Safety and Health Administration (OSHA). There are enormous problems in detecting, monitoring, and assessing health risks in and around the hundreds of thousands of facilities where workers may be exposed to dangerous substances.

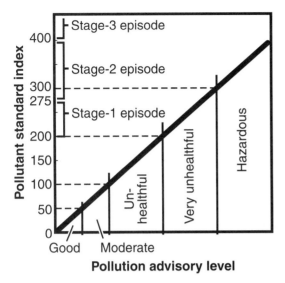

Figure 6.3 The quality of air based on the pollutant standard index (PSI). The PSI is a measure of the actual average concentration of a pollutant in relation to the standard concentration for that pollutant. For example, if the 1-hour average "clean-air" standard concentration for ozone is 0.12 ppmv, then a measured concentration of 0.12 ppmv is equivalent to a PSI of 100, and a measured concentration of 0.06 ppmv is equivalent to a PSI of 50. The quality of air is characterized by the maximum PSI among the pollutants in that air. The chart defines the various categories of air quality corresponding to different PSIs. Pollution "alerts," or "episodes," are called at certain PSI levels: a stage-1 alert is put into effect for PSIs of 200 to 275, a stage-2 alert for PSIs from 275 to 400, and a stage-3 alert for PSIs above 400.

adopted by other states and may become federal guidelines. This is a good example of "trickle-up" environmental policy.

The standards for exposure to pollution can also change over time. For example, the federal standard for ozone has decreased slightly over time from about 0.13 parts per million by volume (ppmv) to 0.12 ppmv. It will probably decrease again in an attempt to catch up with the California standard, now at 0.09 ppmv.

The Pollutant Standard Index

The **pollutant standard index (PSI)**, is a relative scale of air quality that applies to all pollutants. The index is calculated by projecting the measured average concentration of a pollutant onto a scale that ranges in value from 0 to 500. On the PSI scale, the federal standard concentration would have a PSI of

exactly 100. Air with a PSI value of 0 to 50 for all the criteria pollutants would be called *good*. Note that *all* the pollutants (ozone, carbon monoxide, nitrogen dioxide, sulfur dioxide, particulates) must have low PSIs in order for the air quality to be *good*. The other categories of air quality according to the PSI are defined in Figure 6.3.

The ozone concentrations that correspond to the air quality categories and smog episode levels in Figure 6.3 are listed in Table 6.3. Notice that the connection between the ozone concentration and PSI is not a simple linear relationship. Rather, the PSI is a method for standardizing records of air quality and for clearly conveying to the public the state of air quality. An ozone mixing ratio of 0.5 part per million by volume (ppmv) may not mean much to you, but a PSI of 400 suggests a concentration far above the safe level (100).

What happens when the PSI goes too high? A smog alert may be issued. During a stage-1 episode, the authorities recommend that strenuous activities (sports, running, other exercise) be avoided. People, especially children and senior citizens, are advised to remain indoors, where the air quality is generally better (at least on very smoggy days [Chapter 8]). If a stage-2 alert is called, all physical activity should cease (intellectual activities only, please). In addition, industries can be ordered to reduce emissions, and employees can be forced to carpool. If a stage-3

Table 6.3 Ozone criteria (California)

Ozone concentration (ppmv)	Stage	Air quality
0–0.06	—	Good
0.09	Standard	Moderate
0.06–0.12	—	Moderate
0.13–0.20	—	Unhealthful
0.20	Stage 1	Very unhealthful
0.20–0.40	—	Very unhealthful
0.35	Stage 2	Very unhealthful to hazardous
> 0.40	—	Hazardous
0.50	Stage 3	Hazardous
0.60	—	Significant harm

episode occurs (a rare event, since there has not been one in Los Angeles for two decades), everyone can be ordered to stop breathing (only kidding). However, people can be instructed to remain at home while a smog holiday is called.

Monitoring Pollution

Because air pollution is recognized as a hazard to the health and well-being of urban populations, efforts have been made to keep an eye on the levels in affected areas. Monitoring pollution has historically been a task for local authorities. Southern California has one of the most advanced pollution-measurement systems. Here, most of the key pollutants, primary and secondary, are sampled at least hourly at three dozen sites scattered around the Los Angeles basin. The facilities are operated by the Air Quality Management District (AQMD) of southern California (or the Southern California Air Quality Management District [SCAQMD]).

The monitoring of air pollution involves the most advanced technology available in scientific instrumentation. The substances that must be measured comprise only the smallest fraction of air, usually just parts per million. Although the concentrations are small, the pollutants are not harmless, merely difficult to detect quantitatively. The field instruments must also be reliable and robust, working hour after hour and day after day at automated sites. In recent years, the use of satellites for air pollution monitoring has become possible, although the technology has not been applied on a practical scale. In a future scenario, on a sunny Saturday afternoon, a telescope hovering in space over your head will peer down into your backyard and note a smoking barbecue pit loaded with hamburgers. The smog police will arrive soon thereafter to write a ticket and douse the fire.

6.3 Smog Scenarios: A Typical Polluted Day

The evolution of smog follows a predictable pattern: The primary pollutants are emitted beginning in the morning hours when human activity starts in earnest. The primary compounds cook under the sun to form secondary pollutants, which accumulate into the afternoon. Finally, after dusk, with activity waning, the photochemical pollutants can disperse to other regions. The observed variations of the key pollut-

ants, CO, NO_2, and O_3, in the Los Angeles basin are discussed in the following sections.

Meteorological and topographical influences are crucial factors in smog formation in an urban air shed. (See Section 5.5 for a review of these effects in the Los Angeles basin.) In general, the existence of a strong temperature inversion overhead and the marine boundary layer offshore establish the conditions for heavy smog in the basin. The initial development of the smog is tied closely to the sources of primary pollutants from automobiles and industries. The morning rush hour injects tons of primary pollutants into the early-morning boundary layer. A sea breeze (wind blowing inland from the coast) develops during the day as the inland areas are heated by the sun. Sunlight also cooks up the primary pollutants into full-blown smog. The sea breeze carries the layer of smog toward higher terrain, which traps it. Later, during the evening, the land cools and the sea breeze dies down. The inversion layer drops, and pollutants may begin to be advected away in the upper-level winds.

The worst smog episodes occur when the air is stagnant over the Los Angeles basin, trapped below a strong regional temperature inversion associated with high pressure. During such an episode, the pollutants generated one day may linger for several days. The smog accumulates, producing very high concentrations of secondary pollutants. These conditions may eventually lead to an alert and actions to stem the flow of primary pollutants.

The agenda for a smoggy day in Los Angeles can be summarized in the following phases:

1. *Early morning.* Between about 6:00 a.m. and 9:00 a.m., the morning vehicle traffic is the densest, and the emissions of CO, NO, and RH are the greatest. Business and industrial activities begin during this period as well, and continue through the day. In the early morning, the winds are often stagnant, and the inversion layer is very low (sometimes only a few hundred feet above the ground). Marine clouds and fog may be present in the coastal regions, carried inland the previous night with cooler air flowing from the ocean boundary layer. Pollution left over from the previous day may be trapped in narrow layers close to the surface.

2. *Midday.* The primary emissions continue, with a small boost around lunchtime. From about 9:00 a.m. through the early afternoon (~2:00 p.m.), the primary emissions are photochemically trans-

formed into the secondary pollutants (NO_2, O_3, and HC). In the late morning, the sea breeze picks up and transports the coastal pollution inland. Near noon, the intensity of the sun maximizes, and secondary pollutants are generated rapidly. Ozone peaks in the afternoon hours.

3. *Late afternoon.* By this time, ozone concentrations have accumulated to their highest levels. The sun has warmed the land and the sea breeze is quite strong, pushing a cloud of smog far inland against the mountains. At this point, the air laden with pollutants can be lofted through mountain passes onto the high plateau beyond. From there, the pollution can travel hundreds of miles across the desert regions.

4. *Evening.* In the late afternoon and early evening, commuter traffic again builds up, and more primary pollutants are released into the smoggy air. The primary emissions peak between 5:00 p.m. and 7:00 p.m.. However, the sun is lower in the sky, and so insufficient radiation is available to generate much photochemical smog. The primary pollutants themselves accumulate. With one of the key ingredients for ozone formation (sunlight) removed, the ozone concentrations begin to fall. The sea breeze continues to sweep the pollutants inland, diluting their effects closer to the coast.

5. *Late evening.* In the late evening and early morning, the sea breeze dies down, and a weaker land breeze may develop. By this time, radiative cooling has created a low-level temperature inversion. Most of the previous day's pollution is left above this inversion and is dispersed by prevailing regional winds. On the other hand, primary emissions at the surface, which continue during the evening, can accumulate below the nighttime inversion.

The behavior of the principal smog pollutants has been carefully monitored in Los Angeles for many years. The data collected by the AQMD reveal many interesting relationships among the sources of primary pollutants, meteorology, and other factors contributing to smog formation.

6.3.1 CARBON MONOXIDE

Figure 6.4 shows the areas of Los Angeles in which carbon monoxide concentrations most often exceed the clean air standard. It is apparent that CO, a primary pollutant, has the greatest impact in the vicinity of its sources. In Los Angeles, the source of CO is dominated by the large freeways in the western coastal regions and in the San Fernando valley. These sources are directly reflected in the CO data in Figure 6.4. It is interesting that the eastern Los Angeles basin is essentially free of high carbon monoxide concentrations. These areas are less densely developed and experience less traffic than the highly built-up and industrialized areas to the west.

Figure 6.5 shows the diurnal and seasonal variations of carbon monoxide. Figure 6.5(a) reveals a distinct maximum in the CO concentration during the morning rush-hour commute. A secondary peak appears in the late afternoon with the onset of the evening rush hour. The minimum in the CO mixing ratios around noon is the result of several factors. Traffic in general is somewhat lower at midday than in the morning, although this is not the most important factor. The height of the mixed layer is more critical. During the morning, the surface heats up, turbulence is generated, and the temperature inversion capping the boundary layer is forced upward (Sections 5.2.2 and 5.3.3). This mixes and dilutes the CO concentrations at the surface. Later in the afternoon, the solar heating decreases, and the inversion layer drops just as the evening traffic picks up. During the early evening, as the mixed layer collapses, carbon dioxide emissions can accumulate near the surface. Some of the highest CO concentrations of the day can occur at this time.

In the late hours of the night, carbon monoxide concentrations can decline significantly. The effects of land breezes and synoptic winds to dissipate the CO are important. Carbon monoxide is also deposited on land surfaces, although this dry deposition process is quite inefficient (Section 6.2.1).

The monthly concentrations of carbon monoxide in Figure 6.5(b) show an interesting behavior. They are much lower on average in summer than in winter. The cause is again related to the thickness of the mixed layer. In summer, the sunlight is more direct, the heating of the land is stronger, the turbulence generated is more intense, and the mixed layer is deeper. The emitted CO is more dilute in a deeper mixed layer. In winter, about the same traffic occurs as in summer (except possibly for tourists). However, the land heating is weaker, colder air masses pass over the area, and the mixed layer tends to be

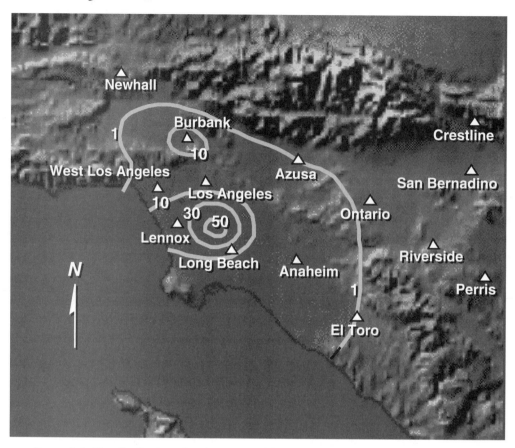

Figure 6.4 The distribution of carbon monoxide pollution in the Los Angeles basin. The contours outline regions subjected to CO concentrations that exceed the federal standard (8-hour average concentration greater than 9.5 ppmv) for a specific number of days each year as indicated by the number on each contour. The data shown are for 1988; the pattern of CO pollution is similar from year to year. (Data compiled by the South Coast Air Quality Management District, 1988)

lower. The CO, accordingly, is more concentrated. More frequently in winter than summer, strong weather fronts can pass by, sweeping away all the pollution with fresh winds. These are those few crystal-clear days when the mountains can be seen from the coast. Averaged over a month, however, the CO mixing ratios are larger in winter.

Carbon monoxide is a primary pollutant. It has an atmospheric residence time against reaction with hydroxyl of at least several days. Moreover, carbon monoxide does not interact strongly with the surface. It follows that the principal way that CO is dissipated is through dilution by mixing or advection. A close relationship between CO concentrations and the turbulent state of the boundary layer can be inferred from the observed behavior of CO shown in Figures 6.4 and 6.5. In summer, the concentrations of carbon monoxide may be lower,

not because less pollution is being spewed into the air (on the contrary, more is probably being emitted), but because the pollution is diluted to a greater extent by vigorous mixing.

6.3.2 Nitrogen Dioxide

Nitrogen dioxide is a secondary pollutant generated mainly from nitric oxide (although there are some direct emissions of NO_2). Figure 6.6 shows the regions in the Los Angeles basin with the highest NO_2 concentrations. These regions are well inland from the areas of the primary emissions of NO_x. These areas are, in fact, nearly coincident with the regions of CO emission. Comparison of Figures 6.4 and 6.6 clearly demonstrates the inland displacement of large nitrogen dioxide abundances. NO_2 is

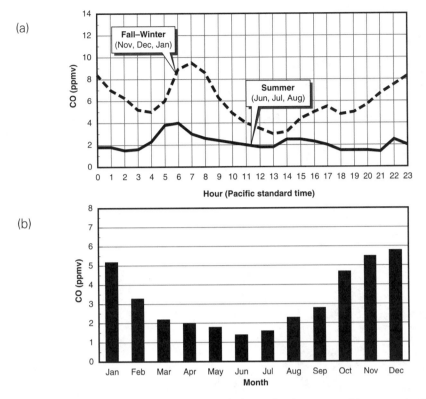

Figure 6.5 The daily and monthly variations of carbon monoxide concentrations in the Los Angeles basin. Both sets of data were collected at Lennox, in the western (coastal) region of the basin, in 1983. (a) The diurnal variation in the CO concentration shows a peak in the early morning caused by heavy commuter traffic. The dashed line gives the daily behavior of CO seen in the fall and winter, and the solid line shows the CO variation typical of the summer. (b) The monthly average CO concentrations indicate that the minimum concentrations occur in the summer. (Data from the South Coast Air Quality Management District, 1983)

produced when NO reacts with hydrocarbons (Equation 6.3). Hence, a period of time is required for NO_2 to form in smog. While the photochemistry is brewing, the air is moving, carrying pollutants with it. In Los Angeles, the air is moving inland on the sea breeze or on the prevailing westerly winds. If NO were monitored as a critical pollutant, its distribution would resemble that of CO (Figure 6.4) more closely than that of NO_2 (Figure 6.6). Indeed, the NO would disappear more rapidly than CO toward the inland areas because it is photochemically transformed into nitrogen dioxide.

Figure 6.7 shows the diurnal and seasonal variations in the concentration of nitrogen dioxide. The daily variation in NO_2 reveals a distinct peak in the mid- to late morning. This peak is obviously delayed from the peak in the concentration of CO seen in Figure 6.5(a). At later times, the NO_2 decreases

because of two specific effects. First, the nitrogen dioxide is diluted by mixing, as carbon monoxide is. Second, unlike CO, NO_2 can be photodissociated (Equation 6.4) and converted to other secondary compounds (for example, PAN). The photochemical reactions of nitrogen dioxide (Equations 6.4 and 6.6) continue to recycle NO and NO_2. Ozone is also a part of this balance. However, the nitrogen dioxide can also be converted into nitric acid by the reaction,

$$NO_2 + OH + M \rightarrow HNO_3 + M \qquad (6.8)$$

Moreover, the formation of PAN and other nitrogen-containing hydrocarbons further lowers the NO_x concentration. These compounds are not easily recycled into NO and NO_2, and so they are effective sinks for NO_x.

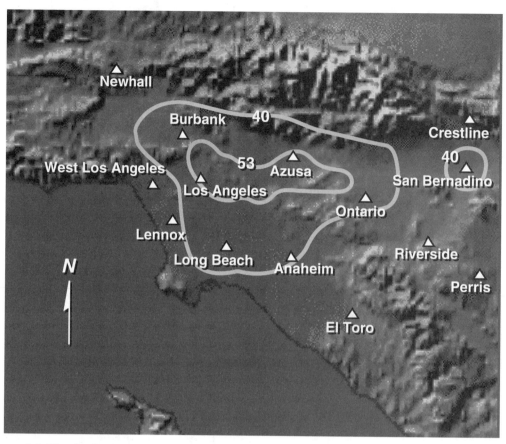

Figure 6.6 The distribution of nitrogen dioxide in the Los Angeles basin. The contours outline regions with yearly average NO_2 concentrations that exceed the mixing ratios in units of parts per billion indicated on each contour. The data shown are for 1988; the pattern of NO_2 pollution is similar from year to year. (Data compiled by the South Coast Air Quality Management District, 1988).

The concentration of NO_2 in the afternoon is maintained by emissions, even though there is a net transfer of nitrogen oxides into terminal compounds like nitric acid and PAN. In the evening, NO_x can be further chemically processed by the reactions

$$NO + O_3 \rightarrow NO_2 + O_2 \qquad (6.9)$$

$$NO_2 + O_3 \rightarrow NO_3 + O_2 \qquad (6.10)$$

$$NO_2 + NO_3 + M \rightarrow N_2O_5 + M \qquad (6.11)$$

$$N_2O_5 + H_2O \xrightarrow{\text{aerosols}} 2HNO_3 \quad (6.12)$$

The last reaction occurs on the surfaces of wet haze droplets in the atmosphere. Notice that overall, the sequence of reactions acts to convert NO_x to nitric acid.

The deposition of NO_2 to surfaces is more rapid than for CO. Typical NO_2 deposition velocities are in the range of 0.1 to 1.0 centimeter per second. At these rates, most of the nitrogen dioxide in a mixed layer 100 meters deep could be removed during one night. However, the chemical conversion of NO_x (NO_2) to nitric acid is even faster in most situations.

The annual variation in nitrogen dioxide shown in Figure 6.7(b) reveals a small seasonal effect. The concentrations tend to be higher in winter than in summer, for the same reason that the CO concentrations are higher. The mixed layer is shallower in winter, and the pollutants are less diluted. Since NO_2 appears later in the day than CO does, the mixed layer is generally deeper than in the early morning. In fact, the difference in the height of the thermal inversion at midday is often not so different from winter to summer.

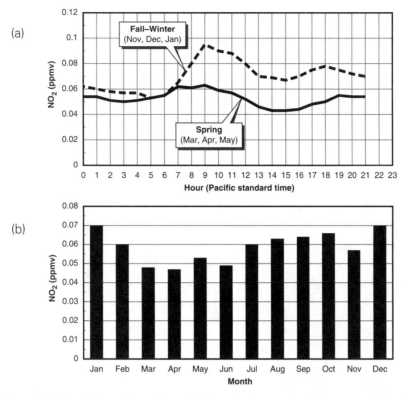

Figure 6.7 The daily and monthly variations of nitrogen dioxide concentrations in the Los Angeles basin. Both sets of data were collected near downtown Los Angeles, in the western-central region of the basin, in 1983. (a) The dashed line shows the daily variation in NO_2 observed in the fall and winter, and the solid line depicts the typical NO_2 variation in the spring. Particularly in the fall and winter, the diurnal variation in the NO_2 concentration shows a peak in the late morning associated with the buildup of oxidants from hydrocarbon photodecomposition in the presence of nitric oxide. The monthly average NO_2 concentrations indicate that the seasonal variation is less apparent than in the case of CO. (Data from the South Coast Air Quality Management District, 1983)

6.3.3 OZONE

Ozone is the component of smog that causes the greatest concern in polluted cites. Ozone is a secondary pollutant, and its production follows that of nitrogen dioxide, through the reactions shown as Equations 6.3 to 6.5. The distribution of ozone in the Los Angeles basin is illustrated in Figure 6.8. Here, the frequency at which ozone concentrations exceed the federal clean air standard are defined over the basin. It is apparent that ozone concentrations are highest in the inland valleys. The greatest amounts are found in cities like Pasadena, Azusa, and San Bernardino, huddled against the mountains to the north. In some of these locations, on more than 140 days in 1988 the ozone concentrations exceeded the clean air standard. Although that record has improved in recent years, the east-ern Los Angeles basin remains one of the most polluted areas in the United States. (See also Figure 6.14, which shows the distribution of stage-1 ozone episodes in the Los Angeles basin.)

Ozone forms later in the day in air masses that are being transported inland. The prevailing winds and the sea breeze carry pollutants from the coastal regions in the late morning and early afternoon. Meanwhile, the smog chemistry is cooking up a batch of nitrogen dioxide and ozone. The worst ozone levels are therefore seen where the smog accumulates at the mountain barriers.

Figure 6.9(a) illustrates the typical diurnal variation of ozone at a receptor site in the Los Angeles basin. The maximum ozone concentrations occur in mid- or late afternoon at any time of the year. The time required to cook up a batch of NO_2 and the strength of the sunlight at midday combine to

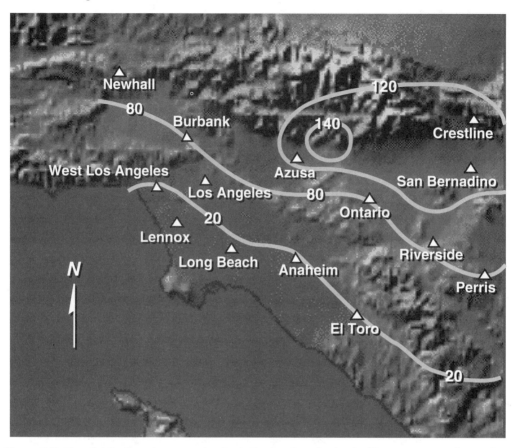

Figure 6.8 The distribution of ozone episodes in the Los Angeles basin. The contours outline regions subjected to O_3 concentrations exceeding the federal clean air standard (1-hour average greater than ppmv) for a specific number of days each year, indicated by the number on each contour. The data for 1988 are shown; the pattern of ozone pollution is similar from year to year. (Data compiled by the South Coast Air Quality Management District, 1988)

generate the highest O_3 abundances in the afternoon. An obvious feature is the much higher ozone concentrations in summer compared with winter. This is also apparent in the annual variation of ozone shown in Figure 6.9(b). The behavior of ozone is completely opposite to that of CO, which peaks in winter. Ozone is diluted by the greater height of the mixed layer in the summer, as is carbon monoxide. However, the greater intensity of solar radiation that drives the photochemistry of smog formation is more than sufficient to compensate for the deeper mixed layer.

The different effects of sunlight compensate each other in determining ozone concentrations. As the intensity of sunlight increases, more ozone is produced by photochemical reactions, and the ozone concentration rises. But on the other hand, intense sunlight increases the surface temperature and decreases the stabilization of the lower atmosphere.

Turbulent activity increases and the mixed layer grows deeper, diluting the ozone concentration. If a strong regional inversion settles in, the mixed layer may not be able to push its way upward. In this case, clear skies, brilliant sunshine, and a strong inversion can lead to the highest abundances of ozone.

At night, ozone concentrations fall to low levels partly because ozone is a reactive oxidant (its oxidizing effect, in fact, is the main cause of health problems associated with ozone exposure). The ozone can react with nitrogen oxides (Equations 6.9 and 6.10) and hydrocarbons left in the smoggy air. Ozone also is rapidly deposited onto surfaces, with a deposition velocity of about 0.1 to 0.5 centimeter per second. Notice that by early evening (Figure 6.9[a]), ozone levels are reduced by a factor of 10 or more. This would be an excellent time to take a run or perform other vigorous activities (which, unfortunately, would conflict with prime-time television

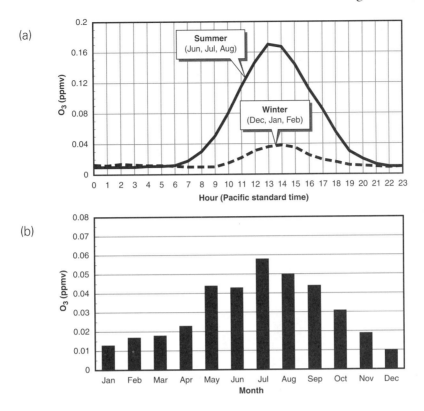

Figure 6.9 The daily and monthly behavior of ozone concentrations in the Los Angeles basin. Both sets of data were collected near Asuza, in the north-central region of the basin, in 1983. (a) The dashed line gives the typical daily O_3 variation measured during the winter, and the solid line gives the O_3 variation typical of a summer day. The diurnal variation in ozone shows a peak in the early afternoon due to the photodissociation of nitrogen dioxide and other oxidants. (b) In the monthly average O_3 concentrations, the well-known seasonal variation, with much higher ozone concentrations in the summer and early fall, is apparent. (Data from the South Coast Air Quality Management District, 1983)

viewing). In fact, it would be even better to exercise very early in the morning before ozone has a chance to form.

6.4 Dissecting Smog

Many of the causes of urban air pollution and photochemical smog should be apparent from the previous discussion. In order to identify potential remedies, however, we must look a little deeper. Smog is a complex problem, with many factors contributing to its impacts. The issue of smog also raises sensitive sociological and economic questions about how to deal with it. To see the technical aspects of the problem in full view, various physical and chemical factors need to be understood. This can be accomplished by carrying out laboratory measurements of

smog reactions and field research in smoggy areas and by analyzing the data obtained using statistical and phenomenological models. The previous section described the characteristics of urban smog observations. In this section, we present further information regarding key processes that influence the distribution and density of smog.

Los Angeles provides a useful laboratory for studying the formation of smog and its properties. Air pollutants have been continuously measured for many years, and major scientific investigations have focused on smog in the Los Angeles basin. From the Los Angeles experience, a number of general properties of smog can also be deduced. Indeed, all the causes of photochemical smog in other cities are apparent in Los Angeles. The principal factors implicated in smog formation may be summarized as follows:

1. *Sources of primary pollutants.* Vehicles—cars, buses, and trucks—are the main initiators of photochemical smog. Industrial emissions of primary ingredients also are significant. The mixture of primary pollutants (that is, the relative amounts of nitrogen oxides and reactive hydrocarbons) can also affect the course of smog evolution. Accordingly, the control of smog must involve the coordinated regulation of all pollutant sources.

2. *Timing of emissions.* The rate at which pollutants are released and the time of day they appear determine the onset and severity of smog episodes. Emissions during the morning are much more effective in generating smog than are emissions in the evening. Generally, the simultaneous injection of primary ingredients produces more concentrated smog.

3. *Distribution of sources.* To be effective in generating smoggy air, the sources of primary pollutants should be concentrated in urban centers or regions (such as the Los Angeles basin). The more distributed the sources are, the less probable that high concentrations of pollutants will build up. Emissions of primary pollutants upwind of a city are more likely to cause smog problems.

4. *Prevailing meteorology.* The most favorable weather conditions for photochemical smog formation are clear sunny skies, warm temperatures, and a regional temperature inversion. These conditions can be found, for example, under stationary high-pressure systems. Unfortunately, these same generic weather conditions are highly desirable in a place that we wish to live and work in.

5. *Regional topology.* The presence of mountain barriers, valleys or lowlands, plateaus, and other extended surface land features affects the flow of air and the dispersion of pollutants in a region. The most likely topographical features leading to smog accumulation are mountain barriers that block or stagnate prevailing surface winds. In coastal regions, sea/land breezes influenced by topography, and the marine boundary layer, also affect smog.

6. *Geographical character.* The distributions of vegetation, the types of vegetation and their hydrocarbon emissions, the rivers and lakes in a region, the moisture content and hydrology of the soils, agricultural activities surrounding an urbanized area, the presence of natural sources of dust, sulfurous gases, and ammonia nearby all influence the formation and intensity of smog in a region.

7. *Season of the year.* The photochemical processes that create secondary pollutants are initiated by sunlight. In the summer, the sun generally remains in the sky longer, and the production of secondary pollutants from primary emissions is more efficient. Moreover, air temperatures are usually warmer in the summer, which boosts the rates of formation of some secondary pollutants.

8. *Urban demographics.* The presence of many old, poorly maintained vehicles, lack of proper emission controls, apathy or ignorance on the part of the public, and tolerance of smog can be factors in persistent excessive pollutant emissions.

The development and impacts of large smoggy air masses are a problem in the San Joaquin Valley in central California, the Ohio River valley in the Midwest, and the Washington-New York-Boston corridor. In the Ohio River valley, for example, the congregation of major industrial cities, including Chicago, Detroit, and Cleveland, provides an extended source of primary pollutants. Power plants, smokestack industries, and heavy vehicular traffic are the main sources of pollution. The region encompasses not only numerous local municipalities, but also many counties and states. This fragmentation of responsibility makes regulating regional air pollution especially difficult. The geography of the midwestern United States helps limit the buildup of smog. Vast areas of flat land and the frequent passage of weather fronts keep the pollution moving eastward. Nevertheless, when winds are in the doldrums, the entire region may be buried under a pall of dirty air.

Smog has different characteristics in different places, owing to differences in such factors. One common factor in many cases is the association of photochemical smog with the emissions of automobiles and other vehicles. This common relationship has led to the close scrutiny of smog formation processes, both in the urban atmosphere and in controlled experiments.

6.4.1   THE EVOLUTION OF SMOGGY AIR

The sequence in which the components of photochemical smog appear during the day is depicted in

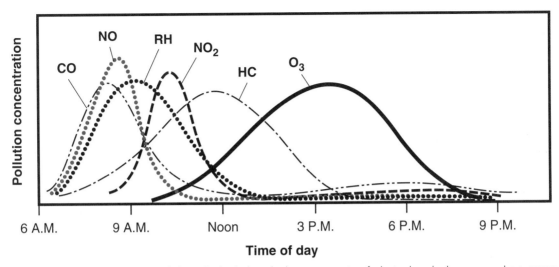

Figure 6.10   The time evolution of the principal chemical components of photochemical smog: carbon monoxide (CO), reactive hydrocarbons (RH), nitric oxide (NO), nitrogen dioxide (NO_2), hydrocarbon by-products (HC), and ozone (O_3). This behavior of the pollutants is observed during a day in a smoggy city. The sequential appearance of the pollutants is related to the timing of the emissions and rates at which CO, NO, and reactive hydrocarbons are transformed into secondary pollutants by photochemical reactions, as explained in the text.

Figure 6.10. The chemical transformation of the primary pollutants proceeds in a regular manner from CO, NO, and reactive hydrocarbons, to NO_2 and secondary hydrocarbons such as PAN, and finally to ozone. This sequence can be seen by comparing Figures 6.5(a) for the diurnal variation of CO in the Los Angeles basin, 6.7(a) for NO_2, and 6.9(a) for O_3. A secondary component of smog, like ozone, is formed as its precursor ingredients are consumed in chemical reactions. The same behavior should be observed under experimental circumstances designed to match the conditions for smog formation.

Figure 6.11 illustrates the result of experimental tests of smog formation under controlled conditions. Such experiments are conducted in **smog chambers**. Smog chambers can be relatively small, but often are as large as a house. The idea is to simulate as closely as possible the conditions that exist in the real atmosphere. The walls of the chamber present a problem, for several reasons. Unwanted reactions can occur on the wall surfaces. By making the chamber bigger, the ratio of the wall area to the volume of the chamber is reduced, and wall reactions become less important. The walls of the chamber also prevent mixing of the gases in the chamber as might occur in the atmosphere by turbulence. So the gases in the chamber may be stirred using fans. Finally, the chamber walls block sunlight. Sunlight is sometimes artificially created with lamps, but it is difficult to reproduce the natural spectrum

and variations accurately. The simplest solution is to construct the chamber out of clear plastic and place it outside in the sunlight. Smog chambers therefore often resemble very large plastic bags.[15]

The behavior of the pollutants in a smog chamber is strikingly similar to that in real smog (Figure 6.11). The secondary hydrocarbons produced in this experiment consist primarily of formaldehyde, acetaldehyde, and peroxyacetylnitrate (PAN). None of these compounds is healthful to breathe (Sections 7.2.2 and 8.3). Notice in Figure 6.11 that the ozone concentration does not decrease rapidly toward the end of the experiment, as it may in urban smog (Figures 6.9[a] and 6.10). The destruction of ozone on surfaces is slower in the chamber than in the real atmosphere. Hence, this feature of the behavior of smog is not closely reproduced in such experiments.

The evolution of smog depends on the initial mixture of nitrogen oxides and hydrocarbons. This effect is shown in Figure 6.12. Here, as expected, increasing anthropogenic emissions of reactive hydrocarbons and nitrogen oxides lead to higher peak ozone concentrations. In other words, the greater the quantity of smog-producing ingredients is, the

15. The bags are very sophisticated, of course. The plastics must be designed to let in all the wavelengths of sunlight, to repel the corrosive pollutants in the bag, and to be flexible and strong. Advanced instruments must also be attached to the bag to measure the light intensity, pollutant concentrations, and other parameters.

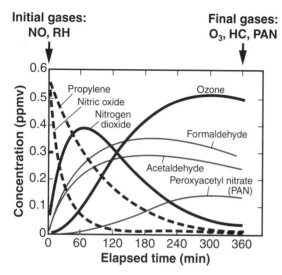

Figure 6.11 The evolution of chemical species in a smog chamber experiment. The initial concentrations of gases in the experimental chamber are the values along the vertical axis at time = 0 (the concentrations are specified in parts per million by volume). The disappearance of the initial NO and reactive hydrocarbon (propylene in this case) is followed by the buildup of NO_2, hydrocarbon by-products, and ozone, just as in polluted air. Controlled laboratory studies can be carried out over a wide range of conditions to determine the fundamental processes contributing to smog formation. Notice that the yield of ozone is approximately equal to the original concentration of the reactive hydrocarbon. (Data from J.N. Pitts, Jr., A. C. Lloyd, and J. L. Sprung, "Ecology, Energy, and Economy," *Chem. Br.* 11 [1975]: 247.)

greater the amount of ozone that can be cooked up. It follows that a reasonable strategy to reduce the ozone in smog is to reduce both NO_x *and* RH emissions. The effectiveness of this and other strategies based on reductions in emissions is illustrated in Figure 6.12. By making a relatively large cut in the release of either reactive hydrocarbons *or* nitrogen oxides, ozone levels can be brought well below the acceptable standard. In either case, however, a minimum level of ozone can be achieved only by cutting emissions. This minimum concentration of ozone is, in part, dictated by natural sources of NO_x and hydrocarbons. Plants, insects, and animals give off hydrocarbons that can form ozone by means of the same reactions that occur in smog. Surely, we do not want to wipe out all these creatures to reduce ozone concentrations further.

In Figure 6.12, the amount of ozone in the atmosphere actually decreases when NO_x emissions are increased, once hydrocarbon emissions have

been brought to a very low level. This response is caused by the reactions of nitrogen oxides with ozone (particularly the reactions defined in Equations 6.9 and 6.10, both of which consume ozone). Thus the diagram reveals the complexity of smog formation and the difficulty of controlling smog. The effort to limit ozone formation in the United States has focused on limiting NO_x emissions. Figure 6.12 suggests that once the emissions of nitrogen oxides have been reduced below about 10 kilograms per square kilometer (10 kg/km^2), reductions in anthropogenic hydrocarbon emissions will have practically no effect on ozone levels. Natural hydrocarbons are then sufficient to produce ozone efficiently from the residual nitrogen oxide emissions. In this case, the limiting factor is the number of nitrogen oxide molecules available to form ozone.

6.4.2 TRENDS IN AIR POLLUTION

The long-term trend in smog conditions in polluted cities has been literally up and down. Air quality began to deteriorate in the middle of this century with a rapid growth in population, industry, and *cars*. Controls on the emissions that produce smog were lax during this period. For example, it was not uncommon for people to burn their trash in the backyard. I can remember raking up huge piles of leaves in the fall and lighting a bonfire to get rid of them. As the levels of air pollution and concern about smog increased, regulations to control smog-forming emissions were enacted. For a long time, however, the growth in emissions was winning. In recent years, however, in some cities like Los Angeles, the overall severity of air pollution has been abating.

Figure 6.13 shows the normal pattern of wind in the Los Angeles basin. The wind field controls the distribution and dispersion of air pollutants from sources (Sections 5.2.3 and 5.5). The relationship is demonstrated by comparing the typical wind pattern in Figure 6.13 with the distribution of ozone stage-1 alerts shown in Figure 6.14. The highest ozone concentrations are found in the areas downwind of the sources of primary pollutants in the Los Angeles basin. The changes in the meteorology of a region can be crucial to the evolution of smog. If the weather becomes more unstable, for example, then mixing is enhanced, and air pollutants cannot reach such high densities. The frequent passage of weather fronts ensures that accumulations of pollution will be

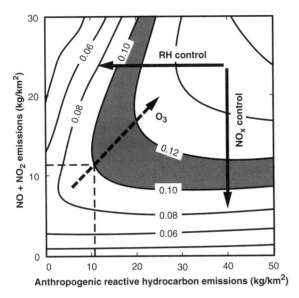

Figure 6.12 The dependence of ozone concentrations in smog on the total emissions of NO_x ($NO+NO_2$) and reactive hydrocarbons (RH) of anthropogenic origin. The contours on the graph represent lines of the constant ozone mixing ration, in parts per million by volume. (The line marked 0.12 represents an ozone concentration of 0.12 ppmv; every point on this line defines an NO_x and RH emission that would yield 0.12 ppmv of ozone after photochemical processing in air.) The shaded zone roughly defines the limit for ozone concentrations in "clean" air. The dashed arrow indicates the general direction in which ozone concentrations increase with increasing RH and NO_x emissions. The solid arrows illustrate the potential decreases in ozone concentrations caused by reducing the emissions of either hydrocarbons or nitrogen oxides. Natural sources of RH and NO_x were considered in constructing this figure. The figure suggests that there is a lower limit to the ozone concentrations that can be achieved by reducing anthropogenic pollutant emissions. (Adapted from W. L. Chameides, R. W. Lindsay, J. Richardson, and C. S. Kiang, "The Role of Biogenic Hydrocarbons in Urban Photochemical Smog," *Science* 241 [1988]: 1473)

blown away. In Los Angeles, for example, a string of cooler, windier summers during the late 1980s and early 1990s has contributed to an overall decline in smog levels in that city.

The wind directions in Figure 6.13 also indicate the general directions in which pollutants escape from the Los Angeles basin. In general, smoggy air is transported through mountain passes to the high desert plateau northeast of the city and to the low deserts to the east and south. Air pollution down-wind of a major urban area like Los Angeles is

therefore sensitive to the local trend in the density of urban smog and to the regional meteorology that transports the air pollution to distant sites.

In Los Angeles, the frequency of stage-1 ozone alerts (ozone concentrations exceeding 0.20 part per million) has decreased steadily since the 1960s. The number of days experiencing stage-1 ozone alerts has been dropping by about 4 days per year, on average, over the past two decades. A careful analysis of hourly AQMD measurements of ozone concentrations over the region shows a consistent trend of reduction in the number of hours during which the federal standard (0.12 ppmv) was surpassed or stage-1 (0.20 ppmv) and stage-2 (0.35 ppmv) concentrations were detected. For example, in 1979, there were 62 hours with at least stage-2 average ozone concentrations measured at 26 AQMD monitoring sites. The number of hours recorded with stage-1 ozone levels was 2649 at these 26 stations, and the federal standard (0.12 ppmv) was exceeded for 10,561 total station-hours.

By 1989, the number of hours with stage-2 ozone concentrations had fallen to zero, with a total of only 5 station-hours at or above stage 2 from 1984 to 1989. Similarly, stage-1 hour-average concentrations were detected only 519 times in 1989, a reduction by a factor of roughly five over the decade of the 1980s. Nevertheless, the federal standard was still being exceeded for more than 4800 station-hours in 1989. The occurrence of unhealthful air was only about a factor of two. It is obviously much easier to reduce the initially high levels of pollution than to continually return the atmosphere to a nearly pristine state. The enormous resources already spent to control pollutants in smog have succeeded in significantly lowering exposure to hazardous air. The problem of eliminating widespread unhealthful air is more intractable, however. The cost of reducing ozone concentrations by each 0.1 ppmv rises dramatically as the average concentrations of ozone fall. It may take much more work to decrease ozone from 0.20 to 0.12 ppmv than it has taken to bring ozone down twice the range from 0.35 to 0.20 ppmv.

Forecasting Smog Episodes

Scientists and officials who work to control air pollution often need to forecast pollution levels. The predictions can be used to plan strategies to limit unhealthful exposure. For example, people could be warned in the morning not to exercise vigorously

Figure 6.13 The general pattern of wind flow in the Los Angeles basin. The pattern is dominated by a westerly flow associated with the prevailing winds and sea breezes. The air flows out of the basin through passes in the northern and eastern mountain barriers. Pollutants thus tend to be pushed eastward and against the northern mountain barriers, where they can accumulate.

that afternoon. Carpooling might be made mandatory. There might be time to forewarn industries about operating restrictions. Schools could be closed in areas subject to the worst smog. A smog-forecasting skill is also useful in interpreting past trends in air pollution. For example, a reliable forecasting model can be employed to correlate the recorded variations in certain smog parameters (like primary emissions) with measured variations in smog levels. Some of the parameters that affect air pollution were summarized in the introduction to Section 6.4. Among those factors are regional weather patterns, time of the year, and composition of the primary pollutants.

Forecasters have developed a simple formula to estimate roughly the greatest concentration of a pollutant that might be expected on a given day:

$$q \approx \frac{c}{u h_m} \qquad (6.13)$$

Here, q is the pollutant concentration, u is the wind velocity measured near the ground, and h_m is the projected height of the mixed layer (or boundary layer [Sections 5.2.2 and 5.3]) that day. The factor, c, is a "constant" determined by trial and error to obtain the best estimates of concentration (the units of c are appropriate for the units in which c and h_m are measured and for the desired units of q; see Section A.2 for an explanation of units). This constant is different for each pollutant (say, carbon dioxide or ozone). The symbol $\approx$ means that only a crude estimate can be made using this relation. Indeed, several important assumptions are implicit in Equation 6.13. Variations in emissions are not taken into account, for example, and the appropriate wind speed and mixed layer height must themselves be estimated from other data, or be divined from the intuition of a gnarly meteorologist with sensitive bunions.

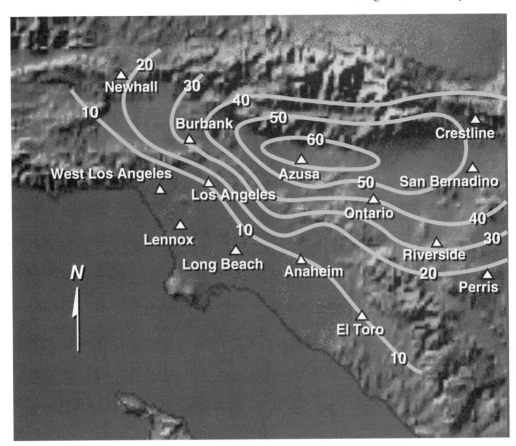

Figure 6.14   The distribution of ozone stage-1 episodes in the Los Angeles basin. These data complement those in figure 6.8, which show the distribution of ozone episodes exceeding the federal clean air standard. The contours outline regions experiencing O_3 concentrations exceeding the stage-1 alert level (1-hour average concentration greater than 0.20 ppmv) for a specific number of days each year as indicated by the number on each contour. The pattern of high ozone concentrations is correlated with the pattern of wind dispersion shown in Figure 6.13. (Data compiled by the South Coast Air Quality Management District, 1988)

The forecasting Equation 6.13 shows the important effects of wind speed and mixed layer height. The greater the surface winds are, the lower the concentrations of pollutants will be. This effect is caused by two factors. First, winds carry, or advect, pollutants from source regions. Generally, the faster the winds are, the more air is being mixed with the pollutants, diluting them over a larger area. Second, faster winds also create more intense turbulence in the boundary layer, further accelerating the dilution. The height of the mixed layer also determines the volume in which the pollutants can be diluted. The greater the height is, the more mixing and dilution there will be.

When measurements of air pollutants are compared with both simple and more complex models for smog, the effect of variations in general weather conditions from year to year can be clearly seen. These variations can then be removed, and the long-term trend in air pollution levels can be deduced. In Los Angeles, for example, the trend has been steadily downward from the early 1970s through the 1980s. When weather effects are included, however, the average levels of smog have appeared to swing up and down over the years, without a clear trend. Such data are shown in Figure 6.15. Notice that the variations in the predicted ozone line closely follow the variations in the observed ozone concentrations. In fact, differences between the measured and predicted ozone levels are typically less than 0.01 part per million when a relatively sophisticated model is used. Accordingly, a significant fraction of the variations in smog may be confidently explained by changes in weather patterns.

An important conclusion to be drawn from Figure 6.15 is that the trend in ozone concentrations in Los Angeles is *downward*. The improving situation can be attributed to the regulation and reduction of primary pollutant emissions in the basin. There is another general lesson to be learned. The detection of long-term trends in a time series of measurements is often masked by the variability that occurs on shorter time scales. Some data processing is usually necessary to identify the trend in the data by the wavelike structures that cover the trend. Such a wave is obvious in Figure 6.15, for example. Several statistical techniques may be used to find the trend. The periods and amplitudes of the waves may be calculated and these variations removed from the data. In other approaches, the data are averaged in various ways to smooth out the high-frequency variations and retrieve the long-term trend. A "regression line"—like the trend line in Figure 6.15—can be directly calculated from a time series of observations by minimizing the average difference between the actual measurements and the trend line.

In all cases, such statistical analyses leave an uncertainty in the derived trend. But statistical analysis alone cannot establish cause-and-effect relationships. We may be able to determine mathematically the long-term trend in ozone levels, but we may not be able to identify the cause. If a logical and consistent explanation is available—such as a reduction in emissions leading to a decrease in average ozone concentrations—cause and effect may be established circumstantially.

The Geography of Smog

In the Los Angeles basin, a distinct spatial pattern in the trend in air pollution has also been noted. During the past decade, the greatest improvements in air quality have occurred in the eastern regions of the basin. The western regions have seen smaller reductions in smog. The explanation stems in part from the fact that the eastern basin historically has suffered much worse pollution than the western basin has. Hence, greater improvement would be expected when emissions are controlled. The eastern basin is downwind of the main pollution-emission sources. As these sources of primary pollutants are reduced, the accumulation of smog at the receptor sites should abate as well.

Another complicating factor is related to the manner in which pollutant controls have been implemented. During the 1960s and 1970s, the control strategy focused on hydrocarbon emissions, as these were most accessible (see Figure 6.22). Limitations

Figure 6.15   The concentrations of ozone in the Los Angeles basin from 1970 to 1988. The ozone concentration averaged over the entire basin and over the summer is given in parts per million by mass. The measurements were taken at the AQMD monitoring stations in the basin. A weather-based model prediction of the average ozone concentration is also shown. For comparison, a linear trend line, which represents the long-term change in ozone after variations associated with weather are removed, is drawn through the measurements (the long-term trend line is based on data that extends back to the mid-1960s). (Data compiled by the South Coast Air Quality Management District, 1988)

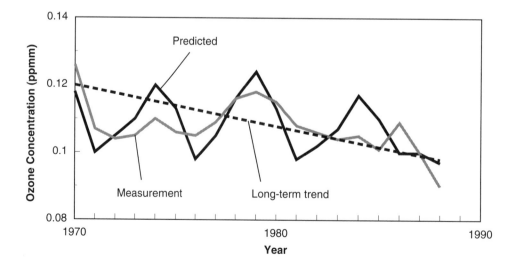

on refinery fumes and gasoline vapors were activated. On the other hand, nitrogen oxide emissions did not abate. The reactive hydrocarbons still generated ozone, only slightly farther downwind of the sources. The photochemical smog reactions were slowed somewhat but were not eliminated. The eastern Los Angeles basin received about the same amount of smog as in previous years.

During the 1970s and 1980s, emissions of nitrogen oxides were also reduced through the redesign of engines and the catalytic converters that scrubbed carbon monoxide from the exhaust stream. The simultaneous reductions in reactive hydrocarbons and nitrogen oxides led to a significant drop in smog over the entire basin, but particularly in the eastern (receptor) regions. The relationship between RH and NO_x emissions and ozone levels is depicted in Figure 6.12.

The Weekend Effect

Oddly enough, in Los Angeles, as the air quality has improved during the workweek, the smog on the weekends has gotten worse. The smoggiest day of the week typically used to be Friday: People racing home early, running errands, and going out to party add to the air pollution woes at the end of the week. Now, however, Saturday has recently surpassed Friday as the worst smog-producing day. The answer seems to lie in the repressed need to drive a car. During the week, we are encouraged to carpool or use public transit. Travel to and from work this way is fine, but trips to the shopping mall at lunchtime are less convenient. So we seem to be driving more on weekends to do all those chores left undone during the week. Saturdays have become hectic forays through crowded streets and aisles to get the groceries, dry cleaning, and a birthday gift for Aunt Sarah.

Weekends are also getaway time: long drives to the outskirts of the city to reach the rutted promised land, four-wheel-drive holidays spent ripping through the countryside, a festival of motorcycles, motor boats, motorized water skis, and snowmobiles. For the kids, there are scooters with engines. And then there's our love of the barbecue! The extra fuel burned and all the dust raised contribute to the worst air quality conditions seen during the week. The level of ozone in many areas is now highest on Saturday or Sunday, even though many industries are closed and truck traffic is reduced. Weekends have suddenly become hazardous to our health.

6.5   Haze and Visibility

Another important form of air pollution is haze, consisting of small particles of dust, soot, and other materials. Haze particles vary widely in size and composition. Figure 6.16 shows how particles are categorized according to their size and composition. This form of chart is frequently used to organize the diverse range of particles and their properties into a concise, orderly display. Note that not all the categories refer to airborne particles, or aerosols. For instance, soils consist of particles with a wide range of sizes. The finest soil particles are clay. When winds blow over dry lakes and riverbeds, dense clouds of clay particles can be raised. Tractors and off-road vehicles also mechanically lift these particles. Clay aerosols have a low fall speed and so can remain suspended for hours or days. On beaches and deserts, the finest particles have been blown or washed away over time, and the coarser sand and gravel usually remain. Strong winds can lift sand particles from the surface, but they quickly fall back to the ground (a typical sand particle drops about 1 meter per second).

In Figure 6.16, the sizes of particles are compared with the wavelengths of radiation (Section 3.2.2). At any wavelength of radiation, the particles that are most efficient for scattering that radiation are about 10 times smaller in radius than the wavelength. Here, the scattering efficiency refers to the total amount of light energy that a *fixed mass* of particles deflects. For a fixed mass of aerosols, the number of particles increases as the particle size decreases (that is, when collected together, the mass of the particles equals the specified mass). For each wavelength of radiation, there is an optimal size of particle for producing the maximum scattering for a given mass of particles. (You get the most "boing" for the buck with particles of optimum size.) In the case of visible radiation, with wavelengths of about 0.4 to 0.8 micrometers (μm), the optimal particle size is roughly 0.1 μm. According to Figure 6.16, the particles in ordinary haze and some smokes and fumes fit into this size range.

Sources of Haze

The sources of haze particles are quite varied. Wind-blown dust was mentioned earlier. Estimates of the total mass of dust raised are very difficult to make and are highly uncertain. Perhaps 200 million tonnes of soil dust particles less than 10 μm in diameter are

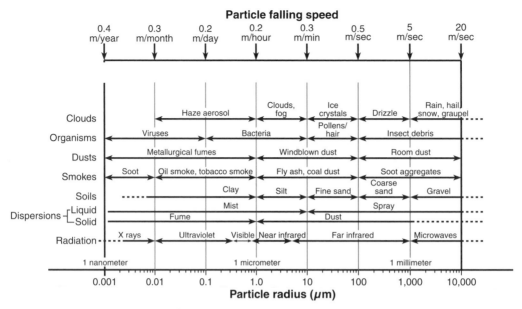

Figure 6.16 The categorization of particles by size and by composition. The particle sizes are indicated along the bottom of the figure. The sizes are compared with the wavelengths of radiation in several spectral bands. The different types of particles are indicated on the left side of the figure. The speeds at which particles of several specific sizes fall through the lower atmosphere are defined along the top (that is, the number of meters that a particle of the specified size will fall during the time interval indicated). The fall speeds correspond to aerosols with the specific density of water, 1 gram per cubic centimeter. Particles with greater densities fall faster, in proportion to their density relative to water.

lofted by winds each year worldwide. Some of these particles are carried long distances in the upper air currents. Dust from the Gobi in China settles over Hawaii. Fine sand grains from the Sahara make their way over Bermuda. The largest source of airborne particles—in terms of the total mass injected into the atmosphere each year—is sea spray. When waves break or bubbles burst at the surface of the oceans, a spray of seawater drops is formed. When these drops evaporate, they leave behind a particle of salt. As much as 900 million tonnes of sea salt spray may be produced each year. These particles are quite large, however, and quickly fall back into the sea. Except near coastlines, where salt aerosols rapidly corrode metal and paint on homes and cars, salt particles are not a significant factor in visibility over land.

The hydrocarbons (terpenes) emitted by plants are another major source of haze. In the atmosphere, terpenes react with ozone and other oxidants to form secondary hydrocarbon species that readily condense into aerosols. Among the most common terpenes are pinene and limonene. A demonstration of this process, which is described in Appendix B (Demonstration 2), can be carried out in the classroom.

The total mass of terpene aerosols formed annually may amount to several hundred million tonnes.

Photochemical haze is also a major source of airborne particles. As much as 250 million tonnes of sulfate aerosol and another 100 million tonnes of nitrate aerosol are formed each year as a result of sulfur dioxide emissions around the world.[16] Most of the photochemical aerosol forms as the result of condensation processes. The condensation sites may consist of preexisting particles of dust, smoke or fume, or liquid droplets of mist or fog. New particles are continually formed by nucleation (Section 6.5.3). The vapors that form these aerosols prefer to be in the condensed state; in other words, they have very low vapor pressures. Among the compounds that like the condensed state are sulfuric acid, nitric acid, ammonia, and a host of organic compounds. Organic

16. The total mass of the aerosols generated from primary gases like SO_2 and NO is generally greater than the total emitted mass of the primary gases themselves. The primary gases are chemically transformed into heavier molecules of sulfate and nitrate, respectively (Sections 3.3.4, 10.2.1, and 10.2.2). Sulfate (in the form of condensed sulfuric acid) is about twice as heavy as the sulfur dioxide precursor gas, and nitrate (as nitric acid) is about twice as heavy as nitric oxide.

aerosols generated from anthropogenic hydrocarbon emissions may add up to several hundred million tonnes per year, although the numbers are very uncertain. Although this is less than the natural source attributed to terpenes, the localization of the anthropogenic organic aerosols can create much more severe local problems. The global natural sources of sulfate and nitrate particulates are roughly 100 million tonnes each. (The biogeochemical cycles and budgets of these compounds are discussed in Sections 10.2.1 and 10.2.2.)

Direct sources of haze from industrial and other activities are particularly diverse. Consider the following: fine dust created by the mechanical handling of coal, cement, and other building materials during application and demolition; fly ash and smoke from coal-fired furnaces; organic aerosols originating from oil refineries and food-processing plants; soot emitted from buses and trucks and oil mists from old automobiles; fumes from smelters and metal works; rubber fragments from tires and powder residues from brake linings; smoke from the burn-off of agricultural wastes and from barbecues; dust raised from roads, including the resuspension of particulates deposited on roadways. To be sure, this list is far from exhaustive. Each person probably has his or her own favorite hateful haze, from the smoke pouring out of a neighbor's barbecue pit or fireplace, to that old Chevy burning oil on the highway.

6.5.1 TOTAL SUSPENDED PARTICULATE

The term **total suspended particulate (TSP)** refers to the total mass concentration of aerosol particles present in the air (that is, the mass of aerosol per unit volume of air, usually measured in micrograms of aerosol mass per cubic meter of air [$\mu g/m^3$]). To determine this amount, all the suspended aerosol particles in a known volume of air are collected and weighed. On the other hand, only particles smaller than about 10 micrometers in diameter can be deposited in the respiratory system—the bronchial tubes and deep in the lungs. The particles with diameters smaller than 10 micrometers (μm) are referred to as **respirable suspended particulate (RSP)**, which are equivalent to PM10. The standards for respirable particulates in smoggy air (Table 6.2) cover a range of concentrations from about 50 to 150 $\mu g/m^3$. Under typical circumstances, most of the mass of the aerosols is contained in the respirable

fraction. If you trapped 1 cubic meter (equivalent to the volume of an average bathtub) of polluted air and filtered out all of the particles in that air, they would weigh, in total, only about 0.0001 grams. The number of particles involved would add up to some *10 billion or more*. That is nothing to sneeze at. (See Section 7.1.2 for a more detailed discussion of the respiratory tract and respirable particles.)

Figure 6.17 illustrates the regional distribution of total suspended particulate in the Los Angeles basin. The areas that most regularly experience TSP episodes correspond to the areas that also experience the most frequent ozone episodes. The similarity in the behavior of ozone and TSP in Los Angeles smog suggests that the two pollutants have common sources (compare Figures 6.14 and 6.17, for example). In fact, much of the particulate matter that composes TSP is the by-product of photochemical reactions that also produce relatively high ozone concentrations.

Daily and seasonal variations in TSP in the Los Angeles basin are illustrated in Figure 6.18. The daily variation shows an increase during the day, with lower concentrations in the evening. The photochemical source of the particulates is implied by the similarity in their behavior to that of nitrogen dioxide and ozone concentrations (compare Figure 6.18[a] with Figure 6.7[a] and 6.9[a]). The advection of polluted air across the basin also plays a role in the diurnal behavior at some locations. The higher early-morning particulate abundances may be caused in part by the cooler temperatures and higher humidities in the morning, which increase the apparent total aerosol mass by the condensation of water vapor on the particles.

The annual variation in TSP follows the pattern for ozone (Figure 6.9[b]). The highest concentrations occur in summer even though the mixed layer is usually much deeper in summer. These observations reinforce the importance of the photochemical generation of secondary aerosols in smoggy air.

The TSP in an urban area has a complex composition that reflects the innumerable primary sources of gases and aerosols that eventually contribute to the suspended particulate. The condensed materials determine the optical properties of the aerosols and, hence, their effects on visibility. Most of the particulate material has low absorptivity and mainly scatters light. Particular components, notably soot from combustion, may cause substantial absorption in some cases. The composition of the suspended matter also

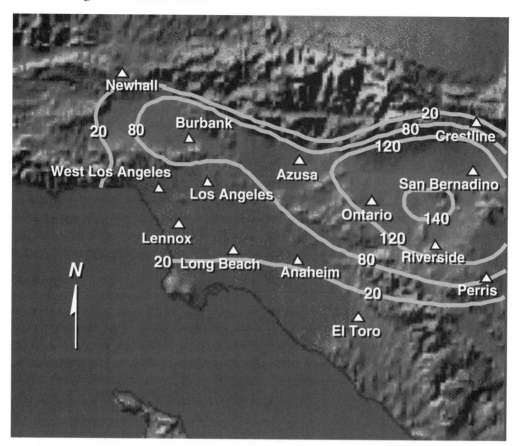

Figure 6.17   The distribution of particulate haze in the Los Angeles basin. The contours outline regions subjected to total suspended particulate (TSP) concentrations exceeding the California state standard (24-hour average greater than 100 micrograms per cubic meter [mg/m³]) for a specific number of days each year, indicated by the shading within the contour (the shading scale is defined at the bottom of the figure). The contours are nested, with the most frequent TSP episodes occurring in the area bounded by the inner contour. The data shown are for 1983, but are similar from year to year. The pattern of haze in the basin shows distinct peaks in both the central urban area and the more rural western region, as explained in the text. [Data compiled by the South Coast Air Quality Management District, 1983]

determines the potential health impacts of breathing the particle-laden air or of ingesting the particles with food and drink. The trace constituents in aerosol particles vary over a wide range. The proximity to specific sources of toxic aerosol materials is critical to estimating the possible exposure and harm.

Roughly 85 percent of suspended particulates associated with human activities originates from fixed sites and 15 percent from vehicles. Near coastlines, sodium chloride (NaCl, ordinary table salt) is a significant component of the aerosols. More commonly, sulfates dominate the trace chemistry of TSP. The following information on the composition of TSP is derived from a limited number of direct measurements in polluted air and does not apply to

all places and times. An average concentration of 10 $\mu g/m^3$ of sulfate, with maximum concentrations of around 100 $\mu g/m^3$, is seen in urban areas. Nitrates contribute roughly 5 $\mu g/m^3$ on average and up to 50 $\mu g/m^3$ at most to the TSP. Organic compounds may add 5 to 10 $\mu g/m^3$ on average. Ammonium is generally a small component of TSP (about 1 $\mu g/m^3$ on average), although in some areas the concentrations may approach 75 $\mu g/m^3$. Trace metals are also common, including iron (average, ~1–2 $\mu g/m^3$; maximum, ~20 $\mu g/m^3$), lead (average, ~1 $\mu g/m^3$; maximum, ~10 $\mu g/m^3$), zinc (average, ~1 $\mu g/m^3$; maximum, ~60 $\mu g/m^3$), and manganese (average, ~0.1 $\mu g/m^3$; maximum, ~10 $\mu g/m^3$). In heavily polluted cities, average peak TSP abundances of

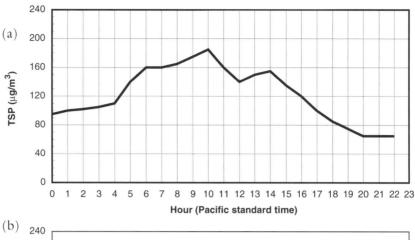

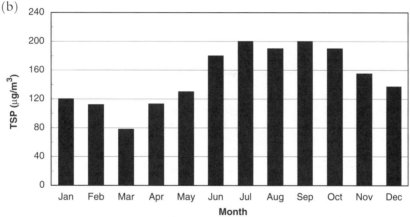

Figure 6.18   The typical behavior of particulates in polluted air using observations from the Los Angeles region. (a) Variation in the total suspended particulate (TSP) loading over the course of an average summer day. The data were taken as 2-hour averages for particulates and were collected during July and August 1977. (b) The annual variation in the total particulate loading near Riverside, California, in the Los Angeles basin. The TSP concentrations are given as average values for each month between 1980 and 1983. [Data compiled by the South Coast Air Quality Management District]

about 50 to 100 μg/m^3 are often detected, with upper limits to the TSP loading of roughly 1000 μg/m^3!

6.5.2   SEEING THROUGH AIR

Visibility is, simply put, the ability to see through the atmosphere. So why should we be concerned about visibility? There are at least two reasons. The first is related to aesthetics. On a smoggy day with low visibility, the atmosphere closes in, enshrouding us in a dense milky cloud of haze. We see shadows and outlines instead of clear, sharp images. Colors are washed out. Scenic backdrops—mountains and seas—are gone. We feel oppressed under the smoky pall. The second reason is related to health. If the air is dense with haze, reducing visibility, the air we breathe is laden with toxic particles. These microscopic bullets are deposited deep in the lungs (Section 7.1.2). They carry irritating acids, chemicals that cause cancer, and heavy metals like lead that can damage the brain and other organs (Chapter 7). Exposure to this haze initiates a number of debilitating respiratory diseases, including asthma and emphysema. No wonder visibility has become a major issue connected with smog.

How far can you see through the atmosphere? Visibility is usually gauged in terms of the ability to discern a perfectly black object against a perfectly white background. A perfectly black object, by definition, absorbs *all* the light falling on it. At some distance, we can barely make out such an object.

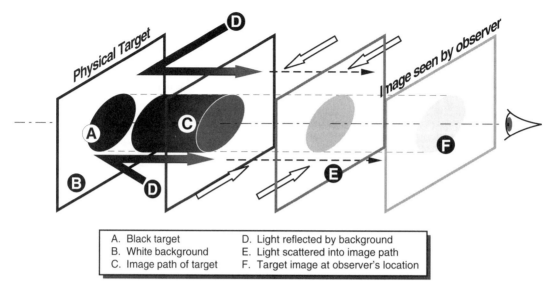

A. Black target	D. Light reflected by background
B. White background	E. Light scattered into image path
C. Image path of target	F. Target image at observer's location

Figure 6.19 The effect of airborne particulates (aerosols) on the visibility of an object at a distance. The radiation scattered by particulates and gas molecules in the direction of the observer's eye masks the light originating from the object and reduces its contrast against the background (an object is discernible only if it has contrast in color or brightness with respect to nearby objects). Generally, the greater the distance is at which an object lies, the more interfacing scattered radiation will accumulate along the path of observation, and the less visible the object will be from the point of view of an observer.

That distance is the defined as the **visual range**, or **visibility length**. Of course, this all is very qualitative. Different people may be able to see the same object at different distances. The exact location of the observer, and thus the path of observation between the object and observer, may affect the apparent visibility. Lighting conditions and the texture of the scene around the object influence the perception of contrast.

Visibility is reduced (or degraded) by scattered radiation (Figure 6.19). When we view an object at a distance, we are attempting to differentiate the light from the object (or lack of light, in the case of a black object) from the illuminated background scene. The light originating from the object (usually reflected light, although it can be radiation emitted by a fire or an incandescent bulb, for example) may be scattered on the way to the eye. The reduction in the direct light intensity from the object tends to reduce its contrast against the background, causing it to fade into the background scene. That seems logical. If all the light originating from an object were scattered before reaching the eye, no information would be available to identify the object. We would not see it.

Air, including the particles suspended in it, scatters radiation and reduces the visual length. At each point along the viewing path between an object and an observer, the quantity of haze particles can be measured, and the intensity of the scattering effect can be calculated. These local scattering properties are usually characterized by a single number, which can be the total mass concentration of suspended particulates, the optical extinction coefficient of the air, or the local visual range (visibility length). Because the optical properties of a gas vary with wavelength, it is normal practice when specifying visibility to use the middle-visible wavelength of 550 nanometers (nm). This corresponds to green light. The effect of color on visual range is relatively small, however, and so will be ignored here.[17]

The mass concentration of TSP has already been discussed. The extinction coefficient for light is described in Section 3.2.2. Basically, extinction refers to the total effect that a material has in scattering and absorbing electromagnetic radiation along a

17. The effect of color is noticeable when Rayleigh scattering is the dominant obscuring effect. Rayleigh scattering is important only on the very clearest days, however. Mountains in the far distance may appear somewhat purple in color because of the dependence of Rayleigh scattering on wavelength, creating the "purple mountain's majesty." In urban smog, the absorption of light by various gases and particles at certain wavelengths may also tinge the air with color (Section 3.2.2).

straight path. The extinction coefficient, ε_p, is specified as an inverse distance, or length, say in units of inverse meters (m^{-1}). The effect of the material on radiation is substantial when the length of the path is comparable to the *inverse* of the extinction coefficient, in the same units. For atmospheric radiation, the extinction coefficient is conveniently defined per kilometer (km^{-1}). Then, roughly, the extinction effect occurs over distances of $1/\varepsilon_p$ kilometers.

The mass density of TSP, ρ_{TSP}, is typically measured in micrograms per cubic meter of air (μg/m^3). With these units, it turns out that the particulate mass density, ρ_{TSP}, and the extinction coefficient, ε_p, can be simply related for urban smog as

$$\varepsilon_p \cong \frac{\rho_{TSP}}{250} \qquad (6.14)$$

The symbol $\cong$ indicates an approximate relationship, for use in making estimates.

The visibility length, l_v, is also related to the extinction coefficient. If visibility is defined in terms of a black object at a distance, the visual length in kilometers is

$$l_v = \frac{3.9}{\varepsilon_p} \cong \frac{1000}{\rho_{TSP}} \qquad (6.15)$$

This relationship can be used once the particulate loading is measured. It follows that as the particulate density increases, the visibility decreases. Doubling the mass of suspended particulates halves the visibility. With typical values of TSP in polluted air of 100 μg/m^3, visual ranges of perhaps 10 kilometers are predicted by Equation 6.15. This is a relatively short distance for viewing.

The effect of recession on our ability to see an object is depicted in Figure 6.20. As the object recedes from an observer, it grows fainter. This effect is enhanced by a larger particulate concentration in the atmosphere and by the elevation of the sun in the sky. If the sun lies in front of the observer in the direction of viewing, the forward scattering of sunlight toward the observer will be intensified. The air appears much hazier. With the sun behind the observer, the sunlight backscattered to the observer is reduced, and the air seems much cleaner. This striking effect is easy to observe when the sun is low in the sky, simply by looking in different directions and noting the change in haziness. The recession effect of

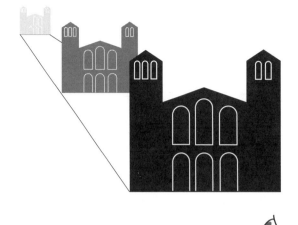

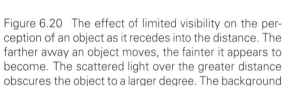

Figure 6.20 The effect of limited visibility on the perception of an object as it recedes into the distance. The farther away an object moves, the fainter it appears to become. The scattered light over the greater distance obscures the object to a larger degree. The background haziness and the object blend together.

visibility is also obvious in a scene with hills or a coastline stretching into the distance. The hills fade into the distance.

If the atmosphere is exceptionally clear (with no suspended dust or haze), one can see over distances up to several hundred kilometers. The greatest visibility is found in dry desert environments, with low winds. The major factor limiting visibility under these circumstances is Rayleigh scattering by air molecules (Section 3.2.2). If there were no air at all, viewing objects at a distance would be similar to viewing objects on the moon, where there is no atmosphere. Objects could be clearly discerned over thousands of kilometers. The fact that we can clearly see the moon and planets on a clear night indicates the potential visual distances in space. Back on Earth, the visibility is greater if a region is elevated above sea level, so that the air density and the Rayleigh scattering are reduced.

In a fog or cloud, the visibility can be very low. An impenetrable wall may appear right in front of you. In a car in the fog or in a blizzard, turning on the headlights can blind you, because the headlamp's light scattered from the water particles is so intense that nothing beyond the front of the car is visible. Even in a thin fog, the visual range may be reduced to 1 kilometer or less. The visibility length is greater in hazy air, ranging up to 10 kilometers. For longer visual lengths, the atmosphere is said to be "clear" to "exceptionally clear."

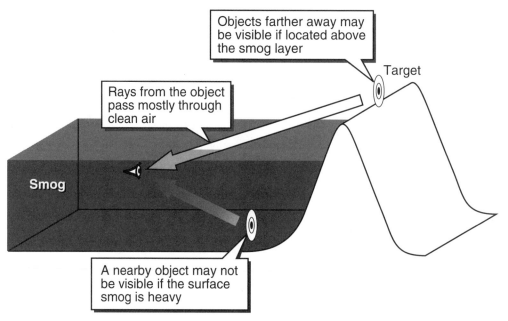

Figure 6.21   Changes in the visibility of a distant object caused by changes in the point of observation. Looking through dense smog over long paths at the ground may completely obscure a distant object near the surface because the visibility is poor. Viewing in the upward (or downward) direction through the same layer of smog produces much less degradation in the viewing. The apparent visibility at a fixed distance can be quite sensitive to the spatial relationship between the observer and the object.

Note that visibility is often difficult to determine, for a variety of reasons, including the fact that it generally varies with the direction of viewing and because visibility-degrading pollutants are not evenly distributed. Also, if the viewing point and the object being viewed are elevated, as in Figure 6.21, it may be possible that one is looking through clean air lying above a layer of polluted air trapped below a temperature inversion.

Figure 6.22 shows the geographical distribution of visibility across the United States. The Rocky Mountain and Great Basin regions have the finest visibility because they have the least pollution. In addition, the air is very dry in these regions, which minimizes the impact of humidity in degrading visual range. In the midwestern and southeastern United States, industrial pollution is heavy, and high humidity in the summer exacerbates the visibility problem. The Ohio River valley shows particularly low visual lengths for these reasons.

Ozone and Visibility

Figures 6.14 and 6.17 imply that high particulate loading and high ozone concentrations are strongly correlated. The co-production of ozone and photo-chemical haze is a reasonable mechanism for this correlation. Simultaneous measurements of ozone concentrations and visibility in the Los Angeles basin clearly reveal this relationship. Typically, when ozone concentrations peak between 0.1 and 0.2 ppmv, the visibility lies roughly in the 10 kilometer range. By the time that ozone abundances have reached 0.3 ppmv, the visibility is typically 5 kilometers or less. Ozone itself has no direct impact on visibility. Rather, it is mainly the particulates generated in smoggy air that degrade visibility.

The haze that accompanies ozone has diverse sources. Primary particles of smoke and dust are significant. In some areas, a film of sooty dust can coat anything left outdoors (including the family dog) in a matter of days. The photochemical reactions of primary organic emissions, which lead to the production of ozone, also generate secondary organic compounds that condense as aerosols. The nitric oxide that also drives ozone production is converted to nitrates on particles. Similarly, sulfur dioxide, which has many sources in common with nitric oxide, is chemically transformed into sulfate aerosols.

Controlling ozone concentrations should also improve visibility. The problem of visibility is more

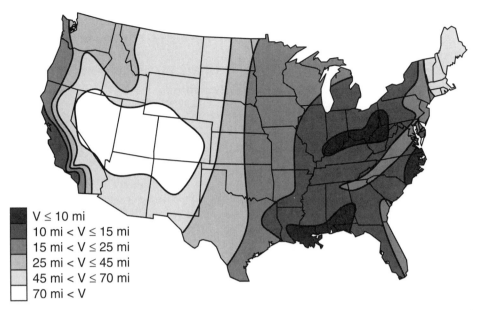

■	V ≤ 10 mi
▓	10 mi < V ≤ 15 mi
▒	15 mi < V ≤ 25 mi
░	25 mi < V ≤ 45 mi
░	45 mi < V ≤ 70 mi
□	70 mi < V

Figure 6.22 The distribution of average annual visibility in the United States. Visual ranges were derived from measurements taken at suburban and rural airports around the United States to avoid excessive contamination from urban air pollution. The data were collected from 1974 to 1976 and averaged over this period. The visibility lengths in *miles* are defined by the shading bar. (Data from Trijonis, J. and D. Shapland, *Existing Visibility Levels in the U. S.: Isopleth Maps of Visibility in Suburban/Nonurban Areas during 1974–1976.* EPA-600/3-78-039. Washington D. C.: US Environmental Protection Agency (1979).)

intractable than that of ozone, however, because visual degradation has a spatial component. One can see far beyond the local atmosphere that is breathed. We discern an enormous volume of space visually. If the visual range in a city falls below a certain value, the air will look soupy. In Los Angeles, for example, the basin is flat and the view can be unobstructed for 50 kilometers or more. If the visual length is 5 kilometers, mountains in the distance will be awash in haze. But even with a 20-kilometer visual range the mountains would be invisible. The local view would improve somewhat, yet the overall impression of suffocating smog would remain in most places. To solve the problem, visibility would have to increase by factors of three or more. Even if pristine air could be imported, visibility would be limited.

In the View of National Parks

The problem of visibility has been obvious in many cities for a long time. One new resident of Pasadena, California (downwind of Los Angeles), is said to have resided in the soupy air for several months before realizing that the San Gabriel Mountains towered just behind his home. It is a rare summer day when the atmosphere in the Los Angeles basin is not

smoky. The street corridors in cities like New York and Chicago also are often filled with gray haze.

The problem has spread to many national monuments. People travel to the Grand Canyon, Yosemite Valley, and the Great Smoky Mountains to *see* their natural beauty. Poor visibility, according to a National Academy of Sciences study, "reduces contrast, washes out colors and renders distant landscape features indistinct or invisible." In recent decades, the visibility has steadily declined to the point that visitors are sometimes faced with dense smog-filled air. Of course, the air pollution is not nearly as bad as in a major urban center. The long viewing distances and otherwise pristine conditions accentuate the impact of any errant haze. Some of the aerosol is indigenous. To wit, the Smoky Mountains were originally named for the "smoky" atmosphere in the area, a background haze produced from terpene hydrocarbons emitted by local vegetation. Similarly, organic hazes in national forests and windblown dust in arid zones can obscure the view.

The degradation of visibility in the national parks is caused by local and remote sources of pollution. In the case of the Grand Canyon, for example, a local power plant—the Navajo Generating Station about

80 miles away in Page, Arizona—has been identified as the source of obscuring haze on certain days, and Los Angeles smog contributes significantly on other days. (The long-range transport of pollutants is discussed in Section 5.2.3.) Studies of the haze problem have concluded that reductions in haze (improvement in visibility) require a regional approach in controlling air pollutants. Local sources of aerosols must be abated, but distant sources must be regulated as well.

Poor visibility does not hurt anyone, at least not in the broadest sense, but it is an aesthetically unpleasant property of a polluted environment. Clear air is clean air. The sight of a purple mountain's majesty is more uplifting than views of a socked-in smoggy city. Health should be the principal concern in areas of heavy air pollution. Ozone and other toxic pollutants in the air must be reduced, but even when that is done, however, visibility will most likely remain poor. Visibility thus may be the next critical issue in air pollution. Poor visibility also emphasizes that we are immersed in particles that can aggravate respiratory problems. Haze can make life in a city much like a yogurt sundae on a plate: short and not so sweet.

6.5.3 ACID PARTICLES AND FOG

A large fraction of the material comprising the aerosols that reduce visibility is in the form of acids, particularly sulfuric acid, nitric acid, and certain organic acids. These compounds are highly corrosive. The local formation and deposition of acidic aerosols and fog can cause health problems as well as erode building materials, including metals and statuary. (The chemistry of inorganic acids [sulfuric, H_2SO_4, and nitric, HNO_3] was described in Section 3.3.4 and is discussed further in Sections 9.2.2 and 10.2.1) Here, our interest is the local conversion of primary pollutants into acids. Sulfur dioxide, for instance, is converted to sulfuric acid by the overall chemical process

$$SO_2 + OH + O_2 + H_2O \rightarrow \cdots \rightarrow H_2SO_4 + HO_2 \quad (6.16)$$

Similarly, nitrogen oxides are converted into nitric acid by the following reaction:

$$NO_2 + OH + M \rightarrow HNO_3 + M \quad (6.17)$$

These chemical processes can occur near sources of SO_2 and NO_x, creating locally concentrated acidity. (In Chapter 9, the problems associated with the long-range [regional] transport and deposition of primary gases and secondary acids are discussed.)

The acids condense on existing aerosols or form new aerosols through a process called **nucleation**. Under certain conditions, a vapor can become dense enough to condense spontaneously. This may happen if the concentration of vapor molecules is increased or if the temperature of the vapor is lowered. The vapor then condenses in the form of microscopic particles, or nuclei (not to be confused with the nuclei of atoms; the aerosol "nuclei" actually consist of dozens of atoms or molecules of the vapor stuck together). The aerosol nuclei grow into larger particles by the continuing condensation of the vapor. The accumulation of acids on aerosols generally causes them to grow to sizes reaching about 1 micrometer (1×10^{-6} m) in diameter. The aerosols generated in this way are secondary particulates.

Other airborne materials may react with the acids in the aerosols to form sulfate and nitrate salts. For example, certain minerals in soil dust can interact with the acid aerosols, producing acid salts such as calcium nitrate, potassium sulfate, and sodium chloride. Ammonia emitted by biogenic activity (decay of organic matter) easily combines with sulfuric and nitric acids to form ammonium salts (Section 10.2.2). The ammonia gas also combines directly with nitric acid in polluted air to nucleate aerosols of ammonium nitrate:

$$NH_3 + HNO_3 \xrightarrow{\text{nucleation}} NH_4NO_3 \, (\text{aerosol}) \quad (6.18)$$

These aerosols then act as condensation sites for sulfates, organic compounds, and other condensable materials, including water.

A Corrosive Urban Soup

The aerosols generated by nucleation and condensation of soluble compounds formed in chemical reactions often respond strongly to changes in the local humidity. Such humidity-sensitive materials include acids and their salts, as well as many organic compounds. These materials absorb water when the relative humidity is high enough. As a result, typical haze particles tend to expand when the humidity rises. Their light scattering efficiency and impact on visibility increase accordingly.[18] Therefore, on a humid smoggy day, the visibility is usually much lower than on a dry smoggy day.

If the haze aerosols are dry to begin with, they will spontaneously form a liquid droplet containing water, at their **deliquescent point** (this process is called **deliquescence**). Deliquescence is defined as the liquefaction of a dry material by the absorption of moisture from the surrounding air. Many salts have a specific humidity for deliquescence. Dry ammonium nitrate, for example, deliquesces (that is, forms a liquid solution of ammonium nitrate dissolved in water) at a relative humidity of about 65 percent. Below that relative humidity, ammonium nitrate aerosols are crystalline. Exposed to higher humidities, the aerosols liquefy spontaneously into haze droplets. The haze is not strongly acidic, of course, because ammonium nitrate is not an acidic compound. But these liquid aerosols can serve as centers for further condensation and chemical reaction.

The acid aerosols and their chemically altered relatives generally fall into the "respirable" suspended particulate category. Such particles are particularly hazardous because they not only are collected deep in the lungs, but also carry corrosive acids and, possibly, toxic metals leached from embedded solid particles composed of fly ash, soot, and metallic fumes.

When the relative humidity in polluted air reaches the saturation point, acid aerosols and deliquescent particles can form fog droplets. The saturation point for water corresponds to a relative humidity of 100 percent. The relative humidity is the percentage of saturated water vapor in an air mass. Water vapor is saturated when the water vapor partial pressure (or the equivalent concentration of water molecules) is equal to the equilibrium vapor pressure over liquid water (Section 12.1.2). When the relative humidity exceeds 100 percent, the vapor is "supersaturated." Any excess water vapor above 100 percent condenses on any available material, including the soluble aerosols. The condensed water becomes fog.

18.   Here we are considering a fixed mass of acid or salt on which a variable amount of water condenses. For a fixed mass of aerosol, there is an optimum size to maximize their scattering effect. However, for any preexisting collection of aerosol particles, the absorption of water by those particles further increases their scattering effect.

Fog Smog Factories

Fog polluted with smog (recall that smog was originally named for the combination of smoke and fog) has several unsavory properties. The polluted fog particles act as centers for the collection and concentration of toxic pollutants. Even after the fog has dissipated (that is, evaporated), the haze that remains is generally much worse than the original particulate. Each fog droplet leaves behind an aerosol particle. Initially, fog droplets form by the condensation of water vapor on soluble aerosols, often through deliquesence. When water evaporates from the droplet, the original aerosol materials are left as a residue particle. In the meantime, the fog droplet has accumulated a variety of compounds from the polluted atmosphere around it and has mediated chemical reactions that increase the mass of the residual aerosol. The fog droplets have transformed smog gases into smog particulates.

The total quantity of unhealthful airborne chemicals may be enhanced by fog. The vapors that condense on fog droplets might otherwise be removed by deposition or photochemical transformation. Acids in particular are more easily formed in foggy air than in dry air. Certain organic compounds can also condense and stabilize on the surfaces of fog droplets. The relatively large surface area of fog droplets is effective in scavenging condensable vapors. The dilute aqueous solution in a fog droplet provides an ideal medium for chemical reactions. Fog is a handy temporary chemical factory, setting up shop for a while to transmogrify primary pollutants and then vaporizing, leaving behind noxious products to be inhaled.

Fog droplets laced with acids are readily deposited on sensitive surfaces. Very small aerosols are not taken up by surfaces as efficiently as are fog droplets and the progeny aerosols of fog. The surfaces of interest include the leaves of plants, limestone and marble exteriors of buildings, corrodible metal structures, and human lung tissue. The aerosols produced by fog are also nicely respirable. In general, haze formed in the aftermath of a fog consists of larger particles that can be drawn deeply into the lung passages and efficiently deposited there (Section 7.1.2).

In southern California, acid fog has been detected that is more acidic than the most acidic rainfall ever recorded. (The broad issue of environmental acidity is addressed at length in Chapter 9.) The production of acid fog in coastal zones, such as along the western rim of the Los Angeles basin, is an important health issue. People who are sensitive to airborne particulates can suffer excessively from acid fog. Earlier this century, the killer fogs of London and other cities acted through the respiratory track. In Los Angeles, the impacts of acid fog are felt most in the coastal areas, although the aerosol progeny of acid fog may affect a much larger area and population.

6.6 Controlling Smog: Everyone's Job

Photochemical smog has become a serious problem not only in the major cities of the developed nations, but also in developing countries struggling to achieve economic growth. In many places, controls on pollution are of secondary interest. For example, in the fragmented eastern European nations, industrial output has been sought almost without regard for the environment. The economic powers are often no more ethical, shipping their most serious pollution problems to the Third World. Automobiles are the primary culprit in urban smog, although regional meteorological factors also play a role. The United States has the largest highway network in the world by far. Even so, the average distance traveled per capita in the United States is about 10,000 miles per year, compared with 12,000 in Denmark, 16,000 in Yemen, and 24,000 in Iraq.[19] More striking, the average densities of cars in Hong Kong (420 per mile of road), Kuwait (220/mile) and Japan (90/mile) are much higher than in the United States (48/mile). In the United States, the number of accidents (120 injury accidents per 100 million miles driven) is also much lower than, for example, in Costa Rica (1300), South Korea (1400), Colombia (1500), and Rwanda (3200). It appears that in addition to air pollution, driving habits could be greatly improved around the world.

The patterns and trends in smog discussed in this chapter can provide important clues to where, from the point of view of clean air, it may be most desirable to live and work. Unfortunately, not everyone can live in a small town far from urban sprawl. But knowledge and understanding of smog can help you to make a better choice of life-style. Knowledge may

allow you to identify cities that have the lowest smog levels and pose the smallest long-term health risks to you and your children. Knowledge can make you aware of the aesthetic problems you might expect to encounter in your new environment. Knowledge is personal empowerment.

6.6.1 REDUCING EMISSIONS OF PRIMARY POLLUTANTS

Figure 6.12 suggests that in order to control urban air pollution, the emissions of primary pollutants, particularly nitrogen oxides and reactive hydrocarbons, must be limited. Indeed, following on the heels of the discovery of the origin of photochemical smog came legislation to restrict emissions from automobiles and other sources. The reduction in emissions of primary pollutants from automobiles over time is depicted in Figure 6.23. During the 1960s and 1970s, emissions per vehicle dropped by 90 percent. Even though the total number of vehicles in use has increased, the overall effect of emission controls has been to decrease total emissions. The mandated reductions in automobile emissions in California are more stringent than anywhere else in the United States.

Despite opposition from automobile manufacturers, the Environmental Protection Agency recently allowed California to accelerate the abatement of exhaust pollutants. California has led the way in demanding lower emissions from car producers. Typically, this has resulted in the rapid development of less-polluting technologies. By 1998, California is asking the automobile industry to include at least 2 percent of "zero emission" vehicles in all California sales. These could be electric-powered cars.

Heaps and Waivers

Older cars emit more pollutants. They also are the most frequent "smokers," or heavy emitters of oil and other fumes, on highways. Compared with modern vehicles, which are designed to minimize air pollution, 1960 Fords and Chevys are an environmental disaster. Even since the mid-1970s, automobile exhaust emissions have been significantly reduced through the application of new technology and improved design. For example, between 1975 and 1991, overall hydrocarbon emissions decreased from an average of about 0.9 gram per mile driven (gpmd) to 0.4 gpmd. Carbon monoxide was cut

---

19. The availability of cheap gasoline in oil-producing countries and the sparse development on huge tracts of open land are obviously important to determining the long-range driving habits in some countries.

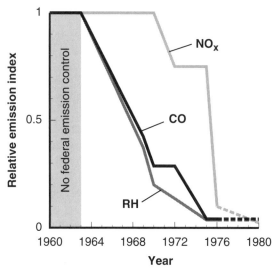

Figure 6.23 The trend in automobile emissions of carbon monoxide, nitrogen oxides, and reactive hydrocarbons since 1960. Pollutant emission indices are given as relative values compared with those of the early 1960s, before the adoption of engine emission standards and controls. The emission index is normally defined as the mass of pollutant released for a specific mass of fuel consumed. Since the mid-1970s, pollutant emission from new vehicles, when they are optimally tuned, has been less than one-tenth of the emissions from vehicles manufactured in the early 1960s. Emission trends have continued downward because of advances in the technology of catalytic converters and increasingly stringent requirements for fuel efficiency. (Data from the Environmental Protection Agency.)

from 9.0 gpmd to 7.0 gpmd. Nitrogen oxide emissions were slashed from 2.0 gpmd to 0.4 gpmd. These technological breakthroughs have led to major improvements in urban air quality.

Nevertheless, the highways are still filled with clunkers and heaps. Amazingly, in California, the number of old cars still registered includes more than 100,000 vehicles of 1958 vintage or earlier, 400,000 cars produced between 1959 and 1967, 200,000 in 1968 and 1969, and so on. In all, nearly 2 million vehicles with emission-control systems designed before 1980 are still operating in southern California. The numbers are staggering. People like to hold on to their cars, like family jewels. They are repaired and patched and driven into the ground. The older a car gets, the more offensive it usually becomes, compared with new vehicles. The logical solution is to remove the worst smokers from the roads.

In southern California, the AQMD is allowing companies to buy and destroy old heaps in exchange

for pollution credit. When an industry buys an old car in California and converts it to scrap metal, the home factory is allowed to emit a little more pollution. The credit is equivalent to roughly 80 percent of the pollutant emissions (of hydrocarbons and nitrogen oxides) estimated by the AQMD for each vehicle operating over a 3-year span. The estimated vehicle emissions take into account usage, gas mileage, and the average pollutant emission factors for each make and model of car. In response to the program, heavy polluters have sought out thousands of old clunkers to gain the right to continue polluting. Instead of instituting emission controls on smokestacks, for example, a firm can buy outdated cars and delay more expensive plant modifications. In theory, the net amount of pollution emissions should decline in proportion to the number of old cars that are taken off the roads. Of course, "new" old cars could be driven in from out of state, creating an interesting market for clunkers in California. The companies participating in the plan include oil refineries, aerospace manufacturers, furniture producers, and utility companies. The AQMD expects that 10,000 to 50,000 cars will be "cashed in" every year. The participation in and effectiveness of the heaps-for-pollution plan have not been evaluated.

Interesting complaints have arisen concerning the car trade-in scheme. Some people have pointed out that the plan may discriminate against low-income workers who can afford only an older make of car and who cannot usually afford the costs of keeping a commuter special in top running form. Collectors who covet antique automobiles complain that destroying the clunkers will dry up their supply of spare parts; they claim that the smokers and rattlers are "a part of history." The firms that actually dismantle the old cars note that the plan will not be profitable to them, since the cars cannot be sold for used parts and must be literally destroyed. Finally, some environmentalists complain that the trade-in plan simply puts off an inevitable industrial retrofit to reduce emissions at the source. The AQMD, they argue, should not be encouraging polluters to continue old habits by buying up old cars.

Another plan developed by the AQMD is to sell **pollution waivers** to industry. The scheme is designed to bring market forces into the air pollution-control arena. The scheme is one element of a long-term goal to reduce emissions of primary pollutants in the Los Angeles basin by 50 percent or more early in the twenty-first century. For example, carbon monoxide emissions, currently close to 5000 tonnes per day, will be reduced to about 2700 tonnes per day by the year 2010 and eventually to about 1300 tonnes per day. Reactive hydrocarbon emissions will drop from about 1400 tonnes per day to 1100 in 2010 and to 180 some time in the future. Nitrogen oxide emissions will decrease from 1100 tonnes per day now, to 900 in 2010, and to 330 beyond that. The waiver plan is called RECLAIM (Regional Clean Air Incentives Market).

According to the waiver plan, a certain quantity of a primary pollutant will be designated as a *unit amount*. A facility releasing that pollutant will be given a number of waivers allowing the emission of that amount over a period of time—say, one waiver for 1 pound of pollutant per year. Individual sources of pollution will be evaluated to establish baseline annual emissions for those facilities, using existing data and supporting measurements. Every facility will be given a specific number of waivers determined by their baseline emissions. Moreover, the facility will be placed on a timetable to reduce emissions. The waivers will be devalued in accordance with the timetable. Slated reductions in Los Angeles, for example, are roughly 6 percent per year for reactive hydrocarbons, 8 percent per year for nitrogen oxides, 9 percent per year for sulfur dioxide, and so on. The waivers (or amount emitted) for each pollutant correspondingly decrease in value at those rates. The market is then set free. Profit motives presumably will move the system toward an optimal economic distribution of pollution abatement costs, and cleaner air.

Imagine a furniture manufacturer that emited hydrocarbons equivalent to 100 waivers. To meet government requirements over the next decade, these emissions must be reduced by 50 percent. Equipment costing $250,000 could be installed to achieve that reduction, which is initially equivalent to 50 waivers. Installation would effectively cost $5000 per waiver ($250,000/50 waivers). Suppose that a nearby refinery has a much higher cost associated with controlling emissions of hydrocarbons— say, $15,000 per waiver equivalent of pollutant. Moreover, suppose that the refinery would like to expand operations by an additional 50 waivers' worth of hydrocarbon emissions. The furniture producer could sell its waivers to the refinery at, say, $10,000 each. The furniture manufacturer would also have to install the emission-control equipment to reduce its hydrocarbon emissions by 50 units. In the deal, the furniture company would make a profit of $5000 per waiver, or $250,000. The refinery could expand its operations without installing new emission-control equipment, saving the equivalent of $5000 per waiver. But, the refinery would be affected by the mandated decrease in overall emissions through the devaluation of the waivers. These reduced-emission goals could be met by phasing out part of the operations, by installing emission-control equipment over time, or by buying up additional waivers each year to cover the shortfall in allowable emissions. In the meantime, the *total* emissions of hydrocarbons would decrease in accordance with the timetable. Everybody wins!

If this example seems complicated, imagine the market chaos that could accompany the bidding for waivers. The waiver plan has an excellent chance of working, but will require time to be implemented effectively. The Environmental Protection Agency has activated a similar plan to reduce sulfur dioxide emissions in the Midwest to reduce acid rain. In recent bidding, a ton of sulfur dioxide for 1 year fetched $122 to $450 from electric utility companies, environmental groups, and brokerage houses. One of the designers of the acid rain marketing scheme was Ronald Coase, who won the 1991 Nobel Prize in economics for developing relationships between economics and law. The market incentive plan and initial auction of sulfur dioxide waivers (called *allowances* in this case) was apparently a big success.

On the other hand, critics argue that waiver schemes merely delay sorely required repairs to the most defective polluting facilities. Certainly, a balance between legislative regulations and economic incentives can be struck to optimize the reduction in pollutant emissions and compliance with abatement schedules.

6.6.2 ALTERNATIVE FUELS

The idea of an alternative fuel to gasoline would have seemed sacrilegious to most people a few years ago. Gasoline is the king of fuels. It is cheap and readily available. There is an enormous infrastructure developed just to produce and distribute gasoline. Practically all cars run on gasoline. Nevertheless, a series of oil shortages, stubborn air pollution, and global climate change (Chapter 12) have led to the reconsideration of gasoline as the premier fuel for transportation. (Issues regarding alternative fuels and recyclable fuels in the context of global climate change are discussed in Section 12.5.) This section looks at the role of alternative fuels in the quest to cut air pollution.

Table 6.4 summarizes the advantages and disadvantages of several alternative energy sources for transportation. It is clear that the number of possible alternatives is growing. But will any of them work?

Improved Gasoline

Gasoline used to be one of the "fuels for fools." Basic gasoline is actually a mixture of perhaps 100 different hydrocarbon species, including suspected human carcinogens such as benzene (Sections 7.1.1 and 7.2.3). Other materials are added to improve the performance of internal-combustion engines. For years, lead was used to boost "octane" ratings of gasoline and prevent engine "knock." The lead is emitted as a particulate in the exhaust. Automotive lead emissions heavily contaminated the environment in the 1950s through the 1970s, after which lead additives in gasoline began to be phased out (Section 7.2.4). Substitute additives have increased the smog-forming potential of gasoline.

Gasoline is a favorite fuel because of its low cost and availability. The cost of gasoline in the United States is only about 6 cents per mile driven. The highest price of gas has been around $2 per gallon (compared with many European countries, where prices are $3 to $4 per gallon). It is worth asking, then, whether gasoline can be rehabilitated. Even if cleaner gas were more expensive, the convenience alone would make it extremely popular. The petroleum industry has therefore undertaken the task of reformulating gasoline, and the results have been mixed.

In winter, when the inversion layer can lie low, carbon monoxide is the problem pollutant in many cities. The cold weather also causes larger CO emissions than usual from cold engines. A new version of ordinary gasoline that is "oxygenated" has been placed on the market in affected areas, including parts of Alaska, California, Colorado, Montana, and New Jersey. The modified fuel is required by law. To make these so-called winter fuels, compounds that are already partially oxidized are added to gasoline. The additives include ethanol and a particular form of ether (**methyl tertiary butyl ether**, [**MTBE**]). Up to 15 percent of the fuel may consist of these added compounds. The oxygenated fuels reduce

carbon monoxide by burning more completely to carbon dioxide in the first place.

Unfortunately, the oxygenated additives increase other pollutants. Some additional formaldehyde is produced in the exhaust, for example. Formaldehyde, with the chemical formula $HCHO$, is a popular building material and embalming fluid. (See Section 8.3 for a complete discussion of its sources and effects.) It is well established that methanol and ethanol fuels generate formaldehyde. In the case of winter fuels, the evidence for significant formaldehyde emissions is not yet in. Many customers have complained of illnesses associated with the use of oxygenated fuels. The smell of winter fuel is different from that of gasoline, being somewhat sweeter. Headaches and nausea are common symptoms that have been reported. One EPA official blames the reports on hysteria. Formaldehyde itself does not produce the symptoms stated, but the MTBE and ethanol may produce lightheadedness if inhaled.

A new "flexible fuel" concept for vehicles is being developed, in which a wide range of gasoline mixtures with various additives could be used in a car. The advantages and disadvantages of such "clean" gasoline mixtures are summarized in Table 6.4. Winter gas is one form of this class of fuels, with the special purpose of reducing winter carbon monoxide concentrations. One could imagine, for example, mixtures of gasoline and methanol ranging between 10 percent gasoline to 100 percent methanol. These **"gasahol"** fuels are relatively easy to manufacture and dispense. Engines are being designed that will automatically adjust the fuel intake rate and timing of the spark plugs to accommodate any such mixture. Among other things, the flexible fuel cars would cost more, perhaps up to $2000 more.

Methanol and Ethanol

Methanol is a simple alcohol that can be produced from natural gas, coal and other fossil fuels, wood, or even garbage. Methanol is also called *methyl alcohol, wood alcohol,* or *wood spirit.* Methane in the form of natural gas is currently the major industrial source of methanol. Accidentally drinking as little as 1 ounce of this highly toxic alcohol can lead to blindness or death.

Ethanol, on the other hand, is eminently drinkable in beverages ranging from wine coolers to vodka. Ethanol is also called *ethyl alcohol* or *grain alcohol.* Pure ethanol is highly combustible (cherries flambeau are incendiary because of ethanol). Ethanol can be

Table 6.4   Alternative Fuels: Pros and Cons

Fuel	Advantages	Disadvantages
"Clean" Gasoline Mixtures	1. Immediate reduction in pollution emissions 2. Fuel-distribution system in place 3. Requires no engine modifications	1. Net environmental benefits uncertain 2. Fuel at least 2–3 cents more per gallon 3. Dependence on imported crude oil 4. No reduction in carbon dioxide emissions
Ethanol	1. Higher octane than regular or premium gasoline 2. Renewable resource (based on corn or sugar cane) 3. Reduces carbon dioxide emissions 4. Lowers CO emissions for gasoline mixtures	1. Less energy per gallon 2. Expensive 3. Causes smog when used with gasoline
Methanol	1. Higher octane than gasoline 2. Overall reduction in hydrocarbon emissions 3. Reduces carbon dioxide emissions 4. Lowers total airborne toxics, except for formaldehyde	1. Less energy per gallon, more frequent refueling 2. Can be dangerous to handle 3. Corrosive 4. Cold engine starts difficult
Natural Gas	1. Abundant supplies 2. Currently inexpensive 3. Lower hydrocarbon emissions than gasoline 4. Reduced CO emissions 5. Small reduction in carbon dioxide emissions 6. Distribution system largely in place	1. Clumsy fuel tank 2. Must refuel every 100 miles 3. Refueling time is greater by 2–3 times
Electricity	1. Quiet 2. Virtually no vehicle emissions	1. Technology is about 5–10 years away 2. Currently, less than 100 miles per charge 3. Maximum speeds 30–65 MPH today 4. Recharge 6–8 hours 5. Environmental impact varies with source of electricity
Solar-generated Hydrogen	1. Renewable energy source 2. Virtually emission-free 3. Does not contribute to global warming	1. Technology is at least 20 years away

Sources: Data from "Fuels of the Future," *Consumer Reports* January (1990) and from other publications describing alternative fuels.

produced by bacterial fermentation of carbohydrates (Section 4.2.1), or it can be synthesized from natural gas and petroleum. Most ethanol used industrially today is derived from fossil fuels rather than from natural fermentation. Ethanol is a toxic compound, although less so than methanol. In small amounts, ethanol induces intoxication, but in larger doses it can be fatal. Because ethanol is a taxable commodity in most countries, industrial production includes contamination with methanol to render the alcohol undrinkable and, therefore, untaxable.

As noted earlier, ethanol is already in wide use as a gasoline additive, to produce winter gas. But it will not be widely used as a replacement for gasoline because it is about twice as expensive to make (Table 6.4). Too bad, because the fuel is natural and recyclable (Section 12.5.1).[20] To fill all the country's transportation energy needs, the mass of plant matter that would need to be grown and fermented is enormous. By one estimate, 40 percent of the U.S. grain yield would produce only 10 percent of the national fuel requirement. And the waste problems would be daunting.

Methanol is the current popular alcohol substitute for gasoline. Table 6.4 lists some of the properties of methanol fuels. Methanol is a high-octane fuel with plenty of pep. According to EPA studies, methanol-fueled vehicles create less air pollution than does gasoline for similar patterns of usage. The reactive hydrocarbons in methanol exhaust are lower in concentration by up to 90 percent, and the toxicity of methanol emissions is lower by 30 percent or so. Mixtures of methanol and gasoline also are attractive because some of the disadvantages of pure methanol are alleviated. With 100 percent methanol fuel, cold-starting engines at temperatures below 50°F can be difficult. Methanol also burns with an almost invisible flame, which presents a safety hazard in some situations. Mixing 15 percent gasoline with 85 percent methanol helps solve these problems. The mileage per gallon with pure methanol is about one-half that with pure gasoline. More trips to the fuel pump

would be necessary. Methanol also damages metal and rubber parts to a much greater extent than gasoline does. Extra care must be taken not to spill the stuff all over the side of the car.

Methanol also generates photochemical smog. In Brazil, where methanol derived by fermentation from sugarcane has been in wide use for many years, air pollution has not abated. The fumes are sweeter smelling than gasoline exhaust, but the smog is just as real. Methanol exhaust also contains more formaldehyde than gasoline exhaust, up to eight times as much. Gasoline fumes do produce formaldehyde through photochemical processes after emission, however. Formaldehyde is a strong eye irritant and has other unfavorable properties (Section 8.3). Catalysts are being sought that will reduce the formaldehyde emissions of methanol-fueled cars.

A study sponsored by the petroleum and automobile industries concluded that methanol and gasoline generate roughly the same amount of smog and that methanol would be much more expensive. These industries have long favored reformulated fuels, such as gasahols. The report has been criticized by California state smog experts on the grounds that substantial overestimates of the cost of producing methanol, and of the quantities of pollutants emitted by methanol-fueled engines, were used in the industry analysis. The alternative fuel debate is just beginning to heat up.

A Few Good Fuels

Table 6.4 lists several other alternative energy sources for transportation systems. Natural gas seems like an attractive alternative in terms of pollution abatement, although, the distribution system and vehicle design would require far more significant reorientation than would be the case for alcohol-gasoline mixtures. The gains in pollution reduction would be significantly greater. Natural gas is also a fossil fuel, like gasoline and most industrially produced methanol and ethanol. This is a drawback in a world where the climate may be warming up as a result of fossil-fuel consumption (Section 12.2.2).

Electric cars have progressed from funny-looking gadgets to authentic vehicles. Still, in 1995, there are no practical mass-produced electric-powered cars on the market. Early claims for electric cars were often inflated and sometimes defied the laws of physics. One amateur car designer (and physicist) screwed together an electric motor and an electric generator and wired them to a few batteries, all in the shell of

---

20. To show how difficult it is to find the perfect fuel for transportation, consider the enormous agricultural enterprise that would be required to grow the carbohydrates for fermentation into ethanol. This operation would require massive fossil-fuel consumption to power tractors and other machinery and to manufacture fertilizers. These activities might have to be switched to ethanol fueled power to make the fuel truly recyclable. The noncarbohydrate fibrous organic waste of fermentation must also be disposed of, becoming in the process another pollutant!

a Volkswagen. He claimed that once the car was cruising on the highway, the generator would replenish the depleted batteries. You could literally drive forever, using the batteries to accelerate up to highway speeds and then switching the generator on to recharge the batteries. A perpetual-motion machine if ever there was one! In fact, it might work *if* you drove downhill all the time.

It appears that electric cars will be available in the near future. Current problems with electric vehicles—including excess size and weight of the batteries, a long recharge time, and lack of convenient facilities for recharging—will be solved (at least to some degree). Currently, the owners of electric prototype cars often have to scurry from electrical outlet to electrical outlet, and the electricity must still be generated *somewhere*. The advantage of electric cars is that the generating plants can be placed in remote locations (like the Navajo Generating Station, which is polluting the Grand Canyon [Section 6.5.2]). The remote placement of power plants allows cities to be cleaner, but may cause problems in the countryside.

Solar-hydrogen fuel is one step beyond electric cars. In this case, solar energy is collected by solar cells and used to generate electricity. The electricity is used to break down water into hydrogen gas and oxygen, as follows:

$$2H_2O \xrightarrow{\text{Solar Energy}} 2H_2 + O_2 \quad (6.19)$$

The hydrogen gas (H_2) formed is collected and placed in storage tanks as the basic fuel. In the vehicle engine, hydrogen is burned with air:

$$2H_2 + O_2 \rightarrow 2H_2O \quad (6.20)$$

The process regenerates water. There is so much water on the Earth that the potential supply of hydrogen gas is unlimited, but the technology to utilize this unlimited supply of energy in a practical way does not exist at the present time.

The search for alternative fuels to replace gasoline must be complemented by other actions to reduce pollution in densely urbanized areas. Many of these actions involve personal sacrifice.

6.6.3 LIFE-STYLES FOR HEALTH AND SURVIVAL

Many of us are trapped in polluted cities. Our jobs are there. So are our families and roots. Most people today would not be able to make a living off the open land. We have become technodependent. We are a society of persons so out of touch with nature that crowded freeways and gasoline fumes seem natural. Even if we managed to escape the urban boundaries, phones, televisions, and microwave ovens would remain our companions. We are also healthier as a population than any previous group. Life expectancy today should swell the number of octogenarians and nonagenarians in the twenty-first century. Yet as the millennium approaches, a pall hangs over our cities and lives, and the world environment is threatened. Given that we are trapped in cities, and certainly on the planet Earth, what behavior makes sense?

To limit the use of polluting fuels in urban zones and over a wider area, a tax on fuel and energy consumption makes sense. Some tax relief for workers in low- and middle-income brackets should also be offered. It has been estimated that a 50 cent per gallon tax on gasoline would reduce total consumption by about 25 percent over 5 years; a $1-a-gallon tax would cut gasoline consumption by almost 40 percent. In fact, energy taxes of some sort are in the works. For gasoline, the taxes may include one or more of the following: a tax on the amount of actual energy created from gasoline combustion; a price increase through taxes to discourage gasoline consumption; a carbon tax on gasoline to encourage less use of fossil fuels; a gasoline tax to compensate indirect public costs associated with gasoline production, distribution, and use; and an oil import tax to discourage dependence on the foreign gasoline market. If all these taxes were imposed at these rates, the per capita annual cost would be only about $120 per year. That is certainly a reasonable amount.

In large diverse urban regions, pollution-control strategies must be thoughtfully designed. The needs of the public must be balanced with the needs of business and industry. The role of agencies, such as the Air Quality Management District in southern California, is critical. They must formulate, apply, and enforce general policies aimed at pollution abatement. Individual decisions and actions are equally important. We can live more modestly, travel less, carpool with co-workers, and work during off-hours to make commuting more energy efficient. We can also cooperate in paying fair energy and pollution taxes. Finally, if the need arises, we can become political activists for the local environment. It is a good cause. What is at stake is our lives.

Questions

1. Discuss the differences between "London" smog and "Los Angeles" smog. What are the critical meteorological conditions and ingredients for each type? If you lived in a city suffering from London smog, what activities and fuels would you demand to have controlled? If you lived in a city with Los Angeles smog?

2. Why do you think there are so many references to air pollution and smog in the literature and poetry of the eighteenth and nineteenth centuries?

3. Explain why you might expect to encounter lower carbon monoxide concentrations in Mexico City during the summer than in Los Angeles during the winter.

4. Why are automobiles (and other vehicles) so prone to generating smog? Consider chemistry and human nature in your answer.

5. Imagine that you have filled a clear plastic bag with a mixture of nitric oxide, oxygen, and a reactive hydrocarbon like propane. There also is a little moisture in the bag. If you keep the bag in the sun for a while, what might be the chemical changes in the composition of the trapped gases? Would there be any visible changes inside the bag?

6. Discuss the reasons that polluted air seems to be much hazier when you look in the general direction of the sun. Consider the fact (Section 3.2.2) that pollution particles scatter sunlight in different directions with different intensity. What experiment are you performing by slowly turning around and scanning the atmosphere in a complete circle during the morning or afternoon? At high noon (with the sun directly overhead), what would you expect to see? Can you now also explain why the most difficult driving conditions occur when you drive into the morning or evening sun with a dirty windshield?

7. The visibility in many national parks is often very poor. What are some of the possible causes of these conditions? Keep in mind the natural effects as well as the human activities. Can you differentiate between natural and anthropogenic effects? (*Hint*: Consider weather as a factor.)

8. Summarize the properties of the alternative fuel sources discussed in the text. Include the sources, performance as a vehicular fuel, and air pollution potential. Which of these fuels would be considered recyclable fuels, providing essentially an eternal source of energy?

Problems

1. Photochemical smog is formed by reactions involving nitrogen oxides and hydrocarbons in the presence of air and solar radiation. If you were to combine 10 molecules of a reactive hydrocarbon with one molecule of NO, how many ozone molecules could you form? Assume that for the hydrocarbon selected, each hydrocarbon molecule can form two ozone molecules under ideal circumstances.

2. Three companies, A, B, and C, emit the same pollutant. Company A emits 100 units per day; B, 50 units; and C, 50 units. The companies are mandated to *reduce* their emissions by 10 percent of these amounts each year. For A, the reduction is 10 units per year, and for B and C, 5 units each year. After 10 years, the emissions from these facilities will be zero. Each company is given one waiver per unit of pollutant. The waivers keep their initial value, but the companies are required to surrender 10 percent of the waiver coupons each year, in pace with the planned emissions reductions. If these companies were to keep strictly to the mandated plan, how many units of pollutant would be emitted over the 10-year phase-out period from the three facilities combined? Company A has a cost of $10,000 per unit to reduce its emissions. B has a cost of $5000, and C has a cost of $20,000 per unit. If pure market forces were at work, how would the waivers be optimally traded among these companies? Who would make a profit? What would be the total pollutant emissions after 10 years? Assume that a company will not invest money to reduce pollutant emissions unless either it has to or it is profitable to do so.

Suggested Readings

Bennett, B., J. Kretzschmatr, G. Akland, and H. deKonig. "Urban Air Pollution Worldwide." *Environmental Science and Technology* **19** (1985): 298.

Blacet, F. "Photochemistry in the Lower Atmosphere." *Industrial Engineering Chemistry* **44** (1952): 1339.

Bleviss, D. and P. Walzer. "Energy for Motor Vehicles." *Scientific American* **263** (1990): 102.

Bridgman, H. *Global Air Pollution*. London: Belhaven Press, 1990.

Evelyn, J. *Fumifugium: or, The Inconvenience of the Aer and Smoak of London Dissipated, Together with Some Remedies Humbly Proposed*. London: Bedel and Collins, 1661.

Fishman, J. and R. Kalish. *Global Alert: The Ozone Pollution Crisis*. New York: Plenum, 1990.

Godish, T. *Air Quality*. Chelsea, Mich.: Lewis Publ., 1991.

Gray, C., Jr. and J. Alson. "The Case for Methanol." *Scientific American* **261** (1989): 108.

Haagen-Smit, A. "Chemistry and Physiology of Los Angeles Smog." *Industrial Engineering Chemistry* **44** (1952): 1342.

Haagen-Smit, A., C. Bradley, and M. Fox. "Ozone Formation in Photochemical Oxidation of Organic Substances." *Industrial Engineering Chemistry* **45** (1953): 2086.

Leighton, P. *Photochemistry of Air Pollution*. New York: Academic Press, 1961.

Lodge, J., Jr. "Selections of *The Smoake of London, Two Prophecies*." Elmsford, N.Y.: Maxwell Reprint Co., 1969.

Middleton, J., J. Kendrick, and H. Schwalm. "Injury to Herbaceous Plants by Smog or Air Pollution." *Plants Dis. Rep.* **34** (1950): 245.

National Research Council. *On Prevention of Significant Deterioration of Air Quality*. Washington, D.C.: National Academy Press, 1981.

Seinfeld, J. *Atmospheric Chemistry and Physics of Air Pollution*. New York: John Wiley, 1986.

Shaw, R. "Air Pollution by Particles." *Scientific American*, **257** (1987): 96.

Sperling, D. *New Transportation Fuels*. Berkeley: University of California Press, 1988.

"Urban Air Pollution: The State of the Science." *Science* **243** (1989): 745.

7

Effects of Exposure to Pollution

The way that pollution affects people is determined by the specific toxic materials they are exposed to. The concentrations of the pollutants and the duration of the exposure are important factors. In this chapter, we explore the physiological and health effects caused by exposure to a number of common toxic pollutants. Toxins may be found in the home, neighborhood, or workplace. Chapter 6 focused on the formation and properties of urban smog. Smog is a local- or regional-scale accumulation of unpleasant compounds at levels that are unhealthful yet not acutely toxic. Exposure to smog may extend over many years before health consequences are evident. On the other hand, some very sensitive persons show signs of ill effects after spending an hour or less in smoggy air.

Standards have been set for the levels of air pollution that can be tolerated before officials will step in to impose measures designed to limit the health hazards (Section 6.2.2). The standards are based on estimates of the risk, or probability, that exposed populations will suffer specific health effects. The risk is usually defined as "acceptable" if the ill effects averaged over the population are kept to a "reasonable" level (for example, 1 chance in 1 million of contracting lung cancer from breathing polluted air for 1 year). In other settings—for example, in the work environment or in a public facility—the standards for exposure may be set according to different rules. It depends on how seriously a certain level of risk is taken—in relation to the importance of holding a job, for example—and how finicky those affected are about protecting their health. In any case, a certain number of people will be greatly affected or even die. They are the statistically unlikely casualties of accepting a "reasonable risk."

Imagine living in a pristine environment, completely devoid of anthropogenic pollutants. The air is always clean and has a natural aroma. Are you free of the risk of health effects caused by toxic pollutants?

No, because the environment produces a host of natural toxins. Smoke from fires ignited by lightning contains many of the same toxic organic compounds as smog. Windblown dust may include dangerous fungi as well as toxins. Volcanic fumes contain deadly hydrogen sulfide and sulfuric acid. Plants give off offensive vapors as a means of self-defense. In most places today, the chance of being poisoned by natural toxins is minuscule compared with the risk of debilitating health effects from human-made pollutants. Nevertheless, there still is a small risk of exposure to natural pollutants. The *additional* risk associated with occupational exposure to human-made toxins and with pollutants encountered in everyday life can be gauged in two ways. First, the risk of ill effects caused by toxic exposure at work and in the home can be compared with the background risk associated with natural toxins. Second, the potential additional risk can be weighed against possible benefits accrued as a payoff for accepting a certain level of exposure to pollutants. The important concepts of "risks" and "benefits," which lead to compromises on safety, are discussed in Section 7.4.2.

The issue of the health effects caused by airborne pollutants has received wide attention in the media. Many of these accounts are basically accurate yet have sensational aspects. A few politically conservative extremists claim that all publicized threats of pollution are exaggerated or even fabricated in certain cases. This paranoid subgroup imagines conspiracies to undermine business and industry—even to destroy freedom and capitalism—hatched by Chicken Little environmentalists. Surely the question of the health risks related to ubiquitous toxins is not simply a question of either science or passion alone. There are value judgments to be made in deciding what might be sacrificed to reduce personal and public risk. This particularly difficult issue is raised in Section 7.5.

7.1 How Pollutants Affect Health

The health effects of pollutants depend on the manner in which they enter the body, the duration of exposure, and the specific physiological responses to each compound. The number of sources and effects of potential environmental toxins are impressive. The china and crystal that you use to serve dinner to friends may be a source of lead (Pb). Lead dissolves in food and drink and is ingested into the digestive system. When absorbed into the body, it can accumulate in certain tissues, particularly red blood cells. Another place in which lead likes to accumulate is brain tissue. High concentrations may cause forgetfulness and eventually irreversible brain damage and loss of intellect. Birth defects have also been associated with lead poisoning.

Excessive exposure to almost any element or compound can prove to be fatal. In daily life, people are continuously exposed to potentially lethal substances, but at low doses. The physiological effects and the threshold exposures for such effects are important to assessing the health risks of environmental toxins. The study of common toxic substances has spanned centuries and resulted in volumes of information. For many exotic modern chemicals, however, a scientific understanding of physiological effects and health risks remains less than satisfactory at the present time.

7.1.1 THE DISCOVERY OF TOXICITY

The study of toxicity in humans has a long history. Newly acquired knowledge of the effects of poisons led to opportunities for mischief and mayhem. The bite of an asp did in Cleopatra, and Socrates drank hemlock to preserve his integrity.[1] More subtle toxic effects have taken more time to discover.[2] In 1775, Sir Percival

Pott, a health professional, deduced that cancer of the scrotum, common in chimney sweeps of the period, was probably induced by some component of the soot that accumulated in chimneys (Percival also noted as a factor in the occurrence of cancer the disinclination of sweeps to take baths). The exposure to high concentrations of soot over long periods of time (together with a lack of cleanliness) was the insightful association that Percival made.

The cause of scrotum cancer in chimney sweeps past, and probably a significant number of other cancer deaths today, is associated with **polycyclic aromatic hydrocarbons** (**PAH**). These compounds are indeed "aromatic," having a strong, often sweet, solvent smell, like gasoline. PAH also is composed of **benzene rings**, a very stable arrangement of carbon atoms. A benzene ring has six carbon atoms strung together like a necklace. The six C atoms are uniformly spaced and form a hexagon (called a benzene "ring"). The simplest form of such a cyclic compound is benzene (C_6H_6) itself. One hydrogen atom is attached to each carbon atom in benzene. Gasoline, kerosene, and other organic fuels contain some benzene. More complex polycyclic organic compounds are composed of benzene rings hooked together in various combinations. The rings fit snugly together like the cells in a beehive. A common carcinogen in animals, and most likely in humans, is **benzopyrene,** which has five rings.[3] Benzopyrene was first isolated in 1933 in the search for a culprit in an epidemic of cancer among workers in coal-tar refineries. It was established even in the late nineteenth century that exposure to fumes and particles in certain petroleum and coal industries carried a high risk of cancer. In fact, several compounds related to benzopyrene that were derived from coal tar proved to be powerful carcinogenic agents.

If you collect the aerosols in smoggy air and extract the polycyclic organic goop from them, that goop is highly carcinogenic (that is, it induces cancerous tumors). In fact, in controlled tests, this residue is 100 to 1000 times more active in damaging cells than is the benzopyrene component alone. The

---

1. Socrates, born in Athens around 470 B.C., was the first of the three great Greeks of antiquity: Socrates, Plato, and Aristotle. A man of modest means, Socrates was known for his intellect and reasoning ability. He developed the art of inductive argument and the revelation of truth through systematic questioning and constructive doubt. In 399 B.C., he was indicted for "impiety" and "corruption of the young," for filling their heads with doubts about the gods worshiped by the elders. He was convicted of these heinous crimes and sentenced to death by the means popular at the time—drinking hemlock (a poison extracted from the herb of the same name which is laced with the deadly alkaloid coiinine). Socrates had an opportunity to escape this fate, but chose to accept the death sentence as the decision of an orderly body of law, however incorrect.

2. A relatively complete history of the identification of toxins in polluted air is provided in *Atmospheric Chemistry* [Barbara

Finlayson-Pitts and James Pitts]. Several anecdotes recounted here are taken from that review. The text is also a rich resource for information on air pollution, although at a generally advanced level of treatment.

3. Benzopyrene represents a class of molecules with similar chemical formulas, but with their atoms arranged in slightly different ways. The molecule that has been most studied as a carcinogen is benzo(a)pyrene, which is found in wood and tobacco smoke, and in diesel engine emissions. In the text, benzopyrene refers to benzo(a)pyrene. (See also Section 8.4.1.)

exact carcinogenic compounds in the goop have never been identified, although their origins include combustion, particularly from coal combustion to generate energy, vehicle exhaust, especially on the soot from diesel engines, and wood-fire smoke emitted from fireplaces. As early as 1954, respirable particles in Los Angeles smog were known to contain strongly mutagenic substances.

The Ames Test

In the 1970s, Bruce Ames developed a quantitative test to evaluate the mutagenic activity of toxic compounds.[4] The **Ames test** uses the common salmonella bacterium to determine the ability of specific materials to attack and mutate the DNA in living cells. A substance that causes significant changes in DNA is referred to as **mutagenic**. The mutations accumulate in subsequent generations of the bacteria and their frequency may be measured. If you have ever suffered from a salmonella infection (usually through tainted food), you will have no sympathy at all for the little buggers in Ames's test. The main advantages of the salmonella bacteria for such tests are their great sensitivity to mutagens and their rapid rate of reproduction. Salmonella are exposed to a specific toxin in a controlled amount. Chemical reactions of the toxin then damage the salmonella's genetic material in the cell.[5] Subsequent generations carry the defects and pass them along to further generations. The number of defects can be counted after a period of exposure and reproduction. A high occurrence of defects indicates a highly mutagenic compound.

The transformation of genetic material can be classified as either mutagenic or promutagenic. Mutagenic materials directly react with and damage DNA. Promutagens do not directly cause such genetic mutations. Rather, reactions of the promutagens

in the cells produce other compounds that can mutate DNA. The effect of a promutagen is therefore indirect. Benzopyrene is a promutagen.

We should differentiate among the terms *toxicity, mutagenicity,* and *carcinogenicity.* A toxic compound causes death by interfering with the organism's normal physiological functions. A poison may cause nerve cells to stop firing, leading to spasm and death. Or a chemical may block the absorption of oxygen by red blood cells. Mutagens or carcinogens, on the other hand, produce specific effects in organisms— the mutation of DNA or the induction of certain tumors in otherwise healthy organisms. Other specific effects can be separately identified. Thus **teratogenic** compounds cause serious birth defects in offspring.

Ames applied his test to a wide variety of materials, both natural and human-made. He noted the propensity for plants to manufacture highly toxic compounds. This is a logical means of protection for a plant, as leaves and fibers laced with poisons are not very appetizing. Ames pointed out that when humans began to cultivate certain vegetables such as potatoes, eggplants, and tomatoes, potent toxins entered the diet. These include **solanine**, the basic toxin in the family of poisons commonly known as *nightshade.* It has been proposed that exposure to these natural poisons and mutagens is largely responsible for the increases in cancer seen in this century. This idea, however, ignores the strong correlation between the widespread production and emission of anthropogenic toxins and the increase in excess cases of cancer. For example, links have been made between the rate of respiratory cancer in people exposed to secondhand tobacco smoke, and benzene (Section 8.4.2). Although the direct connections are difficult to prove, prudence dictates that a warning signal be raised when such a correlation is found. But the idea of dismissing all cancer and toxin-related disease as "natural" is a profoundly flawed approach to environmental health assessment.

Before the Ames test was available, toxicity was (and still is) assessed using animals. The testing procedures were developed during the 1960s. Originally, the experiments were designed to screen compounds for their likelihood of causing cancer in humans. However, because animals do not breed as quickly as bacteria do, the physiological effects had to be accelerated so that they would appear after a reasonable exposure time. Researchers adopted the concept of a **maximum tolerated dose** (MTD) for a species. The MTD is essentially the largest amount of

4. Bruce N. Ames was born in New York City in 1928 and is a professor of biochemistry and molecular biology at the University of California, Berkeley. He developed the well-known Ames test, which is now widely used to detect mutagenic compounds. In one famous early case, this test showed that "Tris," a fire retardant used in children's clothing, was a likely mutagenic and carcinogenic agent. Consequently, Tris was banned. Ames has since taken the position that natural toxins are much more dangerous to the average person than are all the toxins emitted by humans into the environment. Needless to say, Ames's opinions are contested by many environmental scientists.

5. When such a chemical reaction occurs in the laboratory under controlled conditions, it is said to occur *in vitro*—that is, in "glass," or a test tube. If the reaction occurs within a living organism, it is said to occur *in vivo*—that is, in the bodily fluids.

a toxin that can be continually fed, or otherwise administered, to a critter without killing it outright. Mice, rats, hamsters, and rabbits are most often used in these experiments. They reproduce quickly and, in the case of mice and rats, are not objects of great affection to humans.

Toxicological assessments using animal subjects have the advantage that humans are not placed in direct jeopardy. The experimental conditions can also be widely varied. The size of the dose and the length of exposure are easily controlled, along with other factors. Unfortunately, it is usually possible to make only a relatively small number of tests in a laboratory setting, with a restricted research budget and limited time available. The tests are also conducted at extremely high levels of exposure near the MTD. More critical, perhaps, is the fact that the physiological responses of small mammals may be fundamentally different from those of humans. Researchers have, nevertheless, reached important general conclusions from such work. For instance, they have discovered a strong relationship between the toxicity of a compound and its ability to induce cancer. If a compound is toxic at high doses, for example, it is likely to cause cancer at high doses. If low concentrations of a compound prove to be toxic over a long period of time, those conditions are also likely to induce cancer.

Errors in toxicological testing can occur. A noteworthy case involves saccharin, an artificial sweetener. Some years ago, saccharin was found to cause cancer in rats exposed to the MTD. Thereafter, saccharin use was limited, and people were warned about excess consumption. Those not inclined to be concerned about cancer risks complained that one would have to drink hundreds of bottles of artificially sweetened soda each day to get a dose equivalent to that given to the test rats. In addition, it turns out that rats are more sensitive to saccharin than people are. Saccharin causes bladder cancer by reacting with proteins in urine, creating carcinogens in the body. The amounts of these proteins in human urine, however, are lower by a factor of 100 to 1000 than the amounts in rat urine. Accordingly, the risk of bladder cancer in humans who drink diet soda sweetened with saccharin is also lower. The subtle differences between the metabolic processes and physiological responses of rats and humans to saccharin are significant in this case (although saccharin still leaves a bad taste in your mouth).

Cancer and the Clause

Earlier in this century, the trashing of the environment became a source of great concern to conservationists. The despoliation of the environment was not the only problem. It was claimed that the health of millions of people was being placed in jeopardy by exposure to a zoo of carcinogenic chemicals, some accidentally released and some purposely added to food and drink ostensibly to improve it. This concern was vividly expressed in Rachel Carson's landmark exposé of environmental pollution, *Silent Spring*.[6]

During the 1970s and 1980s, the number of deaths caused by cancer in humans increased dramatically. From 1973 to 1981, the number of deaths from lung cancer alone grew by 34 percent. The death rate from melanoma, a virulent form of skin cancer, rose by 30 percent in the same period. Up to 85 percent of lung cancers are associated with tobacco smoking and exposure to second-hand tobacco smoke. Many other cancers are also connected to smoking, including cancer of the mouth, larynx, esophagus, kidneys, bladder, pancreas, and liver. Adding insult to injury, tobacco smoke also causes or aggravates asthma, bronchitis, emphysema, and other respiratory ailments. All these diseases have increased with the worldwide popularity of smoking (Section 8.4).

Skin cancers are related to exposure to ultraviolet radiation from the sun. Melanoma is the most lethal form of skin cancer, although two other forms—basal and squamous cell carcinoma—are far more prevalent. (Additional health effects of exposure to ultraviolet rays are discussed in Section 13.4.2.) Increases in ultraviolet radiation have been connected with depletion of the stratospheric ozone layer (Chapter 13). Some dermatologists

---

6. Rachel Louise Carson (1907–1964) was an American biologist who, through her research in marine biology, became concerned with the toxic pollution of the environment. *Silent Spring*, written in 1962, refers to the indiscriminate killing of birds and other wildlife by widespread applications of pesticides such as DDT following World War II. As a consequence, Carson noted, "[O]n the mornings that had once throbbed with the dawn chorus of robins, catbirds, doves, jays, wrens, and scores of other bird voices there was now no sound; only silence lay over the fields and woods and marsh." Although her most severe forecasts of species extinction never materialized, it was in part because her landmark book launched a full-fledged environmental movement against indiscriminate industrial pollution. Ironically, Carson died of cancer, which she feared as another outcome of widespread pesticide usage, soon after *Silent Spring* was published.

have underscored the rapid rate of increase in melanoma deaths by 30 percent from 1973 to 1981 as virtually an "epidemic" of skin cancer. The connection between these cancers and ozone depletion has not been unambiguously established, however.

In 1958, important congressional legislation was passed to limit our exposure to toxic compounds in food.[7] The Delaney Clause specifically prohibited processed food from containing *even a trace* of any substance that has been shown to produce cancer in laboratory animals or humans. The rule applies only to processed foods and food additives, however. Raw vegetables, for example, can reach market without the same restriction. So, for example, a tomato that might not be pure enough to grind into tomato paste could be sold for a salad. The Supreme Court recently upheld the strict interpretation of the Delaney Clause in a case involving pesticides with names such as mancozeb, phosmet, and trifuralin. The Environmental Protection Agency had informally relaxed the Delaney regulation by not enforcing it; now, that benign neglect has to end, unless the law is changed.

With the limitations inherent in animal testing and the lack of direct data on human response, debates have risen over how to define a "trace" of a toxin. Modern instruments can measure trace compounds in minuscule concentrations. Accordingly, under the strict interpretation of the Delaney Clause, almost everything we eat, drink, or breathe is tainted. Should a "trace" be defined as the amount that could be detected by the most careful analysis possible, or the quantity that might cause an effect with a specific reasonable probability? Many food processors and farmers have been arguing for standards based on the concept of a **negligible risk of cancer** (Section 7.4). Specifically, a negligible risk is defined as one chance in a million of contracting cancer after 70 years of continuous exposure to a toxic compound. Some politicians want to adopt this more liberal rule for chemicals in food, but extend the restriction to *all* foods, raw and processed.

7. James J. Delaney, a feisty congressman from New York, attached his "clause" as an amendment to the Federal Food, Drug, and Cosmetic Act of 1958. That seemingly innocuous amendment has led to significant restrictions on pesticide use, setting in motion an ongoing debate between opponents and proponents of agricultural pesticides.

7.1.2 THE PHYSIOLOGY OF TOXICITY

Toxic compounds are everywhere in the environment. They are dissolved in water; they blow through the air. Most of the time, we are exposed to toxins at very low concentrations or for very short intervals of time. The health effects in such cases may be negligible, or unnoticeable. In order to understand the possible dangers of exposure to a certain toxic compound, it is necessary to know the concentrations and duration of exposure that will begin to cause health effects in humans. This is a difficult task, for several reasons. First, no rational person wants to expose himself, even in a controlled experiment, to a toxin that can make him ill. Second, a particular person's reaction to the toxin depends on their overall health and sensitivity to that toxin. These factors vary significantly from person to person. In many cases, the physiological effects of toxic compounds are determined from accidental exposures. Epidemiological studies of the health impacts caused by chemical spills and industrial accidents can be compiled into a general statistical model of human sensitivity to the compounds involved.

Epidemiology and Pathology

The study of the effects of toxic, mutagenic, and carcinogenic compounds on humans defines the field of **epidemiology**. Typically, a correlation is sought between exposure to a toxin and physiological effects such as specific diseases, sensory anomalies, involuntary responses, other illnesses, and even death. In this way, the "dose-response" relationships are quantified (Section 7.4.1). The data for epidemiological studies come mainly from accidental exposure and controlled laboratory experiments. As we noted earlier, experimentation is usually limited by the availability of willing participants. Moreover, the dosages of toxins that can be administered are usually very low. In small amounts, no effect may be seen during an experiment. However, in larger quantities, the subject might become violently ill or die. Accidents, on the other hand, are uncontrolled and poorly documented events, and so the extent of exposure is difficult to ascertain. The victims may already have died, making an evaluation of the actual lethal dose impossible. In the confusion and disruption following a serious accident, the evolving facts are also obscured.

In order to eliminate some of the uncertainties normally associated with epidemiological studies, cruel experiments have been conducted from time to time on unwitting victims. The National Academy of Sciences recently reported that during World War II, thousands of young servicemen were purposely exposed to mustard gas in order to demonstrate the effects.[8] The project was carried out in secrecy. The men, who were offered weekend passes as an inducement to participate, were never told about the possible long-term dangers or later checked for possible disease pathology. One health scientist called the event a "betrayal and a sad legacy." In the ensuing 50 years, many of the servicemen have suffered bouts of skin cancer, respiratory disease, and impaired vision. Other demonstrated effects associated with exposure to mustard gas are abnormalities in skin pigmentation, chronic skin ulcerations (including ulcers of the cornea), leukemia, psychological disorders, and sexual dysfunction. The need for epidemiological data was so urgent in the 1940s that these tests were carried out even in the face of knowledge, well documented at the time, of potentially devastating health effects.

The epidemiology of exposure to toxins in air pollution is difficult to resolve because the exposure occurs over a long period of time, the conditions of exposure are uncertain, and many factors—some known and some unknown—contribute to the observed effects. Unraveling the extent of exposure to each of many specific toxins requires clever statistical techniques. Large samples must be taken and causative factors carefully identified and quantified. For example, a person's employment history and occupational exposure, eating and exercise regimens, time spent in traffic or on the beach, and smoking and drinking habits all are relevant factors. Medical history and family health patterns also contribute to a person's response to air pollution.

The samples used in epidemiological assessments must be carefully chosen. In some cases, only a small coterie of subjects is available. The statistical certainty of any inferred cause-and-effect relationship is likely to be low. When large samples are used, other factors such as the person's mobility and long-term health care must be considered. Larger samples allow selection of specific characteristics—such as sex, age, weight, and occupation—to be isolated as factors in the assessment. The difficulty arises in identifying all the possible contributing factors. Moreover, the reporting of health effects is often questionable. People sometimes get sick and suffer in silence and despair.

The presence of multiple pollutants in the environment may enhance certain physiological effects. The exacerbation of health impacts due to the interaction of two or more toxins is referred to as **synergism**. In synergy, the combined effect of two toxins is greater than the sum of the effects of the toxins acting independently. One example of an important synergistic effect is the mutagenic activation of certain organic compounds in the presence of nitrogen oxides (the nitrosamines, for example, are carcinogenic substances found in foods treated with additives to improve the color and freshness of red meats).

On rare occasions, people agree to be guinea pigs in experiments designed to determine the toxic effects of particular substances. Convicts sometimes volunteer in return for time off. In a few isolated instances, civilians and military personnel have been surreptitiously exposed to chemicals for the purpose of determining effects on performance. During wartime, prisoners have been sacrificed in sick "scientific" research. Moreover, the results of such experiments, while occasionally useful, are often dubious. Recent laboratory tests, for example, revealed that exposure to nitric oxide (NO) can enhance penile erection. However, do not sniff nitric oxide to become virile! It is a dangerous toxic gas. Physicians may treat impotence with NO by administering it very carefully in low doses. In this case, the benefits of exposure may outweigh the potential health risks.[9]

Another common nitrogen oxide, nitrous oxide (N_2O), is used as a mild anesthetic, primarily in dentistry. Nitrous oxide is also employed as a propellant for spray-on food substances. Low doses of nitrous oxide, also known as "laughing gas," can induce a light-

---

8. Mustard gas is the name for a class of organic liquids containing sulfur and chlorine that act as strong vesicants—that is, induce blistering on exposed skin. Mustard "gas" is actually delivered as a mist of fine aerosol droplets composed of oily fluid. The purpose of these poisonous agents is to disable opponents on the battlefield, without regard for possible long-term effects.

9. Nitric oxide also binds with hemoglobin, much as oxygen and carbon monoxide do (Section 7.2.1). A new artificial blood composed of dispersed hemoglobin molecules, rather than aggregated hemoglobin as occurs naturally in red blood cells, has been tested and found to be relatively ineffective, because the hemoglobin leaks through the walls of blood vessels and scavenges NO from the surrounding tissues. With NO depleted, the blood vessels tend to constrict and blood pressure increases, causing other symptoms. Researchers are attempting to produce a polymerized hemoglobin that will not leak through blood-vessel walls so quickly.

headed euphoria. Continued inhalation can lead to death, however. Recreational breathing of N_2O to get high has already killed a number of teenagers.

Exposure and Dose

We are exposed to pollutants through air, water, and food. The offending compounds we breathe may be in the form of vapors or particles. Toxins found in water and food are often first airborne, carried from their sources on the winds. Exposure to toxins may occur in an **acute dose** (or **acute exposure**—one that is fairly large, but persists for a short period of time) or a **chronic dose** (or **chronic exposure**—one that is smaller, but lasts over an extended period). Acute exposure to pollutants usually occurs in chemical spills, in building fires, and in industrial settings. We are surrounded by toxic accidents waiting to happen. Oil refineries occasionally release clouds of hydrofluoric acid, which is used in the production of high-octane gasoline. Nitric acid—widely used in industrial processes—leaks from time to time. Chlorine gas, another common industrial chemical, has a strong propensity to form a pall of hydrochloric acid when released into the atmosphere.[10] All these acids are highly corrosive and damaging to the lungs. You would know if you were exposed to a high concentration of one of these compounds. Your chest would constrict, and your throat would burn. When you gasped for breath, the deeper you breathed, the worse it would get.

Chronic exposure to pollutants is more insidious. A person may be exposed for years or decades at home or in the workplace before the hazard is discovered. A small manufacturing plant next door may be releasing solvents that can be smelled only occasionally. Gasoline fumes are inhaled when the tank is filled up. Toxic vapors seeping from wall coverings, carpeting, furniture and plastics in a work area produce a microcosmic cloud of unhealthful air. When it is finally realized that vapors inhaled for years may be cancer-causing, it is often too late for many of those exposed.

10. Chlorine is an extremely useful, but hazardous, gas. At high concentrations, chlorine vapor has a greenish yellow color. You can smell it in bleach and near a pool, even when concentrations are less than about 0.2 part per million by volume (ppmv). One ppmv is the occupational limit for 8-hour exposure. At 30 ppmv, coughing, chest pain, and vomiting may presage asphyxiation. Concentrations of 1000 ppmv can be lethal after a few deep breaths. Chlorine was the first in a series of poisonous gases used during World War I.

An individual's exposure to a toxic material is related to the dose of that material they receive. **Exposure** is usually defined by the concentration of a toxin in the local environment and the mode and duration of contact that an individual has with it. Thus, exposure defines the potential for receiving the toxin. The **dose** is the amount actually received, or absorbed, in the body, leading to physiological effects. For example, imagine yourself sitting in a restaurant filled with secondhand cigarette smoke. Assume that the concentration of smoke in the room is relatively constant. Your exposure to the smoke then depends on the concentration of smoke in the air, the amount of air you inhale, and the length of time you remain in the restaurant. The dose, on the other hand, is determined by the fraction of the inhaled smoke that remains in your body. If all the secondhand smoke you breathe is trapped in your nose, throat, and lungs, then the exposure and dose are the same. In this case, the dose can be estimated as

$$D = C_s \dot{V}_b t_e \qquad (7.1)$$

Here C_s is the concentration of smoke in the air (say, micrograms per cubic meter, $\mu g/m^3$), $\dot{V}_b$ is the rate at which air is taken into the lungs (cubic meters per second, m^3/sec), and t_e is the length of time spent in the restaurant (or the exposure time, in seconds). The dose in this case is measured in terms of the total number of micrograms of smoke inhaled, since, $(\mu g/m^3) \times (m^3/sec) \times sec = \mu g$.

Strictly speaking, equating dose with exposure implies that all of the toxin received is retained (in the body). However, some of the toxin may be excreted or exhaled. Accordingly, another factor should be included in Equation 7.1, which defines the fraction of the total potential dose that is actually absorbed, or produces a physiological effect. Often, in determining the effects of a toxic compound, the exposure is stated only in terms of the concentration and duration. The actual dose is never calculated. Hence, the factors that determine the fraction of the potential dose entering the body, or reaching a specific organ, are implicit. Moreover, individual characteristics—such as one's breathing rate in the case of secondhand smoke above—are not explicitly taken into account. Rather, the expected average response is specified for a group of individuals suffering such an exposure under known conditions. This response may be derived from experiments or observations of an exposed population, or can be extrapolated from measurements on animals.

If an individual is exposed to the same toxin from more than one source, the total exposure, or dose, is the sum received from each source over a period of time, or

$$D = D_1 + D_2 + \cdots = \sum_{i=1}^{I} D_i \qquad (7.2)$$

The sum on the right-hand side ranges over all the contributing sources. For the example of secondhand cigarette smoke, one can imagine being exposed by breathing the smoke directly or by ingesting food and drink on which smoke has settled. Clearly, the other sources of exposure to secondhand smoke are smaller than direct inhalation. Exposure by ingestion, moreover, is mainly through the gastrointestinal tract (or gut), while exposure by inhalation can be through either the lung tissue or the gut (the latter owing to the trapping of smoke in the mouth, nasal passages, and throat, followed by swallowing). Determining the pathways and doses of toxic materials can be a tricky business.

Exposure to a specific toxin generally causes predictable symptoms. Individuals may respond to exposure in different ways, but there is an average, or typical, response. A specific dose can be defined that will produce a given symptom in 50 percent of the people exposed—the D_{50}. The dose produces noticeable symptoms in half of the people exposed. If the symptom is *death*, then the dose is called the **LD$_{50}$** (LD stands for "lethal dose"). Similar symptoms may follow a short exposure to a pollutant at high concentrations (acute dose) or a long exposure to low concentrations (chronic dose). Accordingly, standards for exposure usually specify the time intervals that are tolerable for exposure at a fixed toxin concentration. Section 6.2.2 reviews the standards for exposure to common air pollutants. Air quality standards may vary from state to state, from state to federal jurisdictions, and among nations. Such differences often reflect political priorities between human health and a healthy economy.

Different people show different responses to the same dose of a toxic compound. Individual sensitivity may vary from one toxin to another. Many people suffer severe allergic reactions to commonly used chemicals, such as sulfides and phosphates. Others are particularly susceptible to respiratory irritants such as smoke or pollen. We examine next the toxicity of, and potential exposure to, a number of frequently encountered pollutants.

Organs and Toxins

Different toxins affect different parts of the body. Your skin is an organ, an important one that keeps out the rest of the world. Ultraviolet radiation may be considered a toxin. Photons of ultraviolet radiation are the toxic agent affecting the skin. When absorbed, the photons damage and mutate cells deep in the skin. Ultraviolet radiation is a mutagenic and most likely a carcinogenic agent. Other air pollutants affect the skin by acting as irritants and, in cases of frequent exposure, as precursors of cancer of the lungs, respiratory tract, and gastrointestinal tract. The nasal passages, lungs, esophagus, and intestines are, after all, skin on the inside. When pollutants come in contact with the human eye, its sensitive tissues may become irritated. Crying over a chopped onion is aggravating, but red, running eyes caused by smog is a health issue.

When people are exposed to polluted air, they inhale and ingest toxins that may enter the bloodstream. Many toxic materials are soluble in water and stomach fluids and can be absorbed with other substances through the intestinal walls and other membranes. Such compounds include polycyclic organic molecules, metals such as lead and mercury, and sulfates and nitrates. The toxins then travel throughout the body in the bloodstream until they are filtered by the kidneys or liver or are assimilated into tissues (muscle, nerve, bone).

Figure 7.1 shows the areas of the human body where various toxins like to hang out. Among the most potent toxins are the **heavy metals** (not the electric guitar types, but the elemental type). We noted earlier that lead seeks out red blood cells, which act as a repository, as well as the brain, which uses lots of blood. Lead can accumulate in both blood marrow and the nervous system. Mercury (Hg) is another heavy metal that reaches the brain, settling in the stem, and it affects the kidneys as well. Several other elements—including cadmium, chromium, arsenic, and nickel leached from metallurgical fumes and combustion aerosols—can affect organs such as the heart and olfactory canals. The physiological effects of several common airborne heavy metals are summarized in Table 7.1. It is apparent that when many common metals are dispersed in the air at parts-per-million levels, they become rather

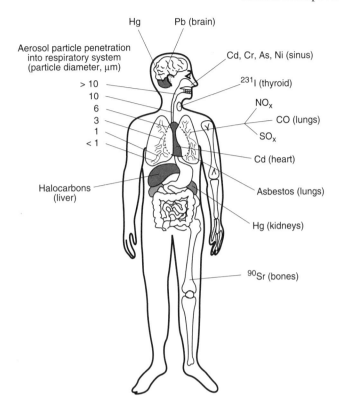

Figure 7.1   Physiological chart of human sensitivity to common air pollutants. The principal gaseous pollutants (NO_x = NO + NO_2, CO, O_3, SO_x = SO_2 + H_2SO_4) affect the lungs and air passages. Most of the other toxins are carried by fine aerosols, or "respirable" particles, that become lodged in the lungs or bronchial tubes, or are swallowed after lodging in the mucous lining of the throat and bronchia. Such toxins include heavy metals (mercury [Hg], lead [Pb], cadmium [Cd], and so forth) and radioactive elements (^{90}Sr, ^{231}I). These compounds tend to accumulate in specific organs or tissues, as indicated in the diagram. Airborne particles composed of solid materials such as asbestos and silicate dust may accumulate in the lungs and cause health problems even decades after exposure.

nasty pollutants. The greatest hazard of exposure occurs during mining, refining, and industrial applications of these metals (Section 7.2.4).

Elements that are "radioactive" also have their favorite places. Radioactive strontium collects in the bones, and iodine in the thyroid gland, for example. (Radioactivity and its effects are described at length in Sections 7.3 and 8.2.)

The Danger in Breathing

A number of airborne pollutants—including nitrogen oxides, sulfur oxides, and aerosols—are continuously deposited in various parts of the respiratory system. Some of these pollutants make their way deep into the lungs, through the trachea into the branching bronchial tubes, which, in diminishing size, direct the air flow into microscopic alveolar ducts and sacs. In the **alveoli**, oxygen, carbon dioxide, and other gases are exchanged between the inhaled air and the blood in capillaries within the sacs' walls. Air is drawn into and forced out of the lungs by the contraction and relaxation of the diaphragm, a muscular membrane that separates the chest from the abdominal cavity.

The depth of penetration of particles into the respiratory tract is controlled by the size of the aerosols and the type of breathing (through the nose or mouth, heavy or easy). The largest so-called **respirable particles**, those that can reach the lungs, are about 10 micrometers (μm) in diameter. Particles larger than this are efficiently captured in the nasal passages, mouth, and throat. Of the remaining respirable particles, different sizes tend to remain in

Table 7.1   Toxic Heavy Metals

Metal	Concentration (ppmm)	Effects[a]
Arsenic (As)	0.5	Cancer of the lungs, liver, and skin; teratogenic; poisonous in large doses
Cadmium (Cd)	0.2	Accumulation in the kidneys, lungs, and heart; symptoms like Wilson's disease; 50 ppmm fatal within 1 hour; carcinogenic
Chromium (Cr)	1.0	Skin rashes, lung cancer (after continued exposure); carcinogenic
Iron (Fe)	10.0	Siderosis, or red lung disease
Lead (Pb)	0.15	Brain damage; red blood cell anemia; paralysis of limbs
Manganese (Mn)	5.0	Aching limbs and back; drowsiness; loss of bladder control; nasal bleeding
Mercury (Hg)	0.05	Central nervous system attack; tremors and neuropsychiatric disturbance
Nickel (Ni)	1.0	Skin rashes, cancer of the sinus and lungs (after continued exposure); exposure to 0.001 ppmm of nickel carbonyl leads to nausea, vomiting, and possible death
Vanadium (V)	0.5	Acute spasm of the bronchi; emphysema
Zinc (Zn)	5.0	Fever, muscular pain, nausea, and vomiting

[a]The effects listed are for the threshold concentrations—in parts per million by mass—required to produce the effects described. The exposure is assumed to occur through inhalation of particulates containing the metal, with the amount indicated effectively absorbed by the body.

different parts of the respiratory system. The larger respirable aerosols are deposited in the upper pulmonary area (the trachea and bronchi). These passages are lined with mucus that capture the particles. Countless tiny hairs, or cilia, move the mucus outward into the esophagus, where the muck can be either swallowed or expectorated.[11] Less than 10% of the largest respirable particles (5 to 10 μm in diameter) reach the bronchi and lungs. For particles of ~3 μm size, 40 to 50 percent reach the lungs. Roughly 30 percent of the 2 μm particles and 20 percent of the 1 μm particles are deposited in the lungs.

Smaller particles (less than about 1 μm in size) easily flow with the rushing air deep into the lung passages. These microscopic particles reach the delicate alveoli, where they can diffuse to the walls.

11. When you breathe in a small hair or feather, it may get caught in your upper bronchial passages. You can feel the tickle as thousands of wiggling cilia work to remove the intruder. The irritation may send you into a spasm of coughing. This is the body's natural way of speeding up the removal process.

Although the efficiency of deposition of these submicron particles is only about 10 to 20 percent, they often carry the most dangerous toxic materials. Indeed, from the point of view of health effects, the most important deposition occurs deep in the lungs where the cleansing mechanisms are not as efficient as in the upper branches of the bronchial tree.

Although some materials can be absorbed through the alveoli into the blood, particulates in general cannot. The solid debris therefore accumulates over time, which reduces the lung's capacity to exchange oxygen with the blood and can eventually lead to serious lung disorders. In polluted atmospheres, the problem is aggravated. For example, in Los Angeles, autopsies of longtime residents reveal lungs stained gray by soot and other fine aerosols. Smokers' lungs, likewise, are discolored by the continual deposition of tars carried by tobacco smoke particles.

A number of specific diseases of the respiratory tract have been associated with air pollution. **Bronchitis** is an inflammation of the bronchial tubes, accompanied by coughing, wheezing, and discharge of sputum. Men are nearly four times as susceptible as women to contracting bronchitis. The disease is usually more severe if it first appears during childhood. **Pulmonary emphysema**, or emphysema, refers to the inelastic stretching and disintegration of the walls of the alveoli (these normally flexible tissues degenerate with abuse, like the rubber in the waistband of shorts that are frequently washed in bleach). After long exposure to airborne pollutants, the deep lung tissues lose their elasticity and become distended, finally breaking down. Among the symptoms of emphysema are breathlessness, tightness in the chest, blueness of extremities, wheezing, and extreme sensitivity to smoke and other pollutants.

Pneumoconiosis is a chronic irritation of the respiratory tract caused by long-term exposure to dust or other particles. Several distinct forms of pneumoconiosis result from extreme exposure to silicate dust (sand), **silicosis**; to asbestos fibers, **asbestosis**; and to coal dust, **anthracosis** ("black lung" disease). Small amounts of silicate and asbestos dusts are capable of producing strong effects. Only 5 *grams* (about one-sixth ounce) of silicate dust are sufficient to cause silicosis. The symptoms of pneumonoconiosis include, initially, shortness of breath and tightness in the chest and, later, severe bronchitis and emphysema.

A normal, healthy person utilizes only about one-twentieth of the lung's capacity for absorbing oxygen, except during strenuous activity. Hence there is a large safety factor built into the breathing apparatus. Once lung tissue is damaged, however, there is no evidence that it can be regenerated. Losing lung capacity is a one-way street, all downhill. The reserve capacity allows one to abuse one's lungs for decades before the cumulative damage becomes obvious.

The properties and toxic effects of a number of common air pollutants are reviewed in the following section. Radioactive materials are treated separately in Section 7.3. (A number of important indoor air pollutants, including radon, formaldehyde, and tobacco smoke, are discussed in Chapter 8.)

7.2   The Toxic Effects of Air Pollutants

The physiological effects of common air pollutants and toxic compounds are determined through toxicological experiments and studies of inadvertently exposed individuals. The important respiratory effects of air pollutants involve the interactions of vapors and particles with the surfaces of the respiratory tract at various depths. Large particles and very soluble gases are collected in the nasal passages and in the throat. Under continuing exposure, the tissues may develop lesions and tumors. In the lung passages, pollutants, including respirable particles, may cause bronchitis, pulmonary emphysema, and pneumoconiosis, including silicosis, asbestosis, and anthracosis.

7.2.1   COMMON INGREDIENTS OF SMOG

Table 7.2 summarizes the physiological responses to a short (1-hour) exposure to CO, NO_2, O_3, and SO_2 over a range of concentrations. The specific responses are explained next. (PAN and some other organic air pollutants are discussed in Sections 7.2.2 and 7.2.3.)

Carbon Monoxide

Carbon monoxide is a tasteless, colorless, odorless gas. When inhaled, it is absorbed in the lungs and enters the bloodstream. Normally, oxygen in the air also dissolves in the blood and attaches to **hemoglobin**, an iron compound, in the red blood cells. That is how we get oxygen to carry out aerobic cellular metabolism. (The evolution of aerobic respiration is discussed in Section 4.2.1.) It happens that hemoglobin also has a strong affinity for CO, forming **carboxyhemoglobin** (CO-Hb). The CO sticks to the hemoglobin and reduces the amount of O_2 that can attach. Accordingly, the quantity of oxygen available to the body is reduced. Thus, as the concentration of CO increases, the availability of O_2 decreases. The result is much like being suffocated.

The effects of exposure to CO are illustrated in Figure 7.2. The amount of carboxyhemoglobin present in the blood depends on the concentration of CO and the duration of exposure to it. As expected, the higher the concentration of CO is, the greater the percentage of blood hemoglobin that is deactivated. For a fixed concentration of CO in air, the fraction of inactive hemoglobin increases with length of exposure. Continuous carboxyhemoglobin levels exceeding 2.5 percent are considered unhealthful. A typical cigarette smoker may have CO-Hb levels of 5 to 10 percent, or more. Such high values of CO-Hb result from a relatively short exposure to very high

Table 7.2   Human response to pollutant exposure

CO	10–30 ppmm?	Time distortion (typical urban)
	~100	Throbbing headache (freeways, 100 ppmm)
	300	Vomiting, collapse (tobacco smoke, 400 ppmm)
	600	Death
NO_2	0.06 – 0.1	Respiratory impact (long-term exposure promotes disease)
	1.5 – 5.0	Breathing difficulty
	25 – 100	Acute bronchitits
	150	Death (may be delayed)
O_3	0.02	Odor threshold
	0.1	Nose and throat irritation in sensitive people
	0.3	General nose and throat irritation
	1.0	Airway resistance, headaches
		Long-term exposure leads to premature aging of lung tissue
SO_2	0.3	Taste threshold (acidic)
	0.5	Odor threshold (acrid)
	1.5	Bronchiolar constriction, respiratory infection

concentrations of CO (exceeding 100 parts per million by volume) in the smoke directly inhaled. Unfortunately, sitting in stagnant freeway traffic for an hour or more can produce similar physiological effects at lower CO concentrations.

The physical responses associated with exposure to CO include headaches and, in the extreme, mortality. Figure 7.2 summarizes the physiological effects for various levels of exposure to CO. At CO-Hb levels that can occur in smokers and occasionally in drivers trapped in heavy traffic, short-term effects may include headache and drowsiness. Poorly ventilated wood stoves, heaters, or barbecues can generate lethal CO concentrations. People have committed suicide by running a car engine in a closed garage, resulting in a fatal exposure to carbon monoxide. One can be exposed to high CO concentrations over a short period of time and recover; but don't try it unless you like to turn blue, fall into a coma, and vomit all over yourself.

Nitrogen Dioxide

Since nitrogen dioxide is relatively insoluble in water, it can be inhaled beyond the moist linings of the mouth, nose, and throat deeper into the lungs. The symptoms observed in people exposed to excess amounts of nitrogen dioxide are summarized in Table 7.2. At concentrations above about 1 to 5 parts per million by volume (ppmv), nitrogen dioxide affects the bronchial tubes and induces difficulty in breathing. Much higher concentrations can lead to painful respiratory failure. Long-term exposure to low levels of nitrogen dioxide have also been found to increase the occurrence of respiratory disease.

The action of nitrogen dioxide involves the reactions of nitrogen oxides on moist tissues to form nitric acid. The reactions are similar in many ways to those occurring in the formation of nitric acid in rain and fog. (See Section 3.3.4 and the discussions in Sections 6.6 and 9.3.) Except at very high concentrations, tissues in the respiratory tract are protected, by the presence of a moist mucous layer, from directly reacting with NO_2. However, nitric acid dissolves in the surface fluids and, as its concentration increases, attacks the tissue as a corrosive compound. One way to avoid breathing excessive amounts of nitrogen dioxide or other water-soluble gases and particles is to filter the air through a wet cloth. The water in the cloth absorbs some of the gas if it is slightly soluble and almost all the gas if it is highly soluble (as most acids are). Adding a little baking soda to the solution that the cloth is soaked in also tends to neutralize the acid, rendering it harmless.[12]

12. For the same reason, people caught in fires are told to place wet towels in cracks beneath doors and to breathe through a dampened fabric. This not only prevents smoke from entering the room, or your lungs, but also filters toxic gases and particles from the smoky air.

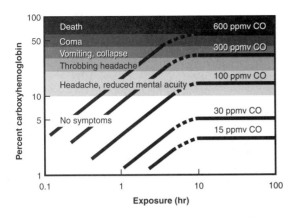

Figure 7.2 The physiological effects of exposure to carbon monoxide. The effect depends on the fraction of blood hemoglobin that is blocked from absorbing oxygen by combining with CO as carboxyhemoglobin, shown as a percentage of total hemoglobin along the vertical axis. The specific effects are defined by horizontal bands, from no symptoms at low carboxyhemoglobin levels to death above about 65 percent. The duration of exposure to CO is shown along the bottom axis. The heavy lines represent the evolution of the human response to specific CO concentrations, which are specified for each curve. For example, when exposed to a CO concentration of 600 ppmv, the carboxyhemoglobin blood fraction increases steadily over several hours to an equilibrium value of about 60 to 70 percent, which leads to coma and death in 10 hours or less. (Data from John H. Seinfeld. *Atmospheric Chemistry and Physics of Air Pollution.* New York: Wiley, [1986]. Based on Figure 2.1, p. 55)

Ozone

Ozone causes a variety of health problems for people exposed to it (Table 7.2). The presence of ozone is easily detected by its sweet, almost sickening, smell. The aroma is especially noticeable if you stick your head outside the door after being in an air-conditioned house with all the windows closed. Exposure to low concentrations of ozone (about 0.1 part per million by volume) can irritate the respiratory tract, particularly those of sensitive people. Concentrations above typical levels in smoggy air (greater than about 1 ppmv), can cause resistance to breathing and headache. Long-term exposure even to low ozone concentrations may damage sensitive areas of the skin, causing premature aging of lung tissue, for example.

Ozone oxidizes tissue. The effect is somewhat like receiving a burn very slowly. The higher the concentration of ozone is, the faster the combustion-like reactions can occur, and the greater the physiological

effect will be. The body responds by lubricating the damaged tissue with fluids and by constricting airflow. Over time, the tissue may be permanently scarred, with a subtle buildup of lesions, as in the case of a severe burn.

Sulfur Dioxide

Sulfur dioxide gas is relatively easily dissolved in water. As a result, it may be absorbed in the mouth and upper respiratory tract (as well as deeper in the lungs). Sulfur dioxide creates a strongly acidic taste in the mouth. The gas also has an acrid smell that can burn the nose at high concentrations. The toxic effects of exposure to SO_2 are summarized in Figure 7.3. Obviously, respiratory congestion is a principal concern in exposure to SO_2. Only 5 minutes of exposure to 10 ppmv or more of sulfur dioxide can cause death.

Like nitrogen dioxide, sulfur dioxide attacks the body through the formation of a strong acid, sulfuric acid in this instance. The formation of sulfuric acid

Figure 7.3 The range of possible effects of human exposure to sulfur dioxide. The exposure time and SO_2 concentration are shown as the variables. The region of excess deaths due to exposure is indicated in the upper-right area of high SO_2 concentration and exposure time. In the hatched area, significant health effects, including severe respiratory difficulties, can occur. Below the hatched region, less serious but nontrivial impacts may be detected. [Data from "Air Quality Criteria for Sulfur Oxides," National Center for Air Pollution Control, U. S. Department of Health, Education, and Welfare (1967)]

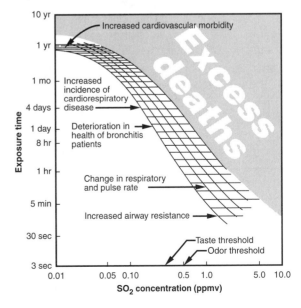

occurs in the upper respiratory tract, unless the SO_2 concentrations are quite high. Then, significant amounts of this gas can penetrate well into the lungs. The acid forms in the moist membranes of the respiratory tract and attacks the sensitive tissues by decomposing the organic molecules. There are grisly reports of murderers attempting to dissolve the corpses of their victims in acid baths. Concentrated sulfuric acid can reduce many organic materials to a charred residue.

7.2.2 EYE IRRITANTS

Certain compounds cause the eyes to tear, swell, and redden. These substances are called **lachrymators**. Among the common lachrymal compounds in smog are peroxyacetylnitrate (PAN), acrolein, and formaldehyde. These chemicals irritate the delicate membranes lining the eye, causing a flow of tears to cleanse the surfaces of the eyeball and eyelids. The potency of lachrymal compounds comes from their ability to sensitize the conjunctive tissues lining the lids and orb. Onions, as many weeping cooks will tell you, are strong lachrymators. When cut, they give off a sulfur-rich organic vapor. A weak solution of boric acid is usually applied as an eyewash to relieve the symptoms of exposure to eye irritants. The fluid cleans the surface and neutralizes any strong chemical agents that may have settled on the eye.

PAN (with the chemical formula $CH_3COO_2NO_2$) is a secondary component of smog formed when certain products of organic oxidation combine with nitrogen dioxide. The precursors of PAN are plentiful in urban air. Acrolein (with the chemical formula CH_2CHCHO) is also an abundant component of urban smog. Acrolein, like PAN, is a secondary organic pollutant. Fires also produce acrolein. This compound, like PAN, is a powerful eye irritant. (Formaldehyde and its health effects, including eye irritation, are discussed thoroughly in Section 8.3.)

Lachrymators are mainly a nuisance in urban life. The long-term effects on the eye of exposure to these particular compounds are most likely to be small. Nevertheless, eye irritation signals the presence of toxic air pollutants in high, and probably unhealthful, concentrations. Concurrent exposure through breathing of these compounds and the accompanying diverse components of smog poses a greater health risk. PAN and other compounds also affect plants and animals. For example, exposure to common levels of PAN in smog can discolor the leaves of plants, causing silvering and bronzing of the leaf surface.

7.2.3 ORGANIC VAPORS

The number of organic compounds that we are exposed to every day is astronomical. We are organic beings. We eat organic matter for nourishment. We emit and breathe organic vapors. Nevertheless, some organic compounds simply are no good for us, and so we should avoid them in any amount. For example, all pesticides and herbicides (bug and weed killers, respectively) are designed to be toxic to living organisms of some type. We may not be closely related to cockroaches (most of us, anyway), but the pesticide that can kill a cockroach can kill you just as well. Today we are surrounded by pesticides: flying-insect sprays, crawling-bug spritzers, flea powders, room insect bombs, garden insecticide dusts, and lawn treatments. Herbicides are less common, but still widespread.

In some areas, pesticides are widely dispersed to eradicate pests such as mosquitoes. In Los Angeles, for example, the insecticide malathion has been sprayed by helicopters over local communities to fight an infestation of Mediterranean fruit flies. One official there was quoted in the *Los Angeles Times* as saying, "Look, we've sprayed something like 750,000 residences. We got 1.5 to 2 million people. How many have we heard from? A few thousand? It's a very small proportion." One advocate of such activities drank a glass of malathion to prove that it is harmless to humans; his fate is unknown to me.

Table 7.3 lists a handful of the common organic compounds that humans release into the environment. All these chemicals are toxic in large doses, some in relatively low doses. One of the worst chemical accidents in history occurred in Bhopal, India, on December 3, 1984. A pesticide plant owned by Union Carbide spilled deadly methyl isocyanate from a storage tank. In the crowded city nearby, up to 300,000 people were exposed to fumes that spread out in a deadly ground plume hugging the surface. Perhaps 14,000 people were seriously affected, and 1800 died within a few days. For the next 2 years, people continued dying at a rate estimated to be as high as 15 per week.

Among the organic toxins that deserve notice are a large class of solvents used in a variety of industrial processes and in common construction materials. For example, benzene is a useful solvent

Table 7.3   Effects of Toxic Organic Compounds

Compounds[a]	Uses and Effects[b]
Acetone	Nail polish and paint remover, cleaning solvent Nose and throat irritation; dizziness at high concentration Smog precursor
Benzene	Solvents, gasoline, and other fuels Respiratory irritation, dizziness Suspected cause of leukemia
Carbon tetrachloride	Cleaning solvent, base for manufacture of chlorofluorocarbons Carcinogenic
Glycol ethers	Component of brake fluid, paints, dyes Kidney damage; nausea at high exposure Fetal toxicity
Methylene chloride	Paint stripper, degreaser Skin irritation; affects heart and nervous system at high concentrations Probably carcinogenic
Methylethylketone (MEK)	Solvent in paints, adhesives, cosmetics Headaches and vision effects at high doses Degradation of memory and reflexes with chronic exposure Possibly mutagenic
Perchloroethylene (PCE)	Metal cleaning, dry cleaning Moderately toxic; liver, kidney, and nervous system effects Probably carcinogenic
Styrene	Plastic and resin production Eye and throat irritation Suspected to be carcinogenic
Toluene	Gasoline additive, detergent production Skin and eye irritation; fetal toxicity Anemia, liver damage, and nerve effects with chronic exposure Smog precursor
Trichloroethylene (TCE)	Cleaning solvent Low toxicity; irritation of eyes and lungs Death at high exposure
Vinyl chloride	PVC plastics and plumbing fixtures Associated with rare forms of liver, brain, and lung cancer Mutagenic to DNA
Xylene	Fuels, lacquers, glues Eye, nose, throat irritation Liver and nerve damage with chronic exposure Smog precursor

[a]The compounds listed are produced in large quantities, up to hundreds of thousands of tons per year. Only a small fraction of the total amount produced is emitted into the environment, usually as fugitive emissions during synthesis, transport, or industrial applications or from products that incorporate these compounds.

[b]The effects listed are for moderate to large doses of these compounds.

for petroleum products, and carbon tetrachloride is widely employed in the dry-cleaning industry. It is estimated that up to 2 million workers are routinely exposed to benzene. The risks include aplastic anemia (related to a dysfunction of the bone marrow, as opposed to anemia related to a deficiency of iron) and leukemia. Solvents used in paints and adhesives, by function, evaporate into the atmosphere. Switching to nonorganic solvents (such as water-based paints) and developing closed systems to recycle volatile substances (as in sealed paint shops) are examples of measures devised to control the release of, and exposure to, organic vapors.

7.2.4 PROBLEM PARTICLES

The transport and deposition of toxic materials carried by airborne particles are a common problem. Toxins on particulates are very efficiently deposited and absorbed in the body. Aerosols laced with toxins may be formed in photochemical smog, emitted from combustion sources or generated by mechanical processes. Lungs, as we already noted, are well-designed particle traps. The aerosols produced in smog and from other common sources (fire smoke and industrial fumes) generally fall into the size range referred to as *respirable*. Where such particles are present, breathing can be hazardous to health. Several of the most threatening particulate-borne pollutants are discussed in this section. Chapter 8 also deals with related issues of aerosol inhalation associated with tobacco smoke, radon, and other indoor particulates. Highly respirable tobacco smoke is one of the most deadly particulates generated by humans (Section 8.4). Regarding radon gas, the danger lies not in the radon itself but in its radioactive "daughters" that collect on "respirable" particles (Section 8.2).

Smoke from Hell

Combustion can produce toxic compounds from common fuels. The analysis by Percival Pott of scrotum cancer in chimney sweeps shows the strong carcinogenicity of smoke. Even smoke generated by burning wood in a fireplace contains small amounts of the strong carcinogenic chemicals benzopyrene, dibenzopyrene, dibenzanthracene, dimethylbenz-anthracene, benzophenanthrene, methylchol-anthene, and dibenzocarbazole.

In recent years, high-temperature incineration has become a possible alternative means for disposing of garbage. The traditional method of dumping refuse in landfills is quickly becoming untenable, as accessible sites are filled and the demand for natural open recreational land grows. Hauling garbage by train to faraway places or sending the waste to sea on huge barges looking for a port of call seems environmentally inappropriate. Burning the trash—reducing it to carbon dioxide, water, and a solid residue—is an attractive concept. Unfortunately, it is very difficult to design a perfectly clean combustion process. The fly ash from existing municipal incinerators has been shown to carry more than 600 organic compounds, among which are several classes of polycyclic organic materials that contain chlorine in their molecules—the polychlorinated dibenzofurans and biphenyls, and the infamous **dioxins**. The dioxins come in many permutations of a basic molecular structure, the most toxic of which is tetrachlorodibenzodioxin (TCDD). Dioxins are also generated when certain common and exotic substances, such as plastics and polychlorinated biphenyls (PCBs), are burned.

Dioxin is a powerful carcinogen in laboratory animals. Indeed, minute doses can induce cancer in hamsters. Two-millionths of an ounce will kill a mouse. Dioxin is absorbed through the skin, in the lungs, or through the gastrointestinal tract. People are apparently less susceptible than hamsters and mice. In 1976, near Seveso, Italy, nearly 37,000 people were exposed to dioxin when there was an explosion in a pesticide factory owned by a Swiss firm. Some of the workers at the accident received high doses of dioxin. Of all the victims, however, only about 100 to 150 developed the skin disorder called *chloracne*. This condition, which appears as a severe surface rash, clears up in time. Very few people died, or showed long-term effects, particularly the expected outbreak of cancer. It seems, in this case, that hamsters are the main victims of TCDD. Nevertheless, some epidemiologists suggest that health problems may surface in the next generation born to those exposed at Seveso.

Following the Seveso accident, a public outcry led to the demolition of contaminated buildings; tainted soil was scraped up, and the offending chemical plant was dismantled. Thousands of tons of debris had to be hauled off and buried. In the process, 41 barrels of dioxin-laced rubbish was lost (not in someone's backyard, one hopes).

Earlier, in 1971, in the small town of Times Park, Missouri, a local entrepreneur purchased used industrial oil and sprayed it on local roads and in equestrian areas to control dust. What he failed to check (or ignored) was the heavy contamination of the oil with dioxin, and so the dust that did rise was laced with it. Dogs, cats, and birds died by the hundreds. Horses fell ill with a rare disease that left them paralyzed and wasted. Needless to say, the town had a public-relations problem. The area was evacuated by government officials. The surface soil was stripped and buried elsewhere. A sleepy Midwestern town was obliterated by concern over dioxin. It could be called the Love Canal of the Midwest.

In the Times Park and Seveso incidents, large areas were contaminated by dioxin. Many families were directly and indirectly affected. The issue of toxic pollutants and their control came to the forefront because of the suffering created by these outrageous events. Laws and regulations regarding hazardous compounds were toughened. But laws are often strengthened only enough to solve an immediate problem, not to forestall future problems.

Diesel engine soot is full of toxic compounds. The soot is formed during the high temperature combustion of diesel fuel, when polycyclic organic compounds in the fuel react to form particles composed mainly of carbon. You can see the black smoke, consisting mainly of soot, spewing from the exhaust pipes of buses and trucks. The surfaces of the soot particles, like the surfaces of fly ash particles, provide a favorable site for toxic compounds to collect. Polycyclic aromatic hydrocarbons collect on the soot. Indeed, tests (for example, the Ames test) show diesel exhaust to be highly mutagenic. Since 1989, the National Institute for Occupational Safety and Health (NIOSH) has stated that diesel exhaust is a potential carcinogen.

The air in large cities is filled with soot. In Los Angeles, where rain is rare most of the year, a steady fallout of soot accumulates everywhere. The soot settling onto a car windshield over a few days will turn a paper towel brownish black. The lungs of longtime city dwellers are invariably stained by soot. Toxins carried by the soot particles into the lungs are perhaps the worst part of the bargain. As noted earlier, the carcinogenic components of urban aerosols, mainly found on soot, are even more active than benzopyrene. A recent analysis shows that the incidence of lung cancer in women living in large Japanese cities has increased by a factor of 10 since the 1950s, closely paralleling the increase in the number of diesel-powered vehicles. Cigarette smoking can be eliminated as a significant contributing factor in this case, owing to the strong social prohibitions against smoking by women. In the United States and other industrialized nations, alarm over the health hazards of diesel soot has led to strict limits on soot emissions from new vehicles beginning in the mid-1990s.

Certainly, any rational person would not want to suck much soot into his lungs. In one published laboratory experiment, a group of convicts agreed to breathe soot from a lawn mower engine to determine how the particles would react in their lungs. What some people are willing to do for a few privileges!

Get the Lead Out

Lead is a highly toxic heavy metal that enters the body through the air and in water and food. Airborne particles have traces of lead. Until recently, this lead originated, in part, from gasoline burned in automobiles.[13] Lead was widely used as a fuel additive (in the form of tetraethyl-lead) to increase the performance of automobile engines. Particles in the exhaust, however, carried the lead into the environment. Today, lead is being phased out of fuels. For most newer cars, you must buy "unleaded" gasoline at the pump. Before the advent of unleaded fuels, however, as much as 500 tons of lead were deposited annually in the coastal waters adjacent to Los Angeles: in runoff following rains, in sewage effluence, and by the fallout of vehicle-exhaust particles. Other important sources of lead include drinking water (in contact with lead pipes and lead-solder joints) and cracking, peeling lead-based paint (that can form dust or be eaten by small children). To control lead from these latter sources, lead pipes and solder and lead-based paints have been eliminated from new construction.

The increase of lead emissions into the atmosphere can be tracked very clearly in ice cores taken in Greenland. During the 1950s and 1960s, the concentrations of lead contained in snow deposited as ice reached peak values 400 times greater than the concentrations recorded 2000 years earlier. The

---

13. Starting in the early 1970s, lead in gasoline has been under continual phased reduction. New automobiles are designed to run on lead-free gasoline. Nowadays, older cars still in use must ping along on unleaded, but environmentally safe, fuel.

increase corresponded to the accelerating use of leaded gasoline. Today, largely because lead has been banned from gasoline in many countries, its concentrations in pack ice have subsided to perhaps 10 times the preindustrial values. Still, in heavily polluted environments, up to 50 micrograms of lead may be inhaled each day, of which 25 to 50 percent is absorbed into the bloodstream. Lead is also ingested with food and water—with amounts ingested up to 300 micrograms of lead a day—of which roughly 10 percent may be absorbed. The present health standards recommend that lead intake be limited to 0.5 microgram per day, and that lead concentrations in blood be less than 15 micrograms of lead per 100 grams of blood (μg-Pb/100 g). To control lead exposure through air pollution, the U.S. ambient air quality standard dictates that average concentrations not exceed 1.5 micrograms of lead per cubic meter of air. Nevertheless, lead intake in urban areas can at times reach 100 times the recommended level, and blood concentrations of 100's of μg-Pb/100 g have been measured in children in cities and workers in industries utilizing lead.

Lead-containing particles are also generated in metallurgical operations. In addition, paints manufactured years ago contain large amounts of lead. As this paint peels and flakes, it creates airborne dust. When particles containing lead are inhaled, settle onto edible food, or are mixed in water supplies, the lead may be ingested and enter the bloodstream. These days, the most serious exposure to lead occurs in occupational situations or in poorly maintained buildings coated with layers of lead-based paint. Nevertheless, everyone is exposed to small amounts of airborne lead.

The disease associated with lead poisoning is called **plumbism**. The general symptoms include paleness of color, moodiness and irritability, a black line near the base of the gums, and abdominal pains ("lead colic"). Advanced cases of plumbism may result in paralysis and loss of nerve function ("lead palsy"), dizziness and confusion, visual disturbances, blindness, deafness, and coma leading to death.

The concentrations of lead measured in the blood of urban dwellers average roughly 20 micrograms of lead per 100 grams of blood (μg-Pb/100 g). The values in rural areas are considerably lower, perhaps 10 μg-Pb/100 g. The physiological symptoms of lead vary with the blood level. At 60 to 70 μg-Pb/100 g, a mild encephalopathy can be seen (that is, a partial loss of brain function). With greater exposure and higher concentrations, the kidneys, brain, and nervous system are seriously affected. Lead tends to accumulate in bone tissue, although over many years, it is eventually purged from the body. Lead contamination of bone marrow inhibits the creation of the oxygen-absorbing agent in hemoglobin, and hence a kind of anemia is induced in victims of lead poisoning. Lead is particularly harmful to young children, because of their higher metabolism and greater air, food, and water intake. In pregnant women, the fetus is also at high risk. It is estimated, for example, that roughly 15 percent of all preschool children in the United States have blood levels exceeding 15 μg-Pb/100 g, the present standard.

The Asbestos Mess

Asbestos is one of the most feared natural materials in the developed world today.[14] This common mineral has been used extensively as a heat insulator and to provide fireproofing. Millions of tons of asbestos have been used in buildings, ships, and other structures for these purposes. Unfortunately, asbestos has a normal tendency to flake and release airborne particles. In many old buildings, layers of asbestos slurry sprayed on ceilings and walls, or asbestos-based tiles on ceilings and floors, release airborne particles when disturbed. Asbestos is also found in the dust from automobile brake pads. Workers exposed to asbestos have a high occurrence of **asbestosis**, a serious deteriorative disease of the lungs. Inhaled asbestos particles become lodged deep in the lungs and remain there for a lifetime. After many years, scar tissue develops around the fibers, impairing normal breathing and spurring dry-coughing episodes. In serious cases of asbestosis, early death can follow as a result of secondary heart disease. In addition, sensitive tissues in the lungs are irritated over time, particularly in combination with cigarette smoking, and can induce cancer of the mesothelial membrane lining the lungs (**mesothelioma**). The incidence of asbestosis has been increasing because

14. Asbestos is a natural mineral that is mined primarily in Canada. Asbestos is easily separated into long fibers that are resistant to fire, do not conduct heat or electricity, and are chemically inert. For these reasons, asbestos is commonly employed in materials that require such properties, including fire and heat insulation. The fibers are often woven into yarn or mixed with binding agents to form solid construction materials. The industries in which exposure to asbestos is greatest include mining, insulation manufacture, and construction.

of the widespread uncontrolled use of asbestos until the 1970s. It is estimated that one-quarter to one-half of all city dwellers have significant, although not life-threatening, amounts of asbestos lodged in their lungs. Asbestos-related diseases can have a long latency period before symptoms appear. Victims often were exposed to asbestos decades before the disease arose.

The risk of contracting mesothelioma from exposure to asbestos fibers has been estimated by the National Research Council. Continuous exposure to an asbestos concentration of 0.0004 fiber per cubic centimeter of air (that is, a total of roughly one millionth of a gram of fibers in an average-size room) carries a lifetime risk of about 160 cases of mesothelioma per million people. The risk increases in proportion to the asbestos concentration. The highest risk is for male smokers—about 300 cases per million—and the lowest risk is for female nonsmokers—about 14 cases per million.

Today, asbestos is treated as toxic waste. Remodeling older buildings containing asbestos requires teams of specialists to remove the asbestos without creating contaminated dust. Thousands of older buildings are still lined with asbestos-based insulation and tiles. In auto shops, I have seen mechanics use air hoses to blow clouds of fine particles from brake drums; that dust is loaded with asbestos. Asbestos is being used less in brake pads today. But enormous piles of asbestos mine "tailings" still remain open, where the wind can lift the fibers. Victims of exposure to asbestos have waged a long legal battle for compensation from the manufacturers and distributors of asbestos and asbestos products, particularly the Johns Manville Company. Recently, the class-action suit was settled, with awards ranging from $2500 to $300,000, depending on the seriousness of asbestos-related disease. That hardly seems like sufficient compensation for a shattered life. Avoidance is probably a better strategy when it comes to asbestos.

7.2.5 PERSISTENT ENVIRONMENTAL TOXINS

Many chemicals manufactured and widely applied for agriculture or other practical purposes may persist in the environment for decades. Figure 7.4 illustrates the buildup and decline of **DDT** (dichlorodiphenyltrichloroethane) in recent decades. The application of DDT as a pesticide caused an environ-

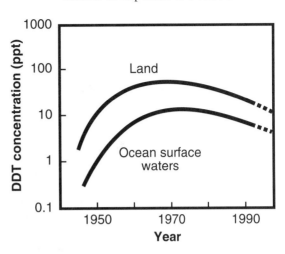

Figure 7.4 Concentrations of the pesticide dichlorodiphenyltrichloroethane (DDT) measured in the environment over a period of time. The concentrations on land and in lake, river, and ocean waters grew during the 1950s and 1960s, until strict controls were placed on its use. Since that time, the concentrations on the land and in surface waters have declined through decomposition and dilution into larger reservoirs. At the present rate, it will take several decades for the environment to recover.

mental nightmare in the 1950s and 1960s. DDT, first fabricated in 1874, was discovered in 1929 to be an excellent insecticide. When sprayed as a dust or liquid suspension, it was particularly effective against mosquitoes that carried a variety of deadly diseases such as malaria, yellow fever, typhoid fever and bubonic plague. DDT also became popular as a means of controlling a host of agricultural pests. One could hardly argue with the value of these uses. The pests, however, had other ideas. Over time, they developed an immunity to DDT. More had to be applied, and the concentration of DDT in the bugs increased. Birds and fish feeding on the DDT-laced insects became the actual victims of the pesticide. Rachel Carson's book *Silent Spring* documents the events surrounding the DDT disaster. Finally, in 1972, the Environmental Protection Agency severely restricted the use of DDT.

DDT demonstrates that persistent chemicals can accumulate to deadly levels in ecosystems. Fish and birds are not the only potential victims. People who eat tainted fish and fowl are themselves poisoned. Although DDT has been largely eliminated from use as a pesticide, its environmental effects will linger beyond the end of the century. DDT, like the lead in gasoline, was used almost indiscriminately before anyone had bothered to ask, let alone answer, the question: Can this compound damage the environment? Many other persistent substances are emitted by human activities that are equally dangerous. Pesticides derived from chlorinated chemicals are often very stable and resist biodegradation. Chlordane is a handy termite killer widely used in the 1960s. Although it is less toxic to humans than some other related pesticides, it is still banned in many countries and today is not as readily available in the United States.

Lead and other heavy metals also persist in the environment and may be concentrated during biological assimilation. Elevated concentrations of mercury in fish are an example of the heavy-metal contamination of natural ecological systems. Moreover, not all persistent pollutants are concentrated in water and soil. Some gaseous pollutants are also stable and persistent in the atmosphere. The entire class of fully halogenated chlorofluorocarbons (CFCs) fits into this category. In this instance, accumulation in the atmosphere has led to significant depletion of the stratospheric ozone layer (Sections 13.5.2 and 14.2.3). It is also worth mentioning the buildup of "greenhouse warming" gases such as carbon dioxide, nitrous oxide and the CFCs in the atmosphere (Sections 10.2.4, 12.2, and 12.3). All these compounds are persistent pollutants that affect the global climate.

Earth as a Toxic Landfill

A sad situation has developed over the past decade. People are more and more willing to dump extremely dangerous and persistent toxic compounds into the environment. As the regulation of dangerous toxins stiffens and the cost of supervised disposal escalates, a growing number of people and businesses are dumping wastes illegally. It may take the form of barrels filled with chemicals dropped into a canyon, a tanker truck drained into an empty field, tainted debris dumped into the ocean, poisonous fluids poured into a storm drain, or toxic gases released under cover of darkness. The waste includes pesticides, herbicides, solvents, petrochemicals, heavy metals, asbestos, fuels, radioactivity (Section 7.3), and so on. Sooner or later someone is likely to breathe the toxic fumes, drink the tainted water, or inhale the poisoned dust. Homes and businesses may be built over the waste, unknown to those living and working there. Love

Canals, small and large, may begin to pop up across the countryside.[15]

Even more deplorable, toxic waste has apparently become a commodity to be traded, or exported, to the Third World. If waste cannot be dumped locally because of stiff regulations, why not send it where the regulations are lax? It would also provide badly needed income for poor countries. Recently, two such dealers were convicted for shipping toxic waste from Los Angeles to Pakistan. They were arrested and tried for violating the national Resource Conservation and Recovery Act. The shipment, which was intercepted in Saudi Arabia, consisted of chemically tainted debris from a fire. The dangerous stuff was neatly packaged in containers with phony identification labels. There is apparently a brisk trade in toxic waste. According to one estimate, as many as 2 million metric tons of toxic debris leave the United States illegally every year. That is quite a bonanza for the rest of the world: a mother lode of illness and cancer.

The Soviet Union as a Toxic Dump

If Love Canal is a nightmare faced by some Americans, imagine the fitful dreams of the former Soviets. Stories are now emerging from the backwaters of the old Soviet Union describing horrible crimes against the environment and horrific consequences for the people living there. Air pollution is so bad it hurts to breathe it; new snow exposed to the air quickly turns gray. In industrial towns scattered across vast regions, the local air, water, and soil pollution is so bad that life expectancy in some heavily polluted areas is a mere 45 years. It never occurred to the past Soviet leadership to *enforce* the strict pollution standards set up to protect workers and the public. In the Soviet Union, environmental protection was just another Potemkin village. Government officials now estimate that 50 million people in 100 cities are

15. Love Canal is a neighborhood in the town of Niagara Falls, New York. In 1978, toxic compounds, including PCBs and dioxin, were detected in the basements of homes in the neighborhood. The residents were found to suffer from abnormal chromosomal damage. It was determined that the Hooker Chemicals and Plastics Corporation had used the area as a dump site years earlier. After covering the toxic waste with dirt, Hooker had donated the land to the city, which allowed a residential development to locate there. Thirteen hundred residents had to be permanently evacuated from the neighborhood. The site remains to be cleaned up despite huge expenditures. "Love Canal" is today synonymous with toxic-waste sites that appear in populated areas.

exposed to pollutants at 15 times the maximum concentrations normally allowed.

Soviet factories pour out mercury, asbestos, polyvinyl chloride, and countless other dangerous chemicals. To dispose of liquid and solid wastes, huge underground cavities have been formed using nuclear explosions. The cavities are then filled with deadly garbage. Radioactivity leaks into the environment along with the buried wastes, contaminating water supplies and soils. Genetic defects and diseases, cancer, and respiratory disorders are rampant. In some cities, only one in six births is normal, and one-third of babies are born with serious disease. The infant mortality rate in the Soviet republics is as high as 40 per 1000 births, compared with about 9 per 1000 in the United States. One can only wonder about the long-term consequences of this indiscriminate poisoning of the earth. The cost of cleaning up this mess is incalculable. Most likely, it will never be fixed, and tens of millions of people in many generations will suffer.

7.3 Radioactivity

One of the most widespread toxins in the environment is radioactivity. Every day we are exposed to natural and human-made sources of radioactivity. Occasionally, a major accident releases deadly quantities of radioactive debris into the atmosphere. More likely, however, radioactive wastes slowly leaking into air, water, and soil provide continuous exposure above the natural background level. Some aspects of radioactivity, the ways in which exposure occurs, and the possible consequences of exposure are summarized in this section.

7.3.1 STABILITY OF THE ELEMENTS

The center, or nucleus, of an atom is composed of protons and neutrons. (The nature of these subatomic particles is described in the introduction to Section 3.1.) Basically, protons have a unit positive electrical charge, and neutrons have no charge (that is, they are neutral). Different elements have different numbers of protons in their nuclei, but all the atoms of a particular element have *exactly* the same number of protons. This number is called the atomic number for that element (refer to the discussions in Sections 2.2.2 and 3.1). Every element has a different atomic number. The atomic number for hydrogen, the simplest element, is 1 (that is, the hydrogen atom has one proton in its nucleus). Helium is the second element in complexity, with two protons (and so has atomic number 2). The same element can have atoms with different numbers of *neutrons*, however. Such atoms—which differ only in the number of their neutrons—are called isotopes of the element (see Section 3.1). All isotopes of an element have the same atomic number.

Particles that carry the same sign of electrical charge (positive or negative) repel one another. This repulsion is a result of the **Coulomb force** exerted by electrical charges on each other. Under the influence of the Coulomb force, particles with like charges experience mutual repulsion, and those with unlike charges, mutual attraction. If it has the same kind of electrical charge, a rubber rod rubbed vigorously with a cloth can be used to charge up two ping pong balls suspended on strings, or two strips of aluminum foil. The balls, or foil strips, will fly apart because of Coulomb repulsion and will remain separated until the charge dissipates. Static electricity is the buildup of electric charge by friction. Sparks jump on a dry day and skirts cling to legs because of static electricity.

You can imagine trying to force protons together to form the nucleus of an atom. The positive charges on the protons cause strong mutual repulsion. As you work futilely to assemble the nucleus, protons keep flying out of the cluster. The more protons that you attempt to add, the greater the repulsion becomes, and the more difficult it is to hold the cluster together. Some sort of atomic "glue" is needed. Fortunately, nature has supplied the glue in the form of neutrons, which bind with the protons and produce a short-range attractive force—the so-called "strong" nuclear force—to overcome the Coulomb repulsion of the protons. The neutrons are the binder for the nucleus. Without the highly unusual strong force, which we never experience directly, no atoms except for hydrogen, with one proton, would be stable. The universe could not exist except as a huge bubble of hydrogen gas—not a very interesting place.

The nuclei of heavier elements (which have more protons) require additional stabilization and more neutrons. Imagine adding protons and neutrons, one at a time, to an atomic nucleus. At first, for each proton added, one or two neutrons could be included to bind and stabilize the nucleus. Above a certain size, however, no matter how many neutrons are added, the nucleus cannot be permanently stabilized. The repulsive Coulomb force becomes dominant. Accordingly, natural elements are generally found with up to only 92 protons (that is, up to uranium [U]).[16] Nevertheless, larger nuclei can be constructed using powerful accelerators that smash together the nuclei of smaller atoms at high speeds. In this way, physicists have recently fabricated artificial elements up to atomic number 111.[17] Such manufactured atoms are all extraordinarily unstable (an atom of element-111, for example, disintegrates into lighter elements in two milliseconds). It is thought that if nuclei with at least 114 protons can ever be fabricated, these super-heavy elements may be more stable. It is doubtful they would be stable enough to have practical uses in everyday life, however.

Nature has contrived to provide us with a certain small number of very stable elements, from which all the things we know and love are made. These elements were formed at enormous pressures and temperatures within the hot cores of stars before the Earth formed. These stable natural atomic nuclei have 84 or fewer protons. The elements with more than 84 protons are inherently unstable and decay by fragmentation into lighter elements. Thus plutonium (Pu, atomic number 94), uranium (U, 92), thorium (Th, 90), radium (Ra, 88) and radon (Rn, 86) all are unstable and eventually decay into lighter elements, including stable lead (Pb) atoms. In addition, many of the stable lighter elements have isotopes that are unstable; such elements include polonium (Po, 84), bismuth (Bi, 83), and thallium (Tl, 81). Because of their instability, these elements (above atomic number 84) are relatively uncommon. They have names that conjure up grim images associated with nuclear radiation, atomic explosions, mutations, and death. Yet in Nature's managerie they are just elements—human exploitation of them has led to the bum rap.

Radioactive Nuclides

When an unstable nucleus decays, or breaks into two or more pieces, it releases the large amount of energy

that originally bound the fragments together. This energy can appear in several forms which, taken together, comprise the **radioactivity**. Radioactivity is, in fact, the products of the decay of unstable atomic nuclei. Radioactivity was first detected in uranium isolated in 1896 by the French physicist Antoine-Henri Becquerel. Soon thereafter, in 1898, the French scientists Pierre and Marie Curie made their famous discovery of radioactivity in radium and polonium: They had noticed that photographic plates were ruined by exposure to these substances. Today, radioactivity can be readily detected using a **Geiger counter**, named for its inventor, Hans Geiger.[18]

For every element, a number of different isotopes can be manufactured by adding neutrons to or subtracting them from the most stable form of the element. A change in the number of neutrons (thus deviating from the optimum needed to stabilize the atom), however, tends to destabilize the nucleus. Such unstable species are called radioactive isotopes, or **radioisotopes**. For example, a normal hydrogen atom has one proton and no neutrons in its nucleus. A neutron can be added to form the stable heavier hydrogen isotope, **deuterium** (**D**). Not all isotopes need be unstable, but if one more neutron is attached to deuterium, the unstable hydrogen isotope **tritium** (**T**) is created. Tritium is radioactive and disintegrates in about 12 years, releasing some of the energy binding its nucleus.

Another example is provided by the isotopes of carbon. The most common isotope of the carbon atom has six protons and six neutrons. The total number of **nuclides** (that is, protons and neutrons) in the normal carbon nucleus is thus 12. This isotope of carbon is given the name carbon-12, or ^{12}C in shorthand notation.[19] Adding a neutron to the nucleus of a normal carbon atom produces a stable isotope, carbon-13 (^{13}C); although this additional neutron in the nucleus destabilizes the atom. The important isotope carbon-14 (^{14}C) disintegrates in about 5700 years (long enough for it to be useful

---

16. The elements with more protons than uranium has are called the *transuranium* elements. These are mainly synthetic, or human-made, elements, although natural plutonium (Pu, atomic number 94) and neptunium (Np, 93) have apparently been detected in minute amounts.

17. In the fall of 1994, a series of experiments at the Society for Heavy Ion Research in Germany successfully combined a nickel atom (atomic number 28) and a bismuth atom (atomic number 83) to form a new, as-yet unnamed element (atomic number 111).

18. Hans Geiger (1882–1945), a German scientist, devised an ionization chamber to detect radioactive decay. The energy released as radioactivity can ionize air molecules or strip electrons from these molecules. Whenever an atom decays in the vicinity of the counter, radioactive "rays" can enter the ionization chamber and produce free electrons, which are collected and measured as a pulse of electricity. The pulse is usually amplified into a "click," the sound commonly associated with radioactivity in the movies.

19. The total number of nuclides (protons and neutrons) in the nucleus of an atom is indicated as a superscript to the left of the chemical symbol for that element. This seemingly odd choice of location in fact leaves room for other numbers and characters that chemists and physicists often attach to chemical symbols.

for dating ancient organic matter).[20] But if a neutron were removed from a carbon nucleus to form carbon-11, that species would disintegrate immediately.

In general, the most stable form of an element has roughly the same number of neutrons and protons. Common isotopes typically have one or two neutrons more or less. The destabilization of a uranium-235 nucleus by the addition of one *extra* neutron is the basis for the "atomic" bomb. In this case, the added neutron causes the nucleus to split, or **fission**, into fragments, releasing a great amount of binding energy.

The radioactive nuclei that have been identified are collectively referred to as radioactive nuclides, or **radionuclides**. It follows that the family of radionuclides consists of all known radioisotopes of the elements. At the present time, roughly 1000 radionuclides have been identified. Of this total, only about 50 are natural, with the rest human-made.

Radioactive Decay and Half-Life

When an unstable radionuclide disintegrates, subatomic particles are emitted. These particles carry a large part of the nucleus's binding energy that is released. Typical radioactive decay produces alpha (α) and beta (β) particles and sometimes gamma (γ) rays. An alpha particle is actually the nucleus of a helium atom, consisting of two protons and two neutrons. This fragment carries a positive charge associated with the protons. A beta particle is just an electron that carries a negative charge. Gamma rays are a very energetic form of electromagnetic radiation with very short wavelengths (Section 3.2). The photons of gamma rays easily penetrate clothing and even the walls of buildings. Careful shielding is needed to prevent exposure.

When an alpha particle is emitted, the process is called **alpha decay**. Since two protons and two neutrons are emitted, the remaining nucleus is "lighter" by *four* nuclides. Moreover, because two protons are removed, the remaining nucleus is that of a lighter element with two fewer protons. We may write a typical alpha-decay process as

$$^{226}\text{Ra} \xrightarrow{\text{decay}} {}^{222}\text{Rn} + \alpha \qquad (7.3)$$

Gamma rays may also be emitted. Note that in this particular process, radium has decayed by emitting an alpha particle, leaving behind an atom of radon, with four fewer nuclides (that is, $226 \rightarrow 222$) and two fewer protons (that is, $88 \rightarrow 86$).

In *beta decay*, a high-velocity electron is ejected from the nucleus, often accompanied by gamma rays. Because the beta particle carries a negative charge, the residual nucleus must be left with an *extra proton*.[21] Obviously, the modified atom is a "heavier" element in the sense that it has a higher atomic number by one unit than the original atom. A typical beta-decay process can be written as

$$^{214}\text{Bi} \xrightarrow{\text{decay}} {}^{214}\text{Po} + \beta \qquad (7.4)$$

Thus although the number of nuclides does not change with beta decay, the atomic number of the decaying species *increases*.

There are four principal decay chains for radionuclides, named according to the initial heavy element: actinium (atomic number 89), thorium (90), uranium (92), and neptunium (93). In each case, the series of α- and β-decay events brings the element closer to lead and thallium. The problem of radioactivity associated with radon gas is related to the uranium-decay chain (an extensive discussion is presented in Section 8.2).

Radioactive decay is measured by the rate at which the decays occur—that is, the number of radioactive-decay events per unit time. The common unit of radioactivity is the **curie** (Ci); a unit obviously named for Pierre and Marie Curie. One curie is equal to 37 billion atomic disintegrations per second (1 Ci = 3.7×10^{10} decays per second). More recently, a unit

20. The carbon-14 dating technique was developed by Willard Libby in 1946, for which he won the Nobel Prize for chemistry in 1960. In this method, the ratio of the number of atoms of carbon-14 to carbon-12 is measured in a sample of organic material. Because ^{12}C is stable, whereas ^{14}C decays over time, the relative concentration of ^{14}C decreases as the sample ages. Cosmic rays, which constantly bombard the Earth, produce ^{14}C in the atmosphere at a continuous rate. That ^{14}C is incorporated into carbon dioxide and is then assimilated by living organisms. Indeed, all living things (including humans) contain in their bodies a small fraction of ^{14}C that is continually refreshed by the recycling of carbon through the biosphere. When an organism dies, the ^{14}C trapped in its tissues immediately begins to decay. The carbon-14 disappears at a specific rate determined by the ^{14}C **half-life** of 5700 years. Thus measuring the residual $^{14}\text{C}/^{12}\text{C}$ ratio of a preserved sample of organic material allows its age to be estimated. Carbon-14 dating was recently used to show that the Shroud of Turin, which supposedly draped the body of the crucified Jesus was actually woven of fabric manufactured between A.D. 1260 and 1390.

21. Since the electron carries away a negative unit of charge, the modified atom has a net positive unit charge associated with the new proton left behind. The charged atom quickly captures an electron from the local environment (which must remain, overall, electrically balanced) and is thus neutralized.

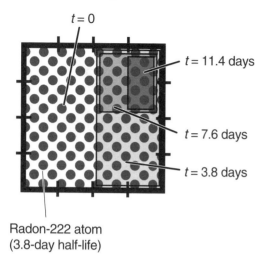

$t = 0$

$t = 11.4$ days

$t = 7.6$ days

$t = 3.8$ days

Radon-222 atom
(3.8-day half-life)

Figure 7.5 The concept of the half-life of a radioactive element. Atoms of the radioactive element radon-222 are shown contained in a sealed box. The half-life of this radionuclide is 3.8 days. After a time equal to this half-life, the original number of radioactive atoms remaining in the box is one-half the original number. After a second half-life has elapsed, the number is halved again, and so on. This is depicted by the sequentially smaller boxes with progressively darker shading.

was named for the original discoverer of radioactivity, Antoine-Henri Becquerel. The **becquerel** is equal to 1 radioactive-decay event per second (that is, 1 Ci $= 3.7 \times 10^{10}$ becquerel).

The **half-life** of a radioactive species is the time it would take for half the existing atoms to decay. The concept of a half-life is depicted in Figure 7.5, and the half-lives of some common radionuclides are given in Table 7.4. Notice that the half-lives range from a small fraction of a second to several billion years!

The decay rate (disintegration events per unit time) of an initial population of radioactive atoms falls continuously over time. That is, the number of decay events occurring per second must drop as the number of remaining radioactive atoms decreases. Using the concept of a half-life described in Figure 7.5, the decay of the total number of radioactive atoms can be calculated as

$$N(t) = \frac{N(0)}{2^p}, \quad p = \frac{t}{t_{1/2}} \qquad (7.5)$$

where $N(0)$ is the total number of radionuclides initially present at time $t = 0$, $N(t)$ is the total number left at time t, and $t_{1/2}$ is the half-life. The **law of radioactive decay** expressed in Equation 7.5 has a

simple interpretation, depicted in Figure 7.5. For each increment of time equal to $t_{1/2}$, the exponent, p (Section A.4) in Equation 7.5 increases by 1 unit and the number of atoms, N, decreases by another factor of 2.

To demonstrate the effect of a decay process involving a half-life, you can use a pile of ordinary objects, such as Cheerios, to play a halving game. Spread out the Cheerios on a table. Now sweep half of them away to one side. Then, in about the same time, sweep away half of those that are left. Then half again. And so on. It will not take very long to remove most of the Cheerios. The number of Cheerios removed at any step is obviously smaller than the number removed at the previous step, although the same *fraction* is removed each time relative to the number left from the previous step. In other words, the *relative* number removed during each time interval is constant. For radioactive decay, this is equivalent to saying that the number of atoms that disintegrate during any fixed time interval is simply proportional to the number that was there to begin with.

7.3.2 SOURCES OF RADIOACTIVITY

There are many sources of radioactivity and other energetic radiation in the environment, but fortunately, most of these sources are quite small. Exposure to energetic radiation may be caused by both natural sources and sources associated with human activities.

Natural Radioactivity

The Earth is continuously bombarded by **galactic cosmic rays**. The galactic cosmic rays consist of energetic particles (mainly small atomic nuclei) that originate in supernovae (spectacular explosions of stars) in our galaxy. As these particles plow through the atmosphere, collisions with nitrogen, oxygen, and carbon atoms generate radionuclides (carbon-14, for example). Radioactive elements are formed at enormous pressures and temperatures in the cores of these same stars that produce galactic cosmic rays. Supernovae spew these elements into the galaxy. When planets like Earth condense, they incorporate the radioactive elements. The decay of these elements within the Earth generates sufficient heat to keep the interior of the planet in a molten state (Section 4.1.1).

Radioactive elements collected during the original formation of the Earth are, over time, exposed at the surface. The presence of natural radioactive elements in minerals leads to sources of radioactive-decay products in the environment. When a heavy radioactive element like uranium-238 decays, the resulting atoms include radium and other radionuclides that like to remain in mineral form. Radium, however, decays into the gaseous radioactive element **radon** (Section 8.2). Unlike most other natural radionuclides, radon vapor can diffuse out of soils and into the atmosphere. A danger arises when radon gas accumulates in buildings and mine shafts. Radioactive radium was once widely used in the dials of watches to make the hands and numbers luminous in the dark.[22] Carrying radium on your wrist poses a certain risk of exposure to radioactivity.

Nuclear Weapons and Reactors

Radioactive minerals are mined for their value as a resource. Uranium is a primary radioactive element with great intrinsic value for generating energy and building nuclear weapons. Direct exposure to natural radioactivity during mining and refining operations is a major problem for the workers in these industries, but other people may be exposed to the end products. For example, radioactive "tailings" from mining operations, if not properly disposed of, can be washed into streams or lifted by winds as dust. Accidental leaks of radioactivity from manufacturing plants, nuclear-power reactors, and repositories may contaminate large areas.

During the 1950s and 1960s, nuclear-weapons explosions at the Nevada Test Site sent large clouds of radioactive dust wafting over nearby towns. The people living there are called *downwinders*. From 1951 to 1963, some 100 nuclear explosions depos-

Table 7.4   Half-lives of Common Radioactive Elements

Radio-nuclide[a]	Common Name	Half-life
^{3}H	Tritium	12.3 years
^{14}C	Carbon-14	5715 years
^{60}Co	Cobalt-60	5.3 years
^{85}Kr	Krypton-85	10.4 years
^{90}Sr	Strontium-90	28 years
^{131}I	Iodine-131	8.04 years
^{137}Cs	Cesium-137	30.2 years
^{234}U	Uranium-234	245,000 years
^{235}U	Uranium-235	700 million years
^{238}U	Uranium-238	4.5 billion years
^{239}Po	Plutonium-239	24,400 years
^{226}Ra	Radium-226	1622 years
^{222}Rn	Radon-222	3.8 days
^{218}Po	Polonium-218	30.5 minutes
^{214}Bi	Bismuth-214	19.7 minutes
^{214}Po	Polonium-214	1.6×10^{-4} secs

[a]Radionuclides are designated by the symbol for the element, with the total number of protons plus neutrons in the nucleus indicated as a left superscript. Since the number of protons in the nucleus of an element is fixed, the number of neutrons corresponding to an isotope can be determined by subtracting the number of protons from the superscript.

ited radioactive dust over the downwinders. Some of them basked in the light of the nuclear fireballs and brushed the fallout from their clothes. The questions and complaints of the downwinders about the effects of the fallout were largely ignored. A more serious alarm was raised when radioactive iodine began appearing in milk and other foods across the country. Atmospheric nuclear testing was banned in 1963.[23] But that may have been too late for many of the

22. The luminous glow is the result of two different physical effects: first, the decay of a radioactive material such as radium that is incorporated in the pigment used to paint the object to be illuminated; second, the excitation of a phosphorescent material that is also mixed in the paint. Phosphors are materials (including the common element phosphorus) that can be energized by absorbing radiation or electricity, for example, and that return to their normal unexcited state by emitting visible light. The most common example of this effect is the eerie glow produced by a "black light" in a dark room. In this case, a phosphor-coated surface is illuminated by invisible radiation from an ultraviolet lamp (our eyes cannot detect ultraviolet radiation, and a UV lamp would appear dark, or black, in a darkened room). In a watch dial, the radioactive particles emitted by radium activate a phosphor that emits enough light to be seen clearly in a dark room.

23. The Nuclear Test Ban Treaty, signed in Moscow on August 5, 1963, required about 8 years of negotiations to complete. During that time, both sides rushed to test hundreds of megatons of nuclear explosives, contaminating the Northern Hemisphere with radioactive iodine, strontium, and cesium. The treaty outlawed nuclear detonations in the atmosphere, in space, and under water but left nuclear testing underground as an option that has been stubbornly pursued to the present.

downwinders, who contracted unusual cancers, perhaps as a result of the exposure to radioactive fallout.[24]

Nuclear reactors are a special case because of their proliferation as a source of electricity. Recently, nuclear power has been boosted as a clean replacement for fossil fuels (Section 12.5.2). This claim is disingenuous. The problem with nuclear reactors is the vast amount of dangerous radioactive residue they create. Tons of radioactive water, concrete, and steel must be disposed of along with the highly radioactive core materials riddled with uranium and plutonium. In the United States and other Western nations, occasional nuclear-reactor accidents release radioactive gases and particles into the atmosphere, which drift in the wind and contaminate large areas. Nuclear-reactor accidents at the Windscale plant in England in 1957 and at Three Mile Island in Pennsylvania in 1979, for example, produced relatively minor releases of radioactivity, but large public outcries. In the case of Windscale, government officials at first tried to hide the details of the accident and later minimized the impacts; radioactive iodine nevertheless contaminated the region, and for several weeks milk sales were banned. Official denial and minimization of danger has not instilled confidence in the public regarding nuclear-reactor safety.

Among countries possessing nuclear capabilities, the former Soviet Union has had the most abominable record of nuclear safety. The Chernobyl nuclear-reactor explosion in 1986 spewed out a dense plume of radioactivity that is estimated to have killed roughly 8000 people, and the casualty list is still growing. Chernobyl capped a long history of problems at Soviet nuclear facilities; the truth has surfaced only since the fall of the Communist regime. From 1948 through 1967, the Mayak atomic plant in the Ural Mountains experienced a series of accidents that contaminated 450,000 people in the area! In one explosion, thousands died of exposure to nuclear waste. A nearby lake used as a disposal site dried up in 1 year, and dust blown from the lake bed contaminated a huge area. The accidents in Russia continue. In 1993, a tank of radioactive debris exploded near

Tomsk and reportedly poisoned 2500 acres of land, although details about the event remain classified. The state of Russia's nuclear facilities, including operational power plants, has been categorized as "perilous" by U.S. nuclear-reactor experts. Ironically, Russia has proposed building up to 31 new power reactors by the year 2010, doubling its electric-generation capacity (and, most likely, doubling its radioactive troubles).[25]

The United States suffers smaller radioactive insults. The Trojan nuclear power plant in Portland, Oregon, for example, was permanently shut down three years ahead of schedule because of safety problems. Large corporations, such as Rockwell International, have been fined or prosecuted for radioactive pollution at sites like the Rocky Flats nuclear-weapons plant near Denver. For years, residents sought information about radioactivity on the site and were reassured that no problem existed. Rockwell recently agreed to pay an $18 million fine for its role in polluting the area, pleading guilty to five felonies and five misdemeanors; the estimated cost of cleaning up the contaminated mess left behind is $1.3 *billion*. A vial of radioactive cesium-137 powder lost on a freeway in California caused a panicky search over several days. The radioactive nightmares and horror stories continue. Stay tuned.

Exposure to Radioactivity

Radioactivity is ubiquitous in the environment and in modern industrial life. There are a number of ways in which people may be exposed to radioactivity (Section 8.2.2). Some of the potential sources of radioactive exposure were described in the previous section. More subtle modes of exposure are discussed next. A variety of units for measuring doses of energetic radiation are defined as well, and the physiological effects of radiation doses are summarized in Section 7.3.3. In general, exposure can occur externally or internally and purposefully, accidentally, or incidentally.

External exposure involves irradiation that originates outside the body. The prompt energetic

---

24.   The fate of the downwinders was documented in Carole Gallagher's *American Ground Zero—The Secret Nuclear War* (Cambridge, Mass.: MIT Press, 1993). The book describes the spread of fallout over most of the United States, and the ghastly medical problems faced by the downwinders and their children during the years of testing and since. Some authorities justify the test program following World War II as necessary to prevent a nuclear weapons gap with the Soviet Union; they argue that the fallout had little effect on anyone.

25.   Adding insult to injury, a nuclear-powered Soviet reconnaissance satellite (*Cosmos*-1402) reentered the Earth's atmosphere on February 7, 1983, and disintegrated in the stratosphere. About 50 kilograms (100 pounds) of uranium-235 were dispersed as fine particles throughout the upper atmosphere, covering the entire globe. The uranium particles settled to the ground over the next few years. Radioactivity from the sky has been an unexpected bonus of nuclear power in space.

radiation from a nuclear detonation is delivered externally. Internal exposure occurs primarily through radionuclides that are ingested (into the lungs or gastrointestinal tract) and subsequently migrate to various parts of the body. The exposure occurs when these radioactive elements decay. Incidental exposure occurs naturally because of the existence of low-level background energetic radiation and radioactivity. Both accidental exposure and purposeful exposure are almost always caused by human activities and represent *enhancements* of normal radiation doses.

Exposure to radioactivity and other energetic radiation can be intentional in medical environments. For example, small quantities of certain radioactive elements are often used for diagnostic purposes: Iodine-131 can seek out tumors in the brain and other organs. Following ingestion and absorption into the bloodstream, a sensitive radioactivity detector (Geiger counter) can be used to locate tumors. In treating some cancers, the radioactivity emitted by cobalt-60 may be used to kill cancer cells localized in tumors. Beta rays (energetic electron beams) may also be used when the depth of penetration into the body must be carefully controlled. Irradiation can be intense, leading to mild radiation sickness, hair loss, and other symptoms associated with exposure to radioactivity. Obviously, the potential benefits of radiation treatment sometimes outweigh the risks of exposure to radioactivity and its side effects.

Medical practice also frequently calls for the use of energetic electromagnetic radiation, particularly X rays, but occasionally gamma (γ) rays. Both X rays and gamma rays produce similar physiological effects, and are used sparingly. Gamma rays are applied to cancerous tumors, destroying the mutant tissues (but perhaps inducing spurious mutations that may later cause problems).

X rays are quite pervasive. Dentists use them, for example, to take a picture of your teeth. The X rays penetrate the flesh of the cheek, as well as the tooth, exposing film held against the inside of the mouth. Of course, tissues are exposed to radiation during the photo opportunity, although the effects of this exposure are thought to be negligible. Modern techniques have reduced the dose of X rays from such an exposure by a large factor compared with the doses administered in similar procedures decades earlier. Nevertheless, the technician who takes the X rays places a protective pad over the patient's body and

genital areas, presumably to limit the dose of X rays absorbed in these areas. In the 1940s, it was common to find a "fluoroscope" in department stores. You could slip your feet into this clever device and see the bones of your feet projected onto a screen. The fluoroscope used an intense X-ray beam to produce the skeletal images. It was a highly entertaining device, particularly for children, who might be fascinated with such a view. Soon, it became apparent that self-exposure to X-ray irradiation was not a pastime to be encouraged, and fluoroscopes disappeared from the scene.

In assessing exposure to radioactivity, one can determine the actual exposure in terms of the total radioactive energy absorbed by the body, or the potential exposure in terms of the concentration of radioactive material in the air that is breathed or in the food and water that is ingested. In the case of a gaseous radionuclide such as radon-222, the concentration in air may be specified by the concentration, expressed in curies. In other words, a ^{222}Rn concentration of 1 curie per cubic centimeter would equal 3.7×10^{10} atoms of ^{222}Rn per cubic centimeter of air. Because the actual concentrations of radon gas are much smaller than this value, a typical unit of concentration is 1 **picocurie** per liter (pCi/liter) of air (which is roughly the same as 37 atoms/m^3, equivalent to about 100 atoms in a good-size closet). (In Appendix A, *pico* is defined as one part per trillion, or 1×10^{-12}, and 1 "liter" is defined as a volume of 0.001 m^3.)

In the case of exposure to radionuclides that are ingested and accumulate in certain organs, and of X-rays and gamma rays (which cause similar effects), the exposure is determined as the total dose of energy deposited per mass of tissue. For a particular radionuclide, the rate at which energy is deposited is the rate of decay multiplied by the energy released per decay event. The rate of deposition can be summed up over time to obtain the total dose. Alternatively, it can be assumed that all of the absorbed radioactive atoms will eventually decay. Then the total dose is simply the total number of radioactive atoms multiplied by the energy released by the decay of a single atom.

Roentgen, Rad, and Rem

Doses of exposure to energetic radiation are measured in units of roentgen, rad, or rem (the singular forms are often used; for example, 100 roentgen, rad, or rem). The **roentgen** is named after the discoverer of X rays, Wilhelm Roentgen.[26] This unit is defined in terms of the ionization of air by the energetic photons of X rays or gamma rays. Ionization occurs when a photon of sufficient energy knocks an electron from an atom or a molecule. The photon must have at least the "ionization" energy of the atom or molecule. X-ray and gamma ray photons have much more energy than they need. In fact, as X rays and gamma rays streak through air, they leave behind a path of ionized air molecules. If the air is dry (that is, consists of only O_2 and N_2), and has a temperature of 0°C and a pressure of 1 atmosphere, then a dose of 1 roentgen will produce about 2 billion positive charges and an equal number of negative charges, in 1 cubic centimeter of irradiated air. Although this may seem to be a rather complex and obscure way to determine a dose of X rays and gamma rays, it is the way health physicists do it. Note that the victim is air.

The **rad** is another way of measuring the dose of energetic radiation in liquid or solid materials, including living tissues. A rad is defined in terms of the energy absorbed by 1 kilogram of a material. One rad is equivalent to the absorption of 0.01 joule of ionizing radiation per kilogram of material (1 rad = 1/100 J/kg). Actually, 1 roentgen is very nearly equal to 1 rad.

To measure the effectiveness of different kinds of energetic radiation in producing damage in living tissues, the **rem** was invented. The rem is used to compare the potentially damaging effects of electromagnetic radiation, such as X rays and gamma rays, with the effects of energetic particles such as alpha and beta particles. Rem is an abbreviation of the phrase "roentgen equivalent man." One rem is the dose of ionizing radiation from any source that produces the *same biological effect* in humans as does 1 roentgen of X rays or gamma radiation. For X rays, obviously, 1 rem = 1 roentgen. Energetic particles are generally more damaging than X rays for the same amount of energy deposited.

26. Wilhelm Conrad Roentgen (or Röntgen, 1845–1923) was a German physicist who received the first Nobel Prize in 1901 for discovering X rays. The first photographs taken by Roentgen using X rays were revealed in 1896. That same year, the first reports of injuries caused by exposure to X rays appeared. Thomas Alva Edison, who was busy designing an X-ray lamp, noticed that his assistant, Clarence Dally, suffered the loss of his hair and the formation of ulcerated sores on his scalp. Dally died of cancer that evolved from ulcerations that later developed on his hands and arms.

As if there were not enough units around to specify exposure to energetic radiation, new units have been introduced to clarify (confuse?) the issue. Hence, we have the **gray** and the **sievert**. One gray is essentially 100 rad, and 1 sievert is 100 rem.

7.3.3 THE PHYSIOLOGICAL EFFECTS OF RADIOACTIVITY

All the energetic particles and electromagnetic radiation emitted by radionuclides can damage human tissues. These radioactive emissions readily penetrate solid matter. Gamma rays are the most penetrating; they can pass right through the body like a rapier. Alpha particles are relatively heavy; they tend to plow through tissue like microscopic bullets. Beta particles, on the other hand, are relatively light in weight and are stopped within a short distance of the surface. With external exposure, beta particles are deposited in the skin. As energetic particles enter the body, they bang into molecules and atoms, initiating chemical reactions by ionizing and dissociating the body tissues. The chemical transformations damage cellular material, particularly DNA molecules. The damage to DNA can accumulate, preventing the tissue from functioning properly and eventually leading to mutations and tumor formation.

Acute (intense short-term) exposure to nuclear radiation over a substantial portion of the body leads to **radiation sickness**, or acute radiation syndrome. Although exposure to intense beta radiation causes severe skin burns, the classic radiation-sickness syndrome is related to whole-body irradiation by X rays or gamma rays, with few external signs of exposure. The initial symptoms include nausea and malaise, weakness and vomiting. These symptoms then subside, and a period of symptom-free latency follows before the main phase of radiation sickness sets in. In one form, severe abdominal pain and diarrhea lead to dehydration, shock, and death within several days. In a second form, which acts through the bone marrow, there is an onset of fever and hemorrhaging after 2 to 3 weeks of latency. Death is caused by an extensive loss of blood.

The lethal acute whole-body dose of energetic radiation is in the range of 450 to 600 rem for an average person (that is, the LD_{50}). Lower doses, which may be given during medical treatments and in localized areas, can cause a temporary loss of hair, a kind of nuclear balding. In a cumulative sense, nonlethal acute doses of radiation also contribute to the long-term health effects of exposure to energetic radiation.

Chronic (low-level, long-term) exposure to radioactivity can induce cancer in humans. One of the primary diseases is leukemia, although breast cancer is also common. The amount of exposure required to initiate cancer is uncertain. It is currently felt that all exposure to radioactivity and energetic radiation is *cumulative* over a *lifetime*. That is, the effects of radiation are determined in proportion to the total dose received since birth. The potential for radiation-induced cancer adds up over the years. There is no threshold exposure for radioactivity. Any exposure is dangerous and should be avoided. As we point out in Section 8.2.2, however, exposure to natural background radioactivity is almost impossible to prevent. The typical natural doses accumulated over a lifetime can result in a significant risk of cancer, particularly as the human life-span is extended. By the age of 70 years, many people will have received a total dose of background radiation equal to about 10–20 rem (refer to Section 8.2.2). This amounts to several percent of the LD_{50} for acute exposure to energetic radiation; a death rate of roughly 10–20 people per thousand might be expected on this basis (also refer to Figure 8.3). Natural radiation, enhanced by localized exposure to radon and other excess sources of energetic radiation such as medical X rays, may ultimately play a role in limiting human longevity. Extraordinary measures would be required to avoid the exposure associated with natural background radiation.

During the years immediately following Roentgen's discovery of X rays,, hundreds of scientists, unaware of the dangers of energetic radiation, received large doses. More than 100 deaths among these pioneering researchers are connected with exposure to X ray. The radioactive luminous dials of watches, noted earlier, used to be painted on by hand. Earlier this century, the painters, mainly women, habitually licked the tips of their brushes to form a fine point. In unusual numbers, they developed lesions in the mouth and on the jawbone, and eventually, many died of bone cancer.

Birth defects are another probable effect of radiation exposure. Radiation strongly affects embryonic tissues, particularly in the nervous system. Many of the offspring of the survivors of the Hiroshima and Nagasaki atomic bombings who were pregnant at the time have suffered an unusually high incidence of mental retardation and reduced cranial capacity. Studies indicate that the greatest sensitivity occurs between the eighth and fifteenth weeks after conception.

Following World War II and extending to 1963 (Section 7.3.2), roughly 200,000 U.S. servicemen participated in the atmospheric testing of atomic weapons. By 1992, more than 13,000 had filed claims for disabilities caused by the tests, and about 1200 were certified as disabled. The actual health effects have been difficult to determine. A new epidemiological study by the Institute of Medicine of the National Academy of Sciences is scheduled for completion in 1998. Meanwhile, the troops want to know what really happened to them. In a propaganda film of the time, an unidentified chaplain soothes the concerned servicemen: "Actually, there is no need to be worried. You look up and you see the fireball as it ascends up into the heavens. It contains all of the rich colors of the rainbow, and then as it rises up in the atmosphere it assembles into the mushroom. It is a wonderful sight to behold."

The Chernobyl Experience

The accident at the Chernobyl nuclear-power plant was a sight not so wonderful to behold. It was perhaps the watershed event in the development of nuclear power (Section 12.5.2). The accident created a public-relations nightmare for the nuclear-power industry and brought the construction of new plants in the United States virtually to a halt. The impacts of the radioactive emissions on the local population of Ukraine produced the greatest nuclear disaster since the atomic bombings of Hiroshima and Nagasaki. The residual effects are still being documented. One fact about radioactive contamination is this: It takes a very long time to go away. The lesson is: Never buy radioactive real estate, even at bargain basement prices.

In parts of Ukraine, cows are forced to swallow large capsules filled with ferrocyanide. No, it is not a bovine mass-murder attempt. The ferrocyanide reacts with radioactive cesium that remains on the fodder the cows have been eating since the Chernobyl accident. The ferrocyanide scavenges the cesium and prevents it from getting into the milk. Among the human casualties of the accident, more than 8000 people have died of radiation exposure and the resultant diseases. By one estimate, more than 35,000 have been rendered incapable of caring for themselves. As many as 1.5 million children had their thyroid glands damaged by exposure to radioactive iodine. In one area of Belarus, the thyroid cancer rate among these children was 80 times normal 5 years after the accident. Epidemiological studies carried out so far suggest that the sensitivity of the young thyroid to iodine-131 exposure had been substantially underestimated.

In other parts of the former Soviet Union, where nuclear-weapons production and testing continued for 40 years with little sense of danger or restraint, stories are beginning to surface that suggest a horrible legacy of the nuclear-arms race. Entire families have been ravaged by cancer. Jaundiced children, with congenital defects and mental disorders, are disabled for life. And people are afraid to face the next generation, afraid to bring forth the radiation-damaged seed that promises more anguish and pain.

7.4 Assessment of Health Risks

The calculation of the health effects that might result from exposure to a toxic pollutant can be a complex and controversial affair. A comprehensive evaluation requires a variety of information concerning the toxin and the exposed individual, including

1. The quantity of a toxic material released into the environment,
2. The resulting concentrations and spatial distribution of the pollutant,
3. The time dependence of the concentrations of the toxin—that is, the potential exposure—at the site of the individual,
4. The uptake and retention of the toxic material—that is, the effective dose—of the pollutant in the individual,
5. The specific physiological and health effects caused by the pollutant as assimilated by the individual,
6. The circumstances of the exposure, including the physiological state of the individual, and collateral exposure to other synergistic pollutants,
7. Other factors that might affect the response of an individual to the pollutant, including medical history and life-style.

The potential risk to one's health from exposure to a toxic pollutant can be determined by a **risk analysis**. The factors just described must be quantified and applied in a sequence of calculations. In general, the definition of the sources and the disper-

sion of the pollutants lead to estimates of exposures and dosages. The "risk" is then defined as the probability that a specific outcome, or health effect, will result from the exposure. If the risk is known, the likely "cost" of incurring the effects can be estimated. If the cost is too high in relation to the "benefit" of carrying out the behavior or activity that creates the risk, the behavior or activity can be modified or terminated to reduce or eliminate the hazard. These somewhat subjective concepts are elaborated next.

7.4.1 Defining the Threat

Imagine that the exposure to a certain toxic pollutant has been defined exactly. Although this is a highly unlikely situation, in some cases, the sources and dispersion of toxins may be calculated accurately. A large amount of data would obviously be needed to carry out a precise computation of exposure. But it could be done. Further, assume that the exposure can be translated accurately into a dose, as defined in Section 7.1.2. The amount of toxin ingested is thus known, and the mode of ingestion is clearly delineated. The evaluation of the risk still requires knowledge of the **dose-response** relationship.

A dose-response relation defines the actual physiological consequences of receiving a specified dose of a toxin. The response is usually expressed as the likelihood of a physical effect's appearing among an average exposed population. For example, the toxic dose that is lethal for 50 percent of normal healthy persons is defined as the LD_{50} (Section 7.1.2). The dose-response relation in other cases gives the probability that a specific effect will occur for a certain exposure under known conditions. For example, exposure to secondhand cigarette smoke while working in a piano bar over a 10-year period may, hypothetically, carry a probability of 1 in 1000 of causing lung cancer. Put another way, if administered to 1000 normal people, the dose of smoke received over that time would be likely to induce one case of lung cancer.

Risk is defined as the potential for, or likelihood of, an adverse effect occurring as a result of exposure to a pollutant. The conditions of exposure are presumed to be known with some degree of certainty, along with the dose-response relationship. Risk is then the probability that a certain exposure will occur, multiplied by the probability that a certain outcome will result. A risk may crudely be estimated from the following equation:

$$R_{ht} = D_t P_h \qquad (7.6)$$

Here R_{ht} is the risk of the outcome, or health effect, h, occurring as a result of exposure over time, t; D_t is the total dose of the pollutant received over that time; and P_h is the probability of the health effect, h, appearing in an average person receiving a *unit dose*, or quantity, of the pollutant. It has been assumed that dose and exposure are equivalent measures of the amount of toxin received. The dose may be specified in units of, say, total rem of energetic radiation received over a lifetime. The probability, P_h, would then be expressed as, say, one chance in a million of contracting leukemia from exposure to 1 rem of energetic radiation over a lifetime. Then the risk of contracting leukemia from exposure to, say, 10 rem during one's life would be calculated as R_{ht} $= 10 \times 1 \times 10^{-6} = 10/1,000,000$, or 10 chances in a million over a lifetime.

In a population of N persons exposed as described, the number that would be at risk of suffering the health effect h over time t, N_{ht}, is given by

$$N_{ht} = N R_{ht} = N D_t P_h \qquad (7.7)$$

In the previous example, the exposure of a population of 1 million people to 10 rem of energetic radiation would cause 10 of them, on average, to come down with leukemia during their lifetime.

Equations 7.6 and 7.7 show that the individual risk from exposure to a pollutant increases as the dose increases (as a result of higher concentrations of the pollutant or longer times of exposure) and as the sensitivity to the pollutant (expressed by the probability of an effect for a unit dose) increases. The number of persons eventually suffering disease also rises in proportion to the size of the exposed population.

Excess Risk

The risk determined above corresponds to exposure to a specific toxic compound under well-defined circumstances. That risk is in addition to all of the other risks one normally faces in life. Hence, we might refer to it as the **excess risk** related to that specific activity or situation. Usually, risks of the same kind—risks of contracting lung cancer from exposure to various pollutants, for example—are combined to give an overall risk. If the various causes are completely independent of one another, the overall risk is simply the sum of the individual risks from each cause considered separately. Thus, the risks of lung cancer from tobacco smoking, from exposure to air pollution, and from medical X rays might all be simply added to determine the total risk.

It is quite common, on the other hand, for different toxins to act together in a synergistic manner, or for one compound, normally relatively harmless, to trigger a dangerous response to another compound. Hence, the health effects of one compound may depend sensitively on the exposure to one or more other compounds as well. This interaction between toxins obviously complicates the problem of determining effective exposure and dose, and excess risk. Thus, in evaluating the threat in a given situation, one must consider the presence of co-carcinogens, or compounds that are responsible for activating the carcinogenic potential of other substances, indirect carcinogens, or noncarcinogenic compounds that break down in the body into substances that are carcinogenic, and synergists, or pairs of compounds that each exhibit enhanced pathogenic activity when both are present. Notice that the risk of exposure to one compound may be either increased or decreased due to the presence of other compounds.

In the remainder of this book, specific risks will be treated as excess risks associated with exposure to an identified toxin. Where co-carcinogenicity, synergy, or other effects are important, these will be discussed separately.

7.4.2 RISKS AND BENEFITS OF POLLUTION

The basic concern with exposure to environmental toxins is the threat posed to health and the quality of life. Insurance is available to protect people from risks that arise in everyday activities. Judicious people carry automobile insurance, fire insurance on their homes, and life insurance to protect their families. Is environmental insurance needed as well? Must the value of one's health be indemnified against the possible effects of exposure to toxic pollutants? (The feasibility of insurance to compensate for damage associated with global climate change is discussed in Section 12.5.) Before insurance can be issued, however, the potential losses must be assigned a market value. This is the purpose of a cost-benefit analysis.

Cost-Benefit Analysis

In determining whether an activity that exposes people to risk will be acceptable to society and at the same time be profitable for the sponsors, a cost-benefit analysis must be carried out. In such an assessment, all the factors creating the risk associated with the activity must be scrutinized to determine the parameters that control the severity of the threat and the actions that could be taken to alleviate the hazards.

The potential cost of exposure to pollution created by an activity, which is related to the attendant risks of specific health effects, can be estimated as follows:

$$C_{ht} = N_{ht}W_h = NR_{ht}W_h = ND_tP_hW_h$$

$$(7.8)$$

Here, the total cost over time, C_{ht}, for a health effect, h, is determined as the number of occurrences of the effect, N_{ht}, multiplied by the valuation, or cost, W_h, of a single occurrence. The total cost of the activity is the sum of the costs of each effect considered separately (Equation 7.8).

Reducing the risk caused by a specific activity reduces the cost to those who would otherwise be affected. This is a cost benefit that can be determined for different scenarios of pollutant exposure and risk mitigation. There are, however, direct expenses required to reduce the risk, and these so-called mitigation costs must be paid by someone. From the point of view of society, an activity could be allowed if there were a net benefit to society. From the point of view of business, an activity could be justified if there were a net profit to investors. For society, if the health risks and costs exceed the social and economic benefits, the feasibility of proceeding with an activity is placed in doubt. For business, if the costs of lowering the risk to an acceptable level decimate profits, such remedies are unacceptable to its investors.

In order to be able to decide whether to proceed, the benefits of carrying on an activity must be quantified. The potential benefits include the creation of local jobs, taxes, and infrastructure, as well as profits for investors that are recycled into the economy. Other real benefits may be associated with the activity. The proposed activity may actually reduce the risks associated with the displaced activities—for example, by eliminating sources of pollution. The calculation of benefits obviously involves mainly economic factors, which may be highly uncertain or changeable.

A case in point is the proposal to build trash incinerators in urban areas. The costs associated with siting and building an incinerator are relatively straightforward to determine. The amounts of toxic emissions from the facilities can also be estimated (although these estimates are often optimistically low), and pollutant doses and health risks can be computed (Section 7.2.4). Benefits accrue to such projects, however. Garbage does not have to be hauled to landfills, saving transportation costs, reducing vehicular pollution, and mitigating the long-term problems arising from vast piles of garbage. Locally, jobs are created and taxes are generated that can be used to support other useful community programs (although jobs connected with the landfill may be lost). A number of vested interests come into play: people who do not want to live near such a facility or fear a decrease in property values around the incinerator; investors who wish to maximize profits and minimize costs; politicians who seek to increase the tax base and provide work, thereby satisfying their constituents. And in the real world, there may be illegitimate factors to deal with, such as profiteering, payoffs, and falsified data.

One difficulty in cost-benefit analysis is determining the value of certain possible outcomes of an activity (aside from the already considerable challenge of estimating the risk and number of casualties). The cost of treating a specific kind of cancer can be estimated using medical records. The price of hospital rooms and surgical procedures can be added up. But that is only part of the true cost of compromised health. What about the pain and suffering of those stricken and their families? The lost work and income? The lowering of societal standards of acceptable living conditions? All these real and perceived impacts must be assigned a value and added to the cost connected with the risk.

The most difficult task in a risk assessment is determining the value of a human life. Are all people equally valuable? In our democratic system and by constitutional tradition, this is the case. But in reality, it is not. Contamination of poor and middle-class neighborhoods is not taken as seriously as pollution in well-to-do areas. The least desirable places to live are usually close to polluting factories or freeways, where property is cheapest. A similar trend is seen in insurance payoffs for accidental deaths. Wealthy and professional casualties fetch the most in airplane disasters, for instance. Such disparities may be related more closely to an heir's access to lawyers than to the intrinsic value of a human life, which should be the same for everyone.

The cost of remediation of potential risks usually includes very tangible actions—for example, the cost of building a facility at a different site or of installing pollution-emission-control devices. These costs limit the profitability of an activity. If the mitigation and remediation costs are too high, the activity will not take place. Then jobs and other benefits are lost. The solution often is to relax pollution standards and increase the local risk of health impacts. Someone will suffer, of course. But that is a decision that must be made in the face of limited resources and the competitive needs of society.

7.4.3 BOX MODELS FOR RISK ASSESSMENT

The box model concept developed in Section 4.1.2 can be applied to the assessment of risk from exposure to air pollution. In this case, the box model is used to estimate the concentrations of pollutants associated with specific sources and sinks. Although the time dependence of the exposure can also be roughly defined with such a model, a steady state situation is the simplest to analyze. Figure 7.6 illustrates the configuration of a box model that can be applied to such problems.

In assessing risk, the concentration of a toxic compound that may cause exposure and the duration of exposure are the most important factors. The "box" in this situation represents a definite spatial volume into which the pollutant is emitted. The volume chosen depends on the problem being addressed. For example, if one is concerned about a toxic leak in the workplace, the volume might consist of a room thought to be contaminated. When assessing the effects of toxins escaping from storage facilities or industries into the local environment, the volume may consist of a ground cloud that expands as it moves away from the source (Section 5.4.2). To investigate the health impacts of smog, the volume may consist of an entire urban airshed (for example, the Los Angeles basin). In the case of global scale air pollution—chlorofluorocarbon emissions and ozone depletion, for example—the entire atmosphere is the "box" of interest.

The parameters that apply to the box model are defined in Section 4.1.2 (see also Figure 7.6 and Section 10.1.2). The basic quantities to recall are the **concentration**, q, measured in mass/volume (for

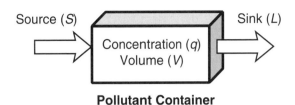

Pollutant Container

Figure 7.6 The "box model" developed in Chapter 4 is applied to analyze exposure to pollutants in air. For this application, the "box" can represent an urban airshed, a valley in which pollutants emitted by a smokestack are confined, or the streets around an industrial plant. The concentration of material in the box, q, is determined by the rate at which the material is added to the box, S, the volume of the box, V, and the residence time of the material in the box, τ. In equilibrium, or a steady state, $S = L$.

example, micrograms per cubic meter, $\mu g/m^3$); the **volume**, V, of the box (measured, say, in cubic meters, m^3); the **source**, S, measured in mass/time (thus, grams per second, g/sec), and the loss rate, or **sink**, L, measured in mass/time (or, grams per second, g/sec). The **residence time** (or lifetime), τ, of a material in the box is controlled by a number of processes discussed in Section 4.1.2. If the size (volume) of the box, the source of material for the box, and the residence time in the box can be estimated, the concentration of material can be calculated in a straightforward manner using the **box model equation**, in this case,

$$q = \frac{S\tau}{V} \qquad (7.9)$$

This form of the box model equation emphasizes the *concentration* of a pollutant in the box. (The version in Chapter 4 emphasizes the total quantity of material in the box, which is treated as a "reservoir.") Figure 7.6 illustrates the simplest box model configuration, having a single source and a single sink (loss process).

Figure 7.7 shows an application of the box model to a practical problem—loss of excess weight. The concentration of fat in the human body is a balance between the rate at which calories are ingested (eating) and the rate at which calories are burned (exercise). To reduce the concentration of body fat, exercise can be increased or eating can be curtailed. Either action would cause some of the stored fat to be used up. On the other hand, dessert is pleasurable, and one may not want to give it up. In Roman times,

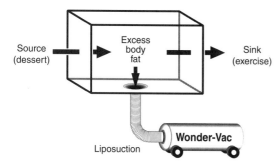

Figure 7.7 Whimsical application of the box-model concept to the problem of weight reduction. Here, the human body is the "box," and the "pollutant" under consideration is fat. The source of the fat is too many desserts, and one of the possible sinks is additional exercise. The easier (or more painful?) way to reduce the concentration of body fat, however, is to have it sucked out by a doctor.

feasting would be followed by forced vomiting. The Romans wanted their pleasure, over and over. Vomitoria were built to accommodate the barfing legions of revelers. Today, plastic surgeons would have you believe that liposuction is the answer. The surgeons carefully vacuum globs of fat from pouches in the stomach, hips, and thighs. The patient may be bruised and purple for a month, and excess skin may sag for a while, but the fat is gone (your purse will be a lot thinner as well). By employing a vacuum, the residence time of fat around your waist can be reduced dramatically (Figure 7.7). The body is a "box," and the liposuction is a "sink."

Roasting a turkey in the kitchen can make the whole house smell nice, as fragrant aromas leak from the oven and fill the adjacent rooms.[27] A house is usually drafty, however. Cracks at windowsills and below doors allow air to flow in and out. This normal "ventilation" of a home renews the air inside every hour or so on average (unless the place is tightly weatherproofed). Ventilation also acts to remove aromas from the house, because fresh air entering the home is (presumably) clean. Accordingly, some of the aroma is continuously being lost. The rate of loss is proportional to the concentration of the aroma in the air multiplied by the rate that the air is exchanged with the outside. A balance is soon established

between the production of turkey aroma (the source) and its ventilation from the house (the sink). In this "steady state" condition, the concentration of the aroma is just large enough that the total amount removed in any time interval just equals the total amount seeping from the oven. The home is a "box" in this example, and aroma is the "pollutant."

If the rate of ventilation is increased, the aroma will be removed faster. For the same source, then, the concentration in the house must be lower. If the rate of ventilation were reduced, the aroma would build up to a higher concentration. In each instance, the time required to adjust the concentration of aroma in the house to a new steady state is roughly the same as the residence time of the aroma in the house. The residence time is determined by the ventilation rate.

Parameters Affecting Risk

The application of the box model requires identifying and quantifying the various sources and sinks of each pollutant, or toxic compound, of interest. For an air pollution risk assessment, the following parameters are important:

Sources. There are a number of ways in which toxic materials can enter an atmospheric region (represented by a box). These processes define the sources of pollutants in the region:

- Direct (or primary) emissions of gases and particles into the region, which provide the major source of pollutants such as NO, CO, RH, SO_2, and soot (Chapter 6),
- Photochemical (and other chemical) processes by which secondary pollutants are generated in the region (for example, the formation of ozone in smog by $NO + RH + O_2 \xrightarrow{\; h\nu \;} O_3$),
- Ventilation, in which winds and turbulence carry polluted air from a nearby polluted region into the region under study (that is, from a source region to a receptor region),
- Resuspension of pollutants from land and the surfaces of vegetation by wind and agitation (for example, dust raised by vehicular traffic).

The processes, in sum, provide the total source for the region, or box. Notice that some of the sources are external to the box and others (photochemistry is an example) can be *internal* to the box. The distinction is not important to our analysis, as long as we define the total source over the region of interest.

27. The aromas of cooking are nothing more than vapors and smoke emitted from foods. By heating meat and vegetables, volatile aromatic compounds are driven into the air. At lower temperatures, these vapors may condense into fine aerosol particles, to which may be added smoke from flames and char. The backyard barbecue is a feast of smoke and good smells, and a major source of air pollution.

Sinks: In general, pollutants may be carried away by winds, deposited on surfaces, or chemically converted to other (perhaps more hazardous) substances. Among the processes of importance are

- Photochemical decomposition in the presence of sunlight and reactive gases such as OH (for example, $CO + OH \rightarrow CO_2 + H$),
- Ventilation, in which winds, convection, and turbulence transport pollutants out of a region or mix the pollutants into adjacent regions (outside the box),
- Dry deposition of gases and particles to surfaces (for example, pollutants sticking on leaves, soil surfaces, and water; plants are particularly effective as scavengers of air pollutants; larger particles can simply fall out of the atmosphere, a process called *sedimentation*),
- Wet deposition, by which soluble gaseous and aerosol pollutants that dissolve in cloud water are deposited on the ground by rainout or are swept up by precipitation through the process of "washout").

These source and sink processes are linked in Figure 7.8. In the box model Equation 7.9, L and S are *summations* over all the possible processes that remove or add a pollutant in the region of interest (the "box"). That is,

$$S = S_1 + S_2 + \cdots = \sum_i S_i \qquad (7.10)$$

$$L = L_1 + L_2 + \cdots = \sum_j L_j \qquad (7.11)$$

The summations extend over all the sources, i, and all the sinks, j. The discussion in Section 4.1.2 described the following relationships between the fundamental box model parameters (which also follow from Equation 7.9, with the steady state condition, $L = S$):

$$L = \frac{qV}{\tau}; \quad \tau = \frac{qV}{L};$$
$$L = S; \quad \tau = \frac{qV}{S} \qquad (7.12)$$

Equation 7.12 represents the different useful ways in which the box model parameters may be rearranged from the basic Equation 7.9. All these forms are equivalent to Equation 7.9, but with a different "unknown" parameter on the left-hand side of the equation to be determined from the "known" parameters on the right-hand side.

Each sink for a pollutant has a related residence time corresponding to the particular process that removes the pollutant from the box. In other words, each loss process in Equation 7.11 has an associated residence time determined by the physics and chemistry related to that process. The loss rate for a particular process, L_j, is related to the residence time for that process, τ_j, in a simple way:

$$L_j = \frac{qV}{\tau_j} \qquad (7.13)$$

It follows, from Equations 7.11 and 7.12, that the average residence time for the pollutant in a box is an average of all the residence times, calculated in a particular way:

$$\frac{1}{\tau} = \frac{1}{\tau_1} + \frac{1}{\tau_2} + \cdots = \sum_j \frac{1}{\tau_j} \qquad (7.14)$$

An important property of this expression is that the value of τ actually lies closest to the *smallest* of the τ_j. The smaller the time constant for a process is, the faster a pollutant can leak out of the box via that process. In other words, it is the *fastest* process that controls the residence time of a pollutant and hence its concentration for a fixed source. This is an important consideration. It implies that the entire suite of processes capable of removing a pollutant need not be analyzed in detail as long as the fastest, or controlling, process is known.

Figure 7.8 A box model for simulating air pollution must include sources and sinks representing processes that produce or destroy or import or export pollutants. The relevant processes are illustrated and discussed in the text. A box model can then be used to calculate average concentrations of pollutants for use in estimating exposure and potential risks to health.

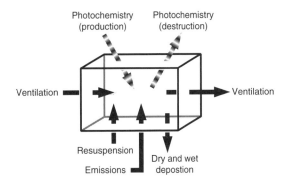

Procedures to Define Risk

This box model can be used to perform simple risk assessments using a well-defined series of steps. First, the problem is characterized in terms of the pollutants of concern, the contaminated region to be evaluated, the sources and sinks of toxins, and the exposure mechanisms and dose-response relations. The box model is used to compute pollutant concentrations for a quantitative assessment of exposure and risk. In many cases, the appropriate model parameters are relatively easy to estimate. The box model equations are then straightforward to evaluate, yielding pollutant concentrations under different sets of reasonable conditions. The parameters may also be varied to reveal the sensitivity of pollutant levels to various physical and chemical factors. Ultimately, an effective means of mitigating the risk might be deduced from such sensitivity tests.

The most uncertain step in the assessment is determining the potential exposure to the contaminants and the likely health effects of such an exposure. Certain general impacts are implied by exposure to specific concentrations of pollutants. These can be inferred by considering the safety standards established for pollutant abundances, as discussed in Chapter 6. For example, exposure to ozone at stage-2 concentrations over many years is likely to cause respiratory damage. More definitive physiological effects, corresponding to both short- and long-term exposures to specific levels of common pollutants, can be deduced from information provided in this chapter and in numerous published studies.

7.4.4 URBAN SMOG: A CASE STUDY

The box model just described can be applied to estimate the levels and effects of smog in the Los Angeles urban area. During the summer, a semipermanent high-pressure system over the eastern Pacific Ocean creates a strong temperature inversion with little rainfall over the entire region. The Los Angeles basin is characterized by a mountain barrier to the north and east. Coastal sea/land breezes and prevailing westerly winds act to concentrate pollutants within the eastern basin against the mountains. In addition, the strong temperature inversion caps the boundary layer and reduces the effective volume of air in which pollutants can be mixed and diluted. Ventilation of the basin is relatively ineffective be-

cause of the barrier and inversion. These factors, among others, contribute to the accumulation of toxic pollutants in the basin.

Assessing the Impacts of Pollutants

The area of the Los Angeles basin is roughly 1000 square kilometers (1000 km^2). Assuming that Los Angeles smog is typically trapped in a mixed layer 1 kilometer deep, the total *volume* of air in which pollutants can disperse during the day is about 1000 cubic kilometers ($V = 1000 \text{ km}^2 \times 1 \text{ km} = 1000 \text{ km}^3 = 1 \times 10^{12} \text{ m}^3$). The Los Angeles basin is a large place, but the quantities of primary pollutants emitted also are large.

The sources, concentrations, and distributions of pollutants in the Los Angeles basin are constantly monitored, and so the total emissions of primary pollutants have been accurately recorded on a daily basis for many years. In the early 1980s these emissions amounted to about 6500 tonnes of CO per day (recall that 1 tonne = 1000 kilograms), 1200 tonnes/day of NO, and 1300 tonnes/day of reactive hydrocarbons (RH). In addition, roughly 650 tonnes/day of particulates are emitted into the Los Angeles basin, with about 350 tonnes/day of secondary particulates generated from emissions of sulfur dioxide. Most of these emissions originate from vehicle exhaust, although roughly half of the reactive hydrocarbons are produced by the evaporation of solvents and emissions from petroleum refineries. Particulates have many different sources, including vehicle exhaust, smokestacks, and road dust.

The photochemical lifetime of CO in the Los Angeles basin is quite long—at least several days—owing to its slow reaction with hydroxyl (OH). The principal sink for carbon monoxide is therefore natural ventilation out of the basin. The average residence time of carbon monoxide in the basin is estimated to be roughly one-half day. Nitrogen oxides (NO, and the NO_2 generated from it) may have a lifetime of several hours. During the day, NO_x is lost principally by reacting with OH and various hydrocarbons. At night, the NO_x can be converted to nitric acid, deposited on the ground, or absorbed in fog. An average residence time of one-sixth day is assumed for NO_x. The primary reactive hydrocarbons are also depleted through reactions with OH and, later in the day, with ozone. Therefore, the residence time of RH is taken to be one-sixth day. Ozone itself readily reacts with hydrocarbons and NO_x and has a substantial velocity for dry deposition. The ozone residence time is, likewise, assigned a value of one-sixth day.

Particulate residence times in a dry environment, such as that characteristic of the Los Angeles basin, are controlled largely by sedimentation and ventilation processes. Considering primary particles, the largest smoke aerosols and particles lifted mechanically from roadways have a relatively short residence time because of their great fall speeds (Section 6.5). On the other hand, some smoke particles, particularly soot, are very small and can remain suspended for many days. These tiny particles compose a minor fraction of the total particulate mass emission. On average, then, a relatively short residence time of one-tenth day is assumed for the primary particles. Secondary particles, composed mainly of sulfates derived from sulfur dioxide, are usually quite small and have very low fall speeds. These fine particles are assumed to have an overall residence time of one-half day.

Pollutant concentrations in the Los Angeles basin can be predicted using the box model Equation 7.9. In order to derive pollutant concentrations in practical units, Equation 7.9 must include a numerical constant to ensure that the units assumed for each parameter are properly converted. Then

$$q = c\,\frac{S\tau}{V} \qquad (7.15)$$

Here, the "conversion" constant, c, has a value that depends on the particular units of the individual parameters.

For convenience, pollutant concentrations are derived in parts per million by volume (ppmv). (See Section A.1 for an explanation of ppmv and related units of concentration.) By comparison, pollutant sources have been specified in tonnes/day, residence times in fractions of a day, and volumes in cubic meters. To make the concentration in Equation 7.15 come out in ppmv, the conversion constant must also relate the pollutant concentration to the total concentration of air molecules. The density of air and the mass of a pollutant molecule relative to an air molecule must be known in order to complete the conversion of units. (The relationship between the concentration and density of a gas is discussed in Section 2.2.1.) Writing out Equation 7.15 in detail for a particular pollutant species, i, yields (note that the density of the atmosphere is taken as 1 kilogram per cubic meter at the surface for this calculation):

$$q_i\,(\text{ppmv}) = \frac{S_i\left(\dfrac{\text{tonne}}{\text{day}}\right) \times \left(\dfrac{1000\ \text{kg}}{\text{tonne}}\right) \times \tau_i\,(\text{day})}{V\,(\text{m}^3) \times \left(\dfrac{1\,\text{kg-air}}{\text{m}^3}\right)}$$
$$\times \frac{30}{M_i} \times 10^6$$
$$= \left(1 \times 10^9 \times \frac{30}{M_i}\right) \times \frac{S_i\tau_i}{V}$$
$$(7.16)$$

The factor of $30/M_i$ in Equation 7.16 represents the ratio of the mass of one air molecule to the mass of one molecule of the pollutant. For CO, $M_i = 28$; for NO, $M_i = 30$; for NO_2, $M_i = 46$; and for O_3, $M_i = 48$. In the case of particulates, this factor can be neglected (that is, by setting $30/M_i = 1$); the concentration is then given in parts per million by *mass* (ppmm). If the factor of $30/M_i$ is ignored for the gaseous pollutants as well, the concentration is again computed in ppmm.

Using Equation 7.16, each of the key primary pollutants may be calculated to estimate their average concentrations. The concentrations of secondary pollutants can also be estimated, as described next. The results of these calculations using the box model approach are summarized in Table 7.5. Examples of the procedure are elaborated for the primary pollutant, CO, and the secondary pollutant, O_3.

Carbon Monoxide: The concentration of carbon monoxide can be computed from Equation 7.16 using the parameters summarized in the text and Table 7.5. Inserting these parameter values yields

$$q_{CO}\,(\text{ppmv}) = \left(1 \times 10^9 \times \frac{30}{M_{CO}}\right) \times \frac{S_{CO}\tau_{CO}}{V}$$
$$= \left(1 \times 10^9 \times \frac{30}{28}\right)$$
$$\times \frac{6500\left(\dfrac{\text{tonne}}{\text{day}}\right) \times \dfrac{1}{2}\,(\text{day})}{1 \times 10^{12}\,\left(\text{m}^3\right)}$$
$$\cong 3.2\,(\text{ppmv}) \qquad (7.17)$$

Estimating the primary gaseous pollutant concentrations is straightforward once the three important parameters in Equation 7.16 have been determined.

Table 7.5   Box model assessment for air pollution in the Los Angeles Basin[a]

Pollutants	Source (metric tons/day)	Residence time (days)	Concentration[a] (ppmm)	Clean air standard (ppmm)
Carbon monoxide (CO)	6500 (Emissions)	1/2 (Ventilation)	~3.2	10 (12-hr average)
Nitrogen dioxide (NO_2)	1200 (Emissions)	1/6 (Photochemistry)	~0.20	0.25 (1-hr average)
Ozone (O_3)	1300 (Photochemistry)	1/6 (Photochemistry)	~0.21	0.10 (1-hr average)
Total suspended Particulate (TSP)	650 (Primary)	1/10 (Deposition)	~0.065	0.10 (24-hr average)
	350 (Secondary, from SO_2)	1/2 (Deposition)	~0.17	

[a]The volume, V, of the Los Angeles basin is taken to be 1000 cubic kilometers, or 1×10^{12} m^3. The equation for concentration is $q \equiv S\tau/V$, where q = concentration, S = source and τ = residence time. The following conversions are used to get parts per million by mass: 1 metric ton = 1000 kg; 1 m^3 air (sea level) = 1 kg.

The computed CO abundance corresponds to an average daytime value.

Ozone: Ozone is generated as a by-product of the oxidation of hydrocarbons in the presence of nitrogen oxides. The total source of ozone by this process must be determined before the ozone concentration can be estimated. The amount of ozone formed, to first order, depends on the total emissions of primary reactive hydrocarbons. The RH emission was specified as 1300 tonnes/day in our case. As a rough rule of thumb, it may be assumed that the photochemical decomposition of a certain mass of RH produces approximately the same *mass* of ozone. The actual mass of ozone generated can be larger or smaller, depending on the availability of NO_x, the types of hydrocarbons composing the RH, and several other factors. After deciding on the best value for ozone production (in tonnes/day), the calculation of its concentration proceeds exactly as for CO in Equation 7.17.

Other Pollutants: In the case of the nitrogen oxides, the constituent of interest is nitrogen dioxide, a major secondary constituent of smog. The sum of NO and NO_2 is treated as a single "constituent": NO_x. In well-developed smog, most of the NO_x is NO_2. Accordingly, for a first-order assessment, the chemical source of NO_2 may be equated with the primary emission source of NO, as long as the mass

factor in Equation 7.16 is set to M_{NO}. Particulate concentrations are computed without the mass conversion factor, as discussed, yielding an average mass mixing ratio for the particulates in the box. Although some reactive hydrocarbons are toxic to breathe, they are not often carefully monitored. Nevertheless, overall concentrations of RH may be readily estimated in the same way as for other primary pollutants. Because RH is actually a complex mixture of organic compounds, it is most convenient to calculate the total abundance in parts per million by *mass* using Equation 7.16 without the mass conversion factor.

The results, collected in Table 7.5, for this simple box model assessment of air pollution conditions are striking. According to these simple estimates, CO and NO_2 levels should be, on average, below the "clean-air" standards established by government agencies. On the other hand, ozone and particulates should exceed the standards. The model prediction is very close to the actual situation in the Los Angeles basin. Notice that the differences between the estimated concentrations and the clean-air standards are relatively small. If the concentration of a pollutant is slightly below the level considered to be unhealthy but exposure is continuous over many years, is one safe from health effects? Cumulative exposure in this case is so close to the threshold for significant long-term impacts that complacency is unwarranted.

7.5 Limiting Risk

The only way to avoid exposure to toxic compounds is to stop living. Society should have the goal of minimizing exposure to hazardous materials. The transport of toxins through the atmosphere is one of the major pathways for exposure. Air pollution is laced with unhealthful compounds. Cigarette smoking is the same as mainlining carcinogens. Fumes and vapors in the workplace generously offer 8 hours of toxic exposure. Common chemicals in the home may lead to even larger doses of dangerous compounds. (See Chapter 8 for a discussion of indoor pollutants.) The avoidance of environmental poisons is a full-time job. Simple actions, however, can prevent or limit exposure. Safety in handling toxic compounds is essential. If people would read and heed packaging labels on pesticides, adhesives, paints, and so on, their exposure to dangerous fumes would be greatly reduced.

At work, any potential source of toxic exposure should be reported and corrected. Acknowledgment of pollutants and potential health risks is the first step in minimizing harm. Yet employees are often hesitant to raise the issue, and employers are frequently reluctant to fix the problem. That is why oversight of the workplace is necessary. The Occupational Safety and Health Agency enforces standards for the concentrations of airborne toxins in work areas, because some companies see safety as an unnecessary overhead expense. It will be a better time, indeed, when one of the universal goals of business is to protect the health and safety of its workers.

The public must also be made aware of the potential dangers created by industries and everyday activities. In California, laws established under Proposition 65 require businesses using hazardous materials to report such use periodically to the public (usually through announcements in newspapers). No one wants to be called a "dirty polluter" today. Business is seeking to clean up its act. At the same time, industry lobbies the government to allow polluting operations to continue, to repeal regulations enacted to control regional pollution, to increase the acceptable levels of exposure to toxins in the workplace, and to delay restrictions on the emissions of hazardous compounds into the atmosphere. Millions of dollars are spent on public-relations campaigns to convince us that an honest effort is being made to protect the environment. We are told that life is better through chemistry, even as we choke on polluted air.

To be sure, great strides are being made in the identification, control, and limitation of toxic pollutants. Everyone is working to reduce the risk. Even so, most people are not aware of the hazards lurking in the corners of their lives. If in doubt, compose a list of potential sources of exposure to pollution, ranging from smog to household chemicals to occupational toxins.

Questions

1. Discuss several ways in which you can be exposed to toxic pollutants carried by airborne particles. Consider all of the pathways by which aerosols may enter the body and their fate along each path. What organs of the body might be involved in each case?

2. Describe the effects that you might experience if exposed to an increasing concentration of sulfur dioxide while working in a closed room. Imagine that the SO_2 is leaking slowly from a tank in the room. At what concentrations of sulfur dioxide might you become aware of the leak? Would you have time to escape before lapsing into a coma? Suppose the leaking gas were carbon monoxide and answer the same questions.

3. Is it reasonable, from the point of view of protecting your health, to argue that exposure to natural toxins is so significant that there is no need to be concerned about the additional exposure to industrial and other human-made toxins, including radioactivity?

4. Describe one or more actual events in which large quantities of toxic materials were spilled or escaped into the environment accidentally. If you were at home and an alarm sounded indicating that a serious toxic spill had occurred nearby, what actions might you take to protect yourself from severe exposure? What common household materials might you keep handy to reduce possible exposure?

5. Briefly state the likely immediate symptoms and long-term effects of exposure to (a) ozone, (b) a lachrymator, (c) diesel soot, (d) X rays, and (e) asbestos.

6. Make a list of all the sources of toxic materials in your home that you are certain about, or can reasonably guess are toxic from their uses. List all of the sources of materials that might be toxic, but which you are not sure about. Describe how you might obtain more concrete information about materials you are uncertain or concerned about.

7. The same as question 6, except consider sources of toxic materials in your school. In your car. In your neighborhood.

Problems

1. You have managed to isolate in a box 400 atoms of radioactive substance A and 100 atoms of radioactive substance B. It is known that A has a half-life against radioactive decay of 1 hour, and B, a half-life of 2 hours. If you were to count the *total* number of atoms of A and B left in the box 2 hours after the start of your experiment, how many would you find? How many after 4 hours?

2. In your workplace, you are exposed to two toxic pollutants: X and Y. The health standard for exposure to X is one part per million by volume (ppmv) 8 hours per day, 5 days per week. The standard for Y is one part per billion by volume (ppbv) 1 hour per day, 5 days per week. Exposure to X at the standard concentration would produce a risk of cancer of 1 in 1 million after 30 years. Standard exposure to B would produce a risk of 1 in 1 million after 30 years. You surreptitiously take a sample of air from the business and have it analyzed for X and Y. It is found that the air contains 3 ppmv of X and 1.5 ppbv of Y. If you want to keep your overall risk of contracting cancer from workplace exposure to 1 in a million or less, how many years could you work here before you would have to quit? Assume that the risk from each toxin increases directly in proportion to the total exposure over time and that the risks from exposure to different substances are simply additive, not synergistic. Which toxin is causing the largest risk to your health?

3. Which situation poses the greatest health risk to you: (a) exposure to secondhand cigarette smoke at the bingo parlor you frequent 4 hours per week, where the smoke concentration is 100 micrograms per cubic meter ($\mu g/m^3$); (b) exposure in your living room, where you spend 15 hours per week and the smoke concentration is 20 $\mu g/m^3$ because your father chain-smokes; or (c) exposure in your friend's room, where you sit 10 hours per week and the smoke concentration is 30 $\mu g/m^3$ because both of you smoke continuously there?

4. Two processes can remove a pollutant (paint fumes) from a closed room. One process involves natural ventilation through cracks around the doors and windows. This process results in a residence time of 10 hours for the pollutant. The second process involves forced ventilation through a system of ducts, which results in a residence time of 1 hour. Calculate the *average* residence time of the paint fumes in the room. Which process is obviously dominant? If the doors and windows could be sealed perfectly, what would be the average residence time? Explain, using physical reasoning and common sense, why one process dominates the residence time.

Suggested Readings

Ames, B. N., W. E. Durston, E. Yamasaki, and F. D. Lee. "Carcinogens Are Mutagens: A Simple Test System Combining Liver Homogenates for Activation and Bacteria for Detection." *Proceedings of the National Academy of Sciences* **70** (1973): 2281.

Block, E. "The Chemistry of Garlic and Onions." *Scientific American* **252** (1985): 114.

Carson, R. *Silent Spring*. Boston: Houghton Mifflin, 1962.

Finlayson-Pitts, B. and J. Pitts, Jr. *Atmospheric Chemistry: Fundamentals and Experimental Techniques*. New York: Wiley, 1986.

Goldberg, A. and J. Frazier. "Alternatives to Animals in Toxicity Testing." *Scientific American* **261** (1989): 24.

Reganold, J., R. Papendick, and J. Parr. "Sustainable Agriculture." *Scientific American* **262** (1990): 112.

Setterberg, F. and L. Shavelson. *Toxic Nation: The Fight to Save Our Communities from Chemical Contamination*. New York: Wiley, 1993.

Tschirley, F. "Dioxin." *Scientific American* **254** (1986): 29.

8

Indoor Air Pollution

When you arrive home at the end of a hard day, kick off your shoes, and plop down in front of the tube with a cold brewsky, the last thing you need to worry about is air pollution. Wrong! Your home, your castle, may contain air more toxic than smog. Sealed up against the outside world, a house is indeed a safe haven from a soupy urban atmosphere. However, because it may be tightly sealed, a house itself can trap pollutants that originate within. Organic chemicals seeping from insulation in the walls, paint fumes, bug sprays, carpet-cleaning solvents, candle smoke, and cooking odors all add to indoor air pollution. Cat litter left festering too long gives off ammonia vapor. The gas burners on the stove emit carbon monoxide. Chlorine gas leaks out of tap water. If a smoker lives with you, secondhand tobacco smoke permeates every room and every fiber. Pets contribute hair and dust. The human inhabitants further degrade the quality of the air by breathing—thereby emitting moisture, carbon dioxide, and other organic compounds—and perhaps by practicing other unsavory but natural habits. If a place becomes noticeably foul smelling, an air freshener may be used to cover up the odor, but that simply adds more chemicals to the problem. If the situation becomes so obnoxious that a window must be thrown open, fresh smoggy air can rush in.

In Chapter 7, we discussed the physiological and health effects of a variety of toxic compounds found in the atmosphere. This chapter covers pollutants that are found principally in homes and in the workplace. There is, of course, overlap between these discussions, and so it is useful to review Chapter 7 before continuing. Asbestos, for example, is found in the atmosphere, but exposure to it occurs mainly in buildings. Some of the complex issues concerning the health effects of asbestos are summarized in Section 7.2.4 and are not repeated here. In this chapter, we identify and clarify several specific

indoor air pollution hazards, including radon, formaldehyde, and tobacco smoke. These are the key indoor pollutants that can seriously affect health and frequently make the headlines. Such pollutants are not occasional threats to small communities of people or isolated neighborhoods. Rather, they are everyday hazards to the large segment of the human population that dwells in homes and apartments, that works in offices and factories, and that finds leisure in bowling alleys and nightclubs. Special attention to these toxic compounds is certainly deserved.

8.1 What Are "Indoor" Air Pollutants?

Indoor air pollution can be considered to consist of any airborne material in a living or working space that may irritate people or affect their health. In this broad definition, **household dust** is an indoor pollutant, made up of fibers, lint, and hair in combination with various sorts of organic debris (see also Section 8.5.1). Our bodies are constantly dropping microscopic flakes of skin and hair and emitting various gases. Pets contribute dander (fine scales of skin, hair, and feathers), larger hair fibers, and dry particulates raised from litter boxes and birdcages. Insects die and decompose into smaller parts. Living in this sea of debris is an army of tiny mites feeding on the rich organic detritus, creating their own waste in turn. The mites, viewed under an electron microscope, are hideous. Freddie Krueger, the horribly disfigured resident of Elm Street, is Robert Redford compared with these guys, who look like a gang of deformed dinosaurs. But the mites are as common as the dust they live in. For the most part, they are harmless, except that some people are highly allergic to them. Fibrous and powdered debris, including countless mites, collecting in the corners of rooms and in air ducts, slowly aggregate into dust particles of various sizes. In dark recesses beneath dressers and beds, the dust may accumulate into fluff balls the size of lemons. More common is the storm of fine dust particles that can be seen swirling in a beam of sunlight piercing a dark room from a window. Cleaning with a broom or vacuum cleaner actually increases this fine dust, even as larger hair balls are captured and disposed of. Entrepreneurs sensitive to the fears of indoor dwellers promise to suck these dust balls—as well as mites, molds, and mildew—from heating and cooling ducts where they lurk, patiently waiting for ventilation systems to send them swirling toward our lungs.

One of the most common sources of allergenic materials in homes may be the beloved pet and the dust created by it. Cats, in particular, are the source of many sneezes. Up to 10 percent of the population may be allergic, to some degree, to cats. Cats release through their skin a certain protein, called **Fel dI**, to which many people are strongly allergic. The protein attaches to hair and dander and is thus carried everywhere in the house. Breathing dust particles contaminated with the protein can lead to an allergic reaction.

Excessive noise may be considered an indoor air pollutant. After all, sound is nothing more than a pressure wave moving through the air. At home, the resident teenager's boom box is often a source of acoustic disturbance and aggravation. In Vancouver, a woman asked to have her husband's snoring recorded after she was diagnosed as partially deaf. He registered a screaming 90 **decibels** (dB)[1] of noise, comparable to a lawnmower or motorcycle, landing him a spot in *The Guiness Book of World Records*. Noise can be taken care of in ways that toxic pollutants cannot (by earplugs, for example). Nevertheless, noise in the modern urban environment is pervasive. People who live in cities, near airports, or next to freeways are continuously exposed to excessive noise. Sound does not have to be deafening to be

a nuisance or a health hazard. Car alarms in the night can ruin a good night's sleep. The roar of trucks and buses on close-by highways can fray nerves. Chronic occupational noise can cause premature hearing loss.

In the home, noise may be a bother. Toxic vapors and particles, on the other hand, can be far more hazardous to one's health.

8.1.1 THE SPECIAL CHARACTER OF INDOOR POLLUTION

Indoor air pollution is unique in several respects. First, many people spend most of their time indoors while at home or at work, and so their exposure to indoor pollution can have a great impact on their health. Second, pollution *in* our homes is a fundamental invasion of private sanctuaries. One does not have to live in a smoggy city to suffer grievously from indoor air pollution. Indoor pollutants are frequently a problem in pristine countrysides. Finally, methods are available to deal with indoor air pollution. Government agencies set standards for exposure to many potential indoor pollutants. If those standards are exceeded, homeowners must usually act on their own behalf to correct the problem. It is literally environmental activism closest to home.

Staying indoors can be an effective way to avoid outdoor air pollutants such as ozone and acidic fog. At least this is true if the ventilation system is functioning properly, thus maintaining relatively clean air inside the building. The interior must also be empty of smokers. In some instances, outdoor pollution may be carried indoors through ventilation systems. Nevertheless, serious sources of pollution are often found inside buildings, including houses. Depending on circumstances, your home environment on a smoggy day may not be the healthiest choice (try the local hermetically sealed mall in that case).

The following list describes a few of the common sources of indoor air pollutants:

- Radon gas seeping into homes from soils through foundations and from water brought into the home through plumbing.
- Vapors escaping from building materials that contain toxic or potentially toxic compounds such as formaldehyde, asbestos, and vinyl chloride.
- Gases and aerosols originating from personal activities such as smoking, burning incense or candles, or using a fireplace.

---

1. The decibel (or one-tenth of a "bel") is a common unit of sound intensity used to rate audio systems and annoying noise levels. The bel is named for Alexander Graham Bell, inventor of the telephone. If a sound-intensity level increases by 10 dB, the amount of power in the sound waves has increased by a factor of 10. A 20 dB increase is therefore a factor of 100 steps in sound intensity; 30 dB is a factor of 1000; and so on. The decibel scale is calibrated against the *least perceptible sound* that a normal person can hear, which is assigned a value of 0 decibels. A sound 10 times louder is rated at 10 dB. Noise becomes physically painful at about 130 dB, which is 10^{13} times more intense than barely perceptible sound!

Table 8.1   Sources of Indoor Air Pollutants[a]

Sources	Pollutants
Soil and ground water	Radon and radioactive daughters
Building materials and furnishings (carpeting, paint, varnish, adhesives)	Formaldehyde Asbestos Vinyl chloride Organic fumes
Personal activities and hobbies	Cigarette smoking Fireplace smoke Solvent and glue fumes
Appliances, cooking, and heating	Carbon monoxide Natural gas Cooking odors Boiler and heater fumes
Household chemicals (bleach, oven cleaner, insect sprays, nail polish, hair spray)	Ammonia Hydrogen chloride Pesticides Organic fumes Aerosols
Electronic equipment and wiring	Organic fumes Electromagnetic radiation
Pets	Hair Feces Proteins Dust
Plants	Pollen Hydrocarbons

[a]The sources are identified, but not quantified. In particular cases, some of the sources are insignificant or absent.

- Gases emanating from appliances, including carbon monoxide and other combustion products, and vapors from stove-top cooking and barbecues.
- Aerosols and vapors escaping from the application or evaporation of household chemicals such as bleach, ammonia, hair spray, perfume, nail polish, and bug sprays.
- Hair and other debris dropping from pets and humans contributing to airborne household dust.

These and some other sources of indoor air pollution are summarized in Table 8.1. (Refer to Table 7.3 for a summary of the toxic effects of organic compounds.) The number of *possible* sources is very large indeed. It has been suggested, for example, that flushing a toilet creates small droplets of water, or aerosols, that may drift throughout the bathroom. What might be imagined to reside on these "latrine aerosols" is revolting. Supposedly, the droplets can settle everywhere—on toothbrushes, for instance.

Accordingly, special devices have been invented to protect toothbrushes from latrine contamination and to sterilize the brush fibers continuously using ultraviolet radiation. The actual health threat from toilet-generated particles settling on a toothbrush is probably negligible. Nevertheless, normal concern with personal health and hygiene leaves one potentially vulnerable to marketing strategies that exploit those concerns. Hence, sterilization devices are suggested for toothbrushes. Some threats are exaggerated, but others are very real indeed.

The Killing Fields

One sits placidly in front of a television or computer screen. Bathed in the dull glow of the screen, the mind drifts in and out of semiconsciousness. There is a benign warmth and safety in the softness of the light. But is the gentle radiation emitted by the tube damaging to one's health? All electrical devices emit stray electromagnetic radiation, or **electromagnetic fields** (**EMF**). (The nature of electromagnetic radiation is discussed in the introduction to Section 3.2.) Such radiation travels through air in the form of electromagnetic fields. Wires, extension cords, radios, appliances, power lines, transformers, electric trains, computers, slide projectors, lightbulbs, electric blankets—all produce EMF. The largest sources of EMF are the high-voltage power lines that crisscross most cities. These transmission lines carry hundreds of thousands of volts of electricity. Stray electric fields are unavoidable, particularly in the vicinity of the transformers that convert the high voltages used for transmission to lower voltages used for power distribution to homes. The phosphorescent screen of a television or computer also emits electromagnetic radiation. The phosphors are excited by a high-energy beam of electrons striking the screen from behind. The electrical currents and voltages necessary to excite the phosphors unavoidably generate stray EMF.

Most of the electrical power consumed in the world is generated at a frequency of 60 hertz. That is, the electromagnetic field associated with the electric current oscillates at a frequency of 60 cycles per second. This is a very low frequency compared with ultraviolet radiation or X rays, for example. Ultraviolet and X rays are made up of highly energetic photons that can be extremely dangerous (Sections 7.3.2, 7.3.3, and 13.4.2). Recall that the energy carried by a photon of electromagnetic radia-

tion is proportional to the frequency of the radiation (Equation 3.19). Low-frequency radiation consists of low-energy photons. These photons cannot break apart the molecules composing living tissue, as can ultraviolet radiation, X rays, and gamma rays; but the potential consequences, if any, of continuous exposure to low-frequency EMF are unknown.

The postulated health effects of EMF exposure range from miscarriages to leukemia, a deadly form of cancer of the bone marrow that inhibits the formation of red blood cells and results in an excess of white blood cells, or leukocytes (leukemia has been associated with exposure to gamma radiation released by the atomic-bomb explosions over Hiroshima and Nagasaki). At one time, the Environmental Protection Agency classified EMF as a "probable" human carcinogen, placing it in the same category as asbestos and dioxins. Later, this classification was modified to "possible but not proven." Some epidemiologists have found a statistical correlation between exposure to EMF and birth defects and miscarriages, as well as cancer. Less serious, but still troublesome, effects occasionally connected to EMF include mood and sleep disturbance, depression, and enhanced sensitivity to certain drugs.

The San Diego Gas and Electric Company was accused in court of negligence in causing an unusual cancer in a child. At only 10 months of age, the girl, living in the shadow of a tangled web of power lines, suffered Wilm's tumor, a rare disease of the kidneys. The intensity of the EMF was 40 times that found in the average home. The girl later recovered when the family moved to another location. In the meantime, the utility company was sued for $1 million. One of the plaintiffs asked whether "we really want to play Russian roulette with our children" by allowing the public to remain exposed to EMF without a full understanding of its effects or proper disclosure by industry to the public of the levels of exposure. Experts from the utility company denied any proven ill effects from stray EMF. The jury sided with these experts in denying negligence by the utility company.

Some epidemiologists now caution that people should minimize their exposure to electromagnetic radiation if possible—without panicking. For example, an electric blanket can be used to warm a bed *before* one gets into it. Major appliances can be avoided when operating. Sure, hide from the washing machine, cringe from the refrigerator. And what about the electric shaver?

Effects of Indoor Toxins

Various indoor pollutants affect health in different ways. (See Sections 7.1 and 7.2 for a discussion of the physiological effects of a number of common toxins.) Carbon monoxide (CO), for example, affects the ability of blood to absorb oxygen. Asbestos lodges in the lungs and produces permanent lesions that may develop into lung cancer. Radioactive elements inhaled or ingested into the body expose the lungs and other organs to energetic alpha, beta, and gamma radiation. Toxic chemicals on smoke particles are deposited in the respiratory tract and may then be absorbed into the blood.

The reaction to indoor pollution exposure can include persistently irritated eyes, nose, or throat; coughs; headaches; dizziness; difficulty in breathing; nausea; and rashes. An increase in the incidence of respiratory infections and higher susceptibility to flu infections are also suspected effects of exposure to indoor air pollution. In some cases, allergic reactions to dust or chemicals found in the home can become debilitating. Even excess water vapor, which builds up in tightly sealed homes, can aggravate respiratory disease. The moisture condenses on cool surfaces, fogging windows and damaging plaster and wood. High humidity can be downright uncomfortable as well.

Exposure to indoor air pollution is often intensified because of the special precautions we take to protect our homes. Bugs crawling over floors and ceilings are unwelcome. Flies and ants are a nuisance. Fleas are a problem in the summer. So all sorts of pesticides are sprayed onto surfaces and into the air. Powders are sprinkled in cabinets and behind sinks. Pest strips are dangled in the kitchen. The living room is fogged, and roach traps are laid. Outside, the soil is soaked with termite and ant killers. The poisoned dust later travels indoors on shoes and on the wind. Occasionally, we manage to eradicate the bugs and, every now and then, the pet.

All good citizens nowadays are conserving energy as best they can. Cracks in doors and windows are sealed to minimize heat loss in winter. Not surprisingly, potentially dangerous pollutants are being sealed in at the same time. Ventilation is the quickest way to rid a home of pollutants that originate indoors. But windows cannot be thrown open in midwinter. Where is the happy middle ground between freezing to death from the cold and choking to death on pollution? Unfortunately, either the sources of indoor pollution must be eliminated or a house must be ventilated, even if moderately. Remember the box model, and imagine that the box is your home (Sections 7.4.3 and 8.2.2).

8.1.2 Indoor Pollution and the News

Indoor pollution is in the news. Radon has been a hot item for years and remains popular among the press. Asbestos is a perennial favorite. Recently, the horrors of cigarette smoking have been revisited—not the direct exposure from puffing itself (that is a proven killer), but the indirect exposure to secondhand smoke, which is debilitating children and adult nonsmokers alike by the tens of thousands. Some of the most critical indoor air pollution issues are discussed in greater depth in the following sections. What about some of the other indoor pollution hazards that make the news? How serious are these threats to health?

Electromagnetic radiation in our homes and workplace is unavoidable. To some extent, the danger from electromagnetic fields seems to be exaggerated. Radiation of such low frequency, with a correspondingly low energy content, cannot directly break down organic molecules. On the other hand, at high exposures, low-frequency EMF cooks meat. That is what a microwave oven does. The risk is thus a matter of dose and dose rate (Sections 7.1.2 and 7.4). Some animals and insects navigate using the Earth's magnetic field. Many living organisms generate a coronal "aura," which surrounds the extremities and may be seen by amplifying the associated electric fields. A few kooks claim that these fields are an indication of internal psychic forces. Unlikely! On the other hand, it is possible that the complex mechanisms operating in living cells may be subject to subtle anomalies when exposed to EMF. Could such anomalies lead to disease? No one wants to be a guinea pig. Concerning common environmental pollutants, one should exercise common sense in limiting exposure to them and to the sensational media coverage that may accompany them.

8.2 Radon: Mother and Daughters

In December 1984, Stanley Watras stepped into his company's Christmas party and set off alarm bells around the world. At the time, he was an engineer working on the new Limerick nuclear-power plant in Pennsylvania. The first alarm he set off, at the portal of the plant, was designed to detect radioactivity on workers leaving the facility. Watras was *entering*. The source of the radioactivity was traced to his home. The radon levels in his living room were so high that simply breathing the air was equivalent to smoking about *100 packs of cigarettes* every day! Eventual death by lung cancer was almost a certainty. In the neighborhood, dream homes overnight became houses of horror. Windows were thrown open in midwinter; children were kept out of basements; and panicked families left town.

Until Watras's unfortunate (fortunate!) encounter with a Geiger counter, radon had been considered to be a problem limited to uranium-mine tailings. Over the years, uranium-mining operations around the country had accumulated huge mounds of leftover diggings. These tailings are still rich in uranium compared with other soils. Later, developers moved in and built homes on the cheap land. Some natural building materials also contain high levels of uranium, and uranium spawns radon. A long time before Stanley Watras was contaminated, workers in uranium mines were known to contract lung cancer at a very high rate. The cause was found to be exposure to radon and its radioactive products. Radon in homes was another matter, however. No one had suspected that large numbers of people might be at risk from the radioactivity exhaled by the Earth.

8.2.1   Poison from the Earth

Radon is a natural radioactive element formed in soils all over the world. Radon itself is a colorless, odorless, tasteless gas. Since it is a gas under normal conditions, radon formed in the ground can seep into the atmosphere. If a home is built over radon-rich soil, radon gas may enter the home and accumulate inside. In some instances, the concentration of radon is great enough to increase significantly the risk of lung cancer. Radon is particularly hazardous in areas where the soil contains large abundances of the parent elements uranium, thorium, and radium. These elements decay by radioactive fragmentation in the soil, eventually releasing radon atoms. Oddly enough, the natural background radioactivity of uranium, thorium, and radium, and the radon that they generate, is relatively harmless. The decay pro-

cesses are slow, and the soil absorbs the radioactive particles emitted (alpha and beta particles [Section 7.3.1]). Radon seeps from the ground at such a slow rate that it is not directly a toxic hazard. But when radon accumulates in a building or mine shaft and undergoes radioactive decay, trouble starts.

The Origin of Radon

Uranium and thorium were present when the Earth formed 4.5 billion years ago. Uranium-238 (the common form of uranium) has a **half-life** of 4.5 billion years; accordingly, about half the primordial uranium remains in the Earth today. (See Section 7.3.1 for a definition of half-life.) Every radioactive element has a specific half-life (Table 7.4). Some radionuclides have very long half-lives (billions of years), and others have extremely short half-lives (much less than 1 second). Of the original radioactive elements that accreted with the Earth, only those with half-lives of a billion years or longer remain in significant quantities.

When uranium decays, one of the intermediate by-products is radium-226 (^{226}Ra). The decay of thorium-232 (with a half-life of some 14 billion years) leads to radium-228 (^{228}Ra). Recall from Section 7.3.1 that the left superscript 226 or 228 on the symbol for the element radium, for example, gives the total number of protons and neutrons in the nucleus of a radium atom, defining the particular isotopes of that element. Radioactive elements are usually tagged with an **isotope** number for identification. The superscript is a simple, compact way to label an atom and should be considered part of its extended chemical name specifying the particular isotope (just as Roman numerals can distinguish among family members with the same name, as in George III).

Both forms of radium generate radon atoms through radioactive decay. The key source of radon, however, is ^{226}Ra formed from uranium. When an atom of radium-226 undergoes radioactive decay (half-life of 1622 years), it emits an **alpha particle**, consisting of 2 neutrons and 2 protons (Equation 7.3). The remaining atom, having a total of 222 protons plus neutrons, is radon-222 (^{222}Rn).[2] Because of this

---

2. Actually, there are two species, or isotopes, of radon: radon-222 (^{222}Rn) and radon-220 (^{220}Rn). These species originate from the two species of radium, ^{226}Ra and ^{224}Ra, respectively. In the latter case, the proximate source of the radon-220 is thallium-228, with a half-life of 1.9 years, which decays into

"birth" relationship between radon and radium, radon is said to be a **daughter** of radium. Why not a son? Radon, being radioactive, continues to decay, giving birth to new daughters. Sons cannot give birth; daughters can. Indeed, the daughters produced by the radon-decay process are the real culprits in the radon story.

Eventually, all radioactive elements decay into stable atoms. These final products of the radioactive-decay process may be referred to as **progeny** of the initial radioactive elements. The progeny of a particular radioactive isotope include all the radioactive and stable species formed during the decay process.

Radon-222 has a relatively short half-life of 3.8 days; radon-220 has an even shorter half-life of about 55 seconds. It follows that the radon in the air around us must have been generated recently and is not expected to linger for very long. The concentrations of radon are naturally limited by its short half-life. Only the radon molecules that, following birth, can diffuse out of the ground within a few days could pose a threat to air breathers. Accordingly, radon-222 is the more important isotope causing human exposure to radioactivity; radon-220 typically decays in the soil, making a negligible contribution to the burden of radioactivity in air. In the following discussions, references to "radon" are to radon-222.

Over several days, the radon vapor decays into a series of other radioactive elements. The two main decay paths for radon-222 and radon-220 are depicted in Figure 8.1. Radon-222 first fragments into polonium-218 by emitting an alpha particle. The polonium-218 (half-life of about 3 minutes) then decays into lead-214, also by emitting an alpha particle. Lead-214 (half-life of about 27 minutes) next decays into bismuth-214, releasing a **beta particle** (high-speed electron). The bismuth-214 (half-life of about 20 minutes) then decays into polonium-214 by emitting another beta particle. The resulting polonium isotope is very unstable (half-life of about 0.00016 second) and decays into lead-210 with the emission of an alpha particle. Lead-210 is relatively stable (half-life of about 22 years). These lead atoms eventually are removed to the ground and, through a series of steps, decay into the stable (nonradioactive) element lead-206.

The processes that create a radon problem make up the first five decay steps for radon-222 in Figure

8.1. At each of these stages of decay, a highly energetic subatomic particle (alpha or beta) is released. These particles can rip through living tissue like microscopic bullets, causing cellular and genetic damage. The impact is most severe when the radioactive elements are in direct contact with the tissue or are absorbed in the tissue or in an organ. Beta particles do not penetrate as deeply as the heavier alpha particles. Nevertheless, both forms of radioactivity destroy living cells.

It is important to note that in its natural form, radon is a gas and so can seep through the pores in soils and escape into the atmosphere. The other radioactive precursors of radon—uranium and radium—are locked in the soil in the form of mineral compounds. These solid and relatively insoluble compounds are immobile in the ground. If pulverized into ultrafine particles, however, the radioactive minerals can be carried as dust by the winds and may be inhaled. Nonetheless, most radioactivity escapes from the ground in the form of radon gas. The polonium, lead, and bismuth that are born as daughters of radon gas also tend to form solid substances. Typically, the daughter elements rapidly attach to any surfaces they come into contact with, including common dust particles in the air. Radioactivity originating from radon is then carried by airborne particles, which may be inhaled or ingested. If the radon is directly inhaled, it is not likely to be absorbed in the body; radon is not very soluble in water or in other bodily fluids and is therefore exhaled. On the other hand, respirable particles easily lodge in the nasal cavities, throat, and lungs. The radionuclides stuck to these particles can then release their energetic radioactive bullets into the tissue nearby. The lungs come under the most intense attack, and lung cancer is the frequent outcome of long-term radon exposure.

The entire decay chain of the radon daughters up to lead-210 occurs over a period of about 50 minutes. In other words, from the time that radon-222 first gives birth to its deadly progeny, all its most dangerous radioactivity is released within an hour. This is just the sum of the half-lives of polonium-218, lead-214, bismuth-214, and polonium-214 (Figure 8.1 and Table 7.4). In 50 minutes, the daughter atoms can stick to dust particles, drift through a room, and be inhaled by the occupants. The polonium-218, which emits highly destructive alpha particles, has a lifetime of only a few minutes. For this reason, almost all the radioactivity associated with ^{218}Po is released in the room where the radon

radium-224, with a half-life of only 3.6 days. On longer time scales, thallium derives from uranium. The major radon source in most cases is radium-226.

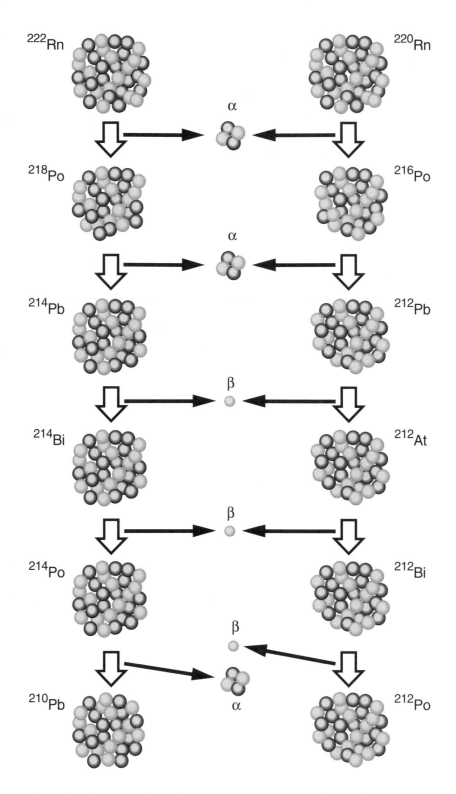

Figure 8.1 The radioactive decay chains for radon in its two isotopic forms. The products at each step of the decay process and the emitted radioactive particles at each step are shown. The chains here are not complete but continue until the stable atoms ^{206}Pb or ^{208}Pb are formed (for the ^{222}Rn and ^{220}Rn chains, respectively). The most common radon isotope is ^{222}Rn.

first decays. The potential "alpha" dose from ^{218}Po can therefore be very large. The alpha dose from the decay of ^{214}Po is somewhat less significant; because they have longer half-lives, ^{214}Pb and ^{214}Bi can be removed from a room by ventilation before giving birth to ^{214}Po.

The characteristics of radon production and decay can be summarized as follows:

1. Uranium-238 (^{238}U), tied up in solid minerals in the soil, decays to form radium-226.
2. Radium-226 (^{226}Ra), also in mineral form, decays to form radon-222, which is gaseous.
3. Radon-222 (^{222}Rn) molecules diffuse through the soil, some escaping into the atmosphere or dissolving in the groundwater.
4. Radon gas seeping into basements or mine shafts or evaporating from tap water accumulates in closed spaces.
5. Radon-222 decays within several days into polonium-218.
6. Polonium-218 (^{218}Po) and the other daughter species that evolve from radon over the next 50 minutes stick onto walls, floors, and dust particles.
7. Polonium-218, lead-214, bismuth-214, and polonium-214—relatively short-lived radionuclides—are inhaled with particles that lodge in the respiratory system.
8. Radioactive decay emits alpha and beta radiation directly into lung tissue.

Radon, the Intruder

Radon comes into the home as an unwanted guest. It sneaks in from the soil under the house or from contaminated building materials in the walls. Radon vaporizes from water running in the sink, shower, or bath; water tapped from geologic formations rich in uranium can have high concentrations of dissolved radon. Figure 8.2 illustrates the major pathways by which radon usually enters a building. The principal ways that radon is removed are through decay into daughters or ventilation from the building. (Ventilation was discussed in Section 7.4.3.) No building is perfectly airtight. A glass jar can be sealed with a screw-on cap, but a home cannot. There are always drafts around windows and doors. A chimney flue provides an opening to the outside world. People coming and going open and close doors. Even openings for electrical and plumbing fixtures allow air to pass through walls. The result is a leaky building. In a typical house, for example, the air inside is replaced with fresh outside air about once every hour. Measured times to ventilate homes vary from less than 10 minutes (a very drafty house indeed) to as long as 4 hours. Most homes have ventilation times that fall in the range of 0.5 to 1.5 hours.

In areas where the climate is always mild, drafty houses are the rule. However, in regions where the weather can be extremely cold or hot, most buildings are usually carefully weatherproofed. Particularly in recent years, buildings have been made energy efficient through the use of weather stripping and insulation. In many office buildings, the air is recirculated more frequently to conserve heating (or cooling) costs. As a result, homes and offices are more airtight, and their ventilation rates are drastically reduced. Although this is wise energy policy, it may be poor health policy. When the ventilation rate is decreased, the concentrations of indoor air pollutants can increase. The relationship is quite direct, and we explain it next.

8.2.2 Radon Exposure and Its Effects

The threat of radon exposure can be quantified. For example, the amount of radon—or, more relevant to the problem, radon daughters—in a home or workplace can be measured. Such measurements can then be used to estimate the potential exposure and to evaluate the health consequences. The analysis is just that of a "risk assessment" applied specifically to radon.

Measurements of radon in homes are now fairly common, particularly in geographical areas where the soil composition points to a high risk of radon production (for example, in parts of the northeastern United States, areas of the Midwest, and in some mountain states). Kits are even available for do-it-yourselfers. A simple technique uses activated charcoal to absorb radon from the air. The charcoal is exposed to the air in a room for a specific period of time (up to 1 week). The exposed charcoal canister is then sent to a laboratory for analysis of the amount of radon on the charcoal, which is proportional to the concentration in the room. Another simple test uses a sheet of special plastic that is hung on a wall. The radioactive radon progeny in the air near the plastic diffuse to the sheet and stick to it. When the progeny decay by radioactive emission, they produce a "track," or damage path, in the plastic. After months of exposure, the tracks are counted, and the

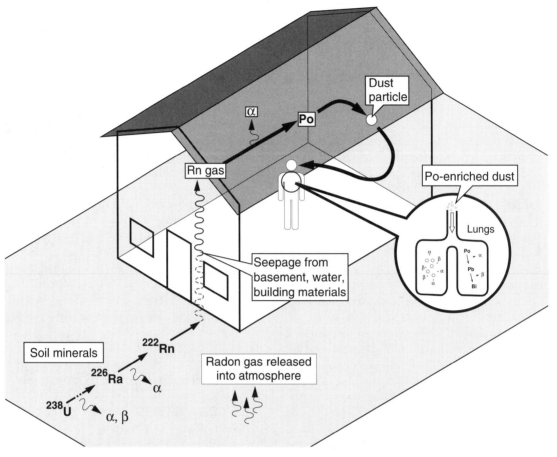

Figure 8.2 The radon problem is depicted for a home placed over uranium- or radium-rich soil. Radon gas enters the basement of the home by diffusing through the soil, or it is released from water used in the home. Building materials may also emit radon. The radon gas, which itself is harmless, decays into daughter species that attach to airborne particles that may become lodged in the lungs. When the daughters undergo radioactive decay, the energetic alpha and beta particles penetrate and damage the lung tissue.

average concentration of progeny is determined. In more complex tests, samples of air may be trapped in a bottle and analyzed for radon daughters. Such techniques require very sensitive analysis.

Gauging Radon Exposure

In Section 7.4.3 we developed a box model to assess pollutant risk. (See Section 7.4 for a review of health risks and their evaluation.) The same model can be used to estimate radon concentrations in buildings and to identify the key factors that control the concentrations. The critical parameters that must be known to carry out such an assessment are the sources and sinks for radon under the conditions of interest.

Radon Sources. The total source of radon is designated as S_{Rn}. This source can be expressed as the amount of radon entering a room each unit of time—for example, as the number of Rn atoms per second. Alternatively, since most rooms are about the same height, the source can be expressed as the quantity of radon entering per unit area of floor space per unit time—that is, Rn-atoms per square meter per second. Finally, because room heights are the same, the source can be expressed as the radon added to each unit volume of the room in each unit time interval— say, Rn atoms per cubic meter per second.

A given amount of radon can also be expressed as a specific amount of radioactivity measured in **curies** (the curie [Ci] defines the rate of radioactive decay, or the number of radioactive decay *events* occurring in a unit time interval [Section 7.3]). The radon-decay rate at any instant is just one-half the radon concentration divided by the radon half-life. In other words, the radon decay rate is directly proportional to the radon concentration. A specific concentration of radon can therefore be defined in terms of its radioactivity measured in curies. Similarly, a source of radon can be expressed in terms of an equivalent source of radioactivity in curies per unit time. Consistent with our discussion, a radon source can also be specified in terms of the radioactivity added to a unit volume of air per unit time—say, curies per cubic meter per second. The *practical* unit used to define the radioactivity of radon is the **picocurie** (pCi). One picocurie equals one-trillionth curie, or 1×10^{-12} Ci. In the following discussion, radon *concentrations* are expressed in picocuries per liter of air (pCi/liter), radon sources in picocuries per liter of air per hour (pCi/liter/hour), and the time constants for various processes in hours. These units are relatively convenient and are often used in published work (however, the number of possible choices and combinations of units is staggeringly large).

The radon sources due to seepage measured in homes and other buildings typically lie within a well-defined range of values. Thus S_{Rn} lies between about 0.01 and 10 pCi/liter/hour. A mean value of $S_{Rn} = 0.5$ pCi/liter/hour is reasonable.

Ventilation provides both a source and a sink for radon. The air outside a building contains a certain amount of radon gas. Radon is continuously escaping from the soil. For the most part, this background gas is diluted by atmospheric mixing to fairly low abundances. For example, a typical background radon concentration is $q_{Rn}^{o} = 0.2$ pCi/liter. Here, the *background* concentration is indicated by the superscript ^{o}. Air that enters a building has a radon concentration equal to q_{Rn}^{o}. The time required for air to be exchanged between a building and the external environment is given by the "ventilation" time, t_{v}. As stated earlier, this exchange occurs every hour or so on average, although the exchange time can be much shorter or longer. Using the principles of the box model, the radon source associated with the ventilation of background gas into a building may then be written as

$$\frac{S_{v}}{V} = S_{v} = \frac{q_{Rn}^{o}}{\tau_{v}} \qquad (8.1)$$

Notice that the equivalent radon source *per unit volume of air*, S_{v}, is defined in Equation 8.1. The *total source* of radon for the interior space is S_{v} multiplied by the volume, V, of that space. At this point, you

may want to review Sections 4.1.2 and 7.4.3 to confirm the relationships among the parameters of a box model. Another way to view this situation is to consider that the box represents a sampling system with a volume of 1 liter, which has been placed in the room of interest.

Radon Sink. Air that is ventilated from a room (or building) provides an important sink for radon. By analogy with Equation 8.1, the equivalent radon loss rate *per unit volume* can be expressed as

$$L_{v} = \frac{q_{Rn}}{\tau_{v}} \qquad (8.2)$$

Here, q_{Rn} is the radon concentration (pCi/liter) inside.

A second important sink for radon is radioactive decay. Obviously, even though radioactive decay removes radon atoms, which are transformed into daughters, these products actually create the potential hazard. The rate of radon decay is related to the radon half-life, written here as τ_{Rn}, which has a value of 91 hours. The decay loss rate *per unit volume* is then

$$L_{d} = \frac{q_{Rn}}{\tau_{Rn}} \qquad (8.3)$$

Note that the quantities S_{Rn}, S_{v}, L_{v}, and L_{d} all have units of pCi/liter/hour.

Radon Concentration. The box model equations show that under steady conditions, the sum of the sources equals the sum of the sinks. Expressed symbolically, this gives the relation

$$S_{Rn} + S_{v} = L_{v} + L_{d} \qquad (8.4)$$

or, considering Equations 8.1, 8.2, and 8.3,

$$S_{Rn} + \frac{q_{Rn}^{o}}{\tau_{v}} = \frac{q_{Rn}}{\tau_{v}} + \frac{q_{Rn}}{\tau_{Rn}} \qquad (8.5)$$

Equation 8.5 is easily inverted to yield the indoor radon concentration, q_{Rn}, in the form

$$q_{Rn} = \frac{q_{Rn}^{o} + \tau_{v} S_{Rn}}{1 + \tau_{v}/\tau_{Rn}} \qquad (8.6)$$

Equation 8.6, in fact, can usually be simplified to read

$$q_{Rn} \cong q_{Rn}^{o} + \tau_{v} S_{Rn} \qquad (8.7)$$

where the $\cong$ sign indicates "approximately, but nearly equal to."

A typical indoor radon concentration can be estimated using Equation 8.7. Since q_{Rn}^o = 0.2 pCi/liter, $S_{Rn} \approx 0.5$ pCi/liter/hour, and τ_v is roughly 1 hour, the predicted radon concentration is about 0.7 pCi/liter. An average value of 1 pCi/liter is usually quoted in studies of the radon problem. Equation 8.7, however, clearly reveals the radon level's dependence on the radon source inside a building and on the ventilation rate of the structure. Extreme values of the radon concentrations can also be estimated from Equation 8.6 for limiting ventilation rates. With very fast ventilation (that is, τ_v approaching zero time), $q_{Rn} \approx q_{Rn}^o$. In this case, the radon is controlled by the environmental conditions outside the building. With no ventilation at all (that is, τ_v increasing to "infinite" time), $q_{Rn} \approx \tau_{Rn} S_{Rn}$. The radon then is strictly controlled by its internal source and radioactive half-life. Both these results are logical. Radon levels are most likely to be elevated in the latter circumstances.

The indoor concentrations of radon daughters can also be estimated using the box model approach. Without working through the details, it turns out that the concentration of each daughter, *expressed as the equivalent radioactivity in* pCi/liter, is proportional to the radon concentration. This approximation is excellent for the conditions under which radon poses a hazard. The relationships between radon and daughter concentrations are

$$q_{Po} \approx q_{Rn} \qquad (8.8)$$

$$q_{Pb} \approx q_{Po} f_{Pb} \approx \frac{2}{3} q_{Rn} \qquad (8.9)$$

$$q_{Bi} \approx q_{Pb} f_{Bi} \approx \frac{3}{4} q_{Pb} \approx \frac{1}{2} q_{Rn} \quad (8.10)$$

For clarity, the isotope numbers have been left off the element symbols in this equation. Equation 8.8 applies to ^{218}Po. Polonium-214 (Figure 8.1) decays so rapidly that it can be lumped together with ^{214}Bi in estimating exposure and risk. In other words, any exposure to ^{214}Bi implies automatic exposure to ^{214}Po, which decays in less than a *milli*second. Equation 8.8 for the concentration of ^{218}Po is the most accurate of the three given. The concentrations of the other radon progeny depend on their rates of decay relative to the rate of ventilation. The "*f*" factors in Equations 8.9 and 8.10, which depend on the ratio of daughter half-life to ventilation time, have been estimated for typical home conditions.

The daughter atoms of radon readily diffuse to the walls and other surfaces in a room and stick there. Once attached to a wall, the daughters can decay relatively harmlessly. However, if a daughter atom becomes attached to a particle, its chance of sticking onto a wall will be reduced by a factor of 10 to 100. The more dust and smoke that is suspended in the air, the greater the likelihood is that radon daughters will attach to particles. Smoke, in turn, leads to an increase in the airborne daughter concentrations and the potential radiation exposure. In a smoke-filled room, one is likely to experience a greater dose of radon-derived radioactivity. Smoke particles provide an ideal vehicle for carrying and depositing radon progeny (and other dangerous substances) deep in the lungs.

Radon Risks

As already noted, the average concentration of radon in a home is about 1 picocurie per liter (pCi/liter) of radioactivity. This means that in an average home, roughly two radon-decay events occur every minute in each liter of air. In outdoor air, the radon concentration is about one-fifth this level, or 0.2 pCi/liter. A common unit that is used to measure exposure to radon is the **working level** (WL). One working level is equivalent to the overall exposure to radioactivity that occurs when the radon concentration is 100 picocuries per liter of air (1 WL = 100 pCi/liter). The working-level exposure includes *all* the potential radiation from radon and its progeny, which is dominated by the alpha emissions of polonium. The working level, however, is defined for an ideal situation without room ventilation or wall effects acting to remove radon daughters. The actual exposure to radionuclides in air is limited by ventilation and plate-out on walls. The effect of such removal can be taken into account by doubling the radon concentration that is needed to produce one working level of exposure. That is, 1 WL = 200 pCi/liter in practical situations.

The working level can be estimated if the radon concentration is known, by applying the following relation:

$$WL \approx \frac{q_{Rn}}{200} \qquad (8.11)$$

If the concentrations of the radon daughters in air are measured directly, they can be used to calculate the working level through the approximate formula

$$\text{WL} \approx \frac{q_{\text{Po}}}{1000} + \frac{q_{\text{Pb}}}{200} + \frac{q_{\text{Bi}}}{300} \qquad (8.12)$$

Since the daughter concentrations may be estimated from the radon concentration using Equations 8.8 through 8.10, Equation 8.12 is actually roughly equivalent to Equation 8.11.

The basic health risk associated with radon exposure is lung cancer. As a measure of risk, the probability of dying of lung cancer after 70 years of exposure to radon is usually stated for various fixed levels of radon. Long-term exposure to radon dramatically increases the chances of developing lung cancer. In a sample of 1000 nonsmoking people living a "normal" life, an average of 10 to 20 people eventually develop fatal lung cancer. This is the "background" lung cancer rate, which has diverse causes, including exposure to air pollution. For people exposed to normal concentrations of radon gas (about 1 pCi/liter), the average additional risk of developing lung cancer is estimated to be in the range of 3 to 13 per 1000. Exposure to radon associated with living in an average home may explain part of the background occurrence rate of lung cancer.

The level of exposure to radon that is considered to be "safe" is 4 pCi/liter, or 0.02 WL. The Environmental Protection Agency defined this safe level based on an assumed lifetime exposure over 70 years in an average home. As many as 8 million homes in the United States may have radon concentrations higher than 4 pCi/liter. The danger of radon is extrapolated from the effects of radon on miners exposed to high concentrations in their work.

The increased risk of lung cancer due to exposure to higher-than-average levels of radon can be compared with the increased chance of lung cancer for smokers. This comparison is shown in Figure 8.3. For example, a person living in a home with radon levels that are 100 times the U.S. average (that is, 100 pCi/liter) has a risk of death due to lung cancer that is nearly the same as a two-pack-a-day smoker. That is, 60 to 210 people per 1000 will likely die from lung cancer at these exposure levels. This is roughly 10 to 15 times the risk for a nonsmoker or for someone exposed to average levels of radon. It is estimated that there are a total of between 5000 and 20,000 lung-cancer deaths in the United States each year due to radon. This makes it the second leading cause of lung-cancer deaths in this country, behind smoking (the total lung-cancer death rate in the United States is more than 150,000 per year). In fact, the combination of smoking and radon exposure is a particularly deadly toxic cocktail (Section 8.4.2). Nearly three-quarters of the radon-associated deaths occur among smokers.

In the uranium and other mining industries, it is well known that miners run an increased risk of contracting lung cancer. The risk is so pronounced, in fact, that in one region of Europe, about 30 percent of all the uranium miners die of lung cancer. Their average level of radon was once equivalent to roughly 1000 picocuries per liter. That astronomical exposure is not even on the charts in Figure 8.3. The death rates in the figure refer to continual exposure over 70 years, a lifetime. At exposure levels of 1000 pCi/liter, life expectancy is much less than 70 years. The threat of mine tailings and the radon they generate was considered so serious that a special law was passed to deal with the problem—the Uranium Mill Tailings Radiation Control Act of 1978. Reassuring to be sure. By some estimates, in the state of Colorado alone there were about 10 million metric tons of tailings to deal with, contaminating nearly 10,000 individual properties. The projected cost of the cleanup in Colorado was $350 million.

Some people do not believe that radon is a serious threat. In fact, a particularly interesting character, an engineer named Galen Winsor, has gone to great lengths to prove this point. Having once worked for General Electric at one of its nuclear facilities, Winsor claimed to have swum in nuclear wastewater, drunk the contaminated water, and eaten powdered uranium oxide. According to Winsor, "If you are as radioactive as I was when I was swimming in the nuclear waste water and eating uranium oxide then the government says you should be classified as nuclear waste and buried 3000 feet underground." Apparently, General Electric believed that Winsor should be avoided, if not buried, and fired him. Winsor's fate is unknown to me, but I wish him luck—he will need it.

The average lifetime doses of radiation that a person receives from radon and from other sources are compared in Table 8.2.[3] These doses have been

3. The doses in this table are given in terms of the equivalent whole-body exposure to energetic radiation that produces the same risk of cancer as the actual pattern of exposure associated with each source. Hence, for radon, which produces radiation exposure that is localized mainly in the lungs, the equivalent whole-body doses are considerably smaller than the localized doses to lung tissue. In the case of radon, exposure to daughters by all pathways for average indoor radon concentrations has been assumed. The actual calculation of the average dose in rem requires considerable information regarding the energies of the emitted alpha, beta, and gamma rays, the probability that the daughters reside in the body at the time of decay, where they reside, the relative biological sensitivity of different tissues, the

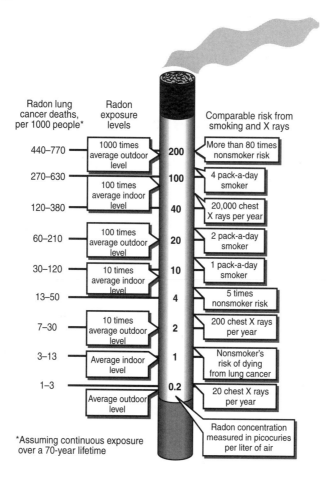

Figure 8.3 The effect of exposure to radon in terms of the incidence of lung cancer. The radon exposure is expressed in picocuries of radon gas per liter (1000 cm³) of air. The estimated lung cancer morbidity per 1000 people for a lifetime exposure (70 years) is indicated on the left-hand scale. The comparative risks of contracting lung cancer from other activities (noted in the boxes) are given for a wide range of radon exposure levels. (Data from the Environmental Protection Agency)

calculated in a way that defines the *relative* risk of cancer from each source. Considering the known mechanisms of exposure to energetic radiation, radon accounts for the largest risk. The other modes of exposure are ubiquitous and therefore unavoidable (discounting medical sources, which are voluntary but usually necessary). Because exposure to radon can be highly localized, the actual doses received by those at risk may be much greater than the average doses indicated in Table 8.2.

variation of exposure with time of day and season, and a number of other factors. Accordingly, such a calculation goes far beyond the scope of the present discussion. Nevertheless, a rough conversion factor of 0.2 whole-body rem/WL for a 70-year exposure has been estimated.

Indeed, as noted earlier, the radon risk is skewed toward those people living in specific geographical regions where radon emissions are high and low ventilation rates in homes are dictated by high energy costs.

From consideration of these data and the documented consequences of occupational exposure to radon, the following conclusions may be drawn: First, exposure to radon on a regular basis and at abnormally high concentrations poses a lifetime risk of cancer that is unacceptably large. This risk is exaggerated in the case of regular smokers of tobacco. Second, radon is generally an avoidable risk. Potential exposure to radon can be readily measured, and actions to reduce exposure are usually

Table 8.2   Sources of radiation exposure and average effective doses

Source	Effective Dose[a]
Cosmic radiation (from space)	~2
Terrestrial gamma radiation (from soils)	~2
Ingested radionuclides: Radon progeny	~4
All others	~3
Nuclear fallout	<0.4
Medical diagnostic X rays	3
Nuclear power plants	~0.4

[a]Units are in rem, over a 70-year period; that is, the number of rem of exposure received by an average person over a full lifetime. A **rem** is a "roentgen equivalent man," the amount of radiation that causes the same biological effect as does 1 roentgen of X rays (see Section 7.3.2 for further details).
Source: Data from P.J. Walsh and W.M. Lowder, "Radon," in *Indoor Air Quality*, ed. P.J. Walsh, C.S. Dudney, and E.D. Copenhaver (Boca Raton, Fla.: CRC Press, 1984), Table 1, p. 145.

straightforward. Even though exposure to energetic radiation from other possible sources may be unavoidable, radon exposure can be controlled to a large degree.

8.3 Formaldehyde

The construction industry in recent years has become fond of using aggregated pressboard and plywood as a primary building material. These construction materials are cost effective because they are fabricated from relatively cheap, low-quality lumber and scrap wood. To add strength to these materials, the sheets and fragments must be tightly glued together. That is where the problem arises. Glues are formulated with organic solvents that evaporate as the glue dries. The solvents end up in the air. If the wood panel is in the wall of your house, those solvents may seep into your living space. The same solvents are used to manufacture excellent heat-insulating materials, and the solvents may escape after the insulating materials are installed. Formaldehyde is probably the most widely used volatile organic component in these building materials. In total amount, it ranked twenty-fourth among all the chemicals produced in the 1980s,

with roughly half used to produce resins for adhesives and related applications.

Formaldehyde offers a classic example of a product with obvious problems ignored by the industries proffering it. Formaldehyde is cheap and easy to obtain and at one time had no environmental baggage to carry. So, sell it and ignore complaints from customers about headaches and nausea; that's the business way. Later, when the nasty facts are made public, deny any knowledge and plead ignorance; that's the business way. When the problem has to be corrected, establish new companies to provide the cleanup services; that's also the business way. The whole formaldehyde mess has been a wonderful business opportunity.

8.3.1 Embalmers' Fluid

Formaldehyde is one of the simplest natural organic compounds, being composed of one carbon atom, two hydrogen atoms, and one oxygen atom (HCHO).[4] It reacts with materials composed of proteins, protecting them from attack by bacteria

---

4. The chemical formula for formaldehyde is usually written HCHO, to indicate that the carbon atom is centrally located, with the hydrogen and oxygen atoms bound only to the carbon.

Table 8.3   Formaldehyde Emissions

Product	Range of emission rates (10^{-6} g per g material per day)
Particleboard	0.4–8.1
Plywood	0.03–9.2
Imitation wood paneling	0.8–2.1
Fiberglass insulation	0.3–2.3
Clothing	0.2–4.9
Carpeting	0–0.06
Paper products	0.03–0.4

Source: Data from R. B. Gammage and K. C. Gupta, "Formaldehyde," in *Indoor Air Quality,* ed. P. J. Walsh, C. S. Dudney, and E. D. Copenhaver (Boca Raton, Fla.: CRC Press, 1984), Table 2, p. 116.

and mold and thus preserving the treated material. This has led to applications of formaldehyde in tanning leather and as a preservative in cosmetics. For obvious reasons, formaldehyde makes an excellent embalming fluid.[5]   Likewise, formaldehyde is highly toxic if ingested or inhaled. A common form of formaldehyde—used in the embalming trade, for example—is formalin, a solution of 37 percent formaldehyde in water. Pure formaldehyde is a gas under normal conditions. Although colorless, it has a strong pickle-like odor that you may recall from biology laboratory in high school when you cut up a frog or starfish.

Wherefore Hides Formaldehyde?

Although formaldehyde is obviously in the air at the local funeral home, where else might it be found? Unfortunately, it is present, in high concentrations, in many homes and other businesses. Common construction materials that emit formaldehyde are urea–formaldehyde foam insulation (UFFI),[6] plywood, and particleboard. Formaldehyde is also given off by certain products containing fiberglass, carpeting, clothing, and some paper products. Examples of measured emission rates of formaldehyde from various products are given in Table 8.3. The emission rates cover a wide range. In a given situation, the actual rates depend on the temperature of the material, whether it is painted or otherwise coated, its exposure to circulating air, the age of the product, and other factors.

To pose a health risk, the emissions of formaldehyde must escape into our living spaces. Formaldehyde trapped in a wall is not immediately dangerous. But if the gas leaks through cracks and openings into our living room, it may become a threat. Even insulation behind walls can emit formaldehyde into a room. Leakage occurs particularly at joints between the walls and floors, doorways and windows, and from electrical outlets and switches. Although a structure might at first glance seem to be airtight, it

5. Embalming, as a means of preserving corpses for transport, viewing, or other purposes, is an art first developed by the Egyptians. Leonardo da Vinci developed an early technique for injecting preservatives into corpses (he personally dissected more than four dozen cadavers in his quest to record the intricate form of the human body). Modern embalming came into vogue in 1775 when Englishman John Hunter placed on display the fully embalmed (and clothed and coifed) Mrs. Martin Van Butchell; her will had stipulated that her husband Martin would control her estate only as long as her body remained unburied, a condition Martin was happy to fulfill by placing her well-preserved corpse on public display. Today, embalming is commonplace in the United States, where the business has been boosted by lobbyists seeking special treatment under the law, but not in Europe.

6. Urea is the primary end product of the metabolism of proteins in mammals and some fish. Accordingly, urea is a principal component of animal urine. The urea is derived from the nitrogen groups in amino acids, which is metabolized to ammonia that the liver then detoxifies by converting to urea. Because urea is rich in nitrogen, it is a common fertilizing agent and an important component of the nitrogen biogeochemical cycle (Section 10.2.2). Pure urea is also used as the starting material for the manufacture of many plastics and drugs. With formaldehyde, it forms a stable fertilizer, and urea–formaldehyde resin, from which adhesives, insulation, and a variety of other products such as buttons and tableware can be manufactured.

is actually likely to be as leaky as a sieve. After all, that's how cockroaches get around.

Tobacco smoke is an important source of formaldehyde, which is generated in the burning ash. Cigarette smoke can contain 40 parts per million of formaldehyde, and a single cigarette generates about 1 to 2 milligrams. Formaldehyde concentrations in rooms with smokers can reach several tenths of a part per million. Such concentrations might cause alarm among those residents if they weren't otherwise engulfed in a cloud of toxins that are much worse (Section 8.4.1).

Formaldehyde Abundances

Concentrations of formaldehyde in homes can vary widely with the season, the construction of the home, and the life-style of the residents (Figure 8.4). When the heating system is turned on early in the winter, it can significantly heat up the walls, floors, and ceilings, causing greater emissions of formaldehyde from the insulation, flooring, and paneling. The fact that the windows are closed at the same time is important, since the ventilation is reduced accordingly. Later, during the coldest part of the winter, the outer walls become quite cold,

and formaldehyde emissions can actually decrease (Figure 8.4). When spring arrives and things begin to warm up, but not enough to open up the house, the formaldehyde rises again. In the summer, formaldehyde emissions tend to increase, particularly during hot, humid weather. Nevertheless, the concentration of formaldehyde indoors may be quite low if the windows are kept open (increasing the ventilation). Sometimes the concentrations of formaldehyde increase when the ventilation is increased, indicating that each case of indoor air pollution may be quite complex in origin and require unique solutions.

The concentrations of formaldehyde encountered in a number of common situations, including industrial settings, are summarized in Table 8.4. The health standards for exposure are also given in the table. Note that ambient (outdoor) air has formaldehyde concentrations of about 0.0004 to 0.08 ppmv (0.4 to 80 ppbv). The higher abundances represent conditions affected by anthropogenic sources of the kind under discussion here, and by urban smog. In smog, many types of hydrocarbons are photochemically decomposed and in the process generate formaldehyde at concentrations reaching 10 to 80 parts per billion (Section 6.4.1). Nevertheless, even in the

Figure 8.4 Measurements of formaldehyde inside a home are shown over the course of a year. The concentrations in parts per million by volume (ppmv) are given for each month of the year. This particular home, located in Tennessee, has been insulated with a urea–formaldehyde foam. Heating the insulation causes the release of formaldehyde, whereas cooling inhibits its release. When the home is well ventilated with outside air, formaldehyde concentrations decrease compared with the levels when ventilation is reduced to retain heat. (Data from R. B. Gammage and K. C. Gupta, "Formaldehyde," in *Indoor Air Quality,* ed. P. J. Walsh, C. S. Dudney, and E. D. Copenhaver [Boca Raton, Fla.: CRC Press, 1984], adapted from Figure 8, p. 123)

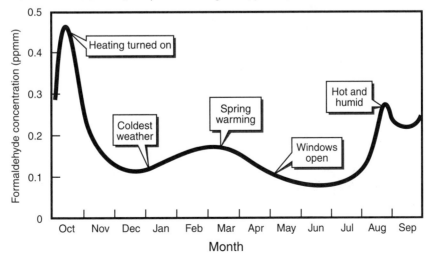

Table 8.4   Formaldehyde Concentrations

Location	Measured concentration range (ppmv)
Ambient air	0.0004–0.08 (0.01 average)
Non-UFFI[a] home	0.01–0.08 (0.03 average)
UFFI home	0.01–3.4 (0.12 average)
Mobile home	0.01–2.9 (0.38 average)
Textile plant	<0.1–1.4
Fertilizer plant	0.2–1.9
Bronze foundry	0.12–8.0
Iron foundry	<0.02–18.3
Plywood industry	1.0–2.5
Hospital autopsy room	2.2–7.9
Standards for exposure	
Indoor air, United States	0.1 (maximum)
Indoor air, Sweden, old structures	0.7 (maximum)
Occupational air, NIOSH[b]	1.0 (30 min. maximum)
Occupation air, OSHA[c]	3.0 (8 hour average)
Occupation air, OSHA	5.0 (maximum)

[a]Urea-Formaldehyde-Foam Insulation.
[b]National Institute for Occupational Safety and Health (United States).
[c]Occupation Safety and Heath Administration (United States).
Source: Data from R.B. Gammage and K.C. Gupta, "Formaldehyde," in *Indoor Air Quality,* ed. P.J. Walsh, C.S. Dudney, and E.D. Copenhaver (Boca Raton, Fla.: CRC Press, 1984), Table 4, p. 119, Table 5, p. 120, Table 10, p. 134.

most polluted air, formaldehyde concentrations are well below the values measured indoors (Table 8.4).

Methane decomposition also contributes to the background concentrations of formaldehyde in rural areas and in otherwise pristine air. Indeed, on a global scale, methane oxidation annually generates several hundred times more formaldehyde than is manufactured worldwide each year. So, why is there an anthropogenic formaldehyde problem? First, formaldehyde generated from methane is distributed throughout the atmosphere; hence, it is already greatly diluted. Second, formaldehyde in the free atmosphere has a very short lifetime, only several hours; thus, it cannot accumulate to any substantial degree. Taking these facts into account explains the very low abundance of formaldehyde—less than a part per billion on average—over the planet.

In homes without urea–formaldehyde foam insulation, formaldehyde concentrations are typically close to outdoor levels. All homes have some sources of fugitive formaldehyde, however, and the average concentrations indoors are three times greater than those outdoors. In UFFI homes, by comparison, the average concentrations are more than a factor of 10 above ambient and are much higher than in homes free of UFFI. In mobile homes, the average is three times greater again than in UFFI homes. In addition, 20 percent of mobile homes tested showed formaldehyde concentrations greater than 0.5 ppmv. In some occupational settings, the average formaldehyde concentrations are 10 times greater than in mobile homes (Table 8.4). It has been estimated that 1.6 million workers are exposed to formaldehyde on the job, with nearly 60,000 exposed for at least 4

Table 8.5   Health Effects of Formaldehyde

Concentration (ppmv)	Exposure time (minutes)	Health effects
0.01	5	Eye irritation
0.05	1	Odor threshold
0.08	1	Cerebral cortex affected
0.2	60	Eye, nose, and throat irritation
0.8	10	Brain alpha-wave rhythm and autonomous nervous system changes
4.0	1	Unbearable without respiratory protection

Source: Data from R. B. Gammage and K. C. Gupta, "Formaldehyde," in Indoor Air Quality, ed. P. J. Walsh, C. S. Dudney, and E. D. Copenhaver (Boca Raton, Fla.: CRC Press, 1984), Table 8, p. 130.

hours each working day. In utopsy rooms, where the embalmed are dissected, the air is laced with formaldehyde. These concentrations may be higher than in a mortuary where formalin is injected into arteries and body cavities to replace natural fluids, presumably with minimum leakage.

A significant fraction of formaldehyde detected in houses has been associated with urea–formaldehyde insulation (UFFI). This material was in widespread use in the 1970s and 1980s. Homes built before 1970 are probably free of UFFI (unless they later had urethane foam blown into the walls). Many consumer complaints—some 1500 or so—concerning irritants from this particular insulation had been lodged with the Consumer Product Safety Commission (CPSC) by 1980. Accordingly, the CPSC banned the use of UFFI in home construction in 1982. But the courts overturned that decision in 1983 on the grounds that the epidemiological evidence from animal studies was not convincing. Nonetheless, UFFI had by that time gotten such a bad reputation that no one wanted to get near it. Several states also acted to ban UFFI. The result has been a virtual cessation of UFFI use in construction. The stuff remains, however, in more than a half-million homes in the United States and Canada.

With the fall of UFFI, other wood products have moved to the forefront as formaldehyde emitters. These sources include particleboard, paneling, plywood in flooring, cabinets and panels, and furniture. The problem is most severe in mobile homes, or trailers. These living spaces tend to be rather cramped, providing minimum air space for dilution. Moreover, trailers typically utilize much more paneling and particleboard products in their structure. The combination of tight space and copious emissions results in high concentrations of formaldehyde.

8.3.2   FORMALDEHYDE'S IMPACTS ON HEALTH

Because formaldehyde is quite soluble in water, up to 95 percent of the vapor that is inhaled can be absorbed in the upper respiratory tract (the nose, mouth, and throat). With nasal inhalation, most of the absorbed formaldehyde is trapped in the nasal tract. When the formaldehyde has been absorbed into an aerosol particle, it can become more "respirable," reaching the lungs. Some of the primary responses to formaldehyde exposure are given in Table 8.5. As might be expected, among the first symptoms of formaldehyde exposure are nose, throat, and eye irritation (Section 7.2.2). Formaldehyde may also produce coughing, nausea, and skin rashes in sensitive people. Indeed, in hypersensitive people, symptoms appear at concentrations well below 0.2 ppmv, as indicated in Table 8.5.

Exposure to formaldehyde in occupational and residential settings has led to numerous accounts of symptoms that, although not characterized well enough to serve as scientific evidence, offer additional insights into its health and psychosomatic impacts. The reported ills include burning or itching eyes and nose, throat irritation, nausea, vomiting, stomach cramps, diarrhea, headache, dizziness, drowsiness, sneezing, asthma, recurring respiratory infections, loss of concentration, disturbed sleep, loss of sense of smell, thirstiness, menstrual irregularities, complications in pregnancy, and low-birth-weight babies. It seems likely that not all these problems are associated with formaldehyde. After all, everyone gets nauseous or drowsy every now and then. What is undetermined in such compilations is how many of the accounts of illness are related to hysteria triggered by the revelation of exposure. Nevertheless, it would be foolhardy to assume that none of these accounts is related to formaldehyde exposure. Similar effects are seen in controlled laboratory experiments.

Formaldehyde can react with amino acids, proteins, and DNA, thereby damaging cells. Although not yet proven to be a direct carcinogen in humans, formaldehyde has been shown to cause nasal cancers in rats at high concentrations (up to 14 ppmv). Rats exposed to formaldehyde and hydrogen chloride vapors had nearly a 50 percent cancer rate after 2 years; formaldehyde and HCl are known to react to form a potent carcinogen. Other evidence identifies formaldehyde as a likely co-carcinogen, enhancing the action of certain well-known carcinogens such as diethylnitrosamine (nitrosamines, which can be generated by cooking meats preserved with nitrites, have been implicated as carcinogens in humans). In some industries, such as fertilizer and textile production and metal processing, workers can be regularly exposed to high concentrations of formaldehyde (at parts-per-million levels). There is evidence from epidemiological studies that these workers have a greater-than-normal occurrence of cancers of the mouth, throat, and nasal cavity, as well as excess disease of the lungs and liver. Moreover, occupational exposure to related aldehydes such as acetaldehyde is implicated in human cancers of the upper respiratory tract. Based on available data, some experts have concluded that formaldehyde is a direct carcinogen in animals and has a carcinogenic potential for humans.

8.4 Tobacco Smoke

Of all the air pollutants that pose a direct health risk to humans, tobacco smoke is the most dangerous and widespread. Tobacco smoke, which is generated by the behavior of people habituated to nicotine, is an insidious pollutant because it can rapidly disperse throughout an enclosed space and expose all of those present. It is not as if smoking (cigarettes, cigars, and pipes) is so pleasant or rewarding. Tobacco smoking is probably the single most deadly addiction afflicting the human species, in terms of the disease and suffering it causes. Most sensible people acknowledge the debilitating effects of heavy smoking; it is difficult to ignore the panting, hacking, gaunt, shriveled chain-smokers around us. Yet, if it is such a destructive addiction, why do so many people continue to smoke, why are so many young people seemingly attracted to smoking, and why hasn't this dangerous but controllable material been outlawed? This section discusses the characteristics and effects of tobacco smoking and some of the issues relevant to the tobacco debate.

How We Got Hooked

In the late nineteenth century, smoking cigarettes and cigars was not a popular habit. The common tobacco plant, *Nicotina tabacum*, is native to South America and Mexico.[7] The indigent peoples of those countries cultivated and used the leaves in the belief that they had medicinal properties. Indeed, the active ingredient in tobacco, **nicotine**, is a strong stimulant and quite addictive.[8] In the sixteenth century, after "discovering" the new world, Europeans adopted tobacco for its "medicinal" effects produced mainly by the nicotine. The worldwide development of tobacco as an agricultural crop soon followed.

It was not until early in the twentieth century that tobacco smoking took off. The automatic cigarette-making machine had been invented in 1880. However, the high acid content of raw tobacco smoke had made it very uncomfortable to inhale the fumes of

7. *Nicotina tabacum* is named for the French ambassador to Portugal, Jean Nicot, who introduced the seeds of the plant to the French aristocracy around 1550.

8. Nicotine comprises roughly 2 to 5 percent of cured tobacco by weight. Some special breeds contain 10 percent or more. Nicotine is a potent addictive drug that has the common odor of tobacco. Laboratory studies suggest that nicotine is at least as addictive as heroin. Inhaled in small amounts, nicotine is a stimulant, and in large amounts it has a tranquilizing effect that appears to induce relaxation. In pure form, nicotine is highly poisonous, inducing vomiting, headaches, convulsions, and paralysis. In fact, nicotine makes a good natural insecticide. Upon exposure to nitric acid, nicotine can be converted to niacin, a useful food supplement.

cigarettes deeply. Tobacco manufacturers, seeing the possibilities for cigarette sales and wishing to satisfy the desires of its customers, developed new techniques for growing and processing tobacco that reduced the acid content, thereby allowing the deep drags that some consider so pleasurable. The two great world wars also advanced the cause of smoking. Recognizing an opportunity to express their patriotism, major tobacco manufacturers began to supply young troops with free cigarettes. At the same time, social barriers keeping women from smoking were broken down. As a result, the legions of lifetime smokers ballooned in the postwar era. By the late 1960s, passengers on airplane flights were treated to a free minipack of four cigarettes and a courteous light from an attendant. There were no nonsmoking sections anywhere at that time, and everyone who traveled by plane might have to inhale dense tobacco smoke for hours. The rugged Marlboro Man, who smoked like a fiend, was everywhere, looking cool (he later died prematurely of lung cancer). Another generation of impressionable youngsters was hooked. With profits soaring, the tobacco industry had no incentive to avoid habituating new generations to smoking, particularly cigarettes.

The aggressive marketing of cigarettes and other tobacco products brought along with it an epidemic of lung cancer and other serious health problems. Health agencies have estimated that up to *several hundred thousand* people currently die each year of diseases related to, or exacerbated by, smoking. That is about six times greater than the number of people that die in automobile accidents, for example. To counter such data and to build a better public image, the tobacco industry has fought back through its industry-supported Tobacco Institute to deny any links between smoking and deadly health effects like lung cancer. Nevertheless, in 1967 the U.S. surgeon general declared cigarettes to be a public health hazard, and in 1972 all tobacco advertisements on television and radio were banned. These actions were based on sound scientific observations concerning the epidemiology of lung cancer, heart disease, and other maladies connected with tobacco smoking. In 1987, smoking was eliminated on domestic airline flights of 2 hours or longer and, in 1989, was banned on every domestic flight in the United States. In two decades, we went from free cigarettes and a courteous light to no cigarettes at all on airliners.

Despite all the bad press, the numbers of smokers worldwide is growing. Cigarettes continue to be sold to children, and particularly to people struggling in developing countries. The addiction to nicotine remains a legal addiction that provides some temporary elevation from the humdrum and tedious, although the long-term personal costs are high. Moreover, there are several ways to become addicted. Cigarettes are the most popular means. But cigar and pipe smoking are also widespread and even more obnoxious to innocent bystanders. Shredded tobacco, specially prepared, can also be sniffed (as snuff) or chewed to extract the nicotine. Snuff was once the tobacco form of choice of aristocrats for obtaining their nicotine fix. Chewing is particularly unsavory, producing copious brown expectorate, not to mention a high incidence of cancer of the mouth.

The addictive nature of tobacco and its nicotine can be seen from statistics published in 1994 by the U.S. Centers for Disease Control and Prevention. According to their survey, only about 8 percent of tobacco smokers who attempt to quit are successful. Of roughly 50 million smokers in the United States, about one-third are trying to kick the habit. The approaches include nicotine patches (which provide the drug without the smoke), hypnosis, and simply "going cold turkey." However, nicotine is so addictive that permanent release from its effects requires unusual willpower. Most of those who quit soon take up the habit again. Local campaigns to discourage people from smoking in the first place have contributed to a decline in tobacco use in the United States. In California, an effort at public education, together with a new 25-cent cigarette tax, reduced the number of smokers from 26.5 percent in 1987 to 22.2 percent in 1990. The Centers for Disease Control have found that the percentage of American adults who smoke has decreased from about 42 percent in 1965 to about 25 percent in the early 1990s and has held steady at that level. Still, the fact that 25 percent or so of the mature population is hooked for life on nicotine, and thus exposed to the other dangerous chemicals in tobacco smoke, is scandalous at a time of unprecedented national awareness of and concern over health. If a quarter of all adults were addicted to cocaine or heroin, for example, society would quickly find a way to end the epidemic, by arresting the producers and dealers.[9]

9. The U.S. Food and Drug Administration has considered classifying nicotine as a drug and subjecting it to national regulations. That could lead to controls, or the outright banning, of tobacco sales to the public. The nature of nicotine addiction is still

Table 8.6   Some components of cigarette smoke

Component	Emission (mg/cigarette) Mainstream smoke	Emission (mg/cigarette) Sidestream smoke
Tar	10.2–20.8[a]	34.5–44.1
Nicotine	0.46–0.92[a]	1.27–1.69
Carbon monoxide	18.3	86.3
Ammonia	0.16	7.4
Hydrogen cyanide	0.24	0.16
Acetone	0.58	1.45
Phenols	0.23	0.60
Formaldehyde	—	1.44
Toluene	0.11	0.60
Acrolein	0.084	0.825
NO_x	0.014	0.051
Polonium-210 (pCi)	0.07	0.13
Fluoranthenes	7.7×10^{-4}	1.6×10^{-3}
Benzofluorenes	2.5×10^{-4}	1.0×10^{-3}
Pyrenes	2.7×10^{-4}	1.5×10^{-3}
Chrysene	1.9×10^{-4}	1.2×10^{-3}
Cadmium	1.3×10^{-4}	4.5×10^{-4}
Perylenes	4.8×10^{-5}	1.4×10^{-4}
Dibenzanthracenes	4.2×10^{-5}	1.4×10^{-4}
Anthanthrene	2.2×10^{-5}	3.9×10^{-5}

[a]Range is for filtered to unfiltered cigarettes.
Source: Data from S. A. Glantz, "Health Effects of Ambient Tobacco Smoke," in *Indoor Air Quality*, ed. P. J. Walsh, C. S. Dudney, and E. D. Copenhaver (Boca Raton, Fla.: CRC Press, 1984), Table 1, p. 160.

8.4.1 COMPOSITION OF TOBACCO SMOKE

Table 8.6 presents laboratory measurements of the various chemical substances found in smoke extracted from cigarettes. The composition is typical for tobacco smokes in general. The smoke compounds are striking for their diversity, abundance, and toxicity. Tobacco smoke has several dominant components, including the tars, nicotine, and carbon monoxide, as well as carbon dioxide and water. The "tars" comprise the black sticky substances (tarlike) in

being debated. Although it has been compared with heroin and cocaine, nicotine produces less euphoria and is apparently a less destructive addictive drug. The tobacco industry is the only special-interest group that continues to deny that nicotine is addictive at all. The fact that the nicotine content of cigarettes is manipulated by cigarette manufacturers suggests to some that tobacco companies have been exploiting its addictive quality for years.

smoke residues that darken cigarette filters and human lungs. These tars are laced with carcinogens. Nicotine, one of the major components of tobacco smoke, is now acknowledged to be an addictive drug like cocaine or morphine. Carbon monoxide is a poisonous gas and an important contributor to air pollution. Carbon dioxide, although dangerous in very high concentrations, inducing asphyxiation, is one of the less harmful components of tobacco smoke.

Notice in Table 8.6 the differences between **mainstream smoke** and **sidestream smoke**. In cigarettes, cigars, and pipes, mainstream smoke is drawn through the unburned tobacco and the filter, if one is present. The tobacco and filter capture some of the smoke components, cleaning up the inhaled smoke somewhat. Later, the materials trapped on the tobacco may be released as the burning tip approaches. When air is drawn through it, tobacco burns fast and hot. Sidestream smoke, on the other hand, is produced as the tobacco burns more slowly and coolly. This smoldering of the tobacco at lower temperatures creates a different initial mixture of compounds in the smoke. In general, the sidestream smoke contains more toxic pollutants than the mainstream smoke. Moreover, the sidestream smoke is not filtered, and hence its composition is not substantially altered as it escapes from the ash.

Mainstream smoke is sucked directly into the lungs at full concentration, whereas sidestream smoke is usually significantly diluted before being inhaled. Dilution is important in protecting nonsmokers from exposure to tobacco smoke. In certain situations, however, the concentrations of sidestream smoke (combined with exhaled secondhand smoke) can accumulate to high levels in poorly ventilated rooms or when many smokers are present.

Toxic Gases and Particles

Tobacco smoke consists of a mixture of gases and particles. The gases consist of the by-products of tobacco combustion in the burning tip intermixed with components of the air. The particles consist mainly of the tars, nicotine, and a variety of organic compounds that have low volatility and therefore condense on existing particles in the smoke stream. Tobacco smoke particles are usually extremely small in size, and all the particles are considered to be respirable. (See Section 6.5 and Figure 6.16 for a general discussion of particles suspended in air.) The ash produced by smoking (up to 25 percent of the dry tobacco mass) is mainly a bothersome waste product of smoking that must constantly be brushed off clothes.

The data in Table 8.6 reveal that tobacco smoke contains a large number of dangerous chemical substances. Only a small fraction of the total number of compounds found in tobacco smoke are listed here (see also Chapter 7 for additional information on specific compounds).

In addition to carbon monoxide, the dangerous gaseous species include hydrogen chloride (which forms hydrochloric acid when mixed in water), acetone (a cleaning agent), acrolein (an ingredient of smog), and formaldehyde (Section 8.3). Other noxious gases such as ammonia, acetone, and nitrogen oxides are contained in the smoke in significant concentrations. Some radioactive polonium-210 also is present. It is one of the radon-222 decay products, which is not shown in Figure 8.1. Polonium-210 is a very short-lived radionuclide (lifetime of about 3×10^{-7} seconds, or 0.3 microseconds) that decays into lead-206.

Many of the carcinogenic compounds in smoke are contained in the smoke particles. These compounds, which consist of large organic molecules built up from cyclic benzene rings (Section 7.1.1), include benzopyrene, a known indirect carcinogen.[10] Indirect carcinogens form carcinogenic byproducts when they are broken down inside living cells; that is, they are precursors of the chemicals that actually do the biological damage. Among the other highly carcinogenic compounds in tobacco smoke are the dibenzanthracenes. Fluoranthenes, benzofluorenes, and many other compounds in Table 8.6 are possible carcinogens, and cadmium has been identified as a carcinogen. They all are unhealthful substances to breathe in at any concentration.

A Stage-XXX Alert

The concentrations of carbon monoxide in mainstream cigarette smoke can be estimated using the

10. Benzopyrene and other carcinogenic compounds in tobacco smoke are also found in smoke from wood fires. Indeed, wood smoke contains even more potent carcinogens, such as dimethylbenzanthracene and methylcholanthene. Not to throw a wet blanket on your next romantic sojourn by the hearth, but try to avoid breathing too much smoke. In winter, wood burning is popular in many towns where under stagnant meteorological conditions, the air can become fouled with toxic carcinogen-laden smoke.

measured quantities of CO generated by drawing on a single cigarette (Table 8.6), the number of long puffs per cigarette (about 20), and the typical lung capacity of an adult (about 1 liter, or 1000 cubic centimeters). These assumptions yield CO concentrations in the mainstream smoke as high as 1000 parts per million by volume (ppmv). This is an extraordinary concentration, exceeding by a factor of 10 the CO concentrations in a stage-3 smog alert (100 ppmv for a 1-hour average) (Section 6.2.2). In other words, inhaling cigarette smoke is like breathing *stage-30* smog! Continuous exposure to such levels of carbon monoxide would asphyxiate the average person in a matter of hours (Section 7.2).

Formaldehyde was discussed at length in Section 8.3. There we noted that tobacco smoke is a significant source, with cigarette smoke containing as much as 40 ppmv of formaldehyde. One can readily estimate the concentrations of formaldehyde that might be caused by smokers in a closed space. According to Table 8.6, a single burning cigarette emits 1 to 2 milligrams of formaldehyde. If this were mixed into an average room with a volume of 30 cubic meters, the formaldehyde concentration would amount to 0.03 to 0.06 ppmv. This is well above background concentrations, even in a polluted environment. Moreover, with several smokers puffing or chain-smoking, the concentrations would be substantially larger. In one experiment, five cigarettes smoked in a room of 30 m^3 produced a measured formaldehyde abundance of 0.23 ppmv—exactly in line with our estimate. These formaldehyde concentrations rival those in mobile homes without smokers.

Now consider the particle concentrations that may be created by a single cigarette. The total amount of particulate smoke in the mainstream and sidestream components in Table 8.6 adds up to more than 60 milligrams per cigarette. In a room of 30 m^3, the average aerosol concentration produced by a single cigarette could be as large as 2000 micrograms per cubic meter ($\mu g/m^3$)! The clean air standard for respirable particulate matter is 50 to 150 $\mu g/m^3$ on a 24-hour average (Section 6.2.2 and Table 6.2). Owing to the ventilation of typical rooms and the adhesion of particles on surfaces, smoke concentrations are not likely to get that high. Remember the box model and the role of ventilation (also see Sections 7.4.3 and 8.2.2). However, with a group of puffers chain-smoking in a closed room, particle concentrations could reach harmful levels (recalling smoky evenings at the kitchen poker table).

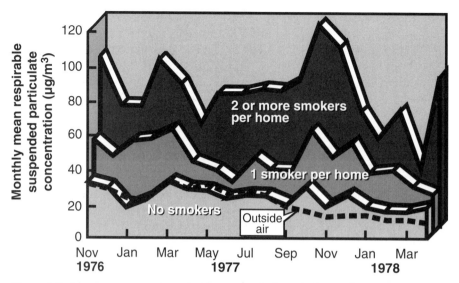

Figure 8.5 Measurements of respirable suspended particulate (RSP) mass concentrations in various social settings as a function of the density of smokers, specified in terms of lit cigarettes per 100 cubic meters of air space. The RSP values for nonsmoking areas are in the lower-left-hand corner. The levels of potential health hazard, as determined from the national ambient air quality standards, are indicated. (Data from J. E. Spengler, D. W. Dockery, W. A. Turner, J. M. Wolfson, and B. J. Ferris Jr., "Long-Term Measurements of Respirable Sulfates and Particles Inside and Outside Homes," *Atmospheric Environment* 15 [1981]: 15)

Figure 8.5 illustrates measurements of respirable suspended particle (RSP) concentrations measured in various indoor situations. First note, in the lower left-hand corner of the diagram, that RSP concentrations observed in rooms without any smokers all lie well within the present air quality standards. Furthermore, there is a well-defined relationship between the density of lit cigarettes (number per 100 cubic meters) and the resulting density of smoke. As the density of cigarettes increases, the particulate load in the air also increases. The data in Figure 8.5 correspond to 38 different locations. For a particular site, a nearly linear relationship would be expected between the densities of cigarettes and smoke, assuming no change in the ventilation rate. It is clear that in many normal settings, smoking can make the air unhealthful and probably uncomfortable to breathe, especially for a nonsmoker. Moreover, logic dictates that in confined, crowded spaces, if smoking is allowed, unacceptably high RSP concentrations are likely.

One recent study found that cigarette smoke released outside buildings or ventilated from buildings contributes to the particulates in smog. In Los Angeles, tobacco smoke accounts for roughly 1 percent of the respirable suspended particles. Although this is a relatively small fraction at present, fugitive smoke particles have a greater impact on visibility (that is, tobacco smoke particles are optimal for light scattering and visibility degradation). As air quality regulations begin to focus on policies to improve visibility in urban areas, the small contribution of tobacco smoke to the fine-particle burden may increase in importance.

Secondhand Smoke

One of the major issues concerning smoking is the **secondhand smoke** it creates. It's one thing for a person to make an individual decision to poison his or her body with the foulest kind of substance. It's quite a different matter for that person to inflict the same poisons on any number of people who may happen to be in the same room, office, restaurant, airplane, or other confined space (Figure 8.5). Even worse, it is unconscionable, and perhaps immoral, for heavy smokers to subject their children, spouse, or other relatives to the unrelenting onslaught of toxic pollution at close quarters.

Secondhand smoke has two major sources. First is the sidestream smoke that escapes from the burning tobacco ash before it can be inhaled. The sidestream smoke is produced at a lower temperature than smoke created by inhaling, and its composition is different from that of "mainstream" smoke (as noted earlier). In general, the sidestream smoke actually has higher concentrations of carbon monoxide, ammonia, and other pollutants. The second source of secondhand smoke is the mucky air that smokers exhale while they are satisfying their addiction. Some of the cigarette (or pipe or cigar) fumes inhaled by smokers are deposited in their noses, throats, and lungs and are eventually swallowed or spat. Hence, the smoke they exhale is cleaner. But even so, the exhaled residual smoke is still laced with carcinogens and toxins. Anyone nearby—a friend or relative or neighbor, or a stranger—may be exposed to these unhealthful secondary fumes.

8.4.2 Tobacco Smoke's Effects on Health

Tobacco smoke is pervasive in our everyday environment. Although we are more susceptible outdoors to exposure to smog, tobacco smoke is a serious health threat indoors. Lit cigarettes may be present at the office, in restaurants, in waiting rooms, or in one's own castle. Accordingly, it is difficult to assess the overall health effects and cost of tobacco smoking. Costs directly accrue to the people who smoke—through degraded health, shorter life-span, and lowered quality of life, as well as the cost of the tobacco itself, which is no longer inexpensive. Costs also indirectly accrue to society at large—through secondary health effects and social expenses related to absenteeism and illness among the legions of smokers. Consider the self-debasement of smokers. It is well known, for example, that they have a dulled sense of taste and enjoy food less; their sense of smell also becomes fouled with tars. Hacking, panting, and shortness of breath are their daily aggravations. Cigarette breath is repulsive to most nonsmokers. And think of all those holes burned in clothing, carpeting, and upholstery by cigarette butts and ashes.

Smoking insults the body in many other ways. The nicotine in tobacco smoke causes blood vessels and capillaries to constrict. This restricts blood flow to the dermis and other tissues. The skin, deprived of a rich oxygen supply, shrivels over time. One of the most recognizable characteristics of long-time heavy smokers is the prunelike mummified skin shrouding their face, neck, hands, and arms. In addition to

being unattractive, premature aging of the skin implies internal damage to this important organ. But serious illness and death are much worse fates.

The Indiscriminate Killer

According to the U.S. Centers for Disease Control and Prevention, in 1990 tobacco smoking was responsible for 130,000 deaths due to lung cancer (most of the deaths from this cause), 180,000 deaths due to cardiovascular disease, and 80,000 deaths from emphysema and other obstructive pulmonary diseases. These are the big three—lung cancer, heart disease, and emphysema. All together, more than 300,000 deaths in one year can be attributed to tobacco smoke, either in whole or in part.[11] Some of the extraordinarily diverse human ailments that have been connected with tobacco smoking (and chewing) are summarized in Table 8.7. These diseases and related health problems are not exclusively associated with tobacco. But there is mounting scientific evidence that tobacco smoke is the primary causative agent for many of the most serious of these problems and is strongly implicated in the others.

Prolonged exposure to tobacco smoke dramatically increases one's chances of developing lung cancer. The obvious link is the carcinogenic tars and vapors inhaled deep into the lungs. As explained in Section 8.4.1, many of the components of tobacco smoke are known carcinogens and mutagens (Table 8.6). Benzopyrene, a particularly nasty carcinogen, has been the subject of extensive epidemiological studies. Given the concentrations of benzopyrene measured in cigarette smoke, it has been estimated that a one-pack-a-day smoker has an excess (additional) risk of contracting lung cancer of 2000 in 1 million per year, or a sure thing in 50 years. For those who smoke two packs or more a day, the excess risk increases to about 3600 per million per year. These are very high risk factors and reflect only the risk of lung cancer. The risks of

---

11. In 1994, Richard Joshua Reynolds III, grandson of the founder of the R. J. Reynolds Tobacco Company, the second-largest in the United States, died at the age of 60 of emphysema and congestive heart failure. Ironically, he smoked the family brands, Camels and Winstons, heavily. Patrick Reynolds, his half-brother, is a staunch antismoking activist. He was influenced by the deaths of his father, Richard Joshua Reynolds, Jr., son of the founder, who died of emphysema when Patrick was only 15 years old, his mother, also a lifetime smoker, who died of emphysema and heart disease, and an aunt who died of lung cancer and emphysema. Patrick's goal is to "...create a society that is totally smoke-free."

Table 8.7   Disorders Related to Tobacco

Organs	*Diseases*
Mouth, nose, and throat	Cancer of the mouth and tongue Cancer of sinus and larynx Loss of sensitivity to taste and smell
Pulmonary tract	Lung cancer, emphysema Coughing and wheezing common Asthma attacks aggravated
Gastrointestinal tract	Cancer of the stomach, colon, rectum, and pancreas
Cardiovascular system	Ischemic, or coronary, heart disease, pulmonary heart disease, congestive heart failure, and strokes Restricted blood supply to internal organs Formation of aneurysms and embolisms
Nervous system	Strokes Accelerated onset of senility
Skeletal and connective tissue	Enhanced osteoporosis Premature aging and wrinkling of the skin
Urinary tract	Prostate cancer

further debilitating health effects are even greater. (See Section 7.4, where risk is discussed.)

Heart disease is widespread among tobacco smokers. Heart disease may prove to be, by a large margin, the greatest tobacco killer of all. Normally, our heart provides a lifetime of trouble-free service; it is an amazingly resilient organ. As we noted earlier, smoking tobacco leads to constriction of the blood vessels and capillaries, which has two major effects. First, the supply of oxygen and nutrients to tissues is reduced. The tissues and organs involved can then atrophy. Second, the heart is forced to work harder to pump blood to all the extremities of the body. At the same time, the heart itself may be starved for blood. The result is more rapid wear and earlier heart failure. This latter effect is referred to as *ischemic* heart disease (caused by inadequate blood supply). Coronary heart disease is the specific case of ischemic heart disease involving the myocardium, or main heart muscle. This major contracting muscle is the pump that keeps blood coursing through our bodies.

Pulmonary heart disease involves the system of arteries and veins servicing the lungs. Clearly, the supply of blood to the lungs—where oxygen is picked up and carbon dioxide is dropped off—must be maintained. The right side of the heart sends blood to the lungs. On this side, the pressure of the bloodstream is low, and the amount of work the heart must do is relatively small. Chronic bronchitis or emphysema can, over time, destroy blood vessels in the lungs, which creates an increasing burden on the right side of the heart, eventually causing heart failure. Smoking is a major risk factor in pulmonary heart disease.

A poor blood supply can also aggravate "hardening" of the arteries, arteriosclerosis, which is often associated with high blood-cholesterol levels. Atherosclerosis is a particular form of this disease that entails fatty deposits on the interiors of medium and larger arteries. Smoking and high fat intake may, indeed, be synergistic risk factors in arteriosclerosis. The degeneration of the major arteries and veins can create aneurysms, or ballooning of the arteries, and embolisms, or clots in arteries and veins. When a blood vessel supplying the brain is blocked by an embolism, or an aneurysm in the brain bursts, a stroke can result. The severity of a stroke depends on the size of the artery, its location, and the extent of hemorrhaging. Atherosclerosis is a common contributor to strokes. The brain is particularly sensitive to the loss of blood supply (ischemia) because of its enormous metabolic demand for energy. Again, smoking is a major risk factor.

Less widely advertised are the other cancers associated with tobacco. Cancer of the mouth and tongue has left many tobacco chewers disfigured for life, if they were lucky enough to survive the disease. Cancers of the gastrointestinal tract also are common, since the tobacco smoke carcinogens deposited in the mouth, nose, and throat are often swallowed. Recent epidemiological studies, for example, have definitely linked smoking with an increased risk of cancer of the colon. These studies are particularly disturbing because they suggest that smoking as a youth—particularly in one's twenties—produces an excess risk of colon-rectal cancer that persists throughout life, even if one later quits smoking.

The Epidemic of Secondhand Illness

What impact does an environment filled with smokers have on nonsmokers? In the United States in 1978, more than 600 billion cigarettes were smoked by about 50 million people. These numbers have not changed much over several decades. That's a lot of cigarettes puffed by a lot of people in a lot of places; indeed, there is almost nowhere to hide from tobacco smoke, which may explain the rising militancy of antismokers and advocates for children's rights.

In a landmark report published in 1993, the Environmental Protection Agency classified secondhand smoke as a human carcinogen, placing it for the first time in the same category as other Group A carcinogens like benzene, asbestos, and radon. The EPA found that secondhand smoke is responsible for approximately 3000 lung cancer deaths annually in the United States. Moreover, each year children under 18 months of age exposed to passive smoke suffer 150,000 to 300,000 respiratory infections, including pneumonia and bronchitis (requiring up to 15,000 hospital visits), and 200,000 aggravated cases of asthma. Another possible effect is the initiation of asthma in otherwise healthy children. Other data show a threefold increase in the occurrence of sudden infant death syndrome in the offspring of mothers who smoke during and after pregnancy. According to the EPA study, at least 20 percent of

all lung cancer in nonsmokers is related to secondhand smoke, amounting to an average lifetime risk of 1 per 1000; this increases to 3 per 1000 for a nonsmoking spouse living with a smoker. Other scientific studies suggest that heart disease associated with exposure to secondhand smoke kills more than 30,000 people annually, although this conclusion is more controversial.

In homes where there are no smokers, the indoor air has about the same concentration of respirable suspended particulate (RSP) as outside air (Figure 8.6). However, in two-smoker homes, the RSP level is three times that in the nonsmoking homes (Figure 8.6). The RSP level also is frequently above the standard for exposure in two-smoker homes. It follows that infants and children and other relatives living in such a home are continually immersed in unhealthful air. There is no seasonal letup either, as there usually is with formaldehyde exposure. Also note that the high RSP concentrations imply that the abundance of the gaseous toxins in the tobacco smoke will be similarly elevated. Obviously ventilation, or an air-cleaning system with a high-efficiency fine-aerosol filter, would help reduce the particulate loading of the indoor air.

Facing the reality of the dangers inherent in exposure to secondhand smoke, public agencies have been appealing to smokers with children to avoid puffing at home. Likewise, guidelines have been drawn up to limit exposure to secondhand smoke at work, in restaurants, and other public places. Nonsmoking pregnant woman, some of whom have been found with as much nicotine in their bodies as if they had regularly smoked five cigarettes a day, have been asked to avoid smokers at all costs, even smoking mates. In Jacksonville, Florida, a judge removed a 7-year-old boy with asthma from his home because his stepfather smoked heavily. Placing the boy in the custody of his father, the judge claimed that he had the legal duty to protect the child's health. In London, a 36-year-old woman became the first British citizen to win compensation for the health effects of secondhand smoke. She sued her employer, arguing that co-workers who smoked had aggravated her bronchitis, and received a $23,000 out-of-court settlement.

No-smoking laws are proliferating to protect the public from secondhand smoke. In California, a statewide antismoking law became effective in 1995. The law applies in all cities except those with *tougher* laws already on the books. Most workplaces and

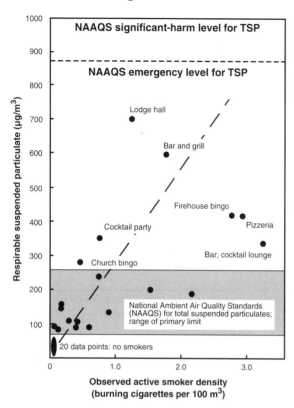

Figure 8.6 Respirable suspended particulates (RSP) measured in homes. RSP is specified as the mass concentration of particles in air that are small enough to reach and be deposited in the bronchial tubes and lungs when inhaled. RSP concentrations are given as micrograms of particle mass per cubic meter of air. The data shown were collected at 80 homes located in six U.S. cities during a period of roughly 18 months. The particulate concentrations, given as monthly average values, are categorized for air from households with no smokers, or one, or two or more resident smokers. Concurrent measurements of the particulate concentration in ambient air outside the home are compared to demonstrate that the higher indoor concentrations of RSP are not associated with external sources of pollution in these cases.

restaurants are designated as nonsmoking sites. Exemptions apply to bars, gambling clubs, truck cabs, hotel lobbies, and small businesses with five or fewer employees, all of whom agree to allow smoking. This broader law follows an earlier regulation banning smoking in more than 20,000 state-owned or leased buildings, including prisons. The California cigarette tax has also provided up to $80 million per year for public education concerning tobacco smoke. Nationwide, antismoking ordinances are catching on everywhere, and hundreds of individual laws have

been passed or are planned. Campaigns are under way to enforce minimum-age laws for buying cigarettes and to outlaw cigarette-vending machines, to ban smoking on all international airline flights, to limit tobacco advertising, and to renew programs to discourage teenage smokers. Are the tobacco companies worried? Regarding the recent gains by antismoking forces, the director of the Tobacco Institute commented that they "tend to get pretty puffed up over small accomplishments." And then there is the counterattack.

8.4.3 SMOKE AND MIRRORS

The tobacco industry has invested large sums of money to convince people that smoking is safe. Stilted arguments may convince a few nonsmokers that tobacco use is OK, but the propaganda primarily reinforces the rationalizations of addicted smokers that their habit is not so harmful. The tobacco industry also employs a team of lobbyists and lawyers to push favorable laws that would encourage access to tobacco and postpone limitations on smoking in public. The profits of the tobacco firms often end up in the coffers of legislators, who must ultimately regulate the industry. More tobacco money is contributed in California than almost anywhere else—in 1992, an average of more than $10,000 per elected official in Sacramento; one legislator received about $220,000! Is this just the normal exercise of civic duty, or are favors expected in return? Several tobacco firms actually sued the EPA in 1993 over its finding that secondhand smoke is a human carcinogen. They disingenuously claimed that the EPA had been selective in choosing the data employed in its assessment, whereas the EPA had rightly rejected faulty or irrelevant data. The Tobacco Institute, which frequently touts half-baked or anomalous evidence or misinterprets research results to support its position on smoking, is no stranger to such a ploy. Lobbying, lawsuits, propaganda—that is all good business practice, of course, but is also irresponsibly contrary to the public interest.

In 1994, tobacco interests sponsored a California ballot proposition that would have created a statewide law to control smoking in restaurants. Sounds good so far. Unfortunately, the measure would have invalidated more than 250 existing local ordinances that effectively ban smoking in eating and working areas. These strong regulations would have been replaced by a much weaker law that would have actually allowed smoking in such places. The measure, according to one analysis, was simply "a fraud upon the public." How brazen to propose such a measure. Are we dealing with self-delusion or unqualified greed? In a related effort, the tobacco industry invested more than $200,000 in a campaign to reverse a restaurant smoking ban in the city of Los Angeles. Tobacco interests heavily supported the "Hospitality Coalition," which advocated smoking opportunities for all people. Although 83 companies and individuals were listed as donors to the coalition, in fact the tobacco companies contributed all but about $4000 to the campaign.

It is ironic to read tobacco-sponsored ads debunking the latest scientific data linking smoking to one disease or another, always tagged with the required disclaimer, faithfully placed as unobtrusively as possible, stating, "Warning: The Surgeon General Has Determined That Cigarette Smoking Is Dangerous to Your Health."

Desperate Denial

The major tobacco companies have denied any duplicity in spreading the nicotine habit. They are, according to their point of view, merely servicing a harmless human pastime. Indeed, executives of many tobacco companies have persistently denied that tobacco smoke is addictive, even in the face of overwhelming evidence that it is. Is this simply a difference of opinion about nicotine or a purposeful effort to deceive the public?

Tobacco money supports research laboratories in which experiments are conducted to demonstrate the innocence of smoking and in which results to the contrary are likely to be suppressed. For years, tobacco company executives denied that the addictive nature of nicotine was known to them, even when company memoranda acknowledged that fact. The tobacco industry has also been accused of carrying on a research program to breed high-nicotine species of tobacco. In one case, a superior tobacco plant was actually developed—dubbed Y-1—with nearly twice the nicotine content of common tobacco. According to tobacco industry spokesmen, this high-nicotine leaf was needed when low-tar cigarettes were developed. That is, because a reduction in tars led to a reduction in nicotine, it became necessary to blend in a higher-nicotine tobacco to maintain "taste." This high-nicotine tobacco is apparently being blended into some present-day brands of cigarettes, supposedly to maintain their "taste" and thereby satisfy the desires of customers.

Claims that nicotine is harmless and non-addictive might be blamed on ignorance. Or it might be psychological denial. Dealers in hard drugs don't tell customers that the stuff is likely to destroy their lives. Street dealers selling heroin to teenagers may simply be responding to the demand of their customers. As far as the dealers know, the stuff might be harmless and not addictive. Would you buy that story?

Banning Smoking in Public

Is it reasonable or possible to ban completely the public use of tobacco? The prohibition of alcohol was attempted early in the nineteenth century but failed to secure wide public support or to overcome the relentless onslaught of opportunists and profiteers. Perhaps no other personal behavior is so widely despised in the United States as tobacco smoking. The health risks are viewed as so serious that non-smokers have become militant in limiting the opportunity to smoke in public. Smokers have fallen into a pattern of denial that risks exist. They are reinforced in this belief by tobacco companies and other special interests who fabricate arguments and fund studies to debunk well-documented threats. Airlines no longer allow smoking and emphasize this by warning people about tampering with smoke detectors in lavatories. Many cities and towns have restricted or banned smoking in public buildings. These defensive tactics are spreading. States have even taken up the cause to clear the air of tobacco smoke. Yet there is no national, let alone international, law to ban smoking where other people might be affected. An outright ban on all smoking, or even on the production of tobacco and its products, seems far off. Nevertheless, the recognition of tobacco smoking as a public menace is a step in the right direction, but only a step.

Making tobacco illegal is probably impractical. It would be much too expensive to enforce such a ban. Smugglers would surely respond to demand. More logically, laws could be passed limiting the access of minors to tobacco products; eliminating smoking in all public buildings, meeting halls, and conveyances; and taxing tobacco at a rate that reflected its detrimental impacts on society. The *direct* public and private health costs associated with smoking have been estimated as $50 billion in the United States alone, or about $2 per pack of cigarettes. Adding to this the other costs related to secondhand smoke and to lost time from jobs and reduced productivity caused by tobacco addiction, the total cost of tobacco to society could perhaps double to $4 per pack. The present federal tobacco tax is roughly a quarter a pack for cigarettes. A fair tax reflecting the actual costs of smoking would probably discourage many people from experimenting with this drug and would encourage many others to shake the habit. Furthermore, youngsters would find it less attractive to pay five or six bucks a pack to gag.

Education is important in limiting the devastation of tobacco. Foremost would be a direct assault on smoking through public health education, as is done with AIDS and other sexually transmitted diseases. However, sharp rebuttals would also be needed to balance misinformation and phony research results promulgated to prove the harmless nature of tobacco smoke. Likewise, glitzy advertising campaigns aimed at potential smokers would have to be offset by public announcements on the dangers of smoking. The forces of profit and greed are at work ceaselessly to maximize addiction to tobacco. These forces can, and must, be repelled.

8.5   Other Indoor Pollutants

A number of other sources of indoor pollution are identified in Table 8.1. A complete discussion of these pollutants is beyond the scope of this book, but we will discuss particular problems involving biological agents and water contamination.

8.5.1   Biogenic Pollutants

Life is everywhere. It is tenacious and achieves a foothold wherever the opportunity arises. In a house, there are countless niches in which organisms can thrive—in the kitchen, the bathroom, potted plants, and pets (Section 8.1). These organisms take the form of bacteria, viruses, fungi, mites, and insects. The detritus of life is also plentiful, including pollen, spores, body fragments, and feces. People, for example, constantly shed skin flakes, some 7 million each minute. Each flake carries an average of 4 bacteria. Mites, ubiquitous in bedding and carpets, graze on the flakes. The bits of dead skin swirl through rooms as dust, carrying organisms to far reaches, much like Noah's ark.

Exposure to biogenic pollutants can cause responses in several ways. First, the organism itself can be pathologic and infectious, particularly certain bacteria and viruses. They may cause, for example, a cold or influenza. Second, the biological material may be an allergen. In such cases, a person develops an antibody to a specific active biological agent, or antigen, when first exposed to that agent. In subsequent exposures, an antibody response is triggered by the immune system. In sensitive people, this response is so exaggerated for certain agents that the body literally attacks itself in attempting to suppress the invading agent (like shooting yourself in the foot trying to scare off a burglar).

Symptoms of exposure to biogenic agents are diverse. Many common allergic reactions produce irritation of the eyes, nose, and throat. Hay-fever-like symptoms can be a response to pollens and fungi. Secondary bacterial infection may follow. Bronchial asthma, with coughing and shortness of breath, may be triggered or aggravated by exposure to allergens in indoor air. Another serious disorder is hypersensitivity pneumonitis, in which the alveoli (air sacs) deep in the lungs are inflamed and may be destroyed. Continuing exposure to the allergens can cause lung scarring and respiratory failure. The symptoms of the allergic reaction are much like those experienced with influenza: fever, chills, and muscle pain, as well as coughing and oxygen deprivation. The "Monday complaints" represent a pattern of allergenic responses that recur when one returns to an environment (such as the workplace) after days of absence. The symptoms can be much like those in hypersensitivity pneumonitis or the 24-hour flu. By the end of the work week, the symptoms may have subsided, but the sensitivity remains.

Pollen is not usually a problem associated with indoor plants. Cultivated flora do not generally generate the very high concentrations of fine pollen grains typical of wind-pollinated plants. For example, hay fever is caused by outdoor exposure to pollen, or exposure to pollen-laden outdoor air that is admitted into the home. Pollen can be excluded to some degree by restricting ventilation or filtering the air.

Arthropods include insects and arachnids, the latter of which include spiders and mites. Such organisms are common in households (even if we don't want them as roommates). The dust mite, as already explained, is a common cohabitant. Its fecal pellets, about the size of a large pollen grain, can be highly allergenic. These critters require high humidity to survive—at least 75 percent relative humidity. Hence, they tend to thrive during the wet seasons or in moist niches in the house. Among the insects, cockroaches deserve special notice. They are so common it is difficult to avoid exposure to their detritus. Antigens carried by organisms consumed by roaches are also found in their fecal droppings. The roach detritus can enter the air as dust.

Overall, biological agents are not as threatening as toxic chemicals in the interior environment. Nevertheless, allergies and asthma, among other disorders, can be extremely debilitating, affecting the lives of millions of people.

8.5.2 INDOOR WATER POLLUTION

Table 8.8 lists some of the pollutants that have been found in drinking water. The major sources for these pollutants in water and their health effects are also described. Radon was discussed in Section 8.2. Water, particularly when it is drawn from wells, can create a radon hazard. Groundwater is in close contact with the minerals that produce radon. Unless subterranean water has an opportunity to aerate, or release its radon gas, before reaching the tap, radon is emitted as the water is used. Lead contamination was reviewed in Chapter 7 (see, for example, Section 7.2.4). Lead in drinking water is particularly a problem in older homes where lead may still be present in the plumbing (there is presently a national ban on lead pipes). Nitrates are dissolved in groundwater and thus are found in drinking water. High nitrate levels (above the EPA standard of 10 ppmm) are found in roughly 600,000 homes, but they are only rarely a health problem, mainly involving infants in whom the nitrate affects the ability of hemoglobin to carry oxygen.

Many of the organic pollutants in drinking water originate from leaks and spills onto soils. These chemicals generally seep slowly through the soil, often taking years or decades to reach the groundwater. Once mixed into groundwater, the contaminants flow through the aquifer and are drawn up with water tapped from the aquifer. These nasty organic toxins may have been stored in tanks that eventually leaked. This problem has plagued the gasoline business for years, with numerous filling stations contaminated by leaks from underground fuel storage tanks. Seepage from toxic-waste sites that have been improperly sealed and from dump sites that were

never designed to handle persistent toxins in the first place but were used for that purpose nevertheless, is another scandalous problem. Accidental chemical spills also occur from time to time. The material may seep into groundwater or freshwater reservoirs before it can be cleaned up, or the spill may never be reported. Purposeful dumping of hazardous materials is, moreover, not uncommon. The high costs of carefully and safely disposing of toxic compounds at special sites makes criminal dumping more attractive. Illegal dumping on land, in storm drains, and in streams and lakes is not unusual. Too often, these fugitive poisons enter homes through the water supply.

Waterborne biogenic agents are not common in treated community water supplies. Unfortunately, as is well known by many travelers to underdeveloped countries, water supplies are not always properly treated for bacteria, protozoa, and other living organisms. Moreover, fresh water in wilderness areas may be contaminated with giardia lamblia. This nasty little protozoan parasite attaches to the intestine and induces giardiasis, characterized by stomach distension, pain, and diarrhea. Sound familiar?

8.6 Indoor Versus Outdoor Pollution

It has now been demonstrated that indoor pollution can be a serious problem. One question that arises is how serious indoor pollution is compared with outdoor pollution? Are there circumstances in which it would be healthier to be indoors than outdoors? Can we avoid smog by staying inside our home or office? When should we throw open the windows and let the "fresh" air in? If we try to conserve energy, will we also end up poisoning ourselves on indoor toxins? In general, these are complex questions without simple answers. A brief discussion follows.

It is noteworthy that indoor pollution has not received as much attention as outdoor pollution. Smog, for example, is such an obvious and widespread problem affecting millions of people in a major city that it would be hard to avoid dealing with it. However, indoor pollution is more of a personal issue. Each case, each home, is different. There are fewer generalities, and cures must be individually designed. The larger societal questions are not as obvious. There is also less information about indoor air quality and toxins. Sophisticated air pollution–monitoring networks that are common in urban areas are not available for

Table 8.8   Contaminants Found in Drinking Water[a]

Contaminant	Description	Sources	Health impacts	Group at risk
Radon	Radioactive gas	Groundwater	Lung cancer	Anyone
Lead	Inorganic chemical Heavy metal	Soft or acidic water in lead pipes, copper pipes connected by lead solder, or brass faucets	Developmental and learning disabilities Low birth weight	Children Fetuses
Nitrate	Inorganic chemical	Wells in agricultural areas	Methemoglobinemia, a blood disorder	Infants under 6 months of age
Secondary pollutants				
Pesticides	Organic chemicals	Runoff and seepage in agricultural areas	In high doses, liver, kidney, or nervous-system damage and possibly cancer	Anyone
Trichloroethylene	Organic chemical	Industrial effluents Hazardous-waste sites	In high doses, nervous-system damage and possibly cancer	Anyone
Trihalomethanes	Organic chemicals	Chlorination of surface water	Possibly cancer	Anyone
Bacteria, viruses, *Giardia*	Microorganisms	Insufficiently disin-fected or filtered water	Intestinal and other diseases	Anyone

[a]Information from *Consumer Reports*, January 1990 pp. 30–32.

homes. Data concerning indoor pollution and epidemiology are sparse, often anecdotal, and generally incomplete. Unlike the situation for outdoor air pollutants, standards for exposure to indoor contaminants are not as well defined or do not exist. It is little wonder that indoor pollution is a public health issue that remains to be focused on, through comprehensive scientific assessments and regulations that can be uniformly applied.

8.6.1   IS IT SAFE TO GO INDOORS?

Figure 8.7 compares measured concentrations of several pollutants in indoor and outdoor air samples (taken at the same time and locale). The samples were taken under different conditions of ventilation and air treatment and show a wide range of air quality. For a normal house without refrigerated air conditioning and with ventilation provided through windows and doors, the average indoor concentrations of many pollutants, such as particulates and NO_x, are similar to those outdoors. Oddly, total oxidants (mainly ozone) can be reduced to about 60 percent of the outdoor values, because reactive species like ozone are particularly sensitive to removal on surfaces with which they come into contact. Air passing through cracks in windows and doors comes into contact with those surfaces and, inside a house, with indoor surfaces as well. Reactive species in the

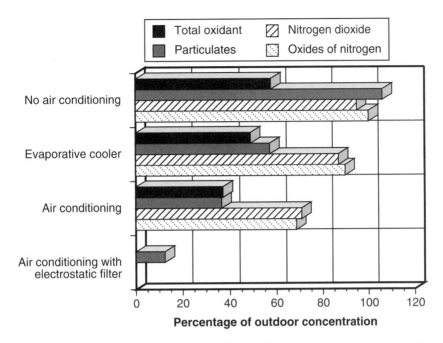

Figure 8.7 Relative levels of several types of indoor air pollutants (for example, total oxidants, airborne particulates) for four specific configurations of air treatment in buildings: no air conditioning and significant ventilation using external polluted air; use of an evaporative cooler, which reduces the need for ventilation; full air conditioning with a further reduction in required ventilation; and air conditioning with recirculated air treated with an electrostatic filter. For those buildings monitored in urban zones, the less often the internal air was exchanged with outside air, the lower the levels of indoor pollutants were. (Data from J. L. Repace and A. H. Lowrey, "Indoor Air Pollution, Tobacco Smoke, and Public Health," *Science* 208 [1980]: 464)

air, such as ozone, are depleted by the surfaces. Oxides of nitrogen, on the other hand, are not removed as efficiently on surfaces and so remain close to outdoor levels.

When an evaporative cooler is used and the house is still ventilated, conditions improve somewhat over those in a closed house. But less ventilation in this case produces a greater contrast between the pollutants inside and outside. The use of a refrigerated air conditioning system (one that does not have to exchange air between the inside and outside but instead transfers heat from the house through cooling coils containing a refrigerant) consistently reduces the level of indoor air pollutants. The effect is most marked if the air is also electrostatically filtered[12] as it is cooled and recirculated.

12. Electrostatic filters are generally more efficient than passive filters. Passive filters capture particles in the air forced through the filter by impaction of the particles on the surfaces of the fibers that make up the filter. The efficiency of the filter, in terms of the smallest size particle that can be captured, is determined by the size of the fibers and the pores between the fibers. In electrostatic filters, the particles are exposed to ions of one electrical charge (negative, say) that stick to and charge the

One factor affecting indoor pollution that is not clearly defined in Figure 8.7 is the rate at which air is exchanged between the indoors and outdoors in each situation. When going from the top to the bottom in Figure 8.7, it should be clear that the rate of exchange decreases significantly. Air that is recirculated indoors becomes progressively cleaner with time because as the residence time of the air is lengthened, the cleaning action of surfaces is more effective. Recirculated air may also be filtered each time it passes through the system. The more interior air that is recirculated, the less "fresh" air that needs to be drawn into the house. The cleansing effect of recirculation, in fact, is limited mainly by the interior sources of pollution. For oxidants like ozone, there are no significant internal sources, and its concentrations can be brought nearly to zero by limiting outside ventilation. For other pollutants like

particles. These electrically charged particles can then be strongly attracted to an electrode or filter of opposite electrical charge (positive in this case). This filter therefore does not rely on impaction alone, but also uses Coulomb electrostatic attraction to draw particles out of the air.

formaldehyde and radon, however, the interior sources are substantial. The concentrations of these pollutants therefore increase as outside ventilation decreases. It follows that internal and external sources of pollution can be balanced through the use of ventilation and filtration to minimize the overall exposure to toxic pollutants.

The data in Figure 8.7 offer another valuable lesson. During severe smog events, one can avoid exposure to highest pollution concentrations by remaining indoors with the windows closed and the air conditioner on. Under these circumstances, most smog pollutant concentrations can be considerably reduced. This is particularly true for ozone (or oxidants), nitrogen oxides, carbon monoxide, and a number of organic compounds characteristic of smog (Section 6.4.1). But it would be a mistake to ignore the interior sources of many of these pollutants. Carbon monoxide is generated by cigarette smoking, gas appliances, fireplaces, and automobiles (that may run for a time in the garage, providing a source of indoor pollution through doorways) (also see Section 7.2.1). Nitrogen oxides are also generated by cigarettes and certain heating or cooking appliances. Considering organic compounds, formaldehyde is an obvious problem, as noted earlier in this chapter.

There also are emissions of various volatile organic compounds (VOC) from household chemicals, plastics, paints, insecticides, and so on (Section 7.2.3). More than 250 VOCs have been measured in indoor air at concentrations exceeding 1 part per billion (not all have been detected at the same time, of course!). Under specific circumstances, you might choose not to stay in the house because of high levels of VOCs—for example, if the interior is being painted, if wood floors are being refinished with polyurethane, or if the house is being fumigated for pests.

Although the number of individual toxic compounds that have been detected in home and work environments is quite impressive, it is unusual for their concentrations to be so high as to pose an immediate health threat. For the most part, it is indeed safe to go indoors. Nonetheless, one must be constantly vigilant and take reasonable measures to improve the quality of the interior environment, where we spend most of our time. We must be particularly sensitive to the exposure of young children to indoor pollution, as they are the most sensitive to toxins and the least able to take corrective action.

8.6.2 Making Indoors Safe

Based on the previous discussions, one can take a number of obvious steps to improve the state of indoor air quality. First and foremost, the sources of indoor air pollution must be controlled. One can easily check on the products used in home improvements, repairs, and retrofitting. Ask the contractor or manufacturer to provide details, and contact the local or state environmental protection agency for an evaluation. Known emitters of pollutants—lead-based paints, urea–formaldehyde foams and adhesives, asbestos flooring or ceiling materials, and so on may be sealed using the proper precautions. Often, sealing is safer than removal, particularly unprofessional attempts that produce more airborne particles and fibers than if the material was left alone. It is also relatively easy to ventilate areas affected by radon seepage by ventilating basements and crawl spaces. If there are smokers in your house, banish them to the porch or backyard to smoke. They may not like the idea, but you will breathe a lot easier. Other simple measures to reduce exposure to indoor toxins and allergens include ceasing to burn incense, washing the family pet more often, and having fewer wood fires in the evening.

The second easy step to reduce indoor pollution is to improve the ventilation throughout your living space. In cooking areas, for example, exhaust fans can be installed. All major appliances should be vented to the outside (logically away from open windows). Exhaust fans in areas that are musty or dusty can also work. Furthermore, dehumidifiers are usually effective in reducing biogenic pollutants.

The third major step in indoor air quality control involves air filtration. Heating and cooling systems are equipped with air filters, which should be checked and changed regularly. High-efficiency filters are now available that can be used to reduce airborne debris in these systems. It might be worthwhile to clean the air ducts as well or to have a professional clean them. When vacuuming, highly efficient dust bags can be used to reduce the recirculated particles. Opening windows while vacuuming can also reduce temporarily high dust levels. Finally, portable or permanently installed air filtration systems can help collect particles from the air and lead to much cleaner spaces indoors. Again, these filters must be properly maintained to be effective.

Energy Efficiency and Health

There is a fundamental conflict between the two important goals of reducing exposure to toxic chemicals in our everyday lives and increasing energy efficiency to preserve resources and forestall global climate change (Chapter 12). In order to mitigate exposure to indoor pollutants, it is often recommended that ventilation be enhanced. This is the easiest remedy when interior sources of toxins cannot be easily removed or deactivated. On the other hand, a higher rate of ventilation causes air conditioning and heating systems to work harder. In both cases, these systems are attempting to modify outdoor conditions. If we let the outdoors in, more energy must be expended to make the necessary modifications. Air conditioning and heating account for a large fraction of all energy usage in developed countries. Accordingly, any decrease in efficiency is counterproductive to global energy conservation.

Contrary to this, major efforts are under way to insulate and seal homes and to persuade businesses to cut down on their energy consumption. Retrofitting and weatherproofing are encouraged by government agencies. Such efforts are logical in a world facing the threat of global environmental change, including global climate warming and stratospheric ozone depletion. In fact, large-scale problems that arise from energy consumption and other human activities are the subject of the third part of this book. Our external environment is affected in this case. The symptoms and impacts are very different. Instead of wheezing, we face sweltering; instead of radon emanating from the ground, we face ultraviolet radiation penetrating from the sky.

These interlocking problems require integrated solutions. As yet, we have not achieved the wisdom to solve such complex issues involving so many entrenched special interests and costing so much to implement. People must set priorities for themselves. Given individual circumstances, is existing indoor pollution more important than projected global climate change? Although expediency calls for priority actions, one should never lose sight of the bigger picture and the longer haul.

Questions

1. Ordinary cooking can produce a number of unhealthful air pollutants, including carbon monoxide, organic vapors, and smoke. Consider how you might decrease your possible exposure to such pollutants when using a gas stove to prepare a meal. What about a barbecue? Is a microwave oven as potentially hazardous as a gas stove in this regard? Try to think of all the ways you might reduce your exposure to pollutants from cooking. Can you imagine any changes in your general eating habits that might be useful in lowering your exposure to pollutants?

2. Describe the conditions in a home and at a home site that could lead to high levels of and exposure to radon gas and its radioactive by-products. Discuss the measures that might be taken to correct each condition. Name several occupations that are likely to be associated with significant exposure to radon.

3. You share an apartment at college with a smoking roommate, who puffs two packs a day at home. How might you minimize your exposure to secondhand cigarette smoke until you graduate? Consider all the possible actions you could take at the apartment and in restructuring your life-style.

4. Formulate the reasons that tobacco smoke presents such a serious health problem. Consider the composition of the smoke, the toxic and carcinogenic properties of the smoke components, and the different modes of exposure to tobacco smoke. Why are young children who do not smoke at greater risk from exposure to tobacco smoke? What public regulations would seem reasonable to minimize exposure to tobacco smoke?

5. Some inhabitants of the planet Nicto smoke roots taken from the taba plant. Smoking the root is apparently addictive. Moreover, those who smoke it suffer 10 times as many rare cancers of the snout, breathing disorders, and third-eye dyslexia as nonsmokers. Health experts on Nicto have carried out a long series of tests and studies and concluded that the smoke of the taba root is highly carcinogenic and causes many related diseases. The taba root industry dismisses these claims as anecdotal and unproven; an advertising campaign aimed at attracting new young smokers has been mounted showing Nictites enjoying the best of life while puffing on taba roots. Based on this sketchy information obtained through

brief communications with Nicto, do you believe that smoking the taba root is harmless to the Nictite's health? Would you smoke it, even if it made you feel good?

Problems

1. A criminal is sentenced to life imprisonment in a small cell. The cell has a volume of 10 cubic meters and an average radon source of 10 picocuries per liter of air per hour (pCi/liter/hour). The air in the cell is exchanged every 4 hours. If the criminal is 20 years old when first incarcerated, determine the probability that he will contract lung cancer by the time he is 90 years old, assuming that no other specific risks for lung cancer exist. [Hint: Refer to Equations 8.1 to 8.7, and use Figure 8.3.]

2. Formaldehyde is being emitted into your bedroom at the rate of 20 milligrams per hour from particleboard in the walls, closets, and floors. The room has a volume of 20 cubic meters, and the air in the room is fully exchanged every hour through leakage. What is the steady-state concentration of formaldehyde in the room in micrograms per cubic meter? In parts per million by mass? Are you worried? What could you do to correct the problem?

3. You are in a room with five smokers. One smoker is smoking two cigarettes per hour; two smokers are each puffing three cigarettes per hour; and two are smoking four per hour. Estimate how many times greater your exposure to tobacco smoke is under these conditions compared with being with only two smokers each puffing two cigarettes per hour.

Suggested Readings

Altman, R. *The Complete Book of Home Environmental Hazards.* New York: Facts on File, 1990.

Hileman, B. "Formaldehyde: How Did EPA Develop Its Formaldehyde Policy?" *Environmental Science and Technology* 16 (1982): 543A.

National Academy of Sciences. *Indoor Pollutants.* Washington, D.C.: National Academy Press, 1981.

Nazaroff, W., and A. Nero Jr., eds. *Radon and Its Decay Products in Indoor Air.* New York: Wiley, 1988.

Nero, A., Jr. "Controlling Indoor Air Pollution." *Scientific American* 258 (1988): 42.

Repace, J., and A. Lowery. "Indoor Air Pollution, Tobacco Smoke, and Public Health." *Science* 208 (1980): 464.

Spengler, J., and K. Sexton. "Indoor Air Pollution: A Public Health Perspective." *Science* 221 (1983): 9.

Turiel, I. *Indoor Air Quality and Human Health.* Stanford, Calif.: Stanford University Press, 1985.

Walsh, P. J., C. S. Dudney, and E. D. Copenhaver, eds. *Indoor Air Quality.* Boca Raton, Fla.: CRC Press, 1984.

9

Acid Rain

Acid rain is a relatively recent environmental problem in many areas of the world. Natural rains that drench forests and fields always contain some acids, but in moderate amounts. The rains that fall downwind of many industrial complexes are laced with acid strong enough to pickle cucumbers. Such "acid" rain affects the chemistry of soils and the forests that grow on those soils and the lakes into which the rainwater drains. The issue of acidic rain is closely related to that of air pollution, although the two problems differ in many respects. Air pollution gives birth to acid rain. The major impacts of acid rain, however, may be far removed from the initial sources of the pollutants. The ozone in smog is a local or, in some cases, a regional problem in areas where the precursors of ozone are released. Acid rain, on the other hand, is a regional and, in some cases, a local problem, often appearing well downwind of the power plants and vehicles that emit the acid precursors.

Acid rain is only one aspect of a broader issue focusing on the acidification of the environment. There are several mechanisms that can deposit acids on exposed surfaces, including

- Acid-carrying raindrops formed from cloud droplets containing acids.
- Acid-bearing snow, sleet, or other precipitation elements that are formed in clouds exposed to acidic gases and particles.
- Acidic aerosols and vapors scavenged from air below a precipitating cloud and thus carried to the ground with the precipitation.[1]

- Acidity deposited by aerosols falling onto or diffusing and sticking to surfaces in contact with polluted air.
- Acid aerosols, vapors, and precursor gases absorbed into or collected by fog droplets settling or drizzling out of the atmosphere.
- Sulfuric, nitric, and other acid vapors absorbed directly from the atmosphere onto exposed surfaces.
- Acid precursor gases such as sulfur dioxide and nitrogen oxides that come into contact with and adhere to exposed surfaces where they react to form acids.

In the worst situations, exposure to both smog and acid rain can occur in the same region. In such instances, heavy local air pollution is involved in the acidification of local fog and rain. Thus, all forms of acid deposition may be active and contribute to the total risk of health and environmental impacts.

This chapter discusses the primary mechanisms of acid formation and deposition, the sources of acidic pollutants, the places most affected by acid rain, the effects of exposure to excess anthropogenic acidity, and the approaches that have been taken to reduce acid precipitation and damage it causes.

9.1 The Tainted Rain

The problem of acid rain—or acid precipitation—and acid deposition has received much attention

---

1. Raindrop scavenging of particles and gases involves collisional and diffusional processes. Drops of water falling through the atmosphere can hit and capture slower-moving aerosols. Gases in air passing over the drop surface can diffuse to and be absorbed into the drop. The deposition of acids or other materials at the ground through these mechanisms is referred to as *wet deposition*. In *dry deposition*, small particles and gases in the lowest air layers are carried toward available surfaces by turbulent motions and thus may stick to soils or vegetation or be absorbed at

water interfaces. Soluble aerosols that can act as cloud condensation nuclei (CCN) may also be incorporated into cloud droplets by nucleation and condensation processes and may later be deposited at the ground by wet deposition. *Rainout* refers to the wet deposition of materials that are entrained into clouds and thus incorporated into precipitation; *washout* refers to the wet deposition of materials below clouds by scavenging mechanisms. (Also see Sections 3.1.3, 6.5.3, and 10.2.1.)

lately because of the significant damage caused in certain regional environments. It appears that acidic rain has already rendered lifeless many pristine mountain lakes and devastated large areas of forest. Part of what makes acid rain a major problem is the enormous area that it can affect. Problems associated with smog are usually localized, although very large numbers of people may be affected. Acid rain, by comparison, covers entire states and even countries. It is a political problem as well as a pollution problem, since the chemicals that cause rain to be dangerously acidic may be emitted in one state or country, but the acid falls on neighboring states and countries. Moreover, the pollutants contributing to the production of acid rain are released in large quantities by power companies, certain heavy industries, and transportation operations. It is not a simple matter to reduce the emissions from such activities without affecting local, regional, or even national economies.

Acid Rain Discovered

In the beginning, the precursors of acid rain were sulfurous gases emitted during the combustion of coal. Coal was known to be a primary cause of pollution even in the thirteenth century, soon after its widespread adoption as an energy source (Section 6.1). Smoke, stench, and blackening soot were the main offenders. By the mid-nineteenth century, coal smoke was creating urban "killer" smog on a regular basis. Scientists of the seventeenth and eighteenth centuries were, indeed, quite familiar with the noxious sulfur emissions associated with coal. But the connection between the acidic emissions and the regional destruction of the environment was recognized much later, after further measurements and analyses.

In 1872, Robert A. Smith published a book in which he coined the term *acid rain*. It had the ominous title *Air and Rain: The Beginnings of a Chemical Climatology*. Smith had earlier measured the composition of rain in and around Manchester, England. He found that in the city proper, the sulfur emissions of coal combustion were primarily in the form of sulfuric acid. Farther from town, in the suburbs, the sulfur was associated with ammonia in the form of ammonium sulfate. Beginning in the 1840s, Smith had made numerous observations around Great Britain from which he noted that the concentrations of sulfate were greater in rain collected near cities, and particularly near coal-burning

facilities. Other scientists began to see changes in environmental acidity in these areas. One recorded the increased acidity in pools of water in peat bogs. In Sweden in the late 1950s, a network of precipitation collectors revealed that the phenomenon of acid rain was regional in scale and that acid deposited in Sweden was coming largely from European sources.

Unlike Europe, where acid rain was recognized as a regional problem by the 1950s, in the northeastern United States and Canada, data were insufficient to characterize the acid rain problem until the 1970s. Up until that time, it had been believed that the regional acidity of precipitation was controlled by natural processes. However, increasing rates of acid deposition in the northeast United States and identification of environmental damage by the acidity set off alarm bells. Monitoring networks were set up to record the patterns of acid deposition. In 1980, the U.S. Acid Precipitation Act was passed, and the National Acid Precipitation Assessment Program (NAPAP), an intensive 10-year study of the problem, was initiated. At about the same time, the United States and Canada began a cooperative effort to limit the cross-border transport of acid pollutants.

One early sign of an acid rain problem was signaled by severe damage to forests over wide areas of Europe. Particularly in Germany, the decline of forests became a major environmental issue. The damaged areas increased from less than 10 percent of the forested regions in 1982 to about 50 percent in 1984. This "Waldsterben," or "forest death," led to dramatic actions to control regional air pollution. It was soon discovered that the combination of high ozone concentrations in smog and high acidity in precipitation worked synergistically to harm coniferous and deciduous trees.

Acid rain today is produced mainly from emissions of sulfur oxides (SO_x) and nitrogen oxides (NO_x) generated by human activities. SO_x consists mostly of sulfur dioxide (SO_2), and NO_x, nitric oxide (NO), and nitrogen dioxide (NO_2). There are both natural and anthropogenic sources of these compounds, which are discussed in Sections 9.3.2 and 9.3.3. The anthropogenic contribution to nitrogen oxide emissions has increased over the past century, owing largely to the introduction of high-temperature internal combustion engines in cars and trucks. Except in certain areas where sulfur emissions have been curtailed, the NO_x contribution to acid precipitation is generally secondary to SO_2.

The chemistry of acid formation is relatively straightforward (Section 9.2.2 and Sections 3.3.4, 6.3.2, 6.5.3, 10.2.1, 10.2.2, 13.5.3). In air, the sulfur and nitrogen oxide emissions are rapidly oxidized to form acids (sulfuric and nitric acids, respectively), which are very soluble in water and hence are efficiently scavenged by clouds and precipitation. In solution, these acids, together with some other compounds, affect the pH (an acidity index) of water and soils. Several mechanisms—including the precipitation of rain and snow and the deposition of fog and vapor—transfer the acidity to contact soil and water. Microorganisms are affected by changes in the pH of their immediate environments as a result of excess acid deposition. The abundances of mineral nutrients are altered, and the food chain can be disrupted. Larger species—fish, mammals, and trees—at the end of the food chain may be damaged as the effects of acidity ripple through the local ecosystems.

9.2   Acidity and pH

How do we measure the acidity of rain? What is meant by acidity, in the first place? And what are the common properties of acid materials? A sour taste is one characteristic of acids, although no one should ever purposefully taste an unknown solution to see whether it is an acid! Acids also turn blue litmus paper red (which is important only in chemistry laboratory). Nature is full of acids, both inorganic (or mineral) acids and organic acids. Our stomachs have an ample supply of hydrochloric acid (HCl), which helps digest the food we eat. This inorganic acid in vapor form, as hydrogen chloride, is also found in the stratosphere, where it is involved in the chemistry of the ozone layer (Section 13.5.2). Urine contains uric acid, which by itself is odorless, colorless, and tasteless. The basic molecules of life—deoxyribonucleic acid, or DNA—are also classified as acids. Most fruits contain acid, such as the citric acid that makes lemons sour. Vinegar contains acetic acid. Car batteries contain concentrated and hazardous sulfuric acid. Bleach is made of hypochlorous acid, which, although less dangerous than battery acid, can still damage skin.

Acids are not the kinds of materials we would normally like to interact with. Acids fume; acids burn. One thinks of horror movies with heroines being lowered into vats of boiling acid. Nonetheless, acidic compounds are everywhere around us and are rather harmless in most circumstances and useful in

many. In the natural world, the atmosphere tends to be slightly acidic, and the oceans slightly alkaline.[2] Precipitation originating from air is also somewhat acidic under ambient conditions. However, when humans begin to use the atmosphere for a waste-disposal system, pouring in sulfur and nitrogen oxides, it is little wonder that the acidity goes up, way up.

9.2.1   THE pH SCALE

Acids readily dissolve in water to form solutions. Such mixtures with water are called *aqueous solutions*.[3] The acidity of an aqueous solution can be defined by its pH. The symbol pH stands for *potenz* (power) of hydrogen (H). It is fundamentally a scale, or yardstick, to measure the acidity or *alkalinity* of liquids. The general characteristics of acids and bases are discussed in Section 3.3.4. Note that the terms *alkaline* and *basic* are often used interchangeably.[4] Moreover, alkaline and alkalinity, and base and basic are the complements of acidic and acidity, and acid and acidic, respectively. Thus, a soluble base that exhibits alkaline properties in solution is referred to as being basic (which is not meant to imply "fundamental" in this context). Just as acids have a sour taste, bases have a bitter taste. The taste is related in part to the nature of acidic and basic solutions and hence to their pHs.

2. Since the atmosphere contains oxygen, compounds released into air tend to become "oxidized." Highly oxidized sulfur and nitrogen compounds usually are strong acids, such as sulfuric acid and nitric acid. The oceans, by comparison, act as a sink for dissolved carbonate minerals, which tend to be alkaline.

3. There are pure liquids, and solutions. A pure liquid is composed of only one substance, for example, distilled water, pure methanol, or mercury (the only pure metal that is liquid at room temperature). A solution is a mixture of two or more substances. The bulk of a solution consists of a **solvent**. Smaller amounts of substances mixed into the solvent are called **solutes**. In **aqueous solutions**, water is the solvent. The solutes can be liquids that are miscible in water (like sulfuric acid), solids that dissolve in water (like salt or sugar), or gases that are absorbed by water (like carbon dioxide).

4. Alkaline originally referred to the basic property of the carbonate and hydroxide salts of the alkali metals, the six elements besides hydrogen in the first column, or Group I, of the periodic table of the elements, particularly sodium and potassium. Another class of metals, the alkaline-earth metals—calcium, strontium, barium, and magnesium in the second row, or Group II—form oxides with alkaline properties; calcium oxide is lime, for example. Moreover, when combined with carbon dioxide or water, calcium oxide forms the basic compounds calcium carbonate and slaked lime, respectively.

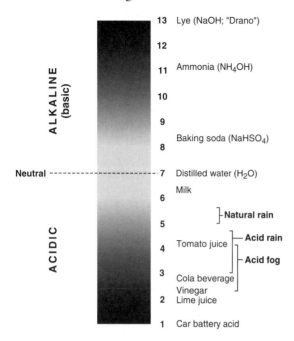

Figure 9.1 The scale of acidity, or pH scale. The more acidic pH values are toward the smaller end (bottom) of the scale, and the alkaline pH values are toward the larger end (top). The pHs of everyday solutions are indicated along the scale. The division between acidic and alkaline solutions is at a pH of exactly 7.0. Solutions with pHs at either end of the scale (strongly acidic or alkaline) are extremely corrosive and dangerous to handle. Note the pH ranges corresponding to natural rain, acid rain, and acid fog.

The pH scale is defined in such a way that the pH of pure distilled water has a value of *exactly* 7.0. This is referred to as the neutral pH. It is the mildest state of a solution with respect to its effects on living organisms. Pure distilled water can be thought of as an especially mild, nurturing fluid. Relative to the neutral pH, acids have lower pH values and bases have higher pH values. Figure 9.1 illustrates the pH scale with examples of the pHs of some common materials.

Substances that we consider mild, like milk and solutions of baking soda that might be taken to settle an upset stomach, tend to have pH values close to neutral. Indeed, baking soda is somewhat alkaline and acts to neutralize stomach acid. Interestingly, many of the liquids that we drink usually are acidic, including milk (only slightly acid), fruit and vegetable juices, colas, and alcoholic beverages. Carbonated sodas are strong enough acids that they can corrode paint finishes. Likewise, some condiments

such as vinegar and lime juice can be quite acidic—too much so to be drinkable. Our stomachs are apparently well designed to handle acids.

Household cleaning agents and chemicals have a definite tendency to be alkaline. Both ammonia and lye are strong chemicals that are difficult just to be around and should never be drunk.

An important fact that emerges from Figure 9.1 is that rain, both natural and polluted, is always acidic. Even pristine precipitation can have a pH of 5 to 6. Acid rain is therefore even more acidic, with pH values of 3 to 5. Worse, acidic fog can have the same pH as vinegar, lower than that of acidified rain. That's all fine, but what does it mean? What is the difference between a pH of 7 and one of 3? Between 5 and 4?

Relation of pH to H Ions

As mentioned in Section 3.3.4, the acidity (alkalinity) of a solution (and thus its pH) is related to the concentration of positive hydrogen ions (H^+) (negative hydroxide ions, OH^-) in the solution. In the case of an acid, the hydrogen ions are formed when the acid dissolves in solution and dissociates into ions. The pH scale, in fact, precisely measures the concentration of hydrogen ions through the following definition of pH:

$$pH = -\log_{10}\left(\left[H^+\right]\right) \qquad (9.1)$$

In this expression, $[H^+]$ is the concentration of the hydrogen ions in the solution. In aqueous solutions, the concentration of H^+ (and other dissolved species) is measured in units of *moles per liter*.[5] These units are important in setting the neutral point of the pH scale. Moles/liter, called the *molar* unit, is abbreviated as M.

5. Recall that 1 liter is equivalent to 1000 cubic centimeters, very roughly the same volume as a British quart. In zoology, a *mole* is a small mammal that burrows in the ground; in chemistry, a *mole* is a specific large number of molecules of a substance. In fact, 1 mole is *exactly Avogadro's number* of molecules (or atoms, in the case of an element) (Section 3.1). Avogadro's number is very close to 6×10^{23} molecules, a huge number indeed. In practical situations, one measures the weight of 1 mole of a substance rather than the number of molecules. The weight in grams of 1 mole of a substance is always exactly its molecular weight in atomic mass units (amu). For example, 1 mole of nitric oxide, NO (N[14 amu] + O[16 amu] = NO[30 amu]) weighs 30 grams; moreover, as expected, it contains 6×10^{23} molecules of NO.

Equation 9.1 also uses a logarithm to the base 10. (See Sections A.1 and A.4 of the Appendix for a description of this function.) Equation 9.1 therefore defines a *logarithmic* scale for pH. This is a log base 10 scale, which means that for each change in pH value of 1, the hydrogen ion concentration and acidity change by a *factor of 10*. However, if the pH increases by +1, $[H^+]$ *decreases* by a factor of 10! Conversely, if the pH decreases by –1, $[H^+]$ *increases* by a factor of 10. This may seem a bit confusing at first, but keep these simple rules in mind: Increasing pH means decreasing acidity, and vice versa; each unit of pH is a factor of 10 in acidity. According to these rules, a solution with a pH of 5.0 is 10 times more acidic than one with a pH of 6.0, and 100 times more acidic than a solution with pH 7.0. Similarly, a solution with a pH of 4.0 is 1000 times more acidic than one with a pH of 7.0.

The hydrogen and hydroxide ion concentrations for a neutral solution are taken to be the concentrations measured in distilled water at room temperature. In pure water, some of the water molecules are always dissociated into H^+ and OH^-. These ions also recombine, and an equilibrium state is maintained between the water molecules and their ions, as follows:

$$H_2O \text{ (aqueous)} \Longleftrightarrow H^+ + OH^- \quad (9.2)$$

where the double arrow indicates that the process proceeds in both directions at equal rates and hence is equilibrated. The ion concentrations have been accurately determined under these conditions. Because electrical charge is also conserved in this process (that is, every time a positive charge is produced, so is a negative charge), the positive and negative ion concentrations must be equal. It turns out that at room temperature

$$\left[H^+\right]_{equil} = \left[OH^-\right]_{equil}$$
$$= 1.0 \times 10^{-7} \text{ M (moles/liter)} \quad (9.3)$$

Hence,

$$pH \text{ (pure water)} = -\log_{10}\left(\left[H^+\right]_{equil}\right)$$
$$= 7.0 \quad (9.4)$$

The "equilibrium constant" for the dissociation of water (Equation 9.2) is the product of the ion concentrations in solution. Obviously, in the case of pure water, the equilibrium constant has the value

$$K_{equil} = \left[H^+\right] \times \left[OH^-\right]$$
$$= 1.0 \times 10^{-14} \text{ M}^2 \text{ (moles/liter)}^2 \quad (9.5)$$

In any aqueous solution, there is so much water present that the equilibrium expressed by Equations 9.2 and 9.5 is always achieved. In other words, the product of the H^+ and OH^- concentrations in *all* aqueous solutions is equal to the same constant—the equilibrium constant for water dissociation. Then from Equation 9.5, the H^+ and OH^- concentrations must be inversely related, or equivalently

$$\left[H^+\right] = \frac{1.0 \times 10^{-14}}{\left[OH^-\right]} \text{ M (moles/liter)}$$

$$\left[OH^-\right] = \frac{1.0 \times 10^{-14}}{\left[H^+\right]} \text{ M (moles/liter)} \quad (9.6)$$

These fundamental relationships allow a universal scale of pH to be established.

The way that hydrogen ion concentrations (and hydroxyl ion concentrations) scale with pH in an aqueous solution is illustrated in Figure 9.2. Note that a neutral solution with a pH of 7.0 has an H^+ concentration of 1×10^{-7} M. This concentration has been adopted as the reference value in Figure 9.2, and all H^+ concentrations are divided by it. Thus it is immediately apparent that a pH of 6 corresponds to 10 times the concentration of hydrogen ions as a pH of 7, and each subsequent unit of decrease in pH corresponds to a factor-of-10 increase in the hydrogen ion concentration. Considering the opposite direction of pH change, each unit of increase on the pH scale implies a factor of ten *decrease* in acidity, or hydrogen ion concentration. However, a decrease in acidity can also be interpreted as an *increase* in alkalinity (or hydroxide ion concentration, OH^-). Indeed, the equivalent scale for alkalinity is exactly complementary to the scale for acidity (Figure 9.2). The two scales match perfectly if one of them is inverted. When the acidity of a solution is known, so is its alkalinity, and vice versa. In Figure 9.2, it follows that each change in pH of 1 unit implies a factor-of-10 change in *both* acidity and alkalinity simultaneously, but in opposite directions.

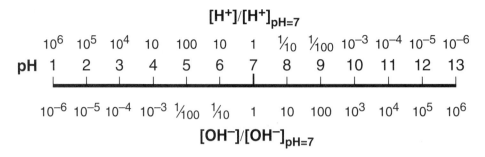

Figure 9.2 The relationships between the pH scale and the concentrations of hydrogen ions (H+) and hydroxyl ions (OH−) in solution. The pH scale is a logarithmic scale: Each unit of pH is a factor of 10 in concentration. Both the reference concentrations of H+ and OH− at a pH of 7.0 are equal to 1×10^{-7} M. As the pH climbs above 7, the solution becomes more alkaline; the OH− concentration increases; and the H+ concentration decreases. As pH falls below 7, the solution becomes more acidic; the H+ concentration increases; and the OH− concentration decreases. For each unit that pH increases, the OH− concentration increases by a factor of 10, and the H+ concentration decreases by a factor of 10, and vice versa.

It seems that acidity is directly associated with the abundance of hydrogen ions (H+) in solution. Similarly, alkalinity is associated with the quantity of hydroxide ions (OH−) in solution. Therefore, it follows that a strong acid produces, molecule for molecule, more hydrogen ions when dissolved in water than a weak acid does. Similarly, a strong base produces, molecule for molecule, more hydroxide ions than a weak base does. It turns out that sulfuric and nitric acids are strong acids, whereas sulfurous and carbonic acids are weak acids. Likewise, sodium hydroxide is a strong base, whereas ammonium hydroxide and calcium hydroxide are weak bases. The difference between strong and weak acids and bases is related to the degree that they dissociate into ions in solution. In general, strong acids and bases mixed with water are completely dissociated; weak acids and bases are not.

The range of hydrogen ion (hydroxide ion) concentrations found in common acidic and basic solutions varies by a factor of 10^{10} or more, that is, a factor exceeding 10 billion. This is a remarkably wide range of variation in an environmental parameter. Accordingly, it is likely that living organisms have learned to recognize and utilize variations in pH to optimize survival. Humans have a highly developed sense of taste that can easily detect fine gradations in sourness and bitterness, or pH. Typically, people prefer tastes closer to neutral than to either extreme. Most organisms can exist only within a rather narrow range of pH that could be either acidic or basic, but not far from neutral. Exceptions include bacteria that live in the gastrointestinal tract or near acidic magma vents. It is likely that the original evolution of life in

the oceans predisposed biochemical organisms to prefer fairly neutral surroundings.

9.2.2 ACIDS IN WATER

Several processes form acids in water under atmospheric conditions and thus contribute to acid rain. Although sulfuric and nitric acid vapors can be produced photochemically in air and later dissolve in cloud droplets, these acids may also be formed directly in droplets from their precursors, SO_2 and NO_x, respectively. This chemistry in the aqueous phase proceeds through a series of stages involving the dissolution and then the oxidation of the acidic precursors. In aqueous solutions, as noted, many compounds readily dissociate into ions. The dissociation process is, in fact, an important one in the overall dissolution and uptake of soluble gases into water droplets. Acid vapors are highly soluble in water mainly because they dissociate so easily in aqueous solution. However, gases like SO_2 and NO_x, which do not dissociate, are not very soluble.[6] At normal atmospheric concentrations—or even at elevated concentrations found in heavily polluted

6. When a gas such as sulfur dioxide, nitrogen dioxide, or carbon dioxide, all of which are soluble in water, comes into contact with a water droplet (or any water surface, for that matter, including the ocean), some of the gas dissolves in the water. The amount that enters the liquid state depends on three factors: the pressure of the gas over the water surface, the solubility of the gas in water, and the amount of water present and accessible. An increase in any of these factors raises the fraction of the gas that dissolves in the water. Also, as the temperature falls, the solubility of a gas, or its ability to form solutions with water, increases.

air—only a small fraction of these gases would be found dissolved in clouds.

Consider the absorption of sulfur dioxide by a water drop. First, the gas molecule enters the water phase, and then it becomes associated with a water molecule, or is hydrated:

$$SO_2(gaseous) \xrightarrow{\text{Dissolution}} SO_2(aqueous) \quad (9.7)$$

$$SO_2(aqueous) + H_2O \xrightarrow{\text{Hydration}} H_2SO_3(aqueous) \quad (9.8)$$

The resulting molecule, H_2SO_3, is the weak acid, sulfurous acid. This can partially dissociate into hydrogen ions (H^+) and bisulfite ions (HSO_3^-):

$$H_2SO_3(aqueous) \xrightarrow{\text{Dissociation}} H^+ + HSO_3^- \quad (9.9)$$

The acidity (H^+) created by the sulfurous acid is generally very small, unless the SO_2 concentrations become extremely high (reaching levels that would kill anyone breathing it). Hence, further steps are required to completely activate the acidic potential of sulfur in the atmosphere.

An oxidizing agent like ozone or **hydrogen peroxide** (H_2O_2) drives the acid activation.[7] Hydrogen peroxide is produced by **hydroperoxy radicals** (HO_2). These radicals are formed when ozone is present in air along with water vapor or organic compounds, and ultraviolet sunlight is available. One simple chemical mechanism leading to HO_2 involves the OH-forming reactions defined in Equations 3.44 and 3.46 (Section 3.3.4) followed by

$$OH + O_3 \rightarrow HO_2 + O_2 \quad (9.10)$$

Hydrogen peroxide then is generated by the reaction

7. Hydrogen peroxide is a common household antiseptic and bleaching agent usually supplied in the form of an aqueous solution. Its powerful oxidizing action destroys bacteria and breaks down color pigments in hair, thus providing a cheap hair bleach. At concentrations exceeding about 8 percent in water, hydrogen peroxide damages the skin by reacting with it. In the atmosphere, hydrogen peroxide can be formed in the presence of ozone and water vapor or when organic compounds are photochemically oxidized. In the troposphere, H_2O_2 is a key oxidant, particularly in aqueous solutions.

$$HO_2 + HO_2 \rightarrow H_2O_2 + O_2 \quad (9.11)$$

Dissolution of H_2O_2 in water occurs as

$$H_2O_2(gaseous) \xrightarrow{\text{Dissolution}} H_2O_2(aqueous) \quad (9.12)$$

Some of the H_2O_2 dissociates into reactive ions:

$$H_2O_2(aqueous) \xrightarrow{\text{Dissociation}} H^+ + HO_2^- \quad (9.13)$$

Mixed with water, hydrogen peroxide is a strong oxidant that can react with the bisulfite ion, generating bisulfate ions (HSO_4^-):

$$HSO_3^- + HO_2^- \xrightarrow{\text{Reaction}} HSO_4^- + OH^- \quad (9.14)$$

Note that the hydroxide ion in Equation 9.14 can recombine with the hydrogen ion in Equation 9.13 to form water (this is one of the reactions implicit in the water dissociation equilibrium expressed in Equation 9.2):

$$H^+ + OH^- \rightarrow H_2O \quad (9.15)$$

On the other hand, the bisulfate ion is clearly acidic, since it readily dissociates to form hydrogen ions,

$$HSO_4^- \xrightarrow{\text{Dissociation}} H^+ + SO_4^{2-} \quad (9.16)$$

The sulfate ion, SO_4^{2-}, by itself is not acidic; the hydrogen ions are the essence of acidity.

The net production of hydrogen ions (acidity) by this series of aqueous processes can be written in shorthand form as the overall reaction between hydrogen peroxide and sulfurous acid:

$$H_2SO_3 + H_2O_2 \xrightarrow{\text{Reaction}} 2H^+ + SO_4^{2-} + H_2O \quad (9.17)$$

An even more compact way of looking at this reaction is to eliminate the water molecule that formed sulfurous acid (Equation 9.8), since it acts only as an intermediate species in the overall reaction. Then

$$SO_2 + H_2O_2$$

$$\xrightarrow[\text{Solution}]{\text{Aqueous}} 2H^+ + SO_4^{2-} \qquad (9.18)$$

It is important to note that the oxidation of sulfur dioxide by hydrogen peroxide in solution creates as many hydrogen ions as does dissolving sulfuric acid directly in water:

$$H_2SO_4 (\text{aqueous})$$

$$\xrightarrow{\text{Dissociation}} 2H^+ + SO_4^{2-} \qquad (9.19)$$

The difference between strong and weak acids is related to their extent of dissociation in aqueous solution. Sulfuric acid, a strong acid, is almost completely dissociated in water, according to Equation 9.19. Similarly, nitric acid is completely dissociated in solution, but sulfurous acid (H_2SO_3) is not. The first step of dissociation (Equation 9.9) does not even proceed to completion, let alone the second step of the dissociation that produces a sulfite ion, SO_3^{2-}:

$$HSO_3^- \xrightarrow{\text{Dissociation}} H^+ + SO_3^{2-} \quad (9.20)$$

Thus, weak acids are reluctant to release their hydrogen ions even when mixed in water. In fact, sulfite ions added to an aqueous solution are likely to capture hydrogen ions (the reverse of Equation 9.20) thereby *reducing* the acidity of the solution.

The reactions that generate acids in the environment can proceed in any water that is available. There are two requirements, however. First, the water must be exposed to the atmosphere, from which reactive gases are drawn. Second, the solution must not be either too acidic or basic, since those conditions preclude many of the necessary reactions. Because moisture tends to condense in cracks and crevices and pits and scratches, all objects small and large are subject to attack by acids that form on their surfaces. Similarly, the surfaces of lakes, ponds, and streams are vulnerable to acidification by virtue of being exposed to the atmosphere.

9.2.3 ALKALINITY: THE ACID BUFFER

Compounds that are basic (alkaline) can act to neutralize acids such as those found in acid rain, lakes, and soils. There are a large number of alkaline compounds, including ammonium hydroxide (NH_4OH), formed by dissolving ammonia gas, NH_3, in water:

$$NH_3 (\text{gas}) + H_2O (\text{liquid})$$

$$\rightarrow NH_4OH (\text{aqueous}) \qquad (9.21)$$

Two other common alkaline substances are sodium hydroxide, or common lye (NaOH), and calcium hydroxide, or lime ($Ca(OH)_2$). All these hydroxide compounds are bases because they have an "OH" group that readily dissociates when the base is dissolved in water (just as acids release hydrogen ions):

$$NH_4OH (\text{aqueous}) \rightarrow NH_4^+ + OH^- \quad (9.22)$$

$$NaOH(\text{aqueous}) \rightarrow Na^+ + OH^- \quad (9.23)$$

$$Ca(OH)_2 (\text{aqueous}) \rightarrow Ca^{2+} + 2OH^- \quad (9.24)$$

The hydroxide ions generated from these bases can recombine with a hydrogen ion derived from an acid:

$$H^+ + OH^- \rightarrow H_2O \qquad (9.15)$$

Amazingly, if a strong, corrosive acid is combined in solution with an equally reactive base, the mixture can end up as harmless water.[8]

The most important base that acts on a global scale to neutralize acids in the atmosphere is ammonia. Ammonia is generated during the bacterial decomposition of organic matter and is a major component of the global nitrogen cycle and the natural food chain. Among the anthropogenic sources, cattle feedlots are responsible for the largest emissions (Section 10.2.2). This source may amount to several million metric tons (tonnes) per year (Mt/yr) in the United States (the estimates range from about 1 to 6 Mt/yr). Fertilizer applications may release another few Mt/yr of ammonia vapor to the atmosphere, although the precise amount is quite uncertain (within a range estimated as 0.5 to 4.5 Mt/yr). Natural lime is not widely available in the environment. On the other hand, it has been spread out in large quantities over lakes in Scandinavia in an attempt to neutralize their excess acidity. If a lake is

8. When this ionic recombination occurs, energy is released. If a concentrated acid and base are mixed together directly, the energy output can be so sudden and large that the mixture literally explodes. Obviously, it is never advisable to mix strong acids and bases.

large and/or the pH is very low, the amount of lime necessary to maintain a normal pH can be quite substantial, however.

Some alkaline compounds do not dissociate directly into hydroxide ions when dissolved in water. These compounds can act as buffers for aqueous solutions, keeping them from becoming too acidic. Calcium carbonate is the most common acid-buffering mineral in the environment. We have already seen that $CaCO_3$ is produced by the dissolution of carbon dioxide in oceans containing calcium ions (Section 4.1.3) and so plays a major role in the carbon cycle (Section 10.2.4). In the atmosphere (clouds, rain, and aerosols), the most common source of calcium carbonate is windblown dust particles. Accordingly, the buffering action of calcium carbonate is largely confined to regions that are relatively dry and windy. The total quantity of dust containing acid-buffering properties lifted into the atmosphere annually is roughly 20 million tonnes, although only about 5 percent of this amount is available and effective in neutralizing atmospheric aqueous solutions. Most of the buffering capacity is associated with the calcium in dust particles raised from road surfaces. Acidity in the western United States and Mexico is most influenced by this buffering effect, although coastal regions much less so than inland areas. In the eastern United States and Canada, acidity is not strongly buffered by wind-borne dust, although soils may still provide buffering capacity.

In soils and lakes, calcium carbonate and other buffers compensate for acid deposition. The process is different from CO_2 dissolution in water, which actually produces weak carbonic acid. Carbonate ions have the effect of absorbing hydrogen ions, as indicated by the following sequence of reactions:

$$CaCO_3(solid) \xrightarrow{\text{dissolution}} Ca^{2+} + CO_3^{2-} \tag{9.25}$$

$$CO_3^{2-} + H_2O \xrightarrow{\text{association}} HCO_3^- + OH^- \tag{9.26}$$

$$H^+ + OH^- \rightarrow H_2O \tag{9.15}$$

Accordingly, a solution placed in contact with calcium carbonate is provided with a source of alkalinity via carbonate ion (CO_3^{2-}) dissolution. These ions readily take up hydrogen ions from the solution, leaving hydroxide ions behind. If acid is added to the solution, excess hydrogen ions released through acid dissociation are neutralized by hydroxide ions, which are replenished by the additional dissolution of calcium carbonate (Equations 9.25 and 9.26). This ability to neutralize hydrogen ions defines the buffering capacity of carbonate minerals. Such buffers can stabilize the pH of a solution by compensating for the acid deposition and release of hydrogen ions. Of course, the buffering action continues only as long as calcium carbonate is available.

There is another side to the buffering coin, however. Works of art and architecture containing calcium carbonate also can decompose when exposed to acidity in air. In fact, marble and limestone—minerals commonly used in construction and sculpture—are composed mainly of calcium carbonate. It follows that building facades and artworks constructed from these materials are at risk of destruction by acid rain and acidic pollution in general. Indeed, this is one of the important objections to the continued emissions of sulfur and nitrogen oxides in populous regions (Section 9.5.2).

Naturally alkaline minerals are present in many soils and act to neutralize most of the acidity arriving on the ground each year. Regions characterized by alkaline soils are therefore less susceptible to damage by acid deposition. On the other hand, areas with soils that are chemically neutral or acidic are stressed when additional acid is deposited by precipitation and other means. Water drainage from nonbuffered soils can also acidify nearby lakes. Poorly buffered soils are found in the northeastern and western United States, upper Great Lakes region, southeastern Canada, Scandinavia, and many other places around the world.

9.3 Sources of Environmental Acids

The sources of acidity measured in aerosols, cloud droplets, and precipitation are varied. Both natural and anthropogenic emissions contribute to acidity. Moreover, both inorganic and organic components are significant. Some of the principal sources of environmental acidity are discussed next. For additional information, refer to discussions of the global biogeochemical cycles of sulfur and nitrogen in Sections 10.2.1 and 10.2.2, and of acidic fog and aerosols in Section 6.5.3. It is worth noting, however, that the primary sources of acid rain and fog, and environmental acidity in general, are emissions of sulfur and nitrogen oxides generated by human activities.

9.3.1 HOW ACID IS ACID RAIN?

Natural rainwater collected in remote areas far removed from any substantial sources of airborne sulfur or nitrogen compounds has a pH of about 5.6. Even the purest rain is slightly acidic. This baseline acidity—the minimum acidity (maximum pH) measured in cloud water and rain, except under unusual circumstances—is a consequence of carbon dioxide in the atmosphere. At atmospheric concentrations (380 ppmv), CO_2 gas is absorbed in water, forming carbonic acid with a pH of about 5.6.[9] This natural acidity is crucial to the weathering of rocks and minerals and drives the Earth's long-term global geochemical cycles. Moreover, this minimum acidity is unavoidable, since carbon dioxide is always present throughout the atmosphere (Section 2.1.2).

In addition to the effect of carbon dioxide, natural sources of sulfur and nitrogen oxides also contribute to the background acidity. For example, sulfur dioxide from volcanic eruptions and nitric oxide generated by lightning are converted into acids that can lower the pH of precipitation. Measurements carried out throughout the world show that the average pH of rain is close to 5.0. The difference between 5.6 and 5.0, or 0.6 units of pH, represents an increase in the hydrogen ion concentration of background precipitation by a factor of roughly 4. This enhancement in hydrogen ions is almost entirely associated with sulfur and nitrogen acids. In remote regions of the world, pH is observed to vary between 5.0 and 5.6. Figure 9.3 shows the measured annual average pH values of rain over North America. Notice that particularly in rural regions far removed from urban zones, the pH falls into the normal range. In polluted areas the pH is typically lower. The two obvious acid strongholds in the United States are in the Northeast and in southern California.

It is not always clear whether rain with a pH in the range of 5.0 to 5.6 has been contaminated with acids from industrial sources. For example, natural sources of sulfur from volcanic vents, or of organic acids from decaying vegetation, could be responsible for many observations in which the pH falls below 5.6. It is very likely, on the other hand, that many cases of pH in the range of 5.0 to 5.6 are caused by anthropogenic pollution. To be conservative, however, only pH values of 5.0 or lower are unequivocally identified with acid rain caused by industrial activity.

Acid rain typically has a pH between roughly 3 and 5. Occasionally, in extreme cases, rain with a pH < 3 is measured. The additional acidity—below a pH of 5—is mainly associated with emissions of sulfur and nitrogen oxides, which are the precursors of sulfuric and nitric acids, respectively. Between a pH of 5 and a pH of 3, the acidity increases by a factor of 100. Values of pH in the vicinity of 5 can be tolerated over time, but values close to 3 would be deadly to vast tracts of the environment. Quite large areas, on the other hand, are subject to rain having a pH between 4 and 4.5. This is the situation in the northeastern United States, where the widespread destruction of sensitive ecosystems has been detected. The pattern of pH shown in Figure 9.3 leads to the inescapable conclusion that massive acidity is tied to dense industrialization. For example, the largest concentration of sources of sulfur dioxide is found in a region extending from the Ohio Valley in the Midwest through the northeast corridor. Not surprisingly, this vast area is also plagued by highly acidic precipitation. Southern California is a region of high nitrogen oxide emissions and, not coincidentally, very low pH rainfall.

Even if the pH of rain is 5 or greater, it still may be polluted. Indeed, it may be dangerously contaminated. For one thing, pH does not measure the amounts of nonacidic toxic materials in the rainwater. This might consist of pesticides, smoke compounds, and heavy metals. Moreover, large quantities of anthropogenic acids in rain may be largely neutralized by alkaline compounds. An example of this situation is found in southern California east of Los Angeles. Huge cattle feedlots emit a dense plume of ammonia that reacts with the acidic air drifting eastward from the city. The result is haze and rain laced with high concentrations of ammonium nitrates and sulfates, but with a pH greater than 5 (Section 9.4). Breathing this chemical haze is certainly unhealthful.

Acids Inorganic or Organic

Considering all the potential sources of acidity for the environment, a number of compounds are found to contribute significantly to the acidity of rain,

---

9. This process is similar to the dissolution of sulfur dioxide in water to form sulfurous acid, as described by Equations 9.7 and 9.8. For a description of CO_2 dissolution in water to form carbonic acid, see Equation 10.22 in Section 10.2.4.

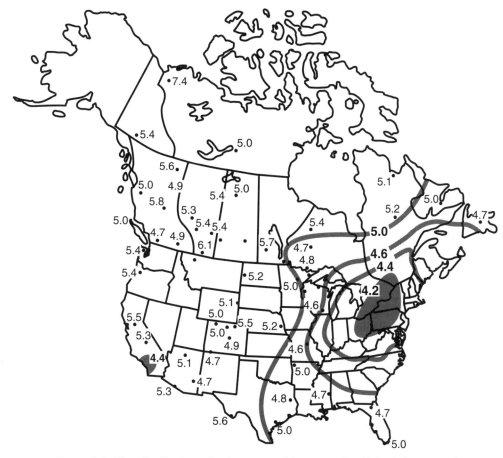

Figure 9.3 The distribution of rainwater acidity over the United States and Canada. The pH values shown at individual monitoring sites are annual averages (means) of all the measurements taken in 1982. The contours of pH were interpolated from these data. Rainwater in the shaded areas had an average pH of 4.2 or less. Note the small area of pH of 4.4 or lower in southern California. (Data from H. L. Ferguson and L. Machta, "Atmospheric Science and Analysis," United States/Canada Working Group 2, *Final Report*, U.S. Environmental Protection Agency, Washington, D.C., 1982.)

clouds, and aerosols. Among the inorganic acids, the following have important roles:

- H_2SO_4: Sulfuric acid is generated from anthropogenic SO_2 emissions and natural sulfur emissions. The latter includes, for example, hydrogen sulfide (H_2S) and dimethyl sulfide (DMS), which are oxidized to sulfur dioxide.
- HNO_3: Nitric acid is derived from both anthropogenic and natural NO_x emissions. Except in certain regions such as California, nitric acid is less important than sulfuric acid in acid rain.
- HCl: Hydrochloric acid, a strong acid, has significant natural and anthropogenic sources and contributes chloride ions to cloud water and rainwater. Hydrochloric acid is always secondary

in importance to nitric acid and sulfuric acid in precipitation.

- H_2CO_3: Carbonic acid is a weak acid formed when CO_2, which is mostly natural in origin (but which has increased substantially because of fossil-fuel burning; see Section 12.2), dissolves in water. Carbonic acid is negligible in acid rain.
- H_2SO_3: Sulfurous acid, another weak acid, is formed when sulfur dioxide dissolves in water. Compared with the more highly oxidized sulfuric acid, sulfurous acid makes a negligible contribution to acidic precipitation.

A variety of organic acids are also found in rainwater, as well as in polluted lakes and streams. These arise principally from biological organisms or their emissions, from the decomposition of organic compounds, or from specific human sources. The organic acids contain the common chemical group – COOH. In total, roughly 25 percent of all the acids measured in precipitation, clouds, and aerosols are organic acids. Among the particular acids detected in rain are the following:

- HCOOH: Formic acid, the most common organic acid found in precipitation, is a by-product of the oxidation of more complex hydrocarbons. Formic acid is also produced by ants as means of defense and imparts its characteristic odor to an ant hill.
- CH_3COOH: Acetic acid is also generated as a by-product of organic photochemical decomposition in air. Vinegar is mainly acetic acid, although atmospheric concentrations are relatively small.
- C_3H_7COOH: Butyric acid has the characteristic odor of vomit and is obviously not very common in rainwater.
- $C_5H_{11}COOH$: Caproic acid, which makes goats smell the way they do, also is not found in very large amounts in precipitation.

There are many kinds of organic acids, which typically have complex molecular structures (although formic acid is relatively simple, with a structure not much different from that of formaldehyde, that is, HCOOH compared with HCHO).

The Composition of Acid Rain

Around the world, samples of precipitation have been collected and analyzed for composition. In regions of North America suffering episodes of acid rain, the collected water is dominated by sulfate ions (SO_4^{2-}), followed by nitrate ions (NO_3^-) and then chloride ions (Cl^-). Carbonate and sulfite ions are insignificant in these conditions. In a typical case, sulfate accounts for roughly 60 percent of the acid anions (the negatively charged ions). Nitrate is typically half of the sulfate. An exception is noted in southern California, where nitrate may equal sulfate in concentration. Chloride falls into third place in concentration, usually well behind nitrate. Exceptionally large chloride ion abundances, however, may be detected near ocean coastlines where sea-salt chloride ions are dominant in the haze that blows off the ocean surface. Nevertheless, sea-salt chloride is not associated with acidity, because its accompanying cation (positive ion) is Na^+, not H^+. The most common cations in precipitation, in addition to sodium, are calcium (Ca^{2+}), magnesium (Mg^{2+}), and ammonium ions (NH_4^+). Typically, ammonium and calcium ions are sufficient to offset most of the nitrate and chloride, leaving most of the acidity (H^+) to be associated with the strongest-acid ions—sulfate.

In Europe, the composition of acid rain is similar to that in the United States, with the expected variability from place to place and time to time. In some areas of China, precipitation can be more intensely acidic but may also contain much higher concentrations of ammonium and calcium ions. These extreme conditions prevail mainly in the industrial zones in the south of China. Downwind of the central deserts, as in Beijing, the acidity may be largely neutralized by alkaline dust.

In remote regions of Canada and Australia, measurements have shown that biogenic emissions of sulfur compounds account for a significant fraction of rainfall acidity, which is associated with the sulfate products (sulfuric acid). The biogenic and anthropogenic sulfur contributions can be separated using isotopic analysis (isotopes are explained in Section 7.3.1). It happens that sulfur in fumes from coal combustion, for example, is isotopically "heavier" than sulfur in gases emitted by bacteria. Hence, the relative sulfur inputs from each source may be determined by measuring the average isotopic weight of the sulfate in the environment. The time variation in the average isotopic weight of sulfate is consistent with stronger biogenic emissions in summer, when bacteria are active. These field investigations also suggest that anthropogenic sulfate deposited on ecosystems can be recycled as acid rain. At first, the sulfate deposited by precipitation is assimilated as a *nutrient*. Later, the sulfur is released into the atmosphere in a reduced form (such as dimethyl sulfide, DMS), which can be oxidized to sulfuric acid. Amazingly, sulfur pollution may be recycled through biogeochemical processes to become acid rain again and again.

9.3.2 SULFUR OXIDES AND ACID RAIN

The presence of sulfur compounds in the atmosphere is a result of both natural and anthropogenic emissions. First, we consider natural sources of sulfur and the processes that cycle sulfur through the

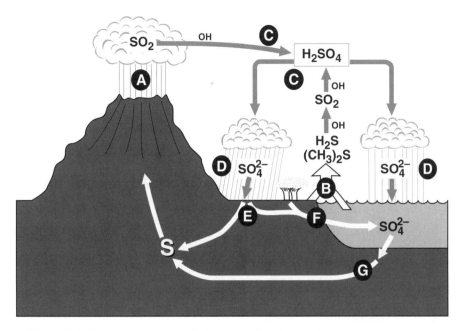

Figure 9.4 Key components of the natural sulfur cycle. The processes that introduce sulfur into, and transport sulfur through, the environment are identified by letters: A, volcanic emissions; B, emissions from living organisms and decaying organic matter; C, oxidation of sulfur emissions to sulfuric acid and other sulfates; D, scavenging of the sulfates; E, seepage into groundwater; F, runoff into oceans; and G, sedimentation in detritus and geologic recycling. The emission of sulfur, its oxidation into sulfuric acid, and the scavenging and precipitation of this acid constitute the natural sulfur-based contribution to the acidity of rainwater.

atmosphere, oceans, land, and the biosphere (consisting of living and decaying organic matter). Additional facts about the sulfur cycle are provided in Section 10.2.1.

The Natural Sulfur Cycle

Figure 9.4 illustrates the principal components of the natural sulfur cycle (which are further quantified in Section 10.2.1). Important sources of sulfur to the atmosphere include SO_2 emissions from volcanic eruptions and dimethyl sulfide emissions from the oceans. Once in the atmosphere, these compounds are chemically transformed into sulfuric acid, which is quickly absorbed in rain, snow, fog, and other condensed states of water. The sulfur, accompanied by acidity, is then deposited on land and ocean surfaces in the form of precipitation and vapor condensation. Sulfur compounds can then be recycled within soils and surface water as an essential nutrient, particularly on land where sulfate concentrations are limited. In fact, sulfate on land is generally depleted because it can be carried to the oceans in water runoff

down streams and rivers and through groundwater seepage. The sulfur is emitted by microorganisms in chemically reduced forms—mainly as sulfides, including hydrogen sulfide, carbonyl sulfide, and dimethyl sulfide. Sulfate is also gradually converted to minerals, such as calcium sulfate, that are not very soluble. Sulfate minerals are deposited in sediments that may eventually be carried deep into the Earth and recycled back into the atmosphere during volcanic eruptions. The sulfur transfer processes summarized in Figure 9.4 involve the atmosphere, soils, living organisms, and the oceans, all of which are important to the recycling of sulfur through the environment.

On occasion, volcanoes are a spectacular source of atmospheric sulfur, exploding with the force of a hydrogen bomb and spewing dense clouds of ash and fumes up to 30 kilometers or more into the stratosphere. In one of these major events, 10 million to 20 million tonnes of sulfur (Mt-S) may be emitted in the form of sulfur dioxide. The sulfuric acid aerosols that later evolve from the SO_2 can blanket the Earth and cause climatic disturbances

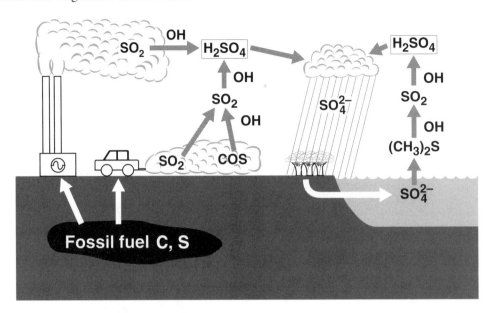

Figure 9.5 Key components of the anthropogenic sulfur cycle. The major source is the mining of fossil-carbon reservoirs for fuel. When the fuel is burned, the sulfur fraction of the fuel (perhaps a small percentage by weight) may be released into the atmosphere as sulfur dioxide and other sulfur compounds. These are oxidized, principally by the action of the hydroxyl radical, into sulfuric acid, which leads to the enhanced acidification of cloud water and rainwater. Note the critical role of OH in the overall chemistry of sulfur.

(Section 11.6.4). Volcanic eruptions this large occur only once every 10 to 100 years. One of the largest recorded historical eruptions—Tambora, Indonesia, 1815—released up to 100 Mt-S. The most recent notable eruptions include Mount Pinatubo (Philippines, June 1991), 10 Mt-S, and El Chichón (Mexico, April 1983), 4 Mt-S.

In an average year, about 20 Mt-S are released as volatile compounds into air through small volcanic eruption vents and fumaroles. These ongoing sulfur leaks have a greater impact on environmental acidity and acid rain than do infrequent major eruptions, for two reasons. First, sulfur injected high into the atmosphere by explosive events is rapidly dispersed throughout the stratosphere, covering the entire planet. Accordingly, the potential sulfur deposition is diluted from regional to global scales. Second, the residence time of sulfur in the stratosphere is 1 to 2 years, causing the sulfuric acid to drizzle out slowly over this span of time. Hence, acidic deposition from major eruptions is never intense (except perhaps near the volcano soon after the eruption, when more serious threats would prevail).

Biogenic emissions of reduced-sulfur compounds (like hydrogen sulfide) from land and shallow bodies of water amount to roughly 30 Mt-S annually (Mt-S/yr). The oceans emit another 40 Mt-S/yr as dimethyl sulfide. Thus, in all, the biosphere, excluding human contributions, may add as much as 70 Mt-S/yr to the Earth's atmosphere. Of course, much of this sulfur ends up as acid in clouds, rain, fog, and aerosols.

The Modern Sulfur Cycle

Humankind has significantly modified the natural sulfur cycle by adding large sources to those already comprising the biogeochemical cycle of sulfur. Industrial activities, in particular, are responsible for emitting huge quantities of sulfur oxides into the atmosphere through fossil-fuel combustion. That sulfur (as well as the carbon in the fuel) had been slowly deposited over tens of millions of years. Normally, the sulfur would have been recycled through sediment uplifting, weathering, and volcanism over additional tens of millions of years. By mining fossil-energy reserves, we are greatly accelerating the rate at which the natural sulfur reservoirs are recycled from the Earth. Figure 9.5 illustrates the major elements of the present sulfur cycle, including the human components. The natural sources and sinks remain, of course. However, in the modern world, power generation, industrial fuel combustion, ore smelting, and transportation are also substantial sources of atmospheric sulfur.[10] These new sources, in fact, are larger in aggregate than the natural sources. As much as 100 million tonnes of sulfur per year (Mt-S/yr) are emitted by human activities directly into the global atmosphere, mainly as sulfur dioxide. By comparison, the global natural atmospheric sulfur cycle involves roughly 90 Mt-S/yr.

As noted earlier, increased emissions of sulfur and acid into the environment are likely to affect the natural emissions of reduced-sulfur compounds by microorganisms. The deposited sulfate provides a nutrient for growth of vegetation, which enhances the cycling of materials through the biosphere and the opportunities for secondary sulfur emission. On the other hand, the increased acidity may affect the pH of local environments and the viability of certain microorganisms. Thus, the food web and nutrient cycles may be perturbed in complex ways.

Anthropogenic Emission of Sulfur Dioxide

In the United States today, about 30 million tonnes of anthropogenic sulfur dioxide are emitted in the form of sulfur dioxide each year. This is equivalent to 15 Mt-S/yr.[11] The U.S. contribution amounts to one-sixth or one-seventh of the total global industrial output of sulfur dioxide. The geographical distribution of the SO_2 emissions over eastern North America is shown in Figure 9.6. The dark squares indicate the most intense regions of emission. In the U.S. Midwest, these correspond to the heavy industrial zone along the Ohio River Valley, and in southeastern Canada, to several areas of concentrated mining and refining activity. The anthropogenic component of the SO_2 emissions comes mainly from the generation of electric power, but also from a variety of industrial processes, including the operation of metal ore smelters (Figure 9.7). The burning of coal is the biggest culprit, since coal can contain from 0.5 to 7 percent sulfur by weight. The higher

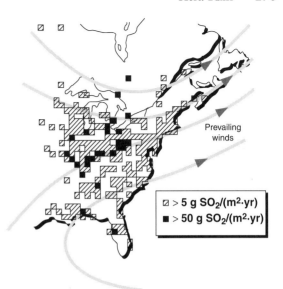

Figure 9.6   The pattern of sulfur dioxide emission in the eastern United States and Canada. The grid squares are 80 kilometers on a side. The solid squares are the sources of the greatest sulfur emissions, averaging more than 50 grams of SO_2 per square meter per year (more than 320,000 tonnes of SO_2 per block annually). The direction of the prevailing atmospheric winds at a height of 1.5 kilometers in July also is shown. The sulfur emissions are thus typically transported to the Northeast and deposited with precipitation in these regions. (Data from R. C. Henry, G. M. Hidy, P. K. Mueller, and K. K. Warren, *"Assessing Sources, Dispersion, Chemical Transformation and Deposition of Atmospheric Sulfur over Eastern North America by Advanced Multivariate Data Analysis"* [Paper presented at the Fifth International Clean Air Congress, Buenos Aires. International Union of Air Pollution Prevention Associations, 1980])

the sulfur content of the coal being burned is, the higher the emissions of SO_2 can be. Because high-sulfur coal is less expensive than low-sulfur coal, high-sulfur coal is preferred in regions with marginally profitable industry. Particularly in poorly managed areas of eastern Europe, China, and the Third World, massive releases of sulfur from industrial facilities are common. Over regional areas, such emissions can cause a factor of 10 to 1000 increase in rainfall acidity above that of natural precipitation.

The streamlines of wind flow over the eastern United States and Canada also are indicated in Figure 9.6. These represent the average directions over which air masses travel across this region (the wind pathways are shown for the month of July in the air that lies above the boundary layer). It is apparent that the sulfur emissions originating in midwestern

---

10.   Fuels used in cars, trucks, and aircraft are relatively free of sulfur. The oil-refining process that yields these fuels removes most of the sulfur in the crude oil. Accordingly, transportation represents the smallest anthropogenic contribution (a small percentage of the total).

11.   A sulfur dioxide molecule, consisting of one sulfur atom and two oxygen atoms, has a molecular weight of 64 atomic mass units (amu). The sulfur atom has a weight of 32 amu, half that of sulfur dioxide. Hence, a given mass of sulfur dioxide is equivalent to nearly exactly half that mass of pure sulfur.

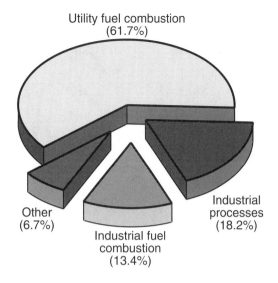

Utility fuel combustion
(61.7%)

Other
(6.7%)

Industrial fuel
combustion
(13.4%)

Industrial
processes
(18.2%)

Figure 9.7 The sources of sulfur dioxide emissions in the United States in 1982. Of the total utility fraction of 61.7 percent, 55.4 percent was from coal combustion and 6.2 percent from oil use. Utilities here refer mainly to facilities producing energy for distribution. For the industrial fuel source, the emission is split nearly 50–50 between coal and oil combustion. Among industrial processes, primary metal refining generates the largest sulfur emissions, accounting for about 9 percent of the total sulfur emission. The "other" category includes transportation fuels, 3.1 percent of the total emission, and commercial fuel burning, 2.6 percent.

states would be transported to New England, crossing over the Adirondack Mountains. Indeed, the deposition of acidity in that region is a major environmental concern (Section 9.5.1).

It has been clear for some time that humans are significantly modifying the global sulfur cycle (Section 10.2.1). On a regional scale, however, the relative disturbances caused by human activities can be even greater. The amounts of SO_2 emitted into the atmosphere (and of NO as well) have increased as the human population has grown over the last century. More people require more total goods and services and consume more total energy. In addition, the advent of the industrial revolution, with its insatiable demand for energy to produce goods and services that society demands, accelerated the consumption of fossil fuels and the consequent emission of sulfur oxides. During the century spanning 1850 to 1950, the main objective of businesses to make profits and the laissez-faire attitude of governments led to the uncontrolled pollution of air, water, and land. Lately, we have become wiser in controlling

emissions from industrial polluters, but the enormous growth in demand for energy and other commodities that require energy to manufacture has driven the rates of pollution upward. Only the "green revolution," which kicked in toward the end of the twentieth century, has begun to stem the tide of environmental pollution.

Figure 9.8 illustrates the growth of sulfur dioxide emissions (and of nitrogen oxide emissions as well) in the United States during this century. Over the entire period, emissions tended to increase with expansion of the population and industry. Between 1900 and 1980, sulfur dioxide emissions roughly tripled. This increasing trend could be traced back another fifty years. Nevertheless, the march upward has not been smooth. Major historical events have had a significant impact on sulfur emissions. During periods of economic depression, emissions generally decrease. In the Great Depression of the 1930s, for example, a precipitous drop in industrial output led to substantially lower emissions of sulfur dioxide. The great world wars that disrupted global civilization through the first half of the twentieth century created an industrial bonanza. During these times, the production of war materials boomed, as heavy industry attempted to keep pace with the massive continuous destruction of warfare. Typically, following each major war, a letdown in material demand led to economic recession, less industrial output, and a reduction of sulfur emissions. These ups and downs in the state of civilization are reflected in the ups and downs of the sulfur dioxide flux in Figure 9.8.

Projections of future sulfur dioxide emissions are clouded by uncertainties in the state of the global economy, population growth, development of clean technologies, and implementation of restrictions on emissions. One scenario for the United States shows present levels of emissions of about 30 Mt-SO_2/yr, increasing slowly to a maximum of about 35 Mt-SO_2/yr in the year 2020. After that, emissions decrease slowly with the introduction of low-sulfur energy generation systems to replace defunct utilities. Another possible scenario has SO_2 emissions decreasing to less than half their present levels by 2020, assuming a rapid introduction of new clean coal-burning technology and a phaseout of existing power plants. The future, at least for the United States probably lies between these two extremes, which differ by a factor of 2 in 2020. Most of the potential reductions in emissions would occur in the eastern United States, as might be expected, since most of the emissions occur there. Accordingly, the acidity of rainfall in the United States may be essentially fixed over much of the next century. It is less likely that it will improve substantially over the next 20 years, particularly if the momentum in regulating emissions is lost. With regard to the rest of the world, it is not certain that global emissions of sulfur will decrease for many decades. There simply is less imperative to spend money to clean the air in many of the most polluted nations. Fortunately, in North America and Europe, where serious efforts to control acid rain are under way, the problem of acidity will not be greatly exacerbated by heavy sulfur emissions on other continents.

9.3.3 Nitrogen Oxides and Acid Rain

Large amounts of nitrogen oxides are released each year into the atmosphere from natural and anthropogenic processes. These emissions can, in regions of concentration, lead to increased acidity of precipitation through the formation of nitric acid. The global nitrogen biogeochemical cycle is discussed in Section 10.2.2. Here, the basic elements of the fixed-nitrogen budget that affect regional acidity are described.

The Fixed-Nitrogen Cycle

The biogeochemical cycle for fixed nitrogen is depicted in Figure 9.9. The important natural and human processes that contribute to this cycle are illustrated. The nitrogen-bearing species shown in the figure include nitric oxide (NO), nitrogen dioxide (NO_2), nitric acid (HNO_3), and nitrate (NO_3^-). These are "fixed" nitrogen species in the sense that they are created by the breakup, or dissociation, of a nitrogen (N_2) or nitrous oxide (N_2O) molecules, releasing a "free" nitrogen atom that can attach to, or be fixed to, oxygen in one of the forms just listed. Both N_2 and N_2O have two nitrogen atoms that are tightly bound together. Considerable energy is required to break these nitrogen–nitrogen bonds. Hence, the concentrations of fixed-nitrogen species are generally much lower than the concentrations of stable nitrogen precursors such as N_2 and N_2O.

The primary natural sources of fixed nitrogen (Figure 9.9) include emissions from soils (about 10 million tonnes of nitrogen each year, or Mt-N/yr),

Figure 9.8 Emissions of SO_2 and NO in the United States during the twentieth century, in millions of metric tons. Rapid changes in the emission rates, particularly for sulfur dioxide, correlate with important historical events that have affected economic activity and growth. In recent years, the U.S. sulfur output has leveled off with the advent of stricter standards for sulfur emissions.

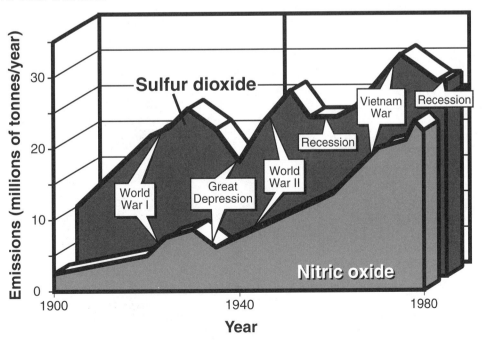

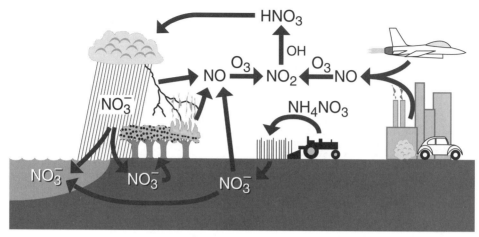

Figure 9.9 Some of the major components of the fixed nitrogen cycle, including contributions from human activities. The emissions of nitrogen oxides ($NO_x = NO + NO_2$) are converted to nitric acid mainly by the action of the hydroxyl radical (OH). Nitric acid is efficiently scavenged by clouds and rainfall. Precipitation, lightning, and biochemistry all have important roles in the fixed-nitrogen cycle.

lightning (roughly 5 Mt-N/yr), and the natural combustion of biomass (about 4 Mt-N/yr). Fixed nitrogen is created by lightning because of the very high temperatures—more than several thousand degrees Kelvin—that are reached in lightning bolts. The nitrogen and oxygen in air can react at these temperatures to form nitric oxide (Section 3.3.4). The source of nitrogen oxides in biomass burning is the fixed nitrogen found as amino acids in organic matter; the nitrogen is first absorbed by plants as a nutrient from soils, usually in the form of nitrate, and is then converted into protein-based material. When these organic materials are burned, the fixed nitrogen is oxidized and released as NO and NO_2. In soils, certain bacteria act to convert nitrogen from the atmosphere into fixed nitrogen, typically into ammonia, NH_3, a reduced compound. Other bacteria can oxidize ammonia into nitrate, NO_3^-, which plants then absorb.

For the natural cycle of fixed nitrogen, clouds and precipitation provide important steps in closing the loop. On the one hand, thunderclouds produce lightning that generates fixed nitrogen. On the other hand, precipitation washes fixed nitrogen from the atmosphere in the form of nitric acid. Acid deposition essentially completes the nitrogen cycle by returning nitrate to soils and surface waters, where it can fertilize plant growth and nitrogen recycling. It is only in extreme cases of contamination of the cycle by acids from human activities that the environment may become overacidified, leading to widespread damage. The natural environment in most places is

sufficiently buffered against natural nitric acid deposition to prevent harm.

Anthropogenic Emissions of Fixed Nitrogen

Figure 9.8 shows the annual emission rate of nitric oxide in the United States since the turn of the century.[12] In recent times, the overall emission rate has been close to 10 million tonnes of nitric oxide per year (Mt-NO/yr), which is equivalent to about 5 million tonnes of nitrogen annually (Mt-N/yr). These emissions have increased by a factor of 4 since 1900. In contrast to the pattern for sulfur emissions in Figure 9.8, the increase in NO emissions has been relatively smooth, with a significant deceleration occurring only during the depths of the Great Depression. The anthropogenic source of NO is tied more closely to transportation than to industrial power generation, which is the reverse situation for sulfur dioxide. Internal-combustion engines work at temperatures that are high enough to generate nitric oxide, whereas many boilers and other industrial burners operate at lower temperatures. The transportation sector of the economy has grown at a fairly regular rate over the last 50 years. After Henry

12. This is given as the equivalent total amount of nitric oxide corresponding to all the nitrogen oxides that are emitted, including nitrogen dioxide and nitric acid. Most of the oxidized fixed-nitrogen emissions are in the form of nitric oxide and nitrogen dioxide, but the exact mixture of these two components is very difficult to determine for most sources. Hence, the equivalent quantity of nitric oxide is usually specified.

Ford[13] introduced the assembly line concept for mass production in 1913, the manufacture of automobiles, trucks, and other vehicles worldwide has expanded steadily until the present. Likewise, the global emissions of pollutants from vehicles have speeded up.

The U.S. emissions of nitric oxide, mainly from vehicles, represent one-quarter of the total worldwide emissions from fossil-fuel combustion—that is, some 40 million tonnes of nitric oxide per year (Mt-NO/yr), which is equivalent to about 20 Mt-N/yr. Aircraft engines account for perhaps 1 Mt-NO/yr (roughly 0.5 Mt-N/yr) of the total emissions to the atmosphere. The seasonal burning of agricultural land has several functions: to remove plant litter, to kill insect pests and weeds, and to recycle minerals in crop stubble. As a result, up to 20 Mt-NO/yr (10 Mt-N/yr or so) may be emitted worldwide from burned fields.

The ammonia in fertilizer is an important anthropogenic component of the fixed-nitrogen cycle. A common form of fertilizer is ammonium nitrate, NH_4NO_3. Ammonium nitrate is formed when ammonia and nitric acid are combined.[14] When spread on soil as a fertilizer, the ammonium nitrate dissolves in water in the soil and dissociates into its ions, ammonium (NH_4^+) and nitrate (NO_3^-). Both ammonium and nitrate ions are sources of fixed nitrogen for agricultural crops. The nitrate can be utilized directly, being absorbed with the water taken up by roots. The ammonium is oxidized to nitrites and nitrates by bacteria in the soil using oxygen from the atmosphere that diffuses into the soil. Although not indicated in Figure 9.9, equal amounts of fixed nitrogen as ammonium and nitrate are contained in ammonium nitrate fertilizer.

The application of fertilizers worldwide releases about 20 Mt-N/yr into soils. Some of this fixed nitrogen escapes into the atmosphere in the form of nitrogen oxides and ammonia (as well as nitrous oxide; see Section 10.2.2). In Figure 9.9, however, the anthropogenic sources of fixed nitrogen that contribute to acid rain are dominated by fossil-fuel combustion processes. For one thing, these latter sources are more concentrated. Moreover, ammonia emissions resulting from fertilizer usage and other agricultural activities can actually neutralize the acidity resulting from nearby sources of nitrogen and sulfur oxides (Section 9.2.3).

9.4   Acid Fog

Acidic particles and fog in polluted urban air were discussed in Section 6.5.3. Los Angeles and other urban areas may, from time to time, have serious problems with acidic fog rather than acidic rain. The fog in Los Angeles typically has a pH value between 4 and 5. In acid fog events, the pH drops to 2 to 3; even a pH value as low as 1.7 has been measured. Acid fog can contain sulfuric, nitric, and organic acids. In southern California, where SO_2 emissions are relatively low, acid fog is dominated by nitric acid. Roughly speaking, there may be three times as much nitrate as sulfate in Los Angeles fog.

Acidic aerosols can readily form in polluted coastal regions where fog is also common. Figure 9.10 shows the stages in the development of such acid fog. Initially, the fog droplets appear over the ocean in humid marine air. They typically condense on aerosols that normally would act as cloud condensation nuclei (CCN). Fog, in fact, is nothing more than a cloud at the Earth's surface. In air over the oceans, CCN are usually composed of sea salt, principally sodium chloride (common table salt). These sea-salt CCN are created when wind blows over the tops of waves, causing them to break, or when bubbles burst at the ocean surface. Sea-salt aerosols are relatively harmless, although the salt does attack painted surfaces and cause metal to corrode (so keep your car covered if you live near the beach!).

As the fog drifts over land, the droplets of water can interact with pollutants emitted at the surface. Acid precursors such as sulfur dioxide and nitrogen dioxide can be absorbed into the fog droplets, forming weak acids. If nothing further happened, acid fog would not be a problem. However, farther inland and later in the day, the fog droplets also absorb oxidants such as ozone and hydrogen peroxide. These oxidants react with the dissolved sulfur and nitrogen oxides, yielding strong sulfuric and nitric acids, respectively. Moreover, acid vapors in the

---

13.   Henry Ford was born to Irish parents in 1863 in Michigan and died there in 1947. Ford dropped out of school at the age of 15 but nevertheless later became fabulously successful as a manufacturer. His first assembly line car, the Model T, sold for about $500 in 1913. By mid-1914, more than half a million Model T's were on the road. Ford is considered to be the father of mass production, which today is used in the production of most consumer goods.

14.   Recall that when ammonia is dissolved in water, it forms a base, ammonium hydroxide, which can react with and neutralize nitric acid, forming a salt, ammonium nitrate.

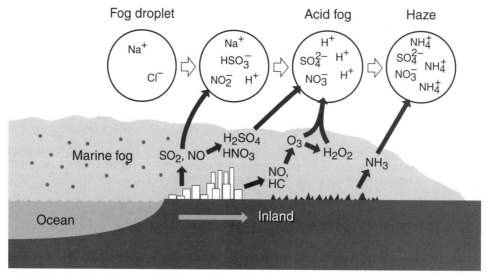

Figure 9.10 The development of acidic fog and haze in a coastal region. The natural sea-salt aerosols and water droplets in a marine fog provide the liquid medium that collects and accumulates acid components of smog. Sulfur dioxide and nitric oxide can be absorbed into the fog droplets and oxidized into sulfuric and nitric acid. SO_2 and NO also are oxidized in the gas phase and then absorbed. Smog reactions between NO and hydrocarbons generate ozone, which creates additional oxidants such as hydrogen peroxide (H_2O_2) that acts on the sulfur and nitrogen components of the fog to produce acids. Ammonia released in rural areas combines with the acids to form ammonium sulfates and nitrates that remain in the form of an aerosol residue when the water evaporates from the fog, thereby creating haze.

polluted air are quickly absorbed into the fog droplets. Not long after landfall, the harmless, salty fog may be transformed into a highly acidic soup.

The acid fog is at its worst if it lingers over coastal regions. Then it has plenty of time to absorb acidic pollutants, acting like a sponge for acids and concentrating them in the fog droplets. The evolution of the fog is not finished, however. If the fog continues to drift inland, it will encounter emissions of ammonia that can neutralize the acids. The fog may also be neutralized by alkaline dust raised from the land surface. At the same time, the fog encounters warmer inland daytime temperatures and begins to evaporate. If the acids have not yet been neutralized at this stage, they will become even more concentrated in the evaporating drops.

Even as the fog moves inland and dissipates, a residual aerosol haze is left behind. These haze particles consist of nitrate and sulfate salts that are formed when the acid fog droplets are neutralized by reactions with ammonia and other bases in the air. This haze is often very dense, causing a profound degradation of local visibility. Haze particles do not completely evaporate, as do water droplets in a clean fog heated by the sun. Indeed, the aerosols that remain after the acid fog dissipates are relatively

large, owing to the accumulation of nitrates and sulfates during their transit from the coastal region through heavily polluted air. These larger particles are also respirable and have significant health implications.

Acid fog can be a problem in any region where fog tends to form in a heavily polluted air mass. Fog can be generated by radiation cooling, marine air, or excess humidity and is common in many areas. Even at a distance from the coast, fogs may be a regular meteorological feature. In California's Central Valley, "killer" radiation fogs occur every year, blanketing highways and causing massive car and truck pileups. In valleys in which pollution may be trapped and fog may form, exposure to acid fog may be likely. Although people can retreat indoors to avoid contact with the acid droplets, plants cannot. Forests may be particularly vulnerable to the effects of long hours drenched in a corrosive acidic fog.

9.5 The Costs of Excess Acidity

The environmental and health costs associated with acid rain and fog, and the primary and secondary air pollutants that cause environmental acidity, are difficult to determine precisely. Among the issues one must consider are economic and aesthetic losses associated with dying lakes, damaged forests, lost agricultural productivity, dissolved artworks, disintegrating infrastructure, and deteriorating human health. Acid rain came to prominence mainly because of its association with the sterilization of pristine mountain lakes. More careful consideration of the causes of environmental decline due to acid deposition points up a number of potential hazards. Calibrating the effects of pollutants generated in local smog on distant lakes and forests, however, is rather difficult. The obnoxious conditions inflicted on city dwellers by modern-day smog are often rather obvious; in contrast, the ambiguous pathologies that are appearing in remote ecosystems are frequently very subtle. Nevertheless, it is worthwhile pursuing a number of lines of inquiry into the possible detrimental effects that acidity may cause in the environment and to human health.

9.5.1 DYING FORESTS AND LAKES

In many forests, the ecosystem is delicately balanced. Trees and other vegetation may be highly sensitive to changes in soil chemistry, among other factors. When acid is deposited on poorly buffered soils, the acidity can leach metals from minerals in the ground. Aluminum, it turns out, is readily mobilized by acids in groundwater; the acids release the aluminum into solution as an ion. In this mobile state, aluminum can move through the soil to roots, where it may be absorbed into the plant tissue. Aluminum is toxic to many plants. At the same time, acidity mobilizes valuable nutrients in the soil—calcium, magnesium, and potassium—which are leached from the soil, creating nutrient deficiencies. These so-called cation nutrients (because they exist as positive ions in aqueous solution) may also be leached directly from leaf surfaces exposed to acid deposition. Contact of acidic cloud and fog droplets with leaves may also cause direct damage to the leaf surfaces or may inhibit transpiration (the breathing action of the stomata). Trees weakened by exposure to acid precipitation, possibly in combination with other pollutants such as ozone, are more susceptible to attack by insects and diseases. Forest regions experiencing problems with acidic precipitation include southeastern Canada, the Adirondack Mountains, Scandinavia, northern Europe, parts of the Sierra

Nevada Range in California, and other high-country areas in the western United States.

Considering the impact of acid pollution on forests and lakes, one must recognize that there are sensitive kinds of organisms and tolerant ones. Just as people often respond in different ways when exposed to air pollution, so do trees and other vegetation and wildlife. Plants that are sensitive to acidity include ash, white oak, azalea, American sycamore, English walnut, Ponderosa and Monterey pine, barley, wheat, and tobacco. Relatively tolerant plants include avocado, birch, dogwood, holly, maple, red oak, spruce, red pine, poison ivy, beets, and rice. Each species needs to be tested to determine its tolerance of acidity. Since individual ecosystems may contain hundreds of individual species—both flora and fauna—the response of an integrated ecosystem to acid assault is much less predictable. For example, an otherwise acid-tolerant species might succumb following acid deposition because its food source is sensitive to acidity. Other species might have embryonic stages that are sensitive, even though adult forms are tolerant. Many fish are sensitive to aluminum ions in water and develop gill disease from continuous exposure. Thus, physical and chemical changes induced by acid deposition can have complex and unforeseen physiological and ecological consequences.

The soils of the western United States are not highly alkaline. Hence, the region is susceptible to damage from acid rain. Even so, the western regions do not yet have a problem as serious as that in the eastern United States. This is mainly a result of the fact that there are fewer clustered sources of SO_2 and NO_x in the west. Facilities that produce the largest emissions of sulfur dioxide, for example, have been dispersed over open desert lands. Nevertheless, because of the areas of heavy pollution, ecosystems in regions adjacent to these areas have been impacted. The effects are particularly serious in high mountain lakes downwind of urban zones. Typically, the watersheds feeding these lakes consist of thin, poorly buffered soils derived from granite rocks. Contaminated snow that accumulates in winter can release a surge of acids into these lakes when the snow pack thaws in spring. This surge can be dramatic, partly because of the tendency for the acids contained in the accumulated acid snow to be leached out quickly as the melting begins.

Several possible interactions may occur when acid precipitation falls on a local environment consisting

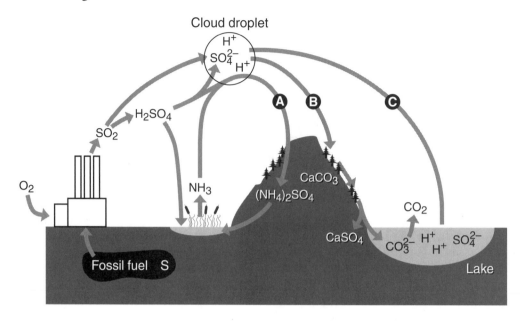

Figure 9.11 Potential interactions of acidity with the environment. The major pathways illustrated are A, neutralization of acids in clouds followed by deposition and fertilization; B, precipitation of acid from clouds followed by neutralization in soils or runoff into lakes; and C, deposition of acids from clouds directly onto water surfaces. The sulfur emissions due to fossil-fuel combustion are emphasized, although the effects of nitrogen oxide emissions would be similar. The excess anthropogenic sulfur acidifies clouds (and, likewise, rain and fog) to abnormally low pH values. The acidity may be neutralized by airborne alkaline substances such as ammonia gas (NH_3), which forms ammonium sulfate as a by-product. The ammonium sulfate is then deposited as a fertilizer (A). If not neutralized, the acid may fall directly onto soils (B). For soils that are well buffered—with calcium carbonate ($CaCO_3$), for example—the acid is neutralized, forming mineral residues such as calcium sulfate ($CaSO_4$). Unbuffered rain may also fall directly onto (or run off the land into) lakes, significantly acidifying them (C). The degree of acidification depends on the buffering capacity of the lake water—for example, its carbonate content.

of forests and lakes, as depicted in Figure 9.11. The chemical effects include fertilization, which is beneficial, and acidification, which is generally harmful. The actual effects of acid deposition depend on the chemical composition of the clouds and rain that are the source of the acidity, and the buffering capacity of the exposed soils, watersheds, and lakes. In Figure 9.11, pathway A illustrates the effective fertilization of a local environment by sulfate deposited from clouds whose droplets have been neutralized by ammonia or other bases. Presumably, the alkaline materials originated locally and were entrained into the clouds by convection and other processes. Path B shows that the acidity in rainfall may be effectively neutralized by mineral buffers in the ground as the runoff passes over and is absorbed into soils. On the other hand, if the soils are not sufficiently alkaline or porous, the runoff could remain highly acidic as it entered streams and lakes. For pathway C, acids are

deposited directly into lakes, where they may be neutralized only by the presence of alkaline buffers in the lake water.

Forests that are exposed both to high levels of acid rain and smog are at even greater risk. The combined action of acidity and ozone seems to be synergistic, that is, producing an overall effect that is greater when they are acting together than when acting independently. Moreover, several other factors may come into play once individual stands of trees are weakened by combined exposure to acidity and ozone. For example, embattled trees are more vulnerable to attack by opportunistic insects and fungi, to extreme temperatures in winter or summer, and to coexisting pollutants like sulfur dioxide and nitrogen dioxide. The interactions of acids with soils create simultaneous stresses on root systems, including aluminum ion poisoning, cation nutrient depletion, and beneficial microbe depletion. The impact of these changes may later be exacerbated by spells of dryness. Synergy between stresses caused by acid rain and smog has been particularly troublesome in Europe's heavily industrialized Rhine Valley region. Here the air is often laden with heavy smog having high ozone concentrations in addition to heavy sulfur pollution and high acidity. In the 1970s and 1980s, large tracts of forest downwind of this source of deadly air were severely damaged. Strict emissions control measures have begun to clean the air and reduce the regional forest destruction.

In the Adirondack Mountains of northern New York State, dotted with picturesque lakes, the ecosystems are relatively fragile. The soils are granitic, with a low buffering capacity. Although damage to many lakes in the region has been noted, the forest damage is less obvious. Up to 20 percent of the Adirondack lakes are no longer habitable for sensitive fish species, and 10 percent cannot sustain even acid-tolerant species. Because of pollution moving in from the Midwest, rainfall in the Adirondack range has an average pH of about 4.3. Worse yet, pH values as low as 2.8 have been detected in the cloud water bathing forests in the region. The combination of acid baths and rains and thin soils has probably contributed to the decline of these forests.

As already noted, lakes are strongly affected by the runoff of acid rain from soils and of meltwater from snow fields. Forests, on the other hand, are damaged by direct contact with acid clouds and fogs and by the leaching of heavy metal ions from soil minerals exposed to acid rain. Soils that are sensitive to acidic leaching are generally poorly buffered. Hence, it is not surprising that the lakes fed by the runoff from these soils also have a low buffering capacity and can be readily acidified. The dual problems of forest and lake decline caused by acid deposition are intimately coupled and have a common origin. In a healthy ecosystem, all elements are healthy. Acid rain is a particular nuisance in fragile ecosystems (like many mountain lakes) because it simultaneously attacks and weakens many elements at once.

9.5.2 A POTPOURRI OF DESTRUCTION

The negative impact of acid rain can extend far beyond the destruction of remote forest/lake systems. Environmental acids devastate local environments, attack agricultural crops, corrode works of art, dissolve structures, reduce visibility, and create human health hazards. In this section, we describe some of these effects and attempt to place the problems they raise in the perspective of their cost to society.

Moonscapes

A dramatic example of an isolated source of pollution (SO_2 in this case) creating an enormous local environmental impact is the Sudbury (Ontario), Canada, ore-smelting operation. Sudbury is one of the largest mining towns in the world, with a population of 150,000 souls living in quaint company homes. Local industry exploits the massive deposits of copper and nickel in the rich earth. Unfortunately, Sudbury's smelting facilities emit 2500 tonnes of sulfur dioxide into the atmosphere every day (about 37 pounds for each resident). In an attempt to reduce the local impacts of this dense pollution, the largest smelter constructed a smokestack 380 meters (1250 feet!) high; nearly one-quarter of a mile. This extraordinary stack allows sulfur dioxide to drift some distance from the town before returning to the ground. But not far enough. Several hundred square miles of habitat just downwind of the smelter have been literally obliterated. Photographs of the area show barren ground devoid of life stretching to the horizon. One is reminded of images beamed from the surface of the moon. The lakes in this region are acidic and totally dead. The denuded land is discolored and uninviting. Sudbury represents the extreme consequences of excessive doses of acidic pollution. It is a, thankfully, localized anomaly in which life over a great area of land has been extinguished for the convenience of dumping waste gases into that infinite trash reservoir, the atmosphere.

No doubt, most of the sulfur emitted from Sudbury drifts far downstream, blending with pollution from many other sources, before being redeposited at the surface. Sudbury sulfur may travel as far as the northeast United States before it falls back as acid rain. Because it is intermingled with sulfur from other sources, the identification of Sudbury as the culprit is clouded. This is just the situation desired by the Sudbury emitters: dispersal of their problem to other towns and counties without cost or liability. The same attitude prevailed in the Soviet Union and Eastern European communist bloc countries for at least five decades. There are likely to be many Sudbury moonscapes to view right here on Earth.

Agricultural Decline

The principal effects of acidic precipitation on agriculture products include reductions in yield, loss of quality, and surface damage. Because the anions in acid rain—primarily the sulfate and nitrate—are essential plant nutrients, crops can be fertilized in nutrient-poor areas. The overall effect on yield for a given crop on a particular soil is the net balance between the impacts of acidification and fertilization for that species. Other stress factors must also be considered, such as water availability, temperature, insect attack, and disease. Finally, agricultural factors and productivity in a region are subject to large variability. One can imagine fluctuations in rainfall and temperature from year to year. Moreover, crop yields may depend on individual crop management techniques, which change over time. Finally, unpredictable events, such as hailstorms, frosts, droughts, or insect invasions, can severely damage crops. Against this background of natural variability and human intervention, the relatively small impacts of acid rain must be identified. In general, major crop perturbations related to atmospheric acidity have not been detected in the United States.

It has been established that the pH of rainwater under most conditions is not low enough to damage crops directly. In California, for example, there are very few recorded instances of significant crop losses from exposure to atmospheric acidity, but exposure to highly acidic fog can damage the surfaces of sensitive fruits, such as some apples. The disfigured produce could not be marketed in today's supermarkets, where only perfect fruit and vegetables are picked up, squeezed, and bought. Nevertheless, such produce could still be used in canned or processed foods, since the nutrient quality is not affected by the surface damage.

The existence of acid precipitation may have more significant indirect implications for agriculture because its precursors are capable of severely damaging agricultural crops. In other words, the presence of sulfur dioxide, nitrogen oxides, and ozone in the atmosphere poses a greater hazard to domestic plants than the acid rain they create. Although the detrimental effects of acid deposition on crops have not been fully investigated (in part because they are expected to be rather small),

exposure to ozone is well known to cause significant reductions in yields of commercially important crops. For example, exposure even to background concentrations of ozone (assumed to be 30 parts per billion by volume) can reduce yields by up to 50 percent for a range of crops, including corn, wheat, cotton, and soybeans.

Kiss Michelangelo Good-bye

In Italy, every city and town has memorable statuary and monuments, buildings covered in marble, and ornate limestone decorations. Italy is waging a war against automobile and industrial pollution to save its irreplaceable antiquities. The largely uncontrolled emissions of automobiles and the chemicals in smog are irreversibly destroying the artistic and architectural heritage of the ancient world. Building materials such as limestone and marble are composed of calcium carbonate. The chemistry of calcium carbonate is important to the global biogeochemical cycle of carbon (Sections 4.1.3 and 10.2.4). It is also important to the degradation of irreplaceable antiquities. Acid precipitation and fog, as well as acidic vapors in smog, are attacking the artistic remnants of classical civilization, not to mention the infrastructure of modern civilization. Sulfuric and nitric acids, in particular, are reacting with the stone and masonry artifacts of ancient times, dissolving, eroding, and disfiguring them.

Although all stone materials are subject to attack, marble and limestone are especially sensitive, since they readily dissolve in acid solutions. Moreover, sulfuric acid converts calcium carbonate to gypsum, or calcium sulfate.[15] Gypsum is used to make "gypsum board," or plaster wall panels. Turning Michelangelo masterpieces into crude plaster is an obscene modern alchemy.

The chemistry of marble and limestone erosion can be summarized in the following reaction:

$$CaCO_3(\text{solid}) + H_2SO_4(\text{aqueous})$$
$$\rightarrow CaSO_4(\text{solid}) + CO_2(\text{gas}) \quad (9.27)$$
$$+ H_2O$$

15. Gypsum, more precisely, is composed of calcium sulfate and water of hydration—two water molecules for each calcium sulfate molecule. In pure form, gypsum is crystalline; another common form is alabaster. The form of calcium sulfate without any water of hydration is called *anhydrite*.

When carbonate rock comes into contact with a solution of sulfuric acid, some of the carbonate enters the solution at the wetted surface. These carbonate ions (CO_3^{2-}) are effectively replaced by sulfate ions (SO_4^{2-}) in the form of dissolved gypsum. At the same time, the carbonate ions are converted into carbon dioxide, which escapes from the solution as a gas.[16] When the remaining water evaporates, solid calcium sulfate is left behind. Deposited as an amorphous surface coating with soot and other materials mixed in, the calcium sulfate appears as a dark, disfiguring crust on the stone.

There are several mechanisms that degrade stone in polluted air. First, the stone material may be dissolved, as just described. In this case, details are worn from the surfaces, faces become unrecognizable, and chiseled text turns unreadable. The surface of the stone can also be discolored by gypsum deposits, as noted. These deposits tend to form on surfaces sheltered from direct rainfall, which has the effect of washing off the dissolved calcium sulfate before it can dry on the surface. A surface buildup can form from the direct deposition of sulfur and nitrogen oxides acid vapors on the stone. This dry deposition process is sensitive to humidity and probably involves water that usually adheres to surfaces. Finally, surfaces exposed to polluted air and precipitation can crack and spall when the accumulated crustal material thickens and breaks apart.

Among the different kinds of stone used in building, granite tends to be the most resilient because of its hardness. Sandstone is generally more sensitive to acid erosion than marble. In experiments on the erosion of marble surfaces exposed to polluted air, about two-thirds of the loss was associated with sulfur acids, and the rest with nitrogen acids. The measured rates for flat surfaces are nearly a factor of 10 smaller, however, than the observed erosion rates of marble columns in polluted cities. Average rates of recession as large as 10 micrometers of depth for each meter of rain have been found.[17] The

rate of loss is related to the shape of the surface as well as its condition and exposure to the elements and acids. In highly polluted industrial areas of Germany, for example, sandstone recession rates as large as 50 micrometers per meter of rain were measured.

The Italians may be losing their battle with automobile pollution and its monumental destruction. The major cities—Rome, Naples, Milan—remain notorious for their traffic jams. Although automobiles have been prohibited from many smaller town centers, swarms of buzzing motorbikes continue to pound the sensibilities and foul the air. Local laws imposing alternate-day driving and no-car days, among other earlier measures, were abolished by the government in 1992 in favor of regional solutions, which never materialized. One major step in Rome, however, has been the banning of VIP caravans racing through the streets; only the president, prime minister, and pope can now have them.

In addition to valuable stonework and statuary, other building materials are degraded by exposure to acid precipitation, fog, and aerosols. Metals are vulnerable to dissolution, pitting, and cracking. Copper, bronze (a mixture, or alloy, of copper and tin), and brass (a mixture of copper and zinc) were used widely in ancient times because of their availability and workability. All three are subject to verdigris, the green-blue deposits of copper salts leached from the metals when exposed to the atmosphere. Green copper-topped buildings are charming, but the decay of the underlying metals can, over time, be substantial, requiring costly replacement. In urban areas, the corrosion rates of metals are closely related to the concentrations of sulfur dioxide in the atmosphere and increase as the SO_2 concentration increases. Zinc, copper, steel, and galvanized steel all are susceptible to dissolution by environmental acids. Conversely, field studies suggest that nitrogen oxides are not a significant factor in metal corrosion. Both the wet deposition of sulfuric acid and the dry deposition of sulfur vapors play a major role in metal degradation, accounting for perhaps half of all metallic corrosion. Measured corrosion rates are highly variable, however, and depend on the conditions of the exposure.

16. When you have heartburn caused by excess stomach acidity, you may take antacid tablets for relief. These antacids are nothing more than a chewable form of calcium carbonate. Hence, as the calcium carbonate reacts with stomach acid, it releases carbon dioxide. That's why you may need to belch up a bubble of CO_2 gas.

17. The mean annual rainfall in the western United States is only about 25 centimeters, and in the eastern United States, roughly 100 centimeters. Europe generally receives 50 to 100 cm/yr, but most of Asia, 25 cm/yr or less. Along the equator, mean rainfall is typically 200 to 300 cm/yr. The greatest recorded

annual rainfall occurred in a remote area of India in 1861, more than 2600 centimeters (85 feet!). In one day in 1952, more than 6 feet fell on Reunion Island. At a rainfall rate of 1 meter/year, a 1-meter thick column could be completely eroded in 100,000 years; its fluting could disappear in a thousand years.

Paints and special coatings are subject to degradation by acidified air. Painted wood and metal surfaces peel and crack sooner. The underlying substrates can also be damaged in the process. Wood is quite sensitive to absorbed acid solutions and, following exposure, is not as good at holding paint. Recently, evidence was collected showing that sulfur-bearing particles deposited on car finishes can etch spots into the finish. The etching occurs if the deposit is allowed to remain on the paint for more than 1 or 2 days. The solution, of course, is to wash your car every day.

Acids corroding everyday building materials—putting aside for the moment the insidious irreversible destruction of cultural objects—is not without its cost to the public. The maintenance and rebuilding of infrastructure in the United States is a focus of political attention today. What would a country be without its highways, bridges, tunnels, skyscrapers, and docks, not to mention its fences, lampposts, sidewalks, fire hydrants, and mailboxes? In many U.S. cities, the infrastructure is dilapidated and getting worse. Reinforced concrete is crumbling as steel reinforcements fail. Roads are pocked with potholes and bridges are near collapse. One estimate places the total cost of infrastructure repairs currently needed at $1 trillion! That is $4000 for every citizen. Moreover, the repairs and bills continue to mount. It is difficult to assess the fraction of these total costs that may be associated with acid deposition. On the other hand, air pollution is certainly a factor in accelerating the decay of infrastructure and accelerating repair replacement schedules. Cleaning the air has many other benefits, it seems.

Haze and Visibility

The formation of haze and the degradation of visibility associated with acidic aerosols and fogs are discussed at length in Section 6.5.3 (also see Section 9.4). The processes leading to haze formation are shown in Figure 9.10. Although acid rain itself does not form haze and actually scavenges haze particles from the atmosphere through washout and rainout, the initial stages of acid formation and condensation are implicated in episodes of low visibility (Figure 9.10). The degradation of visibility is a serious aesthetic issue in urban and industrial areas. Furthermore, negative health effects are usually associated with exposure to acidic aerosols and residual haze. These health effects are discussed in the following section.

9.5.3   HEALTH IMPLICATIONS

In the Ukrainian town of Chernovtsy, children were going bald. Like many towns and cities in the old Eastern-bloc nations, Chernovtsy has had a serious problem with air pollution and acid rain. But Chernovtsy was also experiencing unusually high concentrations of thallium, a highly poisonous metal, in its water supplies and soils. The symptoms of thallium poisoning include nervous disorders and, you guessed it, loss of hair. Thallium is a by-product of ore-smelting operations, ending up in the fumes and dust. It is usually recovered because it is so toxic in the environment. Once used as a rodenticide, however, thallium today has no commercial value. Another property makes it dangerous to release thallium fumes: It readily dissolves in nitric and sulfuric acid solutions. In Chernovtsy, acid rain was carrying a poisonous dose of thallium to the townspeople; a toxic Mickey Finn. Now where did all that thallium come from, anyway? Certainly not from the local industrial facilities nearby. They wouldn't poison the air their own children breathe.

The major health effects of acid rain are indirect and involve the dissolution of toxins into water supplies, whereas the primary effects of acidic aerosols and vapors are direct and involve inhalation. In sensitive regions, the deposition of acid rain on soils and lakes causes the fresh water and groundwater to become more acidic. That in turn allows the water to dissolve more heavy metals from minerals in the soil or lake bottom, or in plumbing and cisterns in water distribution systems or houses. Lead, for example, can be leached from older pipes and fixtures (Sections 7.2.4 and 8.5.2). Only a relatively small fraction of the U.S. population uses surface or groundwater directly, without treatment (this may not be true in other countries—remember Chernovtsy). For these people, estimates suggest that exposure to lead could increase by 5 to 15 percent on average if acid precursor emissions increased by 20 percent. In acid-sensitive areas, however, lead exposure could more than double for some persons. In these cases, lead concentrations in blood could far exceed levels associated with adverse health impacts for chronic exposure. The consequences of lead ingestion include neurological retardation in fetuses and young children.

A second pathway for toxic exposure from acid rain is the dissolution of mercury into fresh water and its bioactivation into highly toxic methylmercury in fish. Accordingly, sport fisherman and Native Americans who consume fish are at risk from increased mercury exposure. Indeed, one assessment assuming the strong acidification of rainwater (to pH 4.5 or less) found an increase in mercury intake by subsistence fishers from a few micrograms per day to more than 200 micrograms per day, above recommended rates. Chronic exposure to mercury can damage the nervous system, leading to weakening of muscles, loss of motor control, and, in high doses, eventual paralysis. Other toxic heavy metals that can be enhanced in groundwater by excess acidity are aluminum, arsenic, cadmium, and selenium.

Since the concentrations of heavy metals in groundwater and lake and stream water are elevated by increased acidity, organisms other than fish also are affected. Plants and animals that are food for people may accumulate the heavy metals. An evaluation of this source of potential health effects shows that it is much less significant than the use of contaminated drinking water and consumption of game fish living in such water.

The direct effects of acidic aerosols and fogs are mentioned in Section 6.5.3. As noted there, the life cycle of acid fog generates respirable aerosols laced with toxic compounds (Section 9.4 and Figure 9.10). These particles are readily deposited in the lungs. Exposure to high concentrations (low pH) of airborne acid solutions can cause respiratory distress and aggravate asthma. It has been found that hydrogen ions associated with sulfuric acid produce the greatest pulmonary response for a fixed amount of acid. The effects of nitric acid appear to be much less severe. Nitric acid accounts for most of the airborne acidity in the southern California area and can account for more than 20 percent of the acidity in the Northeast. One epidemiological study suggests that the occurrence of bronchitis in children was associated with airborne acidity. Exposure to sulfuric acid also suppresses the formation and ejection of bronchial mucus. Hence, other toxins and bacteria deposited in the lungs are not removed as efficiently. Excess illness and death have been clearly associated with the killer smog events of London (Section 6.1.2), and particularly the high concentrations of sulfuric acid aerosols. Even at lower present-day concentrations of airborne acidity, excess mortality and morbidity are expected. But, the presence of ozone, soot, sulfur dioxide, carbon monoxide, and other toxic pollutants in smog complicates the interpretation of these health data.

9.6 Controlling Acid Rain and Fog

It is possible to reduce the acidity of precipitation, fog, and aerosols through reasonable and practical actions. Society and industry must either accept the damage of continuing or increasing acidity in the environment or choose to take actions that would mitigate it. Moreover, it may be cost effective to reduce acidity and reap the benefits. This could be accomplished by any number of schemes to remove acid precursors—primarily sulfur dioxide and nitrogen oxides—from industrial and automotive emissions. The benefits to society are greater agricultural yields, larger timber reserves, less infrastructure maintenance, reduced health costs, and more recreational opportunities.

The possible approaches to control acid rain include schemes to reduce precursor emissions, to disperse the emissions to a greater degree, and to compensate for acidity at receptor sites. Some specific ideas are

- Using low-sulfur fuels in sensitive regions.
- Removing sulfur from high-sulfur fuels, such as coal, before burning them.
- Scrubbing the emissions of large facilities to reduce the amount of sulfur emitted into the atmosphere.
- Dispersing emissions using higher smokestacks to dilute their local impacts.
- Developing large-scale alternative energy sources, such as biomass fuels and solar energy.
- Siting power plants to minimize regional impacts.
- Buffering lakes against acidification by adding lime or other alkaline compounds to the lake water.

How practical are some of these measures? In fact, they all are practical and already in use to varying degrees. Sub-bituminous coal (black lignite) and lignite (peat coal) are low-sulfur fuels that, although having a lower energy content than normal bituminous coal (soft coal), are used where sulfur emissions must be minimized. Anthracite (hard coal) is relatively rare and expensive but is favored for its low smoke production. Sulfur removal from natural fuels can be expensive and is usually done only for special applications, such as aviation turbine fuels. Nevertheless, coal can be processed into a slurry, a liquid, or methane and other flammable gases. In these processes, the sulfur impurities can be reduced significantly, although that adds to the effective cost of the final fuel. The removal of sulfur from coal smoke is also a well-developed technology. Scrubbers oxidize sulfur dioxide in the exhaust stream into sulfate, which is soluble and can be literally washed from the exhaust. Scrubbers are generally bulky and expensive to install and operate. Moreover, the residue sulfates and other smoke components must be carefully discarded.

One of the earliest and cheapest techniques used to disperse sulfur emissions was to build extremely high smokestacks (refer to Section 5.4.1). Some of these enormous stacks tower 1000 feet into the atmosphere. The point is to release the sulfurous smoke as far above the ground as possible so that it takes longer to reach the surface. By that time it would have traveled farther and dispersed to a greater degree. If the stack is high enough, the sulfur may not cause local environmental damage. Typically, the highest smokestacks are built for the largest power plants or smelters with the greatest sulfur emissions. Accordingly, the countryside may still be severely damaged, but over a larger region. Sending the problem downwind is also unneighborly.

The possibilities for alternative energy sources are discussed at length in Sections 12.5.1 and 12.5.2 (also see Section 6.6.2). There seem to be many options here, none of which is an ideal or complete solution. The availability of the cleanest energy sources such as hydroelectric and geothermal power is too limited to meet society's energy demands. Other sources, including solar and wind power, are not cost effective, or they create eyesores. Alternative biofuels result in a new suite of environmental problems. Nevertheless, it would seem reasonable that if a number of these alternatives were combined in a comprehensive energy strategy, a clean global energy system could be constructed.

Siting of sulfur-emitting power-generating plants away from crowded cities is a common practice, and a necessity in most cases. But, pollution is still released from those facilities and often affects large numbers of people. An example of poorly conceived remote siting is the Four Corners power plant in Arizona that supplies energy to Los Angeles, among other places; pollution from that facility is contributing to the degradation of visibility in the Grand Canyon National Park.

Neutralizing acidified lakes and rivers by treating them with alkaline materials is a widespread practice. Scandinavia's governments have established a major program to maintain the pH of their delicate lakes despite acid import from the European mainland. In the United States, a few specific studies of the practicality of liming have been carried out. For example, the National Acid Precipitation Assessment Program estimated the cost of liming all the identified acidified lakes in the Adirondack Mountains in the northeastern United States and pegged the cost at $550,000 to $750,000 annually. On the other hand, an assumed reduction by 50 percent of sulfur emissions in source regions for the Adirondack watershed created $4 million to $15 million in benefits to sport fishers in the region. Moreover, in that case only limited liming was needed. It seems that correcting environmental problems at the source can be much more effective than trying to fix problems at the scene of the damage.

No single step would completely eliminate the problems associated with acidic deposition and precipitation, but a number of practical actions could be taken together, in a coordinated mitigation program, to limit the environmental harm.

Questions

1. Think about the meals you have had over the past week. Which foods do you think were acidic and which basic (alkaline, that is)? Why? Of the different household chemicals you are familiar with, identify the ones that may be acidic or basic.

2. Describe the transformations that a typical sulfur dioxide molecule emitted from a power plant experiences as it moves through the environment. Begin at the smokestack and proceed from there.

3. Gardeners often amend soil to get the pH right for certain plants. If the soil is too acidic, a base must be added, and vice versa. To what kind of soil would you add $Ca(OH)_2$? If the sauce you are making for dinner is bitter to the taste, what common food substance could you add to it to reduce the bitterness?

4. Explain how marble statuary can be damaged by acidic fog. What are the likely physical and chemical mechanisms?

5. The world's oceans are slightly alkaline, with a pH of about 8. If the oceans were acidic to the same degree that they are now alkaline, what would the pH of seawater be? Imagine that the

atmosphere does not contain any carbon dioxide. What effect would you expect this to have on the pH of the oceans (that is, would seawater be more acidic or more basic)? Give the reasons for your answers.

6. Your stomach uses concentrated hydrochloric acid (HCl) to break down food during digestion. Stomach acid can have a pH below a value of 2. Since this is sufficient to dissolve meat, what other special properties must your stomach have? If your stomach used sodium hydroxide (NaOH) instead of HCl in digestion, what specific substances might you take to relieve a case of "alkaline" indigestion?

7. Ammonia (NH_3) is a common alkaline gas that can neutralize acids in aerosols and clouds. To reduce the problem of acidic fog in a city, it has been suggested that large amounts of ammonia be sprayed into the atmosphere from sites upwind of the city. What environmental side effects—good and bad—might be expected as a result of this action? Consider both the effects of gases and aerosols involved in this scheme.

8. Most organisms cannot survive in extremely acidic or basic environments. In fact, most organisms have a rather narrow range of tolerance to pH and generally prefer neutral pHs. Explain why this property of organisms makes sense. In your answer, consider the natural environment in which evolution has occurred.

Problems

1. You have a solution of boric acid to rinse out your eyes. On the bottle, it says that the pH of the solution is 5.0. However, you want to use a milder, less acidic solution, with a pH of about 6.0. One way to obtain the milder eyewash is to dilute the bottled solution using distilled water. Starting with 1 ounce of the bottled solution, how many ounces of water would you add to obtain a solution with a pH of 6.0? If you ever happen to spill acid on yourself, flooding the affected area with water helps dilute the acid and reduce its corrosive effect.

2. At a pH of 7, the hydrogen ion (H^+) concentration in an aqueous solution is very roughly two parts per billion by volume (ppbv). At a pH of 6, how many ppbv of H^+ does the solution contain? If you wanted to make the Earth's oceans uniformly acidic at a pH of 6, how much acid would you need to add to the oceans (refer to Question 5). Assume for simplicity that parts per billion by volume are equivalent to parts per billion by mass in this case (Appendix A.1).

3. You wish to estimate how much acid it would take to acidify the upper layer of the oceans to a pH of 6.0. It is known that the normal pH of ocean water is 8.0. The total volume of water that must be acidified is roughly 3×10^{19} liters. Assume that one mole of acid, which would be equivalent to one mole of H^+, weighs 100 grams. How many metric tons of acid would be needed? Ocean water also holds about 1×10^{-4} moles/liter of dissolved carbonates. How might this affect your calculation?

Suggested Readings

Charlson, R. and H. Rodhe. "Factors Controlling the Acidity of Natural Rainwater." *Nature* 295 (1982): 683.

Corcoran, E. "Cleaning up Coal." *Scientific American* 264 (1991): 106.

Cowling, E. "Acid Precipitation in Historical Perspective." *Environmental Science and Technology* 16 (1982): 110A.

Likens, G., R. Wright, J. Galloway, and T. Butler. "Acid Rain." *Scientific American* 241 (1979): 43.

Mohnen, V. "The Challenge of Acid Rain." *Scientific American* 259 (1988): 30.

National Academy of Sciences. *Acid Deposition: Long-Term Trends.* Washington, D.C.: National Academy Press, 1986.

Roth, P. "The American West's Acid Rain Test." Washington, D.C.: World Resources Institute, 1985.

Schwartz, S. "Acid Deposition: Unraveling a Regional Phenomenon." *Science* 243 (1989): 753.

Smith, R. A. *Air and Rain: The Beginnings of a Chemical Climatology.* London: Longman & Green, 1872.

Part III
Global-Scale Pollution Issues

In this section of the book, we investigate the impacts of air pollution on a global scale. One of the most important contributors to global pollution is the rapidly expanding population of our planet. As the number of people on the Earth increases and the consumption of resources accelerates, environmental pollutants that once may have had only local or regional impacts are now causing global problems. Among the global-scale air pollution issues addressed in the following chapters are the greenhouse warming of the Earth and the depletion of the ozone layer.

Carbon Dioxide and the Greenhouse Effect

The massive combustion of fossil fuels releases CO_2 into the atmosphere. Similarly, accelerating deforestation (particularly the destruction of tropical rain forests) also releases CO_2 into the air. The buildup of carbon dioxide disturbs the Earth's energy balance, resulting in a warming of the global climate. Weather patterns are altered, with extreme weather becoming more frequent, including severe droughts and unusual patterns of precipitation. Desertification (an increase in the expanse of deserts) and a rise in sea level also are possible. Among other consequences, agricultural productivity could plummet; coastal cities could be inundated; and world economies could be disrupted.

Ozone Depletion and the Ozone Hole

Chlorofluorocarbons—chlorine-bearing compounds manufactured for use in refrigerators and air conditioners—are released into the atmosphere in large quantities. Additional chlorine is injected into the upper atmosphere by aerospace activities, such as flights of the space shuttle. The chlorofluorocarbons drift into the stratosphere, where they are broken down by ultraviolet sunlight, releasing the chlorine. The chlorine atoms attack ozone that is normally concentrated in a layer there and destroys a significant amount. With the ozone layer thinned, additional harmful ultraviolet (UV) sunlight can penetrate to the surface. This ultraviolet radiation, in turn, induces a number of health problems in humans, including lethal skin cancers, and creates stress on natural and agricultural ecosystems.

Climate Change Caused by Nuclear War: Nuclear Winter

Some forms of climate change occur over very short time intervals, such as the perturbations associated with volcanic eruptions. In Chapter 4, we discussed the sudden extinction of a wide variety of species caused by a "meteorite winter." In this case, the impact of a large comet or asteroid with the Earth produced a global dust cloud that blocked the sun and cooled the climate worldwide. Over the past 5 decades, a dozen nations have built nearly 60,000 nuclear warheads with enough explosive yield to destroy every major city in the world 30 times over. If a nuclear "exchange" (as a nuclear war is often euphemistically called) were to occur, thousands of large fires would be ignited in cities and oil fields, and unprecedented quantities of soot would be injected into the atmosphere. The soot would block sunlight from the surface, creating an "antigreenhouse effect." Land would cool rapidly and deeply, dropping the world into a "nuclear winter," in which agriculture would be likely to fail and much of the human population would starve. Scientists have speculated that human civilization might not recover from such a trauma.

The Relationship Between Population and Pollution

The current population of Earth—about 5 billion souls and growing—is expected to reach 10 billion or more in the twenty-first century. As the population increases, so will the consumption of natural resources and the level of industrial activity. If past experience concerning the effects of population

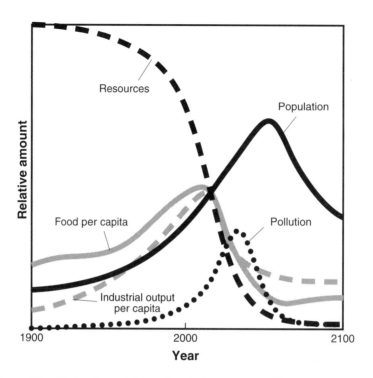

Figure III.1   Projections of future trends in nonrenewable resources, population, food per capita, industrial output, and environmental pollution. The model for this forecast uses simple concepts—for example, that a limited resource (oil is a good example) will be consumed first at an accelerating rate and later at a declining rate until its reserves are depleted. Similarly, the world population that can be supported by a declining resource base must eventually decline. Such conceptual models are useful primarily for illustrating the general relationships among various parameters.

growth and industrialization on the environment is a reasonable indicator, pollution will also increase dramatically. It has generally been the case that greater wealth, although implying greater resource consumption per capita, has also led to concerted efforts to reduce environmental degradation. Thus greater wealth does not necessarily mean more pollution per capita. Conversely, although developing countries use fewer resources per capita than developed countries do, they typically generate more pollution per unit of resources utilized.

Figure III.1 shows hypothetical projections of the future world population, resources, and pollution based on simple economic principles. It is easy to imagine, for example, that nonrenewable resources will decline to a point that industrial output per capita, food production per capita, and eventually the population itself will be forced to decrease dramatically. This outlook suggests that there is a need to develop new energy sources and methods of food

production that consume fewer nonrenewable resources and produce less pollution. Moreover, pressure will increase to recycle scarce materials—metals, plastics, and so on—that would otherwise become pollutants. Some of these conservationist activities are already under way. The scale of the present efforts, however, is dwarfed by the scope of the commitment that will be needed first to improve, and then to preserve, the quality of life planetwide through the next millennium and beyond.

To understand, and eventually correct, emerging global environmental problems like those mentioned, a sound foundation in the science underlying these problems is essential. The following chapters are designed to introduce the fundamental concepts and facts that determine the global state of the environment. Chapter 10 describes the important global biogeochemical cycles of carbon, nitrogen, sulfur, and other elements; the principal chemical compounds involved in each cycle; and the processes that produce, transform, and destroy these compounds. The "reservoir box model" introduced in Chapter 4 is expanded in Chapter 10 to apply to the various chemical cycles. The reservoir approach allows the dynamics of the biogeochemical cycles to be investigated, and then the effects on the reservoirs of various human enterprises can be calibrated.

Chapter 11 provides a comprehensive overview of the Earth's climate system, the basic elements of the planet's energy balance, and a simple climate model. Using this model, we can estimate the effects of a variety of human activities on climate. Chapter 12 is devoted to the greenhouse-warming problem, particularly the impact of carbon dioxide emissions on global temperature. Information on biogeochemical cycles and the climate system from Chapters 10 and 11 is elaborated in Chapter 11. Chapter 13 deals with the stratospheric ozone layer and its apparent depletion by chlorofluorocarbons: the sources and sinks of ozone, the causes of ozone destruction, the "ozone hole," as well as ultraviolet radiation and its effects on humans and ecosystems.

Finally, Chapter 14 introduces the new and important topic of global environmental engineering. If the problems highlighted in earlier chapters intensify and become more threatening, unusual measures may be needed to correct them. Global-scale modification of the natural environment using engineering techniques is already under consideration. The scientific understanding gained in earlier chapters is used in Chapter 14 to evaluate several of the global engineering concepts put forward in recent years. The eventual decision to engineer our environment—or not—may be the most important choice the human species will ever have to make.

10

Global Biogeochemical Cycles

The natural environment, including the composition of the air and seas and the favorable climate for life, is largely a product of the Earth's biogeochemical cycles. These cycles describe how specific elements and compounds—carbon and carbon dioxide, for example—are transferred between the principal global reservoirs—the atmosphere, oceans, land, and biosphere—and the various forms they take in each reservoir. Chemical and physical processes and transformations determine the partitioning of a material among the reservoirs. For example, for a fixed total quantity of carbon in all the reservoirs, a certain fraction is found in the atmosphere, another fraction in the oceans, and so on; these fractions depend on the way the carbon is transferred between the reservoirs, the sizes of the reservoirs, and other factors. The amount to be found in the atmosphere, where it may have the most direct effect on climate, is thus determined by the carbon's overall biogeochemical cycle. Understanding the environment, in this case, requires understanding this cycle.

These cycles are grand indeed. The amounts of material involved are enormous. For example, the carbon cycle transfers more than 100 billion tonnes of carbon per year from the atmosphere to other reservoirs. Although these quantities are large, they are so spread out over the globe that you can barely see the processes at work. As you nurture your vegetable garden in the spring, you are witnessing the transfer of carbon (in the form of carbon dioxide) from the atmosphere to the plants (as organic biomass). Later, when you eat the vegetables and metabolize them for energy, you are participating in the recycling of carbon back into the atmosphere. Humans are a natural link in global biogeochemical cycles, albeit a rather minor link. Unfortunately, our contribution to these biogeochemical cycles through industrial and technological activities, in many instances, is far from trivial and, in some cases, is becoming dominant over the natural cycles.

10.1 The Grand Chemical Cycles of Earth

The elements and materials that we will be concerned with are sulfur, nitrogen, carbon, and oxygen. (Chlorine is discussed in Chapter 13.) These are the most critical cycles for life on Earth, and the environmental issues of concern revolve around these few cycles. The inner workings of these cycles are not simple, however, and we must take a careful tour of them. Each cycle contains a number of key chemical compounds that participate in different parts of the cycle, or in different reservoirs. There are many secondary components for each cycle as well, although the secondary members of each cycle will not be discussed at length in this chapter. (More detailed information on the sulfur and nitrogen cycles can be found in Chapter 6, dealing with smog, and in Chapter 9, on acid rain. The important members of the carbon cycle, carbon monoxide and methane, are treated at length in Chapters 6 and 12, respectively.)

In the sulfur (S) cycle, the key compounds are

- SO_2, sulfur dioxide (atmosphere)
- COS, carbonyl sulfide (atmosphere)
- H_2S, hydrogen sulfide (atmosphere)
- $(CH_3)_2S$, dimethyl sulfide (atmosphere)
- H_2SO_4, sulfuric acid (atmosphere)
- SO_4^{2-}, sulfate ion (oceans, sediments)

In the nitrogen (N) cycle, the key compounds are

- N_2, nitrogen (atmosphere)
- N_2O, nitrous oxide (atmosphere)
- NO_x, nitrogen oxides, mainly NO and NO_2 (atmosphere)

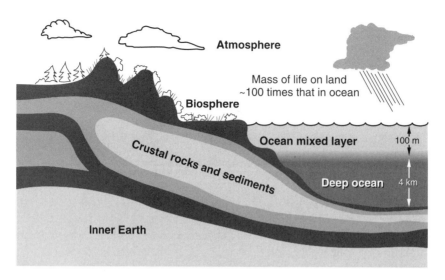

Figure 10.1 The principal reservoirs of chemical materials in the Earth system. The atmosphere is the most tenuous, and the solid earth is the most dense. The oceans are divided into two segments: the upper shallow mixed layer, which is relatively warm with sufficient light for photosynthesis, and the deep, abyssal oceans, which are cold and dark. The biosphere represents that narrow zone near the surface of the land and seas in which life flourishes. In terms of total mass, the solid earth is greatest, followed by the oceans, the atmosphere, and the biosphere, in order of decreasing mass.

- HNO_3, nitric acid (atmosphere)
- NO_3^-, nitrate ion (oceans, sediments, biosphere)

In the carbon (C) cycle, the important compounds are

- CO_2, carbon dioxide (atmosphere)
- CH_2O, organic matter (biosphere, sediments, oceans)
- H_2CO_3, carbonic acid (oceans, atmosphere)
- HCO_3^-, bicarbonate (oceans)
- CO_3^{2-}, carbonate (sediments, oceans)

In the oxygen (O) cycle, the key chemical species are

- O_2, oxygen (atmosphere, oceans)
- O_3, ozone (atmosphere)
- CO_2, carbon dioxide (atmosphere)

The quantities of these elements in the various Earth reservoirs (the atmosphere, oceans, sediments, and biosphere) are discussed in Section 4.1.1 (see also Table 4.1). In each reservoir, the elements are partitioned into the different compounds just listed. Often, only one or two compounds dominate a reservoir. For example, almost all carbon in the atmosphere is in the form of carbon dioxide (although the small percentage of other compounds provides a rich brew for chemical reactions). In the oceans, the principal form of carbon is the bicarbonate ion (HCO_3^-). As the carbon is transferred from one reservoir to another, it is also transformed chemically. So we must be clever in following the carbon through its many disguises as it moves through the environment.

10.1.1 Reservoirs in the Earth System

The principal reservoirs of the elements in the Earth are illustrated in Figure 10.1. (These reservoirs were discussed in Section 4.1.2 and Table 4.1.) The box model described in Chapter 4 offers a simple way to represent the storage and transfer of elements in the environment. That simple model proved to be useful in discussing air pollutants and toxic compounds and could be applied to a number of other practical problems. Here we reconsider the box model and apply it to the reservoirs that control the global biogeochemical cycles.

The relative importance of the various reservoirs for the major chemical cycles of the Earth is summarized in Table 10.1. Each element is distributed differently among the various reservoirs, in patterns determined by the physical, chemical, and biological

Table 10.1   Chemical reservoirs.

| | Reservoir (contribution) | | | |
Chemical	Atmosphere	Oceans	Biosphere	Crust
S (sulfur)	Minor	Large	Small	Large
N (nitrogen)	Large	Minor	Small	Modest
O (oxygen)	Small	Small	Minor	Large
C (carbon)	Minor	Small	Minor	Large

processes that move those materials through the Earth system. The ability of humans to modify the distributions also varies significantly from element to element. The sulfur cycle is most easily disturbed because of the small amount of sulfur occurring naturally in the atmosphere. The carbon cycle, likewise, can be upset by human activities.

10.1.2 SIMPLE RESERVOIR MODELS

The box model developed in Section 4.1.2 can be used to investigate the biogeochemical cycles of interest. Figure 10.2 provides an example of a simple single-box reservoir. In the reservoir model, we are concerned primarily with the total amount of material, Q, in the reservoir and the rate of transfer of material between coupled reservoirs. According to the discussion in Chapter 4, the reservoir equation may be written as

$$Q = S\tau \qquad (10.1)$$

Here S is the source of material for the reservoir, usually specified in millions of metric tons (megatonnes, Mt) or billions of metric tons (gigatonnes, Gt) per year (1 megatonne is equivalent to 1×10^{12} grams, and 1 gigatonne is equal to 1×10^{15} grams). The average lifetime, or residence time, of the material in the reservoir is τ (usually expressed in years). Accordingly, the reservoir amount, Q, is given in megatonnes or gigatonnes. Recall that a gigatonne is a large quantity: One Gt is nearly three times the mass of all the humans on Earth; it also equals the mass of 1 cubic kilometer of water, enough to fill a swimming pool 3 meters deep, 30 meters wide, and 10,000 kilometers long!

Reservoirs in a Steady State

When reservoirs are in a steady state, the parameters S, L, Q, and τ have simple relationships. According to the definition of a steady state, $S = L$ (Figure 10.2). Moreover, if the amount in the reservoir and the source or loss rate can be estimated, then the residence time for the reservoir can be calculated using the relation

$$\tau = \frac{Q}{S} = \frac{Q}{L} \qquad (10.2)$$

The residence time is an important parameter that determines how long it would take to fill or empty a reservoir or to modify significantly the quantity of material in the reservoir at some specific rate of addition or extraction. Another useful interpretation of τ is that it is the average length of time that any individual atom or molecule of the material in question will remain in the reservoir before it is removed or destroyed.

When the residence time and steady state quantity of material in the reservoir are known, the source strength (or drainage rate) for the reservoir may be computed as

$$S = L = \frac{Q}{\tau} \qquad (10.3)$$

If a new source of material for a reservoir, ΔS, is introduced and the residence time for the reservoir is known from other information, then the change in the reservoir amount over a long period of time can be estimated as follows:

$$\Delta Q = \tau \Delta S \qquad (10.4)$$

This result assumes that the loss from the reservoir adjusts to the new source in a linear manner or as a linear response, which is often a good first approximation.

As an example of the application of the reservoir equation, consider the bottom panel of Figure 10.2. In this case, the reservoir amount and source are known. The residence time for the reservoir is thus determined by Equation 10.2:

$$\tau = \frac{Q}{S} = \frac{100 \text{ Gt}}{10 \text{ Gt/yr}} = 10 \text{ yr} \qquad (10.5)$$

Multiple sources and sinks for a reservoir only slightly complicate the picture. Figure 10.3 shows a reservoir with multiple sources and sinks. A reservoir can have any number of sources and sinks, each of which may be independent of the others. However, in a steady state, the total source and sink must be equal:

$$S_1 + S_2 + S_3 + \cdots$$
$$= L_1 + L_2 + L_3 + \cdots \qquad (10.6)$$

or

$$S_T = L_T \qquad (10.7)$$

where (for example)

$$S_T = S_1 + S_2 + S_3 + \cdots \qquad (10.8)$$

Another way to alter the amount of material in a reservoir is to add or eliminate a *loss* process. Then, if the sources for the reservoir are fixed, the quantity in the reservoir will change until a new equilibrium is reached. Equilibrium will occur, however, only if the remaining loss rates vary in proportion to the amount of material in the reservoir, which is usually the case. For example, the rate of loss of a gas from the atmosphere by precipitation depends on the average concentration of that gas in the atmosphere; as the concentration increases, the rate of loss in precipitation increases, and vice versa. The response of the quantity in the reservoir to a change in the loss rate, ΔL, may be expressed through the proportionality

$$\frac{\Delta Q}{Q} = -\frac{\Delta L}{L + \Delta L} \qquad (10.9)$$

Although Equation 10.9 is only an approximation, it is useful in many circumstances.

Figure 10.2   The basic steady state reservoir box model (see also Figure 4.4). The source, *S*, and sink, *L*, are equal in a steady state, or equilibrium. The total quantity of material in the reservoir, *Q*, is constant under these circumstances.

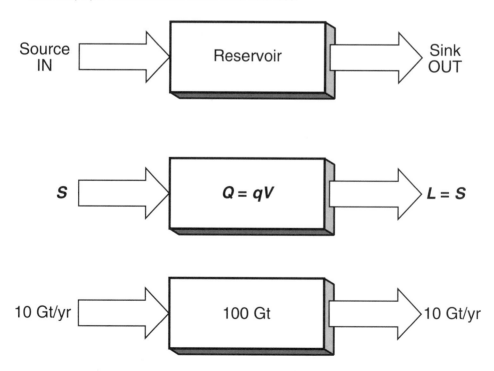

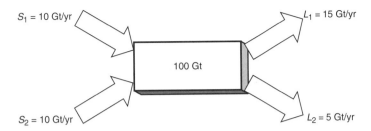

Figure 10.3 A reservoir with two independent sources and two independent sinks. The sources and sinks can represent specific processes. For example, the sources might be associated with biological activity and precipitation. The individual sources and sinks can take on any value but must balance in a steady state.

The total source and sink rates for a reservoir can be used in the reservoir Equations 10.1, 10.2, and 10.3. For example, in Figure 10.3, the following values hold:

$$S_T = 10 \text{ Gt/yr} + 10 \text{ Gt/yr} = 20 \text{ Gt/yr}$$
$$L_T = 15 \text{ Gt/yr} + 5 \text{ Gt/yr} = 20 \text{ Gt/yr}$$

$$(10.10)$$

and hence the residence time can be estimated as

$$\tau = \frac{Q}{S_T} = \frac{100 \text{ Gt}}{10 \text{ Gt/yr}} = 5 \text{ yr} \qquad (10.11)$$

Coupled Reservoirs

Because the Earth is an isolated place surrounded by the void of space, we can consider it a **closed system**. In a closed system, the total amount of any element contained in all available reservoirs is fixed during the times of interest. Accordingly, as materials move between reservoirs, as they must in a dynamic world, the sink of a material for one reservoir must be a source for other reservoirs. It follows that the total mass of any element or material is conserved among the reservoirs even as the materials are continually shifting between them. The reservoirs composing the Earth system are thus coupled together to form a conservative closed system.

Figure 10.4 shows a hypothetical system with three coupled reservoirs. We will assume that each of the reservoirs is in a steady state and that the system as a whole has achieved an equilibrium condition. Because *each* reservoir must be in a steady state in order for the system to be in equilibrium, the sources in Figure 10.4 *must* obey the following relationships:

$$S_{AL} + S_{AO} = S_{LA} \qquad \text{(atmosphere)}$$
$$S_{LA} = S_{AL} + S_{OL} \qquad \text{(land)}$$
$$S_{OL} = S_{AO} \qquad \text{(ocean)}$$
$$(10.12)$$

Equations 10.12 automatically conserve the mass in the system, because the sources and sinks for each reservoir are exactly balanced.

There are a number of ways of expressing the residence times and other parameters for this system, all of which are equivalent. For example, the atmospheric residence time is given by

$$\tau_A = \frac{Q_A}{S_{LA}} = \frac{Q_A}{(S_{AL} + S_{AO})} = \frac{Q_A}{(S_{AL} + S_{OL})}$$

$$(10.13)$$

Note that mass conservation in a closed system reduces the number of free parameters that determine the system's behavior. Thus in Figure 10.4, not all the parameters are independent, which also follows from Equations 10.12 (which can be reduced to only two independent equations by substituting the last in the first).

The reservoir model can be applied to all the key biogeochemical cycles on the Earth. Over relatively short spans of time, the significant reservoirs are the atmosphere, land, oceans, and biosphere. On longer time scales, the lithosphere and sedimentary deposits must be taken into account. Obviously, the Earth system and its chemical cycles are extraordinarily complex. Scientists are far from understanding in detail the global biogeochemical systems. Nevertheless, this simple reservoir approach provides a practical and accurate quantitative description. Many of the

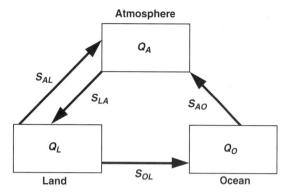

Atmosphere

Q_A

S_{AL}

S_{LA}

S_{AO}

Q_L

S_{OL}

Q_O

Land

Ocean

Figure 10.4 A three-reservoir atmosphere-ocean-land system with specific mass transfer processes between the reservoirs. The quantities of the materials in the reservoirs and the transfer rates, given as sources, are appropriately labeled (for example, the source of material for the atmosphere from the land has subscript *AL*. The type of material is not specified in this example.

leading environmental scientists use the same simple models in their research, and so in pursuing these basic ideas, you are thinking in the same ways as does the scientific elite (who are probably getting paid a lot more than you are to do so).

10.2 Biogeochemical Cycles of the Primary Elements

The reservoir components and equilibrium dynamics of the main biogeochemical cycles on Earth are described in this section. The elements that we are concerned with are sulfur, nitrogen, oxygen, and carbon. It is important to recognize that a number of processes may contribute to the transfer of an element between two reservoirs or to the conversion of an element to different chemical forms within a reservoir. The absorption of carbon dioxide dissolved in ocean waters by means of photosynthesis, for example, involves many species of microorganisms, referred to as *phytoplankton*, and other species that feed on the phytoplankton, called *zooplankton*. At this stage, we are concerned only with the overall conversion rate of inorganic CO_2 to organic matter in the oceans, without getting into the details of the speciation of the organisms. Although the oxygen cycle proceeds through a variety of complex oxygen-containing compounds, it is necessary to differentiate only two critical forms: free atmospheric oxygen (O_2), which we breathe, and oxygen associated with carbon dioxide.

10.2.1 SULFUR

Sulfur is a basic element that, in trace amounts, is essential to life. On Earth, sulfur is found mainly in the form of sulfate minerals. Sulfates are potent fertilizers for plant growth and are sold in great quantities for agricultural application. Yellow deposits of elemental sulfur can form at high temperatures in the Earth or through biological "reduction" (Sections 4.2.1 and 4.2.2). The availability of natural elemental sulfur deposits enabled the early development of gunpowder and fireworks, as well as sulfur drugs. When plant matter is assimilated by certain bacteria, sulfur may be released in the form of toxic and noxious "reduced" sulfur fumes, some of which give the smell of "rotten eggs" to swamp gas and sewage vapors.[1] Huge mounds of elemental sulfur are generated by oil refineries when sulfur is cleansed from gasoline and other fuels. Sulfuric acid is a common industrial chemical, and nearly 100 million tonnes are manufactured each year. The major contributions of human activities to the sulfur cycle are through the combustion and release of sulfur in fuels, particularly coal, and the manufacture and application of sulfate fertilizers. Agricultural and industrial sulfates can find their way into rivers, estuaries, and lakes, where they encourage algal growth. The oceans contain such large quantities of natural sulfate that they are not as vulnerable to anthropogenic perturbations as are freshwater systems.

Sulfur Sources and Sinks

The overall sulfur cycle is illustrated in Figure 10.5. The atmosphere contains the least amount of sulfur (about 5 million tonnes [5 Mt]) of the principal reservoirs, which in total hold more than 5×10^9 Mt. Atmospheric sulfur is primarily in the form of COS (carbonyl sulfide) and SO_2 (sulfur dioxide). The dearth of atmospheric sulfur is directly related to the propensity of this element to form stable, nonvolatile compounds that dissolve in the oceans or precipitate as sedimentary minerals. In addition, the average residence time of sulfur in the atmosphere is very short, about 1 to 2 weeks. The residence time

---

1. You can smell the same odor behind an accelerating automobile with a catalytic converter. In this case, the sulfur in the fuel, which normally would be emitted in an oxidized form such as sulfur dioxide, is "reduced" to sulfates on the surfaces of the catalyst that is placed in the exhaust stream, in order to reduce the nitric oxide to nitrogen gas.

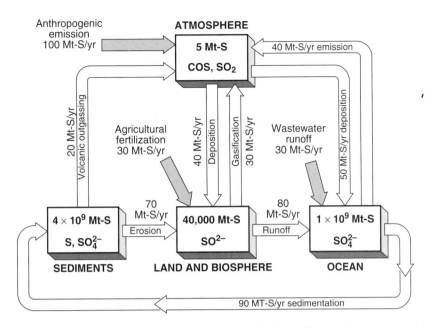

Figure 10.5 The Earth's long-term sulfur cycle. The major reservoirs for sulfur are the oceans and sediments. Sulfur in land soils, surface waters, and living organisms are lumped together in one reservoir. The atmosphere is a minor reservoir for sulfur, although in the form of acid rain and fog (Chapter 9), even a relatively small amount of sulfur can be hazardous to the environment. "Sedimentary" sulfur is separated from sulfur available on land so as to differentiate the sulfur trapped in mineral deposits from that in soils that is available for biological assimilation. The anthropogenic contributions to the sulfur cycle—which are indicated for the atmosphere, land, and oceans—are not included in the sulfur mass balance in this figure.

can be deduced using Equation 10.2, with the total source of sulfur for the atmosphere of roughly 150 Mt-S/yr (including all sources) and the measured atmosphere sulfur burden of about 5 Mt. The short residence time means that anthropogenic emissions of sulfur into the atmosphere are removed very rapidly. The dominant scavenging process is wet deposition, which occurs when the soluble sulfate compounds produced photochemically from COS and SO_2 are absorbed into cloud drops and raindrops. Indeed, the removal of anthropogenic sulfur emissions by precipitation creates the acid rain problem discussed in Chapter 9. Here, we look at this problem from the perspective of the global sulfur cycle.

The basic chemical evolution of sulfur in the atmosphere is described by the following sequence of reactions (Sections 3.3.4 and 9.2.2):

$$SO_2 + OH + M \rightarrow HSO_3 + M$$
$$HSO_3 + O_2 \rightarrow SO_3 + HO_2$$
$$SO_3 + H_2O \overset{M}{\rightarrow} H_2SO_4 \qquad (10.14)$$
$$H_2SO_4 \overset{Rain}{\rightarrow} SO_4^{2-}$$

Sulfur dioxide has a lifetime in the lower atmosphere of only a few days. The reactions of SO_2 lead to the formation of sulfuric acid and sulfates, which are deposited on land and in oceans by precipitation. Sulfates are quite soluble in water and eventually are drained from land surfaces and groundwater into rivers and finally the oceans. Some sulfates are deposited in lakes and, over long periods of time, can form mineral deposits called *evaporites*.

The sulfur entering the atmosphere as a result of biological processes is usually in a more "reduced" chemical state. The sulfur emissions are in the form of sulfides—for example, hydrogen sulfide (H_2S), carbonyl sulfide (COS), and dimethyl sulfide ($[CH_3]_2S$). The sulfides are oxidized upon exposure to the atmosphere, as exemplified by the decomposition of COS:

$$COS + OH \rightarrow HS + CO_2$$
$$HS + O_2 \overset{M}{\rightarrow} SO + OH \qquad (10.15)$$
$$SO + O_2 \rightarrow SO_2 + O$$

The reduced gases emitted by biological activity are thus efficiently converted to sulfur dioxide and sulfates. Most of the sulfides have very short atmospheric residence times, measured in days, with the exception of carbonyl sulfide, which has an atmospheric lifetime of about 1 year. It also happens that COS is the most abundant sulfur constituent of the atmosphere.

The Roles of Sulfur

Sulfur is essential to life. It is contained in proteins and many other organic compounds that regulate life processes. Sulfates are a primary nutrient in the biosphere and may be metabolized to create a wide variety of organic sulfur compounds. When organic matter forms deposits that evolve into fossil fuels (coal, oil, and natural gas), the sulfur remains as a trace component of the fuels. Later, when the fuel is burned, the sulfur is released as sulfur dioxide, causing acid rain and other local environmental problems. If instead the organic debris decays, the organic sulfur may be assimilated by bacteria that emit it in the form of reduced gases.

Sulfur in the atmosphere tends to condense as sulfate particles, or aerosols. These particles have several important functions. The sulfate haze, for example, can effectively scatter or reflect sunlight, increasing the planetary albedo and causing a slight cooling of the climate (Section 11.6.5). The sulfate aerosol particles also act as the primary cloud condensation nuclei (CCN) in the atmosphere. That is, sulfate aerosols readily attract and absorb water vapor, converting to large water droplets when the air becomes very humid. Without such CCN, clouds and rain would not form so easily. The radiative (optical) properties of clouds depend on the number and sizes of the CCN and are thus controlled by sulfates in the air. The cloud optical properties in turn have a major impact on the climate system (Section 11.4.3).

An interesting climate control mechanism involving sulfur was suggested by R. Charlson, James Lovelock, and others. They noted that microscopic phytoplankton in the oceans produce a reduced sulfur gas, dimethyl sulfide, or DMS [$(CH_3)_2S$], in enormous quantities (as much as 35 million tonnes of sulfur are released annually as DMS into the atmosphere from the oceans). Upon exposure to air, the DMS is oxidized into sulfur dioxide, which in turn forms sulfate aerosols. These aerosols modify the reflectivity of low-lying (stratus) clouds that blanket half the world's oceans. The reflectivity also affects the global albedo and thus the equilibrium temperature of the planet. So it may be that microscopic organisms in the oceans have the ability to turn the climatic temperature up and down like a thermostat. If the ocean surface waters become too warm, the phytoplankton pump out more DMS, which increases the reflectivity of oceanic clouds and cools the place down. If the water grows too cold, the plankton slow down their DMS production, decreasing the cloud reflectivity and causing a warming effect. This intriguing biogeophysical climate-feedback mechanism has not been proved. But Lovelock and others point to it as a possible Gaia mechanism, in which the living organisms of Earth act (unknowingly) in their own self-interest to maintain a pleasant climate (Section 4.3.3). Many scientists doubt that the global environment is so robust it can withstand global change, particularly the assault of human pollutants (refer to Section 14.1.1).

Sulfur: The Human Touch

As is apparent in Figure 10.5, human activities are substantially modifying the global sulfur cycle. The atmosphere in particular is being flooded with sulfur dioxide produced by the combustion of fossil fuels. In recent years, the total sulfur emission has actually declined somewhat, as measures to control power plants' sulfur effluents—using sulfur "scrubbers," for example—have been implemented. Nevertheless, as the world population grows, industrialization accelerates, and cheaper sulfur-rich fuels are exploited for energy, sulfur emissions can be expected to rise again.

The direct application of sulfates to land as fertilizer represents a substantial fraction of the terrestrial sulfur budget (fertilizer sulfur accounts for about 30 million tonnes per year [Mt-S/yr]). Most of this sulfur reaches streams and eventually the seas. Added to that is the industrial discharge of waste sulfur compounds directly into waterways, amounting to about 30 Mt-S/yr, and the total runoff to the oceans is double the natural rate. Fortunately, the oceans contain a huge mass of sulfur, and the anthropogenic contribution will remain negligible well into the future. On the other hand, in rivers, estuaries, and coastal waters, where the runoff is concentrated, the anthropogenic effluents can significantly upset the natural nutrient cycles. Here, water can become increasingly anaerobic and poisonous owing to the buildup of hydrogen sulfide and other reduced sulfur

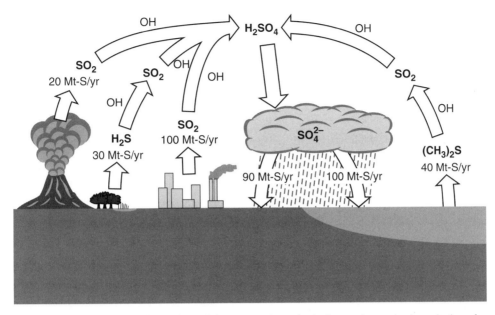

Figure 10.6 The global sulfur budget of the atmosphere, including anthropogenic emissions from coal and oil combustion. The industrial sulfur vapors are estimated to be deposited equally on land and over the oceans, more than doubling the overall natural global deposition rate. The principal chemical species and reactions are indicated for each process.

gases. The full environmental consequences of this sulfur contamination of waterways is not yet determined. Reductions in fisheries might be expected, along with an increase in the stench from once pristine rivers.

The overall global atmospheric sulfur budget subject to the effects of anthropogenic sulfur emissions is summarized in Figure 10.6. It is a rather obvious case of human dominance of an important biogeochemical cycle. The consequences have included—on various temporal and spatial scales—acidic rain and fog, urban haze and low visibility, lung impairment and respiratory death, crop damage and forest destruction, mortality of limnological (freshwater) species, dissolution of stonework and sculpture, degradation of urban infrastructure, and regional climate change. Almost every one of these serious effects is associated with the generation of power needed to run our industrialized society.

The chemical transformations within the sulfur cycle are represented by the processes defined in Equations 10.14 and 10.15 and summarized in Figure 10.6. Note that the most important reactant in the sulfur cycle is the hydroxyl radical (OH), our well-known chemical "scavenger." The decomposition of reduced sulfur compounds and the oxidation of sulfur dioxide to sulfuric acid are driven by OH. If OH did not exist, the concentrations of sulfur gases,

particularly the noxious reduced species like hydrogen sulfide, would build up to intolerable levels.[2]

10.2.2 NITROGEN

The nitrogen cycle is one of the most important and best-known biogeochemical cycles. We all have had the occasion to feed our potted plants a miraculous nitrate fertilizer to induce spring blooms. Nitrogen is a critical nutrient for all lifeforms, providing an essential element for creating amino acids and proteins. The nitrogen cycle is complex, involving the atmosphere, oceans, land, and biosphere. The most important processes are depicted in Figure 10.7. Nitrogen gas (N_2) dominates the nitrogen cycle in terms of the quantity and rate of transfer between the (atmosphere and ocean) reservoirs, although it serves mainly as a source of other trace nitrogen species that have more active roles in the climate system, environmental chemistry, and biochemical processes. Most nitrogen is contained in the atmosphere in the form of N_2, owing to its great stability and volatility

2. Eventually the sulfur gases would be decomposed along a variety of chemical pathways. They might interact with ozone or nitrogen oxides, for example, or be photolytically decomposed by sunlight at slow rates. Sulfur dioxide may also be absorbed by and oxidized in water droplets.

(Section 4.1.1). This is in contrast with the sulfur cycle, in which most of the sulfur is contained in nonvolatile mineral forms, such as sulfates, in the solid Earth and oceans.

Nitrous oxide (N_2O) is not an active gas, either chemically or biologically (although it has a biological source), but it has a number of important roles in the global environment. Nitrous oxide is the major source of NO_x in the stratosphere, where the ozone layer is found (Chapter 13). The nitrogen oxides play a major role in ozone depletion. N_2O is also a

"greenhouse" gas, which contributes to global warming (Chapter 12). In practical applications, nitrous oxide is a useful anesthetic (perhaps better known as *laughing gas*) and has been used as a spray propellant for some nonstick cooking sprays (in part to replace chlorofluorocarbons that were found to threaten the ozone layer [Chapter 13]).

The lifetime (or residence time) of N_2O in the atmosphere is quite long. In accordance with the reservoir model (Equation 10.2), the residence time can be estimated as about 1500 (Mt-N)/15

Figure 10.7   The nitrogen cycle of the Earth in terms of its major components, reservoirs, and transfer rates. The principal form of nitrogen is nitrogen gas (N_2). In the atmosphere, nitrous oxide is the second most common form. "Fixed" nitrogen is found in the form of nitrogen oxides (NO_x and HNO_3) in air, and as nitrate or organic fixed nitrogen (CN) in water and soil. Although the rates of transfer of N_2 between the atmosphere and oceans are extremely high, the overall role in the nitrogen cycle necessary for life is negligible. The dark arrows represent the total source of fixed nitrogen for land (and the total rate of denitrification on land) associated with the action of microorganisms and the application of manufactured fertilizers.

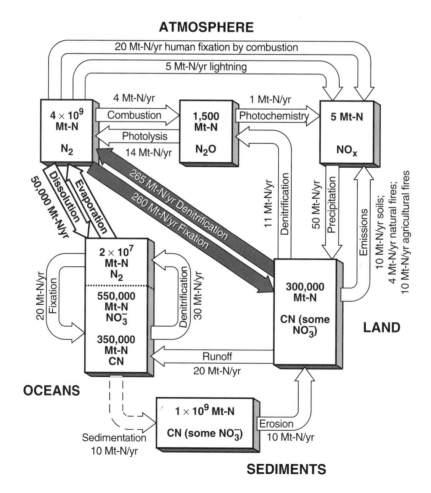

(Mt-N/yr) = 100 years. Thus one might not expect to see substantial changes in atmospheric nitrous oxide over time spans of generations. In the case of nitrogen, the residence time is likely to be hundreds of millions of years. Because the interior of the Earth is relatively depleted of nitrogen at the present time, it is reasonable to assume that no physical or chemical process exists that can recycle it into that reservoir.

Nitrogen Fixation

Nitrogen fixation refers to the processes that split nitrogen atoms from N_2 (or N_2O) to form a series of chemical compounds with a single nitrogen atom (for example, NO and NO_3^-). Nitrogen fixation is distinct from the nitrification process described in Section 4.2.1, in which existing fixed nitrogen in reduced chemical states is oxidized into nitrates. The "fixed" nitrogen species can be assimilated by most living organisms to build proteins and organic tissues. Biological nitrogen fixation is performed by a family of unique microorganisms that live in soils, the *nitrogen-fixing bacteria*. These organisms use nitrogen gas to produce their own fixed nitrogen (required for growth). When the organisms die, the fixed nitrogen in their corpses is converted by other nitrifying bacteria to more generally useful nitrates.

Other specialized microorganisms assimilate the fixed nitrogen and reemit N_2 and some N_2O; these organisms are called the *denitrifying bacteria*. Denitrification is the chemical reduction of nitrates to inert nitrogen gases, for example:

$$NO_3^- + HCN \rightarrow N_2 + CO_2 + OH^- \quad (10.16)$$

where the HCN represents an organic nitrogen-containing material. Equation 10.16 is schematic only; the actual denitrification process involves much more complex chemistry. A small fraction of the total fixed nitrogen is photochemically decomposed in the stratosphere by means of a two-step process:

$$NO + h\nu \rightarrow N + O$$
$$N + NO \rightarrow N_2 + O \quad (10.17)$$

In Figure 10.7 several abiotic as well as biological sources of fixed nitrogen are evident. On land, nitrogen-fixing bacteria contribute roughly 240 million tonnes of nitrogen per year as nitrates (with humans adding about 20 Mt-N/yr in fertilizer). Most of this is denitrified on land and returned to the atmosphere as N_2 and N_2O. Some nitrates also run off into the oceans.

In the atmosphere, combustion and lightning account for a substantial amount of fixed atmospheric nitrogen. There are two distinct combustion sources. During high-temperature combustion (as in automobile and aircraft engines), the oxygen and nitrogen in air can react directly:

$$N_2 + O_2 \xrightarrow{\text{Heat}} NO + NO \quad (10.18)$$

This "nitrogen burning" requires substantial thermal energy to overcome the reaction barrier (activation energy [Sections 3.3.3 and 3.3.4]). The second source of fixed nitrogen in combustion is related to the burning of biomass that contains fixed organic nitrogen. Most of this nitrogen was previously fixed and has been absorbed by growing plants. When the plant matter is burned, the organic fixed nitrogen is oxidized into nitrogen oxides, which are released with the smoke into the atmosphere. The release of fixed nitrogen from biomass by means of combustion has both natural and anthropogenic components. Humans each year torch millions of acres of farmland to remove leftover vegetation debris and recycle mineral nutrients into the soil. The clearing of forests by fire to create pasture accounts for large emissions of smoke and nitrogen oxides each year.

The oceans and sediments hold enormous reservoirs of fixed nitrogen that has accumulated over hundreds of millions of years. Indeed, the oceans are rich in nutrients such as nitrate and sulfate, yet the land is even more productive. The total living biomass on land is about 100 times larger than that in the seas. One explanation is that the land can hold its nutrients longer and recycle them many times before they are washed away or chemically removed. Rain forests offer an example of this efficiency. Although the typical soils in these forests are poor in nutrient value, local ecosystems have evolved to retain critical nutrients such as fixed nitrogen by rapidly reassimilating detrital biomass. There is little organic litter on the forest floor; all the nutrients are in constant use. When the forest is cut down and the nutrients are laid bare on the soils, they cannot be retained as effectively and so are depleted within a few years. The oceans have a similar problem. Most biological activity occurs in the uppermost hundred meters of water called the *photic zone*, where sunlight is plentiful. Without a solid matrix to catch dead organic litter, however, detritus and the nutrients it holds quickly sink into deeper waters where, in the absence of sunlight, biological productivity is minimal. As a result, the ocean deeps are rich in nutrients, but the surface waters are generally barren of life. Exceptionally productive regions are associated with the upwelling of deep water saturated with nutrients. The combination of nutrients and sunlight can result in a bonanza of marine life in these regions.

Ammonia

Ammonia (NH_3) is another important inorganic form of fixed nitrogen. Ammonia is a natural by-product of bacterial consumption of organic matter containing fixed nitrogen. The pungent smell of a compost heap or the cat's litter box is due to bacterial emissions of ammonia. Ammonia is manufactured by industries by combining nitrogen (N_2) and hydrogen (H_2) gases; ammonia is the simplest stable compound of these elements. Ammonia is used as a refrigerant; indeed, before the fabrication of chlorofluorocarbons, ammonia was the primary refrigerant in general use. Ammonia dissolved in water is a common cleaning solution. The main use of ammonia today is in fertilizers. In fact, nitrate, sulfate, and phosphate are combined with ammonia in fertilizers (as **ammonium nitrate**, **ammonium sulfate**, and **ammonium phosphate**).

The largest source of ammonia overall comes from the decay of vegetation in soils. Estimates are crude, but up to 500 Mt-N/yr may be cycled through soil biochemistry in the form of ammonia every year. Most of the ammonia is either taken up from the soil by plants as a nutrient or oxidized by microbes into nitrate. The nitrate is also used by plants or converted by denitrifying bacteria to N_2 and N_2O, which diffuse into the atmosphere. The fixed nitrogen is efficiently recycled through soil ecosystems, being used over and over again by various organisms. The quantity of ammonia that actually escapes from soils into the atmosphere is uncertain, but it could be as much as 50 Mt-N/yr.

The anthropogenic sources of ammonia are dominated by emissions from the excrement of domestic animals. A particularly potent ammonia source arises from cattle feedlots. The pungent, stinging scent of ammonia is always in the air near any area with a large population of these animals. As much as 25 million tonnes of nitrogen as ammonia are produced each year (25 Mt-N/yr) by domestic animals. Human waste contributes perhaps 2 Mt-N/yr, and wild animals, roughly 3 Mt-N/yr. Ammonia vaporization during the application of fertilizers to soils amounts to about 3 or 4 Mt-N/yr.

The total emissions of fixed nitrogen as ammonia into the atmosphere may exceed the emissions of nitrogen oxides. However, ammonia seems mainly to react with and neutralize the acids in fog, clouds, and rainwater. Ammonia easily dissolves in water, and it is very reactive with respect to acidic compounds (Sections 3.3.4 and 9.3.4). Accordingly, almost as fast as ammonia gas is emitted into air, it is absorbed by water and reacts with acids to form salts, including ammonium nitrate and ammonium sulfate. Even in regions of low humidity, ammonia combines directly with nitric acid vapor to form ammonium nitrate aerosols. When ammonia (as ammonium salts) is deposited as an aerosol or in precipitation, it reenters the fixed-nitrogen nutrient cycle.

Ammonia is photochemically unstable in the atmosphere. It can be decomposed by reacting with hydroxyl radicals (OH), eventually forming nitrogen oxides. This source of NO_x is probably small, however. As we already noted, most of the ammonia appears to be scavenged by clouds and aerosols before it can react chemically.

The Roles of Nitrogen

In the environment, nitrogen compounds have a number of important functions. The role of nitrate as a key nutrient has been mentioned. Nitrous oxide affects the global energy balance through the greenhouse effect (Chapter 11). Nitrogen oxides are also a principal ingredient of photochemical smog (Section 3.3.4 and Chapter 6). In some regions, nitric acid is a major component of acidic precipitation (Chapter 9). When nitric acid vapor reacts to form nitrate particles in the atmosphere, visibility is quickly degraded (Chapter 6). The quantity of ozone in the stratosphere is also controlled to a large extent by nitrogen oxides (Chapter 13). Indeed, enhancements in nitrogen oxides related to aircraft emissions and increases in nitrous oxide can lead to ozone depletion.

Nitrogen: The Human Factor

The contributions of human activities to the nitrogen cycle are large and increasing. Figure 10.8 illustrates some of the ways that people influence the nitrogen cycle. Industry fixes about 20 million tonnes of nitrogen each year for use as nitrate fertilizer. Much of the fixed nitrogen applied to fields is washed into streams and lakes, polluting them and encouraging artificial blooms of algae, which deplete the oxygen needed by fish and other organisms.

Some of the nitrate used as fertilizer is denitrified by bacteria to nitrous oxide. Combustion also directly introduces N_2O into the atmosphere (Figure 10.7). Considering the total atmospheric source, most of the nitrous oxide emission is associated with the denitrification of naturally generated fixed nitrogen. As mentioned earlier, nitrous oxide is a potent "greenhouse" gas and is responsible for the depletion of stratospheric ozone. There is concern that increasing rates of fertilization, required to sustain a growing human population in the face of vanishing arable land, will lead to enhanced N_2O concentrations in the future. Indeed, nitrous oxide appears to be increasing slowly (at a rate of roughly 0.1 percent per year). The threat from accumulating nitrous oxide is currently secondary to other potential global hazards.

In many urban areas, the emission of fixed nitrogen in the form of nitric oxide from automobiles is a primary cause of air pollution and smog. One of the principal technological goals has been to reduce the NO_x emissions of automobile engines. Similarly, aircraft engines inject nitrogen oxides into the upper atmosphere, where they can affect the ozone layer. New high-speed commercial aircraft flying faster than the speed of sound may someday be operating in the upper atmosphere (so we can get from Los Angeles to Tokyo in 5 hours rather than 11). The engines will be required to emit very little NO_x. It is no accident that urban areas with dense photochemical smog may also suffer from excess acidity in precipitation and fog. The nitrogen (and sulfur) oxides in smog eventually react to form strong acids (Chapter 9), which are readily absorbed and concentrated in cloud and fog droplets. The goal of reducing the NO_x emissions from internal-combustion engines would alleviate more than one serious environmental problem.

10.2.3 OXYGEN

Oxygen is obviously an element that we can revel in. Oxygen is breath; it is life. More technically, oxygen is required for aerobic respiration, which energizes most of the life on this planet. In addition, the stratospheric ozone layer, which is essential to life, owes its existence to the presence of oxygen in the atmosphere (Sections 4.2.1 and 4.2.3). Most of the elemental oxygen accreted by the Earth remains locked up as mineral oxides in the crust and mantle.

Figure 10.8   The elements of the nitrogen cycle that are affected by fertilizer use. Fertilizers can modify the nitrogen cycle on local, regional, and global scales. Specific processes and the major chemical species involved are shown.

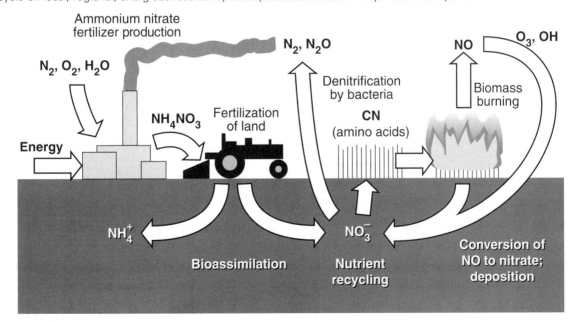

Only a very small fraction of the total, amounting to roughly 1 million gigatonnes(!), is found as "free" oxygen in the atmosphere. The origin of this free oxygen and its long-term fate can be understood by referring to Figure 10.9.

The source of oxygen in the atmospheric reservoir is CO_2. During the early evolution of the Earth, some free oxygen was produced in the atmosphere by the photochemical decomposition of water vapor and the escape of hydrogen to space:

$$H_2O + h\nu \rightarrow OH + H$$
$$H_{Atmos} \xrightarrow{Heat} H_{Space} \quad (10.19)$$
$$OH + OH \rightarrow H_2O + O$$

At high altitudes, the thermal velocities of some hydrogen atoms are great enough to propel them into space. Thus the overall effect of the sequence of reactions in Equation 10.19 is to leave some free oxygen behind in the atmosphere. The rate of hydrogen escape and oxygen accumulation is slow, however.

Later in the history of the Earth, photosynthesis evolved as a biological process and spread throughout the global oceans. A new efficient source of oxygen for the atmosphere was established:

$$CO_2 + H_2O$$
$$\xrightarrow{\text{Sunlight}} \text{'}CH_2O\text{'} + O_2 \quad (10.20)$$

However, the biosphere is a very limited reservoir. Biomass is susceptible to oxidation (by fire, bacterial consumption, and the like) and cannot accumulate:

$$\text{'}CH_2O\text{'} + O_2 \rightarrow CO_2 + H_2O \quad (10.21)$$

But if the organic carbon is buried deep in the Earth, far from the free oxygen in the atmosphere, it cannot react with that oxygen. Organic carbon is continuously buried in sediments as detritus settling to the bottom of seas and lakes, which can evolve into fossilized carbon fuels such as coal and oil. The amount of carbon that is buried in these reservoirs in fact determines the amount of oxygen that can remain free in the atmosphere.

One can imagine that each CO_2 molecule consumed during photosynthesis is decomposed into one C atom and one O_2 molecule. The C atom is part of the organic material formed, 'CH_2O'. The oxygen is effectively released into the atmosphere. If the

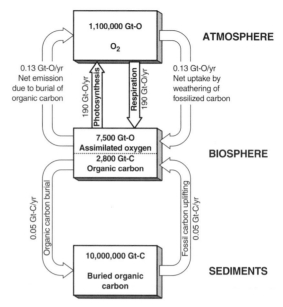

Figure 10.9 The partitioning of oxygen between the atmosphere and the biosphere, and the role of carbon burial in free oxygen production.

'CH_2O' is reassimilated by an oxygen-breathing organism (respiration) or is burned in air, an oxygen molecule will be used up, as in Equation 10.21. Accordingly, the net rate at which oxygen accumulates in the atmosphere must be equal to the net rate of burial of organic carbon in sediments. That rate, which defines the source of oxygen, can be estimated and turns out to be relatively small, about 0.05 Gt-C/yr (50 Mt-C/yr). Since each buried C atom, with an atomic weight of 14, releases an O_2 molecule, with an atomic weight of 32, the source for oxygen is $0.05 \times (32/14) = 0.13$ Gt-O/yr.

Over long time spans, the source of oxygen is balanced by a sink. In this case, the buried carbon is recycled by geologic processes from sediments to the Earth's surface. Exposed to the atmosphere and voracious organisms, the carbon is oxidized into CO_2 by means of Equation 10.21, and thus the oxygen cycle is closed. The average rate of carbon uplifting and oxidation is similarly estimated to be about 0.05 Gt-C/yr. Over long periods of time, the oxygen reservoir is in equilibrium. The total quantity of oxygen in the atmospheric reservoir is therefore determined by the quantity of buried organic carbon. This amounts to about 1×10^7 Gt-C. However, the amount of oxygen in the atmosphere is only about 1.1×10^6 Gt-O (Figure 10.9). A lot of excess oxygen is missing. Of course, the total organic carbon reservoir held in the Earth is difficult to determine; it would take a lot of digging to measure it precisely. Another possibility is that over the eons of the Earth's evolution, most of the excess oxygen was consumed in oxidizing reduced minerals as they were raised from the interior. Reduced sulfides and metal ores are common today, reflecting the original reduced chemical state of the entire Earth system.

In summary, we may conclude that free oxygen in the atmosphere, which is essential to the existence of humans and most other organisms on Earth, is generated by the burial of organic carbon in sediments, primarily in the oceans. The balance between the production and removal of oxygen is controlled by geologic processes involving the cycling of carbon compounds through the oceans and the lithosphere.

The Persistence of Oxygen

The transfer of oxygen between the atmosphere and other reservoirs is dominated by photosynthesis. Nearly 200 Gt-O/yr are processed through the biosphere. At that rate, if photosynthesis ceased, the atmospheric reservoir would be depleted of oxygen in about 5000 years: $(1,000,000 \text{ Gt-O})/(200 \text{ Gt-O/yr}) = 5000$ yr. Although that is well beyond a human lifetime, it still seems precariously close. Fortunately, the time required to burn up oxygen is much longer. The total reservoir of organic carbon that is accessible for oxidation over short times is only about 2800 Gt-C (Figure 10.9). Hence only about 7500 Gt-O can be depleted by oxidizing this carbon—a drop in the bucket. To use up the rest of the oxygen, carbon must be cycled from inside the Earth. This rate is roughly 0.05 Gt-C per year. At that rate, with no new carbon burial, our oxygen supply should last approximately 20 million years! That should put our minds at ease about our precious oxygen supply.

Humans are also mining fossilized carbon to use as fuel. This recovery and consumption of buried carbon represent an acceleration of the natural rate of carbon recycling. Indeed, our current rate of fossil-fuel consumption of about 6 Gt-C/yr is an increase by a factor of $6/0.05 = 120$ over the natural rate of carbon recycling from the Earth's interior. Will this reduce the lifetime of atmospheric oxygen by a factor of 120? Probably not. As we will see, the total reservoirs of accessible fossilized carbon amount to perhaps 10,000 Gt-C. If all this carbon were recovered and

burned, only a small percentage of the atmospheric oxygen would be consumed (Problem 3). Accordingly, oxygen appears to be persistent and reliable despite a heavy assault by human consumption.

10.2.4 Carbon

Perhaps more than any other, the carbon cycle is associated with the climate system and a possible global greenhouse warming of the Earth (Chapter 12). In this section, we review the global carbon cycle in terms of the reservoirs and flows of carbon through the reservoirs. Although considerable information about the carbon cycle has been obtained by scientists over the past 50 years, many of the important details remain uncertain or controversial. Accordingly, this summary will highlight the features of the carbon cycle that are most critical to determining climate and climate change. Keep in mind that as with all the global biogeochemical cycles considered here, the facts concerning enormous and diverse reservoirs and complex transformation processes are limited and imprecise.

The principal reservoirs for carbon in the Earth system consist of sedimentary rocks, fossilized organic carbon, including the "fossil fuels;" the oceans, the atmosphere, and the biosphere. (The important chemical forms of carbon were given in Section 10.1.) The sizes and roles of specific reservoirs are discussed next. To simplify our discussion, we have divided the overall carbon cycle into three components: the long-term cycle involving sediments and the interior Earth, the cycle coupling the atmosphere and land, and the cycle between the atmosphere and the oceans. The first cycle, extending over millions of years, determines the present-day partitioning of carbon between the surface reservoirs and the sedimentary reservoirs and explains to a large extent the current pleasant climatic state of the Earth. The other two cycles are much faster and may therefore be subject to human influence. The integrated carbon cycle is also reviewed at the end of this section.

The Long-term Carbon Cycle

Figure 10.10 illustrates the components of the carbon cycle that operate over very long periods of time, essentially geologic time scales. This carbon cycle is related to the oxygen cycle shown in Figure 10.9, which is one reason for its importance. In the long term, the atmospheric reservoir, which holds only 750 Gt-C, acts mainly to transfer the carbon dioxide emitted by the Earth through volcanism to the oceans, where the CO_2 is dissolved. The oceans perform the function of transforming the dissolved carbon dioxide into carbonates, which can be absorbed by animals to form protective shells of calcium carbonate. Some of these shells eventually settle to the bottom of the oceans and accumulate as sediment. The sediment is later compressed into sedimentary rocks composed of calcium carbonate, one of the most abundant minerals in the Earth, comprising limestone and marble. The oceans also represent a relatively minor reservoir of carbon, amounting to about 40,000 Gt-C. Clearly, the bulk of the Earth's carbon is tied up as sedimentary carbonates, although there is a substantial reservoir of organic carbon, as described later.

The geologic carbon cycle occurs over hundreds of millions of years. Indeed, the time scale for emptying the sedimentary reservoir by geologic processes can be estimated from the information in Figure 10.10; that is, 6×10^7 Gt-C/(0.25 Gt-C/yr) = 240 million years. The cycle is slow because it involves the overturning of the Earth itself. The cycle can be imagined to begin with CO_2 dissolving into the oceans. Carbon dioxide is partly soluble in water (remember that natural rainwater has a pH of about 5.6 because of the dissolved CO_2). The process of dissolution can be written as a series of physical-chemical processes, as follows:[3]

$$CO_2 \text{ (gaseous)} \xrightarrow{\text{Dissolution}} CO_2 \text{ (aqueous)}$$

$$CO_2 \text{ (aqueous)} + H_2O \rightarrow H_2CO_3 \text{ (aqueous)}$$

$$H_2CO_3 \xrightarrow{\text{Dissolution}} H^+ + HCO_3^-$$

$$HCO_3^- \xrightarrow{\text{Dissociation}} H^+ + CO_3^{2-}$$

$$(10.22)$$

The carbonate formed by CO_2 dissolution can be assimilated by marine organisms, small and large, to

3. The dissolution of CO_2 in water is also discussed in Section 4.1.3. The emphasis there is on the aqueous ion equilibrium process that forms bicarbonate and carbonate ions from carbon dioxide. That process is the same as the dissolution process outlined here.

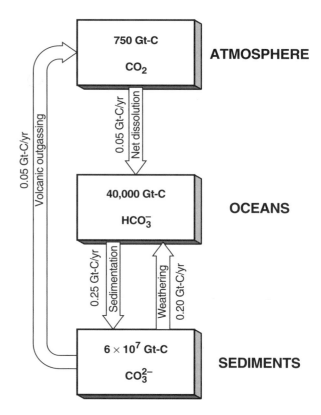

Figure 10.10   The long-term carbon cycle of the coupled atmosphere, oceans, and sediments. The mass of carbon in each reservoir, the processes, and the average long-term rates for mass transfer between reservoirs are shown.

construct skeletons, referred to as *tests*. A clamshell is basically composed of carbonates. Because the oceans hold a huge quantity of dissolved calcium, the favorite carbonate is $CaCO_3$, produced by the association of calcium and carbonate ions:

$$Ca^{2+} + CO_3^{2-} \rightarrow CaCO_3 \quad (10.23)$$

When an organism dies, its shell can fall to the bottom of the ocean and become incorporated into sediments. However, most of the falling shells do not make it into the sediments. The deep oceans, it turns out, can dissolve calcium carbonate back into ions (the opposite of Equation 10.23). In ionic form, it is almost impossible for the carbonate to be deposited as sediment. Accordingly, the most likely place for carbonate sediments to form is in the shallower oceanic regions, particularly around the margins of continents.

The sediments that form are recycled through the Earth. Some of the carbonate materials are subjected to intense pressures and temperatures in the lithosphere. Often, sand is also present in these sediments because it is continually washed into oceans from the continents or is blown as aeolian dust from arid regions long distances over the oceans. Together, at high pressures and temperatures, carbonates and silica react to form silicate minerals and carbon dioxide:

$$CaCO_3 + SiO_2 \xrightarrow{\text{Heat}} CaSiO_3 + CO_2 \quad (10.24)$$

where $CaSiO_3$ represents a generic silicate mineral. Equation 10.24, occurring in the molten regions of the mantle, is called silicate reconstitution (because the silicate is literally reformed from its chemical products). The carbon dioxide released by this process can be emitted through volcanism:

$$CO_2 \text{ (Earth)} \xrightarrow{\text{Volcanism}} CO_2 \text{ (atmosphere)} \quad (10.25)$$

That is, when the heated molten rock, or lava, seeps out of fissures in the Earth or explodes during major eruptions, the carbon dioxide "degasses" from the lava and enters the atmosphere. The lava cools into rocks exposed on the land surface.

The surface of the Earth resembles a jigsaw puzzle in the round. Individual pieces of the puzzle, referred to as *tectonic plates*, are the size of continents and fit together to form the crust of the Earth. The deep ocean floors are actually such plates, and so are most of the major continental land masses and subcontinents. The pieces fit together very neatly indeed and provide a suit of armor for the Earth. The solid plates also move extremely slowly (1 centimeter per year) across the planet under forces exerted by currents flowing in the molten interior. The plates drift like broken ice on a river. Because the entire Earth is completely encased in these plates, any tectonic motion becomes a kind of perpetual slow-motion collision between plates. Incredible masses of rock and sediment crush together at imperceptible speeds, these collisions causing earthquakes and volcanic eruptions. During some of these collisions, sedimentary carbonates are lifted and exposed at the surface. Silicate lava is spewed out by volcanoes and hardens into rocks.

Rocks do not sit there forever, but are slowly eroded by wind and rain. This process of erosion is called *weathering*. Indeed, carbon dioxide in the atmosphere, released from the Earth, accelerates this erosion. Carbon dioxide is absorbed into cloud water and causes rain to be slightly acidic (because it forms carbonic acid). The acid dissolves both carbonate and silicate rocks. Thus the recycling of rocks follows these pathways:

$$CO_2 + H_2O + CaCO_3 \xrightarrow{\text{Rain}} Ca^{2+} + 2HCO_3^- \quad (10.26)$$

$$2CO_2 + H_2O + CaSiO_3 \xrightarrow{\text{Rain}} Ca^{2+} + 2HCO_3^- + SiO_2 \quad (10.27)$$

In the first process, carbon dioxide and water combine as carbonic acid, which, in contact with a calcium carbonate rock surface such as limestone, dissolves the rock into calcium and bicarbonate ions. The ions can flow down streams and rivers into the oceans, where the long-term carbonate cycle begins anew. The erosion of calcium carbonate by carbon dioxide illustrates the buffering capacity of some minerals against the acidity in rain. This effect is important in limiting the potential damage to the environment from acidic rain (Section 9.5). In the buffering process, the mineral reacts with and deactivates the acidic hydrogen ion (H^+). In the case of carbonate rocks in contact with acidified water, the reactions are

$$HNO_3 \xrightarrow{\text{Aqueous}} H^+ + NO_3^-$$

$$CaCO_3 \xrightarrow{\text{Aqueous}} Ca^{2+} + CO_3^{2-}$$

$$H^+ + CO_3^{2-} \rightarrow HCO_3^- \quad (10.28)$$

Nitric acid is used as an example. Sulfuric, hydrochloric, or other acids would have done as well. Notice that the carbonate ion (CO_3^{2-}), scavenges a hydrogen ion and thereby deactivates it. Bodies of water set in a basin of carbonate rock are well buffered against acidity.

In silicate weathering (Equation 10.27), silicate minerals interact with solutions of carbon dioxide, again producing calcium and bicarbonate ions, but also releasing the silica as sand. Eroding mountains thus make beautiful beaches. Beach building and carbon cycling have worked in parallel over many eons and will continue for many more.

The Carbon Cycle: Air and Land

The carbon cycle that couples the atmosphere with the land biosphere is an important component of the overall carbon cycle. The basic elements are illustrated in Figure 10.11. The land biosphere is identified as a carbon reservoir distinct from the land carbonate reservoir. The terrestrial biosphere is dominated by trees and other vegetation. Earth is truly a world of vegetables, with a few opportunistic animals around to nibble on them. The living biomass on land is nearly 100 times as massive as the total living oceanic biomass. Deceased biomass is even more plentiful than living biomass (Figure 10.11). Inanimate organic matter accumulates as plant litter in forests, building up as a layer of humic soil or peat. Eventually, the organic debris is consumed by microorganisms as it decays or in periodic conflagrations, in either case releasing carbon dioxide into the atmosphere.

In the simplified treatment offered here, the land biomass is lumped together as one average "species." Certainly, the total mass of plant matter is much greater than the total mass of animal matter, very roughly by a factor of 100. Plants have quite diverse properties and life cycles. "Annual" plants are seeded or regenerated every year, whereas "perennial" plants have life cycles lasting 2 or more years, and some trees live up to several thousand years. Annuals account for about 20 percent of all the land biomass, including the leafy growth on deciduous trees (that is, trees that become dormant in the winter and drop their leaves, compared with evergreens, which retain their needles for many winters). The vegetative growth of annual plants, including most agricultural crops, is cycled through the biosphere in about 1 year. On the other hand, trees, which comprise about 80 percent of the total biomass, are recycled only every 30 years or so on average. The overall biomass cycling times (from atmospheric carbon dioxide back to atmospheric carbon dioxide) range from less than 1 year to 1000 years, depending on the species. The average cycling time for the whole terrestrial biosphere is close to 10 years.

The global production of organic mass in the environment is controlled by photosynthesis, a pro-

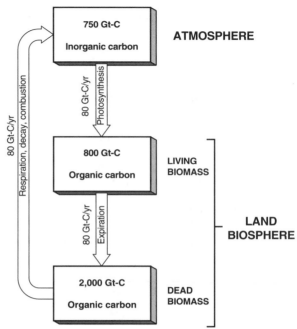

Figure 10.11 The coupled carbon reservoirs of the atmosphere and the land biosphere. The total land biomass is further divided into living and dead material. The processes and average rates of transfer of materials between the reservoirs are also shown. The time scale for the transfer between these reservoirs is on the order of a decade, compared with the time scales of hundreds of millions of years in Figure 10.10.

cess that uses the energy of sunlight and plentiful water and carbon dioxide in the atmosphere to create living matter. (The evolution of life and the appearance of photosynthesis on Earth are discussed in Section 4.2.1.) The complex and essential process of photosynthesis can be summarized as a simple chemical reaction:

$$CO_2 + H_2O \xrightarrow{\quad Sunlight \quad} {}^{\cdot}CH_2O{}^{\cdot} + O_2 \qquad (10.29)$$

In Equation 10.29, 'CH$_2$O' is the unit of organic matter discussed in Chapter 3. Photosynthesis is perhaps the most important life process on Earth. It harnesses the energy of the sun for use by living organisms. It creates 99 percent of all the biomass on the Earth. Photosynthesis generates oxygen used by countless animals for aerobic respiration. It is responsible for the ozone layer, which protects all life-forms from ultraviolet annihilation. It is, in essence, indispensable to all plant and animal life.

The rate of carbon transfer from the atmosphere to the biosphere is roughly 80 Gt-C/yr. Accordingly, the lifetime of the atmospheric carbon reservoir (Equation 10.2) is about 9 years (750 Gt-C/80 Gt-C/yr). Vegetation that dies accumulates as organic debris or humus. The rate that this dead organic matter is converted back to carbon dioxide (through decay and combustion) is also about 80 Gt-C/yr. The presence of huge deposits of organic detritus has led, over billions of years, to the development of armies of bacteria and fungi that can live on the litter. In assimilating the detritus, they oxidize it, generating energy in a process called *respiration*. Respiration causes CO_2 to be released into the atmosphere, as follows:

$$\text{`CH}_2\text{O' + O}_2 \xrightarrow{\text{Respiration}} CO_2 + H_2O \quad (10.30)$$

Inspection shows that Equations 10.29 and 10.30 form a closed cycle. There is no net change in the chemical state of the environment. It is a clean, efficient, and natural cycle. However, these processes have extracted the energy of sunlight to carry out the business of life. Organisms that cannot directly harvest sunlight energy instead harvest the plants that can. Living animal species, including humans, have evolved in this dependent relationship with plants.

Over a year, the rates of photosynthesis and respiration appear to be in close balance. Ecosystems, consisting of numerous interacting organisms exploiting local resources to the maximum, are able to maintain only a limited quantity of living matter on land surfaces. Accordingly, the total area of habitable land determines the total biomass of the Earth. If the area does not change, neither will the total biomass (within some reasonable range of variation, of course). The living organic carbon reservoir on land is thus relatively stable over long time spans. Any small alterations in the global environment are accommodated over a relatively short time of a decade or so. Some species may perish, while others appear to fill new niches in the global ecosystem. If the climate becomes more conducive to vegetation, the forests may spread across the land; if less conducive, the trees may retreat for a time. Thus, over millions of years, the terrestrial biomass ebbs and wanes in rhythm with the beat of the global climate.

Although the average rate of carbon assimilation into organic matter and the average rate of oxidation of organic debris are apparently in balance from year to year, a seasonal pulsation in carbon dioxide in the atmosphere can be measured (Chapter 12). In the summer growing season, CO_2 is rapidly absorbed by growing plants, and the atmospheric concentration dips. In winter, when plants are largely dormant, the microfeeders continue to consume reserves of detritus, leading to a "pulse" of CO_2. The total land areas of the Northern and Southern Hemispheres are quite different. The north holds about 75 percent of the total land mass of the Earth. Likewise, most of the land biomass is found in the north. Accordingly, the seasonal "pulse" in atmospheric carbon dioxide is much more pronounced in the Northern Hemisphere. This pulsation in CO_2 is the heartbeat of a living planet.

Carbon dioxide is the basic food of the biosphere, and land is home to most living matter. If atmospheric levels of CO_2 were to increase, how would the biosphere respond? With more food available, more plants should grow by means of "fertilization." Although seemingly obvious, there is as yet no direct evidence to link significant increases in total biomass with increases in the concentrations of carbon dioxide. It is quite possible that total plant mass is strongly limited by the availability of key nutrients (sulfur, nitrogen) and water (in arid regions, certainly). Changes in weather and climate induced by changes in atmospheric CO_2 (Chapter 12) may also limit the opportunities for plants to take advantage of the extra helping of carbon. Measuring changes in the global biosphere is also very difficult. Imagine weighing all the vegetation on Earth. Even satellite images are difficult to interpret in terms of total biomass. The total biospheric reservoir of roughly 2800 Gt-C has an overall lifetime of about 35 years, determined as

$$\frac{2800 \text{ Gt-C}}{80 \text{ Gt-C/yr}} = 35 \text{ yr}$$

If the rate of carbon uptake by vegetation were to rise slightly—say, an increase of 5 percent, or 4 Gt-C/yr, to 84 Gt-C/yr—the change in the global biosphere would hardly be noticeable, even after a new equilibrium state had been reached at about 2940 Gt-C worldwide. That is a lot of additional organic mass, but it is a very big world!

The Carbon Cycle: Air and Sea

The oceans are a much larger reservoir for carbon than the atmosphere is (Figure 10.12). Most of this

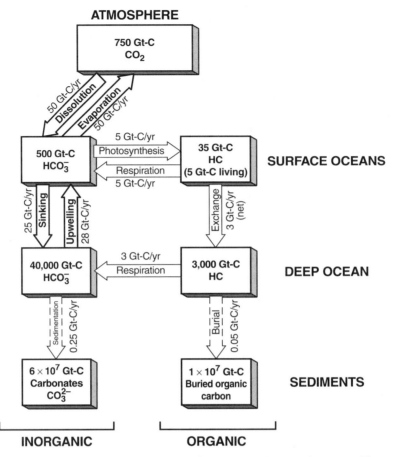

Figure 10.12   The coupled carbon cycles of the atmosphere and oceans. The ocean component is divided into two distinct regions: the surface oceans and the deep oceans. Each region is subdivided into reservoirs for the inorganic (carbonate) and organic (biomass [HC]) carbon components. The sedimentary reservoirs are shown as the primary long-term sink for ocean carbon. The processes and average rates of transfer of carbon between the reservoirs are also given.

is in the form of carbonates, washed down from the land as a result of the rocks' weathering. The carbonates accumulate in the oceans over many hundreds of thousands of years. The total living biomass in the oceans is only about 1 percent of that on land. The oceanic dead biomass (detritus) reservoir, however, is comparable (with a substantial uncertainty) to the land's humic reservoir. This organic nutrient-rich soup residing in the deep oceans provides a bonanza for growth when recirculated to the surface waters.

The inorganic carbon residence time in the surface waters is approximately 10 years. The exchange of carbon dioxide between the surface water and the atmosphere controls the amount of inorganic carbon there. The air-water interface is the principal connection between the atmosphere and oceans, controlling the rates of transfer of water vapor and

other gases between these media, as well as the coupling of atmosphere and ocean dynamics through the interactions between winds and waves.

The living organic carbon in the surface oceans turns over into the deep waters in less than 1 year on average. The complete mixing of surface water into the depths takes longer, about 20 years. Organic matter has a shorter residence time because in the form of fecal pellets, it can sink rapidly to great depths. The organic cycle of the ocean is driven by microscopic plants, the phytoplankton, living in the sunlit surface waters, or photic zone. These simple plants use photosynthesis to convert sunlight to cellular tissue. In the process, they absorb nutrients from the water. The phytoplankton are not alone, however. Waiting to feed on them are hordes of zooplankton. The herbivorous grazers are themselves food for other carnivorous plankton. These, in turn, are eaten by small crustaceans and fish. The marine food chain builds in the size of the feeders up to whales. The whole food sequence rests on the modest shoulders of the phytoplankton—take them away and the food chain would collapse; nearly everything in the oceans would die.

The organic detritus that settles out of the photic zone carries with it valuable nutrients. By thus depleting the surface waters of nutrients, life itself is the principal cause for the relative biological impoverishment of the oceans. Nevertheless, there are enough organisms in the deeper waters to efficiently consume the settling detritus, oxidizing most of it back into carbon dioxide (which, when exhaled, is converted immediately to bicarbonate). The organic carbon cycle of the oceans is therefore similar to the organic carbon cycle of the land and may be represented by the processes in Equations 10.29 and 10.30. A trickle of organic debris reaches the ocean floor and is sequestered as organic sediment (after escaping the hungry jaws of the bottom scavengers). The recycling of this carbon over very long time spans is depicted in Figure 10.10.

The living organisms of the seas have evolved clever means of protection. Both for skeletal strength and as a shield against the environment, numerous species use the plentiful calcium and carbonate ions dissolved in ocean water to form shells and skeletons:

$$Ca^{2+} + CO_3^{2-} \xrightarrow{\text{Shell building}} CaCO_3$$

$$(10.31)$$

These common calcareous (that is, containing calcium carbonate) structures include clamshells and conches. Other microorganisms employ dissolved silica to achieve the same goal, forming siliceous hard parts. Crustaceans such as lobsters and crabs have found an alternative protective material consisting of chitinous exoskeletons, sometimes also impregnated with calcium carbonate. When these organisms die, their remains settle into deeper water. Unlike the soft body parts, which may offer a tasty meal for scavengers, the calcareous parts are not consumed. Very few of the shells, or tests, reach the bottom, however. The deep oceans, it turns out, corrode calcium carbonate. As the tests fall deeper into the abyssal depths, they are dissolved back into calcium and carbonate ions:

$$CaCO_3 \xrightarrow{\text{Dissolution}} Ca^{2+} + CO_3^{2-}$$

$$(10.32)$$

The deep oceans are more apt to corrode calcium carbonate because the water is colder, which affects the chemistry of the ions, as defined by Equation 10.22. The colder the water is, the more CO_3^{2-} that can be tied up as HCO_3^- and H_2CO_3.

In the deep oceans, the carbon dioxide generated from organic detritus by respiration and from calcareous shells dissolved at the colder temperatures adds to the background bicarbonate and carbonate concentrations. As water circulates from the deep to the surface, the excess carbonate is carried along and replenishes the surface reservoir. The overall residence time of water in the deep oceans is roughly 1500 years (Figure 10.12).

The exchange of carbon between the atmosphere and land biomass, or between the atmosphere and surface oceans, occurs in a time shorter than a human generation, whereas the exchange between the surface and deep waters of the oceans spans a time comparable to the age of human civilization. It would be reasonable to wait the former period to allow the environment to heal itself. But we could hardly afford to wait as long as the latter time for a grievous insult of the global environment to be corrected. Unfortunately, existing increases in atmospheric carbon dioxide associated with fossil-fuel use, and any accompanying climatic effects that may follow, are already locked in for centuries (Chapter 12).

The Integrated Carbon Cycle

The integrated global carbon cycle is illustrated in Figure 10.13. The overall cycle is complex and, as of today, plagued by uncertainties. The number of distinguishable individual processes that contribute to the global carbon cycle, including physical, chemical, and biological processes, is enormous. The number of phytoplankton species in the ocean is undetermined. The rate at which fossilized carbon is raised from sediments has not been measured. The response of nearly all species of plants to increases in atmospheric carbon dioxide is unknown. Nevertheless, an important body of information about the carbon cycle has been amassed, as discussed earlier (and in Chapter 12). Also keep in mind that the reservoirs identified in Figure 10.13 represent the minimal number required to describe, with any precision, the Earth's carbon cycle.

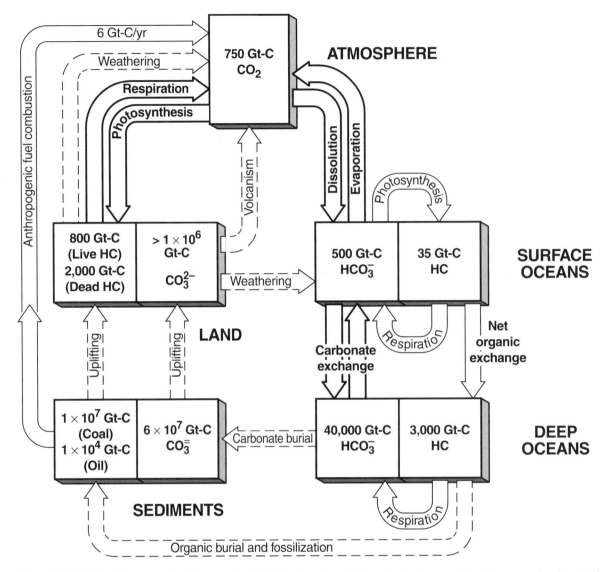

Figure 10.13   The integrated carbon cycle of Earth showing all the principal reservoirs, the mass held in each reservoir, and the processes that connect them. The processes associated with dashed arrows represent the slow, long-term transfer of carbon. The processes shown with solid arrows indicate the rapid, short-term exchange of carbon.

The sizes of the carbon reservoirs are determined by the rates of the transfer processes depicted in Figure 10.13, as described in previous sections. The **primary** (fast) carbon transfer processes can be summarized as follows:

- Atmosphere/land-biomass exchange of carbon dioxide by **photosynthesis** and **respiration**, 80 Gt-C/yr.
- Atmosphere/ocean exchange of carbon dioxide by **dissolution** and **evaporation**, 50 Gt-C/yr.

- Surface-ocean/deep-ocean exchange of carbonate by **sinking** and **upwelling** of ocean water, 25 Gt-C/yr.
- Surface-ocean/water-biomass exchange of carbon dioxide by **photosynthesis** and **respiration**, 5 Gt-C/yr.

The **secondary** (slow) carbon transfer processes include

- Sediment/atmosphere transfer of carbonate as carbon dioxide by silicate mineral reconstitution

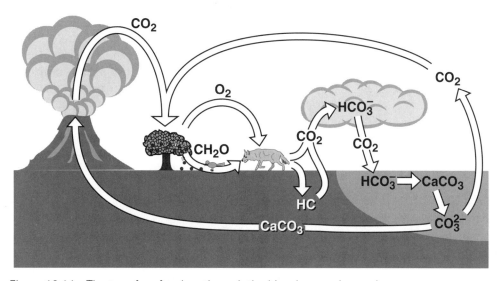

Figure 10.14   The transfer of carbon through the biosphere and geosphere.

and **volcanism**, 0.05 Gt-C/yr.

- Sediment/ocean transfer of carbonate by **weathering** and runoff from land, 0.20 Gt-C/yr.
- Ocean/sediment transfer of carbonate by **sedimentation** and burial of tests, 0.25 Gt-C/yr.
- Sediment/atmosphere transfer of fossilized organic carbon by oxidation during **weathering** of uplifted sediments, 0.05 Gt-C/yr.
- Ocean/sediment transfer of organic carbon by **sedimentation** and burial of detritus, 0.05 Gt-C/yr.

Note that the slower secondary cycles represent small residuals of the primary exchange processes. The secondary cycles cause changes in the reservoirs only after millions of years. For example, the net dissolution (that is, total dissolution minus total evaporation) of carbon dioxide into the oceans represents the average long-term transfer of carbon into ocean sediments. The sedimentary reservoir has a residence time of roughly 240 million years. By comparison, the atmosphere, biosphere, and surface ocean residence times are roughly a decade.

Figure 10.14 depicts the flow of carbon through the Earth system, including biological and geophysical components. Although somewhat whimsical, it is a fact that each carbon atom will, over time, see more of the world than anyone could imagine. Carbon travels across the planet on the winds and through the Earth in sediments. Carbon atoms have been inside countless organisms and served as a meal for many a hungry beast or bug. The number of pathways by which individual carbon atoms may travel

through the environment is countless. Nearly all those carbon "adventures" involve purely natural processes acting according to rules established by eons of evolution.

Minor Carbon Players

The Earth's carbon cycle involves many other compounds besides carbon dioxide. Methane (CH_4) has a mixing ratio of nearly 2 ppmv in the atmosphere (about 0.5 percent of the CO_2 mixing ratio). (The sources and distribution of methane are discussed in Section 12.3.1.) In the atmosphere, methane is oxidized into carbon dioxide and water by the overall process

$$CH_4 + 2O_2 \rightarrow CO_2 + 2H_2O \qquad (10.33)$$

The methane is emitted by bacteria feeding on decaying organic material. In other words, some of the biospheric carbon is recycled as CH_4 rather than CO_2. The total carbon flux as methane amounts to only perhaps 0.4 Gt-C/yr. Moreover, the carbon in methane is eventually transformed into carbon dioxide.

Carbon monoxide has a somewhat larger global source than methane does. In fact, CO is an intermediate photochemical product in the oxidation of methane (the CO later reacts with hydroxyl [OH] to form CO_2 [Section 3.3.4]). Other sources of carbon monoxide include the oxidation of natural airborne hydrocarbons and biomass combustion (forest and grass fires). In total, about 1 Gt-C/yr of CO is cycled between the atmosphere and the biosphere. This hardly makes a dent in the global carbon cycle.

Vegetation also emits a wide range of complex organic vapors into the atmosphere. These include the **terpenes**, such as **limonene**, which gives lemons their distinctive (and pleasant) smell, and **pinene**, which is responsible for the familiar odor of coniferous forests. Bacteria also produce the so-called **nonmethane hydrocarbons** (NMHC). NMHCs consist mainly of organic gases like ethane, ethylene, and propane. These compounds are more complex than methane and are produced in smaller quantities by natural processes. The total mass of nonmethane organic carbon vapors participating in the global carbon cycle is very uncertain, but is probably less than 1 Gt-C/yr.

The Human Impact on the Carbon Cycle

One of the key carbon transfer processes in Figure 10.13 that has not been elaborated is the "anthropogenic fuel-combustion" branch, which links the sedimentary reservoir of fossilized carbon with the atmospheric reservoir of carbon dioxide. The fossil fuels consumed include a significant fraction of methane as natural gas. The current rate of exchange is about 6 Gt-C/yr (Section 10.2.3). The mining and burning of fossil fuels for energy would, at the present time in human history, be classified as a primary net source of carbon for the atmosphere. We have effec-

tively accelerated the natural processes of uplifting and weathering of organic sediments by a factor of more than 100, as noted earlier.

During the combustion and other uses of fossilized carbon, compounds other than CO_2 are released into the atmosphere, often as pollutants. So, for example, carbon monoxide from incompletely burned fuel is a major health threat as an urban pollutant (Chapter 6). Leakage of natural gas is an additional source of atmospheric methane. Of all the anthropogenic carbon emitted into the atmosphere, about 5 to 10 percent may be in the form of CO and CH_4. That emission represents, of course, only a very small fraction of the total global carbon cycle.

Human activity is adding carbon to the atmosphere in other ways. Clear-cutting of forests to obtain lumber or for agricultural use may be transferring another 1 Gt-C/yr into the atmosphere from the land biosphere reservoir. At these rates, atmospheric carbon dioxide levels could double in the next 100 to 200 years. (The fate of this anthropogenic carbon is discussed in greater depth in Chapter 12.) Some of the carbon, released in the form of CO_2, rapidly dissolves in the surface oceans. Most of the rest accumulates in the atmosphere, where it may cause a change in the global climate. The role of "fertilization" and carbon absorption by the biosphere in limiting the accumulation of carbon dioxide is uncertain, as we mentioned earlier.

Figure 10.15   The hydrological cycle of the earth. The reservoirs of water in the oceans are given as total (sedimentary) groundwater and atmospheric water vapor. The principal short-term water-cycling processes also are depicted. (Data from J. W. Maurits la Riviere, *Scientific American* 261 [1989]: 80.)

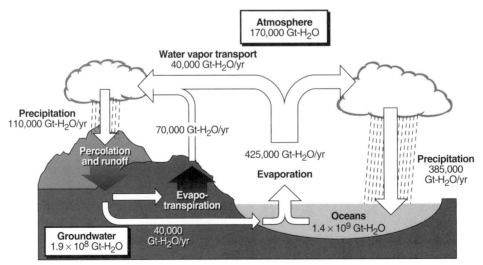

10.3   The Hydrological Cycle

The hydrological, or water, cycle is the most powerful and important cycle of all. The water cycle consists of the humidity we feel in the air, clouds, rain, snow and hail, lightning, streams and rivers, beaches and mud, lakes and oceans (Figure 10.15). Water evaporation and condensation provide a major source of energy, through latent heat release, to drive the winds. Atmospheric water vapor and clouds control the climate to a far greater degree than does carbon dioxide. Without the greenhouse warming caused by water vapor in air, the Earth would be frozen solid (Sections 11.4 and 12.1.2). Water is the medium that breeds and sustains life: Living organisms first evolved in the oceans, and all living things require water for survival.

The hydrological cycle also cleanses the environment. Clouds and rain continually wash pollutants from the air. The runoff of precipitation into rivers and streams carries pollutants and wastes to lakes and oceans. Rainfall percolating through the soil flushes pollutants deep into the ground. Without a hydrological cycle, all land would be a dusty wasteland. Unfortunately, toxins and wastes can be concentrated in the runoff or can accumulate in lakes and groundwater. Eventually, the buildup of pollutants may pose a serious threat to health and the environment.

The contamination of water resources is particularly serious around large cities and industrial zones and in agricultural regions. In large cities like Los Angeles, the first rains after a long dry season flush a cornucopia of exotic wastes into drainage canals and, eventually, the nearby coastal waters. Paint, oil, feces, and solid trash pour out of the city's drainage system. Worse, many industries illegally dump toxic wastes into streambeds. Even pollutants released into the air can settle on the ground and be washed into the streams and groundwater. Water has always attracted polluters. If the discarded waste sinks below the surface, it is out of sight (thus out of mind). If it floats, currents will take it elsewhere. And if the waste is liquid, it can disperse over a large area. But pollution it remains, now generously presented to the population at large for their displeasure.

The transfer rates in the water cycle are enormous. More than 1000 gigatonnes of water evaporate from the oceans each day, most of which falls as rain. The hydrological cycle is particularly powerful in the tropical regions, where abundant sunlight heats and evaporates huge quantities of water. At a few unusual places on Earth, such as the southern coastal regions of Peru, the hydrological cycle is so weak that years or decades may pass with no rain at all. The *global* hydrological cycle is imperturbable in any practical sense. Nevertheless, increased pollution may bring subtle changes, such as changes in the reflective properties of clouds (Section 11.6.5) or major alterations of precipitation patterns induced by greenhouse warming (Section 12.4.3). Further, we have massively engineered the hydrological cycle, to tame tides and floods and harness hydroelectric energy and water for irrigation. We have even tried to modify the weather by making it rain when and where we want. The engineering work in some ways has been successful locally in controlling a wild resource; the price paid is perhaps the loss of that wildness. Regarding the control of weather, we have so far failed at doing it purposefully, even as we seem to be doing it inadvertently. (Engineering the global environment is discussed in Chapter 14.)

10.4   A Global Garbage Dump?

In some ways, human civilization has been treating the global environment like a bottomless trash pit. Cities have filled up pristine canyons and valleys with rubbish. Countless tons of human garbage and sewage have been dumped into rivers and off coastlines. Humans are the only organisms that generate useless rubbish, including not only indestructible solid waste, but also toxic and radioactive materials that poison and despoil the air, soil, and water. All organisms create detritus; it is part and parcel of the natural biogeochemical cycles of the Earth. It should be no surprise that we have long held the attitude that the *global* environment provides a limitless dumping site. The atmosphere and the oceans, in particular, have been seen as enormous reservoirs in which we can dilute our toxic wastes. That view can no longer be accepted. We have seen that each of the major biogeochemical cycles that hold our environment together is under the relentless assault of human activities. Each cycle is beginning to show major changes directly traceable to human activities. The atmosphere is particularly vulnerable because of its small relative mass compared with that of the oceans and continents.

The environmental consequences of all this dumping are now being recognized. The preceding and following chapters of this book detail some of the specific effects: the fouling of the air with sulfuric acid particles, the accumulation of greenhouse gases that excessively warm the planet, the buildup of chlorofluorocarbons that deplete the ozone layer. The consequences for the biosphere, which are not discussed in great depth here, represent the ultimate threat to our healthy future on this planet. Acid rain sterilizes sensitive lakes. Deadly ultraviolet radiation pours through the ozone hole over Antarctica. The quest for food leads to the wholesale stripping of irreplaceable rain forests. Global climate change threatens human societies over the entire planet and stresses natural ecological systems. Toxins in the environment play havoc with wildlife and endanger human health.

Is the Earth a limitless dump site? Can the huge reservoirs we have been describing absorb, process, and recycle human waste at the current and future rates it is being generated? Or will the reservoirs become clogged and poisoned by unnatural, indigestible, and often deadly refuse? Unfortunately, the evidence that we already have in hand points to a global catastrophe on the horizon. The human population is soaring, and the land is being denuded of life-giving vegetation. More and more waste is being poured into the oceans, land, and atmosphere.

The cures for this problem must include recycling nonbiodegradable materials such as glass and metals, reprocessing dangerous waste materials like plastics and solvents, and reusing wood-based materials such as paper and cardboard. Sewage must be treated before being discharged into waterways or landfills. The massive emissions of toxic chemicals into the environment—from smokestacks, discharge pipes, and industrial and agricultural operations—must be controlled. New chemicals and manufacturing techniques must be developed with the specific goal of protecting the environment, and each new product should be assessed for environmental safety before being released for public consumption. We would not allow unlicensed drivers to flood our highways. We cannot allow uncertified wastes to flood our habitat.

Questions

1. What is a "biogeochemical" cycle? How could you describe such a cycle in terms of a simple model? What essential information must you have to make the model quantitative?

2. Can you estimate how much vegetation is required to produce the oxygen you use every day? Suppose that you consume 500 grams (about 1 pound) of food and that your body requires the same mass of oxygen to metabolize the food. Discuss how you reach the solution.

Problems

1. Imagine that your checking account is a "money" box model. The source of money consists of the deposits you make into the account, and the sink consists of the checks you write on the account. The amount of money in the account is the difference between what you deposit plus any interest earned and what you withdraw and any penalties you are assessed. You have $2000 in your account to start (congratulations). If you deposit a paycheck of $1000 each month and earn $20 in interest per month, what is the total value of checks you can write monthly to maintain your $2000 balance? If you get a raise and begin to deposit $1500 a month while maintaining your present life-style, what will happen to your account? Can you be quantitative?

2. Assuming that the world's supply of fossil fuels contains, on average, about 1.5 percent sulfur by weight and that we are burning 6 gigatonnes of fossil fuel per year to generate energy, what is the total mass of sulfur emitted to the atmosphere by this activity? What are the consequences of this sulfur emission? Provide several specific examples. Are there any reasonable alternative uses for this much sulfur?

3. There are about 1 million gigatonnes of oxygen in the atmosphere and about 10,000 gigatonnes of carbon in accessible deposits of fossil fuel. If we can continue to mine this fuel at the rate of 6 Gt-C/yr, how many years' supply would we have (many experts believe we have much less available)? After we have used up all this carbon, how many gigatonnes of oxygen will we have consumed? Explain why the accelerated *rate* of oxygen consumption associated with fossil-fuel use does not threaten the oxygen resource.

Suggested Readings

Berner, R. and A. Lasaga. "Modeling the Geochemical Carbon Cycle." *Scientific American* **260**, (1989): 74.

Bolin, B. and R. Cook, eds. *The Major Biogeochemical Cycles and Their Interactions.* New York: Wiley, 1983.

Charlson, R., J. Lovelock, M. Andreae, and S. Warren. "Oceanic Phytoplankton, Atmospheric Sulphur, Cloud Albedo and Climate." *Nature* **326** (1987): 655.

Ehrlich, P. and A. Ehrlich. *The Population Explosion.* New York: Simon and Schuster Inc., 1990.

Ehrlich, P., A. Ehrlich, and J. Holdren. *Ecoscience: Population, Resources, Environment*, San Francisco: Freeham, 1977.

Holland, H. *The Chemical Evolution of the Atmosphere and Oceans.* Princeton, N. J.: Princeton University Press, 1984.

Newell, R., H. Reichle, and W. Seiler. "Carbon Monoxide and the Burning Earth." *Scientific American* **261** (1989): 82.

Revelle, R. "Food and Population." *Scientific American* **231** (1974): 160.

Singer, S., ed. *Global Effects of Environmental Pollution.* New York: Springer-Verlag, 1970.

Vink, G., W. Morgan, and P. Vogt. "The Earth's Hot Spots." *Scientific American* **252** (1985): 50.

Wilson, E. "Threats to Biodiversity." *Scientific American* **261** (1989): 108.

11

The Climate Machine

The climate system can be thought of as a gigantic machine, whose parts consist of every molecule of the atmosphere, oceans, and land, and every living organism that walks, flies, swims, or crawls. Indeed, it is a complex machine, but one that can be understood by looking at its major parts, one by one, like the parts that make an engine run—the fuel injector, pistons, flywheel, radiator, and so on. Once the functions of the individual parts are known, they can all be bolted together to make the engine go. In the case of the climate machine, the parts include the sun as a source of energy, the atmosphere as the rapidly moving pistons, the ocean as the flywheel, and clouds as the radiator.

In this chapter, we investigate the inner workings of the Earth's climate machine. It is the present climate that makes the Earth habitable. It is the climate that may be changing, threatening to send the environment cartwheeling out of control. It is human civilization that may be forcing the climate into a new state. To understand how the climate can change, we must learn how it depends on factors that are under our influence. We will first define what we mean by climate, and identify the principal factors that control the operation of Earth's climate machine. Then we will develop a simple model that can be used to predict climate change. We will introduce the concept of a "greenhouse effect" and explain the basis of this important climatic process. We will also survey the causes of climate change, both external and internal to the Earth. Finally, we will explore the relationships between society and climate change to place in context the discussions of subsequent chapters.

11.1 Weather and Climate

There is an important distinction between weather and climate. Weather is, in a sense, the meteorological state of the atmosphere that we experience instantaneously at our particular location. If it is raining, we look for shelter. If it is hot, we head for air conditioning. If it is overcast, we may feel sullen or depressed. Climate, on the other hand, is the average weather over an extended period of time in a specific region. The definition is flexible. The period for averaging may be a day, week, month, season, year, decade, century, millennium, and so on. The region for averaging may be a county (Los Angeles, San Diego), a state, a geographical region (southwestern United States, northern Africa), or the entire Earth itself.

Figure 11.1 illustrates how the average temperature might vary on an hour-to-hour basis at a particular location. Typically, the surface air temperature can change by 10 to 20°C from day to night. The temperature changes are caused by variations in sunlight and movements of air masses with different meteorological characteristics (temperature, moisture, clouds, precipitation) over the point of interest. If the temperature over a much larger region were averaged in the same way, however, generally much less variability would be apparent. At a single point, inhomogeneity in meteorology is seen as variability as the atmosphere moves by. Averaging over a large area smoothes this spatial heterogeneity (Figure 11.1).

The global average temperature is the most stable climate parameter. Considering the entire globe at once eliminates the effect of the day–night cycle of sunlight (the diurnal cycle), since the same total area of the globe is always under illumination, even though particular areas experience a diurnal cycle. The global average temperature is a reflection of the total heat energy that the Earth has absorbed from the sun. The atmosphere, land, and oceans have an enormous capacity to store heat. Accordingly, these reservoirs of heat maintain the average temperature over long time intervals despite fluctuations in the global heat input or loss. The result is an average

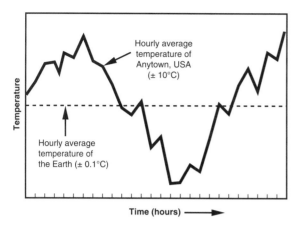

Figure 11.1   The hourly averaged temperature variation (solid line) at Anytown, USA, as measured by the local weather station. The variation in the hourly averaged temperature of the entire Earth, determined by global satellite measurements, is compared. The average planetary temperature does not vary significantly from hour to hour or even from day to day.

planetary temperature that does not vary by more than a few tenths of a degree over centuries and even millennia.

The study of climate, then, focuses on the global balance of energy. The source of energy is the sun. The absorption of solar energy by the Earth, the flow of energy through the Earth reservoirs, and the eventual loss of energy back to space are the processes that concern climatologists. Factors that modify any of these energy processes may be capable of changing the climate. Climatology therefore includes the study of climate variations, both past and future. The climate system is an enormously complex physical-chemical-biological machine, and the science of climatology is still in its infancy—there is no long scientific history as exists for chemistry, physics, and mathematics. Indeed, only a few decades ago, climatology was manifested primarily in the form of the *Farmer's Almanac*, a collection of observations and interpretations of weather events from the recent past. Weather was predicted on the basis of painful bunions and arthritic joints, as well as common empirical wisdom, such as "Red sky at night, sailor's delight; Red sky in morning, sailor take warning." The invention of supercomputers has allowed the science of climatology to move beyond folklore to quantitative forecasting. To be sure, all the proxy climate data (see the introduction to Chapter 4) and common sense of the past have been incorporated into the modern study of climate.

11.2   Energy from the Sun

The sun generates energy by the process of nuclear fusion.[1] At the tremendous pressures and temperatures (perhaps 15 million degrees C) in the cores of stars such as the sun, nuclear reactions can take place. Unlike chemical reactions (Section 3.3), which involve the interactions of the orbital electrons in atoms and molecules, nuclear reactions involve the central nuclei of the atoms. The atoms must be squeezed together extremely violently in order for the nuclei to react, typically by fusing together to form a heavier nucleus (hence the term *nuclear fusion*). But when nuclei do fuse, huge amounts of energy are released. The nuclear fusion of hydrogen atoms (H) to form helium atoms (He), for example, is the basic process that powers the sun. It is also the energy source for the so-called hydrogen bomb, which has awesome destructive capacity.

The energy generated inside of the sun is convected through the fluid interior to the solar surface. The sun does not have a solid surface layer, or lithosphere, or any continents or islands. Rather, it is fluid throughout its depth. The solar atmosphere blends with the apparent solar "surface." Heat liberated from the interior is eventually radiated from the surface and outer solar atmosphere. The **photosphere** is the bright visible disk of the sun, which radiates most of this energy. Although the photosphere appears as the "surface" of the sun, it is actually a layer of tenuous ionized gas, or plasma, about 1000 kilometers thick. The radiation emitted by the photosphere is, to an excellent approximation, blackbody radiation at a temperature of about 6000 kelvin[2] (Section 3.2.1; also see Appendix C).

Just above the photosphere is the lower layer of the sun's atmosphere, the **chromosphere**, which is roughly 15,000 kilometers thick (the radius of the sun, out to photosphere, is about 700,000 kilometers, compared with the radius of the Earth, about 6400 kilometers). The **solar corona**, the outermost

1. The process of nuclear fusion as the source of the energy of the sun and other stars was discovered in 1938 by Hans Bethe (b. 1906), a German physicist (the idea was independently proposed by Carl von Weizsäker). Bethe received the Nobel Prize in physics in 1967 for this discovery. The ongoing research into fusion energy seeks to confine and control this potentially limitless natural source of energy.

2. The actual best-fit Planckian function for solar radiation corresponds to a temperature closer to 5800 K. For almost all discussions of the global climate, however, an assumed temperature of 6000 K is sufficient.

tenuous region of the solar atmosphere, extends hundreds of thousands of kilometers above the chromosphere. The gases in the chromosphere and corona absorb and emit radiation, but do not significantly modify the visible and near infrared regions of the blackbody spectrum (Figure 11.2).

Some of the sun's energy is carried away by the solar wind. The **solar wind** is the outflow of energetic particles—mainly ionized hydrogen and helium atoms—that compose the solar corona and extend it into the interplanetary space. This wind is driven by the thermal escape of atoms from the top of the atmosphere (for example, Sections 2.3 and 4.1.1). The solar wind sweeps past the Earth and interacts with the uppermost air layers. Like a boat moored in a current, the Earth's magnetic field deflects the ionized solar stream, producing a bow wave in the interplanetary plasma. The zone of interaction between the Earth's magnetic field and the solar wind is called the *magnetosphere*. The magnetosphere extends tens of thousands of kilometers from the Earth and is by far the largest region of the Earth's atmosphere. It is also the most tenuous, as the entire magnetosphere weighs only about 1 tonne.

The sun is a turbulent world subject to violent storms. During periods of high activity, the internal convection of the sun causes huge prominences to explode from the surface, injecting dense pulses of fast-moving ions into the solar wind. The intensity of solar activity can be measured by the number of dark cool spots on its "surface." These **sunspots** ebb and wane over a cycle lasting close to 11 years (Section 11.6.2). The spots may be thought of as centers of storm activity. The bursts of energetic particles emitted during solar storms reach the Earth's magnetosphere and are funneled down magnetic field lines at high latitudes, producing the spectacular **aurora** *borealis* (northern lights) and *aurora australis* (southern lights). Thus the Earth is exposed not only to the thermal radiation of the sun, but also to its energetic streams of particles. The thermal radiation accounts for more than 99 percent of the total energy output of the sun, however, and easily reaches the surface over the entire Earth, unlike solar ions, which are completely absorbed in the upper atmosphere.

11.2.1 Solar Illumination

The ability of sunlight to influence weather and climate depends on both its spectrum and the pattern of its deposition at the surface. Figure 11.3 shows that sunlight illuminates the half of the Earth that faces the sun. However, the solar rays hit the planet at different angles, and this affects their ability to heat the surface. At the point on the Earth closest to the sun, the rays are the most direct, hitting the surface from right overhead. At the edges of the planetary sphere, the rays are just tangent to the surface, grazing the edge of the horizon. Figure 11.4 illustrates the relative heating capacity of direct and oblique solar rays. Obviously, the more direct the ray is, the greater the energy deposited *per unit area* of surface and the greater the heating rate will be. Conversely, the more oblique the angle of incidence is, the lower the energy per unit area and the lower the heating rate will be.

The effect of the angle of the sun on heating can sometimes be felt as the day moves to late afternoon and evening twilight. As the sun drops toward the horizon, it may become noticeably cooler compared with noontime. The amount of energy deposited per

Figure 11.2   The spectrum of radiation emitted by the sun. The spectrum is similar to a perfect "blackbody" emission spectrum for an object at a temperature of about 6000 kelvin. The actual spectrum deviates slightly from a blackbody because the solar radiation in different parts of the spectrum originates in different regions of the solar atmosphere. The wavelength of peak intensity is about 0.55 micrometer. The ultraviolet (UV), visible, and near-infrared portions of the spectrum are marked.

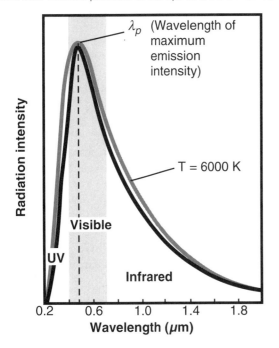

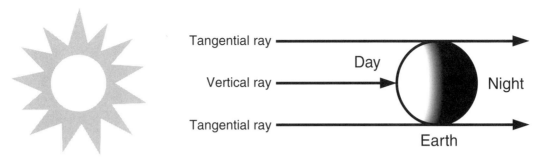

Figure 11.3   The illumination of a spherical planet by the distant sun. The hemisphere facing the sun is always illuminated. The back side of the planet is in darkness. As the planet rotates on its axis, all parts of the planet are exposed to sunshine over the course of a day.

square meter of ground is reduced as the sun sets, and so the local heating also is reduced. If the weather is hot and humid, this effect is less obvious. The temperature has inertia against cooling because of the heat contained in the ground (**sensible heat**) and the potential heat of condensation held by water vapor (**latent heat**). Nevertheless, the angle of the sun is a key factor in regional climate and climate variations.

11.2.2   THE FOUR SEASONS

The changing seasons (winter, spring, summer, and fall) are an important manifestation of variations in the sun's angle of illumination. The Earth rotates on its axis once per day. A point on the surface thus experiences a continuous variation in solar illumination from morning to evening. The total amount of energy received depends on the latitude of the point and the time of year. This dependence can be understood by referring to Figure 11.5.

The Earth's axis of rotation is tilted with respect to its orbit around the sun (the plane of the orbit is called the **ecliptic**). The angle of tilt, or **obliquity**, is about 23.5°. This angle remains constant over thousands of years (but varies significantly over tens of thousands of years [Section 11.6.2]). As the Earth revolves in its orbit around the sun, the axis remains tilted in the same direction. Accordingly, the apparent angle of exposure[3] of the Earth to the sun changes continuously in a range of +23.5° to −23.5°

3. The angle of exposure can be imagined as the angle at which the axis of the Earth intersects the line connecting the centers of the sun and Earth. If the Earth had no tilt, this angle would be 90° (perpendicular) all the time. With a tilt to the axis, this angle varies as 90°± tilt angle, which can be equivalently expressed simply as ± tilt angle.

over the year (which is revealed by a close inspection of Figure 11.5). The tilt angle defines a number of important latitude circles on the Earth. The Arctic Circle, at 66.5° N (that is, degrees of latitude north of the equator), is where the sun dips just to the horizon at midnight during the summer solstice (the longest day of the year in the Northern Hemisphere, around June 22) and rises just to the horizon at midday during the winter solstice (the shortest day of the year, around December 22). The Antarctic Circle, at 66.5°S (degrees of latitude south of the equator), is the equivalent latitude marker in the Southern Hemisphere, and the times of the solar events are reversed. The Tropic of Cancer, a latitude 23.5°N, and the Tropic of Capricorn, at 23.5°S, are the latitudes at which the sun is directly overhead during the summer and winter solstices, respectively.

The Earth's Northern Hemisphere tilts toward the sun in June, July, and August. Hence the sun's rays are most direct at this time of the year (Figure 11.4). The warming effect is therefore greatest then, and we have the summer season. Conversely, in the Southern Hemisphere the Earth is tilted away from the sun (Figure 11.5), and the solar rays are much less direct. Therefore, the Southern Hemisphere has its winter season in June, July, and August. One-half year later, the Earth has revolved halfway around its orbit and tilts in the opposite direction relative to the sun (Figure 11.5). Then, in December, January, and February, the Northern Hemisphere has the least direct illumination and its winter season, while the Southern Hemisphere is most directly lit and enjoys summer.

In between the extreme seasons are spring and fall. In neither of these seasons is the Earth tilted much with respect to the sun. Why, then, are these seasons so different? Spring is a season of warming, from winter to summer, and fall is a season of cooling, from summer to winter. The climate system has reservoirs in which energy can be stored and from which it can be retrieved (Section 11.5). The reservoirs create an inertia in the changes from cold to warm and back and create seasonal lags in temperature change. Nevertheless, both spring and fall are seasons of moderate temperature.

The length of the day varies with latitude and season. In general, summer days are longer, and winter days are shorter. The length of daylight is a function of the Earth's tilt in relation to the sun. If a hemisphere is tilted toward the sun, that half-orb is more fully illuminated, and the days are longer than 12 hours on average. The opposite condition holds in the other hemisphere, where days are shorter than 12 hours on average. At high latitudes, in the extreme situation, the day may be 24 hours long (and so may the night). Above the Arctic Circle (67.5° north latitude) the sun remains above the horizon all day in midsummer and never rises in midwinter.

The tropical regions near the equator are unique because the seasonal variations are small compared with those of other areas of the Earth. You can count on warm temperatures and sunshine in tropical lands. There, the sun always shines fairly directly on the surface, and the daytime is always about 12 hours long (Figure 11.5). Without variations in the angle of incidence and length of daylight, the heating of the sun is fairly uniform all year around, and so is the climate.

The intensity of sunlight at the Earth does vary somewhat over the year because the distance of the Earth from the sun changes. If Earth's orbit were a perfect circle with the sun at the center, the intensity of sunlight would remain constant during the entire year. However, the orbit is slightly flattened, in the form of an ellipse. The deviation of an ellipse from a perfect circle is defined by the **eccentricity of the ellipse**. In the case of the Earth's orbit, the eccentricity is about 0.017, which means that the sun–Earth distance varies by about 1.7 percent from the mean distance, both positively and negatively, during the year, or by about 3.4 percent from minimum to maximum. When the Earth is closer to the sun, it receives more energy and is warmer, and vice versa. This effect is quantitatively described in the next section. The Earth is closest to the sun in winter and farthest from it in summer. This suggests a compensating effect for the influence of the axial tilt on the seasons. Indeed, winters would be colder and summers warmer if the opposite relationship between tilt and eccentricity were to hold. In the past, exactly that situation has occurred and may have contributed to significant climatic variations (Section 11.6.2).

11.3   The Temperature of Earth

The average temperature of the Earth at any moment is determined by the balance of energy held in various heat reservoirs. The total heat content of these reservoirs is controlled by two overall processes: the absorption of sunlight and the emission of thermal radiation.

11.3.1   SUNLIGHT IN, EARTHGLOW OUT

Figure 11.6 compares the spectra of sunlight and terrestrial heat radiation (which might be referred to

Figure 11.4   The difference in heating power between solar rays that impinge vertically or obliquely on a surface. The energy per unit cross-sectional area of the solar beam is constant, but the area illuminated (A_0 and A) varies with the angle of incidence.

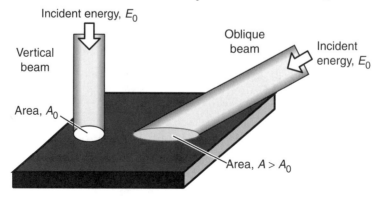

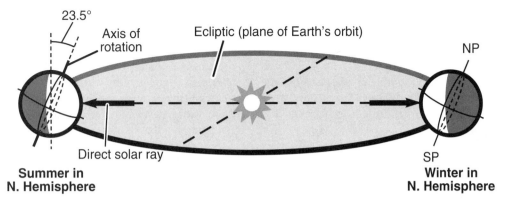

Figure 11.5   The orbit of the Earth around the sun. The plane in which the orbit lies is called the *ecliptic*. The Earth's axis of rotation is not perpendicular to the ecliptic, but is tilted at an angle of about 23.5° from the perpendicular (the obliquity). The Earth revolves around the sun in its orbit one time per year. At different seasons, or quartile distances around the orbit, the Earth tilts in different directions with respect to the sun, as illustrated for summer and winter in the Northern Hemisphere. The distance of Earth from the sun also varies by about 3 percent over the year owing to the slight eccentricity (noncircularity) of the orbit.

as Earthglow, since it represents the emission of energy, much as a light bulb does). The sun is a blackbody with an emission temperature of about 6000 K. The emission peaks at a wavelength of about 0.55 micrometers (as predicted by Wien's law [Section 3.2.1]). The absolute intensity of the radiation reaching the Earth is controlled by the distance from the sun and the size of the sun itself. The size of the sun is essentially fixed and need not be considered a factor in contemporary climate change. However, the distance of the Earth from the sun does vary. The Earth's orbit around the sun is not circular, but elliptical. Although the eccentricity of the orbit is very small, amounting to variations in the mean distance from the sun of less than 2 percent, the variation in distance can be important over a seasonal cycle and on the time scale of ice ages (Section 11.6).

In considering the overall energy balance of the Earth, or other planets, it is helpful to consider the total radiative energy contained in the absorption or emission spectrum of an object. (See the discussion in Section 3.2.1 on the Stefan-Boltzmann law). The total energy emitted by the sun is equivalent to the integral of the solar spectrum in Figure 11.6; that is, the total emission is proportional to the shaded area under the intensity curve for sunlight. Similarly, the total energy emitted by the Earth as thermal radiation is given by the corresponding shaded area in Figure 11.6. These spectrally integrated total emissions are usually expressed as a radiant energy flux in units of energy per unit time per unit area perpendicular to the direction of the source of the radiation.

The units of energy flux are typically specified in joules/sec-meter^2, or watts/m^2.

The total energy flux reaching the Earth from the sun, denoted F_s, depends on the distance from the Earth to the sun, D_{es}, in the following way:

$$F_s = F_{s0} \left(\frac{D_{es0}}{D_{es}} \right)^2 = \frac{f_{es0}}{D_{es}^2} \qquad (11.1)$$

F_s is called the **solar constant** and has units of total radiant energy flux as just defined. The solar constant[4] has a value of about 1400 W/m^2. F_{s0} is the solar constant at the mean (or average) distance of the Earth from the sun, D_{es0}. Notice that the energy flux falls off in proportion to D_{es}^{-2} as the distance from the sun increases (f_{es0} is a constant composed of the other terms in Equation 11.1). This is the general law of the behavior of intensity with distance for spherical radiators, including all stars. It is easy to understand this result if you recognize that the total energy emanating from a fixed source must be spread

---

4. The value of the "solar constant" is somewhat uncertain. In 1906, the Smithsonian Astrophysical Observatory initiated a program to measure the solar constant over a period of years, to establish just how constant it was. But the techniques for measuring solar radiation from the ground required large corrections for atmospheric scattering and absorption and proved to be too inaccurate to actually record the natural variability of the sun. Satellite measurements from space do not involve such corrections. The satellite value for the solar constant is about 1380 watts per square meter, with a variability of about 0.1 percent. However, the satellite data are sparse and span only a few years, leaving open questions regarding the longer-term variability of the sun's energy output.

over an increasingly large (spherical) area as distance from the source increases. In other words, the heating potential per unit area must decrease with increasing distance from the source. The farther away from a campfire that you sit, the less warmth you will feel from the flames. As noted earlier, the relative distance of the Earth from the sun varies by about 3.4 percent during the course of a year because of the eccentricity of Earth's orbit around the sun (see Section 11.6.2).

Albedo and Insolation

An important factor in determining the efficiency of solar absorption by the Earth is the **albedo** of the planet. In brief, the albedo is the fraction of the total radiative energy flux impinging on the planet that is reflected or scattered back to space. The energy contained in the albedo cannot contribute to the climate, and so it is important to quantify this factor.

The different wavelengths of radiation in the solar spectrum are reflected or scattered differently by the atmosphere-surface system. The albedo therefore represents a wavelength-averaged reflectance. The reflectance in general also depends on the radiation's angle of incidence. Accordingly, the albedo for a spherical planet illuminated as in Figure 11.3 must also be averaged over the illuminated disk (and so the different angles of incidence). When all these factors have been taken into account, the average planetary albedo of the Earth is accurately calculated to be about 0.33. This albedo is an extremely useful parameter for calculating the Earth's overall energy balance and climatic state.

The albedo depends in a complex manner on a large number of factors—for example, the distribution of clouds and dust particles in the atmosphere, the amount of snow on the surface, the quantity of pollution in the air, the wetness of the ground, the foam on the oceans. It is impossible to specify all these variables all the time or, in fact, even once! Fortunately, that is not necessary. Satellites have been employed to measure the albedo directly. Averaged over the entire globe, the albedo is found to waver very little over time. Certainly, for the purpose of obtaining a crude picture of the Earth's energy balance, the assumption of a constant average albedo of 0.33 is quite reasonable.

The solar insolation is the amount of solar energy received at the ground. This is not simply the difference between the incident flux (solar constant) and the albedo flux, because the incident radiation can be scattered many times (multiple scattering) by the atmosphere, clouds, aerosols, and the surface. The radiation can also be absorbed in the atmosphere and never reach the ground. It follows that the true solar insolation is usually a fraction of the total energy absorbed by the surface-atmosphere system. In actuality, this fraction is about two-thirds. That is, most of the solar energy absorbed by the Earth is in the form of insolation at the surface; the rest is absorbed into the atmosphere.

For a simple climate analysis, the total energy in sunlight absorbed by the Earth is often treated as the solar "insolation." In addition, the energy deposition is assumed to be distributed uniformly over the entire planetary surface. Because the effective solar energy-collecting area of the Earth is almost exactly

Figure 11.6 Comparison of the spectrum of sunlight reaching the Earth with the spectrum of heat radiation emitted by the Earth (terrestrial radiation, or Earthglow; refer to Appendix C). The wavelength ranges for the two spectra are well separated. Both spectra are caused by blackbody emission: The sun emits at a temperature of about 6000 K, and the Earth, at about 255 K.

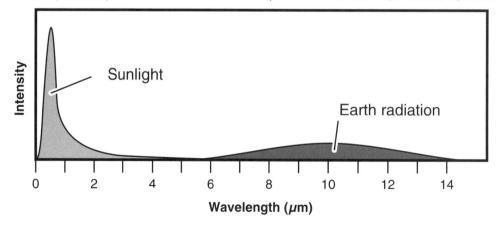

one-fourth the total surface area (Section 11.3.2), the *global average solar insolation*, $\overline{F}_s$, is taken to be

$$\overline{F}_s = \frac{1}{4} F_s (1 - \alpha_e) \qquad (11.2)$$

Here, the solar constant is first reduced by the fraction of the incident radiation that is reflected as albedo and then is reduced by a factor of four, corresponding to the ratio of the surface area to cross-section. The average albedo of the Earth is α_e. Since this is the average fraction of the impinging solar energy reflected to space, it follows that the average fraction remaining to be absorbed (as insolation) is $1 - \alpha_e$.

Most places on Earth do not receive an average amount of solar insolation on any given day or even, in the mean, over an entire year. Nevertheless, the concept of an average solar insolation is useful in certain applications.

Terrestrial Radiation

The Earth emits radiation that can be approximated as blackbody radiation. If we consider the Earth-atmosphere system to be an ideal blackbody, it would exhibit a mean emission temperature of about 255 K. This yields a peak emission intensity at a wavelength of roughly 10 micrometers (Section 3.2). Figure 11.6 illustrates the ideal spectrum of Earth emissions. The area under this spectrum represents the total energy flux emitted by the Earth per unit area of surface, averaged over the entire planet.

The terrestrial emission spectrum is well separated in wavelength from the solar spectrum and lies at much longer wavelengths. This difference is crucial to the working of the climate system, as we will explain later (also see the discussion of radiation in Appendix C).

The terrestrial radiation is emitted by molecules in the atmosphere, by clouds, and by the surface of the Earth. The total energy flux is therefore a complex summation of emissions from different materials in different places. Again, however, by averaging over the whole world, and over time as well, the details become much less significant in describing the overall climate system. For now, we will think of the Earth as an ideal blackbody with a single average emission temperature. Later, we will return to elaborate on this simple and useful representation of the Earth's energy balance.

11.3.2 AN ENERGY BALANCE MODEL

The two principal components of the Earth's climate system must be balanced. This situation is depicted in Figure 11.7. Considering the planet as a whole, the solar energy absorbed must equal the terrestrial energy emitted. Otherwise, there would be a net gain or loss of heat over time, and the temperature of the Earth would drift accordingly. Because the average planetary temperature is known to be very stable over long time spans, the heat source and sink must be nearly equal. This suggests that the climate system may be represented using a simple heat reservoir model.

Figure 11.7   The balance of energy for the Earth. Solar radiation is absorbed on the sunlit side of the planet. Heat (or terrestrial radiation, longwave radiation, far infrared radiation or Earthglow) is emitted from all parts of the world.

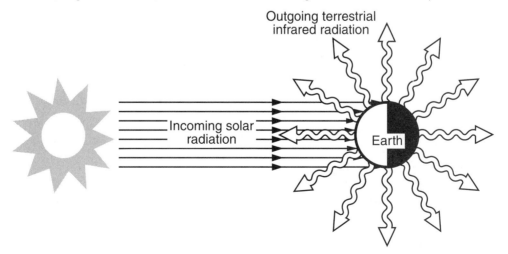

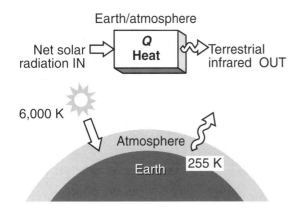

Figure 11.8 The concept of an energy balance box model for the climate. The "box" consists of the atmosphere and the surface elements of the Earth (a land skin layer and the surface oceans).

A Climate Box Model

The total energy contained in the Earth's surface and atmospheric heat reservoirs can be represented by a single box, as in Figure 11.8. The total heat quantity is given by the quantity, Q. (The principles and some applications of box models are discussed in Sections 4.1.2 and 10.1.2.) The source and sink for heat in the model are the absorption of solar energy and the emission of terrestrial radiation, respectively (Figure 11.8). The condition for this simple reservoir model to be in balance, or equilibrium, can be expressed by the relationship

$$\text{Sunlight absorbed} = \text{Heat radiation emitted} \tag{11.3}$$

This relationship demonstrates the operation of processes internal to the climate system (the box) that can absorb, transform, transport, and emit heat. The following sections of this chapter provide details of some of the more important of these processes. The single parameter that connects most of these processes is the temperature of the climate system.

The temperature of a reservoir determines how much heat is held in the reservoir. The connection between temperature and heat content refers to the **heat capacity** of the reservoirs. The heat capacity of a material (water, for example) is the quantity of thermal energy held in a unit mass of the material for each degree of temperature of the material. Air has a heat capacity of about 1 joule per gram of air per °C, and water has a heat capacity of 4 joules per gram of water per °C. If the heat capacity is denoted c_p, then

the change in the heat content of a reservoir is related to the mass of the reservoir, M, and the temperature change by

$$\Delta Q = c_p M \Delta T \tag{11.4}$$

The total heat content of the reservoir may be approximated as

$$Q = c_p M T \tag{11.5}$$

Therefore, once we define the reservoir, the average temperature of the reservoir determines its heat content.

In calculating the average temperature of the climate system using the simple box model of Figure 11.8, the principle of equilibrium (Equation 11.3) can be invoked. It is assumed that all the heat reservoirs in the Earth system can be lumped together using a total mass and an average heat capacity, as follows:

$$Q = \bar{c}_p M_T T \tag{11.6}$$

with

$$M_T = \sum_{\text{Reservoirs, } i} M_i$$

$$\bar{c}_p = \frac{\displaystyle\sum_{\text{Reservoirs, } i} M_i c_{pi}}{M_T} \tag{11.7}$$

The aggregation of the reservoirs in this way is logical when the system is in equilibrium, since all the reservoirs have the same temperature, and the transfer of heat between reservoirs is negligible.

To make progress in determining the equilibrium temperature of the reservoir, however, the heat (energy) source and sink must be appropriately defined. This is done for the following two key components.

Solar Absorption: Input

The total solar energy absorbed by the planet is determined by a number of factors, including the incident solar energy (solar constant), F_s, the size of the planet, and the amount of energy reflected (albedo). In terms of these parameters, the energy source for the climate system is

$$\dot{Q}_{\text{in}} = S = F_s A_e \left(1 - \alpha_e \right) \tag{11.8}$$

Here, the source, S, of heat energy is $\dot{Q}_{in}$, and the solar constant, F_s, was defined in Section 11.3.1. The cross-sectional area of the Earth, A_e, is a well-determined quantity. Planets are spheres. If the radius of a sphere (like the Earth), R_e, is known, the cross-sectional area is simply calculated as $A_e = \pi R^2$. The albedo of the Earth, α_e, is the fraction of impinging solar energy reflected to space. Hence the fraction *absorbed* that contributes to the climate system is $1 - \alpha_e$. Summarizing these results, the rate of total energy input to the climate system is the flux of solar energy impinging on the Earth multiplied by the cross-sectional area that intercepts the radiation, reduced by the fraction of the radiation that is returned to space as the planetary albedo.

It may not be obvious why the cross-sectional area of the Earth is used in Equation 11.8. The entire half of the sphere that faces the sun is always illuminated (Figure 11.3). The lit hemisphere has one-half the total surface area, S_e, of the sphere, which is twice the cross-sectional area. The relationships between areas of a sphere are summarized as

$$S_e = 4A_e = 4\pi R_e^2 = 2A_{Hemi}$$
$$= 2\left[2\pi R_e^2\right] = 2\left[2A_e\right] \quad (11.9)$$

The total energy intercepted by the Earth could be measured in the following way: Imagine constructing an energy detector in the shape of a large disk that can be placed just in front of the Earth. The disk is of just the right size to cast a shadow over the entire planet. That disk would have an area A_e. The energy intercepted by the detector is the solar energy flux (the solar constant) times this area. But the total energy collected is exactly the same as the total energy that is actually intercepted by the Earth at all latitudes. It is certainly not twice this amount.

Another way to look at the deposition of solar energy is to consider the effective heating effect of the sun's rays impinging along the edge of the sphere. Figure 11.4 shows that the heating capacity is diminished in these regions. Hence the average heating rate of the sunlit hemispherical area must be lower on average than the heating rate in the region where the solar rays are direct (the subsolar point). In fact, a precise mathematical evaluation of the total energy deposition over the sunlit hemisphere shows that it is exactly equivalent to the total energy deposited uniformly over an imaginary planetary disk.

Because the Earth rotates on its axis once every 24 hours, the absorbed sunlight is distributed around the entire globe, but the distribution, of course, is not uniform. The low (tropical) latitudes receive the most radiation annually. At higher latitudes (north of the Arctic Circle in the Northern Hemisphere and south of the Antarctic Circle in the Southern Hemisphere), there are seasonal extremes of sunlight, with periods of perpetual darkness in winter and endless daylight in summer. The average energy deposition at high latitudes is less than at the low latitudes. The atmosphere and oceans redistribute some of this heat through currents flowing from the tropics to high latitudes, carrying warmer air and water toward the poles. This flow of heat does not create any net energy, but smooths out to some degree the nonuniform meridional (latitudinal) deposition of sunlight.

Thermal Emission: Output

The energy absorbed from sunlight must be balanced by the escape of energy to space. If no energy were lost, the temperature of the Earth would continue to rise until it vaporized. Although at the current rate of energy absorption this might take several million years, the global climate would become unbearably warm in only a few months. Fortunately, the Earth has a way to shed this excess energy. The Earth does not lose sensible heat (that is, the heat contained in a material by virtue of its temperature and heat capacity) to space because the escape of gases (which contain heat) to space is very slow. However, the Earth can readily emit heat in the form of radiation.

Section 3.2 described in detail the emission of radiation by objects containing heat. The total radiative energy flux (at all wavelengths) emitted by a perfect "blackbody" radiator, per unit surface area of the emitter, is given by the Stephan-Boltzmann law:

$$F_b = \sigma_B T^4 \quad (11.10)$$

The heat radiation energy flux, F_b, has the same units as the solar flux, or constant (energy per unit time per unit area of surface). Moreover, the thermal flux in this case is a function only of the temperature of the object.

We can imagine the Earth as a spherical object suspended in free space emitting heat in all directions, as in Figure 11.7. For the moment, we will not consider the structure of the atmosphere-surface system and its effects on this radiation. Nor will we consider the patterns of solar energy deposition and thermal emission over the sphere. As we noted earlier, the absorbed solar energy tends to be spread fairly uniformly over the planet by means of air and ocean currents. The thermal emission pattern therefore is also fairly uniform over the sphere. It follows that to obtain a crude climate model, we can assume that the thermal emission of the Earth is uniform over the entire surface and may be characterized by a single effective body temperature, T_e, at all points on the surface. According to this model, the total rate of energy emission by the Earth is the thermal blackbody flux corresponding to the effective blackbody temperature, multiplied by the total surface area of the globe, S_e. That is, the rate of energy loss is given by the simple relation

$$\dot{Q}_{out} = L = S_e F_h = 4A_e \sigma_B T_e^4 \quad (11.11)$$

It is important to note that the emission of thermal radiation occurs from the entire surface of the planet, not from a cross-sectional area. Every point on the surface of a blackbody emits the same flux of thermal radiative energy. Although the Earth's surface has nooks and crannies, from any distance away it appears to be a nearly perfect sphere and can be treated as such in our simple climate model. The Earth literally glows like a warm ember in the cold emptiness of space.

The "Effective" Temperature of the Earth

We have now derived expressions for the total rate of solar energy absorption (Equation 11.8) and of Earthglow emission (Equation 11.11). These sources and sinks determine the total amount of energy in the climate heat reservoirs (Figure 11.8). If the climate is to be in balance, the source and sink of energy also must be in balance. In other words, the following equation must hold:

$$\dot{Q}_{in} = \dot{Q}_{out}; \; S = L$$
$$F_s A_e (1 - \alpha_e) = 4A_e \sigma_B T_e^4$$
$$(11.12)$$

Equation 11.12 can be simplified by canceling the common terms A_e, yielding

$$F_s(1 - \alpha_e) = 4\sigma_B T_e^4 \quad (11.13)$$

Note that we could have arrived at the same result if we had equated the *global average* solar and thermal energy fluxes per unit area of the Earth using Equations 11.2 and 11.10.

The validity of Equation 11.13 requires the heat reservoirs of the Earth to be in equilibrium. (The properties of the reservoirs are discussed in depth in Section 11.5.) It turns out that the reservoirs respond rapidly to imbalances in energy sources and sinks. For example, the atmosphere can adjust its temperature by several degrees in a matter of days; the oceans, in months. These time scales are short compared with the variations in the climate system of interest, which span decades, centuries, and longer. We are assured that the climate system is in equilibrium for the conditions assumed in the simple energy balance model.

The heat balance represented by Equation 11.13 is not arbitrary. If the intensity of sunlight increases, for example, the Earth will begin to receive a net input of energy, and the heat content of the reservoirs will increase. According to Equation 11.6, the temperature of the reservoirs must rise in step with the added energy. Now something wonderful happens. As the temperature of the Earth increases, its thermal radiation also increases, not in proportion to the temperature, but to the *fourth power* of the temperature (Equation 11.13). The implication of this nonlinear response is important. Small variations in the **climate forcing** (that is, the solar constant or the albedo, which determine the energy input [Section 11.6]) can be compensated by much smaller variations in the effective radiation temperature.

The climate system has an internal variable—the effective radiation temperature, T_e—that adjusts to bring the energy input and output into balance. This temperature then defines the overall climatic state of the Earth. The value of the effective blackbody temperature can be determined from Equation 11.13 by simple algebraic manipulation, yielding[5]

$$T_e^4 = \frac{F_s(1 - \alpha_e)}{4\sigma_B}$$

---

5. Note that the "fourth root" of a quantity can be calculated by taking the "square root" twice, or the square root of the square root. Recall that when the square root of a number is multiplied by itself, it returns that number. When the fourth root of a number is multiplied by itself three times, it returns that number. For example, the fourth root of 16 is 2; that is, $2 \times 2 \times 2 \times 2 = 16$. But notice that the square root of 16 is 4 (that is, $4 \times 4 = 16$), and that the square root of 4 is 2.

or

$$T_e = \left[\frac{F_s(1 - \alpha_e)}{4\sigma_B} \right]^{\frac{1}{4}} \quad (11.14)$$

Equation 11.14 is the simplest "climate equation" for a planet. Only two factors must be specified: the solar constant, F_s, and the albedo, α_e. The Stephan-Boltzmann constant, σ_B, is very precisely known (Appendix A). For the Earth, both the solar constant and the albedo have also been accurately established, as noted earlier. Substitution of the appropriate parameter values in Equation 11.14 yields the effective temperature of the Earth as 255 kelvin. That is mighty cold—18°C below the freezing point of water (0°C, or 273 K)! If the whole surface were at this temperature, the planet would be a frozen ice ball, devoid of life (certainly as we know it). This situation is depicted in Figure 11.8, in which solar energy is absorbed at the effective spectral temperature of about 6000 K and thermal radiation is emitted at the effective spectral temperature of about 255 K.

Measurements of the surface temperature around the world show that the average temperature of the land and oceans is closer to 290 kelvin, well above the freezing point of water. This is consistent with our everyday experience as well. Although the two temperatures are close—after all, 255 and 290 are not that far apart, considering the simplicity of the climate model used to estimate the Earth's temperature—the difference is indeed a striking one between life and death. Where have we gone wrong in our reasoning about the climate and how to estimate it? If we apply the same reasoning to the other planets in our solar system and attempt to deduce their climates using our simple model, the problem may be revealed.

11.3.3 THE TEMPERATURES OF THE PLANETS

Table 11.1 presents information on several planets in our solar system, including the Earth. For each planet, the relative distance from the sun, in astronomical units (that is, units of the distance from the Earth to the sun), and the albedo are given. The solar constants for each planet can therefore be determined using Equation 11.1. Then the effective blackbody temperatures can be computed using Equation 11.13. These temperatures are shown in Table 11.1, along with the temperatures deduced from measurements of the radiation emitted by each planet. Decades of careful astronomical and spacecraft observations of the planets have provided a fairly detailed picture of the climates of most of the other worlds in our solar system. Surprisingly, the extraordinarily simple climate model we just developed is remarkably good at predicting actual planetary surface temperatures (Venus is the most obvious exception).

The most noticeable characteristic of the planetary climates is that in general the farther a world is from the sun, the colder its climate will be (Venus is an interesting exception). The cause of this behavior is clear from Equations 11.1 and 11.14: The greater the distance from the sun is, the lower the intensity of sunlight and the smaller the solar "constant" for that planet will be. Another striking property emerges from Table 11.1: Planets with very thin atmospheres, such as Mercury and Mars, have a surface temperature that is close to the effective blackbody temperature. Uranus also follows this rule, even though it has no solid surface; in this case, high cloud decks act as an effective surface. Planets that hold significant quantities of carbon dioxide or water vapor, such as Earth and Venus, exhibit surface temperatures that are substantially warmer than their effective blackbody temperatures. Even Mars, with a thin atmosphere composed of carbon dioxide, is noticeably warmer than expected.

The presence of an atmosphere containing carbon dioxide or water vapor, or both, can apparently lead to the surface warming anomalies listed in Table 11.1. The warming effect of an atmosphere is referred to as the **greenhouse effect**. Earth's atmosphere contains substantial amounts of water vapor and some CO_2, which warm the surface about 35°C above the effective emission temperature. Mars, with 10 times the quantity of CO_2 as Earth has, but very little water vapor, shows a modest greenhouse warming of perhaps 15°C. Venus's atmosphere, by contrast, holds an enormous mass of carbon dioxide (nearly 100 times the *total* mass of Earth's atmosphere [Section 11.4.4]), and the greenhouse-warming anomaly is several hundred degrees. The surface of Venus is so hot that some metals would melt there. Thus whereas the greenhouse effect transforms the Earth into a warm and habitable place, which otherwise might have been frozen and barren, the same effect transforms Venus, which otherwise might have been habitable, into a hellish furnace.

Table 11.1   Climates of the planets

Planet	Solar distance (AU[a])	Albedo	Effective emission temperature (box model, K)	Surface temperature (measured, K)
Mercury[b]	0.39	0.30	420	460
Venus[c]	0.72	0.70	300	700
Earth[d]	1.0	0.33	255	290
Mars[e]	1.5	0.20	205	220
Uranus[f]	19	0.80	60	65[g]

[a] Astronomical units; 1 AU is the mean distance of the Earth from the sun, 150 million kilometers.

[b] Mercury has only a tenuous atmosphere because most of the volatile gases have been boiled out of the atmosphere into space.

[c] Venus has a massive atmosphere of carbon dioxide (about 100,000 times the amount of CO_2 as in Earth's atmosphere), with about as much nitrogen and very little water vapor.

[d] Earth's atmosphere is mainly nitrogen and oxygen, with about 1 percent water vapor and some carbon dioxide.

[e] Mars has an atmosphere of carbon dioxide (about 10 times the quantity in Earth's atmosphere), but is otherwise quite thin and dry.

[f] Uranus's atmosphere, at the levels where its radiation originates, consists mainly of hydrogen and helium, with widespread clouds of ammonia and methane.

[g] The "surface" of Uranus actually consists of dense layers of clouds in the upper regions of the atmosphere. Like all the giant Jovian planets (Jupiter, Saturn, Uranus, and Neptune), Uranus has a deep atmosphere of hydrogen that eventually is compressed into a sea of liquid hydrogen.

In the greenhouse effect, the atmosphere acts as a blanket that helps a planet retain the warmth of sunlight. The blanket is leaky, however, full of holes through which heat may still escape to space. The simple climate model derived earlier must be modified to take into account the effects of a greenhouse blanket with holes. This requires a more detailed analysis of the interactions of the atmosphere with solar and thermal radiation.

11.4   The Greenhouse Effect

The simple energy balance model for the climate is illustrated in Figure 11.9. The sun provides energy, which is absorbed by and heats the surface. The warm surface then emits blackbody radiation back into space. When the climate system is in equilibrium, which is nearly always a reasonable assumption over long times, the rate of solar-energy input (absorption) and the rate of thermal-energy output (emission) are equal. The presence of an atmosphere causes two critical alterations of this simple energy

flow: First, the gas molecules absorb and scatter radiation, and second, clouds and small particles scatter, absorb, and emit radiation (Section 3.2.2). We discuss next these processes and their effects on the energy balance and climate.

11.4.1   ATMOSPHERIC BAND ABSORPTION

We have already discussed the fact that the molecules in a gas can absorb radiation at some wavelengths and can scatter radiation at all wavelengths, by means of Rayleigh scattering, for example. If the molecule absorbs a photon of radiation that is energetic enough, the molecular bonds can be fractured, leading to the dissociation of the molecule into products (or photodissociation, for example, the photodissociation of nitrogen dioxide leading to the formation of ozone in smog). Such absorption of radiation generates chemical energy that eventually is converted to heat. Photodissociation usually requires photons of short wavelength (high energy), in the ultraviolet region of the

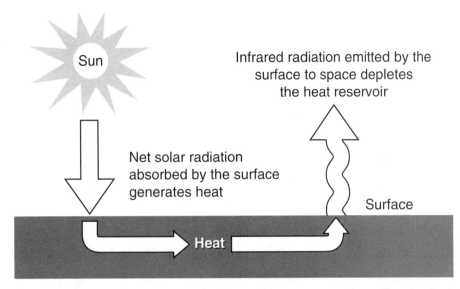

Figure 11.9   The energy balance of a planet without an atmosphere. The absorbed solar radiation heats the surface, which then emits the energy as infrared radiation.

electromagnetic spectrum. If molecules absorb radiation and are not photodissociated, they are generally left in a higher (excited) energy state. Occasionally, the excited molecule simply reemits the absorbed photon by dropping back into its original (ground) state; this is called *resonant scattering*. In other cases, the molecule drops into a different energy state[6] and emits a photon at a modified wavelength; this is called *fluorescence*.[7] Molecules in excited states can also lose their excess

6. Molecules generally have many different energy states in which they can exist. The states are usually identified as electronic, vibrational, and rotational, or combinations of these. The electronic states correspond to various possible configurations of the orbital electrons (Section 3.3.2). The vibrational states arise from the relative motions of the atomic nuclei, which in a molecule can be visualized as small masses held together by springs. The vibrational-state energies associated with the spring action are smaller than the electronic-state energies. The rotational states are defined by the spinning motion of the molecule, much like a dumbell. The vibrational-state energies produce radiation in the near infrared and far infrared spectrums, while the rotational-state energies being smaller produce radiation only in the far infrared spectrum. The complete state of a molecule has an electronic, a vibrational, and a rotational component. Transitions between states of a molecule can involve changes in all three components.

7. The so-called black light used in bars and dance halls is produced by fluorescence or, more precisely, phosphorescence. A sensitive material, often a phosphorus compound, is exposed to ultraviolet radiation, and the radiation is absorbed by electronic excitation. The excited states then relax to lower states, emitting radiation with less energy at a longer wavelength in the visible spectrum. The human eye can detect the visible fluorescence, but not the ultraviolet source. The fluorescence from a sensitized material in a darkened room produces an eerie glowing apparition.

energy during collisions with other molecules. In this case, the energy is converted to the thermal motion, or heat, of the recoiling molecules.

Most molecules experience vibrational and rotational motions (known as *degrees of freedom*) that lead to absorption and emission of radiation in the infrared portion of the electromagnetic spectrum (see Footnote 6). When observed in the laboratory, the absorption appears as a series of sharp spikes, or lines, concentrated in broader spectral intervals, or bands. Other molecules, like carbon dioxide (CO_2) and water vapor (H_2O), have very strong absorption bands. Some molecules, such as oxygen (O_2) and hydrogen (H_2), have very weak bands. These bands typically fall in the range from the near infrared (above about 1 micrometer, μm) to the longwave infrared (up to 50 μm or more). The absorption bands at longer wavelengths are associated with pure rotational energy states (that is, without vibrational components), and the absorption bands in the near infrared involve vibrational-rotational transitions. [Remember that micrometers, microns, and μm are all the same unit of wavelength; see Table 3.1, p. 51]

When radiation is passing through the atmosphere, there is a certain probability that it will be absorbed by air molecules. Figure 11.10 shows the probability of absorption as a function of wavelength throughout the solar and infrared spectral regions. These probabilities correspond to the vertical passage of a photon at a given wavelength through the entire thickness of the atmosphere (for

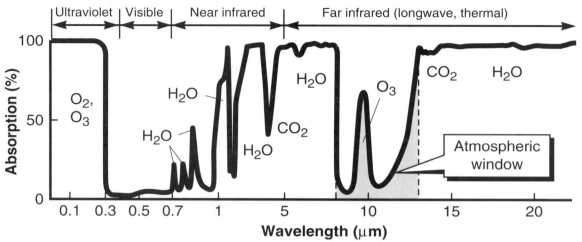

Figure 11.10   The efficiency of radiation absorption by the atmosphere as a function of wavelength through the solar and terrestrial regions of the spectrum (Appendix C). The absorption spectrum is expressed as the percentage of the sunlight incident at the top of the atmosphere that is absorbed by air before reaching the ground. An absorption of 100 percent means that no radiation at that wavelength can penetrate the atmosphere, because of absorption by molecules in the atmosphere. Regions of high absorption associated with specific molecules are identified on the plot. The major spectral ranges are shown: The region from about 8 to 13 micrometers is referred to as the longwave *atmospheric window.* (Data from R.M. Goody, *Atmospheric Radiation: I. Theoretical Basis,* London: Oxford University Press [1964], adapted from data originally presented in Figure 1.1, p. 4)

example, the probability that a solar photon will reach the surface at high noon). At a given wavelength, radiation has the same probability of passing through the atmosphere whether traveling upward or downward. Accordingly, sunlight at a visible wavelength of 0.5 μm is nearly unaffected by atmospheric absorption, whereas at an ultraviolet wavelength of 0.2 μm, it is completely attenuated. Surface infrared radiation at a wavelength of about 8 to 9 μm has little trouble passing through the atmosphere to space, but at 6 μm, escape is not possible.

The longer its path through the atmosphere is, the more likely a photon will be absorbed. (The concept of an optical depth is explained in Section 3.2.2.) The absorption optical depth, $\tau_{a\lambda}$, at a specific wavelength, λ, is the product of the absorption coefficient, the concentration of absorbing molecules, and the length of the path. If the concentration is fixed and the path length increases, the optical depth will become larger, and vice versa. The probability, P_a, that a photon will be absorbed along the path is expressed as the negative exponential of the optical depth:

$$P_a = \exp\left(-\tau_{a\lambda}\right) = e^{-\tau_{a\lambda}} = \frac{1}{e^{\tau_{a\lambda}}} \quad (11.15)$$

Equation 11.15 is known as **Beer's law** of absorption.[8] This rule states that radiation traversing the atmosphere at an angle, as opposed to vertically, is absorbed to a greater degree than vertical rays are.

Inspection of Figure 11.10 shows that atmospheric absorption in the climatically interesting wavelength regions is dominated by water vapor and carbon dioxide. Water vapor absorbs both solar near-infrared and thermal longwave radiation. The important absorption bands are the near-infrared "overtone" (multiple vibrational mode) bands, the 6.3-μm vibration-rotation band, and the pure rotation band at wavelengths longer than 15 μm. Carbon dioxide is active mainly in its 15-μm vibration-rotation band. Ozone also appears in Figure 11.10 in a region without other strong absorption; this is the ozone 9.6-μm vibration-rotation band. Although all these bands are depicted as broad and continuous in Figure 11.10, in reality they are composed of thou-

8.   Beer's law holds whenever radiation at one wavelength, or monochromatic radiation, is being considered, and scattering can be ignored. This is generally the case in the atmosphere at wavelengths shorter than about 0.3 μm (ultraviolet) and greater than 3 μm (longwave infrared). Use of Beer's law requires the evaluation of the exponential function. An easy way to do this approximately is to let $e = 3$. Then if the optical depth is 1, P_a is 1/3; if the optical depth is 2, P_a is 1/(3 × 3) = 1/9; and so on.

sands of individual sharp lines, which can be seen only at very high spectral resolution. For our purposes, only the broad-band properties of the molecular absorption need to be considered.

Atmospheric Windows

The atmospheric absorption spectrum (Figure 11.10) has two prominent "windows," through which radiation can travel fairly easily. Outside these windows, the air is quite opaque. The first window spans the visible spectrum, from about 0.3 to 0.7 μm in wavelength (also extending into the near infrared). This is the solar (shortwave) window. Animal vision evolved to take advantage of the atmosphere's transparency in this region (one need only imagine hunting prey in a dense fog, which is how the world would appear if our eyes worked at a wavelength of, say, 0.2 μm or 6.0 μm).

The second window is in the infrared region, from about 8 to 13 μm. In climate studies, this thermal (longwave) window is the one usually referred to as the "atmospheric window." It happens that the peak intensity of the Earth's thermal emission falls in this window. The window is not wide open, however. Ozone has strong absorption (9.6-μm band) in this region. There is also a weaker background absorption due to the so-called **water vapor continuum**. The continuum is really the cumulative effect of extremely weak absorption in the outlying wings of thousands of water vapor lines in the 6.3-μm and pure rotational bands. All in all, in a relatively dry environment, the longwave window is quite transparent, although in a humid environment the high concentrations of water vapor and accompanying haze particles tend to cloud up the glass.

The greenhouse effect is caused by the relative ease with which solar radiation can reach the surface through the solar window and the relative difficulty that the thermal radiation faces in escaping from the surface. Ideal greenhouse gases are transparent at visible (and near-infrared) wavelengths, yet are opaque at longer infrared wavelengths. Both water vapor and carbon dioxide fit the bill. As we will see shortly, many other gases have similar radiative properties and can act as effective greenhouse gases.[9]

11.4.2 RADIATION EMISSION FROM THE EARTH

What does the radiation emitted by the Earth actually look like? The spectrum of Earthglow is shown in Figure 11.11. A satellite looking at the Earth from a distance well above the atmosphere can measure such a spectrum (using the proper instruments). There are remarkable and important features in this spectrum. First, note that the actual emission spectrum does not resemble a perfect blackbody spectrum (several of these curves are illustrated). But the spectrum does reflect to a greater extent the location of specific bands (for example, O_3 9.6 μm, CO_2 15 μm, H_2O 6.3 μm) that absorb and emit the radiation. It should thus be apparent that different regions of the Earthglow are emitted by different gases *from different parts of the atmosphere*. This conclusion is critical to understanding the detailed structure of the emission spectrum.

There is no reason to believe that all the thermal radiation leaving the Earth is emitted from the same level. In fact, because the atmosphere's absorption varies greatly with wavelength (for example, as in Figure 11.10, but much more so at even finer spectral resolution), radiation at a specific wavelength can escape to space only from an altitude above which the overhead optical depth is less than about one. This emission altitude varies significantly from wavelength to wavelength. The radiation emitted by the gases at the emission altitude has the characteristic blackbody intensity of the temperature of the gases at that altitude. The temperature profile of the atmosphere has a specific structure (Section 2.3.2). The temperature varies substantially with height. It follows that the radiation emitted by gases near the surface, for example, has a quite different intensity, or "effective" emission temperature, from the radiation emitted by gases at the tropopause, or the upper mesosphere, and so on.

The structure of the emission spectrum should now be much clearer. Consider the region around 15 μm, for example. The emitting species in this case is carbon dioxide; we can even identify the inverted image of the CO_2 15-μm absorption band in the spectrum. The altitude of the emission may be estimated as follows: First, determine which of

---

9. A glass greenhouse—after which the greenhouse effect is named—has similar, but distinct, properties. Glass is used because it readily admits sunlight into the greenhouse. The solar energy is converted to heat, just as occurs in the climatic greenhouse. However, in a greenhouse, the heat would escape by convection

(rising) of the hot air into the cooler air outside if the glass did not physically deflect the air currents. In a greenhouse, the hot air is diverted and its heat is prevented from escaping. In the atmosphere, the radiation emitted by the hot air is absorbed and prevented from escaping: similar, but not quite the same, effects.

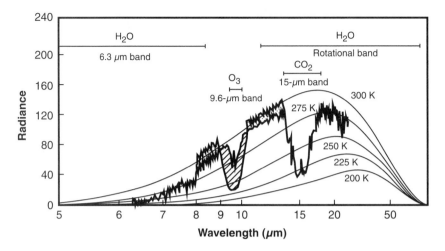

Figure 11.11   The spectrum of radiation emitted by the Earth to space, as measured from a satellite. The spectrum is the jagged line, which defines the intensity of the radiation as a function of wavelength. Several ideal blackbody emission curves (Planck functions) corresponding to temperatures ranging from 200 K to 300 K are shown for comparison. By matching the observed emission spectrum at any wavelength to a blackbody curve, the effective temperature of the Earth's emitting region at that wavelength can be estimated. Indicated above the spectrum are the atmospheric species responsible for the emissions in the corresponding wavelength intervals. The relevant molecular "bands" for each species also are identified. The response of the emission spectrum to the addition of a hypothetical atmospheric absorber in the longwave window region is illustrated by the hatched regions. The measurement corresponds to a cloud-free area of the Earth. (Data from Hanel, R. A., B. J. Conrath, V. G. Kunde, C. Prabhakara, I. Revah, V. V. Salomonson and G. Wolford, "The Nimbus 4 Infrared Spectroscopy Experiment 1. Calibrated Thermal Emission Spectra," *Journal of Geophysical Research* 77 [1972]: 2629.)

the blackbody emission spectra in Figure 11.11 has the same intensity as the CO_2 band emission. It appears that the temperature is close to 220 K. Then compare this value with the temperature profile in Figure 2.6. Temperatures in this range are in the lower stratosphere and middle mesosphere. Either is a possibility, and you would need further information to proceed (the origin of the radiation is actually in the mesosphere). Take, as another example, the region from about 10 to 13 μm. The emission temperature is close to 295 K, based on comparisons with the blackbody curves. Such a temperature is found only near the surface. This is a logical conclusion, since we already know that these wavelengths fall in the longwave window. The emission is thus from the surface, which we can deduce is quite warm.

Another important point from Figure 11.11 is the dependence of the Earthglow's intensity on the absorption strength of the principal emitter. In general, the stronger the absorption band is, the colder is the effective emission temperature will be. Similarly, the greater the atmospheric burden of the absorber is, the colder its emission will be. In both cases, the emission is coming from a height in the atmosphere above which the atmosphere is transparent (at the wavelengths of interest). The stronger the absorption or the greater the amount of absorber is, the higher in the atmosphere and the lower in temperature the infrared emission tends to be. If the radiation temperature is lower, the energy emitted will be reduced, according to Planck's law (Section 3.2.1).

Integral Energy Balance

Refer again to Figure 11.11. A critical feature of this graph is that *the total area under the emission curve is equal to the total energy flux leaving the Earth in the form of heat radiation.* The area of interest is framed by the bottom of the graph (the zero line); in fact, the areas under the blackbody curves in Figure 11.11 obey the Stefan-Boltzmann law for the total heat flux from an ideal emitter. If the Earth emission data in Figure 11.11 represent the average emission spectrum for the globe, then the integral area will equal the absorbed solar energy (averaged over the planet); that is, the energy balance must hold, as in Equation 11.13.

Now we can construct a scenario that demonstrates how the greenhouse effect works. We know that if the solar energy input remains about constant, then the total thermal emission must also remain constant. In other words, the integral area in Figure 11.11 must remain fixed. The curve can be distorted as long as the total area underneath it remains the same. In the previous section, we also learned that the addition of a greenhouse-active gas to the atmosphere can partially "close" the atmospheric window. This generally causes a decrease in the emission temperature in the spectral region where the greenhouse gas absorbs strongly. The situation is illustrated in Figure 11.11 for a hypothetical greenhouse gas that lowers the emission temperature in the hatched region between 9 and 10 μm. Note that the modified emission curve (with the additional greenhouse gas present) falls below the original emission curve. The integral area is smaller, and so less energy is emitted to space. Hence the Earth must be receiving more energy than it is losing, and so, finally, the planetary climate must warm. Indeed, if the surface warms in response to the added greenhouse effect, the surface radiation intensity will *increase*. This response is indicated in Figure 11.11 by the stippled region from 8 to 9 μm and from 10 to 13 μm. In fact, the surface temperature must increase precisely enough for the stippled area to equal the hatched area.

The greenhouse effect is very logical. Greenhouse gases partially close the atmospheric window and trap heat more efficiently at the surface and in the lower atmosphere. To balance this effect, the surface and lower atmosphere must warm and emit thermal radiation at a greater intensity (usually in different spectral regions).

The greenhouse effect is caused when the atmosphere absorbs thermal radiation in spectral regions where it (the atmosphere) is opaque. In accordance with the conservation of energy, the amount of heat absorbed equals the amount reemitted. Roughly half this reemitted heat travels in the same direction as the original thermal radiation, but the other half travels in the opposite direction. Thus absorption and reradiation, occurring over and over, inhibit the direct escape of thermal energy to space. The atmosphere both absorbs and emits radiation with nearly the same efficiency (refer to Kirchhoff's law in Section 3.2.1). That is, the absorptivity and emissivity of air are equal. The fact that the atmosphere absorbs and reemits thermal radiation may be included in the simple climate model developed earlier. In the simplest case, the atmosphere is treated as a single layer of gas with an average spectral absorptivity/emissivity, ε. This approximation represents the entire atmospheric gas layer as a so-called gray-body absorber/emitter with the same (average) emissivity at all wavelengths. In terms of the average emissivity, the equilibrium surface temperature beneath a greenhouse blanket of gas is easily derived by considering the coupled energy balance for the surface and atmosphere (Problem 2). The result is a simple modification of the "effective" temperature derived earlier such that

$$T_s = T_e \left(\frac{2}{2 - \varepsilon} \right)^{\frac{1}{4}} \qquad (11.16)$$

Here, T_s is the actual average temperature of the surface, and T_e is the effective blackbody radiation temperature of the whole Earth defined by Equation 11.14.

The average effective one-layer emissivity of the Earth's atmosphere is roughly 0.75. That is, about three-quarters of the heat radiated by the Earth's surface is effectively absorbed by the atmosphere (Section 11.4.4). Using this value in Equation 11.16 yields an average planetary surface temperature of 288 K, in accord with the measurements. The greenhouse effect of the atmosphere therefore explains the difference between the effective blackbody emission temperature of the Earth and its surface temperature. Indeed, most of the radiation leaving the Earth originates in the upper atmosphere and not at the surface. The surface is distinct from the direct radiative cooling mechanism. The surface thus may be considerably warmer than it would be if it were directly joined to the radiative equilibrium level.

The simple gray-body greenhouse model is not consistent with all the planetary data in Table 11.1. Although it is certainly a reasonable model for the Earth and Mars, with relatively thin atmospheres, it fails for Venus. Even if we assume the maximum emissivity for the Venus atmosphere of 1.0, Equation 11.16 predicts a surface temperature of only a modest 295 K, well below the measured surface

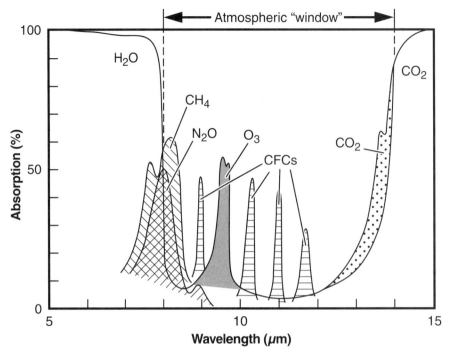

Figure 11.12   The absorption spectrum of the atmosphere in the "atmospheric window" region. The principal absorbing molecules and the relative sizes of their absorption bands are illustrated.

temperature, which exceeds 700 K. Fortunately, it is not necessary to explain this discrepancy in order to understand the Earth's climate. Let us just say that in the case of very thick greenhouse atmospheres, the warming at the surface can be greater than expected from a simple energy balance analysis.

In summary, we may conclude that the energy balance component related to the thermal emission of radiation, although quite complex in its details, must still obey this simple spectral-area integral rule. Moreover, the overall greenhouse effect of an atmosphere like the Earth's is readily explained quantitatively in terms of a simple model consisting of a one-layer gray-body atmosphere overlying a surface.

Closing the Window

The absorption spectrum of the atmosphere in the atmospheric window region is shown in greater detail in Figure 11.12, in which specific bands of important trace constituents are identified. Notice that ozone and the chlorofluorocarbons (CFCs) have strong absorption well within the window, whereas methane and nitrous oxide absorb near the short-wavelength edge of the window. Carbon dioxide is seen to have the effect of narrowing the window

from the long-wavelength side. As these greenhouse gases become more concentrated, their absorption bands in the window become larger. The window slowly begins to close. According to the discussion in the previous section, the Earth's radiation where the absorption is strong is colder and therefore less intense. The surface and lower atmosphere respond by warming, intensifying the emission in the more transparent regions of the spectrum to compensate for the loss of intensity in the window (Figure 11.11).

The greenhouse gases fill the window region of the spectrum, "closing" the window. The CFCs are particularly potent greenhouse gases, for two reasons. First, the absorption bands of these molecules are very strong in the window region. Hence only small amounts of CFCs can produce a relatively large absorption effect. Second, the CFC absorption bands generally fall in otherwise open regions of the window. They do not overlap the other strong bands. When the bands of different molecules lie at the same wavelengths, their effectiveness as absorbers can be reduced. The greenhouse power of CFCs can be gauged by noting that CFC concentrations of about one part per *billion* by volume could produce the same global warming as roughly 300 parts per *mil-*

lion by volume of carbon dioxide. That is, CFCs are more than 100,000 times as effective as CO_2 in causing a greenhouse effect.

Carbon dioxide is not nearly as effective, gram per gram, as are the other greenhouse gases in Figure 11.12. This can be understood by noting that the infrared emission spectrum in the CO_2 15-μm band is already highly saturated (Figure 11.11). Accordingly, further additions of carbon dioxide can act only through very weak absorption in the wings (at the edge) of the 15-μm band and through other weak absorption features (not shown). Because these absorption features are weak, much greater incremental concentrations of CO_2 are required to close the window.

11.4.3 CLOUDS AND RADIATION

Clouds have an important influence on solar radiation. The scattering properties and optical phenomena associated with clouds are described in Sections 3.2.2 and 3.2.3. In addition, clouds have a profound effect on longwave radiation. At thermal infrared wavelengths, in fact, clouds may be treated simply as blackbodies. That is, clouds absorb all the infrared radiation striking them and emit blackbody radiation at the temperature of the cloud. These rules may be applied to almost all forms of clouds: stratus, or lowlying dense cloud layers typically blanketing large areas; cumulus, or convectively active thick localized clouds that appear as separate puffs, often in extensive clusters, and that may grow into towering storm clouds or thunderheads; and cirrus, or high thin ice clouds with great horizontal extent, often wispy in appearance but frequently dense when generated by the outflow of moist air from thunderheads.

Why are clouds blackbodies? It happens that condensed water in the form of cloud droplets or ice crystals is an excellent infrared absorber. Figures 11.10 and 11.11 show that water vapor has absorption bands in the short- and long-wavelength regions of the infrared. Water vapor absorption in the wings of its lines also causes the "continuum" absorption in the atmospheric window (Section 11.4.1). When the molecules of a vapor, such as H_2O, are condensed into a liquid or solid, the individual rotational absorption lines are greatly broadened and tend to fill in regions between the more distinctly defined gaseous absorption bands. The absorption lines are broadened by collisions of the molecules.

The more frequently a molecule is hit by its neighbors, the broader its absorption lines (and emission lines, by Kirchhoff's law) will be. In a liquid or solid state, molecules are so close to one another that they are in continuous collision, and their lines are exceptionally broadened. Naturally, there is some spectral variation in cloud absorptivity (or emissivity), but this is significant mainly in the case of very thin clouds (high, wispy cirrus, for example).

Figure 11.13 illustrates the radiative effects of clouds that contribute to climate. In the solar spectrum, several processes are apparent. First and most important, clouds are excellent scattering agents at solar visible wavelengths and control the Earth's albedo. About 50 percent of the Earth is covered by clouds at any time, and the average reflectivity (or albedo) of these clouds is roughly 0.50. Thus most of the Earth's albedo can be attributed to clouds. Clouds also absorb solar near-infrared radiation (Figure 11.10), which accounts for about half the solar constant. Indeed, about one-fourth the total solar insolation is due to near-infrared absorption by clouds. Finally, through multiple scattering, clouds diffuse the solar radiation emerging from their bottoms. Compared with direct sunlight, this diffuse radiation may be absorbed and reflected in slightly different ways in the lower atmosphere.

The infrared effects of clouds on the energy (radiation) balance are also depicted in Figure 11.13. Because clouds radiate energy as blackbodies, their emission flux, F_H, is proportional to the fourth power of their local temperature (Equation 11.10). Therefore, as a result of the natural temperature lapse rate of the troposphere (Section 2.3.2), clouds located at lower, warmer altitudes radiate energy more intensely than do clouds at higher, colder altitudes. More specifically, dense clouds tend to radiate energy at the temperature of their "surface."[10] The upward longwave flux is therefore determined by the cloud top's temperature, and the downward flux is determined by the cloud bottom's temperature.

10. Clouds do not have a solid outer surface, of course. Flying into a cloud is like being enveloped in a mist. From a distance, some clouds appear to have sharp outer boundaries, almost solid. The sharpness depends on the density of cloud droplets near the edge. The more droplets there are, the greater the optical depth is per unit distance of penetration into the cloud; the less distance is needed to reflect the sunlight illuminating the boundary; and the sharper the boundary appears from a distance. At infrared wavelengths, the optical depth (for absorption) with the same number of droplets is generally larger than the scattering optical depth at visible wavelengths. Accordingly, clouds have even more distinct surfaces at thermal wavelengths.

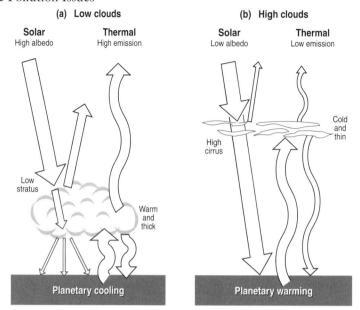

Figure 11.13 The effects of clouds on the flow of radiation and energy in the lower atmosphere and at the surface. Two cases are shown: (a) low clouds, with a high solar albedo and high thermal emission temperature; and (b) high clouds, with a low solar albedo and low thermal emission temperature. The solar components are shown as straight arrows, and the infrared components, as curved arrows. The relative thicknesses of the arrows indicate the relative radiation intensities. The expected impact on surface temperature in each situation is noted along the bottom strip.

Dense clouds, in other words, act just like the ground with regard to thermal radiation—they absorb all incident longwave radiation and emit thermal radiation at the skin temperature.

In the case of cirrus clouds, the situation is more complex. Cirrus are generally thinner clouds than stratus or cumulus, with a relatively low concentration of large ice crystals. Cirrus are not as efficient as stratus are in reflecting sunlight. This is obvious when you view the sun through the feathery cirrus clouds associated with large weather systems. However, cirrus associated with intense cumulus convection are often as dense as stratus. These clouds are certainly blackbody radiators. The thinner cirrus are also reasonable approximations to blackbodies. The large ice crystals in these clouds, although relatively poor sunlight-scattering agents, are excellent infrared absorbers.

Clouds contribute to the greenhouse effect by absorbing the radiation emitted by the surface and lower atmosphere, including that in the infrared window region. Being blackbodies, in fact, clouds close the window completely! If the entire Earth were covered by clouds, there would be no window

at all (in Figure 11.11, for example, cloud effects are absent). Clouds also radiate energy as blackbodies and cool certain regions of the atmosphere. Consider the warm, thick, low-lying clouds in Figure 11.13(a). The clouds and the surface have about the same temperature in this case (although the clouds may be somewhat cooler, of course, depending on their actual height, the temperature lapse rate, and other factors). The thermal flux emitted by the surface is absorbed at the cloud bottom. The cloud bottom emits a flux of the same intensity back to the surface. No heat has been lost from the system so far. However, the top of the cloud also emits heat upward. From the point of view of the energy balance seen from space, the cloud top looks just like the surface. Thus energy is being lost from the lower atmosphere and surface as if the cloud were not there. This energy must be carried to the cloud top, where it can be radiated, and must be replenished continually, or the cloud top would quickly cool off. Atmospheric vertical mixing by means of small-scale turbulence and convection (Sections 5.1.2 and 5.2.3) efficiently transfers heat to the cloud top for emission. Indeed, a self-controlling mechanism is built in. If the cloud top begins to cool down while the cloud bottom remains warm, a thermal instability will develop (Section 5.2.2) that increases the upward convection and mixing of heat to the cloud top.

The dense low-lying clouds also increase the local solar albedo. As we stated earlier, the average albedo of such clouds may be 0.50 or greater. This is twice the average albedo of the Earth and may be five times or more larger than the local albedo over oceans, say. The solar insolation beneath these clouds is reduced accordingly.

We can now summarize the climatic impact of low-lying, warm, dense clouds (Figure 11.13). The solar energy absorbed below these clouds is reduced from the cloud-free case, while the local greenhouse forcing remains relatively unaltered. From simple energy balance considerations (Section 11.3.2 and Equation 11.14, for example), the net effect is a cooling of the local climate.

Next, consider the case of the high, cold, thin clouds depicted in Figure 11.13(b). In this case, the solar albedo can be low. The large ice crystals in these clouds are not efficient in reflecting sunlight back to space. The clouds remain opaque in the longwave infrared, however, and may still be considered blackbodies. The upwelling thermal radiation emitted by the surface and lower atmosphere is absorbed at the cloud bottom. The cloud bottom reradiates the thermal energy back toward the surface. The top of the cloud radiates thermal energy upward, toward space. But the cloud top is much colder than the surface, and the emission to space from the cloud is less intense than the original upwelling radiation blocked by the cloud. From space, the total thermal emission appears to be reduced. Hence the high, cold clouds enhance the local greenhouse effect.

The impact of high, cold, thin clouds can be summarized as increasing the greenhouse component of the energy balance while having a smaller effect on the solar forcing (insolation). The net effect is a warming of the local climate.

A Cloud Thermostat?

According to previous discussion, low-level clouds generally tend to produce a net cooling, and high-level clouds produce a net warming, of the Earth. Overall, the *global* effect of clouds on the Earth's climate is to produce a small net cooling of surface temperatures below those expected in the absence of clouds. An interesting question is whether clouds act as a **thermostat** for the climate. That is, if changes in the forcing of the Earth system warm the climate (say, an increase in the sun's intensity or the addition of greenhouse gases to the atmosphere), then will clouds respond to reduce the potential warming, or enhance it? If the effect of clouds is to moderate the warming, clouds are said to produce a "negative-feedback" effect—negative in the sense that the actual warming is smaller than the potential warming associated with the change in forcing. In this case, the clouds act as a thermostat to limit the degree of warming. If clouds enhance the potential warming, on the other hand, they are said to have a "positive-feedback" effect—positive in that the actual warming is larger than the warming that would otherwise have been caused by the change in forcing. Of course, one may consider similar effects, in the complementary sense, if climate forcing is reduced and temperatures decrease.

The answer to the cloud thermostat question has not been completely resolved. The best existing data suggest that the response of clouds is to enhance the warming caused by the accumulation of greenhouse gases in the atmosphere. Thus clouds are not a moderating thermostat protecting us from climate change, but are actually part of the problem. It is fairly obvious that as the surface temperature warms, say, the evaporation rate of water increases, raising the water vapor concentration of the atmosphere and hence its greenhouse effect. This dependence of water vapor on surface temperature is a critical positive-feedback mechanism for the climate system (Chapter 12 and Section 11.4.3). At the same time, warmer surface temperatures and greater humidity could increase the frequency and intensity of strong convection. The resulting storms would generate high, cold, thin clouds that further enhanced the greenhouse effect.

The existence of a safety valve, or thermostat, in the climate system seems to be likely, because the Earth has enjoyed a very moderate climate for billions of years, but the nature of such a thermostat is unknown. It may in fact involve clouds, but in a more complex manner than we have discussed so far. Moreover, it is well known that the climate system can experience rapid changes (Section 11.7). Clearly, human society should not at this time rely on unidentified or surmised control mechanisms for protection against probable climatic anomalies, either natural or anthropogenic.

11.4.4   THE GREENHOUSE ENERGY BALANCE

All the basic elements of the global energy balance, including the solar and thermal components and the greenhouse effect, can now be assembled as in Figure 11.14 (Appendix C reviews the terminology used in describing the radiation components). Of the total solar energy incident on the Earth (given as 100 units of energy, equivalent to one-quarter of the solar constant, or about 350 watts per square meter, when averaged over the entire surface of the Earth), scattering from the atmosphere, clouds, and the surface reflects about 30 units (the albedo). About 20 units are absorbed in the atmosphere, mainly by water vapor and clouds. The rest of the solar energy, about 50 units, is absorbed at the surface. The total reaching the surface consists roughly of 30 units by direct transmission through the atmosphere, 10 units by scattering from air molecules and aerosols, and 10 units as diffuse radiation emerging from the bottoms of clouds.

Solar energy absorbed at the ground is immediately converted to heat. The surface is also heated by thermal radiation emitted downward from the atmosphere and clouds. Some of the heat (about 30 units) is removed from the surface in the forms of latent heat and sensible heat. Water vapor evaporation requires some energy, which is then carried by the water molecules as potential latent energy (close to 23 units), which is released when the water vapor condenses as liquid water or ice. Thus when a cloud forms, the latent heat of the condensing water vapor is released and warms the adjacent air. To a great extent, latent heat release drives convection in large cumulus clouds and thunderstorms.

Heat can also be transferred from the surface to the atmosphere as "sensible heat" (7 units). Sensible heat is what we may think of as real heat; it is the heat you can feel by touch. In the atmosphere, it is hot air. As the surface warms, the air in contact with the surface absorbs some of the excess heat by means of thermal conduction.[11] This warm air then rises and carries the heat away from the surface (Sections 5.1.3 and 5.2.2). The sensible heat may be converted to the kinetic energy of motion of the air, may warm

other air by mixing, and may be used to drive the adiabatic expansion of rising air parcels.

The balance of the heating of the surface (for the global average situation, and assuming no net heat storage in or loss from surface energy reservoirs) is emitted from the surface as longwave radiation. Only a small portion of this emission (about 5 units on average) penetrates the atmospheric window to escape the Earth. Most of the thermal emission from the surface is absorbed by overlying clouds and greenhouse gases, particularly water vapor. This absorbed longwave radiation, together with the solar radiation directly absorbed into the atmosphere and the latent and sensible heat transferred to the atmosphere from the surface, is effectively converted to heat. The heating is balanced by reradiation of the heat as longwave radiation. The radiation from cloud tops (30 units) and from the greenhouse gases well above the surface (35 units) accounts for most of the thermal energy escaping the Earth. Added to the direct longwave emission from the atmosphere, these exactly balance the net solar-energy input (70 units). The clouds and greenhouse gases also emit longwave radiation back to the surface (nearly 95 units), which contributes to the surface energy balance in the form of the greenhouse effect.

In Figure 11.14, the greenhouse effect can be envisioned in the following way: Solar radiation penetrates the relatively transparent atmosphere and heats the surface. This additional heat must escape the Earth but, in its initial attempt to leave, is largely trapped by the blanketing atmosphere and clouds. Much of the trapped energy is transferred back to the surface, from which it is shed again. However, energy is now also leaking away through various cracks and half-opened "windows" in the atmosphere. The combined emissions occurring as a result of multiple transfers of energy within the climate system maintain the Earth's overall energy balance.

Think of a greenhouse atmosphere in the same way you do a blanket that keeps you warm on a cold night. When you first slip into bed on a cold night, the sheets are icy to the touch, making you shiver. It is not long, however, until your body heat begins to build up underneath the blanket. The warmth is much more comfortable. Each of us is a little energy generator, and the blanket serves to trap and conserve that energy. If you are lucky enough to own a fine down comforter, which happens to be a very efficient insulator against heat loss, you may become downright hot and uncomfortable underneath it.

---

11.   Thermal conduction occurs when "hot" molecules in a highly agitated state of thermal motion collide with cooler molecules and cause them to become more agitated. Thus heat is transferred from warmer to cooler regions within a material, from molecule to molecule, or between materials whose molecules are in contact.

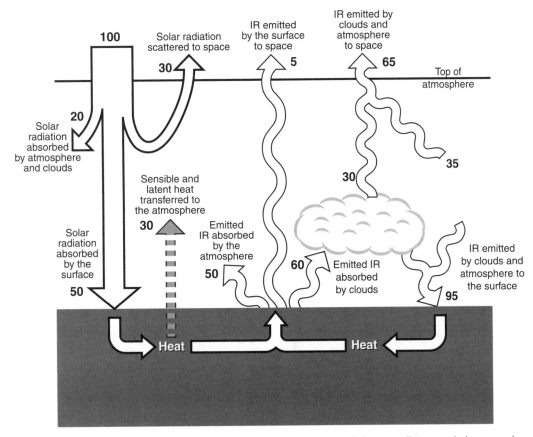

Figure 11.14   The radiative and physical processes that control the overall energy balance and climate of Earth. The shortwave (solar) components are shown on the left, and the longwave (thermal infrared) components on the right. The relative intensities of the radiation components are indicated by the width of the arrows, as well as by numbers representing the relative intensities as a percentage of the global average incident solar energy (that is, one-quarter the solar constant). Also shown (dashed arrow) is the energy transfer associated with latent heat (water evaporation and recondensation) and "sensible" heat (hot air). Energy flux data were generalized from a number of sources.

Kick the comforter off, however, and you have lost your greenhouse protection. Precious body heat is now lost to the room. You begin to shiver again. You pull the comforter over yourself again, but now cut a few holes in it to let some of the excess heat escape. If you make the holes just the right size, you can achieve a perfect temperature for your personal comfort (to the chagrin of whoever shares the comforter).

The Runaway Greenhouse Effect

The Earth is a perfect place to live. The combination of a lucky position in the solar system with just the right degree of greenhouse warming has kept the Earth's climate in a moderate state throughout its history. To be sure, there have been significant long-term climatic shifts, such as the ice ages (Section 11.6.3). Occasionally, there are brief periods of severe climatic anomaly associated with volcanic eruptions (Section 11.6.4). It is also true that to a large extent, we do not understand why the climate has remained so favorable for so long. This amazing fact is reassuring to some (it will never change) and worrisome to others (it could change at any moment). In subsequent sections, we explore in more detail the climate system and climatic change.

Regarding the possible extreme of the greenhouse effect, an example for Earth may be found on Venus. Earth and Venus are sister planets, about the same size and composition and located at similar distances from the sun. However, Venus is about 30 percent closer to the sun than Earth, and that difference is crucial. Being just that much closer to the sun,

Venus suffered a "runaway greenhouse effect," whereas Earth did not (although the Earth was in danger of permanently freezing over [Section 11.6.2]). Probably Venus originally had oceans. The greater solar insolation at Venus would have heated the ocean surfaces to well above 300 kelvin (Table 11.1; the albedo of Venus is not likely to have been greater than 0.7). Hence water evaporated from the Venusian oceans at a much greater rate than it did from oceans on Earth. But this water vapor itself led to an additional greenhouse heating of the surface, which caused further evaporation and warming. Careful calculations, in fact, demonstrate that the warming and evaporation continued until the oceans were dry! The scorching temperatures and drenching atmospheric water vapor concentrations promoted the escape of hydrogen to space (Section 4.1.1) and the eventual depletion of water from Venus altogether.

Without oceans, the volatile carbon dioxide belching from the interior of Venus accumulated in its atmosphere, reaching its present abundance some 100,000 times that in the Earth's atmosphere. The greenhouse window on Venus remained closed, with very few leaks, and the greenhouse effect raged, keeping the surface temperature above 700 K.

The early climate of Venus was unstable. Liquid oceans were not compatible with the more intense solar heating. Carbon dioxide could not be converted to carbonates and stored in sediments and thus remained in the atmosphere. Venus evolved into an oven-baked planet. There is no possibility that the same events could unfold here, even with accelerated carbon dioxide emissions (Chapter 12) and burning of all fossilized carbon. The Earth will not become Venus. There is essentially no chance that Earth will suffer a runaway greenhouse effect (Section 11.6.1).

11.5 Energy Reservoirs: The Climate Flywheel

We can now think of the climate system as a closed system, or box, into which energy flows from the sun and from which heat escapes into space. Of course, the climate system has a characteristic temperature (greater than absolute zero) and so must contain heat. The amount of heat is determined by the size of the heat reservoirs that compose the system and the heat capacities of the reservoirs. (See Section 11.3.2 for a discussion of heat capacity.) The total energy in the climate system is the sum of the energies held in the principal reservoirs. The behavior of the climate system, and particularly its response to changes in the energy input or output, is determined by the sizes and rates of energy transfer between the reservoirs. The concept is identical to the coupled reservoir model for elements in the Earth system, as described in Section 10.1. We will use the model to quantify the climate system.

First, consider simple everyday analogues for the climate system that has been described up to this point—that is, a well-defined physical container that can hold (heat) energy, with both a source and a sink of energy and a mechanism for trapping heat. Now imagine a tea kettle (the Earth) on your stove. Energy flows into the kettle from the flame of the burner (the sun). The water in the kettle (the oceans, say) heats up at first. Then steam begins to stream out of the kettle (heat escape), setting off the whistle. Pretty soon the kettle is whistling away steadily (equilibrium temperature or steady-state climate). The energy in this case is carried mainly by the latent heat of the evaporating water; the actual temperature of the water in the kettle remains roughly constant, pegged close to the boiling point of water at 100°C. If you turn up the burner, the kettle will whistle louder, as more water evaporates and escapes to carry away the extra energy and maintain the temperature near the boiling point.

If you hold the valve on the kettle shut, however, the situation will become more desperate. To force enough vapor out of the kettle, the water must now heat up significantly to increase the water vapor pressure in the kettle. Eventually, the system must reach a new equilibrium at a higher temperature. If you were strong enough (or perhaps stupid enough) to keep the kettle sealed, it would eventually explode! Nevertheless, you have discovered the principle of the "pressure cooker." Pressure cookers are used to cook tough foods at temperatures well above the boiling point of water (normal cooking) in heavy reinforced steel pots you may remember seeing in your grandmother's kitchen. A pressure cooker on the stove is equivalent to the hothouse climate on Venus, where the dense atmosphere holds the heat in.

How often have you gotten into your car on a sunny summer day and been fried on the leather or vinyl seats? Again, you are experiencing the greenhouse effect. Your car is the Earth. Sunlight streams in through the closed transparent windows (the

Table 11.2   Earth's Energy Reservoirs

Reservoir	Volume (km^3)[a]	Mass (Gt)	Temperature (K)	Heat capacity (J/gram·K)	Energy content (J)
Atmosphere[b]	5.0×10^9	5.0×10^6	260	1.0	1.3×10^{24}
Land surface[c]	5.0×10^3	1.0×10^4	290	4.0	1.2×10^{22}
Land subsurface[c]	1.0×10^5	2.0×10^5	280	4.0	2.2×10^{23}
Surface oceans[d]	3.8×10^7	3.8×10^7	280	4.0	4.3×10^{25}
Deep oceans[d]	1.4×10^9	1.4×10^9	275	4.0	1.5×10^{27}

[a] Approximate volumes are specified in order to estimate the reservoir's heat contents.
[b] The atmosphere is assumed to be 10 kilometers thick (roughly one scale height) at density of 1 kg/m^3.
[c] The land surface depth is taken to be 4 centimeters thick for diurnal response, with a density of 2000 kg/m^3. The deep soil layer is taken as 80 centimeters thick, for seasonal variations.
[d] The oceans have a density of 1000 kg/m^3 and average depths of about 100 meters for the surface oceans and 4 kilometers for the deep oceans.

atmosphere) and is absorbed by the seats (the Earth's surface) and converted to heat. But the hot air (thermal radiation) has no way to escape through the closed windows. The temperature continues to rise until the heat escaping through the cracks, or slowly conducted or radiated from the interior, balances the sunlight absorbed inside. Sounds familiar, doesn't it? Now you can use your newly gained knowledge of the climate system to prevent this uncomfortable situation. For example, you can reduce the solar constant (park your car in the shade), reduce the solar insolation (place a white reflective screen in the windows), or eliminate the greenhouse effect (leave the windows rolled down), or some combination of these actions. Understanding the climate is common sense.

11.5.1   Reservoirs for Heat

Energy storage and flow within the climate system are depicted in Figure 11.15. The primary reservoirs for energy should be familiar from our discussion of global biogeochemical cycles in Chapter 10 (see also Section 4.1.2). The sizes, masses, and other characteristics of these reservoirs are summarized in Table 11.2. The heat content of a reservoir, Q, is related, by Equation 11.5, to its mass, heat capacity, and temperature. The principal reservoirs of heat that affect

the climate are the atmosphere, surface oceans, and land surfaces. These reservoirs are joined on relatively short time scales. The processes that transfer energy between the reservoirs, and the rates of energy exchange, are indicated in Figure 11.15. The most massive heat reservoir is that of the deep oceans. There is enough thermal energy there that if it were extracted, it could maintain the atmosphere and surface climate for a thousand years. Unfortunately, mining this energy would not be easy. The deep ocean waters are cold and not readily accessible.

A significant change in the temperature of a reservoir is only a small percentage. That is, for a reservoir at a temperature near 300 K, a temperature change of 10 K is very important. The amount of heat energy that must be added or removed from a reservoir to change the temperature by a certain increment, ΔT, is defined by Equation 11.4. For a 10°C temperature change, a small percentage of the total energy of the reservoir must be transferred. The times required for such effects are readily determined using the relation

$$\Delta t = \frac{c_p M \Delta T}{L} \qquad (11.17)$$

where Δt is the estimated time for the change and L is the total energy loss rate from the reservoir, with other factors being ignored. The information from

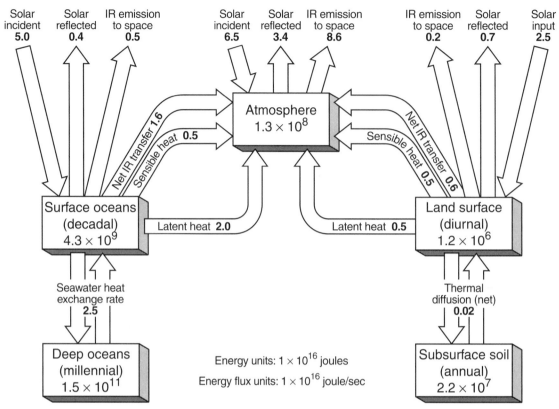

Figure 11.15   A budget for energy in the Earth system related to climate. Each of the major heat reservoirs is indicated by a box, and the amount of energy stored in each reservoir is given. The processes that transfer energy in and out of the reservoirs, as well as the rates of energy transfer, are noted. Energy is expressed in units of 1×10^{16} joule, and energy transfer rates are expressed in units of 1×10^{16} joule/sec (1 watt equals 1 joule/sec). The solar input is divided into components intercepted by the ocean and land surfaces as total direct plus diffuse radiation, and the component intercepted by the atmosphere through absorption or scattering to space. The total solar input is taken to be 14 energy units, which is equivalent to 1400 watts per square meter (the "solar constant") multiplied by the cross-sectional area of the Earth. The source of heat from the interior of the Earth due to radioactive decay is not shown because it is negligible. (Some data adapted from G. W. Paltridge and C. M. R. Platt, *Radiative Processes in Meteorology and Climatology*, New York: Elsevier, 1976. Figure 1.7, p. 15.)

Table 11.2 can be used directly in Equation 11.17. For example, the time required for the surface oceans to cool by 10° if the sun is shut off completely may be estimated as

$$\Delta t = \frac{4.0 \frac{J}{g \cdot °C} \times (3.8 \times 10^7 \times 10^{15} g) \times 10°C}{4.6 \times 10^{16} \frac{J}{sec}}$$

(11.18)

Here, the data are taken from Table 11.2 and Figure 11.15, with appropriate units applied. The net loss of heat from the surface ocean reservoir under these (rather extreme) conditions is, according to Figure

11.15, 4.6×10^{16} joule/sec. The estimated time of cooling is about 1 year (1 year is about 3.2×10^7 seconds). With a small percentage reduction in solar insolation, the ocean surface is therefore likely to cool several degrees per decade.

The land is much more susceptible to rapid temperature change. Applying the same arguments as before, and turning off the sun completely, the time for the land to cool by 10 degrees is roughly

$$\Delta t = \frac{4.0 \frac{J}{g \cdot °C} \times (1 \times 10^4 \times 10^{15} g) \times 10°C}{1.8 \times 10^{16} \frac{J}{sec}}$$

$$= 2.2 \times 10^4 \ sec$$

(11.19)

This cooling time is only about 6 hours! That can happen in one night. If you have ever gone camping on the desert, you have experienced such a fast temperature drop. The ground and surface air can indeed cool off in a few hours. The simple energy balance model is holding up.

11.5.2 ICE: THE COOL RESERVOIR

The energy reservoirs for the climate system are, as you might expect, more diverse the closer we look at the problem. Of course, we do not want to identify every lake and island as a separate potential reservoir, although in a strict sense, each is. On the other hand, there is a major reservoir that we have not identified—glacial ice. The global aggregate of glacial ice and snowpack is referred to as the **cryosphere**. The principal residence for ice on Earth is the Antarctic continent. The Antarctic ice sheets cover an area the size of the United States to an average depth of more than 2 kilometers. That is an incredible volume of ice, amounting to more than 20 million cubic kilometers. Additional volumes of permanent ice are located on Greenland and in a number of mountain glaciers. The Antarctic ice sheets are equivalent in total volume to about 80 meters of depth of the world's oceans. If the ice sheets were to melt, the sea level could rise by roughly that amount. Coastal regions would be inundated. Many of the world's major cities would be flooded.

In the past, massive glaciers have come and gone. During the ice ages (Section 11.6.3), the land in North America, Europe, and Asia was extensively glaciated. The most permanent glaciers, however, have resided on Antarctica, where the ice may be up to 200,000 years old, holding frozen secrets from prehistoric times. For example, air bubbles trapped in the ice have revealed the history of carbon dioxide concentrations over the past 160 millennia (Chapter 12). The ice also exhibits heavy deposits of sulfate and ash following major volcanic eruptions. The temperature of the planet is recorded as well in the oxygen isotopes long ago frozen into the ice.

Ice and snow participate in the Earth's climate system in several ways. As we noted earlier, the water tied up as ice on land represents a substantial fraction of the total surface ocean water. The transfer of this water between glaciers and oceans alters processes that occur in coastal regions, particularly

biological processes that control the carbon cycle (Chapter 10). Ice- and snow-covered surfaces have a high albedo and so increase the reflectivity of the Earth. Imagine what might happen if a substantial fraction of the surface area of the Earth were to become covered with ice. The planetary albedo would increase significantly, and further cooling would result. At some point, the temperature might drop enough for the entire Earth to freeze over, oceans and all. The freezing could be irreversible. The frozen Earth might never thaw, and no life could exist here (at least, not life as we know it). Fortunately, the Earth has never reached that stage of snow and ice coverage that can trigger total freezing. Ice covering the seas, as in the Arctic Ocean, has a strong influence on the energy balance of these regions. A layer of floating ice is, ironically, an excellent insulator against heat loss. The layer of cold ice isolates the ocean water from the atmosphere, preventing efficient latent heat exchange.

The total energy contained in the ice reservoir is about 2.6×10^{25} joules (2.6×10^9 units in Figure 11.15), comparable to the total energy held in the surface oceans. However, the rate of transfer of energy between the ice sheets and the rest of the climate system is much slower than the rate of transfer between the surface oceans and the atmosphere. Thus the effects of the cryosphere must be factored into climate changes occuring over periods of 1000 years or longer.

11.5.3 A COUPLED CLIMATE SYSTEM

The climate system is thus an extremely complex system of energy reservoirs joined together by energy transfer processes. It can be likened to a vast machine with many parts, using sunlight as fuel and spewing heat into space as exhaust. The machine has many moving parts: the winds and clouds, ocean currents, and living organisms, to name a few. The mechanism also has a flywheel—the oceans—to smooth out variations that might cause a rough ride. Like all machines, the climate may require some maintenance. It has hummed along for the past 4 billion years without a checkup (although experiencing a few stalls). But now it may be suffering abuse by overzealous humans. The engine seems to be overheating. We require a reliable diagnosis of the problem, and naturally, we seek the best mechanics available, the modern-day climate scientists. They have studied and probed the intricate parts of the climate machine. They have constructed models of it for computer simulations; the climate machine has been put on a dynamometer and run through its paces. The mechanics are still scratching their heads, though. The mechanism is complicated, and their experience is limited. What are we facing?

In thinking about the climate system, we can distinguish between parts, or "elements," and "processes." The elements represent the actual physical or biological entities that compose the system, and the processes represent the interactions between the elements. In a coupled reservoir model (Section 10.1.2), the reservoirs and materials in the reservoirs are the elements, and the mechanisms that add or remove material or transfer it between reservoirs are the processes. The materials in a reservoir may be transformed into other materials—by chemical processes, for example. In the case of the climate system, energy is treated as an element, as is the sun itself, which represents the source of energy. Energy may exist in different forms, as already noted: thermal,[12] radiative, kinetic, potential. If we wanted to, we could define each form of energy as a separate element of an energy reservoir. Energy is transferred by numerous processes, including the thermal diffusion of heat through a material, the exchange of sensible heat with a material when it is physically transported, and the radiative transfer of energy associated with the emission, absorption, and scattering of electromagnetic waves (or photons) by materials.

Elements of Climate

The number of parts that make up the climate machine is enormous. Of course, if one considers the atmosphere, oceans, and so on as single "elements," or parts, of the system, it becomes much simpler. Think of the atmosphere as, say, the engine, since it converts some of the absorbed solar fuel to the mechanical motion of air and water. But engines themselves have smaller parts. The atmosphere consists of many distinct gases that affect its behavior; it contains aerosols and clouds and various jets and winds; and the temperature changes from region to

region. The shape and distribution of mountains and other landforms, or **orography**, divert the winds in particular directions, like a manifold. The oceans provide heavy doses of latent heat to drive convective cells like huge pistons. Analyzing even one smaller part of the atmospheric engine—water, for instance—requires specifying among other things, the water vapor concentrations as a function of location and time, and the condensed phases of water at various stages in the hydrological cycle, including the distributions of cloud particles. This information, complex as it is, describes only a few of the sub-elements that could be considered. The resolution of elements and their effects might also be refined from global scales, to continent-size regions, to individual clouds, to single water droplets, to water molecules themselves. No one would be foolish enough to subdivide the climate machine into such small parts. But it is possible.

The key components of the climate system are those already discussed: the energy reservoirs, material reservoirs, and biogeochemical cycles described in this chapter and Chapter 10. The chemical compounds critical to climate analysis, from among those discussed in Chapter 10, will be further investigated in Chapter 12. Other specific sub-elements—clouds are an example—were introduced as needed. There is little reason to add details at this point. By now, you should be sufficiently familiar with the most important elements of the climate machine to understand the nature of climate change discussed later in this chapter and in subsequent chapters.

The basic role of solar forcing of the climate system was explained in previous sections. This forcing, the fuel for the climate machine, is an essential element of the climate system. The role of variations in sunlight, climate forcings, and the processes that cause variations in forcing are discussed in Section 11.6.

Processes of Climate

The flow of energy among different elements of the climate system can become quite complex. Details concerning these processes extend beyond the requirements for understanding how the climate system works. According to previous sections, the radiative processes (scattering, absorption, emission) dominate the climate system. Physical processes (and material properties) also are important. The conduction of heat in soils and the capacity of water to store heat are significant. The latent heat properties of water

---

12. Here, thermal energy refers to the energy of motion of a substance's molecules—that is, the energy of thermal agitation. Thermal radiation, or longwave radiation, is just that, radiation. Strictly speaking, thermal energy is composed of both the instantaneous kinetic energy of the molecules' motion and the potential energy instantaneously stored in the molecules in the form of vibrational energy.

are critical. The microphysical behavior of cloud particles and the optical properties of clouds and aerosols are crucial as well. The dynamics of the atmosphere is fundamental to climate and encompasses a range of processes from microscopic turbulence to convection, to cyclones, to global jets and waves.

On a much larger scale, astronomical processes contribute to the Earth's climate and climate change. The sun and the processes that control its output of energy are critical. The brightness of the sun and sunspots are frequently discussed in relation to climate. The orbit of the Earth and small perturbations in it over time have been connected to ice ages. The paths of asteroids and comets through the solar system and the mechanics of impacts of these bodies with the Earth may explain periodic climatic disasters in Earth history. Even the sun's wanderings through the Milky Way galaxy have been proposed as important to the long-term climate.

Chemical processes also play a role in the climate system. The concentrations of greenhouse gases are controlled, to varying degrees, by chemical processes (Chapter 12). Ozone is the most chemically active of all the greenhouse compounds under consideration, and water vapor is the least active. A variety of mechanisms may affect the amounts of climate-active compounds in the atmosphere, including photochemical reactions, biochemical and biogeochemical processes, interactions with clouds, geochemical transformations, and industrial and technological processes. In the last case, exotic and unnatural compounds such as chlorofluorocarbons may be produced and released into the environment, or massive quantities of otherwise natural compounds such as sulfur dioxide and nitric oxide may be emitted.

Biological processes are a key driver of the climate system. Several key greenhouse gases, such as carbon dioxide, methane, and nitrous oxide, are controlled by biological processes. (Chapters 10 and 12 provide more details.) Photosynthesis and respiration are the fundamental life processes on the planet, and they control surface environmental conditions that affect the absorption and transfer of energy in terrestrial and marine components of the climate system. Biological processes have also been invoked as a potential means of modifying or controlling certain components of the climate system—for example, the use of fertilizers to stimulate phytoplankton to absorb excess carbon dioxide from the atmosphere (Chapter 14).

11.6 Causes of Climate Change

The climate is in a constant state of change, from minute to minute and century to century. This statement might seem to contradict an earlier point emphasizing the constancy of climate over long time spans. Indeed, we think of the climate as being relatively stable over the entire period of human civilization, following the last ice age. The apparent inconsistency is resolved by defining the difference between climate variability and climate stability. The long-term reliability of the climate is called *climate stability*. A stable climate is relatively constant; the average temperature of the Earth has varied by only 1° or 2°C over thousands of years. Climate variability is *bounded*. That is, from year to year or century to century, we can depend on the climate's falling within a certain range of temperature. Accordingly, climate stability is an extraordinarily important characteristic of the climate system.

From previous sections, we understand why the climate is stable. The energy input, energy reservoirs, and energy loss rates for the climate system are well defined and fixed over long time spans. Nevertheless, the climate remains variable within a range defined by the stability of the climate system. An analogy can be drawn with the stability of a long flexible pointer held by a lecturer. No matter how hard she tries to steady the pointer, the end will shake continuously around the point of interest. The situation is stable, however: The pointer does not fly off into space, but it merely wobbles back and forth within a small range. That is the stability range for the person and pointer. The lecturer can point reliably within that range (depending on how nervous she might be at the time). But she cannot identify a point more accurately than the variability of the pointer allows. The oscillation of the pointer is controlled by many complex processes, including the steadiness of the lecturer and the flexibility, weight, and length of the pointer.

In the following sections, we review the most important causes of climate change: the nature of climate variability and the sources of this variability, both external and internal to the Earth. Climate also changes in response to variations in the atmospheric concentrations of several greenhouse gases, including CO_2, CH_4, N_2O, and the CFCs. We look at the resulting climatic variability and drift in Chapter 12.

11.6.1 CLIMATE VARIABILITY

Climate variability is the noise in the climate system. It is the change in climate from the average or mean over a time interval of interest. The interval may be a year, a decade, a century, a millennium, or longer. The climate is usually measured by the average surface temperature of the Earth. That temperature is presumed to be known on a time scale that is much shorter than the climatic interval of interest. For example, if we were interested in the climate of the past century, we would want to have, at a minimum, the annual mean temperatures of the Earth during the past 100 years. On the other hand, if we were considering the climate of the holocene (the past 10,000 years or so), average decadal or centennial temperatures would be fine for most purposes. Of course, when averaging the climate over a certain time increment, we lose the variability on time scales shorter than that increment. Often we have no choice because average temperatures, deduced from proxy data (Chapter 4), are sparse and represent long-term averages.

The temperature variations associated with climate variability for recent times are outlined in Section 12.4. Here we discuss the causes and nature of such variability in a more general way, in order to become familiar with the concept of climate variability.

Variations in Forcing

The climate system is "forced" by sunlight hitting the Earth; the climatic response to sunlight depends on the planetary albedo, the greenhouse effect, and a number of other factors. Changes in all these factors cause variations in climate. The importance of climate variability associated with changes in a specific climatic factor depends on the importance of that factor in determining the overall climate. There also should be a match between the rate of change in the forcing factor and the rate of response of the climate system. If these time scales are similar, the climate response may be amplified, or *resonate*. Usually, the response time of the climate system is determined by the residence times of solar energy in the principal heat reservoirs (Section 11.5).

Resonance occurs when a reservoir stores and loses energy cyclically. For example, energy may be delivered to the reservoir only periodically, at certain specific frequencies. Take, for example, a particular climate-forcing factor—the distance of the sun from the Earth. This distance exhibits small variations over time that can be decomposed into simple harmonic, or oscillatory, components. These oscillations can be described by a spectrum of frequencies, or **periods**, and amplitudes. The period is essentially the time span between peak values during an oscillation, much like the separation of the crests in a wave (Section 3.2). The period is thus the repetition time for the variation. The Earth's orbital parameters, as an example, oscillate over periods of about 20,000, 40,000, and 100,000 years. The number of dark spots on the sun (sunspots) oscillates over a period of about 11 years.

If the period of the forcing is shorter than the response time of a reservoir, the reservoir cannot follow the variation from peak to peak as time passes. Daily variations in sunlight, for example, can be tracked by the land-surface energy reservoir and by the atmospheric heat reservoir, but not by the surface-ocean heat reservoir. In other words, the land surface and air temperatures show a distinct diurnal variation, whereas the ocean surface temperature does not. Variations in forcing with a time scale of a year or longer—the annual seasonal cycle in solar insolation, or the near-decadal solar sunspot cycle—can be followed by the surface oceans, but not the deep oceans or glaciers. The forcing associated with ice age cycles lasting thousands of years, on the other hand, can be seen in the deep oceans and in major ice sheets.

Chaos: Internal Variability

Chaos is a term recently applied to describe the seemingly random behavior exhibited by a physical system such as the climate system. Chaotic behavior is not, however, actually random behavior, which in one sense is unpredictable. Rather, chaotic behavior is more constrained than random behavior and is more predictable.

The climate system of Earth is a highly complex machine with a large number of possible "states" in which it may exist at some time. Many of these states are transient, or unstable, and last for only brief spans of time. A small subset of the states are identified as "stable" climatic states, in which the climate seems to persist for long periods. Consider a football player sprinting in short bursts and sitting on the bench between sprints. Sitting on the bench is a comfortable stable state for the player, whereas sprinting is an unstable state that cannot be sustained for more than a few seconds.

Considering all the variables (elements) and transformations (processes) that compose the climate system, the number of *possible* stable and unstable climatic states is enormous. If the climate system were actually a random system, any one of these states might be equally likely to occur. Indeed, the climate would wander hither and yon among this multitude of states. In reality, there appear to be only a small number of climatic states that are actually stable and that have dominated the history of the Earth. The reason for the climate's high stability is not fully understood. While the climate moves among the limited number of stable states in a chaotic manner, the overall climate drift remains bounded during climate shifts. Note that within each stable climatic regime, the climate is still variable over a significant range (in other words, a stable climate state consists of a large ensemble of states with similar characteristics, and the climate drifts among these similar states). The variability in climate generally becomes more pronounced as the spatial and temporal scales under consideration are reduced (that is, regional climate is more variable than global climate, and so on).

Chaos is generated by the interactions among different elements of a coupled physical system. To understand climatic chaos, you would need to analyze the nonlinear mathematical equations that describe the coupled Earth system. That approach is not practical here. Instead, a frivolous example will suffice to illustrate the principle of chaotic behavior.

The crowd of people on a subway platform at rush hour seems to act chaotically. Individual persons mill about for better positions, pushing and shoving to get on and off trains. Nevertheless, the individual motions are not random. Bodies are not careening in all directions at once. Instead, there is order in the overall movement—in particular, the crowd moves into and out of the subway cars in organized streams. If the collection of commuters on the platform were actually behaving randomly, each would be moving independently at every moment, caroming off one another and the walls like gas molecules trapped in a bottle. Fortunately, there is some order to the action: The aggregate movement is bounded; no one ends up on the tracks; and (almost) everyone gets home for dinner. Chaos is thus limited, or directed, behavior that has some aspects of randomness. Chaotic systems, for example, typically have more random smaller-scale internal structures (such as local weather fluctuations within a stable global climate state).

Climatic chaos on a large scale is manifested by fluctuations in the average global temperature over time. Earlier, we noted that such fluctuations have been bounded within a few degrees of the Earth's long-term mean temperature. Over time spans of millions of years, the climate can shift between quite different climate regimes (for example, from a warm climate to an "ice age"). Each regime, although internally chaotic, is also limited in its range of variability. On smaller spatial and temporal scales, climate becomes more chaotic and unpredictable. Locally, such chaotic behavior constitutes weather. Chaotic behavior of the climate system and climate (and weather) predictability are closely related issues.[13] Chaos obviously implies the potential for predictability within certain bounds of variability. For example, forecasting cyclic climatic events such as El Niño in the tropical Pacific Ocean (Section 12.4.2) may be possible with modest confidence. However, for finer predictions of local day-to-day weather changes, internal randomness precludes regular, accurate, long-term predictions. Global average temperature is the least chaotic and most predictable climate parameter; weather is the most chaotic and least predictable.

The climate system can exhibit internal oscillations that are not specifically forced by external factors. These oscillations are caused by linkages between different physical elements composing the climate system. The atmosphere and tropical Pacific Ocean, for example, are mechanically joined in such a way that wind and water currents periodically create large pools of warm surface water over the eastern Pacific Ocean. This is a major factor contributing to the important El Niño Southern Oscillation (ENSO) phenomenon, which periodically triggers anomalous and destructive weather around the Pacific Ocean basin. Naturally, the energy driving El Niño and other internal oscillations of the climate system is derived from sunlight absorbed by the Earth.

Additional internal factors that contribute to climate variability (chaos) include atmospheric

---

13. Accurately predicting the weather has been a dream of meteorologists for the past century. Jacob Aall Bjerknes (1897–1975), a Norwegian-American meteorologist, developed the early theory of weather prediction based on mathematical analyses of wind systems. He also identified the jet streams after studying reports of World War II bomber pilots who had met strong head winds when crossing the Pacific. In 1961, Edward N. Lorenz (b. 1923) discovered "chaos" in atmospheric dynamical systems, establishing an inherent limitation on the predictability of weather and initiating a new field of study focusing on chaotic systems.

convection induced by surface heating, shifting ocean currents, mountain diversions of winds, cloud formation and dissipation, sporadic volcanic eruptions, changes in snow cover and soil moisture, and changes in countless other parameters. Variability results when these factors, acting over different spatial and temporal scales, modify the storage of energy in reservoirs, the transfer of energy between reservoirs, and transformations between different forms of energy (such as latent heat to kinetic energy) within reservoirs. Thus winds can be diverted by the sudden appearance of strong convection over warm moist bodies of water. The convection is not, of course, precisely predictable. The winds then carry moisture to other regions where snow may fall, increasing the surface albedo, which, in turn, reduces the local surface heating. And so on. The predictability of such a lengthening chain of events grows less and less viable. Soon, predictions are lost in a fog of uncertainty.

Although complex and highly unpredictable, the climate system is not completely intractable. We can be fairly sure, for example, that snow will not fall in the Sahara in July or that Himalayan glaciers will remain frozen for the next decade. Climatic uncertainty is bounded, just as is climatic variability. Averaging the climate parameters over time and space reduces variability and enhances predictability. Moreover, if known internal oscillations and external forcings are explicitly taken into account, forecasts of climate change can be made reasonably accurately (Sections 12.4.3 and 12.4.4).

11.6.2 SOLAR VARIABILITY: EXTERNAL FORCING

The climate system is forced externally by physical elements that lie outside the Earth and its atmosphere. The sun provides the basic forcing of the climate system. Solar insolation is one of the principal energy streams for the climate machine. There are other potential external forcing factors: The lunar tides in the oceans and on land may be responsible for geologic activity related to climate change.[14] The impacts of large meteors and comets with the Earth have caused major environmental and climatic perturbations. Over longer times, explosions of nearby

14. The lunar tides are caused by the moon's gravitational attraction to the Earth and are most obvious in the daily tides of the oceans at the seashore. The effect of the moon's gravitation varies around a basic period of about 19 years.

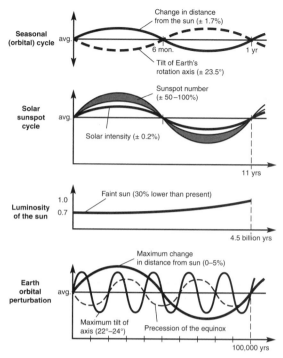

Figure 11.16 Some of the causes of variability in the intensity and distribution of sunlight illuminating the Earth. The origin of the variation is indicated to the left of each panel. The magnitude of each perturbation is sketched along a time axis, with the time scale for the change given; these range from 1 year to several billion years. For the Earth's orbital perturbations, the three principal "Milankovitch" cycles are shown.

stars may flood the Earth with radiation and induce global changes in the environment.

External forcing factors that contribute to changes in the climate are dominated by variations in the sun. The main causes of solar variations are shown in Figure 11.16. The eccentricity of the Earth's orbit around the sun (the degree to which the orbit differs from a perfect circle, or is elliptical) and the tilt, or obliquity, of the Earth's axis of rotation with respect to the ecliptic control the seasonal variations in solar illumination and surface temperature (Sections 11.2 and 11.3.1). If no other variations occurred in the sun, the climate would not vary much over time scales longer than a year (except for chaotic internal variations in the climate system).

Sunspots

Everyone knows there are blemishes, or spots, on the sun's surface, or *photosphere*. A simple astronomical experiment in which a telescopic image of the sun is cast onto a sheet of paper makes these sunspots visible (remembering that you should *never* look directly at the sun). The number of sunspots exhibits a cycle with a period of slightly longer than 11 years, as shown in Figure 11.16. The number of sunspots reflects the degree of solar magnetic activity, which varies from cycle to cycle. The more sunspots there are, the greater the solar activity and the more intense the output of the sun will be. Recent satellite data indicate relatively minor variations in solar intensity with sunspot cycle (roughly 0.1 percent). However, over historical time spans, the number of sunspots has varied more widely. In particular, from about 1660 to 1720, there were very few sunspots; this is called the *Maunder minimum.* At that time, the climate was also unusually cold, about 0.5°C colder than the longer-term mean, the so-called little ice age. There were earlier periods of low solar activity from about 1420 to 1540 (Spörer minimum) and 1280 to 1340 (Wolf minimum). The sunspot data become less reliable as history recedes, however.

Over the years, climate researchers and other scientists have claimed strong correlations between the number of sunspots and other common indicators of climate. Examples include the depth of Lake Victoria in Africa, heavy rainfalls in Europe, the frequency of global lightning, the strength of cyclones over North America, the number of tropical hurricanes, occurrences of thunderstorms over Asia, intensities of aurorae, the number of rabbit pelts collected in Canada, and the quality of wines from France (perhaps only the last is of real interest to most of us). At one time or another, sunspots have been correlated with literally everything under the sun. As you might suspect, many of these correlations are imaginary, based on incomplete or faulty data or overenthusiastic interpretation. Moreover, statistical correlations themselves do not prove cause and effect. In most cases, we have no way to explain the effects of sunspots on particular terrestrial parameters. In the past two decades, sun-weather theories have been less popular.

There are definite effects of sunspot (solar) activity in proxy climate records. Most prominent are the carbon-14 data. Carbon-14, or ^{14}C, is a radioactive isotope of the carbon atom with two extra neutrons (Sections 3.1 and 8.2.1) and a decay half-life of about 5760 years. Carbon-14 is produced when galactic cosmic rays (or fast streams of particles originating in supernovae) bombard the lower atmosphere; the cosmic-ray particles are so energetic that they cause

nuclear reactions when they collide with air molecules. Because the galactic cosmic rays are electrically charged, they have difficulty penetrating the magnetic fields of the sun and the Earth, and so they seldom reach the ground. In general, the greater the sun's activity is, the more effective the shielding of cosmic rays will be. Accordingly, as solar activity increases, the production of carbon-14 decreases. The radioactive carbon is taken up by plants and absorbed in the oceans, providing a proxy record of solar activity in tree rings, fossils, and sediments.

The solar activity deduced from carbon-14 measurements is in good agreement with sunspot observations. The ^{14}C data allow the solar activity record to be extended about 40,000 years into the past. It has been observed that temperature and carbon-14 concentrations are anticorrelated; that is, when temperatures are high, ^{14}C concentrations are low, and vice versa.[15] This is consistent with the anticorrelation of solar activity and ^{14}C. However, the absolute variations in the intensity of sunlight (or the solar constant [Section 11.3.1]) seem too small by almost a factor of 10 to explain the recorded temperature variations.

The "Faint Sun" Paradox

Because the sun is a star, its luminosity has experienced a long evolution. By comparing the sun with comparable stellar objects at different stages of their evolution, astronomers have pieced together the likely history of the sun's brightness. This is illustrated in Figure 11.16. Stars like the sun are born cool and glow brighter over time before burning out (like some late bloomers). The solar constant may have been as much as 30 percent smaller during the early history of the Earth. The implications are significant. From our climate model (Equation 11.14), it follows that with other factors constant, such a decrease in the solar constant would lower the effective radiation temperature of the Earth to about 233 kelvin. Even with the standard water vapor greenhouse effect, the average surface temperature would have been less than 270 K, below the freezing

15. The temperature is obtained using measurements of the long-lived oxygen isotope, ^{18}O, relative to the normal form, ^{16}O. These isotopes are fractionated because of slight differences in the physical and chemical properties of molecules with different isotopic composition. The isotopic fraction of ^{18}O in the H_2O frozen as glacial ice, for example, provides information about the temperature that existed at the time the water condensed as snow that fell onto the ice pack.

point of water. How is it, then, that the Earth did not freeze over early in its history, increasing the planetary albedo and ensuring a continuous frozen state (Section 11.5.2) until the present time? The fact that the Earth is not now frozen when it would be expected to be is referred to as the *faint sun paradox*.

Geologic records clearly show that the Earth has never been completely frozen. Life began its long evolutionary journey more than 4 billion years ago; the planet could not have been frozen then. One possible explanation for this paradox is a greater greenhouse effect. Evidence has been developed that the carbon dioxide concentrations in the early atmosphere were considerably higher than the current 350 parts per million by volume. How high is uncertain, although the cause is logical. Carbon dioxide is a common volatile on the Earth, as it is on Venus and Mars (Section 4.1.1). On Venus, a runaway greenhouse effect during the first billion years or so prevented its oceans from condensing and absorbing CO_2 from the atmosphere (Section 11.4.4); Venus remains a CO_2-laden hothouse planet today (Section 11.3.3).

Earth, on the other hand, is farther from the sun and was originally cool enough for oceans to form, although not all the carbon dioxide liberated from the Earth's interior could be absorbed immediately by the oceans. The evolution of life eventually helped in that task. In the early years of the Earth, when the sun was faint, CO_2 concentrations exceeding current levels but well below those on Venus would be expected. The paradox is resolved (perhaps).

11.6.3   Ice Ages

The "ice ages" have been, and remain, a mystery. During these periods of unusually cold climate, massive glaciers of ice pushed down from the north deep into North America and Europe; the tropics were noticeably colder; the rain forests retreated; and the primitive *Homo sapiens* struggled for survival against the elements. We are currently in an interglacial period of relatively warm climate, the Holocene, which has lasted about 10,000 years. Human civilization and agriculture have developed to their present states during this epoch. The last major glacial cycle preceding the recent warming lasted more than 100,000 years (the Würm ice age). Before that, the Riss (~200,000 years ago), Mindel (~450,000 years past), and Günz (~600,000 year ago) ice ages are

recorded in geologic formations (all four recent ice ages are named for river valleys in Europe). During these major cooling events, the global average temperature decreased by 5° to 10°C. The entire Pleistocene epoch, extending about 2.5 million years into the past, has seen climatic cooling and a series of ice ages of varying duration. The explanation for even the most recent major cooling event about 100,000 years ago is not clear.

In more recent times, during the Holocene, there have been several "neoglaciations." The "little ice age" extending from about 1450 to 1890 is the most recent. Two other prominent but minor ice ages occurred about 2500 and 4500 years ago. In these events, the global average temperature might have dropped by roughly 0.5°C, one-tenth of the decrease for a full-blown ice age. The causes of little ice ages, as mentioned earlier, may be related to solar variability (solar activity and sunspots), although this relationship is very uncertain. Nevertheless, it is natural to ask whether the major ice ages are also manifestations of solar variability, but on longer time scales.

Milankovitch Cycles

The cause of ice ages has been a hotly debated subject for more than a century. We are interested in knowing whether another cycle of strong cooling is imminent (in, say, 2000 years, based on the lengths of previous interglacial times, a mere heartbeat in geologic terms). One idea that has taken hold recently attributes these major climatic variations to the subtle variations in the orbit of the Earth around the sun. This is referred to as the astronomical theory of the ice ages. The gravitational pull of the moon and planets, particularly Jupiter, on the Earth and its equatorial bulge lead to slight fluctuations in the Earth's motion. The orbital perturbations, indicated in Figure 11.16, cause changes in the distance from the sun (eccentricity), in the tilt of the axis of rotation (obliquity), and in the precession of the equinoxes. The precession of the equinoxes is the apparent revolving motion about the sun of the points in the ecliptic where the equinoxes take place. The equinoctal movement results from a combined precession of the Earth's axis and the rotation of its eccentric orbit around the sun. The repetition periods associated with the eccentricity, tilt, and precession are about 100,000, 41,000, and 22,000 years, respectively.

Many of the original ideas linking variations in the Earth's orbital parameters to the ice ages can be attributed to James Croll.[16] Croll calculated the variation in the eccentricity in the Earth's orbit over the past 3 million years using astronomical theory, and he associated changes in eccentricity with variations in the intensity of sunlight at different seasons. He concluded that the occasionally large decrease in winter isolation was the key effect that triggered an ice age. Croll recognized a problem, however: The absolute changes in solar insolation were so small that they would not directly cause the large temperature variations deduced from the geologic record. He suggested that the orbital perturbations were a "triggering" mechanism for other climate parameters. But in the face of mounting evidence against it, Croll's ideas were discounted.

The astronomical theory was revived in modified form by Milutin Milankovitch in 1924.[17] Whereas Croll had quantified the variations in orbital parameters, Milankovitch determined quantitatively the variations in solar insolation with latitude and time. Milankovitch also teamed with two of the leading scientists of the time: Alfred Wegener, a geologist who introduced the idea of continental drift, and Wladimir Köppen, a climatologist who created world maps of climate zones to explain the distribution of vegetation over the Earth. Wegener and Köppen guided Milankovitch's analysis and interpretation of the geologic and climatic records. Milankovitch, under advice from Köppen, also focused on the summer season as the key to the ice ages and emphasized modulations of summer insolation at the higher latitudes. He discovered that only at certain limited times during the history of the Earth had the combination of orbital perturbations acted together to minimize the solar insolation at the latitudes where ice sheets had formed. These minima coincided with the four major periods of glaciation during the past 650,000 years.

In recent years, many independent climatic data sets have been shown to contain components with periods similar to the orbital variation periods of 22,000, 41,000, and 100,000 years. These data have been collected from ice cores and ocean sediments and include variations in oxygen isotopes and pollens for the past million years. They strongly reinforce the astronomical theory of the ice ages. The Milankovitch cycles have been shown to represent the rhythm of the ice ages. Interestingly, the actual cause of the major ice age glaciations remains unresolved. The most prominent change in the total volume of ice in the cryosphere (Section 11.5.2) has a 100,000-year cycle. The change in insolation, however, is too small to explain directly the effect on the ice sheets. The combined internal climatic variability of the atmosphere, oceans, biosphere, and cryosphere is apparently responsible. We simply do not possess adequate knowledge to determine, once and for all, the cause of the ice ages.

11.6.4 VOLCANIC ERUPTIONS

During major volcanic eruptions, huge quantities of gases and particles are blown high into the atmosphere. Dark clouds of ash stretch for hundreds of kilometers from the volcano. Particles falling out of the clouds cover large regions with a fine gray mineral dust. Near these eruptions, day can be turned into night. History has been punctuated by great volcanic eruptions, and we can expect more such eruptions to be a part of our future.

Tambora and Frankenstein

Mount Tambora in Indonesia erupted violently in April 1815, and the following year, 1816, was highly unusual around the world. It is remembered as the "year without a summer." Farmers in the northeastern United States could not grow crops because sporadic frosts throughout the spring and summer killed young plants. In the sparsely settled western United States, the strange weather was recorded as frost damage to tree rings among the hearty

---

16.   James Croll (1821–1890), a self-taught Scottish naturalist, held jobs as a mechanic, carpenter, millwright, insurance salesman, janitor, and maker of "electrical devices for the alleviation of bodily aches and pains." Croll developed the earlier concept of the French mathematician, Joseph Adhemar that the eccentricity of the Earth's orbit produces ice ages. Croll refined the concept by incorporating the precessional effects and published his new theory of the ice ages in 1864. See J. Imbrie and K. Imbrie, *Ice Ages: Solving the Mystery* (Cambridge: Mass.: Harvard University Press, 1979).

17.   Milutin Milankovitch (1879–1958), a Yugoslavian mathematician, refined Croll's theory, by predicting the effects of orbital changes on solar insolation. He carried out the first laborious calculations of variations in solar insolation associated with the combined orbital perturbations over the past 650,000 years. These calculations, which required several months of nonstop computation by hand, were first published in 1924 by two German researchers and formally published in 1938. Today, a supercomputer could solve Milankovitch's problem in a few minutes.

bristlecone pines of that region. In Europe—suffering in the shadow of the Napoleonic Wars—famine stalked the populace as crops failed under wintry skies. Lord Byron wrote his prophetic poem "Darkness" during the grim summer of 1816, and the story *Frankenstein* was penned soon thereafter by Mary Wollstonecraft Shelley. Anecdotal evidence from China testifies to strange weather and poor agricultural output there as well. But no one at the time connected the Tambora eruption with the strange climatic events.

Benjamin Franklin, an American scientist, statesman, and polymath, was the first observer of natural events to describe the possible impact of volcanic eruptions on the climate.[18] Following the particularly severe winter of 1783/1784, Franklin wrote in 1784:

> During several of the summer months of the year 1783, when the effects of the sun's rays to heat the earth in these northern regions should have been the greatest, there existed a constant fog over all Europe, and great parts of North America. This fog was a permanent nature; it was dry, and the rays of the sun seemed to have little effect toward dissipating it, as they easily do a moist fog arising from the water. They were, indeed, rendered so faint in passing through it that, when collected in the focus of a burning glass, they would scarce kindle brown paper. Of course, their summer effect in heating the earth was exceedingly diminished.
>
> Hence, the surface was early frozen.
>
> Hence, the first snows remained on it unmelted, and received continual additions.
>
> Hence, perhaps the winter of 1783–1784 was more severe than any that happened for many years.
>
> The cause of this universal fog is not yet ascertained. Whether it was adventitious to this earth, and merely a smoke proceeding from the consumption, by fire, of some of those great burning balls, or globes, which we happen to meet with in our course round the sun, and which are sometimes seen to kindle and be destroyed in passing our atmosphere, and whose smoke might be attracted and retained by our earth; or whether it was the vast quantity of smoke, long continuing to issue

> during the summer from Hecla, in Iceland, and that other volcano which arose out of the sea near the island, which smoke might be spread by various winds over the northern part of the world, is yet uncertain.
>
> It seems, however, worthy the inquiry, whether other hard winters, recorded in history, were preceded by similar permanent and widely-extended summer fogs. Because, if found to be so, men might, from such fogs, conjecture the probability of a succeeding hard winter, and of the damage to be expected by the breaking up of frozen rivers in the spring; and take such measures as are possible, and practicable, to secure themselves and effects from the mischiefs that attend the last.

In this brilliant piece of scientific reasoning, Franklin essentially defined the volcanic effect on climate. He noted the connection between the harsh winter and the dearth of sunlight the previous summer. This he ascertained using a magnifying glass, which can be used to focus sunlight to a point of such great intensity that paper will ignite when illuminated by it. In 1783, paper would not ignite, indicating a diminishment in the intensity of sunlight at the ground. Franklin also identified the "dry fog" as the culprit. He pointed out, correctly, that normal fogs are dissipated by sunlight as the air warms and the water droplets evaporate. The dry fog was not made of water, however, and so did not dissipate. So Franklin deduced that the haze was associated with volcanic activity in Iceland. In this regard, he perceptively concluded that the volcanic "smoke" could be carried by winds over the entire Northern Hemisphere.

Volcanoes and Climate History

Early in this century, Charles Greeley Abbott was prominent in reviving interest in the role of volcanoes in climate change. He was also involved in setting up a project for the Smithsonian Astrophysical Observatory to measure the solar constant. At that time, measurements had to be made from the ground, as satellite platforms were not available. Because the atmosphere interferes with the penetration of solar radiation to the Earth's surface, corrections were required. To alleviate this problem, the actual measurements were taken at mountain observatories on very clear days. Nevertheless, the ground-based data could pin down the solar constant to within about 1 percent.

When making these measurements, Abbott noticed that during certain otherwise clear days, the

---

18.   Benjamin Franklin (1706–1790) (pseudonym, Richard Saunders) was an American statesman, supporter of the American Revolution, and contributor to the Declaration of Independence and U.S. Constitution. Franklin was a scientist, inventor, author, and diplomat. He dabbled in many fields, inventing the Franklin stove and bifocal eyeglasses, experimenting with electricity by flying a kite in a thunderstorm, inventing the lightning rod, writing *Poor Richard's Almanack*, and acting as a U. S. diplomat in Europe. Franklin helped establish the first public fire departments, lending libraries, and post offices. He is most revered as one of the founding fathers of the United States and an architect of American democracy.

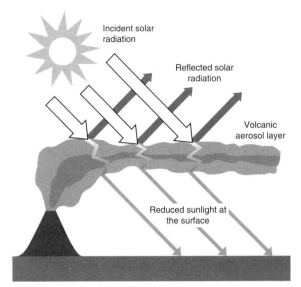

Incident solar
radiation

Reflected solar
radiation

Volcanic
aerosol layer

Reduced sunlight at
the surface

Figure 11.17 The effect of volcanic eruptions on the radiation balance and climate. The layer of particles generated in the stratosphere by a large eruption is seen to reflect sunlight, increase the Earth's albedo, and reduce the heating of the surface.

sunlight was dimmed by a slight haze in the upper atmosphere. The haze was exactly Franklin's volcanic dry fog. Abbott became interested in the nature and effects of these volcanic particles and used his observations in an attempt to correlate solar radiation measured at the ground (presumably affected by volcanic dust) with average surface temperatures and other climatic parameters. In 1883, the powerful eruption of Krakatoa in Indonesia had produced a worldwide particle cloud that caused spectacular twilight glows (Section 3.2.3) and caught the attention of geologists and other scientists. For the first time, a comprehensive study was made of a major volcanic event and its worldwide effects. Using such historical records of volcanic eruptions, Abbott successfully inferred a relationship between volcanic eruptions and climate cooling (by perhaps a few tenths of a degree Celsius). Volcanic eruptions were established as a significant internal source of climate variability.

Figure 11.17 illustrates the basic impact of volcanic eruption clouds on the radiation balance. Major eruptions produce widespread layers of aerosols, or fine particles, in the stratosphere, which efficiently scatter sunlight and enhance the Earth's albedo. The climate cools as a result of the increased albedo (Sections 11.3.2 and 11.6.5). Originally it was believed that the volcanic aerosols were composed primarily of mineral ash particles ejected by the

volcano. Large amounts of ash are certainly emitted and settle over vast regions. But most of the ash falls out within a few months, whereas volcanic aerosols can linger in the stratosphere for 2 or more years. It was the advent of high-altitude aircraft and balloon sampling of the stratosphere in the 1950s and 1960s that led to the discovery of the true nature of volcanic particles. Carl Junge, a German chemist, collected samples of the background stratospheric aerosols in 1961 and found that they were composed of sulfur, in the form of sulfates. Today, the stratospheric aerosol layer is often referred to as the **Junge layer**.

Soon it was also apparent that the long-lived volcanic aerosols that spread over the Earth and created spectacular sunsets and climate change were composed of sulfates, not mineral grains. Volcanoes emit copious amounts of sulfur dioxide, and powerful eruptive volcanoes with towering columns reaching 30 kilometers into the atmosphere inject a large amount of their SO_2 directly into the stratosphere. Much in the same manner that sulfur dioxide emitted by power plants is converted to sulfates in the troposphere, volcanic SO_2 can be oxidized into sulfuric acid vapor in the stratosphere (Section 3.3.4). In this case, the sulfuric acid condenses into a haze of aerosol droplets so small that they do not fall out of the stratosphere for a year or more. During this time, the aerosols are dispersed over the entire globe and affect the global radiation balance. The direct climatic impact of a volcanic eruption is, accordingly, limited by the (relatively) short residence time of the aerosols in the stratosphere.

Hubert H. Lamb, a British climatologist, carried forward the study of volcano-induced climate variability and its effects on human activities. He found that during recorded history, the "year without a summer" and similar major climatic and societal perturbations followed major eruptions. Lamb developed a comprehensive historical database to study these volcanism-climate relationships. He invented the **dust veil index** to calibrate the potential impacts of eruptions. Lamb's work has been extended and reinforced by many scientists. Nevertheless, uncertainties persist concerning the extent of volcanic effects, particularly the patterns of meteorological impacts. The decline in average global surface temperatures the year following a major eruption is roughly 0.5°C or less. A temperature change that small lasting for such a short period of time is difficult to extract accurately from the historical temperature record. There appear to be clear volcanic signals in the temperature record, but the details of the climatic response have not been fully resolved.

The primary effect of a volcanic aerosol layer is to enhance the solar albedo of the atmosphere-surface system. The albedo effect as a contributor to climate variability is discussed in the following section. (In Section 14.3.3, we look again at the effects of volcanic eruptions on the albedo and climate in relation to global environmental engineering.).

11.6.5   THE ALBEDO EFFECT

The variability in climate can be caused by changes in the Earth's albedo. It is easy to see from Equation 11.14 that a change in albedo is equivalent to a change in solar forcing. Because of the greenhouse effect, changes in albedo and solar forcing are not exactly equivalent, but that is a detail at this point. The elements of the climate system that contribute most to the albedo are clouds, surface ice and snow, and aerosol haze. In the previous section, we discussed the albedo effect of volcanic aerosols in causing short-term climate cooling. Almost certainly, variations in the other albedo factors are introducing chaotic fluctuations into the climate.

The extent of and variations in global cloud cover are not well understood. Why is the Earth always about 50 percent covered by clouds? How do the relative distributions of high and low clouds vary, and why? How will clouds respond to global warming or cooling? What is the day-to-day variability in cloud properties worldwide? Clouds play such a central role in the climate system, yet we know very little about them on a global scale. A 1 percent change in cloud cover could be more significant in climatic terms than the variation in solar luminosity over the past two centuries.

The surface of the Earth is not very reflective. The oceans have an albedo of only about 0.1. Regions of dense vegetation have a somewhat higher albedo of perhaps 0.2. Deserts and dry lands can have an albedo of 0.3. Ice and snow, though, have higher reflectivity, with an albedo of 0.7 or more in the case of fresh snowfall. The fraction of land and oceans covered with ice is therefore important factor to the global average albedo. In colder climates, a greater cover of snow and ice increases the surface albedo. In fact, as we noted earlier, the increased albedo reinforces the cooling in a positive-feedback process. In colder climates, the snow and ice margins migrate

toward the lower latitudes and down mountain slopes. This enhances the cooling and promotes further migration. The system is not unstable, however, since even during the most severe glacial periods, the ice margins eventually retreated again. Nevertheless, interglacial periods have been much shorter (approximately 10,000 to 20,000 years) than glacial periods (roughly 100,000 years), which suggests that icy climates are resistant to warming.

When climate cools and the cryosphere expands, cloud cover and land vegetation also change. In ice age climates, deserts spread and the tropics become drier. The complex interconnection among all these elements and the net effect on the albedo are difficult to predict. Models have been developed and tested against observations. The applications of models to the problem of climate change are discussed in Chapter 12.

Smoke, Dust, and Haze

The atmosphere, at various times and in different places, contains substantial quantities of particulates in the form of dust, smoke, and haze. Dust composed of fine soil grains is typically raised from the surface by winds and mechanical agitation, particularly over arid regions. Smoke is a by-product of combustion, including natural wildfires, burning associated with agricultural activities, and emissions from industrial facilities and vehicles using fossil fuels. Haze is composed of other materials that are soluble in water, such as sulfates and nitrates, which have natural and anthropogenic sources. In addition to volcanic eruptions, notable aerosol palls are associated with wind-whipped dust storms, large forest fires, and urban air pollution.

The aerosols can affect the climate both directly and indirectly. Particles directly affect radiation in the atmosphere by scattering and absorbing solar radiation and absorbing and emitting longwave radiation. Of these effects, the scattering properties are generally the most important. When dust loadings are high, the longwave infrared absorption and emission of particles are occasionally quite significant. Aerosols also have an indirect impact on the energy balance because they act as condensation nuclei for cloud droplets. When the concentration of aerosols is high, the number of droplets formed is high, and each drop is relatively small. The opposite effect occurs when the aerosol concentration is low. For the same amount of condensed water in a cloud, smaller droplets produce a larger albedo. Aerosols thus modulate the reflectivity of clouds and the radiative energy balance in cloudy regions. In polluted areas, haze concentrations can be high, and the cloud albedo is enhanced compared with that of unpolluted areas.

The scattering properties of aerosol particles, whether introduced by volcanic eruptions or smokestacks, enter into the global climate equation. The simplest way to account for these effects is through an effective change in the global average albedo. This change in albedo is not always easy to determine, as the albedo effect depends on the quantity of aerosols in the atmosphere, their optical and physical properties, the vertical distribution of particles, and the albedo of the underlying surface. Moreover, because the albedo varies with wavelength and with the angle at which incident sunlight hits the aerosol layer, the changes must be determined for a wide range of conditions. The albedo effect of aerosols can indeed be calculated, but the work may be tedious.

The general behavior of the planetary albedo with changes in smoke, dust, and haze concentrations can be summarized as follows: The addition of scattering aerosols into the atmosphere typically increases the albedo and reduces the surface temperature. The introduction of highly absorbing aerosols in the lower atmosphere (boundary layer) can lower the albedo and warm the surface. However, if the absorbing aerosols are placed in the upper troposphere or stratosphere, they tend to cool the surface, even if the planetary albedo is reduced, through an effect referred to as the *antigreenhouse effect* (Section 14.2.1). These general rules for aerosol effects can be useful for assessing climatic impacts (although care must be exercised, inasmuch as the behavior may be different in specific cases).

Interesting examples of the effects of aerosols on global-scale climate are found on other worlds in the solar system. On Venus, for example, a dense layer of sulfuric acid aerosols continuously shrouds the entire planet (these aerosols are almost identical to the sulfuric acid droplets produced by volcanoes). The aerosol layer increases Venus's albedo to about 0.7, significantly reducing the solar forcing. The extreme greenhouse effect still heats the surface to more than 700 K! Titan, Saturn's most prominent moon, is also blanketed by a dense aerosol haze that prevents our directly observing its surface. The Titan haze is generated by the photochemical transformation of methane into more complex or-

ganic molecules that readily stick together to form particles, sometimes referred to as axle-grease aerosols. The haze is absorbing and controls the albedo and temperature of Titan (whose surface is extremely cold, about 60 kelvin).

Mars is another interesting case. Global-scale Martian dust storms are well documented by observations from Earth and from spacecraft orbiting the planet. The *Viking* instruments that were placed on the Martian surface (to study Martian climate and search for signs of life) also recorded these storms. An event builds up during the week, as dense clouds of dust spread rapidly to cover the Martian landscape. Measurements taken by the *Viking* lander showed that the dust clouds significantly reduced the sunlight reaching the ground and caused a noticeable cooling of the surface temperature. Mars, already a cold place, is getting even colder.

11.7 The Vulnerability of Life to a Changing Climate

The evolution of life and its maintenance depend on a favorable climate. During the long course of evolution, over the past 4 billion years of Earth history, the average temperature of the globe has probably not fluctuated by more than 10°C. There have been periods of significantly warmer climate and colder periods as well. At present, the world seems to be in a fairly optimal climatic state. How long it will stay that way is uncertain. We will explore here the implications of climate change for survival and prosperity. We will look at this problem from three perspectives: first, from the point of view of human evolution and the development of civilization; second, from the perspective of the robustness of the climate system and its apparent tenacious resistance to change; and finally, from the view that catastrophic climate change may be inevitable for worlds governed by the laws of nature.

11.7.1 MODERN SOCIETY AND CLIMATE

The ancestors of humans evolved sometime before the beginning of the Pleistocene era, several million years ago. Modern humans *Homo sapiens sapiens* appeared during the last ice age. These first people, the Cro-Magnons, survived as hunter-gatherers in small groups wandering through large territories seeking sustenance. The land was harsh, but food was abundant, and they persisted, using intelligence as a competitive edge. Early humans left behind tools, decorative relics, and art in the form of wall paintings in cave dwellings. With the end of the ice age, the warming climate changed the landscape considerably. Beginning 10,000 to 20,000 years ago, sea level rose, perhaps by 100 meters or more, as the ice sheets melted. Vegetation, and the animal population it supported, changed steadily. It is believed that, despite their relatively small numbers at the time, humans, by overhunting, hastened the demise of many marginal and faltering species, such as the large elephant-like mastodon. Early humans also apparently used fire as a tool to mold the vegetative landscape for favorable hunting conditions.

The warmer, steadier climate of the Earth allowed humans to experiment with cultivating of wild grains. The horticultural revolution had begun. Soon it was found that grains could be bred, grown, and harvested in a regular cycle, in tune with the seasons. Land was tilled and the first farming communities were established. With the prospect of a reliable supply of food and land to protect, enclaves sprang up. Later, these became cities and city-states, each protected by elaborate defenses. Trade among the cities grew. Technology, art, and culture evolved rapidly on this foundation of sustenance. Our modern technologically driven society is the evolutionary product of these early breathtaking developments.

Despite all the technological inventions of modern society, we still are dependent on dirt agriculture for sustenance. By some estimates, if we had to return to a life of hunting and gathering the natural bounty of forests, fields, and seas, only about 1 percent of the current human population could be supported. Ninety-nine percent of us would be finished. Clearly, we are dependent on organized agriculture for survival.

An Agricultural Achilles' Heel

In 1816, the "year without a summer"—likely caused by the eruption of Mount Tambora in 1815—led to crop failures throughout New England (the most populated region of the United States at the time) and large areas of Europe. Famine was widespread, and many people starved or died of pandemics of disease. In earlier times, changes in climate that were unfavorable, but by no means as extreme as this,

produced terrible results. In the fourteenth century, for example, the climate of Europe became much wetter and damper during the period of climatic cooling that preceded the little ice age. Hubert Lamb described a "time of diseases" then. One particularly nasty illness, ergotism, or Saint Anthony's fire, was caused by rye grains tainted with the ergot fungus. According to Lamb, even a minute proportion of the poisoned grains, baked in bread, would cause the disease. The course of the epidemics was such that the whole population of a village would suffer convulsions, hallucinations, gangrene rotting the extremities of the body, and death. In the chronic stage of the disease, the extremities developed first an icy feeling, then a burning sensation; the limbs then went dark as if burnt, shriveled, and finally dropped off. Even domestic animals caught it and died. And pregnant women miscarried.[19]

The Black Death, a pandemic of bubonic plague, also occurred during this time and killed perhaps one-third of the population of Europe.

Agriculture is the Achilles' heel of modern civilization. Climate is the sword that may cut the tendon, and famine and disease are the specters of defeat. Today's agricultural system is certainly more robust than in times past, supported by the copious application of chemicals (pesticides, herbicides, and fertilizers) and a massive infrastructure (irrigation, machinery, storage, and distribution). Nevertheless, the world as a whole is not well fed. Hundreds of millions of people are starving, and the population continues to increase relentlessly. It is little wonder that climatologists and sociologists are concerned about the next major climatic shift. Depending on the speed and magnitude of the change and its regional implications, billions of people might be adversely affected. Disease would be no stranger. Although modern medicine has many potions to dispense, the recent AIDS epidemic and its resistance to ordinary cures leave an uneasiness about potential future pandemics. Surely they are best avoided.

The Human Component of Climate Change

On the other side of the climate-change coin, we must consider the impact of human activities on the regional and global climates. Since the advent of agriculture on a large scale, humans have significantly changed the landscape. Wide tracts of forest

19. Hubert H. Lamb, *Climate, History, and the Modern World* (London: Methuen, 1982), p. 188.

have been cut down to create fields and pastures. Further harvesting of wood stocks for fuel has denuded large regions. Normal water flow has been diverted and stored to irrigate arid lands and create expansive artificial swamps (rice paddies). Domestic animals have been bred in huge numbers and now roam the land, emitting greenhouse gases into the atmosphere (Section 12.3.1). Fire is continually used to burn off unwanted natural vegetation and crop debris, adding millions of tons of particles to the atmosphere each year. Tens of millions of tons of manufactured chemicals are spread over the land and wash into streams, rivers, groundwater, and estuaries. The local environmental impacts of agriculture can be severe, but even on a global scale, the effects of the spread of human civilization have been felt climatically.

One example is the desertification of large regions in the Middle East and North Africa as a result of overgrazing. The domestic animals eat, trample, and kill the natural vegetation that secures the moisture and topsoil. The clearing of forests—hundreds of years ago in Europe and North America, but more recently in Africa and South America—has changed the Earth's albedo, perhaps by a large percentage. Forests have a low albedo, and pastures—or worse yet, the desertified regions left behind long after agriculture has fled—have a higher albedo. Particularly alarming is the rapid clearing of rain forests in Indonesia and South America, for inefficient subsistence farming or cattle grazing or, more inappropriately, for wood pulp to make cardboard boxes to package VCRs! The rain forests not only are the richest depositories of biological diversity, but also control the hydrological cycle over large areas and so can affect the regional climate of those regions.

11.7.2 Do We Need Climate Insurance?

In Section 4.3, we discussed the robustness of life on Earth in the context of the history of the Earth and the fossil record of life. We introduced two important concepts: the idea that past environments have been guided by the coevolution of life and the climate in a very stable relationship, or homeostasis, referred to as Gaia, and the reality of sudden massive extinctions on the Earth engineered by processes beyond human control. These concepts deserve attention here with regard to possible future climate change and to the implications for the survival of humans and countless other species.

Gaia, the Mother Protector

Gaia, or Mother Earth, has been invoked to protect the environment. The theory of Gaia described in Section 4.3.3, in one sense, seeks to explain the constancy of climate over the ages. The seat of climate stability, in this concept, lies in the response of life to changes in the environment. Rather than simply adapting to every change, obediently following the lead of the physical world, life responds by resisting and modifying every change that threatens its well-being. The result is a more stable world environment hand-tailored by the organisms that must dwell there. The environment is said to be in a state of homeostasis, which can adjust itself internally and is very stable.

We have already seen that biological activity can affect the climate system. Vegetation affects the Earth's albedo. The carbon cycle is controlled by biological processes. Greenhouse gases are emitted by microscopic organisms. All, and many more, of these factors represent couplings between the biosphere and the geosphere that influence climate. Thus life does respond to changes in the physical world and, in some cases, may produce effects that offset the change. The fossil record of life, however, reveals a slaughterhouse of biological debris. Most of the species that have ever lived are now extinct. The extinctions often come in massive waves, as if life were succumbing to irresistible physical change. The biosphere has no collective intelligence, and so perhaps species that evolve and persist are those that, by chance, affect the environment favorably. The delicate balance may be periodically destroyed by an overwhelming physical trauma, such as a large meteor impact. Then in the end, the physical world largely controls the environment and evolutionary pathways.

The Gaia hypothesis provides no evidence for a built-in robustness of the biogeophysical environment or climate. It is not an insurance policy against global change. Gaia cannot be counted on to prevent the putative warming of the Earth, with its attendant disruptions, described in the next chapter.

Punctuated Climate Change and Mass Extinction

It is now evident that the Earth is experiencing one of the most significant extinction events in its long history. Although you cannot see it on a day-to-day basis, over the next few centuries an evolutionary disaster is looming. Many species of fish, mammals, and plants are disappearing. The cause appears to be the expansion of human civilization and technology. Humans are competing more and more with other species for land, water, and other resources. Many animals are senselessly killed for frivolous reasons—the rhinoceros for its horn, which some believe is a sexual aphrodisiac, and the elephant for its tusks of ivory. The development of land and water resources is a more serious long-term hazard. The decimation of the rain forests alone threatens most of the species on the planet. We can freeze specimens and store DNA, but no one believes that the biosphere, if ravaged, could be reconstituted.

The extinction of the dinosaurs 65 million years ago was probably caused by the impact of a comet or meteor with the Earth (Section 4.3.2). The dust cloud could have caused the planet to cool suddenly, destroying the habitat that sustained the dinosaurs and most other species. The fact that such traumatic events have occurred, and the near certainty that similar events will occur in the future, is humbling. Not even Mother Gaia can save the Earth from this eventual cataclysm. Two factors work in our favor, however. First, the date of the next cataclysm may be millions of years in the future. The impact of a large meteor with the Earth is not imminent, not even remotely imminent. The frequency of such events can be gauged by geologic records, and although those records suggest that we may be within a few million years of a major impact, it is just as likely that it would not happen for another 30 million to 40 million years (in the past, there have been long periods without major impacts). On the other hand, smaller meteor events, which are more frequent, could be devastating on regional or continental scales. The effect might be comparable to that produced by detonating 10,000 hydrogen bombs at the same instant in one place.

As time progresses, we may also be able to develop technology for avoiding such collisions when they threaten. It has been suggested, for example, that nations pool their nuclear-weapon stockpiles to use as a huge detonator to deflect any meteor that might collide with the Earth. It would also provide work for the nuclear scientists and administrators who are threatened with unemployment with the demise of the Soviet Union and the birth of a "new world order." There are more logical ways to deal with nuclear weapons (Chapter 14), however, and alternative meteor-deflection schemes will be devised.

Volcanic Winter

Other possible causes of climate change may result in mass extinction. A potential anthropogenic cause, nuclear winter, is not very likely in the present international political scene (Section 14.2.1). A few scientists have suggested that past extinctions were related to extensive volcanism leading to major climatic shifts. During the ice ages, for example, volcanism was more prevalent than it is today. Modern-day major eruptions appear to produce small temporary, but significant, cooling of the global climate. Hence continual, more violent volcanic activity might create a "volcanic winter." This would occur as an intensification and extension of the usual cooling associated with volcanic eruptions. The explanation of strong global cooling caused by volcanism follows the theory for nuclear winter (Section 14.2.1).

The intensity of volcanic activity over time wanes as planets age and cool down. At the current evolutionary stage of the Earth, volcanism is less frequent than it was in earlier eras. Accordingly, the continual violent volcanism required for a volcanic winter is unlikely in the near future. Volcanism may be triggered by other factors, however. The tectonic movement of the Earth's crustal plates could open up cracks that would allow magma to escape. The plates have been deformed by the weight of glaciers. So, for example, if the climate began to cool because of another forcing, ice might build up and trigger volcanism, which would hasten the original cooling. In this case, volcanism would provide a positive feedback for the initial cooling. Although a volcano-cooling mechanism may have played a role in the ice ages, the time scales for such effects are very long.

It appears that the catastrophic change in climate associated with natural external and internal forcing occurs only infrequently. The probability, and hence the risk, associated with such events is very small. By contrast, lesser climate anomalies that are very likely may pose a serious risk to the human populace. If insurance is to be taken out, a policy to cover damages that may be suffered in the predicted global warming of the twenty-first century could turn out to be a wise investment.

Questions

1. Explain what is meant by the "solar constant." Why is the solar constant different for different planets? Is the solar constant for Mars greater or smaller than that for Earth? Explain. Estimate the solar constant for Venus, given that the solar constant for Earth is roughly 1400 W/m^2. Refer to Table 11.1.

2. Can you deduce the meteorological significance of the old saying "Red sky at night, sailor's delight;/Red sky in morning, sailor take warning"? (*Hint*: Consider the nature of the prevailing winds and weather patterns discussed in Chapter 2.)

3. No instrument has ever been sent to the sun to measure its "surface" temperature. Discuss how you could determine this (photospheric) temperature from the Earth. Would you expect the interior of the sun to be hotter or cooler? Why?

4. Define the term *albedo*. List several examples of surfaces that have a large albedo and of surfaces that have a low albedo.

5. In calculating the total solar energy absorbed by Earth, the cross-sectional area is used rather than the area of the globe itself. Provide a reasoned argument for making this choice.

6. Explain how the Earth actually differs from an ideal blackbody emitter of thermal radiation.

7. Before satellites were deployed in space, above the atmosphere, measurements of the solar constant had to be taken at the Earth's surface. The solar constant can be determined by measuring the intensity of sunlight at several wavelengths and "integrating" the spectrum (for example, Section 11.4.2). Describe the effects of the atmosphere that would interfere with such a measurement. (*Hint*: Consider the absorption and scattering effects of the atmosphere in the solar spectrum.) What might you do to minimize these problems? What information would you need to correct for possible errors?

8. Discuss the way in which greenhouse gases added to the atmosphere affect the thermal emission of the Earth.

9. Do all regions of the Earth emit the same amount of longwave energy to space? Explain. What conditions favor larger emissions? Smaller emissions?

Problems

1. Determine the effective temperature of the Earth for assumed planetary albedos of 20 percent and 40 percent, compared with the measured value of 30 percent.

2. Imagine that the Earth's climate system consists of two distinct parts: the surface and a single layer of transparent atmospheric gas above it. Assume that the surface-atmosphere system is uniform around the entire planet and that the average solar insolation (energy input) at the surface is $-F_s(1 - a_e)$ everywhere. The average emissivity (absorptivity) of the atmospheric layer is e; the surface can be treated as a perfect blackbody. Let the entire system be in a steady state, with the energy input equal to the energy loss for both the surface and the atmosphere. Write out the energy balance relationships for the surface and the atmosphere. Derive from these the equilibrium surface temperature and atmosphere temperature. (*Hint*: The energy balance of the atmosphere has three terms: two loss terms due to gray-body emission proportional to the atmospheric emissivity, in both the upward and the downward direction, and one gain term due to the energy absorbed by the atmosphere from the upwelling surface heat emission that passes through the atmosphere on its way to space. The surface has three energy balance terms: one gain due to solar insolation, one loss due to the blackbody emission from the surface, and a second gain due to the absorption of the atmospheric heat emitted in the downward direction.)

Suggested Readings

Abbott, C. and F. Fowle. "Volcanoes and Climate." *Smithsonian Miscellaneous Collections* **60** (1913): 24.

Broecker, W. and G. Denton. "What Drives Glacial Cycles." *Scientific American* **262** (1990): 48.

Budyko, M. *Climate and Life*. London: Academic Press, 1974.

Crutchfield, J., J. Farmer, N. Packard, and R. Shaw. "Chaos." *Scientific American* **255** (1986): 46.

Eddy, J. "Climate and the Changing Sun." *Climatic Change* **1** (1976): 173.

Foukal, P. "The Variable Sun." *Scientific American* **262** (1990): 34.

Haberle, R. "The Climate of Mars." *Scientific American* **254** (1986): 54.

Hansen, J., A. Lacis, R. Ruedy, M. Sato, and H. Wilson. "How Sensitive Is the World's Climate?" *Research and Exploration* 9 (1993): 142.

Imbrie, J. and K. Imbrie. *Ice Ages: Solving the Mystery.* Cambridge, Mass: Harvard University Press, 1979.

Kasting, J., O. Toon, and J. Pollack. "How Climate Evolved on Terrestrial Planets." *Scientific American* **258** (1988): 90.

Lamb, H. *Climate, History and the Modern World.* London: Methuen, 1982.

———. *Climate: Present, Past and Future.* 2 Vols. London: Methuen, 1972.

Lorenz, E. "Climate Determinism." *Meteorological Monographs* **8** (1968): 1.

Lovelock, J. and L. Margulis. "Atmospheric Homeostasis by and for the Biosphere: The Gaia Hypothesis." *Tellus* **26** (1973): 2.

Milankovitch, M. *Canon of Insolation and the Ice Age Problem.* Jerusalem: Israel Program for Scientific Translations, for the National Science Foundation, 1941.

Pearson, R. *Climate and Evolution.* New York: Academic Press, 1978.

Pittock, A. B. "Solar Variability, Weather and Climate: An Update." *Quarterly Journal of the Royal Meteorological Society* **109** (1983): 23.

Prinn, R. "The Volcanoes and Clouds of Venus." *Scientific American* **252** (1985): 46.

Radok, U. "The Antarctic Ice." *Scientific American* **253** (1985): 98.

Ruddiman, W. and J. Kutzbach. "Plateau Uplift and Climatic Change." *Scientific American,* **264** (1991): 66.

Schneider, S. H. "Climate Modeling." *Scientific American* **256** (1987): 72.

Schneider, S. and R. Londer. *The Coevolution of Climate and Life.* San Francisco: Sierra Club Books, 1984.

Trenberth, K., ed. *Climate System Modeling.* Cambridge: Cambridge University Press, 1992.

12

Greenhouse Warming

Imagine walking into a strange, darkened room. You fumble for the light switch, but cannot find it. Stumbling forward, you bruise your shin on the coffee table. Then you cartwheel over a chair, and slam-dance with a lamp. Perhaps it would be wiser to enter this unknown place more cautiously, or with a flashlight. The Earth is the room, our ignorance of how its complex interwoven systems work is the darkness, and enlightened knowledge is the guiding light. Global environmental change appears to be imminent—as evidenced by claims of climate warming—and yet we continue to stumble forward without a strong enough light.

In the previous two chapters, we laid the groundwork for discussing the possible greenhouse warming of the climate. In Chapter 10, we describe the global biogeochemical cycles of carbon dioxide and other elemental components of the climate system. The hydrological cycle, which is central to the greenhouse effect on Earth, is also outlined there. Additional information in this chapter defines the biogeochemical cycles of specific greenhouse-active gases such as methane, nitrous oxide, and the chlorofluorocarbons (the chlorofluorocarbons are also considered in Chapter 13 as they relate to ozone depletion.) Chapter 11 describes the radiative and physical processes that determine the global climate, emphasizing the greenhouse effect, and explains the radiative properties and effects of common greenhouse compounds. Also spelled out is the dominant role of water vapor and the clouds produced by water vapor condensation in controlling the natural greenhouse effect. We also explore the sensitivity of the climate to several variable parameters.

In this chapter, we investigate the effects on climate of changes in the concentrations of trace greenhouse gases, such as CO_2, CH_4, and N_2O. We look at the impacts of anthropogenic activities on these gases, and the environmental implications of such activities. The effects of water vapor are critical. Although water vapor is not directly or significantly affected by human activities, it can be substantially altered through indirect processes. That is, water vapor provides a powerful feedback mechanism that can enhance or moderate the climate impacts of the other greenhouse gases. Finally, we present actions to reduce or avert the potentially hazardous effects of greenhouse warming.

12.1 Greenhouse Gases

A number of trace constituents of air affect the Earth's balance of radiation and climate system (Section 11.4). These gases—partly of natural origin, partly generated by human activities—have been accumulating in the atmosphere at an alarming rate during the past century, but the causes for the increases are not completely understood.

12.1.1 THE GREENHOUSE CULPRITS: A ROGUE'S GALLERY

We are primarily concerned here with changes in trace greenhouse-active gases that may be influenced by human civilization. The specific gases and some of their properties are summarized in Table 12.1. Many other compounds associated with industrial and agricultural activities are also capable of inducing climatic change, and these will be mentioned where appropriate. However, the gases listed in Table 12.1 make up the essential set. In recent years, intense research and debate have focused on these common compounds.

Table 12.1   Greenhouse Gases and Their Properties

Species	Present concentration (ppmv)	Preindustrial concentration (ppmv)	Rate increase (%/yr)	Atmospheric residence time (yrs)
CO_2	365	275	0.4	> 100
CH_4	1.8	0.8	0.8	~10
N_2O	0.31	0.29	~0.2	~150
CFCs[a]	0.001	0	~4	~100
O_3[b]	~0.01–0.2	~0.01	~1	~0.1

[a] The chlorofluorocarbons compose a large class of compounds including CFC-11 ($CFCl_3$) and CFC-12 (CF_2Cl_2). The parameters corresponding to the total burden of CFC-11 plus CFC-12 are tabulated.
[b] Ozone in the troposphere is considered here.

12.1.2   WATER VAPOR: INNOCENT BYSTANDER OR GOOD SAMARITAN?

Water vapor (H_2O), which comprises about 1 percent of the atmosphere, is the principal greenhouse gas in the Earth's climate system. (The radiative properties of water vapor and its role in cloud formation are discussed in Section 11.4; see also Section 10.3 concerning the hydrological cycle.) Water vapor also happens to boost the greenhouse effects of other gases because, as explained in Section 11.4.3, as the surface warms, the amount of water vapor that can evaporate increases. Suppose that carbon dioxide is added to the atmosphere. The CO_2 then creates a greenhouse warming, and more water can evaporate into the warmer air, further enhancing the warming effect. Water vapor produces a positive feedback on the greenhouse effect. Why?

The greenhouse feedback effect of water vapor is related to a common physical property of liquids, their **pressure**. When heated, a liquid tends to evaporate into a vapor consisting of the liquid's molecules. Water left sitting in an open container slowly disappears by evaporation over a period of days. Heated in a pot, the same amount of water vapor can evaporate in minutes. The process of evaporation is greatly accelerated by raising the temperature of the liquid. Raising the temperature increases the vapor pressure of the liquid, and so it tends to evaporate more quickly.

When the Earth is heated up, it evaporates more water, which also cools it down. When you exercise and overheat, you sweat to cool off. Your body

exudes water onto your skin, where it can evaporate (or drip!). Evaporation requires a certain amount of energy, the **latent heat of evaporation**. This is the energy needed to free a molecule from the liquid phase and put it into the gas phase. The latent heat absorbed from your body when you sweat cools you off. The water vapor evaporated from the Earth when it is heated helps the surface keep cool.[1] (See Section 11.4.4 for a more complete description of the surface energy balance.)

The vapor pressure of a liquid is a property of the liquid that does not depend on the amount of liquid or gas present. The vapor pressure depends only on *temperature*, through a simple exponential law:

$$\frac{p_v(T)}{p_v^0} = exp\left(-\frac{L_v}{k_B T}\right) \qquad (12.1)$$

Here, p_v is the vapor pressure at temperature T, p_v^0 is a reference pressure, L_v is the latent heat of evaporation, and k_B is Boltzmann's constant (Section A.3). Equation 12.1 is similar to the exponential Equation 11.15. Note that in Equation 12.1, as the temperature increases, the term in brackets—the exponent—decreases in magnitude, since it varies inversely with temperature. Because the term is negative, the vapor pressure *increases* as the

1. You may have heard of a "swamp cooler." This clever device is used effectively to cool homes in hot dry climates and can be seen on roofs throughout the southwestern United States. The basic principle is similar to sweating. Water is sprayed onto a matrix of metal tubing through which a fluid is circulated. The water evaporates from the tubing and cools the fluid inside by absorbing latent heat from it. The cooled fluid then is piped inside where air is blown over a second set of cooling coils.

temperature *increases*. The dependence of vapor pressure on temperature is so strong that the vapor pressure of water, for example, doubles every time the temperature rises by about 3°C.[2]

A liquid continues to evaporate until the air over the liquid becomes saturated—that is, until the partial pressure of the evaporated gas over the liquid is equal to the vapor pressure of the liquid corresponding to the local temperature. This is the *equilibrium state* of the system. In this state, the amounts of liquid water and water vapor will not change as long as the temperature is held constant. If the air cools slightly at this point, the vapor pressure will decrease, but the concentration and partial pressure of the vapor molecules will remain nearly constant. The air is then **supersaturated**; that is, the pressure of the vapor exceeds the vapor pressure. The **saturation ratio** is defined as the ratio of the partial pressure of the gas in an air parcel to the vapor pressure of condensed gas corresponding to the temperature of the parcel. A supersaturation is a saturation ratio that is greater than 1. In other words, a saturation ratio of 1.1 means that the vapor is supersaturated by 0.1, or 10 percent.

When water vapor is supersaturated, it will condense as fog, clouds, dew, or some other form of water.[3] This explains why clouds usually form well above the ground. The air near the surface is usually undersaturated with water vapor, which is the reason that water can evaporate from the surface in the first place. As these undersaturated air parcels rise by means of convective buoyancy, they cool adiabatically (Section 3.1.1). The water vapor pressure decreases as the air cools (remember that the vapor pressure is a *property* of water and is *independent* of the amount of liquid or gaseous water present). At some height, the vapor pressure becomes smaller than the actual partial pressure of the water vapor in the parcel. Condensation is a result, and a cloud is formed.

The water vapor saturation ratio and **relative humidity** are simply related to each other. Indeed, the relative humidity is the saturation ratio expressed as a percentage. The relative humidity is most useful for characterizing air parcels that are undersaturated—that is, with saturation ratios of less than 1. The relative humidity in undersaturated air can range from very low values of a small percentage in extremely dry air, to 99 percent in air that is uncomfortably humid. Under normal atmospheric conditions, the relative humidity seldom exceeds 101 percent or so, because even a small supersaturation leads to water vapor condensation on soluble aerosol particles or preexisting water drops. A relative humidity of 101 percent and a supersaturation of 1 percent are perfectly equivalent in meteorological terms.

According to recent scientific studies, the water vapor feedback on greenhouse warming can nearly *double* the direct greenhouse effect of most gases. (See the discussion of the temperature feedback parameter in Section 12.4.3.) The role of water vapor and its precise response to global warming, including alterations of cloudiness, are a central issue in the global warming debate. In this sense, water is an innocent bystander. It has always been here to help us when we needed shelter and warmth, incubating life beneath a thick greenhouse blanket. Now water vapor is seen to boost the heating effect of the greenhouse culprits. It is aiding and abetting or otherwise associating with known greenhouse offenders. On the other hand, some climatologists contend that water vapor moderates greenhouse warming and is thus a good Samaritan, stabilizing the climate against excess change. The evidence developed in Chapter 11 points to a water vapor feedback effect that enhances the greenhouse warming associated with other gases. Detailed modeling studies, summarized as a temperature feedback parameter later in this chapter, also show a net warming, not cooling,

2. Imagine what would happen if you heated up a sealed metal can with liquid water in the bottom. Some of the water would evaporate until the pressure exerted by the water molecules equaled its "vapor pressure" at that temperature. Remember that the pressure exerted on the can by the air trapped inside also increases in accordance with the ideal gas law, as the air tries to expand and pushes against the walls of the container (Section 3.1.1). If you continue to heat the can, more of the liquid water will evaporate, and the pressure of the water vapor in the can will increase rapidly. If you are not careful, the vapor pressure will become great enough to burst the container.

3. The difference between a sauna and a steam bath is explained by the idea of supersaturation. In a sauna, the air is hot and dry and far below the point of water vapor supersaturation. When you (cold) first enter, your body heats up by contact with the superheated air, as it would in an oven, and sweat beads from your pores to provide cooling by evaporation; you are contributing water vapor to the dry air in the sauna. In a steam bath, the warm air is already supersaturated with water vapor, which condenses on any cool surface like a mirror or door. When you first enter, the water condenses on your cool skin and releases its latent heat, warming you up. As you warm, you also begin to sweat, but the sweat cannot evaporate in the supersaturated environment and so drips from your body. You are contributing water to the floor.

effect associated with water vapor. Hence for climate assessment, water vapor is assumed to be the innocent bystander who unwittingly becomes involved at the scene of an accident.

12.2   Carbon Dioxide

Svante Arrhenius[4] was the first scientist to recognize the potential climatic significance of CO_2. In 1896 he proposed that the Earth's temperature was controlled by the radiative properties of carbon dioxide (and water vapor) and theorized that the ice ages were caused when CO_2 concentrations were reduced (by some process that remained to be identified). Increases in atmospheric carbon dioxide concentrations had not been detected at that time. Nevertheless, Arrhenius proposed that continued coal combustion might increase the amount of CO_2 in the atmosphere and warm the Earth. It was much later, in 1938, that the first clear indication of increasing CO_2 concentrations was obtained by George Callendar. Even so, it was two decades later, in 1957, that Charles David Keeling began his landmark project to monitor carbon dioxide concentrations accurately over a long time span. Keeling's data showed unequivocally that carbon dioxide has been inexorably increasing since the middle of the twentieth century.

12.2.1   Increasing CO_2: What Is the Cause?

The atmosphere contains, in total, about 750 gigatonnes of carbon in the form of carbon dioxide (Section 10.2.4). The average concentration of CO_2 is currently about 365 ppmv and is increasing at a rate of about 0.4 percent per year (Table 12.1), which is equivalent to about 3 gigatonnes of carbon per year (Gt-C/yr). If that rate of increase continues, the atmospheric concentration of CO_2 will double in about 170 years.[5] This trend reflects recent measure-

ments. To understand why CO_2 has increased, we must know the history of carbon dioxide in the atmosphere. Section 10.2.4 described the basic carbon dioxide reservoirs and the rates of transfer between the reservoirs. We noted that human activities are adding CO_2 to the atmosphere from the fossil carbon reservoirs at a greatly accelerated rate compared with the natural rates of geologic recycling. The specific response of the reservoirs and the potential accumulation of carbon dioxide still must be discussed.

The long-term trend in carbon dioxide concentrations, as determined by measurements of the composition of air trapped in glacial ice bubbles, is illustrated in Figure 12.1. This record can be extended back for tens of thousands of years. However, the comparison of atmospheric CO_2 levels before and after the Industrial Revolution (beginning in the late eighteenth century; the steam engine was invented by James Watt in Scotland in 1751) is instructive. It was precisely at that time that carbon dioxide concentrations began to increase. The use of coal for energy had already become a nuisance (Section 6.1). Its accelerating consumption to fuel the machinery of industry was not only generating energy and air pollution, but also fundamentally altering the global atmosphere. And no one knew that it was happening.

The pre-Industrial Revolution abundance of carbon dioxide was about 275 parts per million by volume, considerably lower than today's 365 ppmv. In terms of total carbon mass, the change has been from some 540 Gt-C to 750 Gt-C, a difference of 210 Gt-C. At the present rate of increase in carbon dioxide, it would take less than 60 years to add that much carbon to the atmosphere. It appears that the preindustrial CO_2 concentrations were relatively stable, at least over historical time scales (the concentration varied significantly during the last ice age, however [Section 12.4.1]). The recent increase is therefore logically connected with the massive burning of fossil fuels.

The more recent, finer-resolution carbon dioxide data collected by Keeling at the Mauna Loa Observatory in Hawaii are given in Figure 12.2. Hawaii was chosen for these and other similar measurements of the composition of the atmosphere because Hawaii has high mountains (like Mauna Loa on the island of Hawaii, on which you can get above most of the clouds and dust of the boundary layer) and because

---

4. Svante August Arrhenius (1859–1927) was a Swedish chemist who developed the theory of ionic dissociation in solution (such as occurs when an acid is mixed with water, releasing hydrogen ions [Section 3.3.4]) in 1884, for which he received the Nobel Prize in chemistry in 1903. Arrhenius had the scientific prowess and imagination to reach beyond his exceptional laboratory experience and apply his knowledge of the properties of chemical compounds to understand the natural environment.

5. The length of time required to double a quantity when it increases by x percent per year is time(years) = $69/x$. Thus for a

growth rate of 0.4%/yr, the doubling time is roughly $69/0.4$ = 170 years.

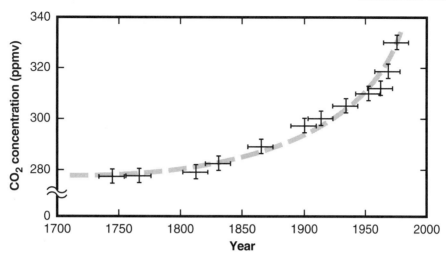

Figure 12.1   The long-term trend in atmospheric carbon dioxide concentrations. The carbon dioxide measurements were obtained from air bubbles trapped in glacial ice sheets. The concentrations are given in parts per million by volume (ppmv). The accuracy of the derived values is defined by the crosses, whose height and length define the range of uncertainty in the concentration and date, respectively, of each measurement. (Data from A. Neftel, E. Moor, H. Oeschger, and B. Stauffer, "Evidence from Polar Ice Cores for the Increase in Atmospheric CO_2 in the Last Two Centuries," *Nature* 315 [1985]: 45. Figure 1, p. 45.)

it is in a very clean environment in the middle of the Pacific Ocean, far removed from sources of pollution. Nevertheless, Keeling slowly came to realize, after tedious years of painstakingly analyzing his data, that even in the remote atmosphere—indeed, everywhere—the air was becoming polluted with excess carbon dioxide. Arrhenius's foreboding of global climate warming due to CO_2 accumulation cast a shadow over the planet. A few perceptive climatologists and oceanographers had already recognized the potential threat. Researchers such as Roger Revelle, a leading oceanographer at the Scripps Institute of Oceanography; Bert Bolin, a climatologist at the University of Stockholm; and Wallace Broecker, an oceanographer working at the Lamont Doherty Geological Observatory, began to study the sources and sinks of atmospheric carbon (Chapter 10) and the roles of the oceans and the biosphere in the uptake of excess carbon.

The Annual Carbon Cycle

The "Keeling curve" for carbon dioxide in Figure 12.2 clearly shows the Earth's biological pulse. The seasonal rise and fall of carbon dioxide are in perfect rhythm with life's basic activity, photosynthesis. In the summer, when the air is warm and sunlight is plentiful, vegetation takes up large amounts of carbon dioxide through excess photosynthesis. In the

winter, when most plants are quiescent, the functioning organisms continue to consume organic debris with a net release of carbon dioxide.

This photosynthesis-respiration cycle is roughly balanced every year and so has little impact on atmospheric CO_2 concentrations. In Figure 12.2, this is indicated by the fact that the annual oscillation in CO_2 is fairly uniform. Nevertheless, the potential rate of carbon uptake by plants is very large. Recall from Section 10.2.4 that about 80 gigatonnes of carbon per year (Gt-C/yr) are taken up by plants through photosynthesis. In balance, roughly the same amount is released by respiration, decay, and combustion. The differences between the times of the year when these processes occur produce the carbon dioxide annual cycle. The cycle has opposite phases in the Northern and Southern Hemispheres, because the seasons are reversed between the two hemispheres. Moreover, the total land area and biomass of the Northern Hemisphere is three times larger than that of the Southern Hemisphere. The annual cycle, as might be expected, is more pronounced in the north. At tropical latitudes, where the seasonal variation itself is minimal, the annual modulation in carbon dioxide is smallest. On the other hand, at high latitudes, where biological activity changes most wildly from perpetual frozen winter nights to endless sunny summer days, the CO_2 cycle is greatest.

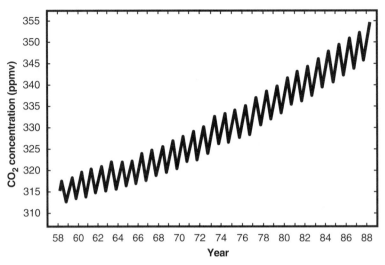

Figure 12.2   Recent variations in atmospheric carbon dioxide. Years are shown on the bottom of the graph; the curve gives monthly average values of CO_2 in ppmv. The seasonal variation associated with photosynthesis/respiration is several ppmv. (Measurements were taken at the Mauna Loa Observatory, Hawaii, by C. D. Keeling.)

The vegetation on land consists of long-lived species (trees and shrubs) and short-lived species (grasses, flowers, and most vegetables). Even some of the long-lived plants have short-lived parts, however; the leaves on deciduous trees, for example, fall off each year. The short-lived species may contain about 20 percent of the total living biomass. The rest, about 80 percent, is tied up in woody material, with an average lifetime of roughly 30 years. A small fraction of this woody biomass—including the majestic giant sequoia and the ancient bristlecone pines, among other species—survives for hundreds or thousands of years! (See Section 10.2.4 for a more complete discussion of the global carbon reservoirs.)

The annual atmosphere-land exchange of carbon dioxide amounts to roughly 1 percent of the total atmospheric carbon reservoir—that is, say, 10 Gt-C/yr. This rate is only about 10 percent of the total rate of carbon uptake by photosynthesis (80 Gt-C/yr, according to the discussion in Chapter 10). Of the total annual uptake of carbon, perhaps three-quarters (75 percent, or 60 Gt-C/yr) is accommodated by short-lived species. The annual cycle of CO_2 is therefore dominated by the life cycle of short-lived plants. The total short-lived carbon reservoir on land, including the associated plant detritus, may total 200 Gt-C, compared with the total land reservoir (living and dead) of about 2800 Gt-C.

Any significant change in the total amount of carbon in the short-lived reservoirs would be much more noticeable than would a change in the long-lived reservoirs. It follows that the long-term trend in atmospheric CO_2 that Keeling detected is not related to changes in the short-term carbon reservoirs. On the other hand, the long-lived reservoirs could, over a century, transfer the amounts of carbon that Keeling discovered. But it is unlikely that the rate of transfer would have been sustained so long or, more recently, would have accelerated exactly in pace with fossil-fuel consumption. Although a component of the atmospheric carbon increase may be related to changes in land biomass, the potential magnitude of the net biomass contribution is usually estimated as less than 1 Gt-C/yr—that is, less than one-third the observed rate of increase.

12.2.2   THE PROBLEM WITH ENERGY ADDICTION

The basic cause of the increase in atmospheric carbon dioxide is fossil-fuel consumption. The problem is rooted in the easy access that humans have had to fossil-fuel stocks and the energy-voracious technology that evolved to take advantage of cheap, accessible energy.

Fossil Fuels: Easy Energy

The first humans did not have fire as a tool. But being resourceful, they soon tamed the power of fire. The original fuels were dead vegetation and perhaps peat. Then, several thousand years ago, coal, lying in open veins, became the choice for making the hot fires needed to forge metals. Coal became more valuable within the past few hundred years, and men were sent into mines to dig it out. Oil, bubbling out of the Earth, became a commodity near the middle of the nineteenth century.[6] Natural gas, leaking out with the oil—initially a bother because it was so flammable and explosive—now is utilized as the cleanest-burning form of fossil fuel. Natural gas is mostly methane. Humans literally stumbled onto coal and oil. As resources, they are cheap and, at one time, were easy to mine. Punch a stick into the ground in some places, and oil would ooze out. Those days are gone. Now we must drill and dig into the bowels of the Earth to get at the stuff. But it is still cheap, at least cheaper than most other fuels.

Figure 12.3 shows the sources of energy used in the United States since 1955. Petroleum, coal, and natural gas—the fossil fuels—dominate the picture. The two cleanest sources of energy, hydroelectric and geothermal, together account for less than 5 percent of the total energy use. Nuclear energy takes up about the same fraction. More than 90 percent is left for fossil fuels, and of this, petroleum is the largest fraction. Although the non-fossil energy sources do not emit carbon dioxide into the atmosphere, they have other problems (Section 12.5).

Sources of Carbon Dioxide

Figure 12.4 shows the sources of carbon dioxide according to activity and political entity. Fossil-fuel combustion is the principal global source of carbon dioxide (~73%), with deforestation estimated to be the second major source (~25%). The production of cement, which involves crushing and baking calcium carbonate, also releases some carbon dioxide. Considering the geographical distribution of CO_2 sources, the United States is the single largest emitter, accounting for about 16 percent of the world total. The European Union (EU), consisting of France, Germany, Great Britain, Italy, and almost a dozen other nations, emits, in total, about the same amount as the United States does. The nations of eastern Europe, once within the Soviet sphere, produce even more CO_2 than the EU does. Obviously, there is no single culprit in carbon dioxide emissions. Several

6. Edwin Laurentine Drake drilled the first productive oil well in the United States at Titusville, Pennsylvania, in August 1859. He died in poverty after failing to patent his drilling techniques and losing his money to speculation on oil fields.

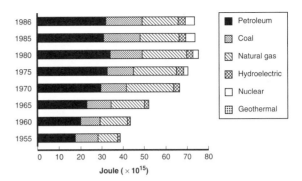

Figure 12.3   Energy production in the United States for selected years during the past half-century. The total energy output in units of 10^{15} joules per year is given (each unit of energy output is roughly equivalent to about 33 megawatts of continuous power). The contributions from various energy sources are indicated. The fossil-fuel sources include petroleum, coal, and natural gas. (Data from Hammerle, R. H., J. W. Shiller and M. J. Schwartz, "Global Warming," Ford Research Technology Assessment Series, September, 1988. Adapted from Figure 10.)

different kinds of fuels used in dozens of countries by thousands of industrial facilities and millions of vehicles cumulatively create a world-class problem.

The states of eastern Europe produced, and continue to produce, enormous quantities of carbon dioxide from coal combustion. Their "planned" economies were spectacular failures. Operations are propped up on a foundation of cheap energy from coal. But energy efficiency is dismal, and much more coal than in Western nations must be burned for the same productivity. Moreover, the "brown" coal often used in the East is highly polluting, as is coal in general. Environmental controls on emissions have been forgotten in the desperate struggle to produce and survive. The eastern European countries have thus evolved an inefficient, polluting energy system that attacks the environment in many ways: unhealthful regional air pollution, poisoned rivers, and massive carbon dioxide emissions.

There is concern that the developing nations will soon become the largest consumers of energy and emitters of carbon dioxide. Large reserves of coal in China, for example, are being developed to support rapid industrialization. Other developing nations with burgeoning populations, like India, will eventually demand massive energy production. Many of these nations are already busy stripping away easily accessible biological and mineral resources without the guidance of long-range planning. The developed

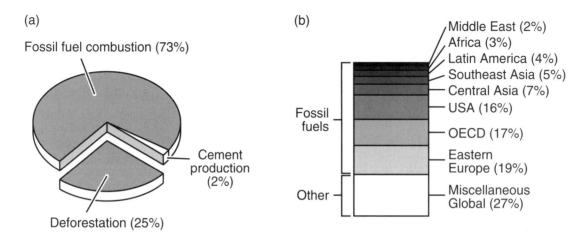

Figure 12.4 Relative sources of global carbon dioxide emissions according to (a) cause and (b) region. The relative magnitudes of the emissions are given as a percentage of the total. (Data from U. S. Environmental Protection Agency)

countries may, at some point, need to offer advanced, energy-efficient technologies and other financial and economic incentives to their less affluent neighbors. Such issues are beyond the scope of this discussion. Nonetheless, unless all nations cooperate, controlling and reducing global carbon dioxide emissions will be difficult. Decreases in emissions in North America, Europe, and Japan could be more than offset by increases in emissions in Asia, Africa, and South America.

Figure 12.5 shows the main sources of two key greenhouse compounds, CO_2 and CFCs. (Here and elsewhere in this book, the related chlorofluorocarbons are treated as one species, unless additional specificity is required. See Section 13.5 for more details concerning the CFCs.) Notice that the estimates for the fractions of the different sources contributing to carbon dioxide emissions differ somewhat between Figures 12.4 and 12.5. In Figure 12.4, fossil fuels are accorded 73 percent of the total source, and in Figure 12.5(a) the comparable figure is 60 percent. In fact, the fossil-fuel source is well documented in terms of the *absolute* amount of carbon consumed. Rather, the uncertainty actually lies in the absolute size of the biomass/deforestation source. The data are meant to reflect the net source for the atmosphere, which includes the effects of biomass destruction by deforestation and the response of the remaining biomass to changes in carbon dioxide and climatological factors. The difference between the numbers in these figures is not critical to our discussion.

The data in Figure 12.5 emphasize the CO_2 source associated with motor vehicles worldwide.

Nearly one-quarter of all fossil-fuel emissions are connected to cars, trucks, and buses. In addition, nearly one-quarter of all CFC emissions (Figure 12.5[b]) originates in air-conditioning units on motor vehicles. The large fraction of total greenhouse gas emissions related to driving is, besides the air pollution it generates, a major reason to improve the fuel efficiency of automobiles and find alternative substances to use in leaky automotive air conditioners.

12.2.3 A GLOBAL RESERVOIR PERSPECTIVE

Section 10.2.4 described the carbon cycle in terms of its global reservoirs and the exchange of carbon between reservoirs on various time scales. Regarding the greenhouse warming due to carbon dioxide emissions, we are interested in recent events. We wish to look forward into history a few decades to a century at most. We thus can neglect the long-term reservoirs for now and focus on the fate of "new" carbon added to the climate system by human activities.

A Contemporary Reservoir Model

Figure 12.6 shows a simplified reservoir model for carbon that applies to the present time. The principal reservoirs of interest are the atmosphere, land biosphere, and surface oceans. The total excess source of carbon for the system is about 6 gigatonnes of carbon per year (6 Gt-C/yr). As noted earlier, most of this is generated by fossil-fuel combustion. Recalling Section 10.2.4, it is obvious that the fossil fuel is mined from deposits of organic carbon laid down

(a)

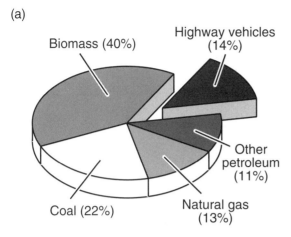

Biomass (40%)

Highway vehicles
(14%)

Other
petroleum
(11%)

Coal (22%)

Natural gas
(13%)

(b)

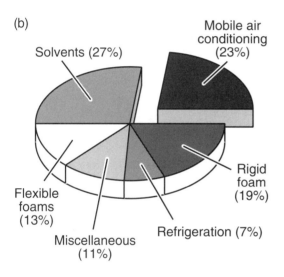

Solvents (27%)

Mobile air
conditioning
(23%)

Rigid
foam
(19%)

Flexible
foams
(13%)

Refrigeration (7%)

Miscellaneous
(11%)

Figure 12.5   Relative sources of the greenhouse gases (a) carbon dioxide and (b) chlorofluorocarbons for the United States. Sources are specified as a percentage of the total in each case. The components associated with motor vehicles are highlighted. (Data from Hammerle, R. H., J. W. Shiller, and M. J. Schwartz, "Global Warming," Ford Research Technology Assessment Series, September, 1988. Adopted from Figure 7.)

hundreds of millions of years ago. The natural cycling time of this carbon back to the surface is perhaps 200 million years. Mining of carbon deposits accelerates this process. The fast surface reservoirs then respond in two ways.

1. The excess carbon dumped into one of the reservoirs (the atmosphere), is redistributed among the closely coupled reservoirs (Figure 12.6) through the fast exchange processes.

2. The fast reservoirs slowly transfer the excess carbon into the longer-lived but less accessible reservoirs (deep oceans and sediments).

For the contemporary problem of global warming, the first process is more important.

The total amount of excess carbon for the atmosphere is roughly 7 Gt-C/yr (Figure 12.6). The 6 Gt-C/yr due to fossil fuels is closely determined by economists. The ~1 Gt-C/yr attributed to deforestation is more uncertain. The estimate is obtained using satellite photographs to determine the areas of forest cleared each year and using measurements of the quantity of biomass in these forests. Most of the vegetation that is cleared is burned. This releases the mineral nutrients held in plant tissues and the soil detritus for use in subsequent crop cultivation. The crops themselves reabsorb through photosynthesis a small fraction of the carbon dioxide released by combustion. All in all, the estimate of 1 Gt-C/yr is sound.

The source of carbon must be accounted for in the accessible reservoirs. The quantity of carbon accumulating in the atmosphere as CO_2 can be accurately calculated from Keeling's (and other) measurements and amounts to some 3 Gt-C/yr. Physical oceanographers have spent a good deal of effort to determine the rate at which carbon is being transferred to the seas. Recent studies show perhaps 3 Gt-C/yr of the excess carbon entering the oceans. Carbon dioxide physically dissolves in water to form carbonic acid and carbonates (Section 10.2.4). The quantity of carbon that can dissolve in the surface oceans depends on the water's temperature, alkalinity, and other factors. Generally, as the oceans warm, less CO_2 can be dissolved. Biological activity and the accumulation of organic matter in the surface oceans are also difficult to modify. As noted before (Section 10.2.4), the amount of biological activity that can be sustained is controlled by the availability of nutrients. Thus unless the supply of nutrients is altered, only a limited amount of carbon can be assimilated by means of photosynthesis. The average ratio of carbon to nitrogen in marine microorganisms is referred to as the **Redfield ratio**, which has a value of about 7 C-atoms/N-atom.

The rate of transfer of CO_2 into the oceans depends on the state of the sea surface and on the amount of carbon already dissolved. In general, as the amount of carbon dioxide dissolved in ocean water increases, the more difficult it becomes to

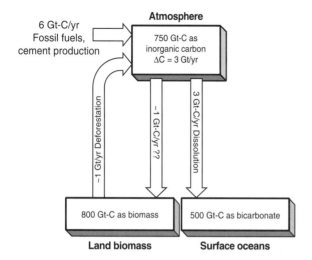

Figure 12.6 The short-term global reservoirs and exchange rates that affect excess carbon generated by modern anthropogenic activities. The measured rate of accumulation of carbon as CO_2 in the atmospheric reservoir is indicated as ΔC.

absorb further atmospheric emissions. In other words, the fraction of CO_2 that remains airborne increases as the amount already taken up by the oceans increases.

Using specific chemical tracers, oceanographers calibrate their estimates of the amount of excess carbon transferred to the oceans. The best of these is radioactive carbon-14, or ^{14}C. During the extensive atmospheric testing of nuclear weapons in the 1950s and 1960s, large quantities of ^{14}C were produced in the form of $^{14}CO_2$. The tagged radioactive carbon dioxide was then transferred to the oceans and the biosphere, just as other CO_2 was. By measuring the amounts in the oceans at different times and their distribution in the world oceans, the processes of transfer to and dispersion through the seas can be deciphered. Carbon dioxide's impact on climate is better understood as a result. Nuclear mushroom clouds, which produced the worldwide fallout of dangerous radioactive isotopes of iodine, strontium, and cesium, at least had one small silver lining.

The Case of the "Missing" Carbon

From the carbon balance in figure 12.6, it is obvious that the atmospheric source of ~7 Gt-C/yr is not balanced by the 3 Gt-C/yr increase measured in the atmosphere and the ~3 Gt-C/yr estimated to dissolve in the oceans. An additional ~1 Gt-C/yr is not accounted for in figure 12.6 (that is, 6 + 1 − 3 − 3 =

1). That's a lot of carbon! It would be hard to sweep 1 *billion* tons per year under the rug. But the world is a big place, and one or more of its enormous reservoirs must be hiding the "missing" carbon. Which one?

It has been suggested that as CO_2 concentrations increase, the biosphere is fertilized, and plant growth accelerates. This would absorb part of the extra carbon in the form of vegetation. The fertilization concept has yet to be adequately quantified. No one is sure if fertilization actually occurs. Different species of plants respond differently to higher levels of CO_2. Some grow faster, and others do not. Moreover, the effectiveness of adding CO_2 to the atmosphere to induce plant growth depends on the availability of nutrients and soil moisture. Temperature is another factor that must be considered, since relatively small variations in temperature can have a large impact on plant growth. Because the increasing amount of carbon dioxide in the atmosphere may alter temperatures and precipitation in currently unpredictable patterns, the net response of the biosphere to increasing atmospheric CO_2 is not so clear.

The land biosphere is the prime candidate as the sink for the excess carbon. Recent detailed calculations using global models for the distribution of vegetation, precipitation, and other factors indicate that the boreal forests of the Earth may be absorbing up to 2 Gt-C/yr. The total biomass in the land reservoirs is 2800 Gt-C (Section 10.2.4). The absorption of 1 Gt-C/yr would hardly be noticeable, at least for many decades. Nevertheless, as the world's forests continue to be stripped away, the amount of possible terrestrial biospheric compensation diminishes. The forests are the repositories of most of the Earth's biomass. Destroying them and not cultivating equally massive stands of new vegetation can, in the long run, only *increase* the atmospheric burden of carbon dioxide. The missing carbon may not be missing for long.

Carbon and other pollutants that humans produce can not disappear spontaneously. Mass is conserved in the Earth system. The carbon we add to the atmosphere must find a resting place. In the case of the "missing" carbon, we do not know where to look because we do not fully understand the global carbon cycle. The pathways that account for the removal of excess carbon dioxide from the atmosphere are critical; more time and diligent research will be required to resolve this issue.

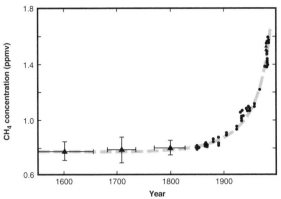

Figure 12.7   Trend in the atmospheric concentrations of methane over the past several centuries. The methane abundances are given in parts per million by volume. The data were obtained from air bubbles trapped in glacial ice and recently cored for analysis. Uncertainty in the older values is indicated by crosses, which define the possible variance in date and quantity. (Data from G. I.Pearman and P. J. Fraser, "Sources of Increased Methane," *Nature* 332 [1988]: 489.)

Nevertheless, we are very close to solving the mystery of the missing carbon. It is not necessary—indeed, it might be a mistake—to wait until the last details are ironed out before proceeding to assess the potential future impacts of CO_2 emissions and atmospheric accumulation.

12.3   Other Greenhouse Gases

In Chapter 11, we looked at the radiative properties of a number of key greenhouse-active gases. Here, we investigate the sources, sinks, and atmospheric concentrations of those gases (Table 12.1). All these gases have much lower concentrations than does water vapor or carbon dioxide, but they do exhibit strong absorption bands in the atmospheric window region (Section 11.4.1). Being unopposed, even small amounts can be significant.

12.3.1   METHANE

Methane has a current mixing ratio of about 1.8 parts per million by volume (ppmv). This is equivalent to about 3.5 gigatonnes (or 3500 megatonnes) of carbon (compared with 750 Gt-C as CO_2).[7] Meth-

ane has not always been so abundant. Figure 12.7 shows the recent history of methane concentration as deduced from ice bubbles. As with carbon dioxide (Figure 12.1), methane concentrations before the mid-nineteenth century were relatively constant, with much lower values than today. A rapid growth in methane abundances seems to keep pace with the recent rapid growth in the human population and the expansion of agriculture worldwide. Methane is currently increasing at the rate of about 1 percent per year. Its concentration has already doubled since before the Industrial Revolution and could double again in the next century.

The Methane Budget

The estimated global sources and sinks of methane are summarized in Table 12.2. Although CH_4 is known to have a number of natural sources, particularly microorganisms that consume organic matter in soils, the anthropogenic sources obviously must be comparable in order to explain the recent historical increase (Figure 12.7). In fact, when quantified, as in Table 12.2, the anthropogenic sources of CH_4 exceed the natural sources. At present, more than 70 percent of the total CH_4 emissions are associated with anthropogenic activities. The total source is dominated by bacterial methanogenesis occurring in swamps and wet soils that become anaerobic, or oxygen poor. Humans have constructed enormous artificial swamps in the form of rice paddies.[8] Rice is a major food crop, and rice paddies are currently a source of methane emissions comparable to those from natural wetlands.

The overall global budget of methane is nicely balanced. Table 12.2 shows that the sources and sinks, determined independently, agree within the range of uncertainty in the numbers (notice that the accumulation of CH_4 in the atmosphere is included

---

7.   Remember that the masses of materials under discussion relate to a particular component of the material. Here, for example, the carbon mass in methane is specified. Other references

may give the total mass of methane, including its hydrogen component. The relative masses of CH_4 and C are 16 and 12, respectively, according to the chemical formula and atomic weights of the elements involved. The relative masses of CO_2 and C are 44 and 12, and of N_2O and N, 44 and 28, respectively. Be careful!

8.   The area of the Earth converted to rice paddies is somewhat less than 2 million square kilometers. This area increased at the rate of about 1%/yr during the first half of the twentieth century, but more recently has grown at one-quarter that rate. Meanwhile, rice production over the same period has increased at the rate of about 3%/yr, more than doubling since the 1940s. The increased use of fertilizers and pesticides accounts for most of the accelerated rice harvest.

Table 12.2   The Global Methane Budget

Source	Rate (Mt-C/yr)[a]		Sinks	Rate (Mt-C/yr)[a]
Methanogenesis			Reactions with	320
Natural wetlands	85	(75–150)	hydroxyl	(260–380)
Rice paddies	80	(45–130)		
Bovines	60	(45–75)	Atmospheric	35
Natural-gas leakage	75	(60–90)	accumulation	(30–40)
Landfills/waste	60	(40–80)		
Biomass burning	40	(35–75)	Soil uptake	20
Termites	15	(5–75)		(15–25)
Oceans and lakes	10	(5–15)		
Clathrates	?	(0–100?)		
Totals	~425	(310–765)		~375 (305–445)

[a] Units are millions of metric tons (tonnes) per year, or Mt-N/yr. One Gt is equal to 1000 Mt. The numbers in parentheses define the range of uncertainty in the sources and sinks.
Source: Data from R. Cicerone and R. Oremland, *Global Biogeochemical Cycles* 2 (1988): 299; and the World Meteorological Organization Global Ozone Research and Monitoring Project.

as a "sink" term in the table). Accordingly, it is believed that no important sources or sinks have been overlooked. Obviously, there is room for improvement in the data. Other data also support the accuracy of the budget in Table 12.2. The carbon isotopes in CH_4 provide clues to the sources of methane. "Old" sources, such as fossil fuels, have different ^{13}C abundances than do "new" sources, connected with biological activity. The fraction of fossil sources represented in Table 12.2 is consistent with the measured ^{13}C in atmospheric methane.

The major sink for methane is its reaction with the hydroxyl radical (OH). This process occurs mainly in the troposphere:

$$CH_4 + OH \rightarrow CH_3 + H_2O \quad (12.2)$$

As a result, the average lifetime of a methane molecule is about 10 years,[9] a relatively short lifetime. If the emissions of CH_4 change, the atmosphere will respond within a decade. This time is actually much longer than the time over which human activities that release methane have developed. Methane in the future can be expected to ebb and wane in pace with human activities.

9. This is obtained using the box model equation $\tau = Q/L$ (Equation 10.2). With Q = 3500 megatonnes and L = 350 megatonnes/year, the residence time is just $3500/350$ = 10 years.

Natural gas consists mainly of methane. The gas is colorless and odorless; your local gas company adds an odor so you can smell leaks. Methane seeps from inside the Earth. Deposits of fossilized carbon often decompose partly into natural gas. The gas easily diffuses through fissures in the ground. Some geologists believe there are enormous deposits of natural gas trapped in rock formations. If we could tap these bubbles of gas, they claim, our dependence on coal and oil could be relieved. But no one has yet found these bubbles. Other researchers have proposed that the permafrost at northern latitudes—regions where the soil never thaws even in summer because the warming rays of the sun are too weak—holds large quantities of methane trapped in clathrates. The *clathrates* are composed of methane molecules held within a crystalline cage formed by water molecules, or ice. Of course, the methane could be released if the ice were melted. But vast methane deposits in the form of clathrates have never been found.

Methane seeps into the atmosphere when fossil fuels are mined. Oil and coal deposits naturally contain methane. The natural gas is often purposefully released or burned at wellheads (flared). Even when natural gas is captured as a resource, some of it leaks into the atmosphere. Apparently, the human component of the natural-gas emission to the atmosphere dominates the geologic component.

Methane and Man (and Beast and Bug)

Termites emit methane. Why? They are not microorganisms. Termites, however, eat cellulose, which animals normally cannot digest. To overcome this problem, termites have evolved a unique gut in which microorganisms that can digest cellulose thrive. These microbes convert cellulose to organic matter that the termites can assimilate—and natural gas that they emit. Termite mounds exude methane. OK, so how many termites are there exuding methane? No one knows for sure. One estimate placed the figure at 100 pounds of termites for every human on Earth. More realistically, termites are a secondary, yet important, source.

Cattle, sheep, and other ruminants also generate methane. There are more than 1 billion head of cattle in the world today. On average, every two families owns a cow. Their population grew by nearly 2 percent a year during the 1950s, 1960s, and 1970s, but has tended to stabilize recently. Cows and other ruminants live by a process called *enteric fermentation*. The plant material that they continually chew on is digested by methane-generating bacteria in their guts. The measured emissions of methane from cattle are significant (Table 12.2). Their droppings, or chips, also reek of methane. Belching cows have been implicated in a climate change! A few researchers feel that our bovine friends have gotten a bum rap. Plans are under way to approach this delicate problem from other directions, to determine once and for all just exactly what gases cows emit.

Over the past few decades, humans have buried enormous quantities of organic waste in landfills. Beneath layers of soil, in the anaerobic environment, methanogens thrive. Methane gas is generated and seeps out. The landfills become natural-gas factories and potential time bombs. Some homes built on landfills have had to be abandoned because of the threat of gas explosions and health hazards. The methane emissions from these garbage burial grounds and other anthropogenic waste-treatment sites amounts to some 60 million metric tons (of carbon) per year. Pound for pound of body weight, humans are more efficient at generating methane than cows are.

12.3.2 NITROUS OXIDE

The atmosphere contains about 1.5 gigatonnes (or 1500 megatonnes) of nitrogen in the form of N_2O,

with an average concentration of 0.31 part per million by volume, ppmv (which is equal to 310 ppbv). The total quantity of nitrogen in the atmosphere is greater by a factor of about 2.6 million. Nitrous oxide, also known as laughing gas, has important roles as a greenhouse-active gas (Section 11.4.1) and as a cause of stratospheric ozone depletion (Section 13.5.2). Careful measurements of N_2O over the past decade show that its atmospheric concentration is rising steadily. The rate of increase is relatively small, amounting to about 0.2 percent per year. At this rate, it would take approximately 350 years to double the concentration of N_2O (see Footnote 5). Nitrous oxide is not an imminent threat, but it has a very long lifetime, roughly 150 years. Accordingly, changes in nitrous oxide are much more permanent. Once we alter the abundance of N_2O, we will have to live with the change for a long time.

The Nitrous Oxide Budget

The global budget of nitrous oxide is summarized in Table 12.3. The only significant sink that has been identified is stratospheric photolysis, which entails the transport of N_2O from the troposphere to altitudes of about 30 kilometers, well into the stratosphere. At great enough heights, above most of the Earth's stratospheric ozone layer, N_2O molecules are exposed to energetic solar ultraviolet radiation that can dissociate them. The process is irreversible, and no nitrous oxide is created by photochemistry in the stratosphere. The stratospheric loss rate is equivalent to 10 megatonnes of nitrogen per year (10 Mt-N/yr). Hence the N_2O lifetime is 1500 Mt-N/10 Mt-N-yr^{-1}, or 150 years, as just noted.

The nitrous oxide budget is in rough balance, well within a range defined by the uncertainty in the estimated sources and sinks; that is, the identified sources offset the known sinks. The sinks are easiest to determine in this case. Stratospheric photodissociation can be calculated fairly accurately with computer models that are used to study the depletion of the ozone layer. The rate of accumulation in the atmosphere (treated here as a sink) has been found from direct measurements.

The source of most N_2O is denitrification in soils, a process fundamental to life (Section 4.2.1). Fixed forms of nitrogen, which are an essential nutrient and a component of organic tissue, are reduced by denitrifying bacteria to N_2O and N_2. Certain plants,

Table 12.3   The Global Nitrous Oxide Budget

Sources	Rate (Mt-N/yr)[a]	Sinks	Rate (Mt-N/yr)[a]
Denitrification in soils	6 (4–8)	Stratospheric photolysis	10 (7–13)
Agricultural combustion	1 (0.5–1.5)		
Fossil fuels	1 (0.5–1.5)	Atmospheric accumulation	3 (2–4)
Biomass	1 (0.5–1.5)		
Oceans	2 (1–3)		
Totals	11 (6.5–15.5)		13 (9–17)

[a] The units are millions of metric tons (tonnes) per year, or Mt-N/yr. One Gt is equal to 1000 Mt. The numbers in brackets define the range of uncertainty in the sources and sinks.
Source: Information from the World Meteorological Organization Global Ozone Research and Monitoring Project.

the legumes, have nodes on their roots in which nitrifying bacteria live. These plants act as natural fertilizers for soils. Other sources of fixed nitrogen are combustion and lightning (Section 10.2.2). The fixed nitrogen, generated in the soil or washed there by rain, is continuously consumed by denitrifying bacteria. Humans intervene in this cycle by applying artificially fixed nitrogen to the land as a fertilizer. As the fertilizer is used for crops, some of it is converted to N_2O. At one time, the fraction of fixed nitrogen in fertilizer that was released as nitrous oxide was assumed to be quite large. Predictions of atmospheric buildup and adverse environmental effects followed. The rate of nitrous oxide conversion and atmospheric accumulation are now known to be quite modest. Nevertheless, the long lifetime of N_2O should counsel caution for the future.

Denitrification can also occur in water that is rich in fixed nitrogen. Sewage outflows typically have high fixed-nitrogen contents and emit nitrous oxide. Anthropogenic combustion, in automobile engines and power plants, also emits nitrous oxide. Fortunately, the amounts appear to be small. One unusual source of N_2O, not listed in Table 12.3, is the production of nylon, primarily for fabrics and hosiery. According to one estimate, the emission rate is perhaps approaching 1 million tons (of nitrogen) per year. We all may have to return to cotton and wool socks to protect the environment, even if they itch more.

12.3.3   CHLOROFLUOROCARBONS

The chlorofluorocarbons (CFCs) are a subset of a larger class of compounds known as **halocarbons**.

The halocarbons are molecules containing one or more halogen atom (of chlorine [Cl], fluorine [F], bromine [Br], or iodine [I]) and derived from organic substances. Many halocarbons are derived from methane (CH_4), for example, by substituting halogen atoms for hydrogen atoms. When only chlorine and fluorine are involved, the resulting compounds are referred to as **chlorofluoromethanes**. When all the hydrogen atoms have been replaced in the original organic molecule, the resulting compounds are referred to as **chlorofluorocarbons**. Two of the most abundant CFCs are CFC-11, or $CFCl_3$, and CFC-12, or CF_2Cl_2. The properties of a variety of important halocarbons are summarized in Table 12.4 (see also Table 13.1). (The properties and chemistry of chlorofluorocarbons and other halocarbons are discussed in detail in Section 13.5.2 from the perspective of global stratospheric ozone depletion.)

Too Smart for Our Own Good

Of all the compounds listed in Table 12.4, only methyl chloride (CH_3Cl) and methyl bromide (CH_3Br) have natural sources. The other compounds are exclusively human made. These substances have a number of uses in industry and everyday life, from air conditioning and refrigeration to dry cleaning and aerosol propellants. One of the most important qualities of the CFCs is their chemical inertness. These compounds can be sprayed in someone's face or leak from a refrigerator with no immediate danger or long-term health hazards. CFCs have no smell, never create a mess when they leak, and harmlessly dissipate into the air without a trace.

Table 12.4   Halocarbon Concentrations and Trends

Halocarbon	Common designation	Mixing ratio (pptv)	Rate of increase (%/yr)	Atmospheric lifetime (yrs)
$CFCl_3$	CFC-11	270	5	76
CF_2Cl_2	CFC-12	460	5	139
CF_3Cl	CFC-13	5	400	
$C_2F_3Cl_3$	CFC-113	65	10	92
$C_2F_4Cl_2$	CFC-114	15	200	
C_2F_5Cl	CFC-115	5	400	
CCl_4	Carbon tetrachloride	110	1	67
CHF_2Cl	HCFC-22	110	7	15
CH_3Cl	Methyl chloride	600	0	1.5
CH_3CCl_3	Methyl chloroform	135	7	8
CF_2ClBr	Halon-1211	~2	15	25
CF_3Br	Halon-1301	~3	20	110
CH_3Br	Methyl bromide	~15	15	1.5

Source: Information from the World Meteorological Organization Global Ozone Research and Monitoring Project.

These compounds, invented by humans, were one of the first miracle compounds offered by modern chemistry to improve the quality of life.

It turns out there are a few drawbacks to using CFCs (Section 14.2.3). The harmless dissipation of CFCs into air in fact leaves a residue of the gas throughout the atmosphere. CFC-11, CFC-12, and a variety of other artificial halocarbons are not biodegradable. Because biological organisms have never been exposed to these compounds, they have not evolved the means to absorb or digest them. By design, CFCs are extremely inert and unreactive. No significant sinks have yet been identified in the oceans, on land, or in the lower atmosphere. The only proven sink involves transport to the stratosphere and photodissociation by solar ultraviolet radiation. This is similar to the dominant sink for nitrous oxide (Section 12.3.2). The breakdown of CFCs by sunlight releases the chlorine atoms into the stratosphere, where the chlorine can attack and destroy ozone. This rather nasty side effect of removing CFCs is discussed at length in the next chapter (see, in particular, Sections 13.5-13.7).

Because of their chemical stability, relative insolubility in water, and lack of biological functionality, the CFCs have very long atmospheric lifetimes. Table 12.4 gives estimates of the lifetimes of the important halocarbons. For CFCs, these lifetimes are about 100 years or more (similar to that of N_2O). Herein lies a trap. Once CFCs are emitted into the atmosphere, they remain there for a century or more. Worse than that, they accumulate over that length of time. In a sense, because CFCs are safe to use in the short term, they are dangerous to use in the long term. We may have been too smart for our own good. (This point is taken up again in Section 14.2.3. Suggestions to correct the situation, aside from reducing the production of CFCs, are discussed in Section 14.3.2.)

It turns out that many of the chlorinated compounds invented in this century have dangerous side effects. Many have been widely restricted in their applications. Notorious are the chlorinated pesticides, including dichlorodiphenyltrichloroethane (DDT) and related compounds (Section 7.2.5). Some of the most toxic compounds known are the dioxins, such as the tetrachlorodibenzodioxins (TCDDs), which can be generated during the incineration of waste materials containing other organic chlorides. Trichloroethylene (TCE), widely used as an industrial cleaning agent, is carcinogenic. TCE improperly disposed of has been detected in some water supplies. Another common dry-cleaning agent, carbon tetrachloride (CCl_4), is extremely volatile and contributes to global ozone depletion. A wide variety of other chlorinated compounds are toxic or contribute to ozone depletion or greenhouse warming, or both. One particularly plentiful industrial chlorocarbon is methyl chloroform, which has an atmospheric lifetime of about 7 years (Table 12.4). The atmosphere and the environment in general are filled with these manufactured compounds.

The CFC Budget

The atmosphere presently contains about 10 million tonnes of chlorine (Mt-Cl), in the form of chlorofluorocarbons, and another 10 Mt-Cl as other halocarbons. The total burden of CFCs roughly corresponds to the *total* emissions since CFC production began to accelerate in the late 1950s. This is consistent with the long lifetimes of these compounds; the atmosphere has been acting as an enormous holding tank for the CFCs leaking from countless small sources around the world over the past four decades. In the 1980s, the total rate of emission of chlorine in the form of CFCs was about 0.4 Mt-Cl/yr. That rate has been decreasing recently and is expected to fall dramatically by the end of the century as a new international treaty limiting CFC production—the Montreal Protocol—takes effect (Section 13.8.1).

The present rate of increase in the atmospheric concentrations of the CFCs is about 5 percent per year. Until recently, the rate has been as high as 7 percent per year. The rate of growth has tapered off because the annual emission rate has dropped. The decrease in CFC emissions has been a direct consequence of concern about stratospheric ozone depletion. The first confrontation between environmentalists and manufacturers of CFCs centered on ozone. Most of the early research into the atmospheric effects of CFCs focused on their chemical impact on the stratosphere. More recently, since the issue of global warming heated up, CFCs have also been implicated as greenhouse-active gases. Indeed, they are so potent in creating a greenhouse effect that even minute amounts in the atmosphere (at the parts-per-billion level) are significant. In fact, a few planetary scientists have suggested that CFCs might be used to modify, or engineer, the climates of cold, greenhouse-deficient planets like Mars, to make them habitable. So far, the inadvertent engineering of the Earth's environment through fugitive CFC emissions has dominated the debate.

12.3.4 Ozone

Ozone is a critical trace constituent of the atmosphere. In the troposphere, the average ozone abundance is fairly low, approximately 10 parts per billion by volume. In the stratosphere, the concentration can be 1000 times as large, or 10 parts per million by volume. In locally polluted tropospheric air, or smog, the ozone abundance can exceed 0.1 ppmv. Even though the global amount of tropospheric ozone is relatively small, large regions of the Northern Hemisphere show concentrations 10 times larger than the global average. These areas are contaminated by heavy pollution from North America and Europe. Included in the pollution are hydrocarbons and nitrogen oxides that lead to photochemical ozone production (Sections 3.3.4 and 6.2). In these polluted areas, ozone concentrations appear to be increasing by about 1 percent per year. Other zones of high ozone concentration are over the tropical Atlantic and Pacific Oceans. These ozone highs seem to be associated with biomass burning in the tropical latitudes; the burning also releases the ingredients for photochemical ozone production.

Ozone as a Pollutant

The atmosphere holds a total of about 3 gigatonnes of ozone, most of which is in the stratosphere. If this ozone were mixed uniformly throughout the atmosphere, from the ground up, the average concentration in air would be about 0.4 ppmv, equivalent to a stage-2 smog alert everywhere all the time![10] Clearly, we are fortunate that ozone forms naturally in the stratosphere and is confined there, with only a small amount of leakage to the surface. The resulting ozone layer is relatively stable, absorbing harmful ultraviolet sunlight before it reaches the surface. (This is discussed at length in Chapter 13.)

Ozone also plays a role as a greenhouse gas. Ozone has a strong absorption feature in the atmospheric window (Section 11.4.1). Changes in ozone abundances in the troposphere and stratosphere each influence the greenhouse effect incrementally. Most ozone is in the stratosphere, which is very cold, making ozone an effective greenhouse agent. Stratospheric ozone is also decreasing because chlorofluorocarbons are increasing (Chapter 13). However,

---

10.   Refer to Chapter 6 for the standards for ozone concentrations in smog and to Chapter 7 for a description of the effects of exposure to ozone in smog.

the chlorofluorocarbons themselves are greenhouse gases. As the CFCs increase, ozone decreases and thus compensates for the total greenhouse impact. The CFC warming effect increases, while that of ozone decreases.

The Ozone Effect

The question with regard to ozone in the atmosphere is: Do we want more or less? As a pollutant in urban areas, we certainly want less (Section 6.6). As a filter for ultraviolet radiation, we certainly want no less than currently exists, and perhaps more. Of course, we want it in the stratosphere, far away from our lungs and eyes. By eliminating chlorofluorocarbons, we maintain and likely increase ozone. We also eliminate the CFC greenhouse effect. These are beneficial results. But we also increase the greenhouse effect of ozone, offsetting some of these benefits. At present, the advantages—in terms of ozone depletion and the greenhouse effect—of reducing CFCs far outweigh the disadvantage of an increased ozone greenhouse effect.

12.4   The Warming Effect of Greenhouse Gases

Climate history may provide an indication of the greenhouse-warming effect of atmospheric gases such as carbon dioxide and methane. If it could be shown, for example, that the concentrations of these gases had changed at the same time the global climate had shifted, the two events perhaps could be connected. For prehistoric climate change, this is not an easy task, as the data are sparse and often ambiguous. For recent changes in the climate, the data are much better, although uncertainties remain. This topic—a possible historical link between climate change and greenhouse gas variations—is a critical element in the current debate about global warming.

12.4.1   CLIMATE HISTORY AND THE GREENHOUSE EFFECT

Section 11.6 discussed the various causes of climate change, both internal and external to the Earth. The long-term history of the Earth's climate is uncertain, but even so, the important role of greenhouse gases is strongly implied. With regard to the "faint sun"

paradox of the early Earth (Section 11.6.2), it has been suggested that carbon dioxide was present in much larger abundances than at present and warmed the climate to a habitable state. Only recently has evidence surfaced for elevated CO_2 concentrations 4 billion years ago in the form of fossilized traces of microscopic organisms that likely required large abundances of CO_2 in order to survive.[11] It has also been known for some time that the extreme heat of Venus's surface is due to the efficient retention of outgassed carbon dioxide in the atmosphere.[12] These exciting revelations show that large increases in atmospheric carbon dioxide are liable to warm the global climate significantly, in agreement with theory and prediction. However, what are the effects of smaller additions of CO_2? Are there any historical data to guide our evaluation of potential global warming from modest increases (say, doubling) in greenhouse gases?

Pleistocene Ice Ages

The Pleistocene epoch of geologic history was a time of significant cooling of the climate from the relative warmth of earlier ages. Over the past 600,000 years, the Pleistocene has been punctuated by severe cooling episodes, referred to as ice ages. During these cold periods, the global average temperature dropped by as much as 10°C and huge glaciers of ice advanced across the continents of the Northern Hemisphere. Warm interglacial periods, with the thawing and retreat of the glaciers, occurred periodically (Section 11.6.3). For only the most recent ice cycle, which began roughly 130,000 years ago, we have a combined proxy record of temperature and carbon dioxide. These data are shown in Figure 12.8.

What is most striking about the data is the nearly perfect correlation between the variations in carbon dioxide and temperature. When the temperature is

---

11.   The results of a preliminary interpretation of the fossil record were discussed by J. William Schopf, a University of California–Los Angeles paleontologist and expert on the early evolution of life and its footprint in fossils and other geologic data.

12.   One of the first scientists to develop the greenhouse theory of planetary atmospheres was Carl Sagan, an astronomer, philosopher, novelist, and inventor currently at Cornell University. Sagan is a modern-day polymath whose contributions to contemporary science and society rival those of Benjamin Franklin and Lord Rayleigh in their own times. Sagan is probably the most articulate spokesperson for science and communicator of science to the public in this century.

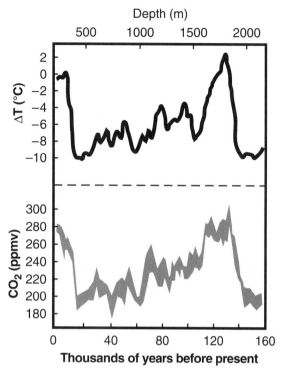

Figure 12.8   Comparison of atmospheric carbon dioxide concentrations and global average temperatures over the past 160,000 years. The CO_2 concentrations are given in parts per million by volume of air. The temperature is shown as a deviation, in degrees Celsius, from the recent average global temperature. The data were obtained using deep ice cores extracted from antarctic glaciers. The CO_2 concentration was determined by analyzing the gas trapped in air bubbles in the ice. The temperature was estimated by measuring the isotopic composition of the hydrogen content of the frozen water. (Data collected and analyzed by J. M. Barnola, D. Raynaud, Y. S. Korotkevitch, and C. Lorius, "Vostok Ice Core: A 160,000 Year Record of Atmospheric CO_2," *Nature* 329 [1987]: 408.)

this conclusion correct? Have we found direct evidence for the critical role of CO_2 in causing climate change?

It is important to note in Figure 12.8 that the maximum variation in CO_2 concentrations ranges from about 170 to 280 ppmv, or a change of about 110 ppmv. This is only slightly greater than the change in CO_2 concentrations that has occurred during the past century (about 80 ppmv) and is much smaller than the change that will have occurred by the time the CO_2 concentration doubles to about 600 ppmv in the twenty-first century (according to some projections). How could such relatively small variations in carbon dioxide have led to such major changes in global climate? The same question has been raised concerning the weak forcing of the climate system by the orbital perturbations of the Milankovitch cycles, which correlate very nicely with the ice ages. No one has a satisfactory explanation.

If one looks more closely at the data in Figure 12.8, another crucial feature emerges. The variations in carbon dioxide do not lead the variations in temperature; rather, they follow the temperature changes. This can be seen only after a careful reanalysis of the data at a finer time resolution than can be shown in the figure. There is an uncertainty of as much as 1000 years in the relative dating of the temperatures and CO_2 abundances using bulk ice-core samples. The reanalysis narrows this uncertainty sufficiently to show that CO_2 is responding to climate change, rather than that the climate is responding to carbon dioxide change. This important conclusion is also logical in light of what we have learned about CO_2. For one thing, as the climate (and so the oceans) warms, the amount of carbon dioxide that can be dissolved in the oceans decreases. More CO_2 will have to be accommodated by the atmospheric reservoir (Section 12.2.3). Because the oceans represent the major reservoir for carbon over periods of 1000 years, the lag in the CO_2 response is understandable.

Although the ice ages were not caused by changes in carbon dioxide, it is likely that CO_2 variations played a role in amplifying the climatic variations. Climate cooling leads to a decrease in atmospheric concentrations of carbon dioxide, because of the increased uptake by the oceans. Lower CO_2 concentrations lead to cooling by diminishing the greenhouse effect. The net result is an enhancement of the initial cooling—that is, a

rising, the carbon dioxide is increasing, and when the temperature is falling, so is the CO_2. Notice that the ice ages consist of many cycles of cooling and warming of differing magnitudes. Most of this variability can be correlated with the Milankovitch cycles in the Earth's orbital parameters (Section 11.6.3). The rest of the variations are probably related to internal chaos in the climate system (Section 11.6.1). From Figure 12.8, it is easy to conclude that ice ages are caused by changes in atmospheric carbon dioxide. Whenever some event reduces the concentration of carbon dioxide in the atmosphere, the climate cools down. When another event releases CO_2, the climate warms up again. Is

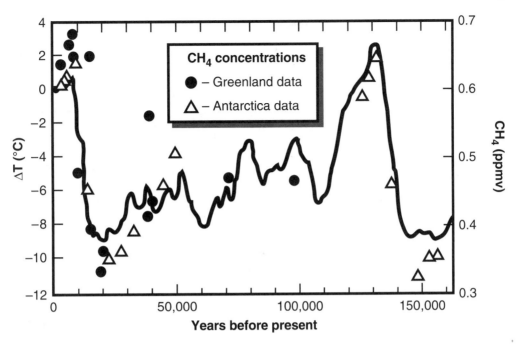

Figure 12.9   Comparison of global average temperatures (from Figure 12.8) with methane concentrations over the past 160,000 years. The methane data were obtained using air extracted from ice cores taken in Antarctica and Greenland (the source for each data point is identified by the symbols). The concentrations are specified in parts per billion by volume of air. (Data from D. Raynaud, J. Chappellaz, J. M. Barnola, Y. S. Korotkevitch, and C. Lorius, "Climatic and CH_4 Cycle Implications of Glacial-Interglacial CH_4 Change in the Vostok Ice Core," *Nature* 333 [1988]: 655. Adopted from Figure 1, p. 656.)

positive-feedback effect involving CO_2. As the climate warms, the CO_2 response enhances the warming. Thus in each case, carbon dioxide acts to increase the warming or cooling trend. This is important because the basic climate forcing that drives the ice ages is not yet known, and CO_2 may be an important player. The likely role of CO_2 in ice age climate shifts strengthens the argument that its current increase will lead to significant global warming. The magnitude of the predicted warming is consistent with a secondary, yet significant, role for CO_2 in ice ages.

Another intriguing set of data has been teased from ice cores taken in Antarctica and Greenland. The data for atmospheric methane concentrations and global temperature are shown in Figure 12.9. The methane results are not as accurate as the CO_2 results because of the much lower concentrations of CH_4. Nevertheless, the data reveal a striking correlation between temperature and methane abundance, similar to that in Figure 12.8 for temperature and carbon dioxide. As in the case of CO_2, it is unlikely that the change in methane caused the variations in temperature.

A more logical explanation, as for CO_2, is that CH_4 is responding to climate change. A plausible mechanism involves biological activity, the major natural source of methane. As the climate cools, biological activity slows down, and methane emission decreases. When the climate warms again, biological activity kicks in, and methane emissions increase. It is assumed that the primary loss mechanism for methane, its reaction with hydroxyl (Section 12.3.1), is relatively invariant. In fact, we might expect to have less hydroxyl during the colder climatic phase (because it is drier), which would act to keep methane concentrations from falling. Since methane actually decreases at these times, an even larger variation in the emission rate between warm and cool periods is implied.

Another possibility is that with a warming climate, methane is released from ice clathrates as they melt. The clathrates presumably re-form when the climate cools again. The clathrate concept, although yielding the proper direction for CH_4 change, has never been verified by the identification of significant quantities of methane trapped in ice. Moreover, the methane, once released, would be destroyed within a decade by

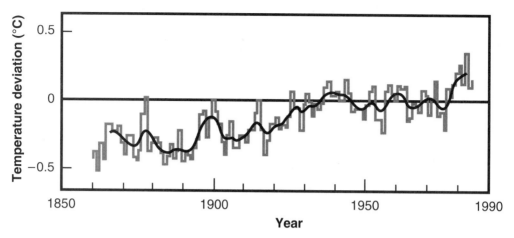

Figure 12.10   Yearly averaged temperature of the Earth since about 1860. The temperatures are presented as deviations from the mean avleu over the past 100 years. The heavy solid line represents a 5-year central running mean, which includes the temperatures 2 years before and after each year in the average. (Data from J. Hansen, D. Johnson, A. Lacis, S. Lebedeff, P. Lee, D. Rind, and G. Russell, "Climate Impact of Increasing Carbon Dioxide," *Science* 213 [1981]: 957.)

reacting with hydroxyl. The depleted methane would have to be replaced, either by biological activity or by leakage from natural-gas deposits.

The combined responses of carbon dioxide and methane to changes in climate detected in prehistoric data from air bubbles suggest a coupled reinforcing climatic role for these gases. Both apparently respond over long time spans in a manner that enhances the background cooling or warming.

12.4.2   RECENT TEMPERATURE TRENDS

Looking closer to the present, we find further evidence for global warming. Figure 12.10 provides a detailed analysis of recent variations in the global surface temperature. The data, collected over a period of about 130 years, represent the most accurate temperature record of the Earth during any period. The data reveal several important features of recent temperature variations. First, and perhaps most important, the general trend in the temperature is upward. The planet seems to be heating up at a significant rate. That trend has continued until the present. A second key feature, however, is the large variability in the temperature from year to year. Notice that the variation in the annual average temperature over a span of several years can be as large as the longer-term variation in the average temperature over a century. That is, for the existing limited measurements, the variability in the data is as large as the trend.

The overall warming trend, if that is what the data in Figure 12.10 actually show, is roughly 0.4°C per century. The year-to-year variability is that large, however. Moreover, there are numerous fits and starts where the temperature begins to change rapidly, and then slows down or reverses. Recall our review in Section 11.6 of the long-term variability of the climate. There, we identified a number of distinct causes of climate oscillation (variability) over long time scales (for example, the orbital perturbations of the Milankovitch cycles). The residual variations that could not be connected with specific forcings were attributed to organized internal noise (or chaos) in the complex climate system.

The variability seen in Figure 12.10 can also be attributed in large part to specific climate forcings on time scales of years to decades. For example, climate variations are known to exist with the following periods (periods of 1 year or shorter, such as the seasonal cycle, have been averaged out of the data in Figure 12.10): the quasi-biennial, or "2-year" oscillation of the tropical atmosphere; the El Niño Southern Oscillation, (ENSO) of the tropical oceans, with a period of roughly 3 years; and the 11-year solar cycle variation. In addition, there are nonperiodic irregular forcings such as major volcanic eruptions that cause significant climate anomalies from time to time (Section 11.6.4).

The most important of the short-time-scale fluctuations is the El Niño oscillation, the dynamical and thermodynamic coupling of the atmosphere and oceans. The mechanism that produces El Niño is not yet understood. What appears to happen is that the prevailing easterly trade winds, which constantly blow across the tropical Pacific Ocean (Section 2.4.2) suddenly subside.[13] This shuts off the upwelling of deep, cold ocean water along the western margin of South America, water that is normally driven by wind friction exerted on the surface of the ocean. The surface waters in the eastern Pacific therefore warm up. As a result, a strong convectively driven atmospheric circulation forms above the warm pool that can start up the winds again. The problem in understanding El Niño is determining what triggers the oscillation. The rocking of the Pacific Ocean from west to east and back again within its basin—like sloshing in a bathtub—seems to set the rhythm. The spread of the warm pool of water and its effect on the circulation of the tropical atmosphere are also critical. El Niño, in fact, generates large regional weather perturbations, having a particularly marked effect on precipitation. In the western Pacific, drought can prevail where rainfall is normally high, and in the eastern Pacific, the opposite conditions occur. Fishing, agriculture, and other economic enterprises are significantly affected by the ENSO.

Even after all the specific known sources of climatic oscillation have been accounted for in the temperature record, small residual temperature fluctuations remain, as well as a long-term warming trend. The small short-term variations represent internal climatic chaos, or global "weather" variations. The warming trend, amounting to perhaps 0.5°C since 1860, has been linked with the recent accumulation of greenhouse gases in the atmosphere. Others disagree, claiming that long-term solar variations lasting for centuries may explain the recent warming trend.[14] Section 11.6.2 discussed the correlation of sunspot activity with climate change. The Maunder minimum in the number of sunspots occurred at the same time as the period of

cooling referred to as the little ice age. Nevertheless, the greenhouse mechanism has the weight of physical evidence in its favor. The solar variability mechanism may contribute to, but probably does not dominate, the current warming trend.

Measuring the Temperature

Not only the cause but also the accuracy of the temperature data in Figure 12.10 can be questioned. The mercury thermometer, invented by Daniel Fahrenheit in 1714, has been the standard instrument to measure temperature since that time.[15] Because it is simple to do so, temperatures have been continually measured and recorded for the past two centuries. That is fortunate because an excellent record of recent historical temperatures now exists. But there are problems. First, the geographical distribution of the measurements is spotty. Temperature records are densest in the more developed regions of Europe, Asia, and North America, for example. Many parts of the world were sparsely populated in the nineteenth century, including the American West. The Europeans' settlement of Asia, Africa, and South America initiated temperature records in widely distributed outposts, although the distribution and reliability of these temperature records are questionable. More disturbing is the sparsity of data for the temperatures of ocean waters, which cover about two-thirds of the Earth's surface. It might seem hardly worthwhile to determine the average temperature of the Earth if the temperatures of the oceans are not known to a reasonably high precision.

There are no permanent floating temperature-recording stations in the oceans, but there are islands and ships. Ships, particularly of the British fleet, have systematically recorded temperatures along their routes of travel since the mid-nineteenth century. These data provide the basis for determining historical world ocean temperatures, along with certain proxy data and, more recently, the introduction of global satellite measurements of

---

13. Easterly winds blow from east to west, and westerly winds blow from west to east. The easterly winds in the tropical latitudes are so regular that sailing vessels could always count on them to move cargo: hence the name *trade winds*.

14. The time duration of these variations would be much longer than a single solar cycle of 11 years, but much shorter than the Milankovitch cycles, which involve changes in solar insolation induced by variations in the Earth's orbit. The centennial and millennial variations in solar sunspot activity and output are caused by internal stellar oscillations that are not well understood but are observed in other stars.

15. An instrument to measure temperature was first invented by Galileo in 1592 and used air rather than a liquid as the working fluid. Galileo's instrument, which measured the expansion of air, although highly inaccurately, defined the concept. In 1654, the grand duke of Tuscany, Ferdinand II, devised a thermometer using a liquid in a sealed glass tube. Fahrenheit improved on the design by using mercury, which has a very low vapor pressure, as the fluid. Unfortunately, excessive exposure to mercury, as might occur in building these instruments by hand, can result in serious brain damage.

surface temperatures.[16] In taking ocean-surface temperatures from a ship, a bucket would be thrown overboard and a sample of water hauled on board, where a thermometer was used to measure the temperature. The procedure was standardized to the extent possible with the conditions existing on ships at sea. Obviously, the depth at which the samples were taken and the change in temperature of the water in the bucket before measurement limit the accuracy of the data. At one time earlier this century, a shift was made from wooden buckets to canvas bags. That produced a systematic change in the recorded temperatures, since the canvas bags cooled at a different rate than did the wooden buckets.

The air temperatures over the oceans have also been measured on ships and from satellites. This information can be translated into ocean-surface temperatures, although there is uncertainty in doing this. Furthermore, measurements of air temperatures over the oceans are relatively sparse for those periods for which additional data are needed.

There are problems with land temperature measurements as well. One serious concern is the "urban heat island" effect. Temperature-recording stations that were originally placed well outside cities may now be surrounded by urban sprawl. It is well known that cities are generally hotter than the nearby countryside (Section 5.4.1). Cities contain more asphalt and concrete that hold the heat, and so they generate more energy locally. The envelopment of measurement stations by urban development can result in an *apparent warming trend* in the temperature record. Other artificial variations in temperature may be caused by changes in the placement of instruments, alterations in the equipment, competence of the personnel, and so on. Once such problems associated with a specific recording station are known, they can be compensated for.

The data shown in Figure 12.10 and in most other records of the "global average temperature" used in climate studies are based heavily on land temperatures. Ocean-surface water and air temperature measurements, particularly those made before the early twentieth century, are considered too unreliable for inclusion. In addition, as we noted,

the ocean temperature measurements are sparse and imprecise. In constructing temperature histories, care is taken to correct for such systematic errors and local effects. Temperature analysts have also attempted to integrate land and ocean data, and modern satellite observations have tried to obtain a more reliable long-term record. The result is an annual temperature history that is probably accurate to within 0.3°C or better. According to Figure 12.10, this is close in magnitude to the actual year-to-year variation and the overall centennial trend in the temperature data.

12.4.3   Forecasts of Greenhouse Warming

The temperature record shows recent global warming. How does this square with the predictions of the greenhouse warming effects of carbon dioxide and other gases that are building up in the atmosphere? Climatologists use a variety of models having a wide range of complexity to forecast the future climate and to determine climates past. Some of these prognostications are discussed next.

Potential Impacts of Greenhouse Warming

The likely climatic effects of increases in greenhouse gases include overall warming of the Earth's climate, a rise in sea level, and disturbances in regional weather and precipitation patterns. These climatic effects imply a range of serious socioeconomic impacts. The reason for concern is emphasized in Figure 12.11. Here, the "four hooks of the apocalypse," representing a sudden growth in the abundance of greenhouse gases, are illustrated. The recent rates of increase in these gases, when compared with historical rates of change, are enormous. There also are new compounds, the CFCs, being introduced that were unknown in nature before this century. This dramatic unprecedented shift in the composition of the atmosphere, caused by anthropogenic activities, is indeed cause for concern.

Surface Temperature

If the atmospheric abundance of carbon dioxide were to double, the estimated increase in the global average surface temperature would range from about 2.5° to 4.5°C. The consequences of such warming have been debated in scientific circles, by governments,

---

16. Satellites do not use thermometers, of course, as they would have to be too long. Instead, they detect the thermal radiation emitted by the Earth's surface in the spectral "atmospheric window" (Chapter 11) and use Planck's law to deduce the emission temperature (Chapter 3).

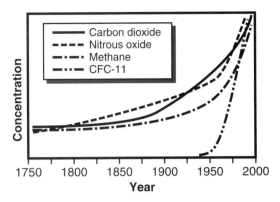

Figure 12.11 The increase over time in the atmospheric concentrations of the greenhouse-active gases CO_2, CH_4, N_2O, and CFCs. The increases are relative and indicate the sudden rapid increases over the past 50 years, which might be referred to as the "four hooks of the apocalypse."

and in the media. It is predicted that a warming climate would lead to a redistribution of flora and fauna worldwide. Some of the consequences would be unusual. One scientist predicted that vampire bats, which feed on the blood of mammals, including humans, could migrate to the southern United States from South America. Other researchers have suggested that warming temperatures could lead to an increase in AIDS cases. At the very least, the crop-growing regions would shift.

The possible warming effect of a greenhouse gas can be specified by its "greenhouse warming potential" (GWP). The GWP of a gas can be defined as the most likely average global temperature increase that would be caused by an increase in the atmospheric concentration of that gas by one part per million by volume. The temperature increase is assumed to be the equilibrium temperature change that would occur after several decades if all other factors remained constant over that period. The GWP of CO_2 is about 0.01°C/ppmv, and those of the CFCs, the most powerful greenhouse agents, are as high as 300°C/ppmv! The potency of the CFCs for warming the Earth is another reason for controlling their production.

Additional aspects of global temperature increases connected to the accumulation of greenhouse gases are discussed in subsequent sections.

A Rise in Sea Level

The impact of future global warming on the sea level is in dispute. But it seems certain that since the last ice age, during the latest period of warmth com-mencing about 20,000 years ago, the sea level has risen about 120 meters in average height. This occurred as the massive glaciers accumulated during the ice-age cycles melted and flowed back into the seas. At the time of the last interglacial period some 130,000 years ago, the sea level appears to have been about 6 meters higher than at present. One possible cause is the collapse of the western Antarctic ice sheet at that time. The major glaciers in Antarctica consist of the eastern ice sheets, resting on land, and the western ice sheets, which extend into the Ross Sea. The western ice shelf is considered to be much less stable than the land-bound glaciers and thus is subject to collapse and melting. The amount of ice involved would cause the average sea level to rise about 5 to 6 meters.

A sudden rise in sea level of 5 or more meters associated with the collapse of Antarctic ice sheets would be a major disaster to coastal civilization worldwide. Low-lying areas would be flooded out-right. Many other areas would be subject to high tides and inundation during intense storms. Countries like Bangladesh, without extensive systems of coastal levees, would be particularly hard hit. Places like Malibu in California would be washed away, and many expensive beach homes would be lost.

There is another cause of the rise in sea level: the thermal expansion of the oceans. When heated, most materials expand. Air, for example, expands according to Charles's law (Section 3.1.1). Roads are built with spaces between concrete sections to allow for expansion and contraction as the temperature changes over the seasons. Water also expands with heating. As a rough estimate, the height of the oceans could increase by 6 centimeters for every degree of warming of the oceans to the depth of the thermocline (about 200 to 300 meters below the surface). Ocean heating and expansion in combination with ice melting would lead to a rise in the sea level.

Recent data indicate that the seas are currently rising. Over the past century, the rise was about 12 centimeters, but during the past 50 years, the rate of rise seems to have accelerated to about 24 centimeters per century. An approximate empirical relationship can be derived from the temperature and sea level data that predict a sea level increase (decrease) of 24 centimeters when temperature rises (falls) by 1°C, with a delay of about 20 years. This **sea level law** is

$$\Delta H_{SL} \approx 24\Delta T \ (\text{cm}) \qquad (12.3)$$

For a global warming of 4°C, the rise in sea level would be nearly 1 meter. The implications for the coastal zones would be significant.

Is the observed rise in sea level due to the warming of the Earth's climate? Or is it a natural variation of some other origin? If the expansion of the seas were not a result of warming, the coincidence would be remarkable. The rate at which the sea level is rising may also accelerate. A few researchers point to two potentially unstable ice masses on the planet today: the western Antarctic Ross ice shelf and the massive Greenland ice pack, which during historical times has been extensively melted (the Vikings had settled there 1000 years ago when the ice had receded). These sources of ocean volume could easily cause the sea level to rise several meters.

An interesting phenomenon takes place when the climate becomes warmer. Initially, the ice sheets get thicker. As the land and sea at the middle and tropical latitudes warm, the rate of water evaporation speeds up (Section 12.1.2). Some of this excess water vapor is transported to high latitudes and deposited as snow. The polar climates are cold enough that the modest warming may not push the high-latitude regions above the freezing point. As long as the temperature remains below 0°C, snow will accumulate, and the glaciers will grow. Evidence exists for this paradoxical increase in ice as the Earth warms. Nevertheless, if the warming continues, the ice will begin to melt, and the oceans will rise. It is therefore also interesting to note that the predicted greenhouse warming is greatest in the highest latitudes.

According to all the facts and possibilities, the most likely result of global warming is a rise in the sea level of perhaps 1 meter, but possibly much less—or more if major ice sheets become unstable. The existing data on past and current changes in sea level support this conclusion.

Drench and Drought

The warming of the Earth will cause the overall rate of water evaporation to increase and, hence, the rate of precipitation to increase. The increase in total precipitation may be 10 percent or larger, although the change could be smaller. No one knows for sure what the total precipitation rate will be in the greenhouse world. But it is very likely to be a world with significantly more rainfall than today in some areas and significantly less in others. At this time, the predictions of regional changes in precipitation are unverified and unreliable. One consistent conclusion is that large areas with currently marginal rainfall could become much more arid. Regions with ample rainfall already have enough—for example, in the tropics. However, the expansion of arid zones in the African Sahel and the American Midwest would require only relatively modest decreases in regional precipitation. Not much more can be said about precipitation changes at this time.

Crop Fertilization

Most plants can be characterized as "C_3" or "C_4" according to the principal enzymes used in photosynthesis. About 95 percent of all vegetation is C_3, with most of the rest C_4. Most food crops are, likewise, C_3 (wheat, rice, potatoes, barley, soybeans, oats, tomatoes, apples, peanuts, and so on), and a few are C_4 (corn, sorghum, sugarcane, millet). The C_3 and C_4 plants respond differently to changes in carbon dioxide concentrations, moisture, and temperature. The changes in these environmental parameters caused by global warming are both connected and uncertain. Accordingly, the response of plants is difficult to predict.

In general, plants will grow more vigorously if CO_2, moisture, or temperature increases; C_3 plants are generally more sensitive than C_4 plants are. Differences arise in the way the plants breathe through the openings in their leaves (stomata) and the relative supplies of light and water for plant growth. Plants that are limited by light are more sensitive to changes in carbon dioxide. Such plants are operating closer to their **compensation point**, at which the uptake of CO_2 by photosynthesis just offsets the release of CO_2 through plant respiration. Higher CO_2 levels can make the utilization of the available photons more efficient under these conditions. The effect is usually greater in C_3 plants. Similarly, when the water supply is limited, plants, particularly the C_3 varieties, show greater sensitivity to CO_2 changes. Water is absorbed by plants through their roots and is lost through the leaf stomata by means of transpiration. To slow the loss of water under dry conditions, a plant may constrict its stomata, which also limits the uptake of carbon dioxide, and therefore the sensitivity to environmental CO_2 changes.

Rises in temperature also generally increase the sensitivity of plant growth to CO_2 concentrations, again more noticeably in C_3 plants. However, a number of complicated physiological feedbacks in plants can modify or eliminate some of these sensitivities. The various developmental stages of plants are controlled by temperature. Thus even if photosynthesis can be accelerated at a higher temperature, the growth of edible mass (fruits, say) may be restricted.

Based on controlled studies, it has been estimated that a doubling of the carbon dioxide concentration alone could increase the growth and yields of C_3 agricultural crops by 10 to 50 percent, and of C_4 crops, by 0 to 10 percent. These benefits would be unevenly distributed around the globe, however. In addition, the accompanying increase in global temperature is predicted to have a negative impact on crop yields in many areas. The higher temperature would lead to more evapotranspiration and would aggravate the loss of water. More rainfall would mitigate that problem, but some areas are predicted to be drier. In many regions, the combination of an increase in temperature and a decrease in precipitation could more than offset any CO_2 fertilization effect.

The most serious potential threat to agriculture, therefore, is the uncertainty and variability of regional weather in a greenhouse-heated world. Unusually hot, dry summer periods, drenching rains, and longer-term climatic shifts all would place agricultural systems under stress. If the average estimated changes in temperature and precipitation were slowly introduced over many decades, agricultural systems, society, and governments could respond appropriately. But in the real greenhouse world there would be sudden shifts and a large year-to-year variability, a "weather lottery" to which agriculture would be hard pressed to adapt.

The Temperature Feedback Parameter

A number of models have been constructed to predict the environmental changes associated with the buildup of greenhouse gases. These models vary widely in the manner in which they represent the climate system. The most sophisticated approaches use a global three-dimensional general circulation model (GCM) to simulate the climate. These models deal with the whole globe and include realistic boundaries for landmasses, the major ocean basins, surface orography (mountains), and the cryosphere. The motions of the atmosphere and all the meteorological parameters are predicted. GCMs allow climate scientists to put the world into a computer, turn the crank, and see the future. In these models,

changes in the concentrations of greenhouse gases and other parameters can be easily reproduced to assess the potential consequences of different emission scenarios. For policy applications, the effects can be analyzed on global and regional scales over long periods of time. (The inaccuracies in such forecasts, which limits their usefulness, are discussed in Section 12.4.4.)

To capture the essential features of climate forecasts, it is not necessary to run a general circulation model on a supercomputer. Those exercises are expensive, time-consuming, and best left to experts. A basic understanding of greenhouse physics and climate responses can be obtained using box models, as we did in Chapters 10 and 11. A useful parameter in discussions of the climate response to variations in forcing is the **temperature feedback parameter**. Recall that in Section 11.3.2, we developed a box model for the climate system and derived the equilibrium temperature for the Earth, expressed through Equation 11.14. Also recall that in Section 10.1.2, we discussed the effect of changes in the sources or sinks of a reservoir on the quantity of material in the reservoir. Now we can apply both of these results to consider the response of the Earth's average temperature to a change in the forcing, either as a result of the external forcing or through a change in the greenhouse effect.

Referring to Equation 11.12, we imagine that there is a *small* perturbation in the energy input or output, expressed as ΔQ, in units of joules per square meter per second (J/m^2-sec), or watts per square meter (W/m^2). Here the perturbation has been averaged over the entire planet. The change could be associated with a change in solar luminosity, the Earth's albedo, or the concentrations of greenhouse gases. In each case, there is a corresponding change in net radiative energy balance, or **radiative forcing**, of the climate system. No matter what the cause, the climate system will readjust to a new equilibrium state with a slightly modified average surface temperature. For example, if greenhouse gases are added to the atmosphere, the blockage of the atmospheric window will have reduced somewhat the thermal emission into space. Then the temperature of the surface and lower atmosphere will increase to compensate by enhancing the thermal emissions in other parts of the spectrum (Section 11.4.2).

The response of the temperature will be "linear" for small changes in forcing. That is, the change in equilibrium temperature will be proportional to the change in forcing. This can be written as

$$\Delta T = f_T \Delta \dot{Q} \qquad (12.4)$$

where f_T is the temperature feedback parameter, with units of degrees Celsius per watt per square meter (°C/W-m^{-2}). For a purely radiative response to a change in forcing (that is, an increase in the thermal emission with no other feedback in the system), f_T is determined to have a value of about 0.25°C/W-m^{-2}. With the positive water vapor feedback on the greenhouse effect, f_T has a value closer to 0.45°C/W-m^{-2}. The positive albedo feedback effect of snow and ice further increases f_T to 0.55°C/W-m^{-2}. The effects associated with the response of clouds (cloud cover, optical properties) may increase f_T to as much as 1.0°C/W-m^{-2} or reduce it to 0.4°C/W-m^{-2}. Although the cloud feedback effect remains uncertain, the combined effects of water vapor and cloud feedback mechanisms can double or triple the warming effect of greenhouse gases such as carbon dioxide, methane, and CFCs.

To estimate a possible range of climate responses, a range of temperature feedback parameters from 0.4° to 1.0°C/W-m^{-2} is reasonable.

The temperature feedback parameter is related to the greenhouse warming potential (GWP) defined earlier, because the greenhouse forcing of a gas, $\Delta \dot{Q}_i$, can be related to the amount of the gas added to the atmosphere, given in terms of its change in mixing fraction, Δr_i. The simplest approximate relationship is

$$\Delta \dot{Q}_i = \dot{q}_i \Delta r_i \qquad (12.5)$$

where $\dot{q}_i$ is the incremental forcing per part per million of gas i added to the atmosphere. Then Equation 12.4 becomes

$$\Delta T = f_T \dot{q}_i \Delta r_i \qquad (12.6)$$

where $\dot{q}_i$ is the forcing parameter for species i, with units of watts per square meter per part per million by volume (W/m^2-ppmv). Typical values for $\dot{q}_i$ are 0.015 for CO_2, 0.53 for CH_4, 2.0 for N_2O, and 280 W/m^2-ppmv for the CFCs. Notice that the greenhouse-warming potential is given by

$$GWP_i = f_T \dot{q}_i \qquad (12.7)$$

Although $\dot{q}_i$ is a well-defined parameter, f_T is uncertain, and thus GWP is uncertain to the same relative degree.

Carbon Dioxide–Warming Scenarios

With the doubling of atmospheric carbon dioxide, from about 300 to 600 ppmv, the maximum potential average global warming given by our simple treatment is close to 4.5°C. Although such a small change seems trivial, it is worth recalling that major climatic shifts are represented by global average temperature changes that are this large. For example, during ice ages, the Earth was only about 5°C colder than it is today. Hence a few degrees of change is a major climatic perturbation. As we explained in previous sections, the sea level could rise as much as 1 meter, and weather patterns would be more variable. We now wish to peer into the crystal ball of global climate modeling to see the future in more detail. What will the world be like, and how will we get there from here?

Figure 12.12 shows a fairly typical forecast of the change in global temperature as a result of carbon dioxide emissions through the next century. Actually, the results shown are taken from a highly sophisticated mathematical representation of the world put into a supercomputer—a general circulation model(GCM). These global-scale models can predict the regional patterns of warming and cooling, as well as of precipitation. The surface temperature predictions in Figure 12.12 have been averaged over the entire Earth—land and seas. Several "scenarios" for the future are offered. Scenarios are developed to provide a range of plausible outcomes of future actions. They usually are based on projections of past behavior, which may be altered by considering potential responses to initial changes. Very rarely does bad behavior continue once brought to our attention.

Scenarios are an essential ingredient in the analysis of climate change and of other environmental problems. Scenarios allow us to explore climate responses as if the underlying actions forcing the system were known. That is, for any given scenario, the uncertainties in the underlying causes are eliminated, and the effects for that particular scenario can be studied unambiguously. This also allows the uncertainties in the Earth's physical representation to be highlighted. Of course, the scenarios could be dead wrong. No one can actually predict the future. That is why a series of different scenarios are usually chosen to encompass the most reasonable projected courses of action. Even then, eventual reality may be different from the chosen scenarios.

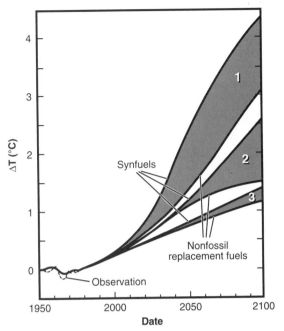

Figure 12.12   Future changes in average global surface temperatures calculated with a global climate model for several scenarios describing carbon dioxide release. The temperature is shown as a deviation from the actual global temperature in 1950. Measured temperatures are superimposed up to 1978. The three scenarios correspond to different assumptions regarding the growth in CO_2 emissions, from an unlikely "no-growth" case (3) to an extreme "fast-growth" case (1). The range of temperature values for each case (indicated by shading) is based on the possibility that some of the growth in energy demand will be met by expanding non-fossil-energy sources, such as nuclear and solar energy (lower boundary of the shaded area), or by relying more heavily on synthetic fuels produced from fossil-carbon sources (upper boundary of the shaded area). (Data from J. Hansen, D. Johnson, A. Lacis, S. Lebedeff, P. Lee, D. Rind, and G. Russell, "Climate Impact of Increasing Carbon Dioxide," *Science* 213 [1981]: 957.)

All of us create scenarios every day of our lives. We review the day's activities each morning; we plan weekend outings and longer vacations; we try to figure out what our friends are thinking or how they might respond to our actions. But events often do not turn out as planned. If we are careful, however, they often will be close enough. Climate scenarios are the same. We realize that none of the scenarios will actually occur, certainly not exactly. But the scenarios provide a means for organizing our thoughts, gauging the outcomes, and taking a reasonable course of action.

The three scenarios in Figure 12.12 assume that the emissions of CO_2 to the atmosphere will remain relatively constant (at about the 1980 emission rate) or will increase substantially during the next century. If no one had ever raised the issue of greenhouse warming caused by CO_2 and other gases, curve 1 in Figure 12.12 would no doubt best predict the future. There would be no reason to stop using cheap fossil fuels or to avoid synthetic liquid organic fuels made from, say, coal. Energy conservation would add to the cost of goods. Forget it. Solar, geothermal, and other alternative energy sources would be too expensive, or exotic, for most people to bother with. On the other hand, if the greenhouse warming theory were proved tomorrow, a worldwide effort would immediately be set in motion to find alternative fuels and energy sources, to conserve energy in all facets of everyday life, and to spread clean technology to all nations. We would be riding curve 3 of Figure 12.12 into the future. It is possible we might even drop below curve 3 and stop global warming completely.

One disturbing fact about the results in Figure 12.12, and about almost all studies of comparable detail, is that a significant amount of warming is already built into our future. It would be impossible to halt all emissions of CO_2 instantaneously; that would be impractical. So more CO_2 will be added to the atmosphere, vast quantities of it, even if drastic corrective measures are taken. The additional carbon dioxide—once equilibrated between the atmosphere and the surface oceans—would remain in the environment for 100 years or more. The climate legacy for the future is being accumulated today and over the next few decades. Unfortunately, like stocks left to accumulate value in a trust for a child's future, no one can be sure what the value of the CO_2 account will be 50 years from now.

If the temperature increases are as large as those indicated for the midrange scenario in Figure 12.12, then some of the serious environmental impacts discussed in previous sections are likely. In addition to the average effects already summarized at a finer resolution, global climate models provide fuzzy images of the future world. For example, the polar regions would warm considerably more than the tropics. That might accelerate the melting of ice and exacerbate the problem of the sea level's rising. Specific areas of drought also appear (although often in different regions for different models). The Great Plains of the United States would grow much drier.

Table 12.5   A Greenhouse Forcing Scenario[a]

Greenhouse gas	Change in concentration (ppmv)	Forcing parameter (W/m²·ppmv)	Change in forcing (°C/W/m²)	Change in temperature (°C)
CO_2	300 to 600	0.015	4.5	2.25
CH_4	1.7 to 3.0	0.53	0.7	0.35
N_2O	0.3 to 0.5	2.0	0.4	0.20
CFCs	0 to 0.005	280	1.4	0.70
Total	—	—	7.3	3.5

[a] The scenario for changes in the greenhouse gas concentrations and climate forcing is based on data from the World Meteorological Organization. A temperature feedback parameter, f_T, of 0.5°C/W-m^{-2}, was used to compute the equilibrium temperature change associated with each forcing.

Parts of Europe, Asia, and North Africa would also become more arid. In California, the snowpack in the Sierra Nevada would thin, and the summer runoff would abate, leading to (even worse) water shortages. By contrast, Australia, India, and parts of South America and Africa would become wetter. Such changes in regional climate would have significant impacts on agricultural productivity. Indeed, parts of Canada and Russia now normally too cold for agriculture might become more temperate in climate and thus better suited to agriculture. Other areas, that are currently productive could become wastelands.

This tableau of the greenhouse world is hazy and ambiguous, like an impressionist's landscape. But there also is substance here. We know too much already to be completely in the dark. We are still squinting at the images, but the light intensity is coming up fast. Each year, through careful research and analysis, scientists are bringing the climate scene into better focus.

The Complete Greenhouse Menagerie

More is changing in the atmosphere than just carbon dioxide. Remember the "four hooks of the apocalypse" in Figure 12.11. Accordingly, to foresee the future, scenarios should account for all the greenhouse gases. Future variations in the concentrations of these gases are highly uncertain. The more variables that are treated in a scenario, the greater the number of possible directions the scenario can take. Because full global climate models based on GCMs are expensive to run, simpler models are generally used to estimate the climatic impacts of mixtures of

greenhouse gases. For such a greenhouse-warming scenario, the increase in climate radiative forcing, $\Delta \dot{Q}$, and equilibrium surface temperature can be estimated using assumptions about changes in greenhouse gas concentrations projected into the future. Table 12.5 provides a typical scenario. In this case, each estimated concentration change has been converted to a radiative forcing using Equation 12.5, and to an incremental warming for that gas using Equation 12.4. The total warming is the sum of the incremental temperature changes.

A more sophisticated projection of global warming is summarized in Figure 12.13, in which the increase in surface temperature is shown as the accumulation of the warming effects of various gases. A specific detailed gas scenario was used to calculate the temperatures.

In Figure 12.13, the greenhouse-warming effect of the trace gases other than carbon dioxide roughly equals the carbon dioxide effect. This result is obviously very sensitive to the trace-gas scenario. In many studies, the long-term non-CO_2 warming effect is 30 to 40 percent of the total. This suggests that care must be exercised when making policies to solve the greenhouse problem. In many existing scenarios, for example, a large fraction of the incremental warming can be attributed to CFCs (indicated as CFC-11 and CFC-12 in Figure 12.13). However, the CFCs and other chlorocarbons are also responsible for degradation of the ozone layer (Section 13.5), and so an international effort is under way to eliminate most uses of CFCs. Toward this goal, the Montreal Protocol and subsequent London amendments were recently ratified by more than 50 nations. This unprecedented treaty calls for the elimination of all

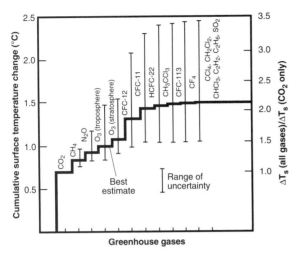

Figure 12.13 Calculated increases in global surface temperatures associated with projected increases in the concentrations of greenhouse gases. The temperature changes correspond to those expected between 1980 and 2030, not the long-term equilibrium temperature changes. Both the absolute temperature change (left scale) and the change relative to the effect of carbon dioxide alone (right scale) are indicated. The total greenhouse-warming effect between 1980 and 2030 is estimated to be about twice the effect attributable to CO_2 alone. Reading the graph from left to right, the temperature effect of each new gas is cumulative, and the effect due to that gas alone is deduced by the height of the step from the previous value to the left. The estimated uncertainty in the calculated temperature perturbation is indicated by the vertical bar, which also is cumulative. The warming connected with CO_2 is about 0.75°C between 1980 and 2030 in this assessment. (Calculations by V. Ramanathan, R. J. Cicerone, H. B. Singh, and J. T. Kiehl, *Journal of Geophysical Research* 90 [1985]: 5547.)

CFC emissions before the end of the century. If this is accomplished, the concentrations of CFCs will begin to decrease by the year 2000. The rate of decay is slow, however, and CFCs will remain in significant quantities for 100 years (Section 13.5.1). But the CFC contribution to global warming will nonetheless remain below a few tenths of a degree Celsius.

In the rising curve of Figure 12.13, a long list of exotic gases trails toward the end. Each gas contributes very little to global warming individually. In total, these compounds account for perhaps 10 percent of the total warming effect. This category of miscellaneous greenhouse gases remains the most enigmatic and potentially significant in the long run. New species are being added to this category every year. The replacement compounds for CFCs, the HCFCs and HFCs, are also greenhouse active. Al-

though the concentrations of these short-lived gases is not expected to grow too large, their cumulative effect will become more important as other greenhouse gases are controlled. Some of the compounds, like CF_4, are so stable that they will remain in the atmosphere for hundreds to thousands of years. Even very small emissions over enough time will become a hazard. Fortunately, environmentalists and government watchdogs are aware of these lurking threats and will be monitoring the situation closely.

12.4.4 UNCERTAINTY IS THE FUTURE

If you were hoping that the best scientists in the world could tell you how warm the Earth will become in the next few decades or where you should move to avoid the consequences of global warming, you are out of luck. No one is yet capable of making such predictions. We may never be able to do it accurately. The reason is that the climate system is extraordinarily complex. Reading of this book should have convinced you of that. Indeed, many of the interrelated physical, chemical, and biological processes pertaining to the atmosphere, oceans, and biosphere are being investigated so as to improve our understanding of the integrated climate system (for example, Section 11.5.3). In earlier sections, where climate model predictions are discussed, the range of uncertainty in those projections is mentioned. Here, specific sources of uncertainty in climate calculations are summarized.

Uncertainty in Observations

Don't always believe what you see. Frequently, data are not what they seem. For example, temperatures measured 100 years ago were based on the same thermometric principles as today, but the places and methods of carrying out the measurements have changed. For example, in the past century, ships at sea regularly measured the temperature of the ocean-surface waters by throwing a wooden bucket overboard and hauling in a sample of water to test. Later, canvas bags replaced the buckets. But these two techniques, so subtly different, yielded slightly different water temperatures. Painstaking research is often needed to uncover such problems. Occasionally, a problem is not discovered or corrected, which can lead to misinterpretation. Rarely will a politician, bureaucrat, or researcher purposely use (or abuse) imprecise data to mislead others.

Uncertainties in observations have several forms. **Imprecision** is caused by instrumental limitations in the sensing of small differences in the parameters measured. All measurement devices have a minimum size that can be recorded, which is controlled by mechanical, optical, and electrical design. Precision therefore reflects the ability to differentiate and record small increments. Measurements may also be inaccurate. That is, experimenters often do not actually obtain the measurements they want because of a flaw in the measurement technique, data-recording system, or interpretation of the raw data. For example, imagine trying to measure the size of a room with a tape measure that has a scale systematically longer than it should be—say, each inch scored on the tape is actually 1.25 inches long. You could measure every distance in the room with great *apparent precision*—to within one-thirty-second of an inch, say—while being inaccurate in the actual distances by several inches. The accuracy of the measurements is limited by the instrument design in this case, not its precision. Accuracy can never be better than precision, however.

If you realize the problem, you can correct your measurements by comparing the lengths on your tape against a "standard" tape that can be trusted, thereby improving their accuracy. This procedure is referred to as a **calibration** of the instrument. Calibration is essential to accurate measurement. Uncalibrated observations have been known to lead even careful researchers down blind alleys. Frequently it is not possible to recalibrate older data because the original instruments no longer exist. It also is difficult to detect systematic errors in data when accurate standards do not exist.

Additional inaccuracy can be introduced by the operator of the measuring equipment. This is especially a problem for remote instruments attended by poorly trained assistants (and sometimes graduate students and feeble-minded professors). Data must usually be interpreted as well. Raw electronic signals from detectors are not useful, and so the data are redefined before they are published. This may amount to "massaging" or, unfortunately, on occasion, "cooking" the data. Too commonly, the users of measurements do not question vigorously enough the procedures to ensure the data's accuracy.

The accuracy required of data depends to a great extent on how they will be used. If one is seeking to identify the proposed trend of global warming over the past century against a highly variable background, the data must be precise and accurate (certainly accuracy of better than a degree would be nice). If, on the other hand, one needs only the average global temperature during this century to compare it with a simple climate model like that developed in Chapter 11, only modest accuracy and precision are required (a few degrees). If the errors in the data are random, confidence in the measured value of a parameter can be improved by remeasuring the parameter several times and averaging the results.[17] This technique can also be used to increase the *statistical* confidence in a series of measurements without improving the accuracy of the individual measurements. If all the recorded values are biased in one direction, the error is said to be *systematic* (for example, our elongated tape yields lengths that are systematically too long). Systematic errors are best reduced by calibration; random errors, by averaging.

In the Earth's environment, the amount of information that could be collected is astounding. It is clearly impossible to measure every parameter of interest everywhere each moment. And it is not necessary, although having only a limited set of data presents new problems. If we wished to determine the average temperature of the Earth and had only one thermometer to use, what would be the best approach? Should we choose an "average" place and leave the thermometer there? Should we move the thermometer as quickly as possible over a measurement grid? In each case, the temperatures would need to be averaged. But should we consider data from certain sites as more or less important than data from others? In dealing with the Earth's climate system, such questions not only are relevant, but also are a primary factor that restricts our true understanding of climate change.

Uncertainty in Predictions

Having dealt very briefly with the inaccuracies in measurements, we now turn to the possible sources of errors in the predictions of climate change discussed earlier. First, climate models depend on

---

17. Although random errors can be considered to occur in an unpredictable manner, they still are bounded and have a well-defined statistical, or average, behavior. Random errors usually follow what is called a "normal" probability distribution, for which the most common error and the range of likely errors—the standard deviation—are well defined.

measurements to function and thus are vulnerable to errors in data. Data are used in several ways in climate models. Measurements of the climate system are used to initialize models, to start them up. Here, data are needed for all the variables of interest. Basic measurements also are required to define the fundamental physical, chemical, and biological processes treated in these models. For example, the absorption band parameters for the greenhouse gases in a range of spectral regions, the physical processes in clouds must be known.

Observations are used to validate climate models. In this crucial exercise, the model is used to predict the climatic outcome for an actual event or period for which direct measurements are available. The predictions are compared with the measurements, and a sense of the reliability of making such predictions is established. Models are not expected to predict all observed variables exactly—indeed, they cannot. The fidelity of the model is defined more in the sense that major climatic features are reproduced and average parameters are forecast accurately.

Climate models have built-in limitations in their representation of the climate system. For one thing, a model has a coarse spatial resolution when compared with the real world close-up. Typical GCM-based climate models divide the Earth into a grid of boxes, each of which may be several hundred thousand square kilometers in area (that is, about the size of the state of Colorado). Over the vast region that is represented by each grid cell, everything is averaged: The land has an average constant height; the soil has an average wetness and is covered with average vegetation; the temperature is the same everywhere; and the sky is filled with average clouds that drop an average amount of rain on all places. Climate models reduce the world to a caricature of itself. But is it an adequate representation? Studies over many years have shown that a coarse GCM grid is sufficient to capture the major movements of winds and weather fronts over long distances, to reproduce the seasonal variations closely, and to predict regions of cloudiness and precipitation reasonably well.

Climate models are nothing more than a set of mathematical equations that describe the basic physics and interactions of the atmosphere, ocean, and land systems. The equations have been put in a numerical formulation suitable for solution on electronic digital computers. The climate model physics at the most fundamental level involves the motions, or dynamics, of the air and seas. The balance of solar and terrestrial radiation is also fundamental (Section 11.3), as is the hydrological cycle (Section 10.3). The interactions between the subsystems and feedbacks within each system join the climate machine together. These are represented by the interdependence of the various model parameters—for example, the dependence of thermal radiation intensity on the fourth power of the temperature.

The mathematical equations that constitute a climate model can be solved quite accurately. The solutions are valid at the spatial resolution of the model and at larger scales. However, there are three general kinds of limitations that introduce uncertainty into the model's results. First, no model actually includes the equations for all the processes known to contribute to climate change. Many of the key elements of the climate system were identified in Section 11.5.3. But no model in existence now, and none that has ever been conceived, treats all of these elements. No computer could handle that many equations. So climatologists must select the most important processes for the problem at hand. Including additional processes leads to refinement of the model. But for all practical purposes, the model can never be truly complete. The sophisticated climate model used to calculate the carbon dioxide-warming scenarios in Figure 12.12, for example, neglects most of the physics of ocean and land processes, and it omits biological processes. Nevertheless, despite these limitations, climatologists are fairly confident of their predictions. Why? Because the world, they feel, cannot be so whimsical that secondary processes would control the course of the climate.

The second problem with climate models is related to their limited spatial resolution. The horizontal distances resolved by current climate models are hundreds of kilometers. Accordingly, events that occur over smaller distances cannot be explicitly represented. These processes must be "parameterized"; that is, they must be described in terms of quantities that are averaged over the larger region. Examples of such processes are cloud formation and precipitation. Clouds range in size from a few meters to hundreds of kilometers in length. Most clouds have structures that are less than 1 kilometer in length. Precipitation from clouds, likewise, is highly heterogeneous. You can be drenched on one side of a street and dry on the other. Imagine, by comparison, a cloud that stretches over an area the size of a state and rains a constant uniform drizzle over the whole region. That is hardly like the real world. Yet for many purposes, it seems to be close enough.

Clouds and precipitation are not the only processes parameterized in climate models. The surface distribution of albedo is typically fixed using grid-average values. Water does not run off in rivers, and there are no lakes smaller than the aggregated Great Lakes. Wind-channeling mountains and valleys are leveled. Urban heat islands are not present. Most current deflecting islands are absent from the oceans. There is no life or biogeochemical cycling of greenhouse gases. The ozone layer is static. The world is an extraordinarily complex place, and representing it through mathematical relationships requires considerable approximation. The crucial principle here is always to retain and resolve the dominant processes.

The third problem with modeling natural events is the possibility that important processes have not yet been identified. If a process is unknown, the chances are small that it will be represented by the mathematical formulation of a model. We cannot say with certainty that such unknown processes exist, for if we knew, they would not be unknown. There is a finite possibility of ignorance, however. Science, which is the most careful, self-correcting, and conservative path to knowledge, has a long history of false theories and discredited ideas. The sun was once believed to revolve around the Earth. More recently, it was proposed that the next ice age, which was imminent, could be avoided by spreading pulverized rock over the continents. We do not hear that idea much today.

The problem of potential unidentified processes is real, but it should not limit our ability to act if the facts point that way. If we were to act only on the basis of certainty, nothing would ever be done. We would live immobile, fearful lives. Risks and benefits must be weighed.

The Water Wild Card

We mentioned that water vapor is a dominant factor in determining the climate of the Earth, through its greenhouse effect and role in cloud formation. Water also participates in all life processes that control the biogeochemical cycles of other greenhouse gases. Water vapor may act as a climatic thermostat (Section 11.4.3). Water vapor is the only greenhouse gas that is not under the influence of anthropogenic change. Some researchers have suggested that water vapor may provide salvation from the projected greenhouse heating of the Earth. It is claimed that the current climate models do not properly account for water vapor processes in the atmosphere, particularly cloud formation and dissipation. We also noted that clouds are not accurately resolved in global-scale climate models. Indeed, clouds are so complex that *no* model on any scale can truly reproduce cloud effects as they are observed. The formation of cloud particles, particularly ice crystals, is not well understood. The origin of lightning is still debated. The pathways for the flow of water through a convective cloud remain uncertain.

On this last point, it can be said that the feedback effect of water vapor depends on its movement through clouds. Consider the following scenario: The climate begins to warm. The tropical ocean surface waters heat up. More water is evaporated, boosting the strength of cloud convection over the warmer oceans. Clouds are pushed higher, and their tops are colder. Because they are colder, the air passing out of these clouds is drier, owing to the strong dependence of water vapor pressure on temperature (Section 12.1.2). If the air in the upper troposphere becomes drier, thermal emissions from the lower atmosphere become more important. These emissions are more intense because the lower atmosphere is warmer. Hence a cooling effect, or negative feedback, is produced. Water is acting as a thermostat to moderate the climate change.

In previous sections, we talked about other feedback effects associated with water vapor and the hydrological cycle. In Section 11.4.3, the feedback effect due to high clouds formed by strong convection was found to be, most likely, positive. In Section 12.1.2, the direct water vapor feedback effect with surface warming was seen to be strongly positive. These positive feedbacks would, accordingly, be offset by this cloud-drying effect. To what extent? Recent analyses of satellite data suggest that the drying mechanism actually may not be working efficiently. Even in cloudless regions downwind of convective systems, the net longwave radiation is reduced by an excess water vapor greenhouse effect. At this time it does not seem reasonable to pin our hopes on a water thermostat to keep the Earth cool.

Uncertainty in Sequelae

The possible consequences following from a series of connected events with a specific cause are referred to as *sequelae*. The warming of the Earth's climate will lead to countless changes in the environment on different scales. As we pointed out, these outcomes are uncertain. The environmental changes will produce a range of social and economic responses. These are even more uncertain. Eventually, the global warming effects will touch individual human lives (not to mention the lives of countless other organisms populating the biosphere). The individual sequelae are unpredictable. The consequences for groups and nations are highly uncertain at best. Accordingly, debates over potential socioeconomic outcomes are left for consideration in other forums.

12.5   Solutions?

How can we solve the conundrum of greenhouse warming? Is there an alternative to playing roulette with the future by continuing to emit into the atmosphere enormous quantities of carbon dioxide and other greenhouse gases? Are we so addicted to energy that we must either find a new cheap source of energy or learn to live with all the drawbacks of the sources we have? Are we energy junkies? Or are we rightfully enjoying the fruits of human labor and intelligence? Is it possible that we have painted ourselves into a corner? Will the quality of life decline in the face of a growing population, diminishing fuel reserves, and increasing pollution? Can a long-term plan for the future be drawn up, even in the face of resistance from countless special interests? Will it be necessary for us to make sacrifices now to preserve the Earth for the future? Or will new technology soon solve the problems we face today? Can we count on nature and society to continue to absorb body blows without a knockout? May we eventually have to deal with the devil in new incarnations to retain our perquisites as CEOs of planet Earth?

These questions are raised here in relation to the specific problem of global warming, but they apply to many of the other issues discussed in this book. We will offer some possible responses. In pursuing these questions, we begin to reach beyond the realm of science into the domain of sociology, economics, and politics. The ice is much thinner here (the warmer it becomes, the thinner the ice seems to get). The sequelae are unknown, as we noted earlier. Hence, we will try to keep to the science and technology of the problem and look at some of the ideas that have been proposed.

Risk and Benefit

Before proceeding to a discussion of possible solutions, it is worthwhile to review briefly the concepts of risk and benefit as applied to climate and other environmental problems. Risk is simply a measure of the potential cost of taking a chance. We face risks every minute in our everyday lives—driving a car, walking down stairs, getting on an elevator, shaving. Even getting out of bed has a risk associated with it; occasionally, you may hurt your back. Most of us are not frozen into inactivity by fear of risk because we perceive the threats associated with common activities as insignificant. Often we try to forget looming risks that we cannot avoid, such as flying frequently or living next to a nuclear reactor. That is rationalization. But try to get fire insurance on your vacation home in a dry woodland. Insurance companies know how to put a price on risk. They charge enough to make a tidy profit.

Simply put, risk is the probability that an undesirable event will occur, multiplied by the cost incurred if it does. Insurance companies place a value on replacing your home and multiply that by the statistical probability of a fire, to figure your annual premium (after adding a margin for profit, of course). Insurance is based on the average outcome of many events over a period of time. Your individual risk is more personal. Once an event has occurred, it is no longer just possible—like one in a hundred. The disaster is real, and the price must be paid in full. Nevertheless, in making many choices in life, you either consciously or subconsciously weigh the average risks or your perception of them. Consider all the types of insurance that many of us carry: life, health, fire, collision, disability, liability, travel, mortgage, title, legal, mental, and so on. We are trying to reduce our risks, or at least the chances that we will lose our wealth.

What about climate insurance? Would you be willing to buy a policy on the future of the Earth's climate? What potential "losses" would you cover? How much would you be willing to pay? Who would carry the policy? These are nearly imponderable questions. First you must determine the probability that significant climate change will occur. Let us say there is a 50-50 chance that the climatologists are right and that the planet will warm significantly in the next few decades. What is the cost to you? What is your personal price tag on a biologically impoverished world? On sweltering summer heat waves? On a new American dust bowl? On swelling masses of starving humans? If you are comfortable, you will not place much value on these outcomes. But if you are compassionate, perhaps you will.

Benefit is on the opposite side of the coin from risk. It is the advantage gained by an action. If you invest money in the stock market, your benefit is the dividend you receive. Of course, risk must be considered. Almost every time we seek a benefit, we must take a risk. Too often we are exposed to hidden risks so that someone else can benefit—for example, from poorly designed vehicles to which cost cutting has been applied so as to increase profits. The benefits of using fossil fuels are numerous and well known. They are cheap, and currently they support a relatively lavish life-style and high quality of life in many parts of the world. But what are the risks? From this book—and your daily newspaper—you know only too well: air pollution, ozone, haze, acid rain and fog, poisoned dump sites, oil spills, odors, and global climate change.

Most of the possible solutions we will describe require costs for implementation. When the price of fuel or electricity rises, most people howl. In terms of risk, these costs could be thought of as an insurance policy. The premium is being used to encourage new technologies or shape market forces to minimize adverse climate change. In many states, we are forced to purchase automobile insurance, to protect the interests of the larger society. Like it or not, you may soon be required to buy climate insurance, in the form of higher prices for essential commodities. As always, we must be wary of the salespeople and shop around for the best deal.

12.5.1 Recyclable Fuels

Some environmentalists believe that the most reasonable method for providing clean energy to society while avoiding the pitfalls associated with fossil fuels (and nuclear power and some other alternative energy sources) is to develop renewable or recyclable organic fuels derived from biomass. These fuels can be produced from vegetation (living or dead biomass) and from natural gas (methane). The most common forms of these fuels in use today are ethanol and methanol, two forms of alcohol. Ethanol, which is the intoxicant in liquor, wine, and beer, can be fermented from a variety of plants. In Brazil, sugarcane is used to produce all the alcohol fuel required for local trans-

portation. In the United States, corn has been suggested as a base crop for ethanol production.

Methanol is most economically produced from the methane in natural gas, or methane is generated from other fossil-carbon sources, such as coal. It can also be distilled from vegetation. Methanol is very toxic to humans; 1 ounce can blind or kill an adult. It is the form of alcohol used in automobile antifreeze solutions. We all have heard stories of alcoholics who have died trying to get high on antifreeze. Although one may drink ethanol, it is still a metabolic poison to the human system and can be fatal in large quantities. To prevent us from getting at alcohol fuel in the future (should it eventually be produced in large quantities), car manufacturers plan to install antisiphoning devices in fuel tanks.

Alcohol, particularly methanol, has been used in blends with gasoline to reduce the overall emissions of pollutants from automobiles. In smoggy urban areas like Los Angeles, this is one approach that has been taken to improve air quality (Section 6.6.2). The advantages and disadvantages of alcohol fuels relative to other transportation energy sources are described in Chapter 6 (see Table 6.4 for a summary). The alcohol fuels have been criticized because they currently are more expensive than gasoline. They also create air pollution, although apparently not as much as gasoline does (Section 6.2).

The potential problems with using alcohol fuels on a large scale are illustrated by the experience in Brazil, where 90 percent of cars run on high-proof ethanol fuel produced from sugarcane. The vehicles are specially manufactured to burn the alcohol efficiently. In fact, an entire infrastructure has been constructed to grow and harvest sugarcane, ferment the cane to produce alcohol, and distribute the fuel around the country. Problems arise because of the waste from the fermentation process, which must be further decomposed for recycling, and the fossil fuel, water, and agricultural chemicals needed to grow the crops. Massive amounts of vegetation must be processed, and the by-products are voluminous. It is estimated that the entire U.S. grain crop might be able to supply one-quarter of the United States' current energy needs. The cultivation of vegetation just for fuel production would expand global agricultural demands considerably.

What is critical about alcohol and related biomass fuels is that they allow the *net* emissions of carbon dioxide to be reduced. The recyclable fuels, which reuse a fraction of the existing carbon dioxide over

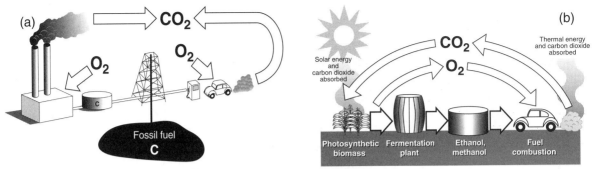

Figure 12.14   A comparison of the chemical and energy cycle of fossil fuels and renewable biomass fuels. (a) Fossil fuels are extracted from reservoirs deep in the Earth. Mining can deplete these reservoirs in a matter of centuries even though it would normally take hundreds of millions of years before the carbon is recycled. When the fossil fuels are burned, new carbon dioxide is added to the atmosphere-ocean system, and the concentrations of atmospheric CO_2 can increase. Fossil-fuel use thus represents an acceleration of the long-term carbon cycle. (b) Recyclable fuels are derived from biomass recently grown. The organic fuel is produced through photosynthesis by extracting carbon (in the form of CO_2) from the atmosphere. The energy of sunlight is captured by the growing plants and stored as chemical energy in organic molecules. At the same time, oxygen is released into the atmosphere, replacing the carbon dioxide absorbed by the plants. When the organic fuel is burned, the oxygen is reclaimed, and CO_2 is emitted back into the atmosphere. Renewable-fuel use represents a short-term recycling of a small fraction of the carbon and oxygen available in the environment.

and over, replace fossil fuels, which generate new carbon dioxide emissions from ancient carbon. Figure 12.14 shows the differences in the carbon cycles associated with fossil and biomass fuels. Note that if biomass is used as the source of the alcohol fuel, there is no net release of carbon dioxide into the atmosphere. Rather, a fixed amount of the CO_2 already present is recycled. In addition, the energy source is eternal; as long as the sun shines, energy can be harvested as biomass produced through photosynthesis. Fossil fuels, the current source of most energy, are projected, at the present rates of consumption, to be available for only a century or so. Then we will have to look elsewhere for energy anyway.

If the methanol in Figure 12.14(b) is produced from natural gas, there will be little impact on the overall emission of carbon dioxide. It makes little difference if the natural gas is burned directly for energy or is first converted to methanol and then burned. Roughly the same amount of carbon is released as CO_2. A partial reduction in CO_2 emissions could be realized if there were a net gain in energy output per fossil carbon atom burned. Then methanol could reduce the net consumption of other fossil fuels, particularly oil. This advantage could be realized only if natural-gas production were greatly expanded. Unfortunately, the proven reserves of natural gas are limited and would soon be tapped. Thus in the long run we are again left with recyclable biomass fuels as an alternative.

It is not as if ethanol and methanol are perfectly clean, biologically pleasant, inexhaustible fuels. They do have polluting side effects, as we observed earlier. They are more expensive. It would require enormous infrastructure investments to bring these fuels into widespread use. But they are cleaner, avoid excess CO_2 emissions, and are inexhaustible. What are the alternatives?

12.5.2   ALTERNATIVE ENERGY SOURCES

The alternatives for generating energy are limited on a finite planet. Consider hydroelectric power, a source of energy derived from sunlight, as are almost all other sources of energy used by society. Sunlight absorbed at the surface is converted to latent heat through the evaporation of water from oceans and soils and through the evapotranspiration of water from plants as they respire. The water rises into the atmosphere where it condenses as clouds and falls as precipitation (Section 12.1.2). The precipitation landing on high ground holds substantial gravitational potential energy. As it flows down rivers toward the sea again, that energy may be harnessed by collecting the water behind dams and funneling it through electric generators (using the same basic principle as the waterwheel). What could be more natural?

Hydroelectric energy has a number of drawbacks. First, it is limited in quantity. Although huge amounts of solar energy are converted to latent heat (Section 11.4.4), the ability to harness this energy requires an appropriate river basin to collect and concentrate the water flow. Such river basins are restricted in number and accessibility. Also, when damming a river, large tracts of land may be flooded and spoiled. Everyone who visits Yosemite National Park in California is in awe of the astounding natural beauty to be seen in the majestic canyons and waterfalls. Very few people realize that a once equally majestic canyon, the Hetch Hetchy, is now completely flooded behind a dam used to generate power and control water flow for irrigation. The head of the Department of the Interior under President Ronald Reagan once suggested that the canyon be drained and reclaimed for recreation and pleasure. Fat chance!

Dams generate electricity, control floods, provide irrigation in arid lands, and offer excellent boating and fishing. But in time, they fill with silt and become useless. Hydroelectric power is one of a number of alternatives to fossil fuels that can be used in certain regions and for specific purposes to relieve the demand for more power plants and to provide a basic reliable source of clean energy. It would be a mistake to believe that hydroelectric energy is the whole answer, however.

Energy Efficiency: Careful Energy

Energy efficiency simply means using energy wisely. Energy should be treated as a limited and valuable resource, not one to be squandered. In the fuzzy first half of this century, before anyone realized the climate might be changing, fossil fuels were wasted. Large automobiles, homes without insulation, inefficient lighting, inadequate mass transit, and many other factors contributed to the massacre. Now we are aware of two impending problems associated with fossil-fuel use:

1. The carbon dioxide emissions from fossil-fuel burning may be causing global greenhouse warming.
2. The known fossil-fuel reserves will begin to run out in the next century, leaving a shortfall in energy production.

Energy efficiency is a logical approach to solving both problems.

We become energy efficient by purchasing smaller cars, putting insulation in the attic of our home, using buses more often, cutting down on our use of hot water, substituting low-wattage lightbulbs, turning down the temperature of the thermostat in winter, and using the air conditioner less in summer. That is, energy efficiency means personal sacrifices. In the United States alone, $300 billion in taxes are spent annually to maintain military forces to keep the peace. How much might we be willing to spend on energy efficiency to keep the climate? To ensure that energy efficiency improves, new technologies for utilizing energy must become a priority for national research. To guarantee the widespread adoption of such technologies, incentives for their emplacement must be established. New technologies must also be shared with developing countries, to avoid natural tendencies to rely on cheaper, less efficient methods of energy generation and use.

Funny things can happen on the way to becoming energy efficient. For example, some energy-generating companies and utilities may complain they are losing business and so raise the cost of energy. We end up paying to maintain a system built to support a wasteful life-style even after we have become more energy conscious. Thus energy efficiency is not just a matter of common sense. It is also a matter of politics and economics. Seeking to do what is logical often leads to illogical actions. Special interests and the status quo can be formidable barriers to change, even beneficial change.

Nuclear Power: Radioactive Energy

Nuclear energy produced by the fission of uranium and plutonium was once hailed as a limitless source of energy that would cost practically nothing to produce. So much for the nuclear enthusiasts who surfaced in the post–World War II "nuclear age." Illuminated by the cold light of reality, nuclear power is expensive, dangerous, and dirty (Sections 7.3 and 8.2). Nuclear energy currently accounts for only about 10 percent of total energy production worldwide. Nevertheless, a few countries have successfully developed a dominant nuclear-energy utility. France is almost solely dependent on nuclear power, with about 75 percent of its electricity generated by nuclear plants.

The shadow of nuclear waste looms, however. In the former Eastern bloc nations, particularly the former Soviet Union, poorly designed nuclear-power plants remain squatting on the landscape like gigantic time bombs. The Chernobyl tragedy may have exposed more than 100,000 people to unacceptable levels of nuclear irradiation.[18] Meanwhile, hundreds of thousands of tons of toxic and radioactive waste continue to accumulate in leak-prone holding tanks near operational and defunct reactors all over the world, waiting for transportation to a final burial site in a place still to be determined. The older reactors, now past their useful lifetimes, need either to be disassembled piece by piece and safely buried far from society or to be left standing as gigantic concrete monuments to the desperate need for energy. Some of the waste from nuclear reactors must be safeguarded for more than 10,000 years![19] Nuclear energy has not been a successful approach to energy sufficiency.

Nuclear energy can, and most likely will, provide a small fraction of the energy needed in the future. The nuclear industry must design a standardized fail-safe reactor system and devise a way to dispose of enormous quantities of radioactive waste safely and cleanly. If these goals were met, there would be no reason to shun nuclear energy. Nevertheless, it is unlikely that tens of thousands of nuclear reactors will be built in the next century to fill the world's energy needs.

Nuclear fission was once touted as the energy source of the future. But it failed to live up to that promise owing to its intrinsic limitations and the side effects of fission. Nuclear fusion is now being touted by some as the energy source of the future.[20] It is claimed that fusion will provide a clean, limitless source of electric power. How soon we forget.

18. On April 26, 1986, near the small town of Pripyat outside Kiev, one of the four nuclear reactors at the Chernobyl power station became unstable during an engineering test. The core of the reactor melted down, spewing an unprecedented quantity of radioactive debris into the atmosphere, killing 250 people at the site, contaminating the local countryside beyond habitation, and sending a cloud of dangerous radioactive debris over Europe. Today, some Russian scientists claim that exposure to the radioactive fallout has been more serious than originally stated. According to conservative estimates, over the next few decades, thousands of people will suffer from cancer induced by the radioactive fallout.

19. The nuclear debris contains uranium-235 and plutonium, which, ounce for ounce, is the most toxic material known on Earth. The most common isotope of plutonium has a half-life of about 24,000 years. To avoid exposure caused by leakage into groundwater or air, the waste materials containing plutonium must be sealed for a length of time longer than its half-life.

20. *Fission* is the breakup of a heavy atomic nucleus into two lighter fragments, and *fusion* is the association of two light nuclei to form a heavier nucleus. The sun generates energy by means of fusion (Section 11.2), and nuclear weapons use both fission and fusion to create an explosion. All existing nuclear-power plants are based on the fission principle.

Despite billions of dollars invested in research, no one has yet been able to produce fusion power above the breakeven point (that is, the point at which the energy output from fusion is at least as great as the energy used to initiate the fusion process). Harnessing fusion energy is like taming a nuclear detonation. Huge machines have been built to control the nuclear fires, which create enough heat to vaporize any known material. Technological breakthroughs are hoped for. Meanwhile, we burn fossil fuels in a primitive ritual of fire and heat.

Several years ago, two scientists claimed they had discovered a "cold-fusion" process occurring in a jar fitted with special electrodes. No exotic containment vessel would be needed to produce cold fusion. Everyone could have a cold-fusion cell in his garage to light and heat his home. Unfortunately, cold fusion was a mistake. It does not work. An easy, painless solution has not been found.

Reaping Wind, Geothermal, and Solar Energy

A number of attempts have been made to tap other natural sources of energy: the wind, the sun, the heat of the Earth, and the tides. None of these approaches has produced energy in sufficient amounts to replace fossil fuels, or even to come close. Nonetheless, all these alternative energy sources provide, incrementally, some relief for our dependence on fossil fuels. Wind energy is collected by windmills, an idea that has been around for a thousand years. What is new is the scale of the effort. In California, for example, hundreds of propeller-driven windmills have been built along ridge lines and in valleys where the winds regularly howl. The result is a nice piece of energy for local communities. The drawbacks include visual clutter by unsightly gangs of ugly structures erected in an otherwise pristine countryside. Geothermal energy is also an oddity at this time. Practical utilization requires easy access to natural hot springs and vents. These springs are also limited in their capacity to heat water and are quenched when too much cool water is recirculated into the vents.

Solar energy holds the greatest potential for clean-energy production. The total solar energy intercepted by the Earth is equivalent to about 140 million gigawatts (GW). (The total output of all the power plants built by humans is roughly 10,000 GW.) This much energy could be collected by an array of solar-energy panels 100 kilometers by 100 kilometers in area—about the size of the Los Angeles basin.[21] There is plenty of desert available in which to locate these collectors, or they could be put in space on gigantic satellites. It has been proposed that the United States construct several hundred large satellites in orbit for the sole purpose of harvesting solar energy. Toward that end, a new, super-space-shuttle would be designed, with 10 times the payload of the present shuttle. Ten flights of the super-shuttle per week would be enough to build and maintain the satellite system. Alternatively, the solar collectors could be constructed on the moon with moon minerals, with the energy beamed back to the Earth in a concentrated form. It all seems very practical, doesn't it?

In everyday applications, solar-energy collectors are relatively inefficient. Moreover, reliable sunny weather is needed. Energy must also be stored for power at night. Space-based solar collectors would create as many environmental problems as they would solve; the space junk alone would be a major hazard. We are probably not ready to begin building cities on the moon while the cities on Earth are decaying. Solar energy must be further developed as a rational alternative to fossil fuels: A carefully conceived combination of energy conservation measures; large-scale development of recyclable-fuel facilities; refinement and expansion of clean-energy technologies, particularly solar and geothermal; and limited construction of new nuclear-energy facilities could provide an alternative to fossil fuels in the future.

12.5.3 Climate Correction: Endangering the Environment

Suppose that we are not able to control our addiction to fossil fuels or to develop practical and economical alternative energy sources. And suppose the worst predictions of the greenhouse climatologists are accurate. Are we doomed? There is one last resort. We can engineer the environment to correct for the problems we have created and set the climate right again. This topic is explored at length in Chapter 14 (see particularly Sections 14.2.2 and 14.3.2). Here we offer a few examples relevant to the carbon dioxide greenhouse problem.

21.   The solar constant at the Earth (Section 11.3.1) is 1400 W/m². The cross-sectional area of the Earth is 1×10^{14} m². Hence the total rate at which energy is intercepted by the Earth (or the total power equal to energy per unit time) is 1400 times 1×10^{14}, or 1.4×10^{17} W, or 1.4×10^8 gigawatts (GW).

Reducing CO_2 Concentrations

Humans are adding carbon dioxide to the atmosphere at a faster than normal rate. This may cause the climate to warm. To solve the problem, why not simply remove the excess carbon dioxide from the atmosphere and return to the natural carbon balance overall? Well, it is difficult to filter carbon dioxide out of air. It can be done by cooling the air to temperatures well below freezing, –78.5°C, at which CO_2 condenses as "dry ice." When exposed to the air, dry ice—nothing more than frozen carbon dioxide—fumes a smokelike vapor, which is nothing more than condensed moisture.[22] Instead of melting, dry ice sublimates directly into CO_2 gas. The problem with freezing CO_2 is that it takes a lot of energy and produces huge amounts of dry ice, which would have to be stored in refrigerators.

There is another solution that we are now familiar with, which is illustrated in Figure 12.15(a). Vegetation can be used to filter CO_2 out of the air by means of photosynthesis. All we need to do is plant the seedlings and watch the trees grow. There have been campaigns to grow more trees to absorb carbon dioxide from the atmosphere. To absorb the 6 gigatonnes of fossil carbon emitted into the atmosphere as CO_2 each year, the average annual rate of photosynthesis would have to be increased by about 5 to 10 percent (Section 10.2.4) for the next century. To prevent the trapped carbon from returning to the atmosphere, all this *new* biomass, which could amount to as much as currently exists on the entire planet, would need to be conserved. All the land surface would need to be turned into forests.

We could go one step further, however, and cut down the newly grown trees when they were mature. The logs could then be buried deep enough to prevent rotting and decay by bacteria. We would be recycling an equivalent amount of carbon originally mined from fossil-carbon reservoirs back into the Earth. Tree burial grounds would become the human equivalent of coal and oil deposits.

22.   The dry ice is so cold that in the air passing over it, a fog condenses from even small amounts of moisture (this moisture would normally condense at a much higher temperature than the freezing point of CO_2). The fog also is heavier than air because it is embedded in pure carbon dioxide (atomic weight 44), which is heavier than air composed of oxygen and nitrogen (average atomic weight of about 29). Accordingly, the fog settles downward and hugs the floor. This eerie "smoke" can be seen pouring over the side of a cooler in which dry ice is used as a refrigerant. The smoke is often used effectively in stage plays.

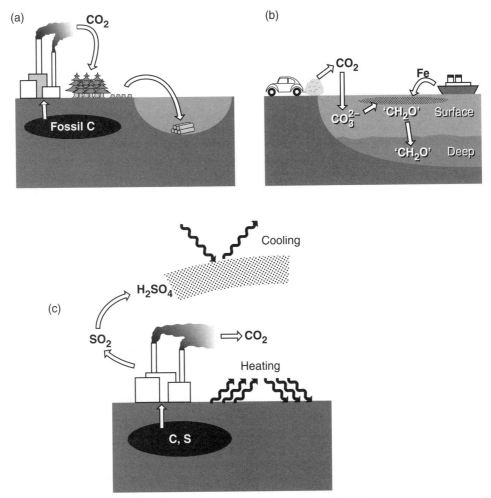

Figure 12.15   Several concepts for limiting the accumulation of carbon dioxide in the atmosphere as a result of fossil-fuel combustion. (a) and (b) The idea is to repossess excess fossil carbon emitted as carbon dioxide as fast as it is released, primarily by using trees to absorb the carbon as wood, which is then buried. (c) The plan is to compensate for the heating resulting from excess carbon dioxide emissions by increasing the Earth's albedo using sulfur particles generated as a by-product of sulfur combustion.

It is fair to ask how many trees would have to be grown and harvested to keep up with the consumption of fossil fuels. Six gigatonnes of carbon—the current fossil-carbon consumption—is about the quantity contained in 1 million square kilometers of forest. To grow and harvest this amount of wood annually would require roughly 10 million square kilometers of land area. Of course, a comparable area would be needed to grow vegetation for a global recyclable fuel system. The burial of wood over the course of a century would require a hole about 100 kilometers wide by 100 meters deep (wood has a density of less than 1 gram per cubic centimeter—it floats on water). Some of the wood could be temporarily used for construction, but it

would eventually have to be buried before it rotted or was eaten by termites (since both processes essentially convert the wood back to carbon dioxide and water).

An alternative to tree growth and burial is shown in Figure 12.15(b). In this scheme, the surface oceans are fertilized with an important nutrient, iron (Fe). In some regions of the world, the surface waters of the oceans, notably the Antarctic seas, are deficient in iron, a deficiency that limits the growth of phytoplankton. By fertilizing the oceans with the proper amount of iron nutrients, microorganisms would be encouraged to grow. In the process, they would absorb excess carbon dioxide from the atmosphere, but to be effective, the assimilated carbon must still be buried. The deep ocean is a perfect burial ground. Instead of dredging the oceans and burying the biomass on land, we would simply let the organic carbon settle into the deep oceans where the residence time is more than a thousand years (Section 10.2.4). In effect, instead of releasing the excess fossil carbon into the atmosphere, we would trap and bury it in the deep oceans.

The major problems facing ocean burial are twofold. First, large tracts of ocean would have to be continually fertilized. A fleet of ships or aircraft would have to be commissioned to spread the millions of tons of iron needed each year. The precise area of the ocean for fertilization is uncertain, but it would probably be comparable to the land area needed for the tree-growth plan. The second problem is that the microorganisms absorbing the excess carbon would not easily sink into the deep oceans. They would be too small, and ocean currents would work against their sedimentation. In addition, other organisms, the zooplankton, would eat the phytoplankton and effectively convert them back to carbon dioxide. In essence, biology would work against the burial scheme as fast as we would make it work for us. Careful calculations suggest that at most, 10 percent of the present fossil-carbon dioxide emissions could be removed by iron-stimulated ocean burial. Some people argue that we might be able to do better by feeding the phytoplankton. Alternative concepts to deep-sixing CO_2 have been proposed, such as pumping it directly into the deep oceans through long pipes.

Compensating for CO_2 Emissions

An alternative to removing excess carbon dioxide from the atmosphere is to allow it to build up, but to compensate for its greenhouse effect. In Section 11.6.4, volcanic eruptions were shown to have a cooling effect on the surface temperature. Following major eruptions, the atmosphere is filled with tiny particles of sulfuric acid, a by-product of the sulfur dioxide emitted by the volcano. These aerosols scatter solar radiation, increase the Earth's albedo, reduce the solar insolation, and cool the climate. Fossil fuels also contain significant quantities of sulfur. In fact, this sulfur is a nuisance, creating haze and acid rain (Chapters 6 and 9). But if the sulfur could be recovered from the fossil fuels and converted to sulfate aerosols in the upper atmosphere, their cooling influence would compensate for the warming effect of the corresponding CO_2 emissions. The situation is illustrated in Figure 12.15(c).

The potential side effects of such an activity on a global scale include the creation of a permanent haze enshrouding the Earth and a possible depletion of the ozone layer. The climate is also not well enough understood to justify such an experiment on a global scale. It might be likened to performing neurosurgery with a screwdriver in the dark.

Greenhouse Common Sense

The three concepts for global environmental engineering depicted in Figure 12.15 are only a small sample of many that have been proposed, some more seriously than others. Do these ideas seem silly? Or perhaps dangerous? Imagine if there were no other choice. This issue is revisited in Chapter 14. It is sufficient to conclude at this point with a brief plan of action to preserve our climate: Encourage the use of natural sources of energy through subsidized technology development and implementation, and reduce the consumption of fossil fuels through stringent energy-efficiency regulations and the sharing of efficient technologies with developing countries.

Questions

1. List the advantages and disadvantages (from an environmental standpoint), of the following energy sources: oil, coal, natural gas (methane), hydroelectric, nuclear, and solar.
2. Summarize the key events and discoveries that led to our current understanding of the carbon dioxide greenhouse problem.
3. Sketch a global reservoir model for methane. Which reservoirs must be included? Specify the amounts of methane in the reservoirs and its principal sources and sinks.
4. Explain the concept of the "net radiative forcing" that results when a greenhouse gas is added to the atmosphere. In this case, what simple assumption is made about the solar input, or forcing? How might feedbacks in the system, particularly those involving water vapor, affect the assumption about solar radiation?
5. Describe the most likely relationships between carbon dioxide and climate during the ice ages.

How might one have affected the other? Then consider the relationships between methane and climate for the same situation.

6. Why are fuels derived from vegetation relatively free of greenhouse side effects? What are some of the potential drawbacks of these kinds of fuels?

7. Outline the basic reservoir principles that underlie the global environmental-engineering concept of growing and burying biomass on land or in the oceans.

8. Discuss all the potential climate effects of emitting CFCs into the atmosphere. Consider their possible direct and indirect effects. (*Hint*: Think about the relationship between CFCs and ozone.)

Problems

1. Imagine that 100 gigatonnes of carbon dioxide are suddenly injected into the atmosphere. Determine where this excess carbon will go in the short term (that is, the first 10 years or so). In other words, determine roughly how the carbon will be partitioned among the easily accessible reservoirs.

2. After an initial partitioning among the fast reservoirs, assume that the excess CO_2 has a lifetime in the atmosphere of 100 years. Methane has an atmospheric lifetime of 10 years. However, methane is (let's say for the sake of argument) 100 times more effective per molecule than carbon dioxide is in producing a greenhouse effect. Neglecting all other factors, estimate the *relative* sources of CO_2 and CH_4 (as equivalent carbon emissions) that would produce the same greenhouse warming after a sufficiently long time to reach equilibrium (or steady state). If available, use the results from Problem 1.

3. Using the simple relationships developed in the text, calculate the global greenhouse warming corresponding to the following trace-gas scenario: Following emission controls, CO_2 decreases from a peak concentration of 400 ppmv to 300 ppmv; CH_4 and N_2O remain constant; the CFCs continue to increase from the current level of about 0.0009 ppmv to a new level of about 0.1 ppmv. What would the world be like in this case?

4. You wish to extract from the atmosphere exactly the same amount of carbon that is added each year from fossil-fuel combustion. You plan to grow trees and bury the trunks. Each mature tree trunk will weigh 2 metric tons and contain 50 percent carbon by weight. It takes 10 years for a tree to grow to this size. How many trees will need to be growing at any one time to compensate for 6 gigatonnes of carbon emissions as CO_2?

Suggested Readings

Ahearne, J. "The Future of Nuclear Power." *American Scientist* **81** (1993): 24.

Arrhenius, S. "On the Influence of Carbonic Acid in the Air upon the Temperature of the Ground." *Philosophical Magazine* **41** (1896): 237.

Callendar, G. "The Artificial Production of Carbon Dioxide and Its Influence on Temperature." *Quarterly Journal of the Royal Meteorological Society* **64** (1938): 223.

Clark, W. "Managing the Planet." *Scientific American*, Special Issue **261** (1989): .

Davis, G. "Energy for the Planet." *Scientific American*, Special Issue **263** (1990): .

Firor, J. *The Changing Atmosphere: A Global Challenge*. New Haven, Conn.: Yale University Press, 1990.

Graedel, T. and P. Crutzen. "The Changing Atmosphere." *Scientific American* **261** (1989): 58.

Gribbin, J. *Hothouse Earth*. London: Bantam Books, 1990.

Hansen, J., D. Johnson, A. Lacis, S. Lebedeff, P. Lee, D. Rind, and G. Russell. "Climate Impacts of Increasing Atmospheric CO_2." *Science* **213** (1981): 957.

Houghton, R. and G. Woodwell. "Global Climate Change." *Scientific American* **260** (1989): 36.

Jones, P. and T. Wigley. "Global Warming Trends." *Scientific American* **263** (1990): 84.

Manabe, S. and R. Wetherald. "The Effect of Doubling the CO_2 Concentration on the Climate of a General Circulation Model." *Journal of the Atmospheric Sciences* **32** (1975): 3.

Mitchell, J., Jr. "Recent Secular Changes of Global Temperature." *Annals of the New York Academy of Science* **95** (1961): 235.

Ramanathan, V., R. Cicerone, H. Singh, and J. Kiehl. "Trace Gas Trends and Their Potential

Role in Climate Change." *Journal of Geophysical Research* **90** (1985): 5547.

Redfield, A. "The Biological Control of Chemical Factors in the Environment." *American Scientist* **46** (1958): 206.

Repetto, R. "Deforestation in the Tropics." *Scientific American* **262** (1990): 36.

Revelle, R. and H. Suess. "Carbon Dioxide Exchange Between Atmosphere and Ocean and the Question of an Increase of Atmospheric CO_2 During the Past Decades." *Tellus* **9** (1957): 18.

Sagan, C. "Croesus and Cassandra: Policy Response to Global Warming." *American Journal of Physics* **58** (1990): 721.

Schneider, S. "The Changing Climate." *Scientific American* **261** (1989): 70.

———. *Global Warming: Are We Entering the Greenhouse Century?* San Francisco: Sierra Club Books, 1989.

———. "The Greenhouse Effect: 'Science and Policy'." *Science* **243** (1989): 771.

13

The Stratospheric Ozone Layer

Ozone is singled out as one of the most important constituents of the atmosphere. It is more important than carbon dioxide or methane. Ozone was the first atmospheric compound to be placed on the endangered list, the first chemical to be preserved by international treaty. The vice president of the United States, Albert Gore, wrote a book discussing ozone, warning of an impending environmental disaster if action were not taken to conserve it. We see, almost daily, stories in the media concerning skin cancer epidemics and "ozone holes." Stratospheric ozone has been, and remains, essential to life; it is imperative that we preserve the ozone layer in a natural state for as long as we hope to survive on this planet. In this chapter, we summarize the characteristics of naturally occurring ozone. The reasons for all the attention paid to ozone and for the alarm over its disappearance should become clear.

13.1 The Ozone Shield

The stratosphere contains a global layer of ozone, a form of oxygen. Ozone protects the biosphere from the deadly ultraviolet rays of the sun. It is generally held that terrestrial life—that is, life on land—was not able to evolve until the ozone layer appeared about 1 billion years ago. The ozone is derived from molecular oxygen, which became common in the atmosphere only after the evolution of photosynthesis in aquatic plants, including such humble species as the algae (Section 4.2.1). Until that time, the atmosphere held very little molecular oxygen, and as a result, little ozone was present. The first plants were themselves protected from ultraviolet radiation by the murky water in which they evolved and were immersed. Nevertheless, these photosynthetic plants thrived on the abundant supply of carbon dioxide and light, and their life processes copiously generated oxygen as a by-product. The ozone layer, thus established, shielded the Earth's surface from ultraviolet radiation during the critical period when life advanced onto land and more complex organisms evolved. We would not be here today were it not for the ozone layer.

Given the scientifically well-established importance of the ozone layer for almost all life on Earth (there are microbes and worms living in mud at the bottom of swamps and near thermal vents in the depths of the oceans), it is not surprising that most scientists who study the atmosphere and the biosphere are concerned that this critical protective shield may be deteriorating. After decades of careful and detailed study, scientists have recently achieved a fundamental understanding of the ozone layer. What they have learned has led some to sound warnings about impending global ozone depletion and the onset of related environmental problems. This chapter tells the story of the ozone layer. First we discuss the origin, distribution, and properties of ozone. Then we describe the role of ozone in filtering ultraviolet radiation, including the effects of overexposure to such radiation. We look at the effects of various compounds—particularly compounds containing chlorine—on the amount of ozone in the atmosphere as well as ozone reductions associated with chlorofluorocarbons (CFCs). Finally, we describe the spectacular ozone "hole" over Antarctica, setting the stage for a discussion of measures that have been suggested to "save" the ozone layer for future generations.

13.2 The Formation and Destruction of Ozone

Ozone is a form of oxygen, the second most common chemical element composing our atmosphere.

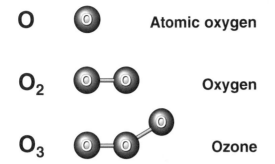

Atomic oxygen

Oxygen

Ozone

Figure 13.1   The forms, or allotropes, of oxygen found in the atmosphere. Atomic oxygen is simply a single free atom of the element oxygen. Molecular oxygen, O_2 (or simply oxygen), is the most stable and common form of oxygen, which we breathe. Ozone is formed by combining O and O_2 to form the bent molecule O_3. Unlike oxygen, ozone is not healthful to breathe.

Practically all the oxygen in air is in the form of molecular oxygen (O_2). However, a small fraction (about one part per million) of the oxygen is in the form of ozone (O_3), which contains three oxygen atoms rather than two (these different forms of the same element are called **allotropes**, in this case, of oxygen). Another form of oxygen found in the atmosphere, atomic oxygen (O), is much less common than ozone (but is nevertheless an important substance in controlling the photochemistry of the atmosphere). The three main gaseous allotropes of oxygen are illustrated in Figure 13.1.

13.2.1   THE PHOTOCHEMISTRY OF OZONE

In the formation of ozone in the upper atmosphere, the first step is the photodissociation of O_2, which generates free oxygen atoms:

$$O_2 + UV \text{ Sunlight} \rightarrow O + O \quad (13.1)$$

Photodissociation occurs when a photon of sunlight is absorbed by an O_2 molecule, imparting its energy to the atoms of the molecule and thus causing it to break apart, or dissociate (Section 3.3.3). The photons in the ultraviolet region of the solar spectrum are quite energetic and can initiate photodissociation. Section 3.2 defined ultraviolet radiation as electromagnetic radiation with wavelengths (l) lying *below* the violet end of the visible spectrum. The process depicted in Equation 13.1 does not occur effectively below an altitude of about 20 kilometers (in the lower stratosphere) because most of the

ultraviolet sunlight at short enough wavelengths (l < 0.25 micrometers [μm], or 250 nanometers [nm])[1] has been removed by O_2 and O_3 absorption at higher altitudes.

The free oxygen atoms produced by Equation 13.1 rapidly create ozone in a **three-body reaction** involving O_2 and a *m*ediator molecule, M (also referred to as a "third-body" [Section 3.3.3]):

$$O + O_2 + M \rightarrow O_3 + M \quad (13.2)$$

Because the atmosphere is uniformly filled with O_2, ozone is generated wherever ultraviolet radiation is incident in the atmosphere (which happens to be anywhere on the globe at altitudes above ~20 km). The photochemical Equations 13.1 and 13.2 in combination produce ozone by the following crucial overall mechanism:[2]

$$O_2 + UV \rightarrow O + O$$
$$\underline{2(O + O_2 + M \rightarrow O_3 + M)} \quad (13.3)$$
Net reaction: $3O_2 \rightarrow 2O_3$

Ozone is continuously produced in the upper atmosphere by this photochemical mechanism. If nothing else happened, most of the oxygen in the atmosphere would eventually be converted to ozone, reaching concentrations intolerable to life (ozone, after all, is a highly toxic air pollutant to living organisms [Section 7.2.1]). Fortunately, a competing photochemical reaction mechanism destroys ozone and naturally limits its accumulation in the atmosphere. The following photochemical reactions occur in the same regions where ozone is formed:

---

1. The units of wavelength that are used in discussing atmospheric photochemistry are angstroms, nanometers, or micrometers (or microns). An angstrom is equivalent to a length of 1×10^{-10} meter, or 1×10^{-8} centimeter. A nanometer is one-billionth of a meter, or 1×10^{-9} m, and a micrometer is one-millionth of a meter, or 1×10^{-6} m. An angstrom is roughly the size of an atom of matter, and a nanometer is roughly the size of a molecule.

2. By an "overall" chemical reaction, or "net" reaction, we mean here that the intermediary steps and products are not explicitly shown. However, the net, or overall, reaction is obtained by summing up *all* the basic reaction steps. In Equation 13.3, for example, the three-body process is included twice; hence the two oxygen atoms produced in the first step are used up in the second step, and no oxygen atoms are left in the "net reaction." The net reaction also conserves the total amount of oxygen (atoms) in the system.

$$O_3 + UV \rightarrow O + O_2$$
$$\underline{O + O_3 \rightarrow O_2 + O_2} \qquad (13.4)$$
Net reaction: $\qquad 2O_3 \rightarrow 3O_2 \qquad (13.5)$

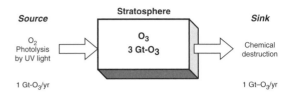

Figure 13.2 A simple box model representation for ozone, or, equivalently, odd-oxygen, in the lower stratosphere, where the main ozone layer lies. The source of ozone is the photodissociation of O_2 followed by the association of O with O_2. Ozone is lost by odd-oxygen recombination. The global rates of ozone formation and destruction in this region of the stratosphere are roughly 1 gigatonne (Gt) per year. At any given time, the stratosphere holds about 3 Gt of ozone.

As indicated in Equation 13.4, ozone can be photodissociated by ultraviolet radiation into atomic and molecular oxygen. In fact, this process is normally much faster than the photodissociation of O_2, which forms ozone. The oxygen atoms generated from ozone in this way, however, almost immediately recombine with O_2 to regenerate ozone by means of Equation 13.2. Every now and then, an O atom encounters an ozone molecule instead of an oxygen molecule and reacts with the ozone as indicated in Equation 13.5. Because of the relative concentrations of O_2 and O_3 in air, a free O atom in thermal motion encounters O_2 a million times more frequently than it encounters O_3.

Whenever an O atom collides and reacts with an O_3 molecule, O_2 is re-formed. The net effect of the sequential occurrence of Equations 13.4 and 13.5 is to convert two ozone molecules to three oxygen molecules (the total amount of oxygen in all forms is, of course, conserved in this process).[3] In scientific jargon, O and O_3 together are often referred to as *odd-oxygen*. This does not mean "strange" oxygen or "eccentric" oxygen. Rather, the term odd-oxygen refers to the fact that both species have an odd number of oxygen atoms (one and three for O and O_3, respectively). Moreover, the one oxygen atom is labile; that is, it is readily transformed between O and O_3 in the stratosphere (Equations 13.2 and 13.4). In fact, each hour almost all the ozone in the stratosphere is recycled through the atomic oxygen form.

The fundamental balance of ozone, or odd-oxygen, is influenced by production through the photodissociation of O_2 (Equation 13.3) and destruction by the odd-oxygen recombination (Equations 13.4 and 13.5). Indeed, a simple "box" model (Section 4.1.2) can be constructed for ozone, as shown in Figure 13.2. Such a model is obviously an oversimplification of the actual ozone layer. In the stratosphere, winds move ozone around the Earth and produce substantial variability in the ozone distribution at any particular time and place. Nevertheless,

the basic behavior of the ozone layer may be understood using the box model.

If all the ozone were suddenly removed from the stratospheric "box," it would take about 3 years to replenish it. This time is determined by dividing the total amount of ozone by the ozone production rate, or source; that is, 3 Gt-O_3/1 Gt-O_3/yr = 3 years. The global source of ozone from O_2 photodissociation (Equation 13.1) is essentially constant over time. The two factors that determine this source—the quantity of oxygen in the atmosphere and the ultraviolet emission of the sun (at least in that part of the ultraviolet spectrum responsible for most of the ozone)—change by less than a small percentage over a millennium. Accordingly, any rapid changes in the global abundance of ozone are more likely to be associated with changes in the ozone sink.

One important consequence of a fixed source of ozone is that an increase in the rates of chemical reactions that *consume* ozone leads to a decrease in the total stratospheric burden of ozone. First, consider that the relationship between the source and sink of ozone is described by the steady-state box model equation (Section 4.1.2). In this case, $S = L$, where S is the source (with units of mass/time) and L is the sink (with the same units). It can also be shown that the O_3 loss rate, L, is roughly proportional to the ozone concentration squared. Such a relationship follows from Equation 13.5. The rate of the reaction that destroys odd-oxygen is, according to simple chemical kinetics (Section 3.3.3), proportional to the product of the atomic oxygen and ozone concentrations. However, we have concluded that O and O_3 are in some sense equivalent species. It turns out that their concentrations are

---

3. Collision and reaction are not equivalent. A collision must occur in order for a reaction to proceed. But many collisions result simply in rebound, as happens when two billiard balls hit. The formation of an activated complex that results in a chemical reaction is discussed in Section 3.3.3.

usually just proportional.[4] If this is the case, the box model equation can be written in the form

$$S = L = l[O_3]^2 \qquad (13.6)$$

Here, $[O_3]$ is the average concentration of ozone in the box (stratosphere), and l is a photochemical coefficient determined in part by the proportionality between O and O_3.

Equation 13.6 can be inverted to find the ozone concentration that would be in equilibrium with a specific photodissociation source, S, for a given chemical state of the stratosphere represented by the coefficient l:

$$[O_3] = \sqrt{\frac{S}{l}} \qquad (13.7)$$

The factor l increases as the concentrations of certain chemical compounds, called **catalytic agents**, increase. As l increases, $[O_3]$ decreases in accordance with Equation 13.7.

13.2.2   THE DESTRUCTION OF CATALYTIC OZONE

Certain chemical agents (which will be indicated for now as the unidentified substance X) can destroy ozone, or odd-oxygen, **catalytically**. The term *catalytic* means that agent X can react with and remove odd-oxygen while X itself is regenerated in the process. Because X is recycled, one molecule of X can destroy many molecules of ozone, perhaps tens of thousands. The basic reaction sequence for such a catalytic cycle can be written as

$$O_3 + X \rightarrow XO + O_2$$
$$\underline{XO + O_3 \rightarrow X + 2O_2} \qquad (13.8)$$
Net reaction: $2O_3 \rightarrow 3O_2$

A related catalytic chemical cycle involves the reaction of an O atom with an XO molecule in the second reaction of Equation 13.8

4. The very fast chemical processes described by Equations 13.2 and 13.4 and summarized in Equation 13.3 control the relative amounts of O and O_3 in the odd-oxygen family. In other words, the ratio of O to O_3 in an air parcel is determined by the values of the local photochemical reaction coefficients, which themselves are functions of the local temperature and pressure and the intensity of sunlight illuminating the parcel. When the local environmental conditions are fixed, therefore, the concentrations of O and O_3 have a fixed ratio or are proportional.

$$O_3 + X \rightarrow XO + O_2$$
$$\underline{XO + O \rightarrow X + O_2} \qquad (13.9)$$
Net reaction: $O + O_3 \rightarrow 2O_2$

As the agent X is recycled, it continues to destroy ozone until some other reaction interferes by converting X to a less aggressive compound. Even small amounts of some catalytic agents are capable of depleting ozone concentrations. In the stratosphere, the following species are known to be important catalysts (X) for ozone destruction: nitric oxide (NO), hydroxyl (OH), chlorine (Cl), and bromine (Br). Notice that each of these agents can be substituted directly into Equations 13.8 and 13.9 to define specific catalytic reaction cycles. Then XO becomes NO_2 in the case of NO, HO_2 in the case of OH, ClO in the case of Cl, and BrO in the case of Br.[5]

The sources of several of these catalytic agents—both natural sources and those associated with human activities—are discussed in Section 13.5 in relation to global ozone depletion.

13.3   The Distribution of Ozone in the Atmosphere

Ozone is not uniformly mixed throughout the atmosphere; its distribution over the planet tends to be highly variable. Urban smog is an example of this heterogeneity. Nevertheless, the global stratospheric ozone layer is dominated by two distinct features: its vertical profile and its pole-to-equator variation. Figure 13.3 demonstrates the vertical distribution of ozone, which holds generally everywhere on Earth.

The ozone layer has a broad maximum in the lower stratosphere, between about 20 and 30 kilometers. This distribution is the result of the photochemical balance between ozone production and destruction. Above 30 kilometers, the production of ozone is rapid, but the catalytic rate of destruction is also very rapid; hence the ozone concentration is low. Below 20 kilometers, the production of ozone is quite small, and ozone can be lost by transport to the troposphere owing to the proximity of the tropopause. Here again, the ozone concentration is low.

5. Two other halogens, fluorine and iodine, are not important stratospheric ozone catalysts. Iodine concentrations are much too low at high altitudes. Fluorine, which has a significant concentration in the stratosphere, does not remain in a reactive form long enough to affect ozone through catalytic reactions.

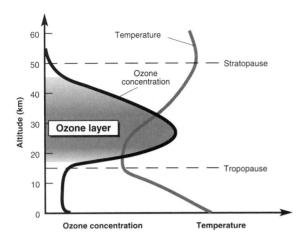

Figure 13.3 The vertical distributions of ozone and temperature in the Earth's atmosphere. The vertical scale shows the altitude, in kilometers. The horizontal scale gives a relative measure of the ozone concentration and the air temperature at each height.

In between these levels, the ozone source is large enough and the ozone loss small enough that substantial ozone concentrations can build up. At the maximum, ozone concentrations are as large as 10 to 20 parts per million by volume (ppmv), and concentrations reach 5×10^{12} molecules/cm^3.

In the troposphere, a background concentration of ozone is also apparent in Figure 13.3. Ozone in the tropospheric contributes about 10 percent to the total global ozone burden and so cannot be ignored. The much higher concentrations of ozone found in polluted cities are not indicated in Figure 13.3. Although locally a major nuisance, this ozone contributes less than a small percentage to the global total.

13.3.1 DOBSON UNITS: OZONE OVERHEAD

The concentration of ozone in air is a useful way to specify its local quantity to determine potential health and environmental effects. The concentration is typically stated as a mixing ratio. For example, in smog, the concentration of ozone is given in parts per million by volume (ppmv). Another useful way to specify the amount of ozone in the atmosphere, to discuss the global effects associated with changes in ozone, is in terms of the total column abundance of ozone, or the **ozone column**. This is the total amount of ozone that is directly over your head, through the entire depth of the atmosphere from the ground to space. The column amount is specifically

defined as the total number of ozone molecules in an imaginary tube having a cross-sectional area of 1 centimeter squared that stretches from the surface through the entire ozone layer to the top of the atmosphere (Figure 13.4). This total column quantity of ozone varies with time and location, but much less so than does the local concentration of ozone at any fixed point in the atmosphere. The column ozone amount is the key to estimating the intensity of ultraviolet radiation that penetrates the atmosphere to the surface.

The ozone column abundance is typically specified (by atmospheric scientists, at least) in **Dobson units**, (DU). George Dobson was a pioneer in measuring the atmospheric ozone, inventing a simple instrument to determine the total quantity of ozone overhead by collecting ultraviolet sunlight scattered by the atmosphere.[6] The **Dobson spectrophotometer**, as his instrument is called, relies on the fact that the extent to which ultraviolet radiation is absorbed depends on the total amount of ozone present between the sun and the measuring device. Rayleigh scattering (Section 3.2.2) guarantees that some of the incident solar radiation will always be deflected toward the instrument. As sometimes happens in science, Dobson's ingenuity was rewarded by having a unit of measurement named after him.[7]

Imagine that we can measure the total amount of ozone overhead by trapping it in an open-ended plastic tube. The tube has a 1-cm^2 cross section and is long enough to reach the top of the atmosphere (a very long tube indeed—some 100 kilometers long!). We carefully raise the tube straight up from the ground, capturing a sample of air from each altitude

6. George M. Dobson (1889–1976) was an English physicist who became interested in meteorology after casually studying the wave motions on Lake Windermere. In the 1920s, Dobson first deduced that the temperature in the stratosphere must increase with height and not decrease, as in the troposphere (Section 2.3.2). He did this by observing meteor trails and noting that the height where they burned up was greater than expected. This suggested that the atmosphere extended to higher altitudes and so had a larger scale height and a warmer temperature (Section 3.1.1). Dobson attributed the warmth to the absorption of sunlight by ozone. Ozone became the focus of his research thereafter. He built his first ozone detector in 1924.

7. The first regular measurements of ozone in the atmosphere were made by Charles Fabry, a famous French physicist who designed numerous optical instruments, beginning in 1912. Lord Rayleigh also studied ozone absorption in the atmosphere, by measuring the spectrum of light from the star Sirius in 1917. In 1918, Rayleigh further noted that the surface air had too little ozone to account for the cutoff in the solar spectrum that Hartley observed.

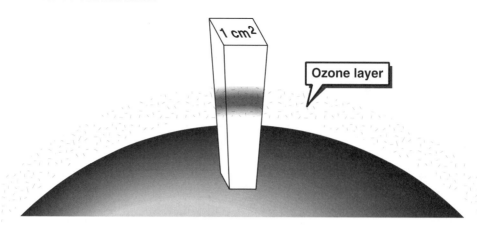

Figure 13.4   The "ozone column" is the number of ozone molecules contained in a column of air 1 square centimeter in area that extends vertically from the surface through the atmosphere, spanning the entire ozone layer. The ozone column is usually measured in Dobson units, where 1 DU = 2.7×10^{16} O$_3$ molecules/cm^2.

as the tube progresses. When the highest altitude is reached, we seal the ends of the tube.

Back at the laboratory, we count the total number of ozone molecules trapped in the tube. If that number were 2.7×10^{16} molecules, we would have measured exactly 1 Dobson unit of ozone. In the Earth's atmosphere, the actual number of ozone molecules in a 1-cm^2 column is about 300 times larger, roughly 800×10^{16} ozone molecules, or 300 DU. If this quantity of ozone were compressed into a layer of pure ozone gas at a pressure of 1 atmosphere, it would be about 3 millimeters thick (about 0.11 inch). In other words, 100 DU is equivalent to a 1-millimeter-thick layer of pure ozone at surface pressure. That is hardly as thick as the skin on an orange.

Figure 13.5 shows the geographical distribution of total ozone measured in Dobson units. The total ozone column amount varies with location. It can be deduced from this map that the *average* amount of ozone in the Earth's atmosphere is roughly 300 Dobson units.

The total ozone map in Figure 13.5 has some remarkable features. Notice that although the total amount of ozone varies over the globe, its distribution is fairly uniform. On average, every point on the globe is covered by at least 250 Dobson units of ozone. No place is left unprotected. Also notice how the ozone contours stretch around the Earth in parallel bands, following the equator. The ozone distribution is therefore said to be "zonally symmetric" and does not vary strongly with longitude.

The total ozone amount depends significantly on latitude. The ozone column increases from the trop-

ics toward the poles. This characteristic of the ozone distribution is important to determining the variation in latitude of ultraviolet (UV) radiation (which penetrates the ozone layer [Section 13.4]). It is well known that with lower amounts of ozone, the intensity of ultraviolet radiation at the ground is greater when the sun is at the same position in the sky (also recall the principle of operation of Dobson's spectrophotometer to measure ozone).[8] Generally, the ultraviolet radiation intensity at the ground increases as the sun rises higher overhead in the sky. Thus the tropical regions of the Earth receive a double dose of ultraviolet photons, because the lowest average ozone amounts are found there (Figure 13.5) and the sun is well overhead all year long.

Figure 13.6 shows the seasonal variation in the average latitudinal distribution of ozone. Several features are noteworthy. At tropical latitudes, the seasonal variation is quite small; the concentrations of ozone are low all year long. The largest seasonal variations occur at high latitudes in the Northern Hemisphere and at middle and high latitudes in the Southern Hemisphere. The largest amounts of ozone are seen in the late winter and early spring in both hemispheres (notice that the ozone maximums are about 6 months out of phase between the hemispheres, like

8.   The instantaneous intensity of ultraviolet radiation depends on the column amount of ozone and the angle of the sun with respect to the "zenith" or overhead, position. At any particular location on the Earth and on any day of the year, the sun moves across the sky during the day in a definite way. The intensity of ultraviolet radiation varies accordingly. The average daily exposure to ultraviolet radiation, therefore, is a function of the day of the year and the latitude.

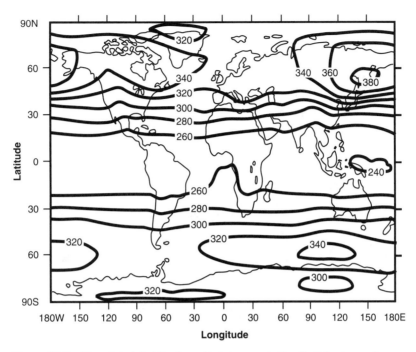

Figure 13.5   The global distribution of total ozone in the Earth's atmosphere. The contour lines form the boundaries of regions in which the overlying ozone amount exceeds the value (in Dobson units) indicated next to the contour. The ozone amounts shown correspond to yearly average amounts of ozone obtained by satellite observations. (Data compiled from global ozone measurements taken by the Total Ozone Mapping Spectrometer [TOMS] instrument [Kenneth P. Bowman and Arlin J. Krueger, "A Global Climatology of Total Ozone from the Nimbus 7 Total Ozone Mapping Spectrometer [TOMS]," *Journal of Geophysical Research* 90 (1985): 7967])

the seasons themselves). There is compensation between the ozone variations in the two hemispheres that keeps the total *global* ozone burden roughly constant. This compensation is in part caused by the transport and redistribution of ozone over the globe and in part by continuous steady photochemical generation of ozone in the sunlit regions of the upper atmosphere.

13.3.2   How Much Ozone Is There?

The average amounts of column ozone are about 300 DU, which corresponds to some 800×10^{16} molecules lying above each square-centimeter of the Earth. The area of the Earth is close to 5×10^{18} cm^2. It follows that the total number of ozone molecules in the Earth's atmosphere (found by multiplying 800×10^{16} ozone molecules/cm^2 by 5×10^{18} cm^2) is 4×10^{37} molecules. Comparing this with the total number of air molecules (that is, 1×10^{44} mol-

ecules), the *average* mixing ratio of ozone in the atmosphere is roughly 0.4 ppmv (that is, $4 \times 10^{37} / 1 \times 10^{44} = 0.4 \times 10^{-6} = 0.4$ ppmv). This concentration of ozone is equivalent to a *stage-2 smog alert*. The average value would apply, of course, only if the ozone were mixed throughout the atmosphere, including the troposphere, by some unusual churning process. Happily, ozone is confined mainly to the stratosphere and is not effectively carried to the ground!

We are rarely exposed to very high concentrations of ozone. In smoggy cities, ozone concentrations occasionally exceed a few tenths of a part per million. Average mixing ratios in the stratosphere, where the global ozone is concentrated, are considerably larger than this. Fortunately, the stratosphere is 10 miles up. So we are safe, at least most of the time. When you fly on a commercial jetliner at 40,000 feet or higher, you may actually be in the lower stratosphere. This is more likely if you are flying a "polar" route over the high latitudes where the stratosphere

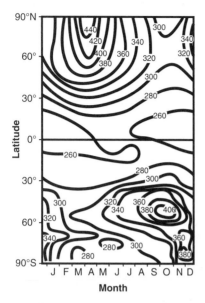

Figure 13.6   The seasonal variation in total ozone. The contour lines correspond to the total amount of ozone in Dobson units as a function of latitude (pole to pole) on a monthly basis. The values correspond to zonally averaged amounts of ozone (that is, averaged longitudinally around the Earth at fixed latitudes). The data also are averaged over several years from 1978 to 1985. (Data from Kenneth P. Bowman and Arlin J. Krueger, "A Global Climatology of Total Ozone from the Nimbus 7 Total Ozone Mapping Spectrometer [TOMS]," *Journal of Geophysical Research* 90 [1985]: 7967)

is closest to the ground (Section 2.3.2). Air from the outside is drawn into the aircraft cabin for ventilation and is compressed to a higher pressure for breathing comfort. The concentration of ozone in this air could be as large as 10 parts per million, far above any smoggy-air value. Ventilation systems on airliners, accordingly, must be equipped with filters to remove the ozone. But sometimes they do not work. The result of exposure to excessive stratospheric ozone can take the form of "cabin sickness," including nausea and respiratory difficulty.

13.4   Ozone and Ultraviolet Radiation

How does ozone protect us? Why should we be concerned about losing ozone from the upper atmosphere, so far above us?

Ozone is beneficial because it absorbs biologically harmful UV radiation emitted by the sun. Ultraviolet light is absorbed by ozone mainly in the stratosphere, where the amount of ozone is greatest.

Although ozone removes most of the harmful UV sunlight before it can reach the surface, a portion still penetrates to the ground. The number of ultraviolet photons that arrive at the surface during a time interval is a function of the total amount of ozone overhead and the elevation of the sun in the sky. The penetration of ultraviolet rays to the ground is greater when the sun is more directly overhead (in the zenith) than when it is lower in the sky closer to the horizon.[9] Exposure to the sun in the early morning and late afternoon is safer than exposure at lunchtime. Most people realize that they are less likely to get sunburned in the afternoon because the burning rays of the sun are absorbed more effectively at this time.

Over many days, weeks, and years, exposure to ultraviolet radiation accumulates as a total **dose**. Dose refers to the total number of photons absorbed by an organism (such as yourself) that cause some specific effect (such as sunburn in humans). When the quantity of ozone is small and the sun is high in the sky, the ultraviolet intensity is greatest, and the dose accumulates most quickly. Any depletion of the ozone layer below its normal abundance can lead to an increase in the ultraviolet intensity and cause the dose to build up more quickly. High ultraviolet radiation doses represent a serious threat to almost all living organisms.

Figure 13.7 gives the altitude to which solar radiation of different wavelengths can penetrate the atmosphere. Only a small segment of the entire solar spectrum is shown. This segment is called the **ultraviolet** region, which can be further divided into several distinct wavelength regions. The **far-ultraviolet** region has wavelengths shorter than about 250 nanometers (nm; 1 nm = 10^{-9} meters), and the **near ultraviolet** region has wavelengths between about 250 and 380 nm. Visible radiation lies above 380 nm. The near-UV spectrum is further broken down into the UV-A, UV-B, and UV-C regions.

The short-wavelength, highly energetic radiation in the far-ultraviolet spectrum is effectively absorbed

9. The attenuation of ultraviolet radiation depends on the "path length" through the absorbing ozone layer. As the sun gets closer to the horizon, this path length increases. As a demonstration, fill a flat pan with water and hold a straw vertically to measure the depth of the water. The straw can represent a ray of light from the sun. Now tilt the straw while holding it against the bottom of the pan. The water line steadily rises along the straw; the path length of water increases as the straw tilts toward the horizon. The amount of water in the straw increases as the straw is tilted. You can trap it and measure the amount by placing your finger over the end of the straw and removing it from the pan.

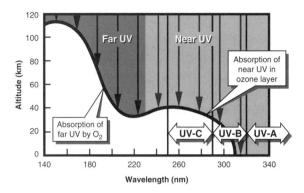

Figure 13.7   The penetration of solar radiation through the Earth's atmosphere. The solid line defines the effective penetration depth of sunlight as a function of the radiation's wavelength; the altitude of penetration is indicated along the vertical axis, with the principal spectral regions of interest noted. The sun is assumed to be directly overhead. The solar radiation at wavelengths greater than about 320 nanometers can easily penetrate to the surface.

by O_2 in the upper atmosphere, at altitudes above 30 kilometers. Ozone absorbs a large portion of the near-ultraviolet sunlight at heights above about 15 kilometers.

13.4.1   REGIONS OF THE ULTRAVIOLET SPECTRUM

Discussions of the health effects of ultraviolet radiation often refer to three distinct regions of the near-ultraviolet spectrum.

- UV-A. Radiation having wavelengths between 320 and 380 nanometers (nm) is referred to as UV-A. Tanning salons use fluorescent lamps that emit UV-A photons to "safely" tan their patrons. Sunlight at the ground has a high intensity of UV-A. Exposure to UV-A radiation is suspected of having detrimental health consequences, including weakening of the immune system. UV-A solar radiation readily penetrates the atmosphere and reaches the surface at nearly full intensity.

- UV-B. Radiation with wavelengths between 290 and 320 nm is called UV-B. This radiation is well established as biologically hazardous. Sunburn is the prompt response of the skin following exposure to intense UV-B radiation. Tanning of the skin produces pigments (**melanins**) in the epidermis that can absorb UV-B radiation, reducing damage to deeper skin tissues. Most of the UV-

B radiation impinging on the top of the atmosphere is absorbed by the ozone layer. A small, but significant, amount of UV-B radiation can penetrate the atmosphere and reach the surface.

- UV-C. Radiation with wavelengths between 250 and 290 nm is designated as UV-C. This radiation is extremely hazardous to people and is broadly biocidal. Occasionally, UV-C is used as a sterilizing agent. Humans and most other organisms have little natural protection against direct exposure to UV-C. This radiation is completely absorbed by ozone in the upper atmosphere.

UV-A radiation passes through the atmosphere relatively unaffected because the most abundant atmospheric gases do not absorb radiation in this spectral region. Ozone has weak absorption here, and nitrogen dioxide absorbs UV-A as well. However, the total column abundances of O_3 and NO_2 are too low to reduce substantially the surface intensity of UV-A. In the UV-B spectrum, on the other hand, ozone absorbs fairly strongly. Moreover, there is enough ozone in the stratosphere to absorb a large portion of the UV-B radiation from the sun at altitudes well above 15 kilometers. Nevertheless, residual UV-B radiation, particularly at the longer wavelengths, leaks to the surface. A sufficiently long exposure to this radiation can result in serious health problems. Ozone strongly absorbs in the UV-C spectral region (in fact, ozone has a maximum **absorption cross section** at a wavelength of 250 nanometers in the center of the Hartley band).[10] At wavelengths shorter than about 250 nanometers, O_2 absorbs most of the solar radiation in the upper atmosphere, above 30 kilometers (Figure 13.7). Far-ultraviolet radiation is therefore completely blocked from reaching the surface. The combined presence of O_2 and O_3 in the Earth's atmosphere is very fortunate indeed! (See Section 4.2.3 for information concerning the evolution of oxygen, the ozone layer, and life.)

Figure 13.8 illustrates the factors that control the amount of solar ultraviolet-B radiation reaching the Earth's surface as a function of wavelength. The

10. The cross section for the absorption of radiation is defined in Section 3.2.2. Basically, every molecule of a gas exhibits a specific area that effectively collects the radiation headed in its direction. The area changes with the wavelength of the radiation. The cross section of a molecule is similar to a parasol held up to the sun and thus intercepts the sunlight over a certain area. If the parasol is black, the light is absorbed; if it is white, the light is scattered.

ozone absorption cross section, which decreases rapidly as wavelength increases in this region, determines how much radiation can penetrate the ozone layer at each wavelength, for a fixed total amount of ozone. When the ozone cross section is large, the absorption is strong, and the penetration of ultraviolet radiation is negligible. This situation occurs at the shorter wavelengths in the ultraviolet. Conversely, when the ozone absorption cross section is small, the attenuation is weak, and the penetration of radiation is nearly complete. These circumstances hold for the longer-wavelength ultraviolet rays. At wavelengths shorter than about 300 nanometers, the radiation intensity at the ground is negligible compared with the solar radiation striking at the top of the atmosphere. At wavelengths longer than 320 nanometers, the radiation intensities are similar.

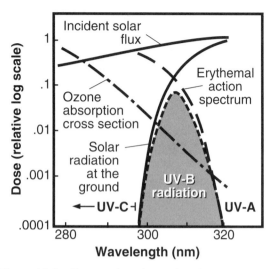

Figure 13.8   Factors that determine the exposure to and effects of ultraviolet radiation received at the ground. The key parameters are the intensity of solar radiation at the top of the atmosphere, the ozone absorption cross section, and the "action spectrum" defining a specific physiological response to ultraviolet radiation. All the parameters are given as relative quantities, with maximum values of 1. This presentation emphasizes the spectral variation in each parameter and the wavelength regions contributing to the surface dose. The light dashed curve defines the dose rate spectrum for producing erythema (sunburn) in humans (the dose rate spectrum is the dose per unit time at each wavelength). Quantitatively (and graphically), the dose rate spectrum is the product of the erythemal action spectrum and the solar radiation spectrum at the ground. The dose rate spectrum for erythema is similar to (but not exactly the same as) the dose rate spectra corresponding to other biological effects of ultraviolet radiation, including DNA damage and the aging of skin. The shaded area under the dose rate spectrum determines the total effective dose rate for producing erythema (again, in the figure the dose rate is a relative value because the other spectral quantities have been normalized to a maximum value of 1). To obtain the absolute dose rate, the spectral parameters must be scaled to their appropriate absolute values; the dose rate would then be multiplied by these scaling factors.

13.4.2   HEALTH EFFECTS OF UV-B RADIATION

Ultraviolet radiation in the B and C bands can be lethal to living organisms. Ultraviolet-A radiation appears to be relatively harmless compared with other natural hazards, including radioactivity, although this conclusion is controversial. The action of ultraviolet radiation on the body is complex and not fully understood. Clearly, UV radiation can affect living cells by attacking the molecules in the cells, particularly DNA and related molecules. Destruction of these molecules results in cellular dysfunction, mutations, and the formation of toxic chemical species. It stands to reason that with continuing exposure, the skin will toughen and age as the cellular scarring builds up. What is not understood is the way in which cellular defects may accumulate over time, eventually triggering the formation of cancerous tumors.

Exposure and Dose

Exposure to penetrating ultraviolet radiation is determined by the **dose rate spectrum**, such as the one shown in Figure 13.8. In particular, the ultraviolet dose rate spectrum defines the number of UV photons absorbed at each wavelength per unit interval of time that produces a specific kind of response, which can be measured. The shaded area under the dose rate spectrum in Figure 13.8 defines the *total* dose rate leading to a specific physiological response following the exposure to ultraviolet radiation. The

response of an organism to radiation exposure at different wavelengths is described by an **action spectrum**, a *specific effect* or response in a *particular organism* exposed to ultraviolet radiation. For example, the **erythemal action spectrum** describes the appearance of sunburn in humans. Other action spectra might refer to the generation of chromosomal damage in cells (mutagenic action spectrum) and the induction of cancer in certain organs. The total dose rate leading to a specific effect is equivalent to the shaded area in Figure 13.8. This area is defined by the product of two curves in the figure: the action spectrum and the solar radiation intensity at the ground at each wavelength.

Figure 13.8 clearly shows that it is radiation in the UV-B spectrum that presents a hazard to health. Notice that the total dose rate (shaded area in Figure 13.8) will go up if the solar-radiation intensity at the ground increases or if the organism exhibits greater sensitivity through either enhanced sensitivity at longer wavelengths or a larger response at wavelengths from 300 to 320 nanometers.

For every 1 percent of depletion in total ozone, the UV-B radiation dose rate at the ground increases by an average of 2 to 4 percent. This estimated enhancement in exposure includes the effect of the sun's motion across the sky during the day. It is well known that ultraviolet (UV-B) radiation is much less intense in the morning and afternoon hours than at midday. At midday, the sun's rays have the most direct route through the atmosphere to the ground. Hence the rays pass through the ozone layer by the shortest route. At other times of the day, the sun's rays are tilted with respect to the zenith (the vertical, or point directly overhead) and strike the ozone layer at an oblique angle. Under these conditions, the distance along a ray through the ozone layer and the absorption by ozone are greater. The UV-B dose rate, likewise, tends to be larger at places on the Earth where the sun rises high in the sky all year round (in the tropics), where the ozone layer is thinnest (also in the tropics); at high elevations, above the clouds, haze, and smog (on mountaintops); and over reflective surfaces (above snow and ice fields, or water).

The accumulated dose of ultraviolet radiation is determined by the dose rate and the duration of exposure. The UV dose builds up over time in much the same way as does the dose from exposure to toxic compounds, discussed in Section 7.1.2. In this case, we can estimate the total dose, D_{uv}, as the product of the dose rate, $\dot{D}_{uv}$, corresponding to the shaded area in Figure 13.8, and the exposure time, t_e,

$$D_{uv} = \dot{D}_{uv} t_e \qquad (13.10)$$

The total UV radiation dose obviously increases over time with any exposure to sunlight. Accordingly, each person has a unique history of ultraviolet exposure and dose. Among the factors that determine the cumulative effects of UV radiation are the dose rates and the areas of the body exposed. For example, a certain total dose received in a short time, resulting in a severe sunburn, is worse than the same dose taken slowly over time, resulting in a tan. Continual exposure of the hands, arms, and face may also lead to localized effects on the skin in these areas.

An important factor in UV-B exposure derives from individual habits and personal biology. Some people like to be tanned all year round. Others work in direct sunshine all day. Many people have fair skin, which is sensitive to ultraviolet radiation. It should be obvious that one's actual exposure to ultraviolet radiation depends on many factors, some of which can be controlled. For example, time spent in the sun can be restricted, and sunscreen can be applied to exposed parts of the body. Other organisms and the ecosystems they inhabit have little control over the dose of ultraviolet radiation they receive. Accordingly, many other species have evolved defensive physiological responses to UV-B radiation, including chemical mechanisms to repair damaged DNA, and the ability to move into the shade when irradiated. As might be expected, natural responses have limited ranges of effectiveness, consistent with the evolution of the species under the influence of relatively small natural fluctuations in background ozone concentrations over long time spans.

In humans, tanning of the skin is a natural response to exposure to sunlight. In the upper layer of skin are cells that, when stimulated by ultraviolet radiation, produce colored pigments, called **melanins**. In humans, melanin is brown to black in color. This pigmentation offers effective protection from the effects of ultraviolet radiation. The color of dark-skinned people is caused by dense natural melanin pigmentation. The skin of people with fair complexions produces melanin, but inefficiently. Freckles, moles, and other dark features on fair skin are the result of localized accumulations of melanin pigments. These dark spots not only offer little protection from ultraviolet radiation, but also may actually be the seats of cancer when they appear following long-term exposure to ultraviolet rays.

Melanin is the most effective natural protection against exposure to ultraviolet radiation. Because melanin collects in the surface layers of the skin, it absorbs UV-B rays before they can penetrate the deeper, more sensitive tissues. Thick hair and clothing also effectively block ultraviolet rays. Melanin absorbs sunlight across the entire visible and ultraviolet spectrum, giving the material its dark color.

More recently, artificial sunscreens have appeared on the market for use as a UV blocker. These ointments consist of molecules that strongly absorb radiation in the UV-B region, or even the UV-B and UV-A regions, but that are transparent to visible radiation.[11] The ointment sticks to the skin and filters out ultraviolet photons before they can enter the skin. Sunscreens slow down or prevent tanning as well as deep exposure to UV radiation.

There are other ways to reduce exposure to ultraviolet radiation. Sitting in the sun only in the morning and afternoon greatly reduces the overall UV-B dose. Before 10:00 a.m. and after 2:00 p.m., the sun is low enough in the sky that the solar ultraviolet rays are strongly attenuated. In earlier days, parasols were used by fair ladies to deflect the intense noontime sun; today, a hat would help. For those who must spend considerable time outdoors, it is better to have a tan than to experience periodic sunburns. Some researchers believe that the risks of building up and maintaining a protective tan are acceptable when compared with the greater harm that can result from tan-fade cycles that may accompany seasonal work outdoors. The issue is controversial and complex. The best rule of thumb is to avoid exposure to the midday sun whenever possible, as failing to do so may lead to some of the health problems described next.

Skin Cancer

The induction of skin cancer by exposure to ultraviolet radiation is an empirical fact that has not yet been quantitatively explained. In other words, the correlation between UV exposure and skin cancer is well established, although the exact reasons for the initial formation of the cancer are still unresolved. Skin cancers include basal and squamous cell carcinomas, and melanoma. Basal and squamous cell cancers are the most common by far, outnumbering melanoma by about 100 to 1. The basal and squamous cell tumors occur mostly on the face, neck, hands, and arms, those areas of the body that are most often exposed to sunlight. Roughly three-quarters of skin cancers are found in these areas. On the parts of the body that rarely see the light of day, basal and squamous cell cancers occur less than a small percentage of the time. The basal and squamous cell cancers are curable in most cases (95 percent or better). Occasionally, if left untreated, they are fatal.

Basal cell carcinomas represent roughly 80 percent of all skin cancers. The initial lesions develop in the basal cells deep within the skin. The tumor first appears as a small pimplelike growth on the skin. Over several months, the protrusion grows very slowly to form a shiny, translucent mass. A central ulcer develops within the lesion, which eventually scabs over. When scratched off, the lesion bleeds and then forms a new smooth surface. The cycle is repeated as the tumor slowly expands. Basal cell carcinomas rarely **metastasize**—that is, spread throughout the body by dispatching cancerous cells into the lymphatic canals and blood vessels. Nevertheless, the basal cell tumor can be very invasive in local tissues and must be aggressively removed surgically or subjected to radiation treatment.[12] Disfigurement may result from the cure, requiring subsequent reconstructive surgery. Therefore, waiting to see if a surface tumor exhibiting the behavior described will heal spontaneously is not a smart move.

Squamous cell carcinomas account for most of the other skin cancers. The squamous cells are platelike and make up most of the skin's outer layers. The squamous cell tumor first appears as a flat red patch on the skin. Because the lesion is red and tends to scale, it may be confused with other skin disorders, such as an allergy or eczema or even a local infection. Squamous cell cancer may metastasize, sending cancerous seed cells to nearby lymph nodes. Squamous cell and basal cell tumors are treated in the same way, although timely treatment is more critical in the case of squamous cell carcinomas to avoid the cancer's spreading.

The frequency of occurrence of skin cancer (excluding melanoma) is given as a function of latitude in Figure 13.9. Note that skin cancers are less common at higher latitudes (that is, as one moves away from the equatorial zone). This drop-off in the observed cancer rate is likely to be associated with the

---

11. Opaque sunscreens were more common in the past. Zinc oxide, for example, is a white paste that can be plastered on the nose to prevent sunburn. Skin-coloring agents have also been popular among fair-skinned tan-wanna-bes. These agents temporarily dye the skin brown. However, they are generally *not* effective as sunscreens and do *not* offer protection from exposure to ultraviolet radiation.

12. Radiation treatment in this case consists of exposing the cancerous tissue to a high dose of short-range energetic radiation, usually from a nuclear decay source such as cobalt-60 or in the form of X rays. The ideal situation is to have all the energy carried by the radiation deposited in the cancer cells and none in the surrounding healthy tissue. Section 7.4.3 discusses the uses and dangers of medical exposure to radiation.

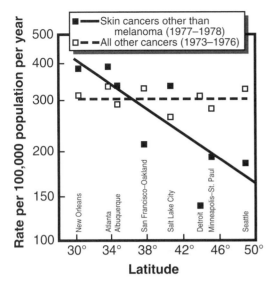

Figure 13.9   The variation of the frequency of cancer occurrence with latitude in the Northern Hemisphere. The occurrence rate is expressed as the number of cases diagnosed per 100,000 fair-skinned people per year. The data were derived from medical records. The occurrence rate of skin cancer, excluding melanoma (which constitutes only a small fraction of the total number of cases), exhibits a distinct trend with latitude. The rate of skin cancer at lower latitudes is a factor of two or three greater than at higher altitudes. The occurrence rate of all other cancers (lung, breast, rectal, and so on) is also shown. That rate is essentially independent of latitude. Many factors that might affect skin cancer rates—for example, life-style, cultural mores—are not included in the data analysis. (Adapted from J. Scotto and J.F. Fraumeni, Jr., *Cancer Epidemiology and Prevention*, D. Schottenfeld and J.F. Fraumeni, Jr., eds., [Philadelphia: W.B. Saunders Co., 1982], pp. 996-1011)

decrease in ultraviolet radiation dose at higher latitudes. Indeed, this behavior can be compared with the occurrence of all other forms of cancer (Figure 13.9), which show no latitude preference, as might be expected. Lung cancer, for example, depends more on how much a person smokes than on where he or she lives.

According to the data in Figure 13.9, the risk that an average fair-skinned person in Atlanta will develop a cancerous skin tumor is about 400/100000 per year, or 4×10^{-3}/yr. Over a life span of 70 years, the total risk is roughly

$$\left(4 \times 10^{-3}/\text{yr}\right) \times \left(70 \text{ yr}\right) = 280 \times 10^{-3}$$
$$= 280/1000 = 0.28$$

In other words, there is a 28 percent chance that a long-time resident of Atlanta will develop skin cancer in his or her lifetime. Notice that some places show higher or lower risks than expected from the average trend line in Figure 13.9. Local meteorological conditions or lifes-tyles may explain the deviations. Thus San Francisco has a lower skin cancer rate because it is cool and foggy there a good part of the year. Sunbathing in the fog is not popular in San Francisco.

Melanoma is a deadly form of skin cancer that can develop within the cells that produce skin pigments. It is believed that melanoma may be triggered by moles or warts that are continually irritated. Melanoma is also linked to exposure to ultraviolet radiation. Figure 13.10, for example, shows the frequency of melanoma at different latitudes in the Northern Hemisphere. As with all skin cancers, there is a general trend for fewer cases to occur at higher latitudes. Although the data are more sparse and scattered than in the case of basal and squamous cell skin cancer, these epidemiological results have led to the general acceptance of a cause-effect relationship between long-term exposure to UV-B radiation and the incidence of melanoma. It may be that persistent mutations, or toxic by-products of UV exposure, accumulate at unusual sites in the skin, such as moles. Later, these sites spawn cancerous cells that grow into tumors.

After the melanoma first appears, it quickly metastasizes, spreading widely through the soft tissues. Melanoma tumors, dark and protruding, can appear almost anywhere. The disease is eventually fatal in about one-third to one-half of all cases (Figure 13.10). Even though melanoma accounts for only about 1 percent of all skin cancers (compare Figures 13.9 and 13.10), the deaths associated with melanoma dominate the overall lethality of skin cancer.

The incidence of melanoma among dark-skinned people is much lower than among fair-skinned populations. The natural evolution of darker skin pigmentation in populations living at lower latitudes and in sunny climes has resulted in a very effective protection mechanism against melanoma and other skin cancers. Those with fair skin who live in these areas, or who like the outdoor life in any climate zone, can reduce their risk of skin cancer by judiciously limiting their exposure to ultraviolet radiation.

Data indicate that the occurrence of skin cancer has increased dramatically since the 1960s. In fact, dermatologists have referred to the trend as a vir-

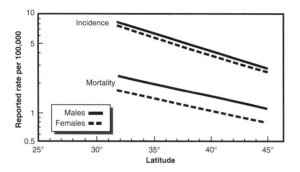

Figure 13.10   The frequency of occurrence (or incidence) of melanoma skin cancer in fair-skinned people as a function of latitude of residence. The occurrence rate is expressed as the incidence of melanoma per 100,000 people per year. The straight lines were drawn through scattered data, as in Figure 13.9. The incidence in men and women also is differentiated. The related mortality (death) rates from malignant melanoma are shown separately as a function of latitude. Roughly one-third to one-half of all melanomas are fatal. The dependence of the incidence and mortality of melanoma on latitude can be seen in these data. (*Environmental Impact of Stratsopheric Flight*, Washington, D.C., Nat. Acad. Sci. [1975]. Adapted from Fig. C2, p. 184.)

tual "epidemic" of skin cancer. Some scientists attribute the increase to the fact that people have been spending more time participating in activities that include exposure to sunlight. Until recently, it was popular—and considered healthy—in many places to have a deep full-body tan, all year if possible. More important, large populations of pale-skinned "northerners" have been migrating to the "sun belts" in the southern and southwestern United States, Australia, and elsewhere. Among fair-skinned people—for example, Scandinavians and northern Europeans—transplanted to sunny environs and outdoor societies such as in Australia and Hawaii, the occurrence of skin cancer is the highest in the world.

Concerns about ultraviolet radiation expressed in the media, particularly in the context of global ozone depletion, have raised warning flags. A tan is no longer a status symbol. Among the fair skinned, pale is acceptable again. People with red or blond hair or freckled skin fare the worst when exposed to the sun. Greater awareness of the hazards of UV radiation has caused many of these people to be more cautious. Moreover, the other nasty side effects of UV-B radiation in addition to skin cancer, particularly premature aging of the skin, have caught the attention of the susceptible masses. Never again will fair

bodies be seen baking under a merciless sun without a prophylactic film of sunscreen.[13]

Additional Health Concerns

Ultraviolet radiation, especially UV-B, can damage the cells and tissues of the skin. The effect is analogous to being burned. The depth of the burn depends on the intensity and duration of the exposure. The body reacts by flooding the damaged skin with blood to ward off infections and induce healing. In cases of extended UV exposure, scar tissue may form and accumulate subtly within the skin tissue, unlike severe heat burns, in which disfigurement is immediate and obvious. All exposed tissues are vulnerable to lasting and cumulative damage. The effects of UV radiation beyond skin cancer that have been identified as clearly undesirable are as follows.

Premature aging of the skin: The penetration of ultraviolet radiation deep into the skin over long periods of time damages connective and supportive tissues. The net result is a weakening of the skin texture, which closely resembles natural aging of the skin, only greatly accelerated. People continually exposed to UV-B radiation in their youth show wrinkles, folds, roughness, and spotting of the skin that become apparent in their forties rather than in their fifties or sixties. Other activities, particularly cigarette smoking, may contribute to premature degeneration of skin tissues (Section 8.4). However, the frequently poor skin texture of persons who have exposed themselves excessively to the sun but never smoked demonstrates the destructive potential of UV-B. The effects of sun on skin are often persuasively demonstrated in the faces of middle-aged farmers and sailors. By their mid-forties, *smokers* who sun themselves regularly often look like the Dick Tracy character Prune Face.

Cataracts and retinal damage: Some researchers have found a correlation between long-term exposure to ultraviolet radiation and cataracts (a clouding of the lens of the eye) and degeneration of the cornea

13.   The amount of sunscreen that should be used is a matter of debate among experts. Some argue that a "sun protection factor," or SPF, of 15 is adequate for most people. An SPF of 15 means that you can reduce your ultraviolet radiation dose by a factor of 15 when you apply the sunscreen as directed before exposure. Other experts recommend an SPF of at least 30, to ensure that skimpy or worn applications will remain effective. Some ultraviolet scientists also prefer sunscreens with a UV-A blocker, since the long-term physiological effects of UV-A radiation are uncertain.

(the outermost transparent film protecting the lens and iris). Excessive exposure to UV-B radiation is also known to cause permanent damage to the retina, the light-sensitive tissue at the back of the eye. In other words, the entire optical structure of the eye is at risk from exposure to UV-B radiation. The UV rays penetrate the optical system along with the visible light that the eye is designed to collect and interpret. Since natural sunlight always contains some ultraviolet radiation, the eye can tolerate a limited amount. But it is not hardened against continual intense exposure to sunlight over 40 or 50 years without the risk of adverse effects. Measures to prevent unnecessary exposure can be taken, such as wearing glasses with UV-blocking lenses and shading the eyes against direct sunlight and glint from highly reflective surfaces such as water, snow, and ice.

UV-A radiation: Ultraviolet radiation with wavelengths greater than about 320 nanometers is not generally considered to be biologically harmful, although recent studies indicate that UV-A radiation may weaken the immune system. Thus following prolonged exposure to UV-A, a person may be more susceptible to infection and disease. This may appear as more frequent colds or influenza or as a general malaise. UV-A radiation, which has longer wavelengths, also penetrates deeper into the epidermis than UV-B does. Accordingly, UV-A may also contribute to premature aging of the skin. Commercial tanning parlors use "safe" UV-A radiation to brown their customers. You are invited to lie in a coffinlike receptacle and receive a full-body dose of UV-A in these establishments. Deciding whether this is an intelligent thing for an educated person to do is your business.

13.4.3   ENVIRONMENTAL EFFECTS OF UV-B RADIATION

Humans are not the only organisms at risk from exposure to ultraviolet radiation. Virtually every species on the planet is susceptible to UV-B radiation. Many organisms have developed protective structures or behaviors (for example, living in the mud at the bottom of a pond greatly limits ultraviolet exposure). Nonetheless, most species still live in sunlight. We summarize next the potentially harmful general effects of ultraviolet radiation on living organisms.

Biocidal Action: Ultraviolet radiation in general is biocidal. That is, it kills living organisms. Microorganisms, such as phytoplankton drifting in the surface waters of the oceans, are particularly vulnerable. Such organisms cannot easily run and hide from the sun. Accordingly, microorganisms have evolved largely in ecological niches that are naturally shaded. They can live in murky water, in the gut of an animal, inside a plant, or buried in the soil. The phytoplankton are unusual because they require sunlight to survive (they are, after all, microscopic plants). They must court the sun and at the same time avoid ultraviolet overexposure. Phytoplankton are a vital component of the marine food chain throughout the world's oceans. These organisms can apparently swim deep enough below the ocean surface to reduce the UV-B dose while still receiving visible light.[14] Perhaps this is a dangerous game, but it is one that phytoplankton have successfully played for 2 billion years.

Damage to plant tissues: Plants in general have no way to hide from ultraviolet radiation. Leaves must be placed directly in the sun to harvest solar energy (although some plants habitually live in the shade). Accordingly, plants have evolved mechanisms to filter out UV rays or to repair the damage they cause. Vegetation acclimated to the tropics or high elevations is naturally less sensitive to solar ultraviolet. Nevertheless, all plants have a threshold tolerance level for UV-B exposure. In low latitudes, the tolerance level may be greater because of evolutionary factors. Increasing UV-B intensities will eventually cause leaf damage, and mortality, in most species. In the case of agricultural crops, useful yields may be reduced as a result of UV irradiation. Even though many food crops are derived from tropical plants, the stress induced by UV radiation can reduce productivity.

Effects on vision: In some insects and mammals, the spectral response for vision is shifted toward the ultraviolet spectrum. Hence changes in ultraviolet radiation intensities can affect the spatial orientation and behavior in these species. In the long term, animal vision may develop pathological diseases, such as cataracts in humans. In 1992, reports surfaced in southern Chile of wild rabbits that had been blinded by ultraviolet radiation leaking through the

---

14.   Ultraviolet-B radiation can penetrate seawater, but only to a depth of several meters. Visible radiation penetrates farther, perhaps to a depth of several tens of meters. This difference is explained by the absorption coefficients of seawater in these two spectral regions. At UV wavelengths, briny water shows much greater absorption than it does at visible wavelengths. The presence of living organisms and detritus further limits the penetration of both visible and ultraviolet rays. A thick mat of rotting algae effectively shades the water below from UV radiation, providing a rich breeding ground for microorganisms.

"ozone hole" (Section 13.7). One could walk right up to these poor creatures and pick them up by the ears. Actually, it seems unlikely that rabbits would be blinded so quickly, although the long-term effect of UV exposure in causing blindness in animals is certainly a real threat.

13.5   Threats Against Ozone

Stratospheric ozone is essential to the survival of most forms of life on the Earth. Concerns have arisen that human activities may be modifying the natural ozone layer. Specifically, observations indicate that chemicals emitted into the atmosphere by modern technology are depleting the ozone shield. Section 13.2 showed that certain chemical "catalysts" can attack ozone. A number of specific catalytic agents (NO, OH, Cl, Br) were identified. In this section, the sources, sinks, properties, and effects of several of the most important catalysts are discussed, with an emphasis on those chemical agents that originate in human activities and threaten the integrity of the ozone layer. In many cases, these agents also have natural sources and sinks.

13.5.1   A LITANY OF THREATS

Since the late 1960s, when interest in the ozone layer blossomed, an impressive number of potential threats to the ozone layer have been identified. It all began with the **supersonic transport** (**SST**) airplane. This marvelous high-speed plane was designed to whisk people all around the world in half the time of subsonic jets. The SST would cruise in the stratosphere, where the air is thinner and friction is lower. Unfortunately, it was discovered that the engines would spew nitrogen oxides directly into the ozone layer. Two scientists, Paul Crutzen and Harold Johnston, independently proposed that the nitrogen oxides would deplete the ozone shield. The SST was shelved. Soon after that, in the early 1970s, the space shuttle program was announced. According to engineering plans, the shuttle engines would spew particulates and chlorine into the stratosphere. Soon, a number of scientists were warning that chlorine would deplete the ozone layer. After considerable debate, the space shuttle was built anyway. But it has never reached full operational potential and thus poses only a minor threat to the ozone layer.

In 1974, during the space shuttle controversy, Mario Molina and Sherwood Rowland suggested that chlorofluorocarbons (CFCs), which also contain chlorine, might cause the ozone layer to thin. CFCs were invented in the 1920s, but had only recently caught on. By 1974 they were in wide use in aerosol spray cans. Not long after the Molina and Rowland wake-up call, CFC-filled dispensers were banned. In quick succession, nitrogen fertilizer was suggested as a danger to ozone; nuclear test explosions were implicated in the loss of stratospheric ozone; and a "bromine bomb" was proposed. This "bomb" consisted of a massive charge of bromine gas that would be deposited in the stratosphere over an enemy's head, creating a hole in the ozone layer and allowing deadly ultraviolet rays to fry him. By 1983, the idea of the unmitigated disaster of a global superpower nuclear war was made even worse by the hypothesis of a "nuclear winter," which included decimation of the ozone layer.

The realization that human activities could destroy ozone and jeopardize all life on the planet logically led to the consideration of natural phenomena that might deplete ozone. Many potential threats were uncovered. A supernova (stellar explosion) occurring near the Earth would pose a rare danger to the ozone layer. No one hid. The impact of meteors large and small would, calculations showed, perforate the ozone shield. No one flinched. These events were spaced millions of years apart. No imminent danger was seen.

Smaller natural perturbations of the ozone layer are associated with activity on the sun. Solar flares emit energetic particles that pierce the upper layers of the ozonosphere. The sunspot cycle also has a small modulating effect on the total amount of ozone in the atmosphere. However, the threat to life on Earth from these fluctuations is minuscule. The real hazard is the continuing human assault on the protective ozone layer, usually inadvertent, but always careless and insensitive. The latest issue involves a new supersonic aircraft, the high-speed civil transport (HSCT). We have come full circle, it seems, from the early days of ozone threats.

13.5.2   CHLORINE

One of the most important catalysts for ozone destruction in the stratosphere is chlorine, which originates in a variety of chemical forms. Biological processes—on land and in the oceans—release chlorine into the atmosphere mainly in the form of **methyl chloride** (**CH₃Cl**). Volcanoes also emit chlorine in the form of **hydrogen chloride** (**HCl**). Methyl chloride supplies roughly 25 percent of the present amount of chlorine in the stratosphere. Measurements of chlorine following major volcanic eruptions (El Chichon in Mexico in 1982 and Mount Pinatubo in the Philippines in 1991 are the most recent examples) have demonstrated that even large eruptions are a minor source of stratospheric chlorine. Chlorine injected into the stratosphere by volcanoes would eventually be deposited in glacial ice sheets, and analysis of the chlorine content of the ice should show a spike in its concentration. However, studies of ice cores reveal no chlorine spikes, even following the largest historical eruptions such as Tambora (Indonesia, 1815) and Krakatoa (Indonesia, 1883).

The human, or anthropogenic, sources of stratospheric chlorine are primarily in the form of industrially produced **chlorocarbons**. This large family of chemicals consists of dozens of specific compounds that contain chlorine atoms and are usually derived from an organic compound such as methane. Many important chlorine-bearing compounds are included in an even larger family of chemicals referred to as **halocarbons**. These are organic molecules in which halogen atoms (in particular, chlorine, fluorine, and bromine) have been substituted for hydrogen atoms. All chlorocarbons are halocarbons, but not *vice versa*.

The common chlorocarbons include **carbon tetrachloride** (**CCl₄**), which is used as a solvent in "dry" cleaning, because it evaporates quickly from cloth (unfortunately causing trouble in the atmosphere). Table 13.1 lists several other important chlorocarbons that contribute to ozone depletion, including methyl chloride and methylchloroform. The most famous chlorocarbons are now the **chlorofluorocarbons** (**CFCs**).[15] These compounds are used as refrigerants, plastic-blowing agents, solvents, and aerosol spray propellants, among other applica-

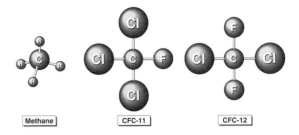

Figure 13.11   The chemical structures of the two most common and important CFCs: CFC-11 and CFC-12. These molecules are closely related to the methane molecule. The CFCs can be manufactured by replacing the hydrogen atoms in CH_4 with chlorine and fluorine atoms.

tions. The two principal CFCs manufactured since the 1950s are **chlorofluorocarbon 11** (**CFC-11**, with the chemical formula **CFCl₃**) and **chlorofluorocarbon 12** (**CFC-12**, with the chemical formula **CF₂Cl₂**). These compounds are purely artificial substances created by chemists and engineers as safe substitutes for other more toxic chemicals, including ammonia and sulfur dioxide. In the early twentieth century, ammonia (and, to a lesser degree, sulfur dioxide) was widely utilized in refrigerators and other cooling devices. The CFCs were designed to provide even better cooling properties and to be safe if accidentally released.

Figure 13.11 shows the structure of CFC-11 and CFC-12 molecules. Each has a central carbon atom with four halogen (chlorine and **fluorine**, in this instance) atoms attached. The chemical symbols for chlorine and fluorine are Cl and **F**, respectively. The CFCs are closely related to natural methane, also depicted in the figure. These CFCs can, in fact, be manufactured by replacing hydrogen atoms in methane with Cl and F.

Table 13.1 provides information on the emissions and properties of a number of chlorine-containing compounds currently in wide use (Section 12.3.3 and Table 12.4). Data are also given for important **hydrochlorofluorocarbons** (**HCFCs**) and **halons**. The halons contain **bromine** (**Br**) and are used principally as fire retardants. The inventories of halons and other brominated compounds in the atmosphere are increasing rapidly because of their superior properties as fire suppressants and agricultural fumigants.

The history of emissions of CFC-11 and CFC-12 in terms of the total mass-produced each year since 1960 is depicted in Figure 13.12. This graph traces

15. These chlorine- or fluorine-based compounds have a number of different names. Chlorocarbons contain carbon (C) and chlorine (Cl), but may or may not contain fluorine (F) and hydrogen (H). Chlorofluorocarbons contain both Cl and F, but generally not hydrogen. Fluorocarbons contain F, but generally not Cl. Chlorofluoromethanes are chlorofluorocarbons derived from methane (Figure 13.11). Occasionally these names are used interchangeably, even though the strict definitions may be inconsistent.

Table 13.1   Chlorofluorocarbons and Other Halogen Compounds[a]

Chemical (symbol)	Emission (10^3 tonne/yr)	Atmospheric concentration (pptv)	Atmospheric lifetime[b] (years)	Applications	Growth rate (%/yr)	Ozone depletion potential[c]
CFC-11 ($CFCl_3$)	260	270	76	Refrigeration, foams, aerosol spray	5	1.0
CFC-12 (CF_2Cl_2)	450	460	139	Air conditioning, foams, refrigeration, aerosol spray	5	3.0
CFC-113 ($CFCl_2CF_2Cl$)	150	65	92	Solvent	10	1.3
Carbon tetrachloride (CCl_4)	75	110	67	Solvent	1	0.35
Methyl chloroform (CH_3CCl_3)	520	135	8	Solvent	7	0.024
Methyl bromide (CH_3Br)	60	~15	1.5	Fumigation	15	—
HCFC-22 (CHF_2Cl)	220	110	15	Refrigeration, air conditioning	7	0.35
Halon 1301 (CF_3Br)	3	~3	110	Fire extinguishers, fumigation	20	—
Halon 1211 (CF_2BrCl)	80	~2	25	Refrigeration, foams	15	—
Methyl chloride (CH_3Cl)	—	600	1.5	—	~0	—

[a] The data on emissions are for 1985, and the data on concentrations and growth rates are for 1990. More recent information indicates similar trends, except for the CFCs, whose emission and growth rates have begun to decline.

[b] The lifetime corresponds to the time over which about two-thirds of the material will be removed from the atmosphere.

[c] The ozone depletion potential (ODP) of a compound is the ratio of the ozone depletion for a unit mass of emission of that compound compared with the ozone depletion for a unit mass of emission of CFC-11.

Sources: Environmental Protection Agency, 1985; World Meteorological Organization Global Ozone Research and Monitoring Project Reports 20 (1989) and 25 (1991).

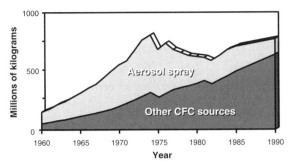

Figure 13.12   The annual production of chlorofluoro-carbons 11 and 12 in millions of kilograms worldwide, based on records kept by the Chemical Manufacturers Association since 1960. The fraction of the total production that is used as an aerosol spray propellant is shown on the graph.

an interesting behavior. The production of these CFCs accelerated exponentially between 1960 and 1974 as industries found more and more uses for these new "wonder" compounds. Then in 1974, scientists Mario Molina and Sherwood Rowland discovered that CFCs could lead to severe ozone depletion. A short time after that landmark discovery, the production of CFCs fell sharply in response to public and political concerns about a potential global environmental disaster. The most frivolous use, as a propellant for aerosol mists, was particularly hard hit.[16] CFC-propelled deodorants and hair sprays were banned in 1978, and mechanical spritzers reappeared on store shelves. Additional restrictions on CFC use were considered, but scientists could not reach a consensus on the merits of the Molina and Rowland theory, and the CFC industry stonewalled on the issue.

As the years passed and public awareness and concern waned, the *total* production of CFCs began to rise again. In 1985, the annual production of CFCs approached that of the early 1970s. Then in 1985, Joseph Farman and his British colleagues announced the discovery of the **ozone hole** over Antarctica. The production of CFC slowed again. By

16.   In an aerosol spray can, the CFCs are stored under pressure in a mixture with the liquid to be sprayed. When a valve is opened, the CFC-liquid mixture is forced out of a nozzle by the pressure of the CFC gas in the can. As the mixture leaves the nozzle, the CFC expands rapidly and disperses the liquid into small droplets—an aerosol mist—or into a foam if the liquid is very viscous. Underarm deodorants, hair pomades, shaving cream, cooking oils, pesticides, and whipped cream all are handily dispensed using this technique. Alternative propellant materials now in use include carbon dioxide, nitrous oxide, and various hydrocarbons.

1988, the evidence of ozone degradation was so convincing that an international agreement to limit future CFC production, the Montreal Protocol, was proposed in 1987 (Section 13.7). So there is clear sailing ahead. Or is there?

The Atmospheric Life Cycle of CFCs

An important characteristic of CFCs is their chemical stability and low toxicity. The CFCs were originally designed to be inert, or unreactive, so that they could be safely used without fear of exposure. These passive compounds were perfect for aerosol sprays, refrigerators, and air conditioners. However, their extreme inertness poses a serious problem. Molina and Rowland first noted it: There is no apparent way that the CFCs, once released, can be removed from the atmosphere. Because of their chemical stability, the CFCs do not react with the usual scavenging agents found in the troposphere, particularly OH (Section 3.3.4). The CFCs are almost insoluble in water and are not efficiently washed out of the lower atmosphere or dissolved in the oceans. Indeed, as Molina and Rowland finally deduced in 1974, the only effective sink for the CFCs is transport into the stratosphere. Here the intensity of short-wavelength ultraviolet solar radiation is sufficient to photodissociate the CFC molecules. This radiation is not usually available in the troposphere, owing to absorption in the overlying ozone layer. The photodissociation of CFCs releases chlorine atoms into the stratosphere. Chlorine can then act as a catalyst to destroy the ozone (Section 13.2.2).

The quantities of CFCs measured in the atmosphere have been increasing steadily since continuous monitoring of these compounds began in the late 1970s (Figure 13.13). From these data, the total mass of chlorocarbons accumulating in the atmosphere each year can be calculated fairly accurately. For the CFCs, the observed increase in total mass over time is roughly the same as the total mass produced by industry (Figure 13.12). This is consistent with the long atmospheric lifetimes of the CFCs (Table 13.1). It takes roughly 100 years for the atmosphere to scrub out CFCs. This time span is much longer than the total period over which CFCs have been manufactured. Accordingly, the atmosphere is still in a state of transition in which the CFCs are being released and accumulating faster than they can be decomposed in the stratosphere. It would take about 100 years of continuous emissions

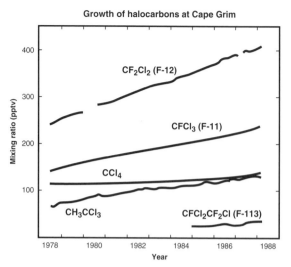

Growth of halocarbons at Cape Grim

Figure 13.13   Measurements of the atmospheric concentrations of several chlorocarbons in the atmosphere from 1978 to 1988. The data were collected at Cape Grim, Tasmania, in the Southern Hemisphere. All concentrations are given in parts per trillion by volume (1 pptv = 1×10^{-12}). The observations include CFC-11, CFC-12, and CFC-113, as well as carbon tetrachloride and methyl chloroform. The measurements have continued to the present time; a steady increase in chlorocarbon concentrations has continued through the early 1990's. Measurements taken at other sites around the globe show similar increases. The concentrations of many other chlorocarbons that are regularly measured are also increasing worldwide. (Data from Fraser and Derek [1989])

for the atmospheric content of CFCs to reach an equilibrium, or steady-state, condition. Until that time, CFC concentrations will continue to increase unless emission (production) rates are reduced.

CFC concentrations in the Southern Hemisphere are almost as high as they are in the Northern Hemisphere. The difference is only a few percentage points. However, almost all the CFCs are emitted in the Northern Hemisphere; the air is transferred between the hemispheres at a relatively slow rate. Several years are required to mix tracers from the northern to southern latitudes across the equator (Section 5.1.3). Nevertheless, the CFCs atmospheric lifetimes are much longer than this interhemispheric mixing time, enough time for the uneven pattern of emissions to be smoothed out before the CFCs are removed. An analogy is filling a swimming pool with a garden hose at one end. The height of the water in the pool can be thought of as the concentration of a CFC in the atmosphere. Water does not accumulate around the hose because the time required to raise

the water surface even a small distance is much longer than the time it takes a wave to move from one end of the pool to the other. In other words, the accumulation time (lifetime) of water (CFC) in the pool (atmosphere) is much greater than the transport time (interhemispheric mixing time) across the pool (equator). Horizontal differences tend to be evened out in either case.

Extensive measurements of the vertical distribution of CFCs in the atmosphere also are available (Figure 13.14). These data show that everywhere on the planet, CFCs are uniformly mixed in the lower atmosphere (troposphere) up to the tropopause (Section 2.3.2). Their mixing ratios drop off rapidly with height in the lower stratosphere above the tropopause, having very small values above about 30 kilometers. This behavior implies that the CFCs are being broken down into their constituent elements, including chlorine, at the higher altitudes. Laboratory experiments with CFCs strongly support this implication. Most important, direct measurements of free chlorine in the stratosphere confirm this basic conclusion, first reached by Molina and Rowland. Logically, this CFC source of chlorine in the stratosphere will deplete the supply of ozone.

The data just described point to a specific sequence of events in the atmospheric life cycle of the CFCs:

1. CFCs are manufactured from basic materials (for example, methane, chlorine, and fluorine). The rate of production has amounted to roughly 500,000 metric tons (1 Mt = 1×10^6 metric tons) of Cl each year.
2. The CFCs are put to use but eventually escape into the lower atmosphere at roughly the same rates at which they are manufactured, with a significant delay between their manufacture and release.
3. These stable compounds, with no substantial sinks, accumulate in the troposphere. Currently, about 10 Mt of chlorine in the form of CFCs reside in the troposphere, and the amount has increased each year in step with continuing emissions.
4. Over time, the natural exchange of air between the troposphere and stratosphere, which is very slow, carries the CFCs above the tropopause. For existing tropospheric CFC concentrations, about 0.1 Mt-Cl/yr is transferred into the strato-

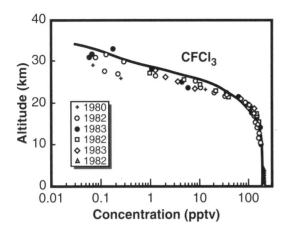

Figure 13.14   Measurements of the vertical distribution of CFC-11 at middle latitudes in the Northern Hemisphere. Data were collected by high-altitude balloon-borne instruments. The concentrations are given as mixing ratios in parts per trillion by volume (1 pptv = 1×10^{-12}). CFC-11 is uniformly mixed in air below the tropopause, at about 15 kilometers. Its mixing ratio drops off rapidly above the tropopause, by a factor of 1000 at an altitude of 30 kilometers. Similar behavior is observed everywhere on the earth and for all CFC compounds. (*Atmospheric Ozone 1985, WMO Report No. 16.* Adapted from Fig. 11-18, p. 634)

sphere in the form of CFCs (note that this is about one-fifth the historical emission rate).

5. The CFCs are exposed to ultraviolet radiation in the stratosphere and break down directly into various products, including reactive chlorine gases. The rate of chlorine production is roughly equivalent to 0.1 Mt-Cl/yr, the amount brought into the stratosphere with the CFCs. A small fraction of the chlorofluorocarbons is added to the amount of CFCs in the stratosphere, which has slowly built up over time.

6. The chlorine released from the CFCs reacts catalytically with ozone. This process continues until the chlorine is transformed by other chemical reactions into less reactive forms, such as HCl.

7. The HCl is transferred downward from the stratosphere into the troposphere by means of exchange mechanisms that balance the upward transport of air that carries the CFCs aloft. HCl is readily absorbed in cloud water and is quickly rained out of the atmosphere. The rate of HCl removal from the troposphere is nearly equal to the rate of chlorine production from CFC photodissociation in the stratosphere.

A Global Perspective on Chlorine

A box model can be used to investigate the atmospheric chlorine cycle. (Concepts and equations for box models are described in Sections 4.1.2, 7.4.3, 10.1.1, and 11.3.2.) The important reservoirs and mass transfer rates corresponding to the CFC component of the overall chlorine cycle are illustrated in Figure 13.15. It is important to note that *instantaneous* values are given in Figure 13.15, corresponding to the current state of the atmosphere. The present CFC reservoirs are not in a steady state, or equilibrium, as can be seen by the imbalance in the mass flow. Accordingly, chlorine (contained in CFCs) is continuing to accumulate in both of the atmospheric reservoirs: the troposphere and stratosphere (Figure 13.13). The system will reach a state of equilibrium when the source and sink have equalized. Figure 13.15 shows that this could happen if

• The tropospheric source remains constant while the tropospheric concentration increases by a factor of about five above current levels. This assumes that the transfer rate of CFCs from the troposphere to the stratosphere is proportional to the amount in the troposphere.[17]

• The tropospheric source decreases by a factor of five to about 0.1 Mt-Cl/yr while the tropospheric concentration remains constant.

• The tropospheric source decreases and the CFC concentration increases in such a manner that the equilibrium reservoir model equation (Section 10.1.2) is satisfied. For example, if CFC emissions were reduced by 50 percent, a new equilibrium state could be achieved with about two and one-half times the present concentration of CFCs.

The overall chlorine cycle of the atmosphere is illustrated in Figure 13.16. The cycle is quite complex and involves many physical, chemical, and biological processes. Indeed, in the overall global chlorine cycle, the CFC component is now the easiest to understand, because both the major sources and the

17. The rate at which air is transferred between the troposphere and stratosphere is not directly affected by the amount of CFCs mixed in the air. Hence, the net rate of CFC transfer should be proportional to its concentration in the air. If the CFC concentration is twice as large, twice as much will be moved with the same amount of air. Complications can arise if the CFCs begin to modify the global energy balance and circulation, affecting troposphere-stratosphere exchange rates, or if the photochemistry of air in the stratosphere is significantly modified by the buildup of CFCs. Such detailed considerations are beyond our scope.

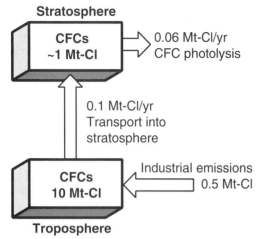

Figure 13.15   The atmospheric budget of chlorine in the form of chlorofluorocarbons is shown as two joined global "boxes," or reservoirs, which represent the troposphere and stratosphere. The amount of chlorine in each reservoir is given in millions of metric tons (or tonnes) of chlorine (Mt-Cl). The instantaneous mass flows are specified in units of Mt-Cl/yr (millions of metric tons of chlorine per year). The tropospheric source of CFCs is associated with industrial production and release, and the stratospheric sink is related to CFC photodissociation into chlorine and other by-products.

major sinks are well defined. The natural components of the chlorine cycle are much less understood. Recent work has quantified the methyl chloride source as contributing about 0.6 part per billion by volume of chlorine to the atmosphere. Methyl chloride also reacts rapidly with hydroxyl radicals and is removed primarily in the troposphere before it can be transported to the stratosphere. Volcanoes are even less important as a chlorine source, as we noted earlier.

Chlorine Chemistry and Activity

The chemical aspects of the atmospheric chlorine (and CFC) cycle are shown in more detail in Figure 13.17. Note that chlorine tends to *partition* itself among several species: Cl, **ClO** (**chlorine monoxide**), HCl, and **ClONO₂** (**chlorine nitrate**). Most of the chlorine in the stratosphere (~99 percent) is in the last two forms under normal conditions (but see the following section on the ozone hole). The key **active chlorine** constituents are Cl and ClO, which react catalytically with ozone. The region of maximum ozone depletion by chlorine catalysis (indicated by the shaded region of the ozone profile in Figure 13.17) corresponds roughly to the region of

highest ClO concentrations. Hydrogen chloride and chlorine nitrate are the main chemical reservoirs for non-CFC chlorine in the stratosphere and do not react with ozone.

The chlorine chemical cycle begins with the photodissociation of chlorofluorocarbons, releasing chlorine atoms. For CFC-11, the process is

$$CFCl_3 + h\nu \rightarrow CFCl_2 + Cl \qquad (13.11)$$

The chlorine atoms then interact with ozone, or odd-oxygen, through a catalytic reaction sequence (Equation 13.9), as follows:

$$
\begin{aligned}
O_3 + Cl &\rightarrow ClO + O_2 \\
ClO + O &\rightarrow Cl + O_2
\end{aligned}
\qquad (13.12)
$$

Net reaction: $O_3 + O \rightarrow 2O_2$

During this process, most of the chlorine atoms are converted to ClO. Eventually, the chlorine is further partitioned into HCl and ClONO₂ by a number of reaction cycles, including the following:

$$
\begin{aligned}
Cl + HO_2 &\rightarrow HCl + O_2 \\
HCl + OH &\rightarrow Cl + H_2O
\end{aligned}
\qquad (13.13)
$$

and

$$
\begin{aligned}
ClO + NO_2 &\xrightarrow{M} ClONO_2 \\
ClONO_2 + h\nu &\rightarrow ClO + NO_2
\end{aligned}
\qquad (13.14)
$$

All the chlorine species are cycled into the various forms according to these reactions (and others not shown). Chlorine atoms have a strong affinity for hydrogen atoms and readily form HCl (the first reaction in Equation 13.13, for example, and several similar "hydrogen abstraction" processes). The HCl molecule is very stable and remains intact for a long time before being recycled into chlorine atoms. Similarly, ClONO₂ is more chemically stable than ClO. It follows that a short time after Cl is released from CFCs, free chlorine is repartitioned mainly into HCl and ClONO₂, neither of which reacts with ozone.

The natural partitioning of chlorine into relatively inactive forms (HCl and ClONO₂) is very fortunate indeed. If all the chlorine currently being released from chlorofluorocarbons into the stratosphere remained in active forms (Cl and ClO), the concentra-

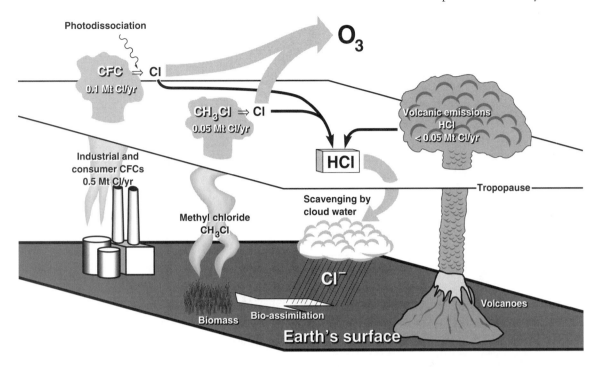

Figure 13.16   The global atmospheric chlorine cycle. Chlorine sources include methyl chloride (CH_3Cl) from biological activity, hydrogen chloride (HCl) from volcanic activity, and CFCs and other chlorocarbons from industrial activities. The chlorine "source gases" enter the stratosphere when air is transferred naturally across the tropopause. In the stratosphere, methyl chloride and CFCs are broken down photochemically, releasing chlorine atoms that catalytically attack ozone. Eventually, this free chlorine is converted to HCl. Finally, HCl is transported back across the tropopause, where rain scavenges it and deposits the chlorine on the surface as chloride ions.

tions of these catalysts would be roughly 100 times larger, and *the ozone layer would be completely eroded*! Mother Nature is watching over us. Or is that Mother Gaia's shadow?

13.5.3   Nitrogen Oxides and Ozone Change

The nitrogen oxides, or **NO_x**, consist of a family of related chemical species that include nitric oxide (NO) and nitrogen dioxide (NO_2). The atmospheric cycle for the nitrogen compounds is similar in many ways to the chlorine cycle. The principal features are illustrated in Figure 13.18. In this case, **nitrous oxide (N_2O)** is the main source gas for NO_x, analogous to CFCs in the chlorine cycle. In the stratosphere, ultraviolet radiation in the presence of ozone results in the production of NO from N_2O. NO (or, NO_x) then participates in a catalytic reaction cycle with ozone (Equation 13.9), thereby destroying some of the ozone. The NO_x, however, is converted by other reactions to an inert reservoir compound, **nitric acid** vapor (**HNO_3**), analogous to the conver-

sion of Cl and ClO to HCl and $ClONO_2$ in the chlorine cycle. The nitrogen is removed from the stratosphere mainly in the form of HNO_3.

The nitrogen cycle begins with the decomposition of nitrous oxide, which proceeds as follows:

$$O_3 + h\nu \rightarrow O^* + O_2$$
$$N_2O + O^* \rightarrow NO + NO \qquad (13.15)$$

Notice that ozone actually initiates the formation of nitrogen oxides by spawning excited oxygen atoms (O^*) (Section 3.3.4).

Nitrogen oxides affect ozone through the important catalytic reaction cycle

$$O_3 + NO \rightarrow NO_2 + O_2$$
$$\underline{NO_2 + O \rightarrow NO + O_2} \qquad (13.16)$$

Net reaction: $O_3 + O \rightarrow 2O_2$

This catalytic cycle (Equation 13.16) is similar to the chlorine cycle (Equation 13.12), with NO

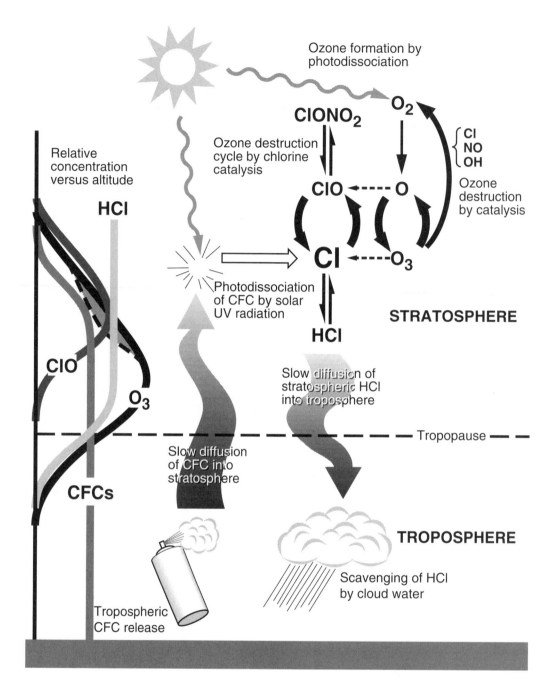

Figure 13.17   The atmospheric chlorine cycle, emphasizing the photochemical process and interactions between chlorine species and the ozone (odd-oxygen) cycle. Some of the physical processes illustrated in Figure 13.16 are included here. The vertical profiles of several key chemical constituents are shown to illustrate their spatial relationships to the vertical structure of the atmosphere and to one another. The shaded portion of the ozone profile indicates the altitude in which chlorine is most apt to deplete ozone.

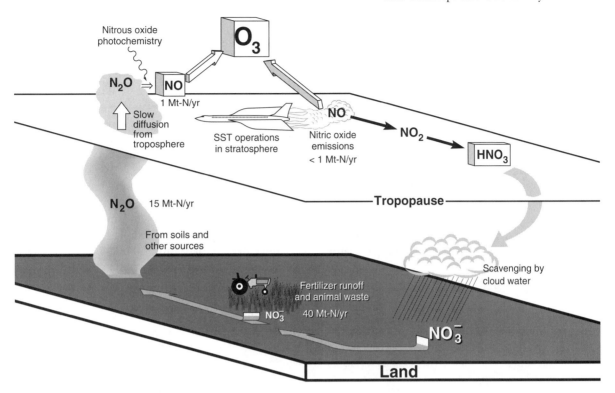

Figure 13.18   The global cycle of nitrogen oxides (NO_x). All mass flow rates are given as millions of metric tons (or tonnes) of nitrogen per year (1 Mt-N/yr = 1×10^6 tonnes per year). The principal source of NO_x in the stratosphere is nitrous oxide (N_2O), which is generated through biogenic activity. N_2O is transported into the stratosphere from the troposphere—analogous to the transport of CFCs—where it is photochemically decomposed into nitric oxide (NO). Nitrous oxide has no other important sinks in the atmosphere. The NO catalytically reacts with ozone. NO is efficiently converted to nitric acid vapor (HNO_3), however, which is chemically inert with regard to ozone. The nitric acid is transported downward into the troposphere and is scavenged by precipitation, analogous to the role of HCl in the chlorine cycle (Figure 13.16). The main anthropogenic sources of stratospheric NO_x occur indirectly in the form of N_2O generated by combustion and fertilizer denitrification. The direct anthropogenic sources include NO_x injection by high-flying aircraft (including supersonic planes operating in the stratosphere).

substituting for Cl. The nitrogen oxides are further cycled through nitric acid as follows:

$$NO_2 + OH + M \rightarrow HNO_3 + M$$
$$HNO_3 + h\nu \rightarrow NO_2 + OH \qquad (13.17)$$

There is a close parallel between the chemical processes of nitric acid and chlorine nitrate (Equation 13.14), and the existence of an acidic nitrogen reservoir (HNO_3) is analogous to the acidic chlorine reservoir (HCl).

Figure 13.18 shows a high-flying aircraft emitting nitric oxide into the stratosphere. Aircraft, like automobiles, generate NO (NO_x) as a by-product of high-temperature combustion. As we noted, NO_x is a powerful catalytic agent that destroys ozone in the stratosphere. By contrast, in the lower atmosphere, NO_x emissions can lead to ozone *production* in smog

(Section 6.2.1). If heavy aircraft traffic were to materialize in the stratosphere and if the engines were not specially designed to limit NO emissions, then significant ozone reductions would be expected. In the late 1960s and early 1970s, forward thinkers had proposed building a large high-speed aircraft to fly in the stratosphere—the so-called supersonic transport (SST). The major airframe manufacturers were excited about the business prospects. But no one looked very carefully at the potential environmental consequences. It is not surprising that the American project was killed when chemist Harold Johnston pointed out that the nitrogen oxides emitted by any substantial fleet of such planes would severely deplete the ozone layer. Johnston had applied the simple chemical arguments just described to shoot down the SST. Undaunted, the French, British, and Soviets continued to develop a small number of SSTs, which proved to be economically impractical.

Recently, the idea of flying large numbers of supersonic airplanes in the stratosphere has been revived in the form of a high-speed civil transport (HSCT). The National Aeronautics and Space Administration (NASA) is doing basic research on the HSCT and its effects on the stratosphere. The nitrogen oxide emissions from the HSCT engines are being reduced significantly in response to earlier criticisms of the SST. Supersonic airplanes also generate a sonic boom at the ground. This is a shock wave produced when an object moves through air faster than the speed of sound. The crack of a bullwhip is a very weak sonic boom. Flying mainly over uninhabited land and open oceans is one of the proposed solutions (although the number of organisms disturbed by the booms will probably be enormous). There are also new questions regarding the effects of aircraft emissions on the occurrence of stratospheric ice clouds that lead to ozone depletion. The effects would be similar to those observed in the "ozone hole" (Section 13.7). The final verdict on the HSCT is not yet in.

13.5.4 BROMINE AND THE HALONS

Bromine is a less common halogen than chlorine. In the atmosphere, chlorine is about 100 times more abundant than bromine. Bromine is also less common in industrial processes and applications. Nevertheless, bromine surpasses chlorine in its ability to destroy ozone through catalytic reactions, because a larger fraction of bromine than of chlorine in the stratosphere remains in chemically active forms (that is, Br and **BrO** [**bromine monoxide**]). The natural sources of bromine are dominated by **methyl bromide** (**CH$_3$Br**), a by-product of biological activity in the oceans, similar to methyl chloride. Methyl bromide is also manufactured in large quantities (about 60,000 tonnes per year). A smaller amount of methyl bromide is generated by burning biomass and using leaded gasoline (which contains bromine as an antiknock additive).

Methyl bromide is widely applied as a fumigating agent for soil pests, including termites. Consequently, a large fraction of the CH$_3$Br volatilizes from treated soil (about 50 percent). Nearly 15 percent of all manufactured methyl bromide is used in California for agricultural fumigation. Industry advocates have

placed the worldwide cost in crop and trade losses of a ban on methyl bromide at $50 billion. Methyl bromide is highly toxic, and at least one person has died after entering a treated residence. At the present time, about two-thirds of the total CH$_3$Br source is natural, and one-third is related to human activities. The fate of methyl bromide will soon be decided by the international community, using the Montreal Protocol as a framework.

A variety of organobromides are also manufactured for use as fire retardants. One subset of this class of compounds is referred to as the halons. Some information on the halons and methyl bromide is given in Table 13.1. The simplest halons are fully halogenated methane molecules, similar to the CFCs except that one or more bromine atoms replace chlorine and fluorine. Typical of fully halogenated molecules, the halons are chemically quite inert, with long atmospheric lifetimes. At the present time, the halons account for about 30 percent of the total atmospheric burden of bromine gases. However, the increasing use of halons as fire retardants will soon threaten the ozone layer. Recognizing this threat, timetables have been established under the Montreal Protocol to phase out all halon emissions by 1996.

Bromine is released from methyl bromide and various halons in much the same way as chlorine is released from methyl chloride and CFCs. The free bromine atoms participate in a catalytic cycle:

$$O_3 + Br \rightarrow BrO + O_2$$
$$\underline{BrO + O \rightarrow Br + O_2}$$

(13.18)

Net reaction: $O_3 + O \rightarrow 2O_2$

Bromine has another interesting reaction cycle that does not require sunlight to work. Sunlight generates the oxygen atoms that complete the catalytic cycles based on Equation 13.9 by means of the photodissociation of ozone. But bromine can also react as follows:

$$O_3 + Br \rightarrow BrO + O_2$$
$$O_3 + Br \rightarrow BrO + O_2$$
$$\underline{BrO + BrO \rightarrow Br + Br + O_2}$$

(13.19)

Net reaction: $2O_3 \rightarrow 3O_2$

In this case, two bromine atoms participate in the

catalytic cycle, destroying two ozone molecules each time around.

Bromine is partitioned into HBr and $BrONO_2$, through reactions similar to those for chlorine. Unlike chlorine, however, most of the bromine can remain in active forms, Br and BrO. Atom for atom, stratospheric bromine is about 10 times more effective in depleting ozone as chlorine is.

13.6   Forecasts of Global Ozone Depletion

Using the general ideas and quantitative results developed in previous sections, we can begin to forecast potential ozone depletion associated with human activities. Such activities include the manufacture and use of CFCs, and the flying of high-altitude aircraft. To make detailed predictions, scientists use sophisticated computer models, which might treat dozens of chemical constituents and hundreds of reactions. But all that is just an elaboration of the simplified models and mechanisms described in this text. Some results from these detailed studies are discussed next.

The stratosphere currently contains about 3.6 ppbv of chlorine, which would be close to 0.6 ppbv if all the synthetic chlorine were removed. The chlorine that is present has apparently caused a 2 to 3 percent global average reduction in the total amount of column ozone.[18] This reduction is based on years of observations of the ozone layer using satellite-borne and ground-based instruments. Because the change in ozone is small and the amount of ozone fluctuates naturally from year to year, estimates of the average depletion and its connection to CFCs remain somewhat controversial. A decrease of 2 to 3 percent in ozone means an increase of 4 to 6 percent in the average dose of UV-B radiation at the ground. Over time, the excess ultraviolet radiation will cause thousands of additional cases of skin cancer. These additional cancers will appear as a statistically significant increase in the incidence of cancer among a large population, and most probably in older people.

---

18.   The reduction in the total amount of column ozone (in Dobson units) is of primary interest here, inasmuch as this controls the penetration of UV-B radiation to the surface. Ozone reductions vary with location and time of the year. For convenience, these variations may be averaged, and the mean change in ozone over the entire planet and for all seasons provides a single number as the fundamental measure of risk.

13.6.1   SCENARIOS AND PROJECTIONS

The real threat of ozone depletion lies in the future. We have measured the current state of the ozonosphere. The degradation so far is tolerable. What will happen in the future? To see the future, we need a crystal ball. Our crystal is not made of silicon dioxide quartz, however, but highly purified silicon in the form of computer chips. The chips are wired together in intricate patterns to carry out the most complex calculations that humans have ever made. The purpose is to determine the fate of the ozone shield that protects all life. To work, these computer simulations require information concerning the past, present, and future behavior of all the parameters that affect ozone. The key parameters include the composition and state (temperatures, motions) of the background atmosphere that holds ozone, the coefficients of the chemical processes that directly or indirectly influence ozone, and the variations of the catalytic trace gases that attack ozone.

Projections of future atmospheric chlorine concentrations associated with CFC emissions are given in Figure 13.19. Controlling the future production and release of CFCs is the subject of a recent, unprecedented international treaty, the **Montreal Protocol** of 1987 (Section 13.8.1). The initial agreement calls for a reduction in the worldwide production (and hence release) of CFCs by 50 percent from 1986 levels by 2000. More than 33 nations have signed the treaty. Amazingly, even with this large cut, *the total amount of atmospheric chlorine will continue to increase*, more than doubling during the next century (Figure 13.19). Recall the discussion in Section 13.5.2 and Figure 13.15. If the current source of CFCs and chlorine were halved, the quantity of chlorine in the atmosphere would still need to increase by roughly a factor of 2.5 in order to bring the sink into balance with the source (Figure 13.15).

The amount of chlorine added to the atmosphere by a particular compound and its impact in terms of ozone reductions may be summarized in terms of a single parameter—the so-called ozone depletion potential (ODP). The ODP for a compound is defined as the ratio of the reduction in column ozone that would result from the emission of a unit mass (say, 100,000 tonnes) of that compound, to the ozone reduction that would result from the same mass emission of CFC-11. The ODP thus measures the ozone-reducing capacity of a chlorocarbon

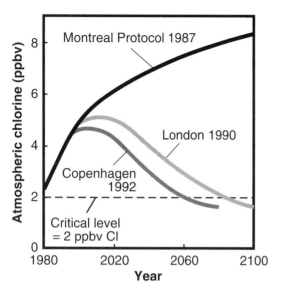

Figure 13.19   Projections of the total concentration of chlorine in the atmosphere (in parts per billion by volume, ppbv) corresponding to several scenarios for future emissions of CFCs and other chlorocarbons. The baseline emission rates are for 1990. The scenarios are based on the Montreal Protocol of 1987 and the London (1990) and Copenhagen (1992) amendments. The original treaty called for a 50 percent reduction in CFC production worldwide by 2000. Later amendments accelerated those cuts and added additional compounds to the hit list, including carbon tetrachloride and the halons. The largest reductions in emissions are necessary to bring the stratosphere back during the next century to the near-natural chlorine levels of the 1960s. The "critical level" for stratospheric chlorine, above which the ozone drops to an unacceptable level, has been pegged at 2 ppbv of Cl. (Data from the World Meteorological Organization Global Ozone Research and Monitoring Project, 1992)

relative to CFC-11. The ODP for chlorofluorocarbon 11 is exactly 1.0 (since both the two compounds being compared are CFC-11). Table 13.1 lists the ODPs for a number of other chlorine-containing compounds. The ODPs are determined using advanced computer models with prescribed emissions of chlorocarbons and comprehensive ozone photochemistry (the models have limitations, however). In general, CFC-11 is the most potent ozone-depleting agent, gram for gram, although CFC-12, CFC-113, and carbon tetrachloride are in the same league.[19]

19.   The ODP is obviously related to the increase in stratospheric chlorine associated with the emissions of a specific chlorocarbon. The larger the ODP is, the greater the increase in atmospheric chlorine loading for a given mass emission will be. The curves in Figure 13.19 are thus controlled by the combina-

If the emissions of CFCs are controlled more stringently in the future (according to the London and Copenhagen amendments to the Montreal Protocol [Table 13.2 and Section 13.8.1]), the concentrations of chlorine could be stabilized at 5 ppbv or less and then reduced to ~1 ppbv by the end of the twenty-first century (Figure 13.19). During the worst period during the first half of the next century, global ozone depletions could reach 5 percent, and skin cancer rates could rise by 10 percent or more. Because of the long residence times for CFCs in the atmosphere, chlorine increases and ozone decreases associated with CFCs can persist for 100 years (Figure 13.19). If for any reason the CFC phase-out deviates from the Montreal Protocol's timetable, additional chlorine would accumulate in the atmosphere and remain there for 100 years. That is hardly an honorable legacy for future generations.

Figure 13.20 shows a detailed forecast of possible decreases in total ozone associated with a buildup of CFCs. The reductions in total ozone are given as a function of time at different latitudes. An important general point emerges from this scenario: Ozone depletion is closely related to latitude, with the greatest decreases in ozone occurring at the highest latitudes. In a way, this is fortunate because the expected increase in average UV-B doses for a specific reduction in ozone would be less at high latitudes than at low latitudes. The calculated ozone depletions for 1990 in Figure 13.20 are consistent with observations, as determined in a recent global ozone data assessment. Projections into the twenty-first century highlight the potential threat of the ozone depletion problem. Ozone reductions could double and last until the end of the next century even if limited action is taken to curtail the production of CFCs worldwide. The complete elimination of CFCs, halon, and other chlorocarbon emissions by the millennium would avoid the unacceptable ozone decreases depicted in Figure 13.20.

The computer models that are used to make forecasts of ozone depletion are imperfect. The effects of hundreds of chemical reactions producing and destroying ozone and the influences of the winds dispersing ozone over the entire planet must be calculated for a century. A large enough computer does not yet exist to make such calculations with all the details included, and so practical approximations

tion of industrial emission rates and ODPs for the various chlorine-bearing compounds of interest.

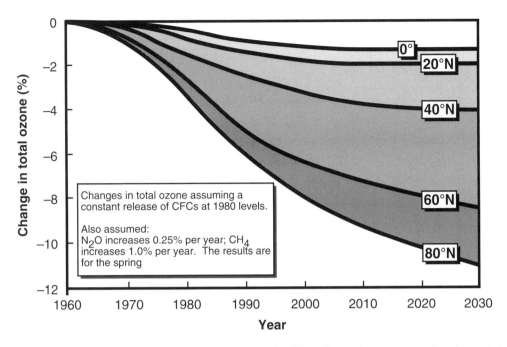

Figure 13.20   Scenario for the depletion of the ozone layer by CFCs. Predictions were made using a state-of-the-art computer model of the ozone layer photochemistry and dynamics. The ozone projections are given as a percentage decrease in the total amount of column ozone over time, beginning in 1960. The reductions of ozone at several different latitudes in the Northern Hemisphere are plotted as separate lines. The emissions of CFCs are assumed to be constant over time in these calculations, and other gases that affect ozone are assumed to change; details of the emission scenario are summarized in the inset.

have been substituted. The Earth's ozone layer is typically reduced to a vertical slab extending from the South Pole to the North Pole and reaching from the ground to the stratopause. This two-dimensional representation takes into account all latitudes and altitudes of interest. However, many of the intricate longitudinal motions that shape the ozone layer are lost. Although such approximations may seem drastic, the two-dimensional representation is considered to be quite reliable for predicting the rough magnitude and latitudinal distribution of ozone perturbations. Much greater uncertainties are associated with the treatment of stratospheric particulates and their "heterogeneous" chemical reactions in these models. The "ozone hole" is an example of an effect that cannot be predicted without considering particulates. Heterogeneous chemistry is discussed in the next section.

Projecting the emissions of gases that will affect ozone in the future is a bit dicey because the number of such compounds is very large. Consider, for example, the lists in Tables 13.1 and 13.2. In addition, ozone is influenced by other compounds such as nitrous oxide, which generates nitrogen oxides in

the stratosphere (Section 13.5.3), and methane, which affects the partitioning of chlorine compounds and participates in several other ozone chemical processes. Methane, for example, can react with nitrogen oxides in the lower stratosphere, generating ozone, as in photochemical smog. Methane that is oxidized in the stratosphere also provides a major source of water vapor for the middle and upper stratosphere. The overall methane oxidation scheme can be summarized by the process

$$CH_4 + 5O_2 \xrightarrow{\text{OH, NO, } h\nu} 2H_2O + 2O_3 + CO_2 \quad (13.20)$$

The number of ozone molecules formed in this process depends on several factors and is thus variable. At least one or two ozone molecules are likely to be produced in the lower stratosphere.

To make reasonable projections of ozone change, variations in all the gases that affect ozone should be considered. The atmospheric concentrations of nitrous oxide and methane have been steadily increasing since the Industrial Revolution (Sections 12.3.1 and 12.3.2). The ozone change scenario in Figure 13.20 accounts for the recent growth rates for these gases and extends those trends into the future. Because such projections cannot be guaranteed, usually a range of scenarios must be run. The more parameters that are uncertain, the more possible scenarios can be dreamed up. The number of plausible cases begins to soar as the future advances ahead of us. In the end, however, reality will differ from predictions, and so we should at least be prepared for the worst outcomes, and aware of the best, should the actions now in progress be fulfilled.

13.6.2 SIGNATURE OF THE CULPRIT

The effects of different catalysts on ozone can often be identified by an atmospheric "handwriting" ex-

Figure 13.21   The relative changes in the ozone concentration as a function of altitude for two specific catalytic agents : chlorine (Cl) and nitrogen oxides (NO_x). The ambient ozone profile is shown for comparison with the perturbed profiles. Each catalyst has a distinct impact on ozone. The NO_x signature shows an ozone *increase* in the lower stratosphere, similar to the effect of NO_x on smoggy air. Even so, the overall amount of column ozone may still be reduced by nitrogen oxides if the emissions extend into the upper stratosphere. Note that the profiles of ozone depletion illustrated here correspond to middle and low latitudes, where the effects of stratospheric particles on ozone chemistry are minimal.

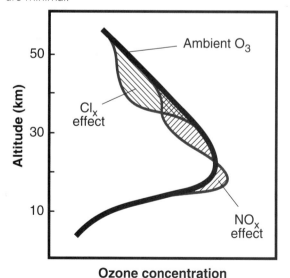

Ozone concentration

pert. Each catalyst interacts with ozone most efficiently in a particular region of the atmosphere. Accordingly, specific catalysts leave their "signature" on the ozone layer. The typical vertical distributions of ozone depletion caused by chlorine and nitrogen oxides are depicted in Figure 13.21. Chlorine tends to attack ozone primarily in the altitude ranging from 30 to 50 kilometers (Figure 13.17). Nitrogen oxides, on the other hand, affect ozone mainly below about 30 kilometers. In fact, NO_x injected into the lower stratosphere can actually *increase* stratospheric ozone. This response is caused by a smoglike interaction between NO_x and methane, a light hydrocarbon present in the lower stratosphere (Section 6.2.1). It follows that aircraft operating just above the tropopause—as the existing subsonic commercial air fleet does—may actually enhance the ozone layer.

Excess nitrous oxide (N_2O) generated by agricultural activities and anthropogenic combustion may be another threat to the ozone layer. The N_2O would be decomposed by photochemical processes at altitudes near 30 kilometers (Section 13.5.3). The NO_x generated would strongly deplete ozone in this region (Figure 13.21). If N_2O concentrations in the atmosphere were to double, the amount of ozone could be reduced by 10 to 15 percent. Although doubling of N_2O now seems unlikely until the twenty-first century (Section 12.3.2), nitrous oxide is a potent ozone-depleting agent. Moreover, the atmospheric lifetime of N_2O is about 100 years, similar to that of many chlorofluorocarbons. Any change in nitrous oxide levels and the accompanying ozone depletions would persist for a century. There are good reasons to be cautious about N_2O emissions in the future and to watch carefully their behavior in the atmosphere.

Forecasts of future ozone depletion are controversial for several reasons, not the least of which centers on uncertainties in the scenarios used. In addition, ozone depletion expected from catalysts already in the atmosphere has been difficult to detect. If calculations of the present state cannot be confirmed, predicting the future seems even more remote. The current impact of chlorine on global ozone is small—only a small reduction in ozone averaged over the planet. This reduction also is difficult to measure precisely against natural fluctuations in the ozonosphere. Commercial aircraft traffic has apparently added some ozone to the lower stratosphere, further obfuscating the chlorine effect. An apparent chlorine *signature* (Figure 13.21) has been detected in some of the ozone measurements, but the evidence is not yet ironclad.

The lack of a clear global signal of reduced ozone connected with CFCs has led some authorities to argue that the ozone layer is robust against *any* human activity. They contend that we simply do not understand the complex atmospheric system well enough to make reliable forecasts. Moreover, they point to putative natural forces that might alter the ozone layer to a greater extent than CFCs do—volcanic eruptions and solar variations are frequently invoked—and argue that the human contribution to ozone change is trivial. It follows that no corrective or preventive measures are needed. The "status quo" and "business as usual" prevail!

That narrow attitude was recently shaken by the discovery of a sharp decrease in the abundance of ozone over Antarctica. The highly unusual depression is widely referred to as the *ozone hole*. The ozone hole can be found only over the Antarctic continent in the southern (austral) spring. The discovery of the ozone hole, the reasons that it forms above the South Pole, and the implications for ozone everywhere else in the world are discussed in the next section.

13.7   The Ozone Hole

The possible environmental impacts associated with stratospheric ozone depletions of 5 or 10 percent are serious enough. Imagine that suddenly overhead, there is a hole in the ozone layer the size of a continent with 50 percent or more of the ozone missing. That scenario would have made an excellent plot for a horror movie. Unfortunately, such a hole has appeared. Not on the silver screen, but over the South Pole.

13.7.1   Discovery

Joseph Farman and his British colleagues had been traveling to Antarctica each year for more than a decade. They spent the winter there in isolation and perpetual darkness. The British scientists had a long-established base at a place called Halley Bay. During those annual visits, Farman used a simple Dobsonlike instrument to measure the thickness of the ozone layer over Halley Bay when the sun reappeared in the spring. He recorded these data, year after year,

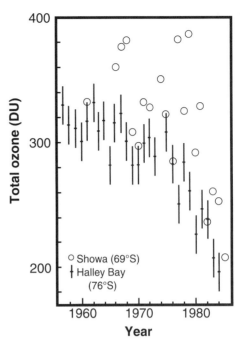

Figure 13.22   Measurements of the total amount of ozone in Dobson units over two locations on the Antarctic continent. The Halley Bay data were collected by J. Farman and his colleagues. The Showa data were gathered by a Japanese research team. For each year, the average amount of ozone for the entire month of October is given, and the average is taken over daily measurements. The main difference between the two stations where the observations were made is the latitude of the sites. Halley Bay is considerably closer to the South Pole. (Data from J. Farman, B. Gardiner, and J. Shanklin, "Large Losses of Total Ozone in Antarctica Reveal Seasonal Interaction," *Nature* 315 [1985]: 207 for Halley Bay)

carefully and methodically. Then in 1984, Farman noticed something strange about the ozone data. When he looked at the amounts during the month of October, he saw that the ozone thickness had started to drop precipitously. His original data are reproduced in Figure 13.22. Note that winter in the Southern Hemisphere occurs during summer in the Northern Hemisphere. Similarly, the austral (southern) spring corresponds to the boreal (northern) fall. That is why Farman's springtime ozone measurements were taken in October.

Farman and his co-workers published their startling discovery in 1985. They speculated that the sudden onset of ozone loss was connected with the accumulation of CFCs in the atmosphere. The CFC-ozone depletion theory was well established at the time, but had not been proved by direct observations.

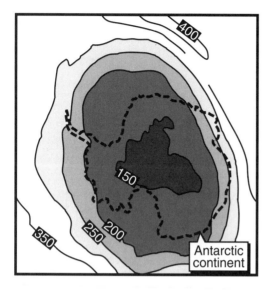

Figure 13.23   A map of total ozone over the Southern Hemisphere on September 29, 1987, obtained with the TOMS (total ozone mapping spectrometer) satellite instrument. Contours correspond to different amounts of ozone, given in Dobson units. The Antarctic continent is outlined, and the "ozone hole" is seen to encompass the entire continent. This view is typical of the situation in September and October, when the largest ozone depletions were recorded. The images are generated by processing satellite measurements of scattered ultraviolet radiation. From space, you would not be able to see the ozone hole as it is depicted here.

Farman's paper soon launched an intensive scientific inquiry into the cause of this **ozone hole**.

Satellites had been monitoring the global distribution of ozone since the late 1970s. By the early 1980s, ozone was disappearing rapidly over Antarctica each spring. This ozone depletion had actually been recorded by the satellites. The ozone values were so startlingly low, however, that the analysts studying the processed data thought they surely must be incorrect and so disregarded the results. Following the publication of Farman's report, they reanalyzed the data. Those satellite images now provide the most striking views of the ozone hole (Figure 13.23). The data show that ozone amounts plunge in September and October each year, but recover to normal values in November and December. The ozone levels then remain more or less stable until the following spring (September). The depletion extends over most of the Antarctic continent, as confirmed by measurements at locations such as Halley Bay and Showa.

The ozone hole has continued to deepen since 1984, reaching record low levels of about 120 DU in 1990. That is roughly one-third of the total ozone amounts measured in the 1960s. Not only has the minimum value of ozone been decreasing over the years, but the area of the "hole" has expanded. Figure 13.23 shows a satellite image of the ozone hole taken in 1987. The contour of total ozone corresponding to 150 DU covers a large portion of Antarctica. The ultimate size of the hole is limited by meteorological conditions in the Southern Hemisphere (Section 13.7.2). Nonetheless, large pieces of the hole have broken away in the late spring and traveled over Australia and southern Chile. Large enhancements in ultraviolet radiation have been detected beneath these "miniholes." Reports of blind rabbits in southern Chile and other bizarre incidents have been loosely connected with these miniholes, although such severe effects seem unlikely (Section 13.4.3).

13.7.2   THE POLAR VORTEX

The ozone contours in Figure 13.23 are symmetrically distributed with respect to the South Pole. This suggests that the ozone hole is associated with a regular feature of the Southern Hemisphere's wind system, the wintertime **polar vortex**. The southern stratospheric polar vortex is a powerful circulation system that develops every fall and winter. In late fall and early winter, the air in the stratosphere over Antarctica begins to cool rapidly, particularly after the sun drifts below the horizon. The air cools by radiating thermal energy into space. Because of this cooling, a thermal gradient is set up between the cold polar region and warmer air at lower latitudes (closer to the equator). The thermal gradient causes midlatitude air to flow poleward as air subsides over the pole. However, the rotation of the Earth creates a Coriolis force (Section 2.4) that accelerates the air sideways (longitudinally), forcing it into a spinning motion around the pole. As the polar region continues to cool, the rotation picks up speed. The swirling motion that results is shown in Figure 13.24.

The polar vortex is mechanically similar to other common vortices in fluids. A tornado is a violent and dangerous example of an atmospheric vortex, with winds reaching speeds of 800 kilometers per hour! A hurricane is a much larger vortex, with peak winds of

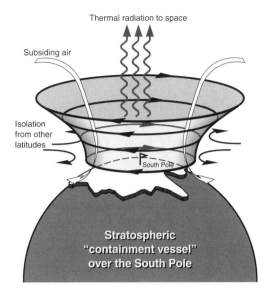

Thermal radiation to space

Subsiding air

Isolation from other latitudes

South Pole

Stratospheric "containment vessel" over the South Pole

Figure 13.24   The main circulation of wind over Antarctica in winter and early spring. The polar vortex is composed of strong west–east winds in the stratosphere that extend to latitudes of about 65° south. The winds also extend vertically through the entire depth of the stratosphere. Air cannot easily move across the vortex boundary in either direction, and so the interior of the vortex is effectively isolated from the rest of the atmosphere. However, air can descend slowly within the vortex as the trapped air mass cools by emitting radiation to space.

perhaps 200 kilometers per hour. The swirling action produced when you flush a toilet is a kind of vortex. Similarly, the funnel of air dipping into the drain when you empty a sink or bathtub represents vortical circulation. In each case, the spinning motion keeps the rotating fluid from rushing into the center of the vortex. The rotation also strongly inhibits any motions across the wall of the vortex, as the inside of the vortex is effectively isolated.[20]

20.   You can strengthen the vortex in a sink by swirling the water in one direction before opening the drain. It helps if the drain is in the center of the basin; otherwise, the swirling action is likely to disturb the symmetry of the vortex and cause it to dissipate. The general rotational motion, or vorticity, in a fluid is concentrated at a drain vortex according to the physical law of **conservation of angular momentum**. According to this law, a spinning object speeds up as it approaches its axis of rotation. For a simple demonstration of this effect, tie a light weight to a string and spin it on the string, using your finger as an axis. As the string winds around your finger and shortens, the weight will rapidly accelerate. Spin the weight in the opposite direction, and it will slow down as the string lengthens. A bola works on the same principle, with three tethered iron balls accelerating as they entwine the legs of an animal (or the neck of an opponent). Ice-skaters employ the same principle to produce a blurring spin by pulling in their appendages. A vortex in a sink will spin faster and

In the case of the polar stratospheric vortex, the spinning motion encloses an area the size of Antarctica. The winds in the polar jet forming the walls of the vortex reach maximum speeds of 300 km/hr. The huge plug of air within the spinning vortex is referred to as a **containment vessel**. In this vessel, polar chemistry can occur in isolation from the rest of the atmosphere. The isolated air also becomes extremely cold, occasionally falling to temperatures below 180 K (–93°C). The cooling is aided by the fact that the polar vortex remains in darkness during the entire winter, receiving no heat from solar illumination. The vortex, moreover, is dynamically isolated from warmer air at higher latitudes, and so it receives no sensible heat by horizontal transport. A stabilizing feedback mechanism is set up; as air in the vortex grows colder, the spinning motion is accelerated, and so the air inside becomes more isolated.

A polar vortex is also set up in the Northern Hemisphere in winter. However, the northern winter vortex is much weaker and less stable than the southern vortex. This difference is the result of geography. The Southern Hemisphere has relatively symmetrical geographical features surrounding the South Pole. The Antarctic continent and southern Pacific Ocean are fairly uniform in longitude. The Southern Hemisphere also has a larger fraction of ocean surface compared with the Northern Hemisphere, which leads to more uniform seasons. As a result, the southern vortex spins around its pole with less drag and fewer disturbances.[21] In the north, however, the geography is more complex, with land areas, mountain ranges, and seas scattered around the pole. The northern vortex is buffeted by weather systems and cannot build up the same head of steam or get as cold.

Both the southern and northern polar vortices break down in spring. This is caused by the return of

grow larger as you swirl the water around the drain, adding vorticity.

21.   Weather develops in the troposphere as large waves superimposed on the uniform circular polar winds. The waves may become unstable and spawn storm systems. These tropospheric weather systems also disturb the stratosphere by pushing up the tropopause, which separates the stratosphere from the troposphere. When the tropopause is distorted in this way, the stratospheric polar vortex is deflected and can become unstable itself. The response of the vortex is similar to that of a spinning top. On a flat surface, a top tends to stabilize at a point and spin for a long time. On a rough or uneven surface, the top hops from point to point or wanders over the surface following the low terrain. The punctuated movement causes the top to wobble more and increases frictional drag, which slows it down faster.

sunlight, with a fundamental change in the dynamics of the stratosphere. In fact, the polar vortex reverses direction during the spring as the polar air, illuminated by the sun, warms up. During the reversal, vortex air sometimes mixes rapidly with midlatitude air. These springtime vortex breakups occur earlier and more frequently in the Northern Hemisphere than in the Southern Hemisphere. Occasionally, the vortex collapses spectacularly in a **"sudden warming"** event, in which a massive infusion of warm midlatitude air all but destroys the circumpolar flow. The northern polar vortex is more susceptible to these events than the southern vortex is; accordingly, it is both less stable and shorter lived.

The dynamical differences between the polar vortex in each hemisphere are critical to the development or absence of the ozone hole. The southern vortex is stable and cold every year. The northern vortex is typically less stable and warmer and less persistent. An ozone hole is clearly present in the Southern Hemisphere spring. No hole has been detected in the Northern Hemisphere (but see Section 13.7.5).

The structure and properties of the polar vortex are consistent with the observed size, geographical location, and timing of the southern ozone hole. But the meteorology alone cannot explain the ozone depletion. Atmospheric dynamics specialists have performed numerous studies showing that winds by themselves can produce only a small fraction of the ozone deficit that is measured. In addition, the onset of the hole in the early 1980s is not consistent with any noticeable climate shift in the stratosphere. If meteorology were the only factor causing the ozone hole, it should have been present throughout the century and earlier. But we have no evidence pointing to an ozone hole before the 1980s.

One early theory for the ozone hole proposed that sunspots were the cause. Sunspots are associated with solar storms that can modulate the atmosphere of the Earth is subtle ways. In the early 1980s, the number of sunspots and solar activity were on the upswing in their 11-year cycle—perfect timing. One effect of solar activity is to increase the bombardment of the upper atmosphere with protons (the nuclei of hydrogen atoms). This proton rain is often very intense, especially during **solar proton events**. The protons produce nitric oxide as they crash through the air above about 50 kilometers altitude. As we already pointed out (Section 13.5.3), nitric oxide reacts with and destroys ozone in the middle and upper stratosphere. Logically, the NO generated at high altitudes by solar protons could be carried downward and react with ozone.

Sunspots were a novel idea, but wrong. Measurements taken directly in the ozone hole have demonstrated that the concentrations of nitrogen oxides are, in fact, greatly *reduced* in the regions of the largest ozone depletion. Moreover, the ozone hole did not recover when the number of sunspots and solar activity decreased. Two nails are enough for this coffin—two very big nails. The mystery of the ozone hole remained.

13.7.3   POLAR STRATOSPHERIC CLOUDS

In 1978 a satellite was launched carrying an instrument that looked at the edge of the Earth (the "limb") to measure aerosols in the stratosphere.[22] The instrument scanned the horizon at high latitudes, well into the polar regions in both hemispheres. It detected the ubiquitous background stratospheric sulfate aerosols; that was expected. But during the first austral winter, the satellite picked up a much stronger signal, and only at high latitudes in the Southern Hemisphere. These unexpected aerosols clouds were too dense and massive to be composed of sulfate aerosols. They were named the **polar stratospheric clouds** (**PSCs**). PSCs were discovered almost a decade before the first alarms about the ozone hole. Research on these mysterious clouds initially proceeded with no connection to stratospheric ozone.

Polar stratospheric clouds form because air trapped inside the winter polar vortex can cool to extremely low temperatures. In fact, as stratospheric air cools, two distinct types of PSCs can appear. One type is closely related to another rare cloud formation, the **nacreous clouds**, or **mother-of-pearl clouds**. Mother-of-pearl clouds are named for their beautiful irridescent colors, caused by the selective scattering of the different wavelengths of sunlight from small spherical particles. These clouds are composed mainly

---

22.   Looking toward the Earth from a satellite in space, the atmosphere appears as a thin luminous halo bordering and just above the solid disk, or edge, of the planet. The effect is highlighted when the sun is behind the Earth. This piece of the atmosphere, ringing the horizon from the point of view of the space observer, is called the *limb* of the planet. Optical measurements are often made in the limb because there is no surface directly along the line of observation to cause interference by scattering or emitting radiation.

of water ice, much like the common cirrus clouds seen as wisps in otherwise clear skies. However, nacreous clouds, unlike cirrus clouds, can reside in the stratosphere.[23] The polar stratospheric clouds that are related to nacreous clouds—that is, composed of water ice—are referred to as *Type-2 PSCs*. This clever name derives from the fact that Type-2 PSCs form at *colder* temperatures than do the other, more common type of polar stratospheric clouds, referred to as *Type-1 PSCs*.

The Type-1 PSCs are composed of frozen *nitric acid* and water. Clouds of nitric acid are not found anywhere else in the solar system. In the Earth's atmosphere, temperatures must fall below about 195 kelvin for Type-1 cloud particles to condense. There also must be sufficient nitric acid vapor present for Type-1 PSCs to form. The appropriate conditions for Type-1 PSCs—low temperatures and high nitric acid vapor concentrations—are confined primarily to the lower stratosphere in the winter polar vortex.[24] If air temperatures fall below about 190 kelvin (83°C below the freezing point of water), clouds of pure water ice may begin to condense. These Type-2 PSCs are a major form of water ice clouds in the stratosphere, although mountain lee wave (mother-of-pearl) clouds are frequently seen, and occasionally, aircraft contrails may be observed.[25]

23.   Nacreous clouds are rare because the stratosphere is very dry, and so air temperatures must be extremely low before water will condense as ice. Typically, nacreous clouds do not form until temperatures drop to 180 to 190 kelvin. Under such conditions, the ice particles in nacreous clouds are quite different in size and shape from the ice crystals in cirrus clouds, which form at warmer temperatures closer to the freezing point of water (273 kelvin). Nacreous clouds are most often observed in the lee waves of mountain ranges. Air rushing over the mountain deflects stratospheric air upward, causing it to rise and cool adiabatically; the deflected air parcels then oscillate up and down beyond the mountain. Water clouds may condense in the cold troughs of these waves. The clouds appear to be fixed with respect to the mountain and usually have sharp, well-defined edges. On occasion, such clouds are mistaken for hovering UFOs (unidentified flying objects), since they seem to remain fixed in the air.

24.   The stratosphere over the tropics has temperatures that are cold enough to form PSCs; however, there is usually too little nitric acid vapor for the clouds to condense. The upper mesosphere is also cold enough for these clouds to form (Section 2.3.2), but it contains no nitric acid vapor at all.

25.   The temperature in the upper mesosphere may drop so low that water ice clouds can also condense there as "noctilucent," or "night-luminous," clouds, so called because they usually appear as glowing waves against the night sky. The coldest temperatures in the mesosphere occur over the polar regions during the summer season, when temperatures as low as 130 kelvin (more than 140°C below the freezing point of water!) are possible. Hence, noctilucent clouds are most frequently observed

The sequence of cloud formation in the wintertime polar stratosphere can be summarized as a series of events unfolding as the air cools. When air temperatures drop below about 195 to 200 kelvin, the background stratospheric sulfate aerosols become highly supercooled, and some of them freeze. At temperatures lower than about 192 to 195 kelvin, nitric acid ice condenses on the frozen sulfate aerosols to form Type-1 polar stratospheric cloud particles. The Type-1 PSCs grow much larger in size than the sulfate aerosols, by absorbing the abundant nitric acid vapor in the local environment. If temperatures continue to fall below about 185 to 190 kelvin, water ice can begin to condense on the Type-1 particles, forming Type-2 ice crystals with Type-1 cores embedded in them. Type-2 particles are much bigger than Type-1 particles, because water vapor is much more abundant than nitric acid vapor in the stratosphere. If the Type-2 ice crystals fall out of the stratosphere, they will remove the nitric acid trapped in them. This process is known as **denitrification**, and is important to the chemical evolution of the ozone hole.

The PSCs of both types have been extensively investigated with balloon- and aircraft-borne instruments and from satellites. Their general physical properties and distributions are reasonably well established. The Type-1 and Type-2 clouds are found in the polar stratosphere in both hemispheres during the winter. But in the Southern Hemisphere, Type-1 and Type-2 PSCs can literally fill the polar vortex in the winter, whereas in the Northern Hemisphere vortex, PSCs are more sporadic and less dense, and Type-2 PSCs are quite rare. The clouds generally lie between the altitudes of 15 and 25 kilometers. This is almost precisely the same altitude range over which polar ozone rapidly disappears in the spring!

Chemistry in the Clouds

Atmospheric scientists first linked PSCs to the ozone hole because they seemed to be hanging out together. It was theorized that some unusual chemistry must involve the PSCs. Soon thereafter, laboratory studies revealed that certain chlorine gases, which were normally inert in the gas phase, reacted vigorously on ice surfaces. Careful analysis showed that

in summer at high latitudes (near or below the Arctic Circle in the Northern Hemisphere, since the clouds are too faint to be detected against the continuously illuminated sky of the perpetual Arctic summer day).

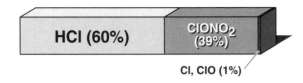

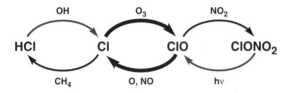

Figure 13.25   The partition of chlorine gases between inert reservoir species (HCl and ClONO$_2$) and catalytically "active" species (Cl and ClO). The percentage of the total chlorine in each form is indicated (here we consider only "liberated" chlorine derived mainly from CFCs, but from other chlorocarbons as well). Also shown are the key reaction pathways that shuffle the various chlorine species to establish the partition.

the clouds can effectively activate the chlorine derived from CFCs, making it far more reactive toward ozone.

As discussed in Section 13.5.2, stratospheric chlorine is normally partitioned so that only a small fraction—about 1 percent—is in active forms that catalytically attack ozone. Figure 13.25 makes this point by illustrating the partition of chlorine among HCl (hydrogen chloride), ClONO$_2$ (chlorine nitrate), ClO (chlorine monoxide) and Cl. The most important reaction pathways that join these chemical species also are shown in Figure 13.25. Some of the specific reactions were described in Equations 13.12 through 13.14. The central cycle involving Cl and ClO represents the fast catalytic cycle that can destroy ozone. Not all the other details are important at this point.

PSCs alter the normal state of affairs in several ways. First, the cloud particles act as a local sink for some compounds. Nitric acid is strongly affected because it condenses to form the cloud particles. The PSCs also provide surfaces on which unusual chemical processes can take place. Chemical reactions that occur on particle surfaces are referred to as **heterogeneous reactions** because both gases and solid surfaces are involved. Laboratory studies have shown, for example, that HCl and ClONO$_2$ react very efficiently on ice surfaces. This critical reaction is depicted in Figure 13.26. The chlorine species produced by this reaction is **molecular chlorine gas**

(Cl$_2$). The reaction can be summarized in chemical notation as

$$HCl + ClONO_2 \xrightarrow{\text{PSC Ice Surface}} Cl_2 + HNO_3 \quad (13.21)$$

The molecular chlorine evaporates from the surface of the ice because it does not have a strong affinity for ice, although the nitric acid remains frozen to the ice surface.

The immediate effects of this reaction are to convert HCl and ClONO$_2$ to Cl$_2$ vapor and to remove gaseous nitrogen oxides by converting them to condensed nitric acid. The reduction in nitrogen oxides inhibits the recycling of activated chlorine back into the chlorine nitrate reservoir (Figure 13.25). The NO$_x$ catalytic cycle is also slowed, but that effect, which might benefit ozone, is secondary. Also, the nitric acid condensed on polar stratospheric cloud particles may be removed as the particles fall into the troposphere. The resulting denitrification can intensify and prolong the ozone depletion by inhibiting the re-formation of chlorine nitrate through reactions involving nitrogen oxides produced by the photodissociation of HNO$_3$.

Cl$_2$ does not react directly with ozone, but chlorine gas has a green tinge, which means that it absorbs visible sunlight. The Cl$_2$ is then readily broken apart into chlorine atoms by photodissociation,

$$Cl_2 + h\nu \rightarrow Cl + Cl \quad (13.22)$$

The free chlorine atoms react vigorously with ozone. The overall effect of heterogeneous chemistry on PSCs is to transform the inert chlorine reservoir species, HCl and ClONO$_2$, into active chlorine species, Cl and ClO. Indeed, measurements of chlorine constituents in the ozone hole reveal that chlorine is almost fully activated in the springtime polar vortex. The concentrations of ClO reach one part per billion or more, 100 times greater than the concentrations found in the rest of the stratosphere. This drastic repartitioning of chlorine in the presence of PSCs is shown in Figure 13.27.

A high concentration of ClO alone is not sufficient to explain the sudden depletion of ozone in spring. It turns out that a new chemical species still needed to be discovered. Mario Molina, codiscoverer of the CFC-ozone problem, took the essential step when he proposed that chlorine monoxide could form a **dimer**—that is, a double of itself—in

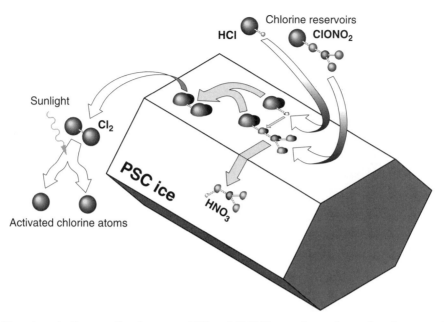

Figure 13.26   The steps in the reaction between HCl and ClONO$_2$ on the surface of an ice crystal. The atomic structure and conformation of the molecules and the configuration of the reacting species and products on the ice are depicted. The products of the reaction are chlorine gas (Cl$_2$) and nitric acid (HNO$_3$). The Cl$_2$ molecule leaves the ice surface, but the HNO$_3$ molecule remains bound to the ice.

the following simple way:

$$ClO + ClO + M \rightarrow Cl_2O_2 + M \quad (13.23)$$

The Cl$_2$O$_2$ is then broken apart by sunlight.

The formation and destruction of the ClO dimer actually define a very potent ozone catalytic cycle, which proceeds through the following steps:

$$Cl + O_3 \rightarrow ClO + O_2$$
$$Cl + O_3 \rightarrow ClO + O_2$$
$$ClO + ClO \xrightarrow{M} Cl_2O_2$$
$$Cl_2O_2 + h\nu \rightarrow Cl + Cl + O_2$$

$$(13.24)$$

Net reaction: $2O_3 \rightarrow 3O_2$

Although this catalytic cycle is somewhat more complicated than the others already discussed, the effect is the same. Here, two chlorine atoms conspire to destroy two ozone molecules through the intermediary formation of the ClO dimer. The chlorine atoms are recycled in the process. *The dimer mechanism is essential to explaining the ozone hole.*

Bromine can also play a role in forming the ozone hole. In this case, chlorine and bromine cooperate in

a catalytic cycle. The reaction sequence is

$$Cl + O_3 \rightarrow ClO + O_2$$
$$Br + O_3 \rightarrow BrO + O_2$$
$$ClO_2 + BrO \rightarrow Cl + Br + O_2 \quad (13.25)$$

Net reaction: $2O_3 \rightarrow 3O_2$

The catalytic cycle Equation 13.25 involves one chlorine atom and one bromine atom. The cycle strengthens the catalytic activity of both halogens and is said to be *synergistic*, or self-reinforcing. This cycle contributes to, but is not essential to, the formation of the ozone hole.

13.7.4   OZONE DEPLETION: THE HOLE STORY

An extensive campaign of scientific measurements in the winter polar stratosphere was waged between 1987 and 1992. These field data have been exhaustively analyzed and combined with laboratory measurements and modeling predictions to formulate a complete and consistent theory for the formation of the ozone hole. In fact, the hard scientific results uniformly support the general hypothesis

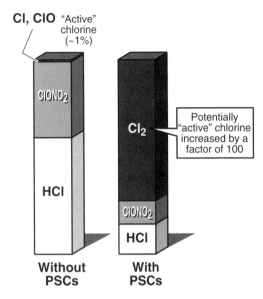

Figure 13.27 The partition of chlorine among its various chemical forms in the normal stratosphere (left) and in the polar winter stratosphere with PSCs present. In the latter case, almost all the chlorine from the inert reservoirs, HCl and ClONO$_2$, has been preactivated into Cl$_2$ gas. With the first sunlight in spring, the Cl$_2$ is fully activated into Cl and ClO by means of solar photodissociation.

outlined in previous sections.[26] The sequence of events that creates the ozone hole can be summarized as follows:

1. Over decades, CFCs accumulate in the global troposphere and are slowly carried into the stratosphere.

2. Under illumination by ultraviolet solar radiation above roughly 20 to 30 kilometers, the CFCs are decomposed into chlorine atoms (Cl) and chlorine monoxide (ClO).

3. The free chlorine partitions itself through normal chemical processes primarily into hydrogen chloride (HCl) and chlorine nitrate (ClONO$_2$), with only a small fraction (~1 percent) remaining in the catalytically active forms, Cl and ClO.

4. In early winter, the polar stratospheric vortex winds up as the sun drops below the horizon and the stratosphere cools strongly.

5. The vortex entraps ozone and other gases, including the partitioned chlorine species, and

26. Note that in the current account of the ozone hole, not all the contributing chemical actors and physical stages have been introduced. That larger play with a more complex plot is not much different in its action and outcome from the story just told.

isolates them from the rest of the atmosphere.

6. The dark isolated vortex continues to cool by radiating the residual heat into space as infrared radiation; this further strengthens the vortex and seals its walls to form a containment vessel.

7. As the vortex air cools below about 195 kelvin, Type-1 polar stratospheric clouds (PSCs) begin to condense from the nitric acid vapor trapped in the vortex.

8. The inert chlorine reservoir species HCl and ClONO$_2$ stick to and react on the Type-1 PSC ice surfaces and are converted to molecular chlorine (Cl$_2$) gas.

9. If temperatures drop below about 185 to 190 kelvin, Type-2 PSCs begin to form, leading to "denitrification" of the lower stratosphere.

10. The "heterogeneous" chemical processing continues throughout the winter, during which the inert chlorine reservoirs are largely converted to "preactivated" molecular chlorine vapor, and nitrogen oxides are removed.

11. When sunlight first returns to the polar vortex in early spring, the molecular chlorine is photodissociated and fully activated into Cl and ClO.

12. The extraordinarily high concentrations of active chlorine in the form of ClO (~100 times the normal abundance) rapidly destroys ozone through halogen catalytic reaction cycles, which are accelerated by the formation of the chlorine monoxide dimer (Cl$_2$O$_2$) and the presence of bromine monoxide.

13. Local ozone concentrations decrease in a matter of weeks by 90 percent or more at some altitudes, with the amount of total column ozone decreasing to less than half its initial value.

14. As time progresses, the polar vortex begins to warm up, becoming less stable and starting to leak as it slows down; the vortex may collapse in a "final warming" event.

15. During the period of vortex breakdown, ozone-rich stratospheric air is transported from the middle latitudes into the polar region, and the ozone hole rapidly "fills in"; nitrogen oxides and nitric acid also are replenished.

16. During late spring and summer, the normal chemical composition of the stratosphere is reestablished, and the polar region becomes primed for the next ozone hole.

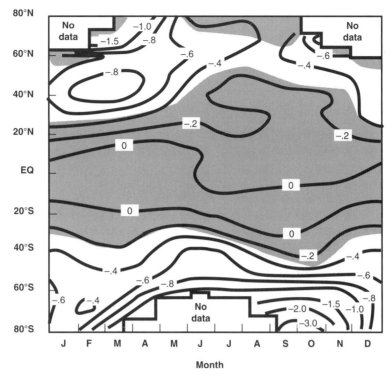

Figure 13.28   The global distribution of ozone reductions measured by the TOMS satellite system over an 11-year period. The results are shown as contours defining the rate of change in total ozone overhead, in percent per year. The values are plotted as a function of latitude, and the time of year is indicated by month along the bottom of the graph. The rate at each latitude represents a geographic average in longitude around the earth (according to Figure 13.4, the longitudinal variation is expected to be small). The rates are also averaged over the first 11 years of the TOMS data. Since the graph displays average *annual* reductions in total ozone, the ozone change (in percent) over the entire period, 1979 to 1990, can be estimated by multiplying these numbers by 10. (Data from R. Stolarski, P. Bloomfield, R. McPeters, and J. Herman, "Total Ozone Trends Deduced from Nimbus 7 TOMS Data," *Geophysical Research Letters* 18 [1991]: 1015)

This series of events that causes the ozone hole to form is amazingly complex. It was fortunate that more than a decade of scientific research preceded the first appearance of the hole. With some entrenched knowledge and subsequent ingenuity, the perplexing riddle of the ozone hole has been solved rather quickly—before global damage appeared. So far, the luck of *Homo sapiens* has held.

13.7.5   A GLOBAL OZONE DISASTER?

In contrast with that of the Southern Hemisphere, the meteorology of the Northern Hemisphere is not as conducive to the formation of an ozone hole. Scientific missions to the Arctic winter stratosphere have found chemical perturbations similar to those detected over Antarctica, and local ozone reductions of up to 15 percent, but no large-scale ozone hole as

in the south. The formation of an ozone hole requires that certain chemical conditions be satisfied. For example, the persistence of polar stratospheric clouds over a wide area seems to be a necessary condition. In the Arctic, PSCs are not nearly so common as they are in the Antarctic. The unstable vortex in the north does not cool as deeply. But if the Arctic meteorology were to change, favoring more frequent PSC formation, ozone holes might begin to appear in the Northern Hemisphere as well.

Figure 13.28 shows the distribution of global ozone reductions between 1978 and 1990. In the tropical latitudes and for all seasons, a change in total ozone has been essentially undetectable. By contrast, substantial average rates of ozone decline have been observed at higher latitudes in both hemispheres. In the Southern Hemisphere, large ozone depletions associated with the ozone hole are obvious during September and October at latitudes poleward of about 60 degrees. What is surprising are the large ozone depletions recorded in the northern springtime at high latitudes, and at lower latitudes throughout the winter half-year in both hemispheres. Even in the midlatitudes of the Northern Hemisphere (30°–60°N), decadal reductions in total ozone of 6 to 8 percent are seen in the winter. These reductions are larger than would be expected from some model predictions (for example, Figure 13.20).

The causes of the accelerated ozone decreases at high latitudes have not been determined. It is likely that the ubiquitous stratospheric aerosols mentioned earlier are responsible. These aerosols are much less massive than polar stratospheric clouds. They also are fairly unreactive under normal stratospheric conditions. Recent experiments, however, indicate that these aerosols can become highly reactive when cooled below about 200 K. The sulfate aerosols, moreover, are always present at all latitudes, unlike PSCs. Thus, like PSCs, stratospheric sulfate particles may be able to activate chlorine at a fast enough rate to reduce ozone concentrations substantially. The scientific case for aerosol activation has not been proved. But if it turns out to be correct, then major volcanic eruptions in the future—which increase the stratosphere sulfate aerosol burden 100-fold—may pose a temporary hazard for ozone over large regions, particularly at high latitudes.

Pictured in Figure 13.29 is the global variation in the thickness of the ozone layer. Important perturbations in total ozone are also identified. To the extent that the physics and chemistry underlying ozone depletion are currently understood, it seems highly unlikely that a "global ozone hole" will form. Polar stratospheric clouds cannot form over the entire planet. Furthermore, the amount of chlorine in the atmosphere should soon begin to decline under the Montreal Protocol. These facts are hopeful.

Figure 13.29   The global distribution of total ozone and several important anomalies in the ozone layer's thickness. Particularly noteworthy is the "ozone hole" over the South Pole, which occurs in September and October (austral spring) and is characterized by 50 percent reductions in total ozone. An "ozone dent" has also appeared over the northern high and middle latitudes in February and March and involves ozone reductions in the range of 5 to 10 percent.

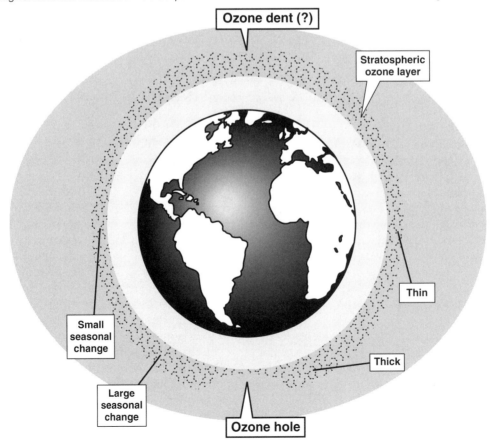

A more worrisome possibility is that the gases and particles injected into the stratosphere by a future fleet of high-altitude aircraft might further activate chlorine over the Northern Hemisphere. The upper atmosphere also has begun to cool under the influence of accumulating carbon dioxide from fossil fuels (Section 12.2.1). The combined effects of aircraft emissions and stratospheric cooling may produce conditions more favorable to PSCs, and the result might be large and unexpected ozone depletions over heavily populated areas. The suggestion of such threats to the ozone layer brings an atmospheric scientist's blood to a boil.

13.8   Solutions and Actions

Critics of the CFC-ozone depletion theory were quieted by the discovery that CFCs cause the ozone hole. Chlorine released by chlorofluorocarbons is the culprit, unequivocally. In the stark light of the scientific evidence, even CFC manufacturers—who for years argued against the theory and resisted the development of alternative chlorofluorocarbons—have joined in finding a new course. International political awareness of and concern about the ozone problem have expanded as well. This concern is expressed in the Montreal Protocol. All life on the planet and all nations are endangered by ozone depletion. The actions that can be taken to help solve the ozone problem are obvious:

1. Phase out or severely limit the production of CFCs and other chlorocarbons that have long lifetimes in the atmosphere. In particular, eliminate CFC-11, CFC-12, carbon tetrachloride (CCl_4), and a few other less commonly used substances (Table 13.1).
2. Design alternative compounds to fill the important needs for refrigeration, air conditioning, solvents, and so on. Substitutes for CFC-11 and CFC-12 are already being manufactured and marketed. One of these compounds, **hydrochlorofluorocarbon 22 (HCFC-22)** is considered as a temporary replacement for CFC-11 and CFC-12. The hydrochlorofluorocarbons contain chlorine, but have short atmospheric lifetimes because the molecules contain at least one hydrogen atom that can react with OH in the troposphere. Plans are under way to produce an even safer compound, **hydrofluorocarbon 134a**

(**HFC-134a**), which contains *no* chlorine. The HFCs also contain hydrogen and therefore react in the troposphere. This prevents their accumulation in the atmosphere and forestalls any significant contribution to greenhouse warming (Section 12.3.3). Unfortunately, some forms of HCFCs and HFCs cause health problems and so cannot be used. A few researchers have suggested that the products of decomposition of HCFCs and HFCs may themselves be long lived and contribute to the depletion of stratospheric ozone. Although these contentions remain unproved, they do deserve attention.

3. Phase out the use of bromine-containing compounds, as well as chlorine, that endanger the stratospheric ozone layer. The most obvious candidates for a phase-out are the long-lived halons and the excessively used pesticide methyl bromide. As chlorine compounds are removed from the atmosphere, bromine compounds could assume an increasingly important role in global ozone depletion (although their effect is moderated by reduced chlorine levels).
4. Monitor carefully the concentrations of potentially dangerous compounds in the atmosphere, and the ozone layer itself, over a sufficient period of time to establish a baseline against which future changes can be accurately determined. This important activity would allow future policymakers to take early action based on sound scientific evidence before the ozone layer is significantly damaged.

13.8.1   The Montreal Protocol

The Montreal Protocol on Substances That Deplete the Ozone Layer, first proposed in 1987, provides a framework for phasing out CFCs and other halocarbons. The treaty has been accepted by all of the developed countries that manufacture CFCs in large amounts. By 1992, more than 50 countries had ratified the treaty. The original agreement has also undergone two major revisions, in 1990 in London, and in 1992 in Copenhagen. The amendments greatly accelerate the phase-out of CFCs and add HCFCs, halons, methyl chloroform, and carbon tetrachloride to the list of chemicals *non grata*. The latest timetable for reducing halocarbons in the atmosphere is summarized in Table 13.2. The production of CFCs, methyl chloroform, and carbon

Table 13.2 Terms of the Montreal Protocol[a]

Montreal Protocol of 1987	
Chemicals	Timetable
CFC-11, 12, 113, 114, 115	Reductions from 1986 production levels 20% by 1995 50% by 2000

London Amendments of 1990	
Chemicals	Timetable
CFC-11, 12, 113, 114, 115	Accelerate 100% reductions to 1996
CFC-13, 111, 112, 211, 212, 213, 214, 215, 216, 217	Reductions from 1989 production levels 20% by 1993 85% by 1997 100% by 2000
Halons 1211, 1301, 2402	Reductions from 1986 production levels Freeze at 1986 level in 1992 50% by 1995 100% by 2000
Carbon tetrachloride	Reductions from 1989 production levels 85% by 1995 100% by 2000
Methyl chloroform	Reductions from 1989 production levels 30% by 1995 70% by 2000 100% by 2005

Copenhagen Amendments of 1992	
Chemicals	Timetable
CFCs, carbon tetrachloride, methyl chloroform	Accelerate 100% phase-out to 1996
Halons	Accelerate 100% phase-out to 1994
HCFCs[b]	Initial cap in 1996 35% reduction by 2004 100% by 2030
Methyl bromide	Initial cap in 1995; further action to follow

[a] This schedule represents the global objectives for eliminating the production of most artificial chlorocarbons. Individual countries have established timetables that may accelerate the phase-out.

[b] The hydrochlorofluorocarbons are considered to be temporary, or bridging, compounds replacing CFCs until safer chemicals are developed.

Source: World Meteorological Organization, 1992.

tetrachloride will be completely stopped in the European Union countries, the United States, and several other nations by 1996. The HCFCs, which are accepted as a temporary substitute for CFCs until even safer chlorine-free hydrofluorocarbons (HFCs) are developed, will be phased out by 2030.

The importance of the accelerated reduction in CFC production is underscored by the predictions in Figure 13.19. According to the conditions of the original Montreal Protocol, a 50 percent reduction in CFC emissions by 1998 would actually allow the stratospheric chlorine content to more than *double* over the next century! The consequences for global ozone, and particularly for ozone at the northernmost latitudes, would remain perilous at best. Moreover, reductions by up to 85 percent would merely maintain the *status quo* for chlorine for the next 100 years (Figure 13.19). It has become clear that reductions in CFC emissions by 95 percent or more are necessary to allow the stratosphere to return to a more or less natural state by 2100. In addition, such an action would keep the chlorine level in the stratosphere below the critical level of one to two parts per billion by volume. This amount currently seems to be causing a substantial loss of ozone in the Northern Hemisphere.

Some Third World countries, looking toward the day when their societies will demand universal refrigeration, air conditioning, and access to the luxuries of high technology, have been reluctant to embrace a treaty that forecloses many options. China, for example, with the largest potential for CFC consumption of any single nation, has asked the West for assistance in developing the infrastructure to manufacture and utilize ozone-safe compounds. These would displace CFC-11 and CFC-12, which can be produced cheaply using older technology. The Western nations have been reluctant to share their new technology, although they are moving to help the Chinese develop environmentally acceptable technologies.

The Montreal Protocol is a landmark treaty. It demonstrates that in the face of a plausible threat to the global environment, nations can cooperate to mitigate the threat. Before this agreement, hopes for broad international cooperation to solve environmental problems were in doubt. It will be crucial to monitor the evolution and, more important, the implementation of the Montreal Protocol. The world faces other serious environmental threats—global warming comes to mind—that will eventually have

to be resolved through a political process similar to, but far more complex than, that leading to the global ozone treaty.

13.8.2 SAVING THE EARTH'S OZONE LAYER

Human ingenuity created the ozone depletion problem, and human ingenuity can solve it. Clever chemists invented chlorofluorocarbons as the ideal substitutes for toxic gases that posed a hazard in our everyday environment. The CFCs were perfect until the side effects were discovered (Section 14.2.3). Nowadays, concern about losing the ozone shield is so pervasive that almost everyone has gotten into the act of finding cures. But sometimes the cure is worse than the disease. Schemes to fix the damaged ozone layer using exotic technologies are, for the most part, misinformed—and missing the point (Section 14.3.3). The best way to prevent the depletion of ozone by CFCs is to *eliminate the production and use of CFCs*. The Montreal Protocol provides a framework for achieving that goal.

In the meantime, ingenuity has surfaced. It is fascinating to recount the story about the aerospace manager who discovered that CFC-based solvents could be replaced with *lemon juice*. He had been puttering around the kitchen after work searching for an effective solvent to clean microelectronic circuit boards. One evening, he tried a lemon. The citric acid in the juice squeezed from the fruit proved to be as effective as some CFCs. His company is using the technique as a cheaper, safer alternative to CFC solvents used in the defense industry. An important goal of society is to reduce exposure to ultraviolet radiation and thus the occurrence of skin cancer during the period until the ozone layer is fully healed, some time by the mid-twenty-first century. At the University of Arizona, a team of researchers recently found that a hormone that allows frogs and lizards to alter their skin color can also induce deep tanning in white males. The researchers produced a synthetic form of the hormone, alpha-melanotropin, and injected it into a group of volunteers over a 2-week period. One week after this treatment had ended, the subjects had indeed become tan, a deep uniform tan that lasted for 3 to 4 weeks. Some of the men were accused by acquaintances unaware of the experiment of taking long and undeserved vacations. Perhaps they were on a mission to save the ozone layer.

Everyone, every sensible person, should be concerned about the possible loss of the ozone layer, but we should not panic. We must be informed and knowledgeable. Most of us are burdened with a variety of everyday concerns. We cannot dedicate our lives to protecting a tenuous layer of gas 10 miles over our heads. What can we do? We can use products that pose no threat to the ozone layer, express our concerns to political representatives, teach our children what we know about the environment, and be willing to make modest sacrifices to sustain the protective ozone shield.

Questions

1. Why is ozone considered a dangerous pollutant, even though it shields humans from harmful ultraviolet sunlight?
2. Can you name any other compounds or materials that may be considered helpful to humankind, but that can also be harmful when misused or released into the local (or global) environment?
3. Reconstruct the steps that lead to the formation of the ozone layer. What are the key photochemical processes that control the abundance of the ozone layer?
4. Why are the concentrations of chlorofluorocarbons (CFCs) continuing to increase in the atmosphere, even though the amounts produced each year have begun to decrease? Can you construct an answer to this question using the simple "box model" concept, when the system is not in a steady state, or equilibrium?
5. Why doesn't the ozone layer come down all the way to the Earth's surface? What physical and chemical factors might be at work?
6. If the concentration of ozone in the stratosphere were reduced by a large fraction, would the stratosphere become warmer or cooler? Why do you think so?
7. Reconstruct the physical and chemical scenario that leads to the formation of the Antarctic ozone hole. Begin the trail in early austral winter, when the polar vortex is first established. Include in your scenario the meteorological players and the supporting roles by polar stratospheric clouds and chlorofluorocarbons. Explain why the ozone hole appears in the spring, but disappears soon after.
8. Compare the amounts of ozone in smoggy air with the concentrations in natural stratospheric air, and then explain why one could not simply solve the ozone depletion problem by carrying smog compressed in bottles to the stratosphere and releasing it.

Problems

1. You can measure the amount of ozone in the atmosphere by measuring the intensity of sunlight at certain wavelengths in the visible spectrum. The reduction in the light intensity from that at the top of the atmosphere is proportional to the amount of ozone between the instrument and the sun. Suppose you are monitoring the intensity of sunlight at one wavelength when the intensity suddenly increases from its usual value of 0.950 to a value of 0.975. These values are relative to the fixed intensity at the top of the atmosphere, which you obtain from a satellite measurement and which does not change over time. What could you conclude about a possible change in the ozone layer? Would you be worried, and if so, why?
2. Which of the following hypothetical chlorofluorocarbons poses the more serious threat to the ozone layer, as far as you can tell from the data given? CFC-A: Release rate = 1×10^{26} molecules per second (globally); lifetime = 10 years; four chlorine atoms per molecule. CFC-B: Release rate = 1×10^{24} molecules per second (globally); lifetime = 1000 years; two chlorine atoms per molecule.
3. The air in your airplane cabin is drawn in from the outside. The plane is flying in the lower stratosphere where the ozone mixing fraction is 1 ppmv. The air is compressed by a factor of three in pressure when it is drawn into the cabin. It is also heated from 200 K to 300 K. The heating destroys 50 percent of the ozone in the stratospheric air. Are you likely to get sick? What is the mixing fraction of ozone in the cabin?

Suggested Readings

Crutzen, P. "The Influence of Nitrogen Oxides on the Atmospheric Ozone Content." *Quarterly Journal of the Royal Meteorological Society* **96** (1970): 320.

Dobson, G. *Exploring the Atmosphere*. Oxford: Clarendon Press, 1963.

Dotto, L. and H. Schiff. *The Ozone War*. Garden City, N.Y.: Doubleday, 1978.

Farman, J., B. Gardiner, and J. Shanklin. "Large Losses of Total Ozone in Antarctica Reveal Seasonal ClO_x/NO_x Interaction." *Nature* **315** (1985): 207.

Fishman, J. and R. Kalish. *Global Alert: The Ozone Pollution Crisis*. New York: Plenum, 1990.

Gribbin, J. *The Hole in the Sky: Man's Threat to the Ozone Layer*. New York: Bantam Books, 1988.

Grobecker, A., ed. *The Natural Stratosphere of 1974*. Climatic Impact Assessment Program Monograph 1, DOT-TST-75-51. Washington, D.C.: Department of Transportation, 1975.

Johnston, H. "Atmospheric Ozone." *Annual Reviews of Physical Chemistry* **43** (1992): 1.

———. "Reduction of Stratospheric Ozone by Nitrogen Oxide Catalysts from Supersonic Transport Exhaust." *Science* **173** (1971): 517.

Molina, M. and F. Rowland. "Stratospheric Sink for Chlorofluoromethanes: Chlorine Atom-catalyzed Destruction of Ozone." *Nature* **249** (1974): 810.

National Academy of Sciences. "Causes and Effects of Changes in Stratospheric Ozone: Update 1983." Washington D.C.: National Academy Press, 1984.

———. "Causes and Effects of Stratospheric Ozone Reduction: An Update." Washington D.C.: National Academy Press, 1982.

———. "Environmental Impact of Stratospheric Flight: Biological and Climatic Effects of Aircraft Emissions in the Stratosphere." Washington, D.C.: National Academy Press, 1975.

———. "Halocarbons: Effects on Stratospheric Ozone." Washington D.C.: National Academy Press, 1976.

Rowland, F. "Chlorofluorocarbons and the Depletion of Stratospheric Ozone." *American Scientist* **77** (1989): 36.

Scientific Assessment of Stratospheric Ozone, 1989: Volume 1. Global Ozone Research and Monitoring Project, Report 20. Geneva: World Meteorological Organization, Switzerland, 1990.

Solomon, S. "The Mystery of the Antarctic Ozone 'Hole'." *Reviews of Geophysics* **26** (1988): 131.

Stolarski, R. "The Antarctic Ozone Hole." *Scientific American* **258** (1988): 30.

Stolarski, R. and R. Cicerone. "Stratospheric Chlorine: A Possible Sink for Ozone." *Canadian Journal of Chemistry* **52** (1974): 1610.

Stratospheric Ozone [special issue]. *Ambio, Journal of the Swedish Academy of Sciences*, October 1990.

Toon, O. and R. Turco. "Polar Stratospheric Clouds and Ozone Depletion." *Scientific American* **264** (1991): 68.

Wofsy, S., M. McElroy, and Y. Yung. "The Chemistry of Atmospheric Bromine." *Geophysical Research Letters* **2** (1975): 215.

14

Global Environmental Engineering

The daunting environmental problems—local, regional, and global—discussed in the previous chapters must be solved, or at least controlled, if human civilization is to advance and prosper—if people everywhere are to achieve an acceptable standard of living and comfort. Most of the identified problems are associated with the widespread application of technology, particularly for the production of energy. Such technologies are deeply ingrained in economies and ways of life. Constituencies may seek to regulate the most offensive activities, but often these regulations are circumvented. Over the long haul, alternative sources of energy will need to be found. But what can be done in the meantime to preserve a decent quality of life? This chapter considers the emerging issue of **global environmental engineering** (**GEE**), which seeks technological cures to solve intractable environmental problems or to preserve as the status quo a degraded state of a declining environment.

GEE might be looked on as the next logical step in the coevolution of human intelligence and technology (Section 4.4). This coevolution has created a profound codependence between society and technology. In seeking solutions, it is difficult to evolve in reverse, to recede to an earlier state. The answer always seems to lie ahead in new technology. That, in turn, leads to deeper dependence. Is technology like heroin? Or Valium? Are we headed for a painful siege of withdrawal or a stuporous afternoon at the mall? Should we be so optimistic, complacent, or shortsighted as to presume that a livable environment can be maintained in the face of increasing pollution through increasing doses of technology?

and the quality of life have fallen backward only during episodes of global warfare. Achievements in science and technology have surged. Everyday conveniences abound, and sophisticated helpmate devices, like refrigerators and washing machines, are now taken for granted. One problem that cannot be ignored, however, is pollution of the environment as a by-product of population growth and technology. Garbage littering roadways and waterways is too visible to overlook; smog blanketing cities is too thick to see through. Subtle changes in the ozone layer and in the climate promise an uncertain future.

As a newspaper headline declared recently, "Tinkering with the environment is tempting." It is often seen as much easier to compensate for harmful behavior than to modify or stop the behavior. Smoking is a bad habit. But rather than suffer the discomforts of nicotine withdrawal, many people would rather puff on "low-tar" cigarettes and use a breath freshener. The long-term damage is rationalized in terms of short-term pleasure or profit. If an antidote for the effects of chlorofluorocarbons (CFCs) on stratospheric ozone could be found, it would be much easier to continue manufacturing refrigerators that use CFCs than to redesign refrigerators to run on more complex and expensive compounds requiring new manufacturing techniques. The antidote itself might be expensive and cause tertiary environmental problems, but how much easier it would be to stay with the old way of life. If the ozone layer is depleted, new crops can be genetically engineered to survive the increased ultraviolet radiation. If aggressive pathogens emerge to ravage the crops, stronger pesticides can be developed. If those pesticides kill birds well, that may just be the cost of human survival.

14.1 What Is Global Environmental Engineering?

In this century, the wealth and health of the human species have steadily increased. Standards of living

14.1.1 LIVING THERMOSTATS: NATURAL COMPENSATION

Nature has evolved complex systems that exhibit self-control. Many natural systems are internally

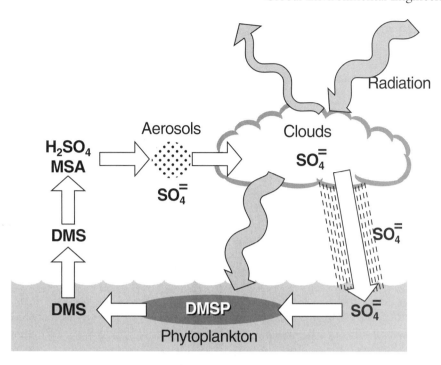

Figure 14.1   Example of a natural climate control mechanism. Dimethyl sulfide (DMS) released by phytoplankton in ocean surface water enters the atmosphere and is oxidized, forming sulfate aerosols that alter the reflective properties of marine stratus clouds and hence modulate the solar insolation at the ocean surface where the phytoplankton live. The chain of events leading back to phytoplankton is impressively complex and poorly understood. (Adapted from R. Charlson, J. Lovelock, M. Andreae, and S. Warren, "Oceanic Phytoplankton, Atmospheric Sulfur, Cloud Albedo, and Climate," *Nature* 326 [1987]: 655)

controlled by physical, chemical, and biological processes that limit the number of variations the system can accommodate. The climate system, for example, has a number of built-in feedback mechanisms, involving oceans and clouds, that help damp large climatic swings (Section 11.5). Groups of organisms coexisting in ecosystems are balanced by the availability of nutrients and by relationships between predator and prey.

An example of a naturally occurring mechanism that may influence the climate is illustrated in Figure 14.1. The mechanism involves the compound dimethyl sulfide (DMS), which is produced by phytoplankton in the oceans' surface waters. The sequence of events and their impact on the overall climate, triggered by the production of DMS, are quite complicated. DMS seeps from the ocean into the lower atmosphere—the marine boundary layer. That fact has been ascertained by measurements of DMS taken in air over regions where phytoplankton are active. The DMS is subsequently oxidized to form sulfates. This is known from laboratory studies and analyses of marine atmospheric chemistry. The sulfates form new aerosols, a process that has been

demonstrated by observations of particles over the oceans. These new aerosols affect the properties of marine stratus clouds that condense on the aerosols. This effect is less certain. Unusual behavior of marine clouds has been observed following the passage of ships: The smokestack emissions create long-lived "tracks" in the clouds. The appearance of ship tracks suggests that DMS emissions may have a similar effect on marine clouds.

The *climate* connection to dimethyl sulfide is still far away. The DMS-generated aerosols can modify the reflectivity, or albedo, of marine clouds (Section 11.6.5). In particular, the affected clouds can become more reflective. This modification has been noted in satellite observations of clouds over the oceans with ship tracks embedded in them. According to the discussions in Sections 11.6.4 and 11.6.5 (also see Section 14.3.2), it follows that an increased albedo tends to cool the climate. Thus a possible connection between the production of dimethyl sulfide by phytoplankton and a change in climate can be established.

There are two important questions that remain unanswered, however. Is the effect of DMS produced by plankton large enough to be important on a global, or even a regional, scale? And is the feedback loop closed; that is, does the climatic change caused by the DMS in turn affect the phytoplankton and their rate of production of DMS and hence the aerosols, and so on back to the climate?[1] In other words, is there a continuous cycle of cause-and-effect that may either amplify or diminish the climatic signal? Scientists simply do not yet know the answer to this crucial question, although it is likely that the DMS-climate connection is very weak.

The DMS-cloud relationship, which represents a rather small part of the global climate system, demonstrates the extraordinary complexity of the natural world. Myriad physical, chemical, and biological factors must be understood before quantitative predictions are possible. When a new technology inadvertently throws one process out of kilter, entire systems can be disturbed. The means chosen to correct the problem should rely on knowledge of the entire system. But most frequently, that is not the case.

Alternative and Corrective Technologies

Technology has inarguably upset natural checks and balances in a number of important systems. Since technology created these problems, it is reasonable to consider whether technology can provide solutions. There are two approaches that seem worthwhile to pursue: alternative technologies and corrective technologies. **Alternative technologies** should offer nonpolluting substitutes for currently polluting activities. **Corrective technologies** should provide complementary means to fix problems associated with other essential activities. Alternative technologies replace undesirable products and activities with more desirable ones. Corrective technologies attempt to compensate for, or mask, the original problem.

---

1. Positive and negative "feedback" are important to determining the behavior of complex or coupled systems. Think of a *positive-feedback loop* as reinforcing or in "phase." Psychologists use positive feedback—praise or a reward—to reinforce desired behavior. A *negative-feedback loop* is usually more stable; it strongly limits the possible excursions that the system can take. Negative feedback is commonly used in electronic circuits to ensure stable output signals. In the climate system, positive feedback amplifies small perturbations, and negative feedback dampens perturbations, like shock absorbers on a car.

It's a Big World After All

It is relatively easy to dream up schemes for improving the environment or compensating for pollution. The scientific basis for such schemes must be verified, of course, and all possible side effects—both good and bad—must be identified. The world population requires reassurance. Even putting these issues aside, however, another crucial question must be considered in all concepts for altering the global environment: Is the scheme even practical in terms of engineering technology and total cost? The enormous scales of these problems are not often understood by the polluters or the proponents of solutions.

Think of the numbers. The sun continuously deposits roughly 100 million gigawatts of power (the same as 100 billion megawatts) on the Earth. A single large power plant generates something like 1 megawatt of power. Humans collectively produce about 10,000 gigawatts (10 million megawatts), or 0.01 percent of the solar input (which explains why the energy dissipated as heat by civilization is not contributing significantly to planetary warming). Roughly 0.1 percent of the absorbed solar energy is converted by plants to chemical energy stored in biomass. That energy is released when the biomass decomposes or is burned. To fill all of society's present energy needs, about 10 percent of the existing biomass potential-energy production would need to be harnessed. Alternatively, solar-energy collectors with a total area of at least 10,000 square kilometers would be needed in orbit. That area, although not much larger than a small state, would require unprecedented activities and expenses to construct in space.

The atmosphere weighs 5 quadrillion metric tons (or tonnes); one part per billion by mass of the atmosphere amounts to 5 million tonnes. The ocean weighs 300 times as much as the atmosphere and contains heat energy roughly equivalent to 500 years of total solar input. The lower atmosphere has a volume of more than 5 billion cubic kilometers, and the stratosphere is four times larger. The surface area of the oceans is more than 300 million square kilometers. The living organisms on our planet weigh almost 1 trillion tonnes, about 200 tonnes for every living person. The ozone layer weighs 4 billion tonnes and is continually being renewed (roughly once every month).

By comparison, a large truck can carry 10 tonnes; a jumbo jet, 100 tonnes; and a large ship, 1000 tonnes. A home takes up 100 square meters; a city, perhaps 100 square kilometers. It would take all the people currently on Earth 1 million years to breathe all the air in the atmosphere. Humans and their most impressive engineering projects and structures are puny in comparison with the constructs and scales of the natural world. Yet in a number of ways, humans are damaging the global landscape by undermining or destroying critical vulnerable links and components. Like microscopic parasites that invade and weaken the heart muscle, humans are infiltrating and compromising the life-sustaining tissues of the biosphere. Can vital functions be maintained indefinitely? Or will the Earth someday need artificial organs to survive?

14.1.2   PLANETARY ENGINEERING

As a human enterprise, global environmental engineering has much in common with another technological objective: the modification of other planets to make them habitable for humans. The goal of **planetary engineering** is to alter the surfaces and atmospheres of nearby objects in the solar system to mimic the environment of Earth. Future generations might even "terraform" planets in other star systems throughout the universe. To start out, however, only three objects in the solar system have the correct size and composition to construct a livable world (aside from the Earth, where the quality-of-life rating is slipping). These objects are Mars, Venus, and Titan, the largest moon of Saturn. The Earth's moon—the closest object to us and therefore the most accessible—is too small and its gravity is too weak, to retain an atmosphere. Moonites would be forced to live in space suits and domed towns. The other possible places to hang the human shingle are so remote and inhospitable that enormous investments and long-term commitments would be necessary to ensure successful terraforming projects.

Concepts for planetary engineering have arisen from the debris of global scale environmental damage on Earth. It is widely recognized that human activities are modifying the composition of the Earth's atmosphere and climate. If global-scale changes can be produced inadvertently here on Earth, why not purposefully on another world? Obviously, planetary environments *can* be altered significantly and possibly can be fine-tuned. But before embarking on

projects to make other worlds habitable, we should perhaps concentrate on preserving our only safe haven in the solar system.

In most of the planetary engineering projects that have been proposed, the same principles can be applied as in the case of global environmental engineering. In particular, the radiative balance of a world can be changed by modifying the solar insolation (with sun shades), planetary albedo (with aerosols), or atmospheric greenhouse effect (with carbon dioxide and other gases). When making the necessary modifications, the composition of the atmosphere must be maintained within certain bounds (Table 14.1). Such limits to the basic composition of the environment pertain to the preservation of life as it has evolved on the Earth. If distant worlds are to host humans and other species, those engineered environments must conform to standards established here on Earth.

Mars is a frozen world, with an average surface temperature (~220 kelvin) more than 50°C below the freezing point of water (0°C). Venus is a hothouse world whose surface temperature (~730 kelvin) is about 360°C above the boiling point of water (100°C). Titan is absolutely gelid, making Mars appear balmy by comparison, since Titan's surface temperature (~95 kelvin) lies roughly 180°C below the freezing point of water.

To change Titan into a productive and living world would certainly pose a grand challenge to the human intellect. The intensity of sunlight reaching Titan is only 1/100 of that at the Earth. With such a weak source of light, even photosynthesis would be problematic. Nevertheless, Titan is thought to be more amenable to planetary engineering because of the large masses of greenhouse-active gases condensed on its cold surface. By artificially heating the surface, these gases might be released and provide a strong positive feedback on the initial surface warming. The most efficient way to warm up Titan and evaporate its abundance of greenhouse gases could be to heat the surface directly using energy generated by nuclear fusion. The fusion furnaces would use hydrogen isotopes isolated from compounds frozen on the surface.

The intensity of sunlight on Mars is weak (Mars is much farther from the sun than Earth is [Table 11.1]), but more significantly, the Martian atmosphere is too thin to create a greenhouse warming. On Venus, the solar intensity is strong (actually about twice the intensity as at the Earth), but more

Table 14.1   Limits to Planetary Habitability

Parameter[a]	Limits for survival[b]	Comments
Temperature of the planetary surface (~15°C)	0 to 30°C	Most species cannot survive below freezing or above −30°C for prolonged periods of time, for various physiological reasons
Total atmospheric pressure (1 atmosphere)	> 0.01 atmosphere	For most plants, assuming an air like mixture of gases
	> 0.5 atmosphere	For humans, in air, based on response to high altitude
	< 5 atmosphere	For humans, owing to narcosis (suffocation) from exposure to nitrogen and other gases
Oxygen (O_2) concentration (~0.2 atmosphere)	> 0.001 atmosphere	For plants, to perform respiration
	> 0.13 atmosphere	For humans, to avoid hypoxia (lack of sufficient oxygen)
	< 0.30 atmosphere	For plants, to avoid excessive flammability
Nitrogen (N_2) concentration (~0.8 atmosphere)	> 0.001–0.01 atmosphere	For plants to ensure sufficient nitrogen fixation
	> 0.30 atmosphere	For humans, to produce adequate total atmospheric pressure
Carbon dioxide (CO_2) concentration (0.000365 atmosphere or 365 ppmv)	> 0.00015 atmosphere (150 ppmv)	For plants, minimum concentration for photosynthesis
	< 0.01 atmosphere (10,000 ppmv)	For humans, to avoid toxicity associated with long exposure

[a] The ambient values for Earth are shown in parentheses.

[b] The limiting values are rough figures corresponding to the existence regimes of common flora and fauna. <, "less than"; >, "greater than."

Source: Information from C. McKay, O. Toon, and J. Kasting, Nature 352 (1991): 489.

important, the Venusian atmosphere is dense with greenhouse gases. On Mars, the challenge would be creating a stable greenhouse atmosphere thick enough to hold in the dim sunlight that does arrive. On Venus, the trick would be to cool the surface and at the same time remove the greenhouse gases that might trigger a new runaway greenhouse effect (Section 11.4.4). For Mars, the solution might lie in the frozen soils and ice caps, which hold large amounts of carbon dioxide and water. For Venus, the answer might involve metallic minerals near the surface that can react with carbon dioxide to form carbonates. For each of these planetary-engineering schemes, the scale of operations would be immense. In the case of Venus's rocks reacting with CO_2, for example, the surface over the entire planet would need to be mined and processed to a depth of about 400 meters!

In the futuristic plans for planetary engineering, we should include the eventual likelihood of genetic engineering of new species. One could imagine revolutionary new microbes that could live in the concentrated sulfuric acid clouds of Venus, eating carbon dioxide and converting it to graphitic carbon for use in creating an "antigreenhouse" effect (Section 14.2.1). Plants might be designed that could thrive on low levels of sunlight. One futurist has gone so far as to suggest that rather than engineering planets to suit people, we should genetically engineer people to suit the available planets. Any volunteers for cosmic surgery?

14.2   Technological Traps

Since the Industrial Revolution, society has amassed a number of basic technologies. The specific applications range from transportation and communication to energy production and medicine. We enjoy technological wonders such as television and air travel. Humans can now be rebuilt part by part (up to a point). A turkey can be cooked in half an hour.

How many people actually understand these technologies? Only a handful of scientists and engineers are familiar with the inner workings of a television set or microwave oven or nuclear-power plant. How much do we need to know? Are we sure that these technologies, which we take for granted, are safe? The industries that develop and distribute these technologies reassure us of their safety. Even so, regulatory bodies and watchdog agencies have been established to keep an eye on things. Are the scientists and engineers themselves smart enough to recognize potentially hazardous technology? If the past is any measure of skill in this regard, the answer is "Not always." Countless collapsed bridges, crashed airplanes, and sunken ships attest to the limited human ability to forecast and forestall disasters associated with technology. Even the most specialized and expensive technologies are not immune from engineering flaws; the space shuttle and the Chernobyl power plant are examples (Sections 7.3.2 and 7.3.3). On the whole, society is relatively safe; at least on the surface, that appears to be the case. The real threats of technology arise from subtle traps not yet "sprung" that lie along the path of progress.

In the following sections, we look at several well-known technological traps that have already been sprung.

14.2.1   NUCLEAR WINTER

The threat of nuclear war has diminished in recent years with the breakup of the Soviet Union and the democratization of the Eastern bloc. New strategic-arms treaties have promised to reduce the superpowers' arsenals by a factor of two or more in the next decade. So everything is OK. Right?

Into the foreseeable future, thousands of nuclear warheads will remain in the hands of more than a dozen nations. The political stability of some of these nations is in doubt. The weapons caches are powerful enough to destroy modern civilization, city by city, 10 times over. In addition, the danger of a nuclear winter following the massive use of these weapons in warfare remains real, although much less likely since East-West rapprochement.

Nuclear winter is the name of a new phenomenon associated with the mass detonation of nuclear weapons. Nuclear winter is the deep, short-term cooling of the Earth's climate. Dark smoke generated by massive conflagrations ignited by nuclear blasts would prevent sunlight from penetrating to the ground, leading to a rapid drop in land temperatures. Agricultural crops, which are sensitive to meteorological fluctuations, would be devastated by unprecedented weather anomalies. Crop destruction would be further aggravated by the loss of transportation and other infrastructure support. In addition, concurrent environmental stresses, very likely including large ultraviolet radiation doses beneath a depleted ozone layer, would compromise plant growth. Those people that survived the initial nuclear exchange would be faced with a lack of food and water and health services, even as they were enervated by widespread radioactive fallout and a variety of other serious environmental stresses. The world after a nuclear war would probably be dominated by mass starvation and epidemics.

How could the world leadership have allowed us to get into this mess? Why has the future of humanity been placed in jeopardy? In the rapid advance of scientific frontiers during the first half of the twentieth century, physicists could hardly avoid stumbling across the secrets of the atomic nucleus and the methods for releasing the enormous energy stored there (Section 7.3.1). The design of nuclear explosives is actually quite straightforward, even if the devices are expensive to build. During World War II, a team of scientists were brought together at Los Alamos, New Mexico, to design and build the first atomic bombs.[2] Soon after the first successful

---

2. Many of the greatest physicists and chemists of this century participated in the effort to develop the atomic bomb, including Leo Szilard, Enrico Fermi, Hans Bethe, George Kistiakowsky, and

detonation of a nuclear device in 1945, two of these new and devastating weapons were dropped on the Japanese cities of Hiroshima and Nagasaki. The human species lost its innocence. Mass destruction, once restricted to natural events like earthquakes and floods, could now be manufactured and delivered in small packages.

Following World War II, "national security" and "missile gaps" were invoked, on both sides of the Atlantic Ocean to justify the senseless expansion of nuclear arsenals. Politicians and bureaucracies could hardly resist wielding such a powerful stick. Businesses were not inclined to pass up opportunities to reap enormous profits at the expense of taxpayers. Minor dictators could see the value of such compact weapons of mass destruction for threatening neighbors and the world at large. The politics of nuclear weapons thus motivated a potentially self-destructive balance of terror. After 40 years and trillions of dollars spent, the world is left with tens of thousands of useless and dangerous warheads, vast areas of radioactive contamination, and hundreds of thousands of scientists, engineers, and bureaucrats trained exclusively to build, maintain, and justify nuclear weapons.

The Aftermath of Nuclear War

In a nuclear burst, the fireball created by the explosion emits radiation like a blackbody emitter (Section 3.2.1).[3] The effective temperature of the fireball is

---

Edward Teller. J. Robert Oppenheimer was the leader of the **Manhattan Project**, as it was called. Oppenheimer and most of the other physicists later opposed the push to expand the nuclear arsenal and design new weapons of mass destruction, particularly the "hydrogen bomb." Oppenheimer was later stripped of his security clearance because of false accusations questioning his loyalty to the United States. Albert Einstein did not participate in the Manhattan Project, but was instrumental in convincing President Franklin Roosevelt to start the project, emphasizing that Germany might be seeking to build nuclear weapons to use against the Allies.

3. Nuclear weapons generate their energy from nuclear reactions, unlike conventional weapons that derive energy from the chemical reactions of "high explosives" such as TNT. Fission weapons use the same nuclei-splitting reactions as nuclear-power plants do, only in a highly controlled manner. The so-called hydrogen bomb derives most of its energy from nuclear fusion reactions, similar to those that drive the sun. To initiate a fusion explosion, a fission detonation is used to create the necessary temperatures and pressures. The weapons that destroyed Hiroshima and Nagasaki would now be mere triggers for hydrogen bombs. The power of a nuclear weapon is measured in kilotons (KT) or megatons (MT). A 1-kiloton weapon has roughly the same explosive power as 1000 tons of TNT! A 1-megaton weapon is

---

about 7000 kelvin, similar to the radiative temperature of the sun. The fireball light is emitted for only a few seconds as an intense **thermal pulse**. Flammable materials—paper, cloth, vegetation, fuels, and so on—close to the explosion (within about 50 kilometers for a 1-megaton [MT] detonation) can be ignited by the thermal pulse. As a result, nuclear explosions can initiate massive fires. At Hiroshima and Nagasaki, fires engulfed the areas destroyed by the atomic bombs. In Hiroshima, an unusual **firestorm** arose from the rubble of the city and consumed all combustible matter over an area of 10 square kilometers. The fierce storm generated swirling winds and temperatures high enough to fuse metal and glass.

When oil, plastics, asphalt, and many other common materials are burned, the smoke produced is exceptionally black and sooty. The individual particles of soot are typically less than 1 micrometer in size (too small to be seen by the naked eye). By comparison, fires in vegetation produce a much lighter-colored smoke, usually brownish or even white. The sooty smoke, which is always associated with city fires, has a much greater impact on sunlight than does the lighter vegetation smoke. A cloud of sooty smoke appears black because it is absorbing the impinging radiation (Section 3.2.2). However, it has also been observed that this smoke has a much smaller affect on radiation at longer wavelengths, in the thermal infrared spectral region. These unique properties of soot—strong solar absorption and weak thermal absorption—are some of the factors responsible for the nuclear-winter effect.

Following a large-scale nuclear exchange, a dense layer of smoke might accumulate in the atmosphere. It has been estimated that as much as 10 million to 100 million metric tons of sooty smoke might be generated from urban fires. Even if spread over an entire hemisphere, this would be enough soot to dim the sun at noon everywhere. Wildfires often blot out the sun hundreds of miles downwind. During the Persian Gulf War of 1991, soot from oil-well fires created nighttime during the day in Kuwait. Conflagrations ignited in hundreds of cities in a nuclear war would have a quite different character and on a greater scale. The sun might be blotted out over entire continents for weeks. Huge soot clouds would encircle the globe.

---

equivalent to 1 million tons of TNT, about the total amount of all the explosives used in World War II. Some individual nuclear weapons are 10 MT, and the Soviet Union tested a 50-MT weapon.

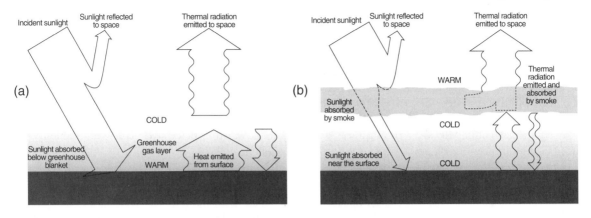

Figure 14.2   Comparison of the (a) natural "greenhouse" warming effect caused by gases in the lower atmosphere and the (b) "antigreenhouse" cooling effect produced by absorbing smoke particles in the upper atmosphere. In the normal greenhouse effect, gases and clouds trap thermal radiation created by solar heating near the surface and enhance the warming of the ground. In the antigreenhouse effect, the smoke layer blocks sunlight, reducing the heating of the surface while still allowing thermal radiation from the lower atmosphere to escape to space, cooling the surface. In the antigreenhouse effect, a large-scale temperature inversion is formed in the upper atmosphere because sunlight absorbed by smoke strongly heats the air even while the surface is cooling sharply.

In 1816, Lord Byron, wrote the poem "Darkness." It was the same year Mary Wollstonecraft Shelley wrote her novel, *Frankenstein*. It was a glum and depressing year following the eruption of the Tambora volcano (Section 11.6.4). Byron's poem is a premonition of global disaster occasioned by the dark clouds of nuclear winter:

> The bright sun was extinguish'd ... and the icy earth
> Swung, blind and blackening in the moonless air;
> Morn came and went—and came, and brought no day,
> And men forgot their passions in the dread
> Of this their desolation; and all hearts
> Were chilled into a selfish prayer for light. No love was left;
> All earth was but one thought—and that was death,
> Immediate and inglorious; and the pang
> Of famine fed upon all entrails.

"Darkness"

The smoke clouds generated by nuclear fires would disrupt the energy balance of the planet. The amount of sunlight reaching the surface would be minimized. As a result, surface temperatures could drop significantly. Figure 14.2 illustrates the radiation balance for the normal atmospheric greenhouse effect and the modifications that a dense layer of soot would cause. Because of the smoke, sunlight that normally penetrates to the surface would be absorbed and reflected by the smoke particles. But the longwave infrared energy emitted by the surface and lower atmosphere could still escape because smoke particles are not as effective at absorbing this radiation. This combination of sunlight depletion and thermal leakage would create an unusual *antigreenhouse effect*. The surface and lower atmosphere would be strongly cooled, or refrigerated. Meanwhile, the smoke layer itself would be sharply heated by the absorbed sunlight. This unprecedented pattern of continental-scale heating of the upper troposphere and cooling of the lower troposphere and surface would create a stable temperature structure, or inversion (Section 5.3). Under these circumstances, vertical mixing and turbulence would be suppressed, thereby isolating the surface and preventing it from being warmed effectively by heat transfer from warmer air layers above. Deep convection and precipitation would be inhibited. Accordingly, the nuclear winter-induced temperature inversion would limit the removal of soot from the warmer upper layers by mixing and washout, thus extending the residence time of the soot in the atmosphere.

The presence of a soot layer in the upper atmosphere and the heating of that soot by the sun represent a positive-feedback system. Heating of the soot would cause the atmosphere to stabilize, lengthening the residence time of the soot and thus allowing the heating to continue for a longer time. In addition, the heating of the soot would actually cause parcels of the soot to rise, much like hot-air balloons. The soot would thus rise higher and last longer. This additional positive feedback is referred to as a *self-lofting effect*.

Acute Global Climate Change

The nuclear-winter theory most likely will never be tested (if it were and if the theory were correct, the world would never be right again). However, like other environmental issues, nuclear winter can be simulated using a model. Today, there are available advanced global climate models running on advanced computers. Smoke emissions corresponding to a hypothetical nuclear exchange have been inserted into these model atmospheres, and the evolution of a nuclear winter has been predicted. The sources of soot in a nuclear war are defined by the available weapons and targets. These are concentrated in the United States, Western Europe, and the Soviet Union. The possible scenarios for a nuclear war have been argued endlessly and, of course, will never be settled to the satisfaction of all strategists. For the simulated nuclear winters, several scenarios have been fabricated (all on paper, with minimum damage to the participants).

The changes in surface temperatures predicted for a typical nuclear-winter scenario are given in Figure 14.3. The continental interiors beneath the nuclear-generated smoke clouds would become very cold.

The coastal regions and islands, however, would be relatively immune to the cold (if not the dark), as these areas are warmed by nearby ocean heat reservoirs. The normally mild climates of western coastal zones and islands are the result of ocean warmth carried by winds. Hence frigid weather would also be less likely in coastal regions during a nuclear winter. But most other places would be vulnerable to deep cooling.

If the soot were allowed to disperse in the model without absorbing sunlight (that is, as a passive atmospheric tracer), most of the soot particles would be removed by rainout within a few weeks. Regional and global climates would be only slightly affected; a nuclear winter would not grip the land. However, if the soot were allowed to absorb solar radiation (as it really does) and heat up the atmosphere, self-lofting of the heated smoke clouds would result. In the model, soot is rapidly transported from the lower troposphere into the upper troposphere and stratosphere. A widespread temperature inversion forms in the model, suppressing deep convection and precipitation and stabilizing the soot against its removal.

Substantial quantities of smoke can be carried on heated winds from the Northern Hemisphere into

Figure 14.3   Calculated surface cooling caused by the emissions of soot into the troposphere as a result of nuclear war. Patterns are shown for the temperature differences, or anomalies, relative to a "control" calculation without soot emissions. The temperatures are 7-day averages taken 20 days after the start of the "war." Most of the temperature anomalies are caused by strong cooling due to the antigreenhouse effect. The cooling is greatest over large continental land masses, where temperatures may drop by more than 25°C within several weeks. Over the oceans and in coastal regions, the cooling effect is greatly moderated by heat transfer from the oceans. These surface temperature perturbations exceed all known climatic anomalies since the last ice age, more than 10,000 years ago. (Calculations were made by G. Glatzmaier and R. Malone at Los Alamos National Laboratory, 1986)

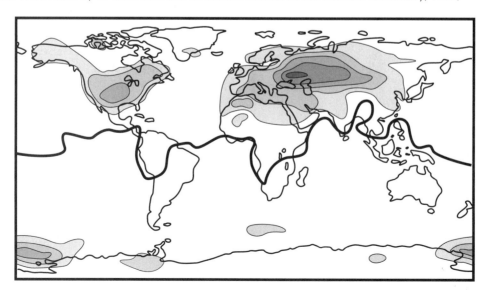

the Southern Hemisphere stratosphere in a matter of weeks, creating a global-scale climate problem. The putative climatic changes would include rapid land cooling by 10° to 20°C, particularly in the vast agricultural zones of the Northern Hemisphere. The anomalies would be so intense that entire crops would almost certainly be lost almost everywhere. It is likely that losses would also occur in subsequent years. The impact of the projected food shortages would be devastating. One comprehensive assessment predicted up to 3 billion human casualties of hunger and disease during the first year. The future of civilization beyond that point would seem grim, with little infrastructure remaining to support a long-term recovery.

This is global Armageddon rising up to consume a model Earth residing in the memory chip of a computer. This numerical Armageddon, although harmless, teaches valuable lessons about rational restraint in pursuit of peace. Humans, plagued by greed and madness, have contrived smaller versions of Armageddon that also hold lessons concerning the environmental aftermath of warfare.

Kuwait and Saddam's Revenge

Imagine more than 500 oil wells burning in an area the size of Los Angeles, with turbulent plumes of dark sooty smoke boiling into the atmosphere. Imagine a black sheet of dense smoke filling the sky, turning noon into midnight, nightfall all day long, day after day. Imagine the air thick with petroleum fumes and acrid smoke; imagine a soot fall of black oily particles settling everywhere, staining everything they touch. Imagine the ground dark and desolate under the suffocating pall of smoke, crops withered beneath sunless skies, discolored by soot. Imagine lakes and rivers of shimmering oil soaking the land and fouling the waters. Imagine an oily black rain—like that falling after the atomic bombing of Hiroshima—splattering the landscape and contaminating fodder and soil. This is not a description of hell. It is a picture of Kuwait and other areas of the Middle East in the wake of the Persian Gulf War of 1991. Near the end of that war, Iraqi leader Saddam Hussein unleashed one of the most violent purposeful assaults on the environment in human history.

The events that unfolded in Kuwait had never occurred anywhere else at any other time. Iraqi forces systematically sabotaged some 800 oil wells,

causing fires at about 530 wellheads. Sooty smoke from the oil fires darkened the skies over an area of about 75,000 square kilometers. Land temperatures cooled as much as 15°C below normal. Reports from as far away as Turkey and Afghanistan described greasy "black rain" falling over large areas. Added to this misery was the largest oil spill ever (perhaps 10 million barrels of crude oil sloshing in the northern Persian Gulf).

The Kuwaiti disaster was certainly a horrific demonstration of the misuse of technology against humanity and the environment. It could have been worse. As it turned out, the oil fountains at wellheads burned very efficiently, generating less than one-tenth of the soot that might have occurred under other circumstances. The oil was also contaminated with brine, which left the soot particles coated with salt and ready to be washed out by the first rainfall. If more soot had been generated and if it had been in its usual state of high resistance to washout, the black clouds could have spread much farther. The soot might have affected the Asian monsoons and might have led to climatic anomalies similar to those following large volcanic eruptions (Section 11.6.4). Compared with a putative nuclear winter, however, the Kuwaiti fires and soot clouds were small potatoes. The world can be thankful for that.

14.2.2 CARBON DIOXIDE

The anthropogenic gas that contributes most to global climate warming is carbon dioxide, which is generated by fossil-fuel combustion. It is perhaps ironic that the fuel that drove the engines of the Industrial Revolution also fueled the degradation of the global environment. The science of global warming is described in detail in Chapters 11 and 12, and so there is no need to cover this ground again. But we offer a few comments in reference to technological traps.

Exhaust from the "Engine" of Industrialization

Most of the conveniences enjoyed by modern society were derived from massive investments in energy production. Early civilization benefited from the discovery of coal as a cheap and efficient fuel for heating and cooking. Later, coal was used to produce steam to drive various mechanical devices and eventually electric generators. Free-flowing oil proved to be a boon to industry near the turn of the century. Oil refined into gasoline led to a boom in transportation. Coal and oil also proved to be a bane to the environment. Smog and spills have caused havoc with local environments. These regrettable side effects have been manageable in some cases; in other instances, smog and oil slicks were the price paid for power. In time, more and more "essential" uses were found for fossil-fuel energy and the products that can be made from these materials, including fabrics and plastic.

Civilization and its citizens have become completely dependent on fossil energy sources, like junkies on heroin. That would be fine if the sources of the drug were unlimited, and the side effects of using it were minor. Neither condition holds. Even though the supplies of oil and coal are vast, the accessible reservoirs will probably be depleted during the twenty-first century. Even before that, the recovery and refining of fossil-fuel reserves will grow much more expensive as the depth and quality of the fuels drop over time. Yet civilization is hooked on the stuff.

Early industrialists who profited from coal and oil use never questioned its value to society. Despite serious air and water pollution—deemed acceptable as a trade-off for modern products and conveniences—fossil-fuel exploitation raced ahead at full speed. There was no suspicion of global-scale effects. If scientists had stepped forward at the time warning of possible uncertain effects on the Earth's future climate, industry would have brushed them off as alarmists. Svante Arrhenius's early ideas concerning carbon dioxide and climate were not immediately connected with the need to control fossil fuels. No one really wanted to see a potential problem with such a large cash cow.

Should the producers and users of fossil fuels have been responsible for recognizing obvious threats to the climate? Would they have modified their activities in the face of enormous losses of profits, even if they had been convinced that a change in climate was likely? Past experience with industry and business suggests that it would be naive to assume even modestly beneficent acts on their part. Rather, the task of enforcing environmental standards falls on the shoulders of ordinary citizens and civil servants. Although the mess with air pollution and carbon dioxide is, to a degree, the result of individual self-interests, the information necessary to make conservative decisions to protect the long-term quality of life was never made available to the public. Time and again in history, critical information denied common awareness—intentionally by those who profited from public ignorance—has resulted in long-term environmental tragedy. In the case of fossil fuels, the future hazards may be monumental indeed, although the actual effects remain uncertain (Section 12.4.4).

The Benefits of Air Pollution

In the case of fossil-fuel consumption and the smog that accompanies it, a silver lining has been found—in fact, two silver linings. Smog, it turns out, can absorb ultraviolet radiation and it can cool the greenhouse warming. It is somewhat ironic that the ozone in smog may limit the ultraviolet radiation leaking through a damaged ozone shield. Over the past two decades, as stratospheric ozone has declined a few percentage points on average over the globe, ultraviolet radiation at the surface has not increased in response. In some cases, measurements of ultraviolet radiation in urban regions indicate lower intensities. The moderation of the UV light is related to the ozone and other absorbing components of smog. Paradoxically, the air that chokes us also shades us from irradiation.

Is it a reasonable compromise to suffer bad air quality in order to avoid harmful ultraviolet rays? Hardly. A quick reading of Chapters 6 and 7 should convince anyone even slightly concerned with his or her health that smog must be eliminated, or at least be minimized. The point is that smog needs to be reduced, *and* the ozone layer needs to be protected. There can be no compromise on either issue. The idea that these problems offset each other—no harm, no foul—is nonsense. Indeed, reductions in stratospheric ozone can intensify smog. The increased flux of ultraviolet radiation accelerates smog reactions, cooking up more ozone near the ground. The problems of smog and stratospheric ozone reduction are connected. Both problems must be corrected, not encouraged.

It has recently been discovered that the particulates generated by sulfur dioxide emissions and biomass burning reflect sunlight and lower the temperature of the Earth (the so-called albedo effect [Section 11.6.5]). Thus sulfate aerosols and vegetation smoke particles are effective scattering agents that reduce the amount of solar energy absorbed by the planet (Section 14.3.2). This compensating property of fossil-fuel combustion (in other words, the cooling effect of the particulates generated, which offsets the warming effect of the carbon dioxide emissions) represents a treacherous sleight of hand. The combustion-generated aerosols are present as long as the fuels are burned; when the burning stops, the aerosols disappear in a matter of weeks. Carbon dioxide, on the other hand, remains in the atmosphere for hundreds of years (Sections 10.2.4 and 12.2). The warming potential of the CO_2 is masked as it becomes more concentrated. The warning signals of climate change are suppressed. Action to correct the problem is delayed. Eventually, when the fuels run out, the cooling effect of the aerosols will disappear, and the warming effect of the carbon dioxide may appear full blown.

The intricate relationships among the physical, chemical, and biological effects of large-scale technology are not widely appreciated by most lay people, policymakers, or scientists. Scientists may overlook the connections because of their natural academic tendency toward narrow specialization. Most physical scientists, for example, are not familiar with biological principles, and vice versa. But technology sets traps between academic disciplines. There is no easy solution to this contrivance of intellect to focus on details rather than to view the "big picture." Some ideas about protecting the environment, society, and civilization from technological harassment are addressed in Section 14.4.

14.2.3 Chlorofluorocarbons

The environmental controversies involving chlorofluorocarbons (CFCs) are discussed at length in Chapter 13, in Section 12.3.3, and elsewhere in this book. The concentration of CFCs is a perfect paradigm for the pitfalls of new technologies, particularly those compounds intended for widespread use: Beware the wolf in sheep's clothing.

"Miracle" Compounds

Like so many "miracle" compounds, the chlorofluorocarbons had to be invented. Automobiles and refrigerators and television sets are not mined from the ground or harvested like fruit from trees. They must be manufactured from natural raw materials. Moreover, design and fabrication schemes must be worked out before any can be made. CFCs, it turns out, are much simpler than the refrigerators and air conditioners they are used in. They are by no means harmless, however.

Chlorofluorocarbons were invented in the early 1930s as a safe replacement for common refrigerants of that time, including toxic ammonia gas. The CFCs have superior properties as a coolant in air conditioners as well. They are non-toxic and can be breathed without harm. Hence CFCs can also be used as a propellant for underarm deodorants, hair sprays, and other compounds used in personal hygiene. Further, CFCs are so inert chemically that they can be used in a variety of industrial processes that require a nonreactive buffer gas. For example, plastic and rubber foams can be blown using CFCs. Finally, because they are relatively cheap and easy to make, the common CFCs have been widely adopted for every possible use.

Environmental Hangover

The environmental problems that eventually surfaced from the widespread use of the miracle CFC compounds are now legendary. The depletion of the ozone layer (Section 13.5) and greenhouse warming (Section 12.3) are the two main global environmental issues of this century. Like a drunk on the morning after a binge, we are still woozy from the effects of CFCs. We must give them up, yet we are not sure how we will live without them. Although it is clear that to continue using CFCs would lead to severe damage, we experience the discomfort of withdrawal in the form of roll-on deodorants and higher prices for cars. Some damage has already been done to the vital environmental "organs" of the Earth. Signs of cirrhosis of the ozone layer have appeared, although it has not yet failed as a vital filter of toxic rays. The planetary temperature has risen, but the world has not become feverish yet. Unfortunately, the symptoms will persist well into the twenty-first century. So far, just a mild hangover, not delirium tremens. Just a legacy of stress and cancer for the next five or more generations.

14.3 Technological Cures

Technology has caused many of today's most serious environmental problems, and the countryside is strewn with technological traps that may snap shut at any moment. Can technology provide solutions as well? It makes sense. If ozone depletion causes skin cancer, medical techniques can be developed to remove the cancerous lesions produced by excess ultraviolet radiation. If global warming causes the sea level to rise, levees can be built using modern engineering techniques. If nuclear weapons are proliferating around the world, a defensive shield can be constructed in space to ensure national security. Are these ideas feasible? What other technological patches might be worth pursuing? Is there a planetary prophylactic to protect against environmental degradation?

14.3.1 PREVENTING ARMAGEDDON

Ever since the first nuclear device was exploded, alarm bells have been ringing. The scientists who invented nuclear weapons immediately realized the jeopardy in which civilization had been placed. Most of the scientists began to lobby against the production and deployment of weapons of mass destruction. Later, after those pleas had been ignored and huge arsenals had been collected, they worked for disarmament and the abolition of nuclear weapons. A few of the inventors took another tack. Rather than dismantle the nuclear systems coveted by a superpower, they thought it might be possible to build other systems that would make such weapons "impotent and obsolete." Enter the Strategic Defense Initiative (SDI).

Star Wars

If there ever was a bankrupt technological concept, "Star Wars" is it. The idea is to place a shield in space to stop enemy missile attacks using satellite-based weapons. The technological problems of constructing a reliable system to operate for decades in a space environment and to perform flawlessly the first and only time it is ever used are now agreed to be insurmountable. Indeed, such a system could never be tested properly. More to the point, the proposed weapons technologies either would not work or would be vulnerable themselves. Despite early optimism and a few shady promises by proponents, lasers that could fire X-ray beams at enemy warheads have been shown to defy the laws of physics. A backup concept—chemical lasers the size of small ships drifting through space—is less than impractical; it is ludicrous. And "brilliant pebbles," small high-speed "guided bullets," are not much better than BB guns

against a concerted missile attack. Petulant technologists have shrugged off criticisms of the proposed high-tech devices, fibbed to presidents, and wasted enormous resources to pursue this phony concept.

By 1993, the total price tag for Star Wars was about $30 *billion*! The project is continuing, and the costs are accruing. For that money—roughly $100 for every U.S. citizen so far—there is not a single useful product to show, let alone a "shield." The money has been wasted in an orgy of spending on oversold and overvalued technology.

In the original Star Wars concept, the United States would be preserved intact in an all-out nuclear war with the Soviet Union, by destroying Soviet missiles and warheads in flight. That goal, embraced by a misinformed president as a moral alternative to mutually assured destruction (MAD), was, if not mad, at least a little loony in the face of 10,000 Soviet warheads. Sights were lowered to preserving enough of the U.S. economy so that the country could prevail over the Soviet Union in the aftermath of a nuclear war. Somalia would likely have more economic and military viability than the United States after an all-out nuclear attack. When the Soviet bloc finally crumbled in the late 1980s, the Star Wars objective was further reduced to the protection of U.S. cities from nuclear attack by Third World powers such as Libya and Iraq. Imagine long-range nuclear missiles launched by Libya at the United States! More logically, a small nuclear bomb would be smuggled into the country and detonated. Behind all the smoke screens and lame excuses, tens of billions of dollars have been wasted on Star Wars.

During the Persian Gulf War, the Iraqis launched a number of Scud missiles at various targets. On the defense, U.S. Patriot missiles scored several hits, although apparently many misses as well. Star Wars advocates took this spotty record as "proof" of the "defensive shield" concept. Woe to us. Tens of billions more dollars will likely be spent to fend off the Scuds that will never come.

The Star Wars fiasco is a prime example of the manipulation of facts, use of secrecy, and lobbying and special interests applied to subvert the best interests of society and, ultimately, the global environment. Money wasted on such technological nonsense enervates the economy (there are few useful "spin-offs" from weapons research) and skews the priorities of governments.

Meteor Defense

The advocates of Star Wars have a new mission: To prevent Earth from being destroyed by a large meteor. There is compelling evidence that a 10-kilometer-diameter meteor collided with the Earth 65 million years ago, wiping out the dinosaurs and most other species of that epoch (Section 4.3.2). The agent of extinction probably involved major climate disturbances triggered by the explosive impact, worldwide dust clouds, global fires, and searing acid rain (Section 11.7.2). Even a much smaller meteor impact could wreak havoc on human civilization. The frequency of meteor impacts (that is, the number occurring in any fixed time interval) increases as the meteor size decreases. Small meteor impacts hit the Earth more frequently than large ones do. An object (a comet or asteroid) the size of the meteor that killed the dinosaurs hits the Earth only once every 30 million years. However, an object the size of a football field may hit every 10,000 years.

The Tunguska meteor exploded over a remote area of Siberia on the morning of June 30, 1908. The detonation flattened 2000 square kilometers of dense forest, blowing full-grown trees over like matchsticks. The closest observer, 60 kilometers away, was blown off his feet. If the Tunguska meteor had fallen over New York City, the casualties would have numbered in the hundreds of thousands, and Manhattan would have been leveled. Such an event happens every few hundred years. Whenever a comet enters the solar system on its way around the sun, it might be deflected by Jupiter or Saturn onto a collision course with Earth. Although comets will pass within several hundred thousand miles of the Earth in the next century, it is highly unlikely that one will actually strike the surface. Fortunately, near misses do not count.

Nevertheless, Star Wars proponents have scared up their own bogeyman—meteors. They point out that the collision of a meteor with Earth is inevitable. Indeed it is. They mention that even a small hit on a city would be devastating. Indeed it would be. They note that a bigger impact could change the climate and create a "meteorite winter," leading to agricultural failure and worldwide famine. Shades of nuclear winter. What they fail to underscore, however, is that the probability of an event of any consequence happening during the next few centuries is vanishingly small.

Despite incredible improbability, the deployment of a space- or moon-based nuclear-tipped meteor-terminator is recommended. Carrying a warhead of up to *1 million* megatons of nuclear explosive power, the terminator would sit and wait for an invading asteroid or comet.[4] At the right moment, perhaps centuries in the future, this supermissile would be launched to pulverize the interloper. On Earth, we would hope and pray that the guidance system had remained sound. The tens or hundreds of billions of dollars would have been well spent.

A few years ago, a cold war was conjured up to justify obscene expenditures of public funds to build useless weapon systems that now must be dismantled at additional great cost. Today, a cold rock is cast as the enemy in another fuzzy scheme to spend tax revenues. The next thing you know, the civil defense advocates of the past will reappear urging everyone to build a personal meteor shelter.

14.3.2 COOLING DOWN THE GREENHOUSE

The threat of global changes in climate associated with greenhouse warming has fostered a cottage industry in technological cures. After all, a practical scheme could forestall the climatic chaos that may follow warming *and* likely turn a handsome profit for the inventor of the scheme. This potent wedding of philanthropy and profit has spawned a slew of climate-sensitized entrepreneurs and technologists. The basic physical, chemical, and biological principles that allow such schemes to blossom are described in other sections of this book. We discuss next some creative applications of these principles (also see the summary of some current ideas in Section 12.5.3).

Fortifying an Anemic Ocean

The oceans represent the largest reservoir of carbon dioxide that humans have access to in a relatively short time. (In Sections 10.2.4 and 12.2.3, the global cycle of carbon dioxide is described in detail.) The oceans naturally absorb excess carbon dioxide from the atmosphere, but this process takes hundreds of years to occur. Why not speed it up? Indeed, if "carbon burial" in the oceans could be accelerated,

---

4. The largest nuclear device ever detonated was equivalent in explosive power to about 60 million tons (60 megatons [MT]) of TNT. In principle, there is no limit to the size of an explosive device based on nuclear fusion. The trigger for such a device would be a series of nuclear fission explosions the size of the atomic bomb dropped on Hiroshima.

the need to curtail CO_2 emissions would disappear. Normally, living organisms in the oceans constitute a "carbon pump." They incorporate carbon into their bodies, and when they die, the carbon sinks to the bottom with them.

The carbon uptake in the oceans begins when carbon dioxide dissolves in the ocean water to form carbonate compounds (Section 10.2.4. Equations 10.22, 10.23, and 10.31). Microscopic plants (phytoplankton) in the oceans absorb the dissolved CO_2, much as terrestrial plants absorb CO_2 from the air. Through photosynthesis, this carbon dioxide is converted to organic matter (Section 4.2.3). The phytoplankton are then eaten by zooplankton, which are eaten by larger organisms, eventually leading to food for whales. The marine food chain is anchored by phytoplankton. When the larger organisms die or defecate, organic "detritus" is generated. The detritus can be eaten by bacteria and recycled as carbon dioxide, much as organic debris is recycled on land. Otherwise, the carbon-rich material settles into the deep oceans, from which it will not return for centuries.

The biological productivity of the surface oceans is generally limited by the availability of trace nutrients, especially fixed nitrogen, phosphate, and iron. There is plenty of carbon dioxide, sunlight, and water, of course, to carry out photosynthesis. Plant growth in particular is restricted by deficits in nutrients (the limiting factor for zooplankton and other aerobic feeders is oxygen, whereas bacteria usually consume everything in sight under most conditions). Nutrients are absorbed in fixed ratios compared with carbon; these fixed proportions are called the **Redfield ratios**. The Redfield ratio of fixed-nitrogen (N) to carbon (C) atoms, for example, is about $N/C = 1/7$. This is roughly the elemental ratio of nitrogen to carbon found in living organisms and the amino acids from which they are built.

Different areas of the world oceans exhibit deficits in different key nutrients. In the southern Pacific Ocean near Antarctica (the Southern Ocean), the waters are relatively poor in iron but rich in other nutrients. The Southern Ocean is, however, a region where cold water is sinking to form abyssal "bottom water." Carbon brought down there is carbon removed for a long time.

A clever marine scientist recognized the possibility of burying carbon in the Southern Ocean if it were fertilized with iron. The phytoplankton in this area are limited in productivity mainly by a lack of iron.

That is, the waters are anemic. Iron could be added in a soluble form using ships or aircraft to spread it over a sea area of several million square kilometers.[5] One soluble iron compound, ferric chloride, is cheap and plentiful. About 1 million tonnes of iron per year would be needed to remove 1 gigatonne (1 billion tonnes) of carbon. The Redfield ratio for iron (Fe) to carbon (C) is about $Fe/C = 1/1000$. Hence, the amount of carbon removed is just the amount of iron spread over the water divided by the Redfield ratio for iron. This estimate assumes that all the iron is eventually tied up in biomass that sinks into the deep ocean.

Ships laden with iron would ply the southern Pacific Ocean spraying iron and deep-sixing carbon dioxide. The idea is simple and elegant. It uses basic biological, physical, and chemical concepts. Forget the thousand ships laden with iron, vast ocean tracts unnaturally fertilized, billions of dollars spent. The solution, although large on an engineering scale, is approachable. Unfortunately, it would probably not work. Oceanographers carrying out detailed simulations of the carbon cycle under the conditions existing in the Southern Ocean have tentatively concluded that most of the carbon would not sink, but would be recycled by mixing before it could be "buried." The potential removal of carbon dioxide by this method might amount to one-tenth the total present-day source from fossil fuels. That is enough to make a dent, but only a dent.

One group of Japanese researchers has proposed another method for burying carbon dioxide in the oceans. The CO_2 would be compressed into a liquid and pumped to the ocean bottom. Being denser than seawater, the liquified carbon dioxide would stay put. Moreover, the overlying pressure of the water would keep the CO_2 from vaporizing and rising as effervescence. So far, no carbon dioxide has been sunk this way. A proposal similar in concept to those just mentioned uses wood as the carbon vehicle. The trees would be grown, the logs would be harvested, and the wood would be buried in an environment where decay was very slow. To keep up with the output of carbon dioxide, several million square

5. Iron cannot be dumped into the ocean in its pure metallic state. Nails would sink immediately. In order to be useful to living organisms, the iron must be transformed into a soluble species of ferric or ferrous iron. Iron exposed to air and water rusts, producing these soluble forms. Metal buried in soil eventually decomposes by rusting and is mineralized. Rusting takes time, however. The decomposition process can be accelerated with acids, and the iron crystallized for convenient application.

kilometers of forest would have to be under continuous cultivation and cutting for a century or more. It could be done, of course, but it would pose extreme new problems. Perhaps a better solution is to allow the carbon dioxide to accumulate and compensate for its effect on the climate.

Smoke and Mirrors: The Albedo Effect

The basic planetary energy balance, which determines the global climate over long time scales, is outlined in Chapter 11. One of the key factors controlling the energy balance is the albedo, or reflectivity of the Earth. The albedo, in turn, is controlled by a number of parameters, including the conditions of the land surface, the cloudiness of the sky, and the amount of smoke in the air. (Specific relationships between the albedo and climate are described in Sections 11.6.4 and 11.6.5; see also Section 3.2.2.) In particular, the albedo can be affected by changes in the particulate loading of the atmosphere. The more particulates that are present, the hazier the air and generally the more reflective the atmosphere will be. The albedo would be reduced only in the circumstance that highly absorbing particles, like soot, were present in large amounts.

There is a substantial body of evidence illustrating the effect of a change in albedo on climate, caused by volcanic eruptions. The eruption of Mount Tambora (Indonesia) in 1815 provides the most spectacular example of the potential climatic impact of volcanic eruptions in historical times (Section 11.6.4). Nonetheless, the Tambora event is not well documented because the event was quite remote and the geophysical data collected at the time were quite crude.

Relatively small volcanic eruptions, such as Mount St. Helens (which exploded in Washington State in 1980), do not cause global effects. In the case of St. Helens, a plume of ash spread over the western United States, but did not go much farther. On the other hand, larger recent eruptions, such as El Chichón (Mexico) in 1982 and Mount Pinatubo (Philippines) in 1991, have had a major global impact. Both Pinatubo and El Chichón emitted a large amount of sulfur dioxide, along with ash and other debris. Sulfur dioxide that is injected into the stratosphere is converted to sulfuric acid aerosols (Section 3.3.4), and these spread over the entire globe, causing spectacular purple twilights (Section

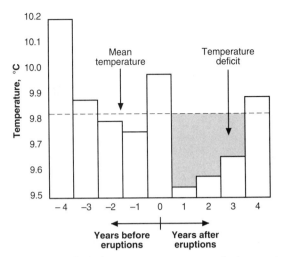

Figure 14.4 Global average temperatures before and after seven major historical volcanic eruptions (1815, 1822, 1831, 1835, 1846, 1902, and 1963). The number of years before (negative) or after (positive) the year an eruption occurred is indicated. The global temperature averaged over all seven eruptions is shown for each year in the relative chronology. There is, on average, an apparent temperature deficit of several tenths of a degree Celsius during a span of 1 to 3 years following such eruptions. (Data from S. Schneider and C. Mass, "Volcanic Dust, Sunspots, and Temperature Trends," *Science* 190 [1975]: 741)

3.2.3). In addition, less sunlight reaches the Earth's surface, and the climate cools slightly. Although the extent of cooling is uncertain, a global average temperature decrease of 0.5°C is expected the year following a major volcanic eruption. The volcanic aerosols disappear from the stratosphere over a period of several years, and the climatic anomaly fades just as quickly.

The effects of notable historical volcanic eruptions on the global temperature are illustrated in Figure 14.4. Variations in the magnitude of the effects caused by volcanic eruptions can be attributed to differences in the materials emitted by each volcano, the height of the volcanic injections, and the latitude of an eruption. If the eruption plume extends into the stratosphere, for example, the sulfuric acid aerosols that are formed can persist long enough to disperse over a large area of the globe. The stratosphere is dynamically stable, like a large temperature inversion (Section 2.3.3). Storm clouds do not penetrate from weather systems in the troposphere below. Rain and snow do not form there. Hence the removal rate of volcanic debris from the stratosphere is quite slow. Small particles have a

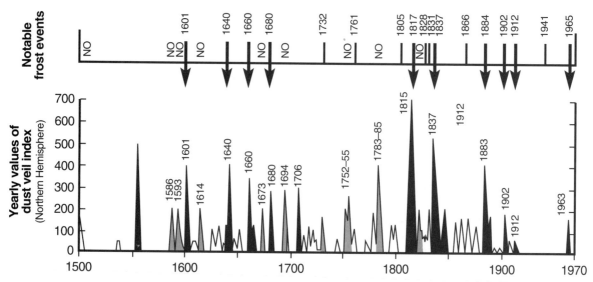

Figure 14.5   The record of frost damage in bristlecone pine trees in the southwestern United States, correlated with major volcanic eruptions. The upper panel identifies the years in which frost damage is evident in tree ring growth. Arrows indicate those years when a major volcanic "dust veil" also existed in the Northern Hemisphere, as suggested by the optical data summarized in the lower panel in the estimated "dust veil index" (DVI). The DVI roughly corresponds to the optical depth of the volcanic aerosol layer in the stratosphere. Years of unseasonable frost damage (upper panel) are indicated without arrows if no substantial dust veil was noted (lower panel). The symbol NO in the upper panel indicates a year in which a notable dust veil appeared, but not frost damage. Volcanic events corresponding to years with frost damage are shown as solid-filled peaks in the lower panel. Other major dust veil events are shown as gray-filled peaks. (Data from V. LaMarche and K. Hirschboeck, "Frost Rings in Trees as Records of Major Volcanic Eruptions," *Nature* 307 [1984]: 121)

residence time of one to several years. That is long enough to allow a solar-energy deficit to build up, but not long enough to cause a long-term climate anomaly.

The fact that volcanoes can change the climate is supported in a number of geologic, biological, and historical records. In one particularly important set of data, damage to tree rings indicates years with extreme weather and climate anomalies. Figure 14.5 gives the record of frost damage to the ancient bristlecone pines of the southwestern United States. The striking point of this study and of many other data is that climate can be manipulated relatively easily within certain small bounds. The short-term climatic variability associated with volcanic eruptions, the El Niño phenomenon in the equatorial Pacific Ocean (Section 12.4.2), and solar variability (Section 11.6.2) all demonstrate that tweaking the climate system is possible. The magnitude of the average year-to-year temperature changes for natural perturbations is small—1 degree or less—but the fact of a climatic response to specific forcing is clear.

Climate forcing associated with changes in solar-energy input, caused by variations in the sun's brightness or the reflectivity of the Earth, have similar climatic implications. Both kinds of phenomena can be studied using the energy balance box model of Section 11.3.2. Indeed, the effect of a change in the planetary albedo on average surface temperatures over a long time period can be estimated using the simple climate Equation 11.14. This relationship can be rewritten in a form suitable for calculating small temperature changes from the norm (the "climate-change" equation):

$$\frac{\Delta T_s}{T_s} = -\frac{1}{4}\frac{\Delta\alpha_e}{(1-\alpha_e)} \cong -\frac{\Delta\alpha_e}{3} \qquad (14.1)$$

Here, the normal surface temperature, T_s, decreases ($\Delta T_s < 0$) as the albedo, α_e, increases ($\Delta\alpha_e > 0$). Since the average surface temperature is close to 300 kelvin, Equation 14.1 can also be expressed approximately as

$$\Delta T_s \cong -100 \times \Delta\alpha_e \qquad (14.2)$$

Hence a change in the average planetary albedo of 0.01 (from the current albedo of about 0.33, that is, a 3 percent change in the albedo) can lead to a surface temperature change of about 1°C.

If the albedo increases, the surface will cool. If the albedo decreases, the surface will warm. In either case, the shift toward a new climatic state would take decades or longer to evolve, because the ocean heat reservoir would take a long time to equilibrate (Section 11.5). The greenhouse effect and any internal adjustments in the climate system could further modify this result. Nevertheless, Equations 14.1 and 14.2 are useful for making first-order estimates of global climate changes related to long-term variations in the albedo.

Several possible schemes for intentionally altering the albedo of the Earth are described next. In the particular cases we use, our goal is to cool the planet to compensate for an increase in the abundances of carbon dioxide and other greenhouse gases. Neither the details of the climatic response on regional scales nor the evolution of the response over time is considered in any depth. Instead, the objective is to create artificially a first-order compensating effect for the projected global warming of several degrees Celsius associated with greenhouse gases. The fact that much is being neglected should immediately raise a warning flag.

The Sulfate Shield

The ozone shield protects the Earth from harmful solar ultraviolet radiation. It happens that a "sulfate shield" also exists that may protect the Earth from climate warming. Unfortunately, it is an inefficient prophylactic. The sulfate shield actually consists of the aerosols in the lower atmosphere. In Section 14.1, one source of these aerosols—dimethyl sulfide (DMS)—was discussed in the context of a natural climate feedback system. Another component of the sulfate shield consists of the aerosols generated in polluted air. Our old nemesis, polluted haze (Section 6.5), may be acting as a climate "thermostat" to limit greenhouse warming. The sulfate particles originate as sulfur dioxide emitted mainly during the combustion of fossil fuels, which are the primary source of atmospheric sulfur, equaling or surpassing most natural sources (Sections 9.3.2 and 10.2.1). The sulfur emissions undergo chemical conversion to sulfates and end up on haze particles or in acid rain.

Figure 14.6 shows the geographical distribution of sulfate aerosol effects. The sulfate aerosol, which is dominated by the human consumption of fossil fuels, is concentrated in the northern midlatitudes. Unlike those in the stratosphere, aerosols in the troposphere have a relatively short residence time in

Figure 14.6   The distribution of sulfate aerosols in the lower atmosphere. The contours provide a relative estimate of the climatic forcing effect of the particles, the effect being greatest where the aerosols are denser. Most of the sulfate effect is due to anthropogenic emissions of sulfur dioxide over the Northern Hemisphere's continental landmasses, particularly at middle latitudes. (Data from R. Charlson, J. Lovelock, M. Andreae, and S. Warren, "Perturbation of the Northern Hemisphere Radiative Balance by Backscattering from Anthropogenic Sulfate Aerosols," *Nature* 326 [1991]: 655)

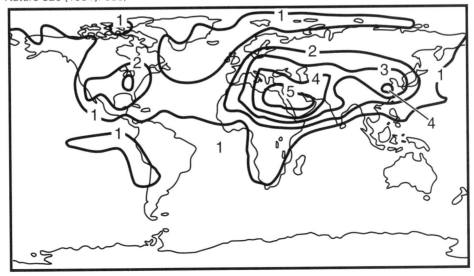

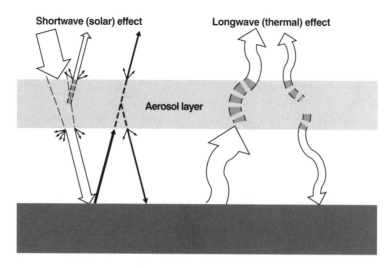

Figure 14.7   The effects of a layer of sulfate aerosols located in the stratosphere on the fluxes of solar (shortwave) and thermal (longwave) radiation. Sunlight is affected mainly by aerosol scattering, which increases the reflected component, thus enhancing the albedo and reducing solar insolation. Thermal radiation is affected primarily by radiation, warms the stratosphere somewhat, and strengthens the greenhouse effect. The effect on solar radiation is typically much larger than the effect on infrared radiation.

air. Indeed, these aerosols near the surface are depleted in a matter of hours or days. The aerosols formed over the eastern United States travel over the North Atlantic Ocean, but rarely make it as far as Europe. Similarly, the concentrations of sulfate aerosols over southern Europe and eastern Asia reflect the local sources of sulfur emissions in those areas. Recall that these pollution particles are thought to provide a "benefit" in reducing ultraviolet radiation at the surface and cooling the planetary greenhouse effect (Section 14.2.2).

The tropospheric aerosols reduce global temperatures in two ways.

1. The aerosols directly reflect sunlight and enhance the planetary albedo.
2. The pollution particles, like those generated from dimethyl sulfide (Section 14.1), cause clouds to become more reflective, further enhancing the albedo.

Both these tendencies to increase the albedo are fairly weak, however, owing to the difficulty in altering cloud reflectance.

The fact that tropospheric aerosols are short lived means that they would be less useful as climate moderators in environmental engineering schemes. To produce a compensating albedo effect on greenhouse warming, huge amounts of sulfur would have to be released into the atmosphere, as much as

several hundred million tonnes per year. Such actions, filling the air with respirable sulfate particles (Section 7.1.2), would not be viewed as a general benefit to human health. And the impact on visibility would be devastating.

The 3 Percent Solution

The role of stratospheric aerosols in controlling the global radiation balance and climate is depicted in Figure 14.7. It happens that the sulfate particles have a much stronger effect on visible radiation (at short wavelengths) than on infrared radiation (at long wavelengths). This allows the aerosol layer to cool the surface, because of two effects.

1. Less warming sunlight reaches the surface.
2. The thermal longwave radiation emitted by the surface in the "atmospheric window" spectral region (Section 11.4) is not efficiently trapped to enhance the greenhouse effect.

The net effect, therefore, of adding sulfate aerosols (or any other small scattering particles) in the stratosphere is to cool the surface, similar to the antigreenhouse effect of smoke described in Section 14.2.1 in relation to nuclear winter.

Figure 14.8 shows the cooling effect of stratospheric sulfate aerosols. The aerosols are defined in

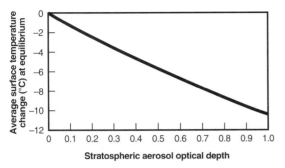

Figure 14.8   The surface cooling effect of a layer of stratospheric sulfate aerosols. The aerosol optical properties are defined in terms of the optical depth of the layer for sunlight in the middle of the visible spectrum (at a wavelength of 0.55 micron). The surface temperature decreases correspond to the new equilibrium condition, or steady state, of the climate system after fully adjusting to the modified radiation forcing. The new equilibrium state is not reached for a decade or longer, owing to the thermal inertia of the oceans. (Data from J. Pollack, O. Toon, C. Sagan, A. Summers, B. Baldwin, and W. Van Camp, *Journal of Geophysical Research* 81 [1976]: 1071)

terms of the optical thickness of the particulate layer. (See Section 3.2.3 for a definition of "optical thickness," or "optical depth.") The scattering efficiency of the aerosol layer increases as the optical depth increases. The albedo of the aerosol layer and the cooling effect of the aerosols increase with scattering efficiency. Over a relatively wide range of aerosol optical depth, the relationship between the optical depth and surface equilibrium temperature change is quite linear. That is, if the optical depth of the sulfate layer is doubled, the decrease in surface temperature will also double. From Figure 14.8 it is apparent that an optical depth of 0.1 can lead to a surface temperature decrease of about 1.5°C.

The optical depth of an aerosol layer usually specified at a specific wavelength of radiation, say the mid-visible wavelength of 0.55 micron, or 550 nanometers (nm). The optical depth varies with wavelength; typically, the optical depth decreases slowly as the wavelength increases, roughly in an inverse relationship to wavelength. That is, if the wavelength doubles, the optical depth is halved. The potential decrease in surface temperature depends on the wavelength dependence of the optical depth and on the length of time that the aerosol layer is present. If the aerosol properties are fixed over a long period, then the surface temperature change will achieve a new steady state after several decades. This new state

represents the *equilibrium* climate perturbation corresponding to that aerosol layer. If the aerosol properties change over time (that is, if the particle sizes, the optical thickness of the aerosol layer, or other parameters defining the particles vary), then the perturbation in surface temperatures at equilibrium will be affected accordingly. Recall that major volcanic eruptions create optical depths of ~0.1 to 0.2 but that the global surface temperature decreases only about 0.5°C. Volcanic aerosol layers are too short lived to achieve an equilibrium state of maximum surface cooling. That is, the cooling is transient and smaller than the potential cooling effect of a permanent aerosol layer.

How can the stratospheric sulfate layer be thickened? Volcanic eruptions do this naturally by injecting large amounts of sulfur dioxide directly into the stratosphere. Lifting 10 million to 30 million metric tons of sulfur dioxide to stratospheric heights in aircraft each year has been suggested as a means of mimicking volcanic eruptions. This would require something like a thousand jumbo-jet flights every day. The flights would need to continue for as long as the threat of global warming persisted. Since carbon dioxide may linger in the atmosphere for a century or more, an artificial aerosol layer would have to be maintained over that span of time. To be effective, the sulfur dioxide must be dispersed throughout the stratosphere. The aircraft would have to cover most points on the globe, flying at all latitudes in both hemispheres. The aerosols generated in dense sulfur dioxide trails just behind the aircraft would likely be too large for optimal climate modification. Moreover, new planes would need to be designed to fly at much higher altitudes than the present jumbo jets can (up to at least 20 kilometers, compared with a ceiling of about 14 kilometers for existing commercial large-body airframes).

The quantity of aircraft exhaust emitted into the stratosphere itself would be unprecedented. Nitrogen oxides and water vapor from the engines could lead to an unacceptable depletion of ozone through direct chemical attack (Section 13.5.3). Indeed, in the past, stratospheric aircraft fleets have been banned by Congress for just this reason. The threat is particularly serious in the Northern Hemisphere, where corrections for global warming would be sought.

Figure 14.9 depicts an alternative concept to compensate for global warming. The carbon-based fossil fuels that are widely used to generate energy

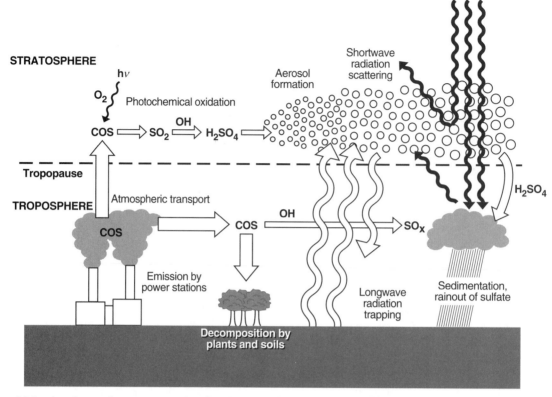

Figure 14.9 A scheme for compensating for the greenhouse-warming effect of carbon dioxide by forming a semipermanent layer of stratospheric sulfate aerosols using the sulfur in fossil fuels. That sulfur, which contributes to local and regional pollution problems, can be converted to carbonyl sulfide (COS) before being emitted into the atmosphere. Carbonyl sulfide is more stable in the atmosphere and, over time, is transported to the stratosphere. There, COS is photochemically transformed into sulfuric acid, which condenses into aerosols that modify the radiation and climate.

and that emit carbon dioxide into the atmosphere as a by-product also contain trace amounts of sulfur. Following combustion, the sulfur is released primarily in the form of sulfur dioxide. On average, fossil fuels contain a small percentage of sulfur by mass. Coal, in general, contains the most sulfur, oil somewhat less, and natural gas the least amount. The sulfur is a nuisance. (Sections 6.1.2 and 7.2.1 [in addition, Sections 7.1.2 and 7.4.4] describe the health hazards of sulfurous smog.) Sulfur emissions also create haze that degrades visibility (Section 6.5). In addition, sulfur emissions create acidic rain in regions of the world where energy production is highly concentrated (Sections 9.3.2 and 9.5.2). To avoid these problems, sulfur is removed from petroleum during processing or is scrubbed from smokestack effluents of power plants. The removal of the sulfur is expensive. The cleanup is forced by the serious nature of the pollution that is generated. If sulfur were removed from

fossil fuels and used to offset greenhouse warming, the multiple benefits to society could be enormous.

That is the concept sketched in Figure 14.9. Carbonyl sulfide (COS) is a common "reduced" form of sulfur. Along with hydrogen sulfide and dimethyl sulfide, carbonyl sulfide is one of the most important sulfur compounds in nature. COS is produced by bacteria in anaerobic environments and can be absorbed by plants. Combustion processes also generate some carbonyl sulfide, and it may be formed as a chemical product when carbon disulfide (CS_2) is photochemically decomposed. Carbonyl sulfide is the most abundant sulfur-bearing gas in the atmosphere, having a relatively uniform mixing ratio of about 0.5 part per billion by volume (ppbv) throughout the lower atmosphere. The lifetime of COS in the atmosphere is uncertain, but appears to be at least 1 year. That is an important property, which allows COS to drift far from its sources before being destroyed.

When COS is emitted at ground level, it can be transported over long distances and may travel between the hemispheres. Because of this dilution, COS does not generate local sulfate haze. Nor does it significantly acidify precipitation on regional scales. In other words, converting the sulfur in fossil fuels to COS largely eliminates the local and regional environmental impacts of the sulfur emissions.

What about the global effects of the COS emissions? The total amount of sulfur emitted by fossil-fuel combustion worldwide approaches 100 million tonnes (Mt) of sulfur annually (Section 9.3.2 and 10.2.1). But after being converted to sulfuric acid (Section 3.3.4), that quantity could only marginally increase the acidity of rainfall around the world. In fact, by spreading around the sulfuric acid, widespread environmental damage is avoided even as regional acidity is mitigated.

Most of the carbonyl sulfide emitted in the troposphere is destroyed there—up to 80 percent. The rest, about 20 percent, is transported into the stratosphere, where it is transformed into stratospheric sulfate aerosol particles. The observed concentration of COS falls off with height above the tropopause, owing to photochemical decomposition at high altitudes. The background stratospheric aerosol layer is a consequence of the sulfur liberated from naturally occurring COS. A fraction of the excess COS generated from fossil fuels would therefore add to the normal aerosol layer.

Nearly 50 million tonnes of sulfur (S) could be converted to COS annually (equivalent to about 100 Mt of COS). The conversion of 20 percent of this COS to stratospheric sulfate aerosols would be equivalent to injecting roughly 20 million tonnes of sulfur dioxide into the stratosphere,[6] like having a major volcanic eruption every year. Because this artificial sulfur injection could be maintained over many years, an equilibrium climate cooling of several degrees Celsius would be expected.

The advantages of the COS scheme can be summarized as follows:

1. The solution already lies in the fossil fuels that are causing the problem. The 3 percent sulfur content of the fuels provides the source of reflective aerosols for mitigating greenhouse warming.

2. The cost of converting fugitive sulfur emissions to COS would be cheap compared with the cost of drastically reducing CO_2 emissions associated with energy production using fossil fuels.

3. The technology for converting SO_2 to COS is quite simple, involving basic thermodynamics and catalytic chemistry. Moreover, the economy and society would not be significantly disrupted during the changeover to COS emission.

4. The excess COS would be widely dispersed and diluted by winds around the planet, eliminating most of the local and regional pollution effects, including sulfate haze and acid rain, connected with sulfur emissions from fossil-fuel combustion.

5. The stratospheric aerosol layer would be formed in a natural way, without the need for aircraft flights or other forms of mass intervention. The cooling mechanism would be similar to that following volcanic eruptions.

6. The thickness of the enhanced aerosol layer and its duration over time could be closely controlled by regulating the rate of COS emission.

7. Because sulfur would be removed from fuels, or combustion products, before being emitted into the atmosphere as COS, the sulfur could be retained, thereby improving regional air quality in any case;

8. If the COS emission were halted for any reason, the atmosphere would return to its initial state within a few years (because the atmospheric lifetime of COS is ~1 year).

There seem to be many advantages to this concept. How could it fail? The disadvantages have not been mentioned, and they are not trivial. For one thing, carbonyl sulfide, like most sulfides, is highly toxic. The atmospheric COS would be concentrated by a factor of 100 or more if the climate mitigation scheme were implemented. Concentrations of COS would approach 0.1 part per million by volume (ppmv). Ozone in this amount is considered to be a health hazard. Carbonyl sulfide in this concentration certainly is. Thus, the high concentrations of COS required for mitigation would pose a worldwide health hazard, not just locally but everywhere at all times. Near the COS emission sites, the concentrations could be considerably

6. For each S atom in SO_2, there are two O atoms. The mass of S is 32 amu, and the mass of O is 16 amu. Hence each SO_2 molecule weighs twice as much as an S atom. Ten Mt of S is equivalent to 20 Mt of SO_2. For COS, the arithmetic is slightly different. A carbon atom has a weight of 12 amu. A COS molecule therefore weighs 12 + 16 + 32 amu = 60 amu, which is slightly less than the 64 amu of SO_2.

larger and therefore more deadly. Moreover, when COS decomposes, it corrodes metals and causes stomatal damage in plants. Adding insult to injury, sulfides like COS have a powerful odor. For example, the smell of rotten eggs is due to hydrogen sulfide (H_2S). Any system designed to produce COS would generate an overwhelming stink over a large region. As in the case of SO_2, COS emissions would need to be diluted by using very high smokestacks, for example, or by mechanically mixing emissions with clean air.

The major problems with massive COS emissions would be related to the aerosols created by the excess sulfur. Indeed, the disadvantages in this regard are common to all solutions that propose creating a stratospheric sulfate aerosol layer to mitigate climate warming. An aerosol layer thick enough to cool the climate by several degrees would cause the skies overhead to be milky white, not blue. This effect is related to the scattering of sunlight by the aerosols and is much like the effect of haze on visibility in smoggy air (Sections 3.2.2 and 6.5.2). No more blue skies, ever, anywhere, just an oppressive global pall of haze. Still, that may be only a minor issue concerning aesthetics.

The stratospheric sulfate aerosols also are implicated in the global depletion of ozone. Such particles have been shown to cause ozone destruction, much like the polar stratospheric clouds responsible for the ozone "hole" (Section 13.7.4; see also Section 14.3.3). After the eruption of Mount Pinatubo in 1991, the total amount of stratospheric ozone worldwide declined by up to 10 percent between 1992 and 1993. That depletion healed as the volcanic aerosols were removed from the stratosphere. In the COS emission scenario, however, the aerosol layer would be semipermanent, being renewed continuously for a century or longer. Thus while the greenhouse effect was being fixed, the ozone layer would be threatened. The ozone degradation would persist as long as the artificial aerosol layer was present.

The sulfate aerosols themselves would scatter and block some of the dangerous ultraviolet radiation leaking through the depleted ozone layer. However, the scattering effect of the aerosols is insufficient to prevent a net enhancement of UV-B radiation at the ground. This situation would not be acceptable without large doses of sunscreen.

The objections to massive COS emission to correct the greenhouse warming effect can be summarized as follows:

- The toxicity of sulfides
- The smell of sulfides
- No more blue skies
- Stratospheric ozone depletion, with ultraviolet spring replacing greenhouse summer

The 3 percent sulfur solution for climate mitigation shows a common outcome encountered in dealing with environmental problems: Even the simplest ideas become increasingly complex the deeper you probe. Even straightforward technologies can generate nasty side effects. Easy or convenient solutions quickly grow into bigger headaches. Specialists with quick answers turn out to be charlatans.

Fourth of July

The Fourth of July celebration last year: hot dogs, fireworks, rockets bursting in air. Smoke and flares and warm flat beer, a cool summer night awash in the glare. We could be celebrating the Fourth of July all year long if one group of technicians had their way. They propose to turn down the greenhouse warming effects of carbon dioxide by filling the stratosphere with dust. That is not a novel idea, as the previous sections demonstrate. The method of delivering the dust to the stratosphere, however, is rather unusual. It would be lofted using 16-inch artillery shells!

It seems that the navy has a number of large ships equipped with very large guns. The guns fire enormous shells that are usually filled with explosives directed at targets miles away. Most of the time the guns are silent. Why not, it is argued, use those guns to save our way of life? Instead of high explosives, the shells can be filled with dust and a small explosive charge. The propellant would be powerful enough to loft the shells as high as 15 kilometers, into the lower stratosphere. Like microscopic volcanic eruptions, each shell would add a little dust to the stratosphere. Eventually, a dense layer could be built up and maintained by continuous bombardment. We have the artillery; we have the dust. Why not?

For one thing, the global scale of the problem comes into play. To be effective, perhaps 30 million metric tons of dust would need to be injected into the stratosphere every year.[7] Suppose that each artil-

---

7. Mineral dust would generally be less effective than sulfate aerosols in creating climatic anomalies. For one thing, it is not easy to produce and widely disperse mineral grains with sizes much smaller than 1 micrometer. Small grains tend to stick together, as in a powder, and are difficult to separate. Sulfate aerosols, on the

lery shell could carry and disperse 100 kilograms (about 220 pounds) of fine dust particles. That amounts to roughly 300 million shells fired each year. That amounts to about 10 shots every second of every minute of every day for the next century. If 1000 guns could be made ready for the task (worldwide, the number of such guns is perhaps a few hundred), they would need to be fired every minute or so forever. The manufacture of shells would be a problem. Shrapnel falling from the skies would be a problem. Noise would be a problem. However, since the guns would be mounted on ships, they could be kept at a distance from populated areas and moved around to generate a more uniform dust layer. As long as the guns were not pointed straight up, any duds would fall harmlessly into the sea.

The technical and infrastructure problems with this ludicrous scheme are so profound it is hard to believe that anyone would seriously embrace it.

Sunshades, Balloons, and Boogie Boards

If an increase in planetary albedo will fix the climate, there are a variety of ways to manage it. After all, the albedo is related to reflectivity. Everyday experience tells us that white objects reflect more light than black objects do. A mirrors reflects almost all the light that falls on it. A number of schemes based on this simple principle may be devised to cure a change in climate. One idea proposes placing huge solar shades in space. These space "parasols" would reduce the amount of sunlight impinging on the Earth. Only a small reduction would be needed (somewhat less than the increase in albedo required to produce the needed temperature compensation).[8] Hence the area on the ground that would have to be shaded is roughly 3 million square kilometers. Unfortunately, to create this equivalent shading effect on the Earth from space would require a sunshade about 100 times

larger. Such a large size is needed because it must be placed at a great distance from the Earth, in a gravitationally stable position referred to as a *Lagrangian* point. In this position, the shade would cast only a partial shadow on the Earth, and an effect similar to a partial eclipse of the sun. Under these circumstances, the shade must be considerably increased in size to produce the same effective reduction in solar insolation as a shade near the Earth would produce.

Several researchers are exploring the idea of using balloons to increase the albedo. Balloons with shiny metallic coatings reflect sunlight very effectively. Imagine constructing a fleet of such balloons, filling them with helium, and letting them loose in the atmosphere. As they drifted around the world, the balloons would act like little clouds, reflecting sunlight away from the Earth. The balloons would be much smaller than a typical cloud, which can be hundreds of meters to several kilometers in size. We imagine the balloons to be only about 1 to a few meters in diameter. The balloons could be made to float in the lower stratosphere. Scientists already fly balloons there. So far it sounds easy enough.

How many reflecting balloons would you need to compensate for greenhouse warming? The answer is disconcerting. The cross-sectional area of the Earth is close to 120 million square kilometers. Accordingly, the effective area of shading required is crudely 4 million km^2 (or roughly 3 percent of the cross section).[9] Assume that the cross-sectional area of a typical balloon is about 4 square meters, or 7 feet in diameter (note that a spherical balloon does not reflect into space all the sunlight hitting it, depending on the time of day, varying amounts of the reflected light are reflected toward the surface). It follows that at least 1 trillion balloons will be required. A *trillion*

---

other hand, can be generated photochemically. Since they are nucleated from the vapor phase, their sizes are initially much smaller than a micrometer. The larger size of the dust grains has two disadvantages. First, the grains fall more rapidly to lower altitudes, where they are readily removed from the atmosphere. Second, the larger-size particles are less efficient, for a fixed amount of mass, at increasing the albedo (Sections 3.2.2 and 6.5.2). Hence to produce the same increase in albedo, considerably more dust mass, compared with sulfate mass, would need to be injected.

8. The average albedo of the Earth is close to 0.33. To produce a 3-percent decrease in solar insolation, the albedo must increase by 0.03, which is close to 10 percent of the present albedo (Equation 14.1).

9. The area that must be covered by balloons is actually much larger than the average effective cross-sectional area of the Earth that must be shaded, 4×10^6 km^2. The area beneath the balloons is increased by a factor of two, because the *surface* area of the sunlit hemisphere is twice the cross-sectional area of the planet, and by another factor of two because the dark hemisphere, which rotates into daylight every 12 hours, must also contain balloons. Try an experiment. Take any spherical object—a globe, a basketball, or a grapefruit—and cut out a piece of cloth or aluminum foil equal in size to the cross-sectional area of the object. Now divide the material into pieces, and stick these randomly on the surface of the object. What fraction of the total surface appears to be covered by the material when you look at the object from different directions? Note that on average, the half of the sphere you can see at any time is about 25 percent covered. Working backward, to have an average shadow of 4 million km^2 projected onto the daylit hemisphere, an area four times as large, 16 million km^2, must have balloons floating overhead at any instant.

balloons! That is equivalent to nearly 200 balloons for every human on the planet. These balloons, moreover, would not be permanent. If carefully designed, they might, on average, last 1 year in the upper atmosphere. In other words, 1 trillion balloons must be launched every year. If everyone alive launched one balloon each morning (like taking a vitamin pill), the reflective shield could be maintained.

The problems with such a plan are overwhelming to say the least. What would the sky look like with a trillion balloons floating around? Occasionally the balloons would cluster together over a region, accidentally creating a solar eclipse. The surfaces of the balloons would strongly affect gases, like ozone, in the stratosphere. It is likely that there would be major chemical perturbations. Spent balloons falling from the sky at a rate of several billion per day would foul the countryside, waterways, and the oceans. If longer-lived miniblimps could be designed—to last 10 years, say—they would tend to collect at high latitudes, forming massive superballoon clusters that would black out these regions all year around. Putting motors on the balloons to push them around would create other problems.

OK, instead of balloons, a cheaper and faster scheme has been proposed: the "boogie board" solution. Styrofoam is a white, reflective material that floats on water. The oceans have a naturally low albedo, less than 0.1. By floating styrofoam on the oceans, the planetary albedo could be increased to 0.5 or more. It is logical to ask how much of the ocean must be covered by styrofoam to achieve climate protection. The required reflective area is similar to, but larger than, the area that must be shaded by balloons—equivalent to a small percentage of the cross-sectional area of the Earth. The oceans cover only about two-thirds of the surface area of the planet, and clouds normally blanket half of the ocean area at any time. Accordingly, the fractional area of the oceans that must be covered by styrofoam would need to be about three times as large, or close to 10 percent.[10]

Think of the oceans awash in styrofoam. Life in the seas would be devastated. Alternatively, styrofoam continents (something like two Antarcticas) could be constructed and moored in the major oceans. The continents could serve a number of useful purposes, such as providing space for resorts along a vast coastline. Guests could be entertained by fireworks shot by naval guns to fill the stratosphere with dust. One vision of the engineered world of the future is depicted in Figure 14.10.

14.3.3   FIXING THE OZONE SHIELD

The role of chlorofluorocarbons (CFCs) in reducing stratospheric ozone was outlined earlier (Section 14.2.3). CFCs already contaminate the environment, and the CFCs already released will linger well into the twenty-first century. The Antarctic ozone "hole" and serious worldwide ozone depletion are among the identified consequences. Although the Montreal Protocol (Section 13.8.1) has been adopted, which proposes to eliminate CFC production before the turn of the century, it will be decades beyond that time before the ozone layer recovers significantly. What if we cannot wait that long? Suppose that ozone becomes even more depleted, as now appears to be happening in the Northern Hemisphere? What technological alternatives should society have at its disposal to preserve the ozone layer? Are there any corrective schemes that make sense? If one were found, would industry then argue that CFCs could be safely manufactured again?

Ideas for saving global ozone have surfaced as the ozone crisis has deepened. Next we look at a few of these schemes.

Lasers Against CFCs

The problem with CFCs is their long lifetime in the lower atmosphere. There are no known significant sinks for CFCs in the troposphere. Photodecomposition in the stratosphere determines the loss of chlorofluorocarbons. All the CFCs that are released must eventually be processed and destroyed in the stratosphere, which suggests a rather obvious solution: Introduce a new sink for CFCs in the lower atmosphere. This simple idea has led to a number of proposals.

One of the most thoughtful ideas is to use laser beams to break down CFC molecules. The action

10. Clouds having a relatively high albedo always cover about 50 percent of the Earth. Styrofoam lying under these clouds would not be effective in increasing the albedo. Accordingly, twice the ocean area must be covered. An additional increase by a factor of 1.5 would be needed because the oceans account for only two-thirds of the Earth's surface (1.5 is the inverse of 2/3). A simple multiplication of the two factors shows that about 10 percent of the ocean surface would need to be paved with styrofoam.

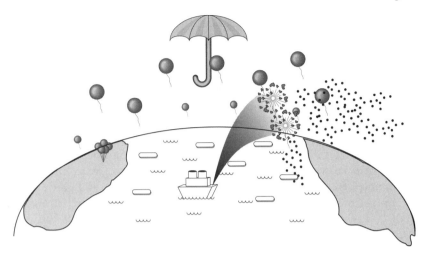

Figure 14.10   The world as it might look with active global environmental engineering to control the climate. The temperature correction schemes include lofting artillery shells filled with dust into the stratosphere from naval vessels, releasing a trillion large balloons into the upper atmosphere, and floating styrofoam on the world's oceans. All these schemes would seek to enhance the planetary albedo, thereby reducing the solar forcing of the climate. An alternative concept, whimsically illustrated in the form of a large umbrella, would place large sunshades in space to decrease the sun's illumination. The engineering scale and cost of these proposals would be astronomical.

would be much like that occurring in the stratosphere, where energetic ultraviolet solar radiation does the job. Unfortunately, ultraviolet radiation does not travel very far in the lower atmosphere, limiting the lasers' effective range. Moreover, radiation that is capable of destroying CFCs would also attack a wide range of substances. Heavy artificial doses of UV in the troposphere could create all sorts of strange chemical side effects, including a choking global smog. What is needed is a magic bullet that destroys only a particular CFC molecule and nothing else. That bullet would be a photon of radiation specially tuned to each CFC. Such a selective bullet does not exist, however.

A clever scientist has come up with an answer: Use more than one bullet, or photon, in rapid sequence. The photons would excite a CFC molecule by a series of steps to an energetic state that readily dissociated. Think about climbing a ladder to reach the top of a building from the ground. If the ladder had only one rung at the top, you might have difficulty getting there in one step. It would be much easier to move up the ladder in small steps, rung by rung. You might also envision spacing the rungs in a particular way that you could negotiate but that a child, say, could not. As an analogue, you might imagine raising a molecule level by level from its lowest "ground" state of energy to a higher energy state from which it could fly apart, or photodissociate (Section 3.3.3). The molecule, in

this case, would be moved rung-by-rung from the ground to the top floor. This could be done in stages using a series of "small" photons each of low energy. Every compound would have a specific pattern (actually, a number of definite patterns) of photons that could take it to the top. The rungs of the ladder would be spaced differently for each compound, however. A sequence of photons that could dissociate one compound generally could not dissociate another. A highly selective mechanism for photodecomposition therefore would be available: **multiphoton dissociation**.

Multiphoton dissociation does not occur in daylight because the intensity of sunlight at each wavelength (or energy) is too low. A laser is capable of producing an extremely high intensity of light at one specific wavelength.[11] For the greatest effect, the laser beam would be pulsed, with the light emitted in short bursts lasting a millionth of a second or less.

11. Electromagnetic radiation having a single well-defined wavelength is referred to as **monochromatic** radiation. In nature, radiation from matter typically has a characteristic spectrum, or distribution, of wavelengths. The narrower the spectrum is, the closer the radiation will be to being monochromatic. When the amount of energy carried by a beam of radiation is fixed, the intensity of the radiation at any wavelength is inversely proportional to the width of the spectrum. Monochromatic radiation has a narrow spectrum and a high intensity at the central wavelength. A laser generates radiation with an extremely narrow spectral range by stimulating the emission of a single specific spectral line of a material.

Such a laser beam shining on a parcel of air could produce conditions ideal for multiphoton dissociation. The laser must be tuned to the proper wavelength to select the molecule of interest.

For this scheme to work, the entire atmosphere must be processed by laser beams over a time span that is short compared with the current atmospheric lifetime of CFCs. A period of 10 years is commonly assumed, although a shorter time span would be preferable. In that time, all 5 billion cubic kilometers of air in the lower atmosphere must be blasted with intense laser radiation. Designs have been worked out. Between 1000 and 10,000 sophisticated lasers would be required. The total energy requirements of several gigawatts (GW) to tens of gigawatts have been projected. So what's a few watts between friends? The lasers themselves do not actually exist. The costs of building such machines several meters in diameter to generate megawatt beams of radiation, with the specialized optics needed to deflect and guide the rays, would be high. Even at that, the proposed laser systems would barely keep up with the present emissions of CFCs. To reduce CFC concentrations to negligible amounts in a reasonable time, a project 10 times greater in scale would be called for.

The idea of laser spotlights swinging through the skies all over the world forever is not particularly reassuring. The intensity of the beams would exceed that of sunlight by at least a factor of 1000. Birds flying through a beam would be fried. This solution for the CFC problem, which even makes a little sense technically, makes no sense in the global context.

Charging up the Stratosphere

Another approach to saving ozone is to deactivate chlorine after it is released from CFCs in the stratosphere. This solution might allow CFC production to continue by compensating for its effects within the stratosphere itself. Figure 14.11 offers a plan to nullify the CFC's effect on ozone. The idea is based on the fact that chlorine atoms carrying a negative electrical charge do not react with ozone. Recall from Equation 13.12 that chlorine atoms participate in an important "catalytic" reaction cycle that destroys ozone (Sections 3.3.4, 13.2.2, and 13.5.2). If this reaction could be inhibited, ozone depletion by CFCs would be halted. Indeed, studies of the chemical processes show that whereas the neutral chlorine-ozone reaction occurs rapidly:

$$Cl + O_3 \rightarrow ClO + O_2 \qquad (14.3)$$

the equivalent "electrified" version of this reaction does not:

$$Cl^- + O_3 \xrightarrow{\text{NO WAY!}} ClO^- + O_2 \qquad (14.4)$$

The challenge then is to charge up the chlorine.

Positive and negative electrical charges are relatively easy to separate. On a dry day, when you shuffle across a carpet and reach for the door, a painful electric spark may jump from your finger to the knob. Shoes rubbing against carpet fibers pick up extra negative charge (electrons) from the carpet fibers, leaving the fibers with an equal positive charge. The electrons travel through your body (which is an excellent conductor of electricity) and jump from your hand to the metal doorknob as you reach for it. From there, the electrons drift through the air to rejoin the positive charges.[12] A **Van de Graaff generator** acts on the same principle.[13] The Greeks discovered that amber (fossilized tree resin) becomes activated when rubbed and attracts bits of paper, styrofoam, feathers, or other similar objects. The Greek word for amber is *elektron*, and so the negative electrical particles later associated with this phenomenon were called *electrons*.

A Van de Graaff generator could be flown on an airplane. The positive charge it generated could be transferred to water droplets by bringing the drops into contact with an electrode. If the drops were

12.   Opposite electrical charges attract each other, and like charges repel, according to Coulomb's law (Section 7.3.1). Electrical charge also likes to move to the outside surface of an object, particularly a good conductor. When a charged object is brought near a conductor, an "image" charge of opposite sign is produced in the conductor. This image charge creates a voltage that may be strong enough to cause a sudden discharge. A charged object exposed to air slowly leaks its charge to the air, forming currents of electrically charged molecules, or ions. The ions readily attach to dust particles, which slows them down considerably. In air, drifting ions represent a weak current that carries electrical charges back to their original source. By nature, positive and negative charges are always seeking routes to recombine. That is why most electrical devices, particularly those working at high voltages, must be carefully insulated.

13.   Robert Jemison Van de Graaff (1901–1967) was an American physicist who invented the electrostatic "generator" named after him. He was influenced by Marie Curie's work and began studying atomic physics. He saw the need in experimental physics to generate high voltages to manipulate charged particles. His generators have produced up to 13 million volts simply by rubbing a fabric on hard rubber and collecting the charge on a sphere.

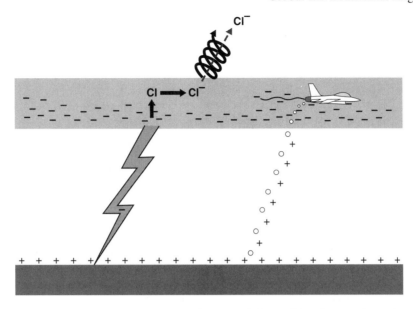

Figure 14.11   A plan to preserve the ozone layer against attack from chlorofluorocarbons, by attaching negative electrical charges to the chlorine released from CFCs, thus chemically deactivating the chlorine. A fleet of jumbo jets would fly through the stratosphere trailing wires that emitted a negative electrical charge into the air. Water carried by the aircraft would be released as artificial rain to carry unwanted positive charges to the lower atmosphere and surface. The charged chlorine atoms would then spiral up the fair-weather electric-field lines and be accelerated into space.

released from the bottom of the plane, the positive charge would be carried downward into the lower atmosphere. The negative charge would remain on the plane, however. As the negative charge accumulated, electric potentials would develop making it difficult for the drops to fall. The solution would be to trail a wire behind the plane. Electrons would flow into the wire and, because of the high potential developed by the accumulating charge, would easily leak into the air around the wire. This leakage of electrons is referred to as a **coronal discharge**. Such discharges, called **Saint Elmo's fire**, can be seen during a storm as glowing light emanating from pointed objects and wires or dancing from the tops of masts on ships. At high voltages, electrons escaping from a surface are accelerated and bang into the surrounding air molecules, exciting them to emit visible radiation.

Chlorine is a very **electronegative** element. It has nothing to do with attitude. Chlorine just likes to acquire and hold onto electrons. Free electrons can be soaked up by free chlorine atoms. The plan outlined in Figure 14.11 to deactivate chlorine might work. Right? Wrong. There are so many problems with this scheme that it would require an entire chapter of this book to explain all of them coherently. Instead, we will just sketch a number of the major objections.

1. Chlorine is not the most electronegative substance in the stratosphere. For example, nitrates and sulfates, which are more abundant, attract electrons more strongly than chlorine. The charging solution would not even have a chance to start up.

2. Less than 0.01 percent of all the chlorine in the stratosphere is in the form of chlorine atoms. The rest is in the form of HCl, ClO, and other compounds. As soon as the free chlorine atoms were charged up, they would immediately be replaced from other chlorine compounds by efficient photochemical processes that operate to maintain a balance among Cl, ClO, and HCl. Indeed, for the scheme to work, all the chlorine in the stratosphere would eventually need to be electrified. That prospect represents an extremely dangerous transformation among chlorine species, similar to the conditions that exist in the ozone hole (Section 13.7.3).

3. Water droplets can carry only a limited electrical charge before they literally explode from Coulomb repulsion. Assuming that the charge-carrying capacity of the drops is optimal and that only 0.01 percent of the chlorine must be electrified, about 10 billion tonnes of water would be necessary to separate the required electrical

charge.[14] Jumbo jets carrying 100 tonnes each would have to fly *100 million sorties* to lift that much water into the stratosphere.

4. The electrical charge separation created—with negative charge in the stratosphere and positive charge in the troposphere—would generate powerful electrostatic fields. The charges would drift in these fields and recombine. The entire system could be discharged within a day by means of atmospheric conduction. Accordingly, the water runs would have to continue unabated at the rate of 100 million trips per day essentially forever. These flights would destroy the ozone layer.

5. If the goal of charging up 0.01 percent of the chlorine were achieved—which, in any case, would have absolutely no ameliorating effect on ozone depletion—the voltage difference between the ground and the stratosphere would be so large that lightning bolts 10 miles long would fly through the atmosphere. Everyone's hair would be standing on end, and corona discharges would be dancing everywhere.

The end result of any attempt to conserve the ozone layer by charging chlorine using artificial rain would most likely be the total destruction of ozone and the end of the world as we know it. This idea for saving the ozone layer was put forward by an intelligent physicist—who apparently does not understand how the atmosphere works. Even respectable scientists make mistakes.

Filling the Ozone Hole

Measures have also been offered to prevent ozone depletion on regional scales. In particular, we describe next a concept to "heal" the ozone hole. (See Section 13.7 for background information concerning this phenomenon.) The basic motivation for intervention rests on the possibility that the ozone hole may become troublesome before the Montreal Protocol has a chance to significantly reduce atmospheric chlorofluorocarbon concentrations (Section 13.8.1). Figure 14.12 depicts the evolution over a year of the ozone hole and a few key parameters

14. The estimate assumes that the water drops are 1 millimeter in radius. This is a minimum size to ensure that the drops would fall into the troposphere before evaporating. Larger drops would be even better, but would carry less electric charge per ton of water.

associated with it. The main action comes in the austral spring, when sunlight returns to the high southern latitudes. Chlorine, which has been activated over the winter on polar stratospheric clouds, is quickly converted to chlorine monoxide (ClO). The ClO participates in destroying the ozone. The ozone depletion continues until ClO is redirected into the normal chlorine reservoirs, mainly hydrogen chloride (HCl).

The high ClO concentrations and deep ozone depletions are maintained by the stability of the southern winter polar vortex (Section 13.7.2). One thought is to destabilize the vortex, which would abruptly end the conditions favorable for ozone depletion. Alternatively, the ClO concentrations could be suppressed. Recall that the critical chlorine catalytic cycles driving ozone destruction entail the chlorine atoms reacting with ozone (Sections 13.5.2 and 13.7.3 and Equations 13.12 and 13.24). Chlorine monoxide is generated in the process. The chlorine-charging scheme discussed earlier proposed deactivating the chlorine atoms by adding an electrical charge. In the concept considered here, the Cl atom would react with light hydrocarbons such as ethane (C_2H_6) and propane (C_3H_8), converting chlorine to HCl. These common hydrocarbons react vigorously with chlorine and do not produce any unusual or long-lived by-products.

The reactions of interest are

$$\begin{aligned} Cl + C_2H_6 &\rightarrow HCl + C_2H_5 \\ Cl + C_3H_8 &\rightarrow HCl + C_3H_7 \end{aligned} \quad (14.5)$$

The chemical products on the right-hand side of the reactions would interact further with chlorine. The details are not crucial. The net effect would be to convert Cl and ClO to HCl. Thus the chlorine species would be forced toward a normal partition (Figure 13.27).

In Figure 14.12, the plan would be to add the selected hydrocarbons to the polar vortex in late winter to inhibit the sudden increase in ClO at first light. These hydrocarbons, which are normally gaseous, would be spread across the vortex using aircraft (oops, we've heard a lot about problems with aircraft). The hydrocarbons are stable in polar darkness and would disperse throughout the vortex as winter progressed. When sunlight first appeared in spring to activate chlorine, the hydrocarbons would begin to do their work. The action is depicted in Figure 14.13.

The ozone hole-plugging scheme seems relatively straightforward. Light hydrocarbons like ethane and

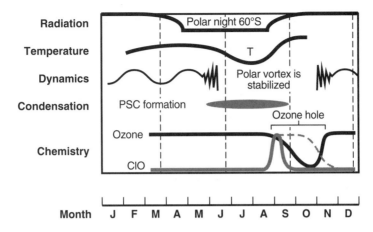

Figure 14.12   The conditions under which the ozone hole forms over Antarctica. The critical period for ozone depletion occurs during a short period of time in September and October. The ozone loss is associated with the high concentrations of the free chlorine radicals (ClO) sustained by the persistent wintertime polar vortex. The idea is to add hydrocarbons to the polar vortex at this time, to quickly suppress the concentrations of ClO before catalytic chlorine reactions severely reduce ozone levels. (Information from R. Cicerone, S. Elliott, and R. Turco, *Science* 254 [1991]: 1191)

propane are available, cheap, and, aside from the danger of explosion, fairly easy to handle. Propane fuel is already in wide use in the form of liquefied petroleum gas (LPG). Calculations suggest that to prevent ozone depletion in austral spring, about 20,000 tonnes of liquefied propane or ethane would have to be lifted into the polar stratosphere. A small fleet of specially fitted aircraft could make the flights over a 3-week period; 10 planes might be enough. This seems a rather small price to pay to fill the ozone hole should it become a serious hazard. There is some doubt the plan would work, however.

The ozone hole is quite extensive. It covers an area as large as the Antarctic continent, and is more than 10 kilometers in vertical extent. No aircraft is presently capable of flying over the entire range of altitudes in the vortex where ozone is normally depleted. For the scheme to work, the hydrocarbons must be more or less uniformly distributed throughout this vast region. Here, nature is uncooperative. The atmospheric region of interest has relatively weak mixing. In the stable vortex, little turbulence is present (Section 5.2.1). It is uncertain, therefore, whether the hydrocarbon vapors would be spread uniformly enough to cure the ozone hole.

Under the unusual atmospheric conditions that characterize the southern winter polar vortex, HCl derived from active chlorine through the action of injected hydrocarbons would not be secure. In fact

if polar stratospheric clouds were present (Section 13.7.3), the HCl could be quickly recycled into active chlorine. In the process, the hydrocarbons would be consumed. Under some not unlikely circumstances, the addition of hydrocarbons might actually exacerbate the ozone-depletion problem. Polar stratospheric clouds have been observed to persist at high latitudes far into the spring, and these lingering clouds might cause the injected hydrocarbons to turn against the ozone. The result could be a prolonged, aggravated depletion of the ozone layer.

We should emphasize that the decomposed polar ozone represents only a small percentage of the global ozone layer. Filling the ozone hole is really global environmental engineering on a minor scale compared with some of the other ideas discussed in this chapter. Nevertheless, all the elements of surprise, uncertainty, and awesome scale can be found there.

Zeolites, Bacteria, and Other Exotica

Many of the ideas concerning remediation of the ozone layer are simply crackpot. Others, although physically reasonable, are unfeasible because of the global scale or high cost. A few ideas may be reasonable and feasible, but carry too much uncertainty. Is there a winner of a thought, a gem of an idea, lurking

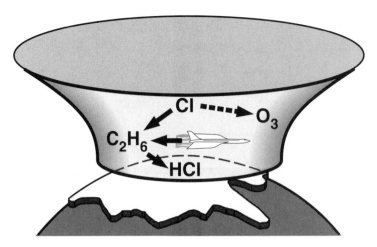

Figure 14.13   A scheme to plug the Antarctic ozone hole by releasing light hydrocarbons (in this case, ethane [C_2H_6]) from high-flying airplanes. The hydrocarbon would react with ozone-active chlorine atoms to form HCl, which does not react directly with ozone. The hydrocarbons would eventually be oxidized into carbon dioxide and water, in amounts too small to disturb the global stratosphere.

out there? One cannot eliminate the possibility of a foolproof scheme to stop the loss of ozone while allowing the production of chlorofluorocarbons. It is a long shot. But given the enormous stakes, an army of researchers are thinking hard (some, unfortunately, not hard enough). It is inevitable that a variety of ideas have surfaced, and a few of these are mentioned here.

Think of a zeolite as a microscopic sponge. It is a material having many nooks and crannies that can absorb chemicals from water or gases from air. When a zeolite is exposed to water, for example, it can filter out undesirable chemicals. When it is exposed to air, pollutants can be absorbed inside its labyrinthine structure. Natural zeolites consist mainly of silicate minerals, often of volcanic origin. Pumice is a fine, vesiculated, glassy substance with a huge effective surface area. Special zeolites can also be engineered to attract and hold specific compounds, or families of compounds. Zeolites are used in petroleum refining, in which certain hydrocarbons are absorbed and separated; in the dehydration of gases, in which water is scavenged; and in water softening, in which minerals are adsorbed.

It does not take a great stretch of imagination to try to manufacture a new zeolite that can absorb chlorofluorocarbons. The zeolite could be ground into fine particles and sprinkled into the atmosphere. After scavenging CFCs, the zeolites would settle to the ground, carrying the CFCs with them. The CFCs would remain harmlessly trapped in the zeolite honeycomb.

Such a zeolite dose not exist, however. Whether it is possible to design and mass-produce a zeolite that preferentially absorbs chlorofluorocarbons remains to be demonstrated. At a minimum, 10 million to 100 million tonnes of the stuff would need to be spread over the Earth to soak up the chlorofluorocarbons already mucking up the atmosphere. The long-term stability of the zeolites would be a serious issue. If the zeolite particles heated up or decomposed, the CFCs would be released. However, because we don't have an actual zeolite material that targets CFCs to experiment with, this nifty idea must be filed away for future reference.

A biological researcher recently performed an interesting experiment. He scooped mud from the bottom of a pond, and in the laboratory, he bubbled CFCs through the mud. Lo and behold, a smaller amount of CFCs escaped in the bubbles than he had pumped in. Microbes living in the mud were apparently eating the CFCs. Or were they? The researcher took another sample of mud and heated it to kill all the microbes. Bubbling CFCs through the sterilized mud produced a smaller loss. The difference could be attributed to microorganisms consuming the chlorofluorocarbons.

Because CFCs are not natural compounds, living organisms have not evolved in their presence and so have not learned how to use CFCs for food (Section 12.3.3). On the other hand, compounds with similar molecular structures, like methane and methyl chloride, are widely used by living organisms. Moreover, bacteria are known to evolve rapidly when exposed to novel environmental conditions. Perhaps, living at the bottom of lakes and ponds are organisms waiting to feast on CFCs. As the concentrations of CFCs in the environment—in air, water, and land—continue to increase, microorganisms may appear that can metabolize these inert gases. Better yet, microbes that like CFCs could be genetically engineered and released into the environment. Selective breeding and genetic engineering have resulted in microorganisms that can eat oil spills and toxic wastes. Why not CFCs?

Other possible explanations for the experimental observations were noted earlier. For example, chlorofluorocarbons may have been absorbed on the fine clay particles that compose pond mud. Heating may have altered the properties of the mud and slowed the uptake of CFCs. Bacteria may have had nothing to do with the missing CFCs. Clearly, more work is needed to isolate the actual effects of microorganisms on chlorofluorocarbons and related compounds.

Plans to carry ozone manufactured at the ground into the stratosphere have been proposed. Ozone, condensed for shipment, might explode, however. Another project would pump into the stratosphere smoggy air, containing high concentrations of ozone, from polluted cities. Two serious problems might be solved at once. But the means for pumping smog 15 miles high is left unresolved, and the impact of the filthy smog on the pristine stratosphere is not mentioned. And so it goes. One after another, ideas surface, are poked at, and sink. So many wild ideas, so little time!

14.4   A Rational Approach to Environmental Management

The application of technology to solve environmental problems is often accompanied by problems, especially when human intervention must extend over global distances and span decades and centuries. Some of the examples we cited reflect a lack of basic understanding about, and sensitivity to, the natural environment. Ideas to deep-six CO_2, grow and bury millions of trees, inject chemicals into the stratosphere, and build huge parasols in space all seem a bit ludicrous. It is apparent that caution must be exercised by those empowered to make decisions regarding environmental intervention, serious mistakes leading to a significantly reduced quality of life in the future.

Science and Nonsense

All right. The environment is threatened; indeed, it is slowly declining before our eyes. What can be done to prevent a disaster? Consider the following observation about human nature, made seven decades ago, that in a way frames the current debate over the environment: "The optimist proclaims that we live in the best of all possible worlds; and the pessimist fears this is true."

Some people feel that we should do nothing and let events take their course. A few idiots argue that laissez-faire has never led to a global disaster. Put aside child labor and sweat shops, urban blight, and Third World poverty. With regard to the environment, simply careening downhill toward a brick wall hoping for a soft landing is no solution. Neither can we "go back to the caves," as some refer to ecological retrenchment. There are not enough caves to go around anymore.

A cadre of clever pseudotechnicians and political hacks have appeared denying that global environmental problems exist in the first place. They find numerous flaws in the generic "disaster" theories of environmental degradation. Thus we hear about the "hole" in the ozone theory or the "cooling down" of the greenhouse-warming effect. These polemicists are outside science. Rather, they exploit the scientific method itself, which at its best doggedly questions accepted ideas and seeks tests of existing hypotheses. By emphasizing the small differences of opinion that often arise among competing researchers and highlighting irrelevant observations that seem to conflict with scientific consensus, the polemicists piece together a phony case. Their nonscientific approach is worse than nonsensical; it is dishonest.

Who are these people? They range from practicing scientists acting as "the loyal opposition," to extremists on the political right and left, and religious zealots. Unfortunately, some of the loyal opposition rarely submit their scientific "results" for peer review; they would rather publish authoritative editorials in newspapers. An example at the more extreme end of the spectrum is Lyndon LaRouche, an occasional political wanna-be who advocates "Star Wars" and universal fusion power and supports aggressive actions to sell these ideas. A more typical conservative society, organized under the name of the author Ayn Rand, believes that most environmental concerns are unfounded.[15] Society members

15. Ayn Rand (1905–1982), a Russian émigré to the United States, was a writer who believed in laissez-faire capitalism and the full freedom of individual persons. Her major novels, *The*

offer evidence to "prove" that CFCs do not deplete ozone and that the greenhouse-warming effect is a deception. Hundreds of similar organizations generate pamphlets and posters and are willing to comment on scientific issues at the drop of a hat. Most of the information dispensed by these groups is distorted or plain wrong. In a related phenomenon, countless public-information groups have surfaced carrying names invoking environmental concern that are organs for special-interest groups that wish to exploit the environment. Examples are the Desert Conservation Institute, sponsored by the mining industry; the Information Council of the Environment, run by coal and utility interests; the National Wetlands Coalition, which seeks to open wetlands to development; the Environmental Conservation Organization, supported by land improvement contractors; and the Wilderness Impact Research Foundation, which seeks to "...educate the public about the damage that wilderness causes to society, the economy, and even wildlife."[16] Again, beware the wolf in sheep's clothing!

Many people exposed to media hype have come to believe that technology will always be available to solve environmental problems. Technology has been known to perform miracles. Behold the atomic bomb. Behold television! During the late 1950s, Walter Cronkite hosted a television show called *The Twentieth Century*. Each week new technological wonders expected to be common in the next few decades were unveiled: limitless nuclear energy, nonpolluting cars magnetically levitated and automatically piloted on tracks, cures for cancer. Few of the wonders promised for this century have actually materialized; instead, the technologies have turned against us. Nuclear energy generates a nightmare of radioactive waste. Automobiles foul the air in cities around the world. Chemicals meant for better living fuel an epidemic of cancer. Television itself devolves into an inane medium of mass marketing.

The examples in this chapter demonstrate that concepts to tinker with the environment are relatively easy to dream up. Their implementation is more difficult, however, usually involving dumb-

foundingly expensive construction projects. Worst of all, dangerous and unforeseen side effects abound. Many of the people who are devising and advocating technological cures are amateur geophysicists. We hear, for example, ideas for preserving the ozone layer from well-meaning engineers and physicists who do not understand how the ozone layer works (unlike the readers of this book). Under cursory inspection, a seemingly creative idea becomes a harebrained scheme. No person or organization has the breadth of expertise required to decipher the full complexity of the natural world, with its interacting biological, physical, and chemical systems. A consortium of experts is needed. Even then, important linkages may be overlooked, and side effects may be ignored. What can be done to ensure that profound, perhaps irreversible, damage is not incurred when dealing with environmental problems and that an enormous sum of money is not wasted chasing phantom technological solutions? How can environmental degradation be limited or reversed?

Behavior and Ethics

If the world as a whole is to prosper and remain livable, then persons, industries, and nations must develop higher standards of behavior and ethics in dealing with the environment. People must be willing to make sacrifices over the short term to build an environmentally sound infrastructure for the long haul. The environment is not just a fad for today. It is the foundation supporting long-term human survival. Business and industry must develop an agenda for environmental action. The leaders of commerce must *believe* that the environment is worth saving. Faking concern over the environment as a public-relations ploy is unethical and dangerous in these times. Nations must abandon selfish imperatives and join a global effort to preserve key elements of the environment. World civilization must be brought into closer harmony with the natural world. Formulating more equitable distributions of resources worldwide, sharing nonpolluting "clean" technologies with less-developed countries, and negotiating international environmental treaties are important steps.

If this all sounds like "one-world" gibberish, think again. The environment is no longer a matter of concern only for vegetarians and flower children. The global habitat is on the minds of presidents and vice presidents, scientists, baby boomers, beef eaters,

---

Fountainhead and *Atlas Shrugged*, express the conflict of individualism with collectivism and conformism in terms of her philosophy of objectivism, which asserts that "moral" precepts are objectively valid.

16. Information on these societies originally appeared in 19913 in *The Observer*, a publication of the Audubon Society of Whittier, California, in 1993.

and Middle America. It has become a matter of survival—very likely the survival of generations not yet conceived. Every day, we face personal choices in behavior and life-style that will shape the future. It is easy to point an accusing finger at business and its leadership when assessing damage to the environment. Surely, greed is the engine for environmental destruction. Yet it is the public that benefits from, and enjoys the products of, business. Thoughtless consumption encourages industrial activities that pollute. Vocal disapproval of such activities and boycotts of pollution-generating products would stem the tide of environmental destruction. Each of us may realize that that is the right personal choice, but we may be too busy right now to participate. Perhaps the costs appear too dear to pay. More to the point, people like us have little power or influence on the course of events. Nonsense, nonsense, nonsense!

Education is the key to saving the environment. High schools and colleges should be places where students learn how the environment works. Everyone should gain a basic understanding of the natural world, which ultimately feeds and nurtures our species. Each of us should be aware of the need for a healthy environment. Sensitivity to environmental deprivation should be taught universally. Information is a powerful stimulant. Environmental activism should become a normal part of our lives. An appreciation of the benefits of a clean, healthy environment would make the cost of conservation palatable. Every person would have a role in nurturing that lovable global organism called the biosphere.

Individual citizens can have a positive influence on the state of the environment if they express themselves and are heard. But unless leaders of countries and corporations believe that the environment is a cause worth championing, it will be difficult to halt the downward slide of the quality of life worldwide. The more desperate the situation becomes, the more attractive technological fixes will become. Leaders who understand the environment are less likely to be misled by seemingly attractive proposals by technologists. It is reasonable to expect the world's leaders to become familiar with the technical issues concerning the environment at a level comparable with the descriptions in this book. It would be irresponsible to expect from leadership any less vigilance today with regard to the environment than was expected in the recent past with regard to nuclear weapons and their potentially devastating effects on civilization.

Treaties and Laws

The Montreal Protocol controlling the production of chlorofluorocarbons to save the global ozone shield is an astounding milestone in international environmental law (Section 13.8.1). For the first time in history, the Earth's collective population has recognized a serious threat to the global environment and has acted to fix it. The solution that was finally adopted is *not* a technological cover-up. The source of the ozone depletion problem—chlorofluorocarbons—is being eliminated. This mandates large and permanent changes in a major global industry under unprecedented international oversight and regulation. Personal life-styles will be changed by the treaty (which will affect the costs for refrigeration, air conditioning, dry cleaning, and so on [Sections 12.3.3 and 13.5.2]). Luckily, in this instance, the changes are likely to be relatively painless. The offending industry is small. New, ozone-safe compounds are available to replace the older ozone-depleting chlorofluorocarbons. Nevertheless, over the next decade, the transition to a CFC-free world under the treaty's guidelines will require sacrifices, ingenuity, and huge investments of capital. Despite the costs and inconvenience, world leadership is in essential agreement with the decision to proceed.

Imagine the complexity of the treaty that will be drafted to control the emissions of greenhouse gases, including carbon dioxide. To forge such an agreement, all the creativity and ingenuity of the human species will be required. More important, a consensus of leadership is necessary. Such a consensus must be built on sound science. The geophysical, biological, and chemical basis of global climate change has been under intense development for several decades and is documented in a mountain of technical reports and assessments. Polemicists lingering at the margin of science hoot at this work, calling it biased and selective. They seek, and often attract, the attention of policymakers. Technological "fixers" feed at the outskirts, promising answers for money. An army of lobbyists representing myriad special-interest groups argue against forceful actions. As in any legal proceedings, "experts" on both sides of each issue are paraded before the world court. The media often seek sensationalism rather than balance in pitting the arguments of fringe elements against the conclusions reached by the central body of science.

Laws are imposed on all aspects of life to control a broad range of unacceptable activities and behaviors. The environment must be protected by laws, just as other facets of our lives are, and for the same reasons. To be effective, laws must be fair and equitable. They also must be enforced. The Montreal Protocol is fair and equitable, but it remains to be seen how well it can be enforced. A treaty to control greenhouse gases is still in the formative stages, yet the battle to win the hearts and minds of world leaders is well under way.

Technology for the People

Technology and its underlying science and engineering should be applied to provide life-enhancing experiences while preserving or improving the quality of life. Too many technologies that have been established by industry and government are dirty and dangerous. The world badly needs clean, safe technologies, which are exactly what a new breed of engineers and scientists, sensitive to the fragility of the environment, are seeking: nonpolluting energy sources, renewable fuels, safe waste disposal, toxin-free foods, pristine air, drinkable water, fumeless transportation. Rather than just the "cheapest" and "easiest," another adjective has been added to the design lexicon, the "cleanest." Technology is being designed with the environment in mind from the outset, not merely as an afterthought; technology that is sensitive to the ecology of the land and water; ultimately, technology for people.

In applying clean technology for the purpose of conservation, we are practicing preventive medicine. Pollution of the body over time—say, through exposure to tobacco smoke—leads to health problems such as lung cancer. Pollution of the environment during this century has created serious life-threatening problems for people and for countless other living organisms as well. Clean technology can help eliminate these problems in the future. Imagine a gas leak in your home. You can throw open the windows to let the gas out. You might even live with the windows open for a while without fixing the leak. Of course, you would get wet when it rained and cold during the winter. Better yet, why not fix the leak and close the windows? Stop the pollution at the source. That is the philosophy of clean technology. But unfortunately, it is not usually the philosophy of environmental engineering.

In referring to the possibilities for engineering the environment of Earth and the other planets, Carl Sagan and his colleague James Pollack; pointed out that

> ... a short-term imperative for planetary engineering exists for only one world in the solar system, our own. Careless or reckless applications of human technological genius have put the global environment at risk in several different ways. The Earth is not a disposable planet ... The first step in engineering the solar system is to guarantee the habitability of the Earth.

—J. Pollack and C. Sagan, "Planetary Engineering," in *Near Earth Resources*, University of Arizona Press, 1993.

From what has been said in this chapter, I hope it is evident that the application of technological schemes to fix environmental problems is generally a mistake. Nonetheless, the use of technology as a tool in environmental remediation must not be abandoned. In some cases, the environmental risks associated with the small-scale application of a particular technology are well understood. In such cases, small-scale tests could be performed to determine potential safe uses of the technology. Nevertheless, in almost every instance of environmental pollution discussed in this book, the first and most logical approach to remediation is to identify and eliminate the *source* of the pollution. Technology applied to mask or correct undesirable environmental conditions while leaving the cause undiminished should be second, third, fourth—or last—on the list of remedial options.

Questions

1. Discuss the difference between a "positive-feedback loop" and a "negative-feedback loop." The brake pedal and accelerator pedal in a car are designed to be worked alternately, using the same foot. Would you consider this design to have positive or negative feedback? Some people drive with one foot on the brake pedal and the other foot on the gas pedal. Why could this be dangerous?

2. Summarize the physical and biological foundations for the concept of reducing atmospheric carbon dioxide by fertilizing the oceans with iron. What other schemes involving living organisms could you dream up to bury carbon in the deep oceans (consider the fact that many

marine organisms construct hard shells or skeletons composed of calcium carbonate)? How might you imagine using genetic engineering of marine organisms to enhance the ocean's carbon biological pump?

3. In the ocean iron-fertilization scheme, phytoplankton are fed extra nutrients to accelerate the absorption of carbon dioxide in the oceans. Suppose that the same planktonic species emitted dimethyl sulfide (DMS) as they grew. What would be the overall effect on global climate of ocean fertilization under these circumstances (describe the relevant processes and effects in a general way)? Would the DMS emission cause a positive or a negative feedback on the global temperature change associated with carbon burial?

4. Discuss three schemes to cool the Earth's climate using the *albedo effect*. In your answer, describe the material that will be used, the manner in which the albedo will be affected, and the possible side effects of the proposed activities. Also outline the possible quantitative aspects of the projects, in terms of the needed amounts of materials, the necessary infrastructure to carry out the work, and the relevant size and time scales.

5. Discuss two geotechnological-engineering concepts that have been proposed to prevent chlorofluorocarbons from damaging the stratospheric ozone layer. Explain why these schemes might be impractical. If someone had suggested replacing all the chlorinated fluorocarbons (CFCs and HCFCs) with chlorine-free compounds that contain bromine in place of chlorine, would you be relieved of your anxiety about ozone depletion? Explain.

6. Someone has bubbled chlorofluorocarbon vapors through mud collected at the bottom of a lake and found that the concentrations of CFCs were reduced. Assuming that a particular kind of bacteria in the mud have consumed and destroyed the CFCs, how might this discovery be used in a geoengineering project to save the ozone layer? Discuss in a general style how you would use these bacteria. What problems might you encounter in making this scheme operational? Consider issues related to the biological aspects of this idea and the requirements to reduce CFCs on a global scale.

Problems

1. Someone has proposed that we transport ozone into the stratosphere on jumbo jets to fill in any holes that are found. The process would be like filling potholes in a street. According to current estimates, perhaps 5 percent of the total normal amount of stratospheric ozone would need to be replaced every month. Calculate how many jumbo jet flights, each carrying 100 tonnes of stabilized liquid ozone, it would take each day to keep the ozone layer full in this case. Is this a practical solution?

2. You have decided to use sand particles to reflect sunlight from the top of Venus's atmosphere in the hopes of cooling its climate to a habitable level. Someone has calculated that you will need a layer of particles with an optical depth of 2 at visible wavelengths. She has also calculated that if the sand grains have a radius of 0.5 micrometer, you will have to build an aerosol layer with a mass loading of 1 gram for every square meter of (planetary) surface to obtain that optical depth. The results will vary with the size of the particles. The optical depth, τ, of the layer is proportional to the mass loading, M, divided by the particle radius ($\tau \propto M/r$). (a) If the surface area of Venus is 400 million square kilometers, how many metric tons of 0.5-micrometer particles will be required to create a global aerosol layer of the proper optical depth? (b) If you can only make dust grains that are 1 micrometer in radius, how many tonnes of these would be needed to achieve the same optical depth? (c) You can use the space shuttle to transport the particles from Earth to Venus. Each shuttle flight will carry 100 tonnes of dust. How many flights will be needed to construct the aerosol layer in (a)? (d) If each shuttle round-trip to Venus takes an average of 6 months to complete, and the dust layer needs to be replaced once a year, how many shuttle vehicles will have to be flying at any given time?

Suggested Readings

Broecker, W. *How to Build a Habitable Planet.* Pacific Palisades, Calif.: Eldigio Press, 1985.

Brown, L., C. Flavin and S. Postel. *Saving the Planet.* New York: Worldwatch Environmental Alert Series, W. W. Norton and Co., 1991.

Cicerone, R., S. Elliott, and R. Turco. "Global Environmental Engineering." *Nature* **356** (1992): 472.

———. "Reduced Antarctic Ozone Depletions in a Model with Hydrocarbon Injections." *Science* **254** (1991): 1191.

Crutzen, P. J. and J. Birks. "Twilight at Noon: The Atmosphere after a Nuclear War." *Ambio* **11** (1982): 114.

Harwell, M. and T. Hutchinson. *Environmental Consequences of Nuclear War, Volume II. Ecological and Agricultural Effects.* New York: Wiley, 1985.

Martin, J., R. Gordon, and S. Fitzwater. "Iron in Antarctic Waters." *Nature* **345** (1990): 156.

McKay, C., O. Toon, and J. Kasting. "Making Mars Habitable." *Nature* **352** (1991): 489.

Oberg, J. *New Earths.* Harrisburg, Pa.: Stackpole Books, 1981.

Pittock, A., T. Ackerman, P. Crutzen, M. MacCracken, C. Shapiro, and R. Turco. *Environmental Consequences of Nuclear War, Vol. 1, Physical and Atmospheric Effects.* New York: Wiley, 1986.

Pollack, J. and C. Sagan. "Planetary Engineering." In *Near Earth Resources*, ed. J. Lewis and M. Matthews. Tucson: University of Arizona Press, 1993.

Sagan, C. and R. Turco. *A Path Where No Man Thought: Nuclear Winter and the End of the Arms Race.* New York: Random House, 1991.

Turco, R., O. Toon, T. Ackerman, J. Pollack, and C. Sagan. "The Climatic Effects of Nuclear War." *Scientific American* **251** (1984): 33.

———. "Nuclear Winter: Global Consequences of Multiple Nuclear Explosions." *Science* **222** (1983): 1283.

Appendix A

Scientific Notation, Units, and Constants

This appendix reviews the "scientific notation" used in the text; the units commonly used in physics and chemistry to quantify length, mass, and so on; and the manipulation of these units. We discuss some of the physical constants and the essential mathematical operations that are used in the text.

A.1 Scientific Notation

Most scientific discussions use numbers that are either very large or very small. Examples of such numbers can be found in every chapter of this book, including the total number of molecules in the Earth's atmosphere, 10^{44}; the fraction corresponding to a mixing ratio of one part per trillion by volume (pptv), 10^{-12}; the amount of carbon emitted into the atmosphere each year from fossil-fuel combustion, about 10^{19} grams; and the wavelength measure for ultraviolet radiation, roughly 10^{-7} meters. If we had to write out the first number, it would read: 100,000,000,000,000,000,000,000,000,000,000, 000,000,000,000. That is, the digit 1 followed by a string of 44 zeros! The second number would read: 0.000000000001. These longhand numbers are clumsy and impractical.

Basically, any large or small number can be expressed as the product of two terms. The first term is a **prefactor** of order unity (that is, a number with a value between 1 and 10) that gives the precision, or accuracy, of the original number. The second term is a **power of 10** (that is, 10^p, where p is the **exponent** or "power" of 10). The power of 10 defines how many times the number 10 is to be multiplied by itself. For example, $10^4 = 10 \times 10 \times 10 \times 10 = 10,000$ (that is, 10 multiplied by itself four times). As an example of the use of this notation, consider the large number 1,100,000 (which is named one million one hundred thousand). In scientific notation this number may be written 1.1×10^6.

Notice that to arrive at this form, we must count the digits from the position where the decimal point is originally found (or would be placed) in the number toward the left until only one non-zero digit remains to the left. The decimal point is fixed here, and the resulting figure is the prefactor. The number of digits counted to the left becomes the power, or exponent, of 10. The exponent is positive if we have counted from right to left, but is *negative* if we have counted from left to right. In either case, after moving the decimal point, the remaining figure (with one digit remaining to the left of the decimal point) is the prefactor (usually with the unnecessary zeros lopped off). Try this procedure with the number 0.0000000035. In this case, we count from the original decimal point to the right nine digits, leaving the prefactor 3.5 (all the zeros to the left are dropped) and thus yielding the number in the form 3.5×10^{-9}.

The usefulness of scientific notation is easily appreciated for arithmetic with large and small numbers. Imagine that you want to calculate the total number of molecules in the Earth's atmosphere for a species that has a mixing ratio of 1 pptv. This would be written literally as 100,000,000,000,000, 000,000,000,000,000,000,000,000,000 × 0.000000000001 = 100,000,000,000,000,000, 000,000,000,000,000. How much easier is it to write $1 \times 10^{44} \times 1 \times 10^{-12} = 1 \times 10^{32}$. All we do is multiply the prefactors and add the exponents! Consider the examples given later, and enjoy.

APPLICATIONS OF SCIENTIFIC NOTATION

To illustrate the use of scientific notation, we offer some examples. The number 550, for example, can be written in scientific notation as 5.5×10^2. That is, $550 = 5.5 \times 10^2$. In this case, the normal decimal representation is more convenient than the scientific representation. Naturally, the simplest notation is adopted for specific numbers in the text of the book. For some numbers, the choice can be a matter of taste. Take, for example, the number of people on the Earth—approximately 5 billion. This may be written out as 5,000,000,000 or more compactly as 5×10^9. You should become familiar with both representations.

If we consider the number of molecules in Earth's atmosphere, there is little dissension. Which number is more convenient to write and which value is more immediately clear: 1×10^{44} or 100,000,000,000, 000,000,000,000,000,000,000,000,000,000,000? In addition to compactness and immediate recognition of magnitude, numbers expressed in scientific notation are much easier to manipulate in arithmetic operations. Some examples of these operations follow. *Multiplication*: The rule for multiplying numbers in scientific notation is to *multiply the prefactors* and *add the exponents* (or the powers of 10). For example,

$$1 \times 10^5 \times 1 \times 10^8 = (1 \times 1) \times 10^{8+5}$$
$$= 1 \times 10^{13}$$

$$1 \times 10^{-3} \times 1 \times 10^6 = (1 \times 1) \times 10^{-3+6}$$
$$= 1 \times 10^3$$

$$1 \times 10^{-2} \times 1 \times 10^{-4} = (1 \times 1) \times 10^{-2-4}$$
$$= 1 \times 10^{-6}$$

$$(2,500,000,000) \times (200)$$
$$= (2.5 \times 10^9) \times (2 \times 10^2)$$
$$= (2.5 \times 2) \times (10^9 \times 10^2)$$
$$= 5 \times 10^{9+2}$$
$$= 5 \times 10^{11}$$

Division: The rule for dividing numbers in scientific notation is to *divide the prefactors*—that is, to divide the numerator prefactor by denominator prefactor—and *subtract the exponent* of the number in the denominator from the exponent of the number in the numerator. For example,

$$\frac{1 \times 10^4}{1 \times 10^3} = \frac{1}{1} \times 10^{4-3} = 1 \times 10^1$$
$$\left(or \quad \frac{10,000}{1,000} = 10 \right)$$

$$\frac{1 \times 10^4}{1 \times 10^5} = \frac{1}{1} \times 10^{4-5} = 1 \times 10^{-1}$$
$$\left(or \quad \frac{10,000}{100,000} = 0.1 \right)$$

$$\frac{1 \times 10^{-6}}{1 \times 10^{-5}} = \frac{1}{1} \times 10^{(-6)-(-5)}$$
$$= 1 \times 10^{-6+5} = 1 \times 10^{-1}$$

$$\frac{6 \times 10^4}{2 \times 10^3} = \frac{6}{2} \times 10^{4-3} = 3 \times 10^1$$
$$\left(or \quad \frac{60,000}{2,000} = 30 \right)$$

$$\frac{5 \times 10^{-3}}{2 \times 10^5} = \frac{5}{2} \times 10^{-3-5} = 2.5 \times 10^{-8}$$

LARGE AND SMALL NUMBERS

Tables A.1 and A.2 summarize the notation and identifying names of the most common large and small numbers used in scientific discussions. These numbers may be applied to any specific quantity by attaching an appropriate suffix indicating the units of measure involved.

USING MIXING RATIOS

Table A.3 lists the common mixing ratios, or mixing fractions, that are used to characterize gases in the atmosphere. The principal ratios, or fractions, used are parts per million, parts per billion, and parts per trillion (abbreviated ppm, ppb, and ppt, respectively). These mixing fractions can be expressed in parts *by volume (or number)* or parts *by mass* of air. In the former case, for example, one might specify parts per million by volume, or ppmv. For smaller fractions it would be ppbv and pptv. For mixing fractions specified as parts by mass, the corresponding abbreviations are ppmm (for parts per million by mass), ppbm, and pptm, respectively.

"Parts by volume" refers to the *fraction of the total number* of molecules in a unit volume of air of a certain kind of gas—say O_2 or CO_2 or H_2O. The *ratio* of the number of such molecules to the total number of molecules in the volume is the fraction of interest; hence the term *mixing ratio*. An

Table A.1   Large Numbers

"Normal" notation	Scientific notation	Standard name
1	1×10^{0}	One
10	1×10^{1}	Ten (*deca*-)
100	1×10^{2}	Hundred (*hecto*-)
1000	1×10^{3}	Thousand (*kilo*-)
10,000	1×10^{4}	
100,000	1×10^{5}	
1,000,000	1×10^{6}	Million (*mega*-)
10,000,000	1×10^{7}	
100,000,000	1×10^{8}	
1,000,000,000	1×10^{9}	Billion (*giga*-)
1,000,000,000,000	1×10^{12}	Trillion (*tera*-)

Table A.2   Small Numbers

"Normal" notation	Scientific notation	Standard name
0.1	1×10^{-1}	Tenth (*deci*-)
0.01	1×10^{-2}	Hundredth (*centi*-)
0.001	1×10^{-3}	Thousandth (*milli*-)
0.0001	1×10^{-4}	
0.00001	1×10^{-5}	
0.000001	1×10^{-6}	Millionth (*micro*-)
0.000000001	1×10^{-9}	Billionth (*nano*-)
0.000000000001	1×10^{-12}	Trillionth (*pico*-)

Table A.3   Mixing Fraction Definitions

One part per million	1 ppm	1×10^{-6}	one out of each million
One part per billion	1 ppb	1×10^{-9}	one out of each billion
One part per trillion	1 ppt	1×10^{-12}	one out of each trillion

easy way to understand the mixing ratio is to imagine grabbing a sample of air in a bottle. You count the total number of individual molecules in the bottle (which would amount to about 2×10^{22} molecules in a quart-size container!). Next, you count the number of specific molecules of interest. If it were carbon dioxide, for example, you would find that the quart bottle held about 7×10^{18} molecules of CO_2. The ratio of these two numbers is $7 \times 10^{18}/2 \times 10^{22} = 3.5 \times 10^{-4}$, which is the same as 350×10^{-6}, or 350 parts per million (by number or volume).

The relationship between volume and number may not be perfectly clear. In fact, the "mixing ratio by volume" and "mixing ratio by number" are equivalent for air. According to the ideal gas law and law of partial pressures (Section 3.1.1), the volume occupied by every molecule in a gas mixture is exactly the same. A molecule of oxygen and a molecule of carbon dioxide take up the same "space" in a container. The molecules themselves are so small that they do not fill a significant part of the volume, and there is no reason to assign more of the remaining space to one molecule or another. Now imagine that all the molecules of a single component of the mixture have been isolated in one section of the container. If they were constrained at the same pressure as that of the original mixture, the fraction of the total volume occupied by that component would be exactly proportional to the fractional number of molecules of that component in the original mixture (we are assuming that the temperature is fixed during these impossible manipulations).

"Parts by mass" is interpreted in a similar way as "parts by volume." For a specified volume of air, the *fraction of the total mass* associated with a certain gas is defined as its mass-mixing ratio in that volume. The volume- and mass-mixing ratios are, of course, related. The mass mixing fraction of a certain component is its volume-mixing fraction multiplied by its molecular weight and divided by the average molecular weight of air. More directly, the mass-mixing ratio is the ratio of the density of the component to the density of air in the parcel of interest. The following algebraic relations hold:

$$r_{i,m} = \frac{\rho_i}{\rho_A}$$

$$= \frac{m_i n_i}{m_A n_A} = \frac{m_i}{m_A} \frac{n_i}{n_A} = \frac{m_i}{m_A} r_{i,v} \quad (A.1)$$

where $r_{i,m}$ is the mass-mixing ratio of a gas species i in air, $r_{i,v}$ is its equivalent volume-mixing ratio, and ρ_A and m_A are the density (mass/volume) and average molecular weight (atomic mass units, amu) for air, respectively.

In this book, we use volume or number and mixing ratios or fractions wherever possible. Some books and research papers use mass-mixing fractions. They are easily converted to volume fractions when the molecular weight of the gas is known.

A.2   The Metric System: Units and Conversions

All physical parameters have units by which they are measured. For example, **time**, **length**, and **mass** are basic physical parameters that, unless given specific values, are not particularly useful in quantitative work. Similarly, temperature and electrical charge are basic physical parameters. (See Section 2.1.3 for a further discussion of temperature.) The units for each parameter provide a point of reference that defines the magnitude of the parameter. For example, time may be specified in seconds, hours, days, weeks, months, years, decades, centuries, millennia, and so on in smaller intervals such as milliseconds and microseconds. Accordingly, any specific interval of time can be expressed as a specific number of units of time. As long as the basic unit of time itself is well defined by some standard or reference, any given interval can then be precisely specified in those units. In almost all scientific systems of units, seconds are the basic measure of time. Likewise, kelvin (K) is the preferred unit for temperature.

A system of units is the collection of basic units for all the physical parameters of interest. For example, the **mks metric system** of units consists of lengths given in meters, mass in kilograms, and time in seconds. A related system of metric measure called the **cgs** system specifies lengths in centimeters, mass in grams, and time in seconds. Throughout this book, an attempt is made to conform to the *Système Internationale d'unités* (SI) designation for all units. The SI units are essentially the mks metric system units. The Congress of the United States legalized the use of the metric system in 1866. This acceptance was reinforced by the United States' signing of the Treaty of the Meter in 1875. Before that, only measurements in the British system (feet, pounds, quarts) were legally accepted. Even so, to this day Americans remain addicted to the more difficult and scientifically abandoned British system of units.

Other physical parameters deriving from the basic parameters are given units consistent with the basic units. For example, volume has dimensions of length times height times width. All three dimensions are actually length, so volume has overall units corresponding to length cubed, or l^3 (later we explain "cubed"). Thus volume has units of meters cubed (m^3) in the mks metric system and centimeters cubed (cm^3) in the cgs metric system. There is an important distinction between the terms *dimensions* and *units* as they are used here. Dimensions define the way that a quantity depends on basic physical parameters, and units define the scale by which the dimensions are measured. Thus length (l) is a basic dimension, and meter is a unit of length. Similarly, mass (m) is a basic dimension, or physical attribute, and kilogram is a specific measure of the amount of mass. Time (t) is also a basic dimension, and seconds are one of the possible corresponding units.

Force depends on the basic parameters as mass times length divided by time squared, or $m \times l/t^2$. Note that the dimension of mass as indicated by the abbreviation, m, in the dimensional expression, should not be confused with meter, which is a unit of length, l. Likewise, the dimension of time, which is abbreviated as t in dimensional expressions, should not be confused with the metric ton, or tonne, a unit of mass (also abbreviated t). One must be careful to distinguish between a dimension and its unit of measure, and between a dimensional expression and its equivalent combination of units. For example, in the mks metric system, the combination of units corresponding to the dimensional expression for force, $m \times l/t^2$, are kilogram-meter/second-second, or kg·m/sec^2. The newton (N) is defined as 1 kg·m/sec^2.

Energy can be defined as force times distance, which depends on the basic physical dimensions as $m \times l^2/t^2$, with corresponding metric units of kilogram-meter-meter/second-second, or kg·m^2/sec^2. The joule (J) is defined as a 1 newton-meter (N-m), or 1 kg·m^2/sec^2.

More complex parameters, such as a radiant energy flux, can also be defined in terms of the basic dimensions. For example, any energy flux may be defined as energy per area per time, which can be decomposed into a dependence on basic parameter dimensions as, $(m \times l^2/t^2)/(l^2 \times t) = m/t^3$. We

have several equivalent combinations of metric units for the energy flux: $J/m^2{\cdot}sec$ = kilograms/sec^3 = watts/m^2 (where only the second form is given in basic units). The watt (W) is the specific unit of power in the metric system, where 1 watt is 1 J/sec.

COMMON UNITS OF MEASURE

We next compare the most frequently employed units of measurement for length and mass in different systems of units. Note that one of the basic physical parameters (and units) that is not treated explicitly in this appendix is **electrical charge**. Electrical charge is important to the development of electromagnetic theory, which has not been discussed in the text. Accordingly, the unit of charge and the units deriving from it, are not defined further.

Length

1 meter (m) = 100 centimeters (cm)= 1000 millimeters (mm)
1000 m = 1 kilometer (km)
1 m = 1×10^6 micrometers (mm); 1 mm = 1×10^{-6} m
1 m $\cong$ 39 inches (the symbol $\cong$ means "very closely, but not exactly, equals")
100 m $\cong$ 110 yards
1 inch $\cong$ 2.5 cm
1 mile $\cong$ 1.6 km

Volume

1 liter (l) = 0.001 cubic meter (m^3) = 1000 cubic centimeters (cm^3)
1 liter $\cong$ 61 cubic inches $\cong$ 1.06 quarts

Mass

1 kilogram (kg) = 1000 grams (g)
1 metric ton (t) = 1 tonne = 1000 kg = 1×10^6 grams (g)
1 kg $\cong$ 2.2 pounds
450 g $\cong$ 1 pound
1 tonne $\cong$ 2200 pounds = 1.1 English ton (or ton)

Among the many derivative units that can be defined, such as the newton, joule, and watt, are the following useful units. Note that pressure is defined as force per unit area.

Pressure

1 pascal (Pa) = 1 newton/meter^2 (1 N/m^2; or 1 kg/
  m·sec^2)
100,000 Pa = 1 bar (roughly the average pressure at
  the Earth's surface)
1000 millibar (mb) = 1 bar
1 mb = 100 Pa
1 atmosphere (atm) ≅ 1013 mb
1 atm ≅ 1 bar ≅ 14.7 pounds/inch^2

MANIPULATION OF DIMENSIONS AND UNITS

Dimensions of parameters and their corresponding
units can be treated like variables in an equation. The
symbols representing the dimensions of length (l),
mass (m), and time (t) can be manipulated in any
expression in which the original physical quantities
are multiplied or divided. We did this when manipulating the basic dimensions of derived quantities
such as force and energy flux. We can write out
similar relationships for the units corresponding to
the dimensions. Thus, $m \times l/t^2$ is equivalent to
kg·m/sec^2. In more complex combinations of dimensions or units, we can cancel dimensions or units
that appear in both the numerator and denominator
and collect dimensions or units that are multiplied.

One of the most important rules regarding the use
of units is that whenever they are combined in any
expression, *the units must always be consistent*. One
should never, for example, mix together mks and cgs
units; or use seconds in one part of an expression and
years in another, unless it is clearly established that
such usage will not cause problems. Accordingly, if
an expression with two or more parameters is being
evaluated, you have two choices: (1) Make sure all
the parameters are in the same units (mks, say) or (2)
correct the units of the individual parameters, or the
final expression, by multiplying by appropriate **conversion factors**.

The principal guidelines for the conversion of
units can be demonstrated using a simple example.
Suppose you want to evaluate the distance that an air
mass had traveled in a certain time and you know the
velocity (speed) of the parcel. You will need to
evaluate the expression

$$d = v\tau \qquad (A.2)$$

where d is the distance, v is the velocity, and τ is the
time. The velocity has units of length per unit time,

or l/t. The basic dimensions are consistent because
Equation A.2 can be written in dimensional form as,
$l = (l/t) \times t = l$. Notice that the dimension of time
cancels out on the right-hand side of Equation A.2.

If velocity is measured in meters per second and
the time is measured in seconds, then clearly the
distance will be in meters. This is a consistent set of
units. However, if the velocity is given in m/sec and
the time is given in hours, the units will be inconsistent. We must convert the parameters to consistent
values, by using a conversion factor. In this case, we
note that 1 hour = 3600 seconds and so write
Equation A.2 as follows:

$$d \,(\text{meters}) = v\left(\frac{\text{meters}}{\text{second}}\right) \times \tau\,(\text{hours})$$

$$\times \left(\frac{3600 \text{ seconds}}{1 \text{ hour}}\right) \qquad (A.3)$$

$$= 3600 \, v\tau$$

The converted equation can now be evaluated using
the original unconverted values of the velocity and
time parameters. The conversion factor that has been
applied in this example is the last term in the middle
section of Equation A.3. Notice that this term changes
the units of hours in the numerator to units of seconds
by multiplying by the number of seconds per hour.

Another example follows from the box model
equation for the concentration, q, of a material in a
reservoir (see Equation 4.1, for example):

$$q = \frac{S\tau}{V} \qquad (A.4)$$

where S is the source, τ the residence time, and V the
volume of the reservoir. Imagine that the source is
given in tonnes per day, the residence time is given
in hours, the volume is specified in cubic kilometers,
and the concentration is in grams per cubic centimeter. Clearly, each of the parameters determining q
needs to be converted to the appropriate units.

Instead of illustrating this procedure step by step,
the various conversions are shown collectively in a
modified version of Equation A.4:

$$q\left(\frac{g}{cm^3}\right) = \frac{S\left(\frac{\text{tonne}}{\text{day}}\right)\left(1 \times 10^6 \, \frac{g}{\text{tonne}}\right)\tau\,(\text{hours})\left(\frac{1 \text{ day}}{24 \text{ hours}}\right)}{V\left(km^3\right)\left[\left(\frac{1000 \text{ m}}{1 \text{ km}}\right)\left(\frac{100 \text{ cm}}{1 \text{ m}}\right)\right]^3}$$

$$= \frac{S\tau}{V} \left[\frac{\frac{1 \times 10^6}{24}}{(1000 \times 100)^3} \right] = \frac{S\tau}{V} \left(\frac{1 \times 10^{-9}}{24} \right) \quad \text{(A.5)}$$

In this massive equation, the same simple form of the original equation is retained. Each of the parameters on the right-hand side of the equation is operated on by a specific conversion factor. The conversion factors simply relate the original units to the desired units, based on the relationships between the common units of measure listed earlier. Some of the conversions require more than one step. For example, to get from kilometers to centimeters, we first convert kilometers to meters, and then convert meters to centimeters. This could be done in one step if the relationship between kilometers and centimeters were already known—that is, $1 \text{ km} = 1 \times 10^5 \text{ cm}$. The final overall conversion factor in Equation A.5 combines all the individual conversion steps by straightforward multiplication and division.

A.3 Physical and Mathematical Constants

In physics, chemistry, and mathematics, certain numbers are special. These special numbers are usually constants that make general relationships between parameters into exact relationships with the appropriate units. For example, the thermal energy, E, of a molecule in a gas is proportional to the temperature, T, of the gas. But the energy is given quantitatively by the relation

$$E = \frac{3}{2} k_B T \quad \text{(A.6)}$$

Here the physical "constant," k_B, fixes the amount of energy. It was originally determined by the physicist Ludwig Boltzmann, and so is called **Boltzmann's constant**. Many simple and useful relations in science contain a fundamental constant such as Boltzmann's constant. Some of these, which are referred to in the chapters of this book, are listed next.

PHYSICAL CONSTANTS (AND THEIR COMMON SYMBOLS)

The following constants have associated units of measure for the mks metric system of units.

Universal Constants of Physics and Chemistry

Atomic unit mass, $m_0 = 1.67 \times 10^{-27}$ kg/amu: The atomic unit mass is equivalent to the mass of a proton or neutron in an atomic nucleus. If the total number of protons and neutrons in a nucleus is known—that is, the number of atomic mass units, or "amu"—the total mass of the nucleus will be the number of amu multiplied by the unit mass, m_0. For convenience, the weights of atoms and molecules are stated in amu rather than kilograms. The weight of 1 mole, or Avogadro's number, of atoms or molecules is equal in grams to the number of amu; that is, $m_0 N_A = 1 \times 10^{-3}$ kg/amu = 1 g/amu.

Avogadro's number, $N_A = 6.02 \times 10^{23}$ molecules/mole: Avogadro's constant defines a "mole" of a substance; it is exactly N_A molecules—or smallest molecular entities—of that substance. The mass of a mole of a substance is equal to the atomic weight of the substance expressed in grams. For example, the atomic weight of ozone (O_3) is 48, equivalent to the combined atomic weight of three oxygen atoms. One mole of ozone therefore is the same as 48 grams. In other words, 6.02×10^{23} ozone molecules weighs 48 g.

Boltzmann's constant, $k_B = 1.38 \times 10^{-23}$ J/K: Boltzmann's constant relates temperature to the thermal energy of motion of single molecules. To emphasize this point, the units of k_B can be written as J/K·molecule. Boltzmann's constant is the most frequently used physical constant in descriptions of gases and their mechanical and thermodynamic behavior.

Gas constant, $R_g = 8.31$ J/K-mole: The "universal" gas constant is closely related to Boltzmann's constant, since both connect temperature with energy. In this case, the energy is per *mole* of gas, rather than energy per molecule. As you might suspect, the gas constant can be defined in terms of Boltzmann's constant and Avogadro's number as $R_g = k_B N_A$. This equivalence is easy to check.

Gravitational constant, $G = 6.67 \times 10^{-11}$ N·m²/kg²: The gravitational constant quantifies the gravitational force that one body of a given mass exerts on another body of known mass. The law of gravitation applies to *all* objects regardless of size. The smallest objects exert some gravitational force on the largest, although only the gravity of very large objects, such as the Earth, is obvious.

Planck's constant, $h = 6.63 \times 10^{-34}$ J·sec: Planck's constant relates the energy of a photon of radiation to the frequency of the radiation. (Frequency and wavelength are connected through Equation 3.18.) Only the frequency or wavelength of light needs to be known in order to determine the energy of a photon of radiation at that frequency or wavelength (assuming that the radiation is traveling in free space or in air where the speed of light is constant).

Speed of light, $c = 3.00 \times 10^8$ m/sec: The speed, or velocity, at which light, or *any other radiation*, moves in a vacuum is exactly determined as 3×10^8 m/sec. Moreover, the velocity of a photon is independent of wavelength (or frequency) and holds for all photons. The speed of light in other materials (air, glass, water, and so forth) is slower, although the velocity in air is almost exactly the same as in a vacuum.

Stefan-Boltzmann constant, $\sigma_B = 5.67 \times 10^{-8}$ W/m^2·K^4: The Stefan-Boltzmann constant relates the power emitted by a blackbody, per unit surface area of the blackbody, to the fourth power of its temperature (Section 3.2.1).

Wien constant = 2900 μm·K: The Wien constant can be derived from Planck's constant for blackbody radiation and relates the temperature of the blackbody to the wavelength at which the radiant energy is the most intense (Section 3.2.1).

"Constants" of the Physical World

The following physical "constants" have been used in the text of this book or may be of general use to those studying the atmospheric environment. Note that in several cases, the "constants" are not truly constant, but may vary by some small amount. In many applications, particularly at the level of detail required in this book, they may nevertheless be treated as true constants.

Angular speed of rotation of Earth, $\Omega = 7.29 \times 10^{-5}$ radians/sec: The Earth rotates on its axis at a nearly constant "angular velocity." The unit of angular motion is the "radian"; there are exactly 2π radians in a single rotation about an axis (later we define π). One radian equals about 57 degrees of angle. The Earth rotates through 2π radians, or 360 degrees of angle, in somewhat less than one day, or 86,400 seconds.

Atmospheric pressure at sea level = 1013 mb = 1.013×10^5 N/m^2: This is the average pressure of the atmosphere.

Density of air at sea level = 1.225 kg/m^3: The average density of the atmosphere at sea level is determined by the average surface pressure and temperature.

Gas constant for dry air (R_g^*) = 287 J/K·kg: The gas constant for dry air can be obtained by dividing the "universal" gas constant by the average molecular weight of air in amu, or about 29 g/mole, and multiplying by 1000 to convert from grams to kilograms. The gas constant says that each kilogram of air holds 287 joules of thermal energy for each degree kelvin of its temperature.

Gravitational acceleration (g) = 9.81 m/sec^2: The gravitational *force* exerted by the Earth on an object at sea level is simply the *weight* of the object. The weight is the mass of the object multiplied by the gravitational acceleration, $W = mg$. Often, an object's "weight" and "mass" may be treated as interchangeable parameters; the equivalence derives from the constancy of the gravitational acceleration, which relates them.

Mass of the Earth = 6.00×10^{24} kg: The masses of the major components of the Earth system are discussed in Sections 2.3.1 and 4.1.1.

Radius of the Earth = 6.37×10^6 m: This is the average radius of the Earth. The Earth tends to be somewhat flattened at the poles rather than perfectly spherical; the greatest radius is at the equator because rotation on an axis creates a centripetal force that causes the equatorial regions to bulge.

Solar constant (F_S) = 1390 W/m^2: The solar constant is discussed in Section 11.3.1.

Sun-Earth distance = 1.50×10^8 m: The mean distance of the Earth from the sun corresponds to the orbit of the Earth around the sun. The difference between the maximum and minimum distances is about 5 percent of the average distance. Accordingly, the distance between the sun and the Earth is only roughly constant.

MATHEMATICAL CONSTANTS

π: A mathematical constant that is exactly equal to the ratio of the circumference of a circle to its diameter, π also gives the area of a circle when multiplied by the radius of the circle squared. (See Section A.4 for a definition of "squared.") There are exactly 2π angular radians in one rotation or revolution about a point or axis: $\pi = 3.1416$.

e: The base of the **natural logarithms**, or *ln*, is $e^{\ln(x)} = x$ (Section A4). Also, e is the base for the exponentiation in the exponential function, exp; $e = 2.718$.

A.4 Mathematical Operations

The principal mathematical operations used in this text are taking the "square," and its complement, taking the "square root," using the natural exponential operation, and algebraically manipulating simple equations. These operations are summarized next.

SQUARES AND SQUARE ROOTS

The "square" of a number is simply the number multiplied by itself, so the square of 2 is $2 \times 2 = 4$. The square of 3 is $3 \times 3 = 9$, and so on. The square of any number can be written in the form

$$s(n) = n \times n = n^2 \qquad \text{(A.7)}$$

where $s(n)$ is the "square of n," and n is the number to be "squared." The exponent shown as a superscript of n is the "power" to which n is raised—here, the second power. The second power of n is the same as the square of n. The number n may be a very large number, a fraction, or any physical or mathematical constant.

The operation that is opposite to squaring a number is taking its "square root." The special mathematical symbol, $\sqrt{}$, is used to indicate a square root. The square-root operation determines the quantity that, if multiplied by itself, will yield the original number. For example, we know that $2 \times 2 = 4$; therefore, 2 must be the square-root of 4, or $\sqrt{4} = 2$. There is nothing mysterious here, just a little arithmetic. The square root of a number can be written in shorthand notation as

$$r(n) = \sqrt{n} \; ; \; r \times r = n = n \times n = n^2 \quad \text{(A.8)}$$

It follows that n is the square of r. The square root may also be written as a the fractional one-half power of a number, such as: $r = \sqrt{n} = n^{1/2}$. Further relationships may then be written, for example:

$$\sqrt{s(n)} = \sqrt{n^2} = (n^2)^{1/2} = n^{2 \times 1/2} = n^1 = n$$

$$\text{(A.9)}$$

This relation shows that the square root of the square of n is n. A key point is that when a number is raised to a "power" in two or more consecutive operations, the final effect can be obtained by multiplying to-gether the separate powers into one number. As you might expect, if you undo a mathematical operation you just performed with an exactly complementary operation, you will end up at the starting point.

HIGHER POWERS

We are not restricted to "squares" in multiplying a number by itself. It would be just as easy to multiply a number by itself three times; or four times, and so forth. The cube of a number is that number raised to the "third power," or multiplied by itself three times. Thus the cube, or third power, of 2 is $2^3 = 2 \times 2 \times 2 = 8$. The cube of 3 is $3^3 = 3 \times 3 \times 3 = 27$. And so on. The "cube root" is also defined as for the square root: The cube root is the one-third power of a number that, when multiplied by itself three times, returns the original number. Hence the cube root of 8 is $8^{1/3} = (2 \times 2 \times 2)^{1/3} (2^3)^{1/3} = 2$. One only need perform a complementary operation to obtain a root (although this operation is not always so easy).

These arguments may be extended to the "fourth power." The fourth power of 2 is $2^4 = 2 \times 2 \times 2 \times 2 = 16$; for 3, it is $34 = 3 \times 3 \times 3 \times 3 = 81$. The "fourth root" is defined in a similar way as for the other roots. The fourth root is the one-fourth power of a quantity. In the case of the number 16, the fourth root is 2, found as $16^{1/4} = (2^4)^{1/4} = 2$. One trick (you will perhaps not find so entertaining) is to find the fourth root of a quantity by taking the "square root" twice, or the square root of the square root. This is illustrated for the number 16:

$$16^{1/4} = (16^{1/2})^{1/2} = \sqrt{\sqrt{16}} = \sqrt{4} = 2 \quad \text{(A.10)}$$

Notice that the square root of 16 is 4 (that is, $4 \times 4 = 16$), and the square root of 4 is 2. The one-fourth power is equivalent to applying two one-half powers consecutively, since $\frac{1}{4} = \frac{1}{2} \times \frac{1}{2}$.

EXPONENTIALS AND LOGARITHMS

Exponential functions are commonly used in physics and mathematics. Many fundamental processes can be described by "exponential" behavior. The exponential of a number, n, is defined by the mathematical expression

$$\exp(n) = e^n; \; \exp(-n) = e^{-n} = \frac{1}{e^n} \qquad \text{(A.11)}$$

The number, n, is the "exponent" in this relation. The mathematical constant, e, is called the "base of the natural logarithms" (its value was given earlier). Notice that when an exponential function is inverted, only the sign of the exponent changes (the last two relations in Equation A.11). Any number, n, can be placed in an exponential function. Moreover, the exponential function is *always positive*, even when the exponent is negative. The exponential of "zero" is exactly "one": $\exp(0) = 1$.

The logarithmic function is the complement of the exponential function. Taking the logarithm of an exponential yields the exponent. The logarithmic function is represented by the symbol ln.[1] Then the following relations hold:

$$\ln[\exp(n)] = \ln e^n = n; \; \ln e = 1 \quad (A.12)$$

We can calculate (or "take") the logarithm of any *positive* number. Although the logarithm of a negative number is not defined, the value of a logarithm may be negative if the original number is smaller than 1. The logarithm of "one" is exactly "zero": $\ln 1 = 0$.

If x is a number larger than zero, then its logarithm is $m = \ln x$. Because the exponential and logarithmic operations are complementary, it follows that

$$\exp(\ln x) = e^{\ln x} = e^m = x \quad (A.13)$$

Accordingly, any positive number can be written as an exponential of a base number, which is the base of the logarithms (Footnote 1).

The exponential and logarithmic functions are related to the scientific notation discussed earlier in this appendix. It turns out that $\ln 10 = 2.3$. Then the following relations hold:

$$\exp(2.3n) = e^{2.3n} = (e^{2.3})^n = 10^n \quad (A.14)$$

In other words, taking an exponential is similar to dealing with powers of 10 in scientific notation. When multiplying or dividing numbers, exponents may be added or subtracted, respectively, as in scientific notation. For example:

1. Logarithms that are calculated using the mathematical constant e as the base are called *natural* logarithms. This is sometimes written as $\ln_e$, although the symbol ln represents the natural logarithm. Logarithms can be taken with respect to any base as long as the base has a positive value. The most common alternative base for logarithms is 10 (that is, $\ln_{10}$), which is consistent with the representation of large and small numbers in powers of 10, or in scientific notation.

$$\exp(n) \times \exp(m) = \exp(n + m);$$
$$e^n e^m = e^{n+m};$$
$$10^{2.3n} \times 10^{2.3m} = 10^{2.3(n+m)} \quad (A.15)$$

The division of exponential functions, like the division of numbers expressed in powers of 10, is equivalent to subtracting the exponents:

$$\exp(n) \div \exp(m) = \exp(n - m);$$
$$\frac{e^n}{e^m} = e^{n-m};$$
$$10^{2.3n} \div 10^{2.3m} = 10^{2.3(n-m)} \quad (A.16)$$

Although the terminology and notation used to describe exponentials and logarithms may seem a bit obscure, these functions are indispensable to mathematicians and scientists. They are an important part of the shorthand language of science.

ALGEBRAIC EQUATIONS

We have been using equals signs throughout this appendix to indicate equivalence between two numbers or quantities or functionalities. Such "equations" are nothing more than a statement that two specific quantities are equal. Trivial equations of the form $x = x$ (for example, $2 = 2$), convey no information; they are obvious identities. But more general equations that connect different parameters or functions can convey useful information in a compact and convenient form. Thus the equation $x = y$ simply states that the quantity x is equal to the quantity y. It should be clear that this statement can be inverted to say y is equal to x. The two statements are perfectly compatible, given the equality. In fact, all the equations in this book can be interpreted in either direction.

If either x or y or both are functions of some sort, the equation may be an "algebraic" equation. For example, we may have the relation

$$x = y^2 \quad (A.17)$$

This equation represents a straightforward statement that x equals the square of y, which establishes the unequivocal relationship between x and y. Indeed, from our previous discussion, we can easily deduce that y is the *square root* of x. That is, the equation can be inverted, or solved for y in terms of x. In this case, the procedure is simple: We take the square root on each side of the equals sign and switch sides to get

$$y = \sqrt{x}, \quad \text{or} \quad y = x^{1/2} \qquad \text{(A.18)}$$

In any equation, we can apply the same operation to both sides of the equals sign without changing the equality. In this example, we took the square root. We could also have divided each side by the same constant or taken the fourth root of each side, or taken the logarithm, or carried out any number of operations sequentially. Usually, the operations are chosen to simplify one side of the equation or to isolate one of the variables or parameters on one side of the equals sign. In the preceding example, we isolated y on the left-hand side of the equation (which is usually the side chosen for the single variable).

We can take a concrete example from Chapter 11, in which Equation 11.12 was derived from the energy balance model of the Earth:

$$F_S(1 - \alpha_e) = 4\sigma_B T_e^4 \qquad \text{(A.19)}$$

In this equation, we could have isolated, or solved for, any of the parameters or variables. Typically, one variable is to be determined from all the other information given. In this instance, it is the temperature, T_e. This can be accomplished by performing the following sequence of operations on each side of the equation: Divide by 4; divide by σ_B; take the one-fourth power of each side; reverse the sides of the equation. Having performed these steps, you will have derived Equation 11.14:

$$T_e = \left[\frac{F_S(1 - \alpha_e)}{4\sigma_B} \right]^{\frac{1}{4}} \qquad \text{(A.20)}$$

Equations are fun to work with because they are concise, precise statements of facts about the world. So do not be intimidated by these wonderful little artifacts of science.

INEQUALITIES

Not all algebraic relations are exact. Equations are exact when they are connected with an "equals" sign (=). But we sometimes wish to express a conditional relationship between two parameters. For example, it may be important to know when one quantity is "less than" another or "greater than" it. Such conditions are expressed in a concise way by means of special mathematical symbols. The symbol < is literally interpreted as "less than." If the condition $x < 0$ holds, then the value of x is always "less than" zero. It can be $-1, -2, -10^6$, or any other negative number. Related symbols are >, which means "greater than"; ≤, which translates to "less than or equal to"; and ≥, which means "greater than or equal to." The inequalities express a certain degree of uncertainty in a relationship and, at the same time, a degree of certainty. Someone may admit to being older than you but may not say by how much. In that case, you could mathematically write

their age > my age

You are using a mathematician's shorthand.

Appendix B
Demonstrations of Common Natural Phenomena

In this appendix, we describe several simple experiments in order to demonstrate a few of the basic physical and chemical principles discussed in the text. It is not recommended that students perform these experiments except under the supervision of a trained instructor. A few of them require special equipment or chemical compounds that are not readily available to students and that can be dangerous to handle.

Demonstration 1: Light Scattering by Small Particles

BACKGROUND

The experiment shown in Figure B.1 demonstrates the basic characteristics of light scattering by fine particles. Airborne particles (aerosols) have a wide range of sizes. The relative efficiency with which aerosols scatter sunlight depends on the size of the particles relative to the wavelength of the light radiation of interest (Sections 3.2.2, 6.5.2, and 14.3.2). Very fine particles, with sizes much smaller than the wavelength of light, cause Rayleigh-scattering effects. Larger particles (including cloud droplets) cause Mie-scattering effects. In the atmosphere, Rayleigh-scattering produces the blue color of the sky and the redness seen in the twilight. Mie-scattering is responsible for the white appearance of clouds. Conversely, particle sizes can be estimated by observing the characteristics (color, intensity, and polarization) of the scattered and transmitted radiation.

In the experiment described here, Figure B.1, a colloidal suspension is used in place of an aerosol. A **colloidal suspension** consists of fine particles or droplets suspended in a fluid. For instance, when you try to mix oil and vinegar together to make salad dressing, vigorous shaking produces a colloid of small oil droplets drifting in the vinegar base. The harder you shake, the smaller the droplets of oil will become. Oils does not mix easily with vinegar or water and tends to form colloids rather than true solutions, or mixtures of a solute in a solvent. Fluids that are readily mixed together in solutions are said to be **miscible**. A small amount of milk added to a large beaker of water will form a colloid, or emulsion, not a solution. Milk is not highly miscible in water because of its high fat content. The milk instead breaks up into small globules, or droplets, that are analogous to aerosols suspended in the atmosphere. Indeed, fine atmospheric aerosols represent a form of colloidal suspension.

The small droplets of milk scatter light quite effectively, which can be observed by illuminating

Figure B.1 The demonstration of light scattering by small particles. An emulsion of milk and water is very effective in demonstrating scattering effects.

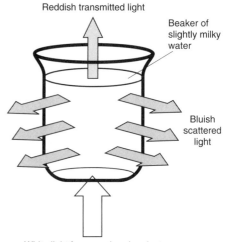

Reddish transmitted light

Beaker of slightly milky water

Bluish scattered light

White light from overhead projector

the beaker with light. If the milk breaks up into small enough particles, the scattered light will have a bluish cast, and the transmitted light will be reddish. If these spectral characteristics are observed, then Rayleigh scattering is occurring in the milk-water suspension. Furthermore, the milk droplets must be smaller than the wavelength of the incident light. If no coloration is apparent—that is, if the emulsion appears white—then Mie scattering is dominant, and the droplets of milk are larger than the wavelength of visible light.

EXPERIMENTAL PROCEDURE

1. Place a clean beaker full of clear water onto an overhead projector. It is preferable to use a large or tall beaker so that the path lengths are long enough to produce sufficient scattering and reduce the amount of transmitted light. Note the lack of scattering in the water and the whiteness and brightness of the transmitted light.

2. Add about a teaspoon of milk to the few liters of water in the beaker. As the milk breaks up, the water becomes visibly clouded. Carefully examine the scattered light for any bluish coloration. Then examine the transmitted light by looking at the projection screen. As we noted, the bluish tint of the scattered light and the reddish tint of the transmitted light indicate Rayleigh scattering and suggest that the milk has dispersed into very small particles.

Demonstration 2:   Gas-to-Particle Conversion in Smog

BACKGROUND

Aerosol particles are generated both from pollutants in photochemical smog and from natural organic vapors emitted by vegetation. The aerosol particles form a haze, which can restrict visibility in regions where pollutants and organic compounds have been emitted. One class of secondary aerosols in pollution is produced when organic vapors are oxidized by ozone (Sections 3.1.2, 3.3.4, 6.2.1, and 6.5).

Ozone is a major component in photochemical smog, and it is also readily generated in the laboratory using ultraviolet light at sufficiently short wavelengths. The UV radiation breaks up oxygen molecules in ambient air into oxygen atoms. The oxygen atoms then rapidly combine with other oxygen molecules to form ozone. Organic vapors in air come from the incomplete combustion of fossil fuels, from fuel vaporization, and from natural emissions of vegetation. Ozone easily reacts with a large number of organic compounds. Upon reacting with ozone, some organic compounds form new hydrocarbon molecules that are "sticky." That is, the sticky hydrocarbons have very low vapor pressures and tend to condense, or aggregate, into small particles. Eventually, these organic particles become large enough to scatter significant amounts of light, which is manifested as a noticeable haze in the air. In rural forested areas where there is no urban smog, a haze may still be apparent, which comes from natural organic compounds through gas-to-particle conversion processes.

Unlike those found in an urban environment, where anthropogenic sources of hydrocarbons are dominant, the hydrocarbons found in rural areas are composed mostly of terpenes, which are complex aromatic organic gases emitted by many plants. For example, pinene is a terpene that causes the odor associated with pine trees. The terpenes react with ozone to produce compounds that condense as aerosol particles. The relatively small size of the aerosols formed in this way—that is, by molecular aggregation—is evidenced by the tendency of natural organic hazes to have a bluish cast, which indicates Rayleigh scattering by small particles.

EXPERIMENTAL PROCEDURE

1. Place an ultraviolet lamp in a clean glass beaker with a glass cover (Figure B.2). This UV source must emit electromagnetic radiation at short enough wavelengths to photodissociate oxygen molecules (the "black light" commonly available in novelty stores is inadequate for this experiment). *Note that all ultraviolet lamps can be dangerous at close range.* Ozone is produced in the beaker through a process similar to that occurring in the stratosphere, where ultraviolet sunlight dissociates the oxygen in air to generate the ozone layer.

2. After a few minutes, turn off the UV light, and carefully remove it from the beaker. Try to avoid breathing the ozone that will leak out of the beaker. At this point it is useful to note that the

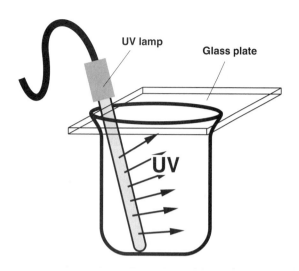

Figure B.2 Setup for the haze-production demonstration. The UV lamp generates a high concentration of ozone in the beaker. A glass cover should be used to prevent the ozone from escaping (ozone is a toxic substance!).

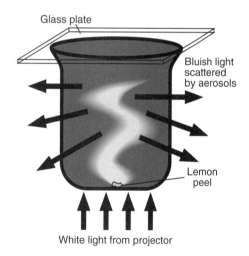

Figure B.3 The effect of the ozone reacting with a piece of lemon peel dropped into the beaker. A dense haze of organic aerosols is generated in a matter of minutes.

beaker is clear and does not appear to contain any particles, even though it is filled with ozone. To create a hydrocarbon haze, drop a freshly shaved piece of lemon or orange peel into the beaker. The piece need be no larger than a fingernail. The peel of a citrus fruit emits terpenes, particularly limonene. Limonene gives the fruit its distinctive and pleasant odor. The li-

monene is also the source of hydrocarbon vapor for the formation of haze in this experiment.

3.   Almost as soon as the peel is dropped into the beaker, the limonene will react with the ozone, and the product hydrocarbons will coagulate into visible aerosol particles. The aerosols can be highlighted by shining a light through the beaker (for example, by placing the beaker on an overhead projector during the experiment, with appropriate baffles to create a beam of light). With the proper lighting, a smoky haze can be seen rising from the peel in wisps and plumes (Figure B.3). These swirl along through the air currents in the beaker. Eventually, the entire beaker will be filled with the organic "smoke" as a uniform haze.

Demonstration 3:   Atmospheric Pressure and Water Vapor Condensation

BACKGROUND

The experiment shown in Figure B.4 demonstrates the weight of the atmosphere and the force of atmospheric pressure (Sections 2.1.3, 2.2.2, 2.3.3, 3.1.1, and 12.1.2). A trick is used to create a substantial pressure decrease inside a closed container. Water is heated and evaporated in the container. The water vapor displaces air from the container, which is then sealed. As the container cools, the water vapor recondenses into liquid water, and the pressure inside the can drops. The demonstration thus also shows how the amount of water vapor in air is sensitively controlled by temperature. Water vapor always attempts to achieve equilibrium with its saturation vapor pressure by condensing or evaporating. At its boiling point of 100°C, the vapor pressure of water is equal to 1 atmosphere.[1] At room temperature, the vapor pressure of water is less than 1 percent of atmospheric pressure (which explains why even

---

1.   The definition of the **boiling point temperature** of a liquid is the temperature at which the vapor pressure of the liquid equals atmospheric pressure (1 atmosphere). At the boiling point, the internal vapor pressure of the liquid is sufficient to form bubbles in the liquid. The bubbles rise to the surface, bringing the liquid to a "boil." Different liquids have a different boiling point, which is a basic property of the substance. At high altitudes, where the air pressure is lower, the boiling point is correspondingly lower; that is, the vapor pressure of the liquid becomes equal to the surrounding atmospheric pressure at a lower temperature.

the most humid air contains less than a few percentage points of water by weight). Hence the pressure exerted by water vapor in a container can be controlled over a wide range by varying the temperature of the container. Thrusting it into an ice bath would allow the container to be cooled further, for example, causing more water vapor to condense and depressing the water vapor pressure in the container below the room temperature value. If the container originally were completely filled with water vapor, having expelled all the air, the pressure in the container could be made much smaller than the pressure of the atmosphere.

Changes in pressure caused by vapor condensation and evaporation are often experienced in everyday life. If hot food is placed in a plastic container, sealed, and stored in the refrigerator, the container may collapse as the food cools and moisture in the container condenses. The container lid may be sucked in and difficult to remove because the reduced vapor pressure inside makes it difficult to open. The more rigid the container is, the more difficult it can be to remove the top. Recall the discussion of the behavior of gases in Section 3.1.1. If the volume of the container is held fixed as the vapor is cooled, the pressure difference from the atmosphere will be greater than if the volume shrinks, as would happen if the container could distort. A similar effect occurs when a can of paint stands around for a long time and the oxygen in the can reacts with the solvents in the paint to form a film. Opening the can may produce a hiss of air being sucked into the vacuum that is left. At the other extreme, heating a liquid in a closed container is always dangerous, as the increasing vapor pressure will eventually rupture the container, therefore, the need for safety valves on devices operating at high temperatures. Similarly, in the experiment described next, the container should never be sealed when heat is being applied to it.

EXPERIMENTAL PROCEDURE

1. A new metal jerrican (paint thinner can) can be used as the container for this experiment. Obviously, the can should not contain paint thinner or have ever been used to store any flammable solvent. Fill the bottom of the can with water. After the water inside the open can has boiled for several minutes, most of the air molecules in the can will be displaced by evaporating water mol-

Figure B.4 Configuration to demonstrate the force of atmospheric pressure and to investigate the vapor pressure of water. The container used to hold the water should be sturdy and clean.

ecules. Since the can is open, the pressures inside and outside are equalized at 1 atmosphere. The boiling pressure of water is also 1 atmosphere. Ideally, all the air can be expelled, but in reality, some air will remain in the can.

2. Remove the can from the heat and quickly cap it. To hasten cooling, the can may be immersed in cold water or ice water. As the vapor inside cools, water vapor condensation quickly depresses the pressure in the can to 10 percent or less of atmospheric pressure. Soon the atmospheric pressure exerted relentlessly on the outside of the can overcomes its mechanical strength and crushes it (Figure B.5). If the metal walls of the can were thick enough to resist the force of atmospheric pressure, the can would retain its shape. However, even a rather sturdy metal container is easily crumpled by a force of up to 14 pounds per square inch of surface (the force of atmospheric pressure). The can may resist for a while as it first begins to cool, but soon it will be crumpled by an invisible force—the tenuous air around it.

Demonstration 4:   Acid Rain Formation

BACKGROUND

The formation of acid rain involves a sequence of chemical processes in which primary air pollutants, such as nitric oxide (NO), are converted to acids. In this experiment, nitric acid is produced and detected

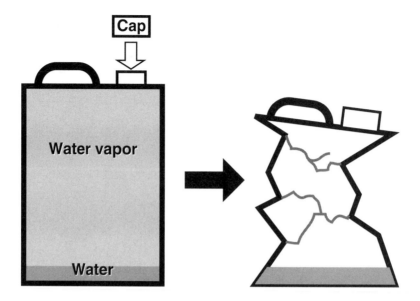

Figure B.5 Results of cooling a metal can filled with water vapor at its boiling point.

(Sections 3.3.4, 6.2.1, 6.5.3, 9.2, and 9.3.3). An intermediate product of nitric acid formation, nitrogen dioxide (NO_2), is seen as a reddish brown gas in the experiment. Nitrogen dioxide is readily absorbed in liquid water to form nitric acid (HNO_3). The acidity of water can be demonstrated by the use of an "indicator," **methyl red** in this case, which turns reddish in color when the pH drops below a threshold value of about 5.0 (Section 9.2). This experiment illustrates how water can be acidified by exposure to NO_x. In the atmosphere, however, other processes can acidify rainwater or even snow (Sections 6.5.3, 9.3, and 10.2.1). For example, aerosol particles formed in the atmosphere downwind of metal ore smelters, which emit large amounts of sulfur dioxide, often contain large amounts of sulfuric acid. When these particles are scavenged by water drops or ice crystals, they too can become highly acidic.

EXPERIMENTAL PROCEDURE

1. Prepare a 1- to 2-liter flask of nitric oxide (NO) by first evacuating all the air from the flask (a good vacuum pump is required) and then refilling the flask with pure NO to a pressure of about 100 mm-Hg. Have available 250 milliliter (ml) of distilled water (pH = 7.0), and 150 ml of a premixed solution of methyl red in distilled water. This mixture will be yellow in color, indicating a neutral pH of the water.

2. First, note that the flask of pure nitric oxide is perfectly clear (unless some air remained in the flask) [Figure B.6(a)]. Next, slowly open the valve on the flask, allowing some ambient air to enter. Close the valve. As the air mixes with the nitric oxide, the NO will react with the oxygen (O_2) in the air to form nitrogen dioxide (NO_2). This gas will assume a distinctive reddish brown color [Figure B.6(b)].

3. Remove the cork from the flask, and quickly add the distilled water (not letting the nitrogen dioxide escape). Close the flask and shake it vigorously. The reddish brown nitrogen dioxide will disappear as it is absorbed into the water, and the flask will become clear again. As the nitrogen dioxide is absorbed into the water, nitric and nitrous acids are produced [Figure B.6(c)].

4. In order to demonstrate the acidity of the water in the flask, transfer some of the clear liquid from the flask into a column of the methyl red indicator solution, which is initially yellow because the pH of the solution is > 5.0. However, as the clear solution from the flask is added to the column, the yellow indicator turns bright pink [Figure B.6(d)]. The color change of the indicator shows that the pH of the liquid in the flask is well below 5.0. The pH-neutral distilled water added to the flask has been highly acidified, just as clouds have been that form acid rain.

The chemistry that occurs in the beaker is slightly different from the chemistry forming acids in the

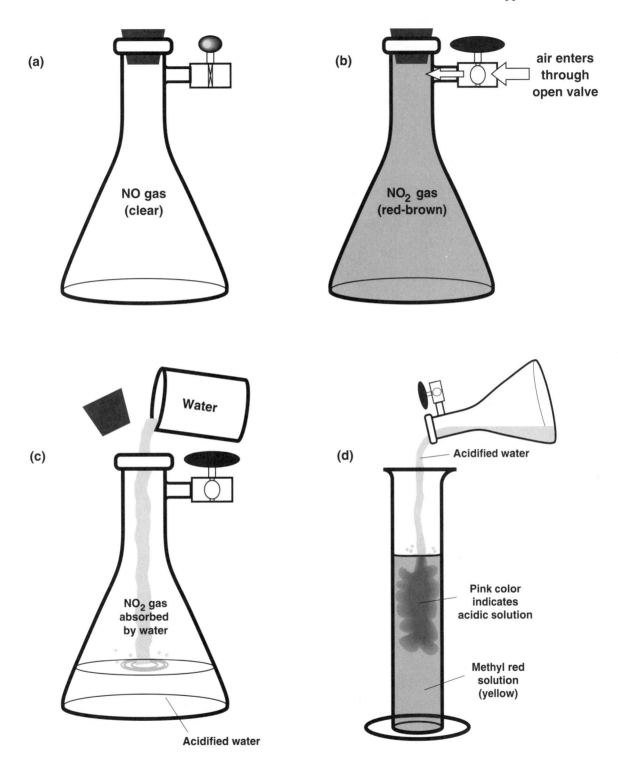

(a) NO gas (clear)

(b) NO₂ gas (red-brown)

air enters through open valve

(c) Water

NO₂ gas absorbed by water

Acidified water

(d) Acidified water

Pink color indicates acidic solution

Methyl red solution (yellow)

Figure B.6 Setup for the demonstration of the formation of acidic rain. The flask must be completely evacuated and backfilled with nitric oxide (a), leaving a sufficient vaccuum for air to be drawn into the flask through a valve, as shown in the next panel (b). Nitric acid is formed when nitrogen dioxide is mixed with water (c). The presence of nitric acid can be shown using a pH "indicator," in this case, a solution of methyl red (d).

atmosphere. In air, particularly polluted air, radicals such as hydroxyl (OH) are abundant. These radicals oxidize nitrogen and sulfur oxides into nitric and sulfuric acids (Sections 3.3.4, 6.3.2, 6.5.3, 10.2.1, and 13.5.3). In this experiment, ions in an aqueous solution perform the same function. Acid is quickly produced in this case because of the high concentration of nitrogen oxides in the flask; the amounts are *10 million times greater* than in polluted air. Things often need to be accelerated in a demonstration. The reaction that forms acids in the experiment is

$$NO_2 + NO_2 + H_2O \rightarrow HNO_2 + HNO_3 \tag{B.1}$$

The initial formation of nitrogen dioxide is also somewhat different from the processes occurring in smog. In polluted air, reactions of nitric oxide with hydrocarbons—denoted RO_2 in the text—or with ambient ozone, lead to the formation of NO_2. In the experiment, the nitric oxide concentrations are so large that a slow reaction between NO and molecular oxygen is effective. Thus

$$NO + NO + O_2 \rightarrow NO_2 + NO_2 \tag{B.2}$$

These differences are not really significant with regard to the origin or properties of the acids in the environment.

Appendix C
Radiation Nomenclature

Radiation in the environment is referred to in so many different ways that it may be confusing to nonspecialists. Figure C.1 below identifies the various regions of the solar spectrum that are important in various applications, such as determining the energy balance of the climate system—Chapter 11— or describing the photochemistry of ozone—Chapter 13. The various regions named are used throughout the text, where they are explained from time to time for clarity. Details concerning the nature of radiation in the environment are given in several different contexts in Sections 3.2, 11.2, 11.4 and 13.4; also see *radiation* and related entries in the index. The standard units for measuring the wavelengths of radiation are summarized in Table 3.1 on page 51 of the text.

The complete spectrum of radiation includes both high energy waves with very short wavelengths (**X rays** and **gamma rays**) and low energy waves with very long wavelengths (including **microwaves, radio waves** and other common electromagnetic fields,

or **EMF**). For most environmental problems, however, it is the radiation lying between these extremes that is most important. This radiation includes the ultraviolet, visible, and infrared regions. Often this central region of the full spectrum of radiation is divided into **solar** and **terrestrial** components (Figure C.1), which may be further subdivided or re-named for specific applications. For example, in climate studies, reference is frequently made to **shortwave** and **longwave** radiation, while in photochemical investigations **ultraviolet** and **visible** radiation are the key components. Figure C.1 depicts the various wavelength ranges that apply to the common spectral regions of interest, and indicates how these regions are related.

The solar spectrum is the sum of the **ultraviolet, visible**, and **near infrared** regions. Moreover, the ultraviolet spectrum includes the **UV-A, -B** and **-C** regions, as well as **far ultraviolet** radiation at shorter wavelengths. The visible spectrum includes the **primary colors** of light—blue, green, and red (B, G,

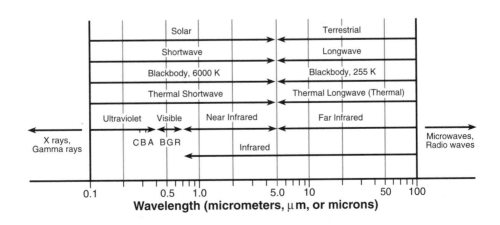

Figure C.1 The different spectral regions of environmentally important electromagnetic radiation are illustrated. The divisions shown represent the commonly used categories of radiation.

and R in Figure C.1). **Solar** radiation is equivalent to **shortwave** radiation. The spectrum of **infrared** radiation actually has three subregions, denoted near, middle and far infrared. For simplicity, however, we subdivide the infrared into two regions—the **near infrared** at relatively short wavelengths and the **far infrared** at longer wavelengths (refer to Figure C.1). The infrared spectrum overlaps both the solar and terrestrial radiation regimes. **Terrestrial** radiation, consisting mainly of far infrared radiation, is equivalent to **longwave** radiation. **Blackbody** radiation may correspond to either short or long wavelength regions, these being differentiated by high and low emission temperatures, respectively. Blackbody radiation emitted at short wavelengths is frequently called **thermal shortwave** radiation, and that at long wavelengths, **thermal longwave** radiation.

Table C.1 summarizes the relationships between the commonly used spectral regions of radiation. The definitions and equivalencies given in the table are approximate, although sufficiently accurate for the purposes of the present exposition. These spectral ranges, in fact, are somewhat fuzzy, can overlap to a degree, and may vary depending on the specific application.

Table C.1 Relationships between radiation components

Spectral region	Spectral equivalence
Solar	Ultraviolet + Visible + Near infrared (+ X rays) = Shortwave
Terrestrial	Far infrared (+ Microwaves) = Longwave
Infrared	Near infrared + Far infrared
Ultraviolet	Near ultraviolet + Far ultraviolet = UV-A + UV-B + UV-C + Far ultraviolet
Shortwave	Solar = Near infrared + Visible + Ultraviolet
Longwave	Terrestrial = Far infrared
Blackbody, 6000 K	Solar = Shortwave = Thermal shortwave
Blackbody, 255 K	Terrestrial = Longwave = Thermal longwave
Thermal shortwave	Solar = Near infrared + Visible + Ultraviolet = Blackbody, 255 K
Thermal longwave	Terrestrial = Longwave = Far infrared = Blackbody, 6000 K
Visible	Shortwave - Ultraviolet - Near infrared = Blue + Green + Red
Near infrared	Solar - Visible - Ultraviolet = Infrared - Far infrared
Far infrared	Terrestrial = Longwave = Infrared - Near infrared
Thermal	Terrestrial = Longwave = Far infrared = Blackbody, 255 K

Note: The terms in brackets represent extensions of the spectral composition beyond the wavelength limits given in Figure C.1. In the infrared region, the middle infrared has been incorporated into the near and far infrared regions.

Index